Fundamentals of Chemical Engineering Thermodynamics

Prentice Hall International Series in the Physical and Chemical Engineering Sciences

Visit informit.com/ph/physandchem
for a complete list of available publications.

The Prentice Hall International Series in the Physical and Chemical Engineering Sciences had its auspicious beginning in 1956 under the direction of Neal R. Amundsen. The series comprises the most widely adopted college textbooks and supplements for chemical engineering education. Books in this series are written by the foremost educators and researchers in the field of chemical engineering.

Make sure to connect with us!
informit.com/socialconnect

 |  |

Fundamentals of Chemical Engineering Thermodynamics

Themis Matsoukas

PRENTICE HALL

Upper Saddle River, NJ • Boston • Indianapolis • San Francisco
New York • Toronto • Montreal • London • Munich • Paris • Madrid
Capetown • Sydney • Tokyo • Singapore • Mexico City

Many of the designations used by manufacturers and sellers to distinguish their products are claimed as trademarks. Where those designations appear in this book, and the publisher was aware of a trademark claim, the designations have been printed with initial capital letters or in all capitals.

The author and publisher have taken care in the preparation of this book, but make no expressed or implied warranty of any kind and assume no responsibility for errors or omissions. No liability is assumed for incidental or consequential damages in connection with or arising out of the use of the information or programs contained herein.

The publisher offers excellent discounts on this book when ordered in quantity for bulk purchases or special sales, which may include electronic versions and/or custom covers and content particular to your business, training goals, marketing focus, and branding interests. For more information, please contact:

    U.S. Corporate and Government Sales
    (800) 382-3419
    corpsales@pearsontechgroup.com

For sales outside the United States please contact:

    International Sales
    international@pearson.com

Visit us on the Web: informit.com/ph

Library of Congress Cataloging-in-Publication Data

Matsoukas, Themis.
  Fundamentals of chemical engineering thermodynamics : with applications to chemical processes / Themis Matsoukas.
     p. cm.
  Includes bibliographical references and index.
  ISBN 0-13-269306-2 (hardcover : alk. paper)
  1. Thermodynamics—Textbooks. 2. Chemical engineering—Textbooks. I. Title.
  QD504.M315 2013
  660—dc23
                                                               2012025140

Copyright © 2013 Pearson Education, Inc.

All rights reserved. Printed in the United States of America. This publication is protected by copyright, and permission must be obtained from the publisher prior to any prohibited reproduction, storage in a retrieval system, or transmission in any form or by any means, electronic, mechanical, photocopying, recording, or likewise. To obtain permission to use material from this work, please submit a written request to Pearson Education, Inc., Permissions Department, One Lake Street, Upper Saddle River, New Jersey 07458, or you may fax your request to (201) 236-3290.

ISBN-13: 978-0-13-269306-6
ISBN-10:   0-13-269306-2

Text printed in the United States at RR Donnelley in Kendallville, Indiana.
Third printing, March 2016

Executive Editor
Bernard Goodwin

Managing Editor
John Fuller

Production Editor
Elizabeth Ryan

Packager
Michelle Gardner

Copy Editor
Alyson Platt

Indexer
Constance Angelo

Proofreader
Gina Delaney

Cover Designer
Alan Clements

Compositor
Laserwords

To my mother, Vana,
TM

Contents

Preface	xiii
Acknowledgments	xvii
About the Author	xix
Nomenclature	xxi

Part I Pure Fluids — 1

Chapter 1 Scope and Language of Thermodynamics — 3
1.1 Molecular Basis of Thermodynamics — 5
1.2 Statistical versus Classical Thermodynamics — 11
1.3 Definitions — 13
1.4 Units — 22
1.5 Summary — 26
1.6 Problems — 26

Chapter 2 Phase Diagrams of Pure Fluids — 29
2.1 The PVT Behavior of Pure Fluid — 29
2.2 Tabulation of Properties — 40
2.3 Compressibility Factor and the ZP Graph — 43
2.4 Corresponding States — 45
2.5 Virial Equation — 53
2.6 Cubic Equations of State — 57
2.7 PVT Behavior of Cubic Equations of State — 61
2.8 Working with Cubic Equations — 64
2.9 Other Equations of State — 67
2.10 Thermal Expansion and Isothermal Compression — 71
2.11 Empirical Equations for Density — 72
2.12 Summary — 77
2.13 Problems — 78

Chapter 3 Energy and the First Law — 87

- 3.1 Energy and Mechanical Work — 88
- 3.2 Shaft Work and PV Work — 90
- 3.3 Internal Energy and Heat — 96
- 3.4 First Law for a Closed System — 98
- 3.5 Elementary Paths — 101
- 3.6 Sensible Heat—Heat Capacities — 109
- 3.7 Heat of Vaporization — 119
- 3.8 Ideal-Gas State — 124
- 3.9 Energy Balances and Irreversible Processes — 133
- 3.10 Summary — 139
- 3.11 Problems — 140

Chapter 4 Entropy and the Second Law — 149

- 4.1 The Second Law in a Closed System — 150
- 4.2 Calculation of Entropy — 153
- 4.3 Energy Balances Using Entropy — 163
- 4.4 Entropy Generation — 167
- 4.5 Carnot Cycle — 168
- 4.6 Alternative Statements of the Second Law — 177
- 4.7 Ideal and Lost Work — 183
- 4.8 Ambient Surroundings as a Default Bath—Exergy — 189
- 4.9 Equilibrium and Stability — 191
- 4.10 Molecular View of Entropy — 195
- 4.11 Summary — 199
- 4.12 Problems — 201

Chapter 5 Calculation of Properties — 205

- 5.1 Calculus of Thermodynamics — 205
- 5.2 Integration of Differentials — 213
- 5.3 Fundamental Relationships — 214
- 5.4 Equations for Enthalpy and Entropy — 217
- 5.5 Ideal-Gas State — 219
- 5.6 Incompressible Phases — 220
- 5.7 Residual Properties — 222
- 5.8 Pressure-Explicit Relations — 228
- 5.9 Application to Cubic Equations — 230
- 5.10 Generalized Correlations — 235
- 5.11 Reference States — 236

5.12	Thermodynamic Charts	242
5.13	Summary	245
5.14	Problems	246

Chapter 6 Balances in Open Systems — 251

6.1	Flow Streams	252
6.2	Mass Balance	253
6.3	Energy Balance in Open System	255
6.4	Entropy Balance	258
6.5	Ideal and Lost Work	266
6.6	Thermodynamics of Steady-State Processes	272
6.7	Power Generation	295
6.8	Refrigeration	301
6.9	Liquefaction	309
6.10	Unsteady-State Balances	315
6.11	Summary	323
6.12	Problems	324

Chapter 7 VLE of Pure Fluid — 337

7.1	Two-Phase Systems	337
7.2	Vapor-Liquid Equilibrium	340
7.3	Fugacity	343
7.4	Calculation of Fugacity	345
7.5	Saturation Pressure from Equations of State	353
7.6	Phase Diagrams from Equations of State	356
7.7	Summary	358
7.8	Problems	360

Part II Mixtures — 367

Chapter 8 Phase Behavior of Mixtures — 369

8.1	The Txy Graph	370
8.2	The Pxy Graph	373
8.3	Azeotropes	380
8.4	The xy Graph	381
8.5	VLE at Elevated Pressures and Temperatures	383
8.6	Partially Miscible Liquids	384
8.7	Ternary Systems	390

| 8.8 | Summary | 393 |
| 8.9 | Problems | 394 |

Chapter 9 Properties of Mixtures — 401
9.1	Composition	402
9.2	Mathematical Treatment of Mixtures	404
9.3	Properties of Mixing	409
9.4	Mixing and Separation	411
9.5	Mixtures in the Ideal-Gas State	413
9.6	Equations of State for Mixtures	419
9.7	Mixture Properties from Equations of State	421
9.8	Summary	428
9.9	Problems	428

Chapter 10 Theory of Vapor-Liquid Equilibrium — 435
10.1	Gibbs Free Energy of Mixture	435
10.2	Chemical Potential	439
10.3	Fugacity in a Mixture	443
10.4	Fugacity from Equations of State	446
10.5	VLE of Mixture Using Equations of State	448
10.6	Summary	453
10.7	Problems	454

Chapter 11 Ideal Solution — 461
11.1	Ideality in Solution	461
11.2	Fugacity in Ideal Solution	464
11.3	VLE in Ideal Solution—Raoult's Law	466
11.4	Energy Balances	475
11.5	Noncondensable Gases	480
11.6	Summary	484
11.7	Problems	484

Chapter 12 Nonideal Solutions — 489
12.1	Excess Properties	489
12.2	Heat Effects of Mixing	496
12.3	Activity Coefficient	504
12.4	Activity Coefficient and Phase Equilibrium	507
12.5	Data Reduction: Fitting Experimental Activity Coefficients	512
12.6	Models for the Activity Coefficient	515
12.7	Summary	531
12.8	Problems	533

Chapter 13 Miscibility, Solubility, and Other Phase Equilibria   545
13.1 Equilibrium between Partially Miscible Liquids   545
13.2 Gibbs Free Energy and Phase Splitting   548
13.3 Liquid Miscibility and Temperature   556
13.4 Completely Immiscible Liquids   558
13.5 Solubility of Gases in Liquids   563
13.6 Solubility of Solids in Liquids   575
13.7 Osmotic Equilibrium   580
13.8 Summary   586
13.9 Problems   586

Chapter 14 Reactions   593
14.1 Stoichiometry   593
14.2 Standard Enthalpy of Reaction   596
14.3 Energy Balances in Reacting Systems   601
14.4 Activity   606
14.5 Equilibrium Constant   614
14.6 Composition at Equilibrium   622
14.7 Reaction and Phase Equilibrium   624
14.8 Reaction Equilibrium Involving Solids   629
14.9 Multiple Reactions   632
14.10 Summary   636
14.11 Problems   637

Bibliography   647

Appendix A Critical Properties of Selected Compounds   649

Appendix B Ideal-Gas Heat Capacities   653

Appendix C Standard Enthalpy and Gibbs Free Energy of Reaction   655

Appendix D UNIFAC Tables   659

Appendix E Steam Tables   663

Index   677

Preface

My goal with this book is to provide the undergraduate student in chemical engineering with the solid background to perform thermodynamic calculations with confidence, and the course instructor with a resource to help students achieve this goal. The intended audience is sophomore/junior students in chemical engineering. The book is divided into two parts. Part I covers the laws of thermodynamics, with applications to pure fluids; Part II extends thermodynamics to mixtures, with emphasis on phase and chemical equilibrium. The selection of topics was guided by the realities of the undergraduate curriculum, which gives us about 15 weeks per semester to develop the material and meet the learning objectives. Given that thermodynamics requires some minimum "sink-in" time, the deliberate choice was made to prioritize topics and cover them at a comfortable pace. Each part consists of seven chapters, corresponding to an average of about two weeks (six lectures) per chapter. Under such restrictions certain topics had to be left out and for others their coverage had to be limited. Highest priority is given to material that feeds directly to other key courses of the curriculum: separations, reactions, and capstone design. A deliberate effort was made to stay away from specialty topics such as electrochemical or biochemical systems on the premise that these are more appropriately dealt with (and at a depth that a book such as this could do no justice) in physical chemistry, biochemistry, and other dedicated courses. Students are made aware of the amazing generality of thermodynamics and are directed to other fields for such details as needed. A theme that permeates the book is the molecular basis of thermodynamics. Discussions of molecular phenomena remain at a qualitative level (except for very brief excursions to statistical concepts in the chapter on entropy), consistent with the background of the typical sophomore/junior. But the molecular picture is consistently brought up to reinforce the idea that the quantities we measure in the lab and the equations that describe them are manifestations of microscopic effects at the molecular level.

The two parts of the book essentially mirror the material of a two-course sequence in thermodynamics that is typically required in chemical engineering. The focus of Part I is on pure fluids exclusively. The PVT behavior is introduced early on (Chapter 2) so that when it comes to the first and second law (Chapters 3 and 4), students have the tools to perform basic calculations of enthalpy and entropy using steam tables (a surrogate for tabulated properties in general) and equations of state. Chapter 5 discusses fundamental relationships and the calculation of properties from equations of state. It is mathematically the densest chapter of Part I.

Chapter 6 goes into applications of thermodynamics to chemical processes. The range of applications is limited to systems involving pure fluids, namely power plants and refrigeration/liquefaction systems. This is the part of the course that most directly relates to processes discussed in capstone design and justifies the "Chemical Engineering" in the title of the book. It is one of the longer chapters, with several examples and end-of-chapter problems. The last chapter in this part covers phase equilibrium for a single fluid and serves as the connector between the two parts, as fugacity is the main actor in Part II.

The second part begins with a survey of phase diagrams of binary and simple ternary systems. It introduces the variety of phase behaviors of mixtures and establishes the notion that each phase at equilibrium has its own composition, and introduces the lever rule as a basic material balance tool. Many programs probably cover some of that material in the Materials and Energy Balances course but the topics are central to subsequent discussions so that a separate chapter is justified. Chapter 9 extends the fundamental relationships, which in Part I were applied to pure fluids, to mixtures. This chapter also introduces the equation of state for mixtures and the calculation of mixture properties from the equation of state. Chapter 10 is a short chapter that establishes the phase equilibrium criterion for mixtures and applies the equation of state to calculate the phase diagram of a binary mixture. Chapters 11 and 12 deal with ideal and nonideal solutions, respectively. Chapter 13 goes over several topics of phase equilibrium that are too small to be in separate chapters. These include partial miscibility, solubility of gases and solids, and osmotic processes. The last chapter in this part, 14, covers reaction equilibrium. The focus of the chapter is to establish the fundamental relationships, which are then applied to single and multiphase reactions. Standard states are discussed in quite some detail, since this is a topic that seems to confuse students.

Overall, a great effort has been made to balance theory with examples and applications. Examples cover a wide range, from direct application of formulas and methodologies, to larger processes that require synthesis of several smaller problems. It has been my experience that students are more willing to accept what they perceive as abstract theory if they can see how this theory is tied to practical industrial situations. Realistic problems are rarely of the paper-and-pencil type, and this brings up the need for mathematical/computational tools. The choices today are many, from sophisticated hand calculators to spreadsheets and numerical packages. The textbook takes an agnostic approach when it comes to the type of software and leaves it up to the instructor to make that choice. Typically, the problems that require numerical tools are those involving calculations with equations of state. Some problems lend themselves to the use of process simulators but, by deliberate decision, there is no specific mention of these simulators in the book. As with the other computational tools, the choice is left to the instructor. In my experience, the best approach with problems that require a significant number of computations is to assign them as projects. Picking problems with industrial flavor not only motivates students in engineering, it also offers convincing justification for the practical need for theory and numerical methodologies.

—Themis Matsoukas
University Park
May 2012

Acknowledgments

This book is the product of some 20 years of teaching thermodynamics at Penn State, but it would not have happened without the help, direct or indirect, of several individuals, colleagues, students, and friends. I must start by thanking Andrew Zydney, my colleague and department head, who encouraged me to spend my 2010–2011 sabbatical finishing this textbook. Without Andrew's nudging, the book would have remained in the perpetual state of a draft-in-progress. I want to thank my colleagues, Darrell Velegol, Scott Milner, Seong Kim, Rob Rioux, and Enrique Gomez, who trusted drafts of the book, in various stages of completion, for use in their classes. Among those who reviewed the near-final draft, I want to thank Michael M. Domach and Tracy J. Benson, who offered comments that were especially insightful and encouraging. The number of students, whose feedback over the years shaped this book, is too large to list. But special thanks must go to my Fall 2011 class of CH E 220H, the honors section of Thermo I, on whom fell the unenviable task of debugging the first finished version of this book. They did this admirably. Of all the students in that class special mention goes to Steph Nitopi, Twafik Miral, Brian Brady, and Ashlee Smith, whose attention to detail was nothing but extraordinary. Bernard Goodwin and Michelle Housley of Pearson Education guided me through this book project—my first one—with a light hand and flexible deadlines. I want to thank Elizabeth Ryan, John Fuller, and Michelle Gardner, who, as members of the production team, brought my vision of this book to life. My wife Kristen, and daughter Melina, helped in many ways, big and small, in carrying this project out. And lastly, I feel grateful to LuLu, our little dog, who sat approvingly on my armchair next to me through countless hours of typing.

About the Author

 **Themis Matsoukas** is a professor of chemical engineering at Pennsylvania State University. He received a diploma in chemical engineering from the National Metsovion Polytechnic School in Athens Greece, and a Ph.D. from the University of Michigan in Ann Arbor. He has taught graduate and undergraduate thermodynamics, materials, and energy balances, and various electives in particle and aerosol technology. Since 1991, he has taught thermodynamics to more than 1000 students at Penn State. He has been recognized with several awards for excellence in teaching and advising.

Nomenclature

In general, properties are assumed to be molar unless indicated differently. A corresponding extensive property is indicated by the superscript ^{tot}, for example, $H^{\text{tot}} = nH$. In dealing with mixtures, a variable without subscripts or superscripts (e.g., H) is the molar property of the mixture; a property with a subscript (e.g., H_i) is the corresponding property of the pure component; a property with an overbar (e.g., $\bar{H}_i$) is the partial molar property of the component in the mixture.

List of Symbols

A	Helmholtz free energy, $A = U - TS$
a_i	activity of component i
F	generic thermodynamic property
f	fugacity of component
G	Gibbs free energy, $G = H - TS$
$\bar{G}_i$	partial molar Gibbs free energy
G^E	excess Gibbs free energy
$\bar{G}_i^E$	excess partial molar Gibbs free energy
G^R	residual Gibbs free energy
$\bar{G}_i^R$	residual partial molar Gibbs free energy
$G°$	standard Gibbs energy of formation
H	enthalpy, $H = U + PV$
$\bar{H}_i$	partial molar enthalpy
H^E	excess enthalpy
$\bar{H}_i^E$	excess partial molar enthalpy
H^R	residual enthalpy
$\bar{H}_i^R$	residual partial molar enthalpy
$H°$	standard enthalpy of formation
ΔH_{vap}	enthalpy or heat of vaporization
ΔH_{mix}	enthalpy or heat of mixing
K	equilibrium constant

k_B	Boltzmann constant
k_{ij}	interaction parameter in equation of state
k_i^H	Henry's law constant for component i
$k_i^{H'}$	Henry's law constant for component i based on molality
L	liquid fraction of a vapor-liquid mixture
m	mass
n	number of moles
N	number of components
P^{sat}	saturation pressure
Q	heat
R	ideal-gas constant, $R = 8.314$ J/mol K
S	entropy
$\bar{S}_i$	partial molar entropy
S^E	excess entropy
$\bar{S}_i^E$	excess partial molar entropy
S^R	residual entropy
$\bar{S}_i^R$	residual partial molar entropy
S°	standard entropy of formation
ΔS_{vap}	entropy of vaporization
ΔS_{mix}	entropy of mixing
$\Delta S_{\text{mix}}^{\text{id}}$	ideal entropy of mixing, $\Delta S_{\text{mix}}^{\text{id}} = -R \sum x_i \ln x_i$
U	internal energy
V	volume; also used to denote the vapor fraction of a vapor-liquid mixture
W	work
w_i	mass fraction
x_i	generic mole fraction or mol fraction in liquid (two-phase system)
y_i	mole fraction in gas (two-phase system)
z_i	overall mole fraction in two-phase system

Greek symbols

β	volumetric coefficient of thermal expansion
γ	activity coefficient
κ	coefficient of isothermal compressibility

μ		chemical potential
ν		stoichiometric coefficient
ξ		extent of reaction
π		number of phases (in Gibbs's phase rule)
ρ		density coefficient
ϕ		fugacity coefficient
ω		acentric factor

Subscripts

i	component or stream
L	liquid
V	vapor
gen	generation
sur	surroundings
univ	universe

Superscripts

tot	total (extensive) property, for example, $H^{\text{tot}} = nH$
R	residual
ig	ideal-gas state
igm	ideal-gas mixture
id	ideal solution
E	excess
sat	saturation
vap	vaporization
°	standard state

Part I

Pure Fluids

Chapter 1

Scope and Language of Thermodynamics

Chemical processes involve streams undergoing various transformations. One example is shown in Figure 1-1: raw materials are fed into a heated reactor, where they react to form products. The effluent stream, in this case a liquid, contains the products of the reaction, any unreacted raw materials, and possibly other by-products. This stream is pumped out of the reactor into a heat exchanger, where it partially boils. The vapor/liquid mixture is fed into a tank, which collects the vapor at the top, and liquid at the bottom. The streams that exit this tank are finally cooled down and sent to the next stage of the process. Actual processes are generally more complex and may involve many streams and several interconnected units. Nonetheless, the example in Figure 1-1 contains all the basic ingredients likely to be found in any chemical plant: heating, cooling, pumping, reactions, phase transformations.

It is the job of the chemical engineer to compute the material and energy balances around such process: This includes the flow rates and compositions of all streams, the power requirements of pumps, compressors and turbines, and the heat loads in the heat exchangers. The chemical engineer must also determine the conditions of pressure and temperature that are required to produce the desired effect, whether this is a chemical reaction or a phase transformation. All of this requires the knowledge of various physical properties of a mixture: density, heat capacity, boiling temperature, heat of vaporization, and the like. More specifically, these properties must be known as a function of temperature, pressure, and composition, all of which vary from stream to stream. Energy balances and property estimation may appear to be separate problems, but they are not: both calculations require the application of the same fundamental principles of thermodynamics.

The name *thermodynamics* derives from the Greek *thermotis* (heat) and *dynamiki* (potential, power). Its historical roots are found in the quest to develop heat engines, devices that use heat to produce mechanical work. This quest, which was instrumental in powering the industrial revolution, gave birth to thermodynamics as a discipline that studies the relationship between heat, work, and energy. The elucidation of this relationship is one of the early triumphs of thermodynamics and a reason why, even today, thermodynamics is often described as the study of energy

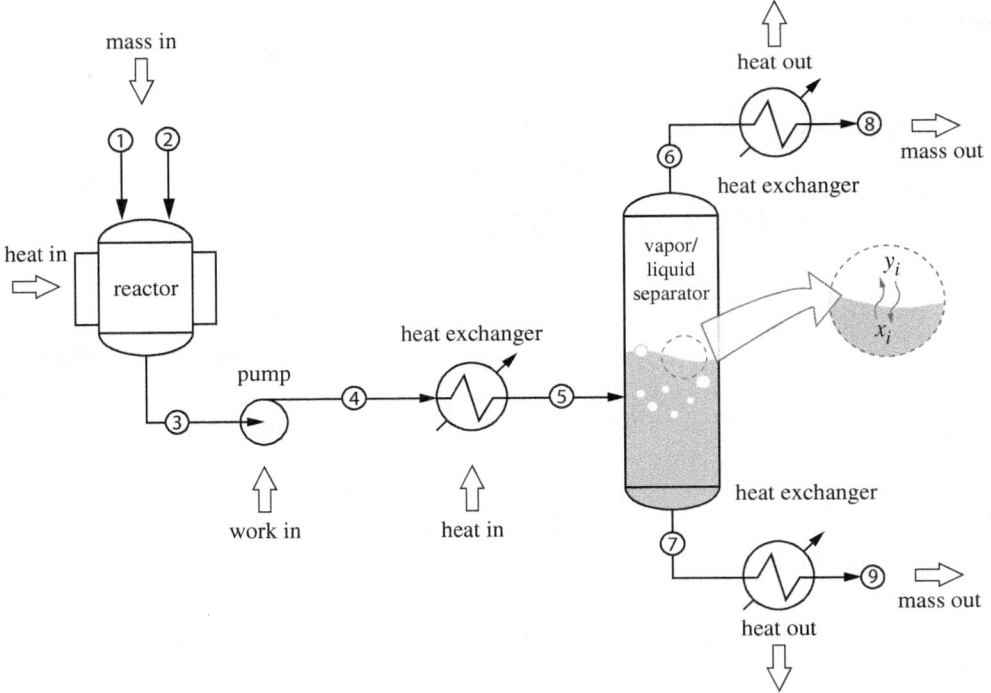

Figure 1-1: Typical chemical process.

conversions involving heat. Modern thermodynamics is a much broader discipline whose focus is the equilibrium state of systems composed of very large numbers of molecules. Temperature, pressure, heat, and mechanical work, as manifested through the expansion and compression of matter, are understood to arise from interactions at the molecular level. Heat and mechanical work retain their importance but the scope of the modern discipline is far wider than its early developers would have imagined, and encompasses many different systems containing huge numbers of "particles," whether these are molecules, electron spins, stars, or bytes of digital information. The term *chemical thermodynamics* refers to applications to molecular systems.

Figure 1-2: J. W. Gibbs.

Among the many scientists who contributed to the development of modern thermodynamics, J. Willard Gibbs stands out as one whose work revolutionized the discipline by providing the tools to connect the macroscopic properties of thermodynamics to the microscopic properties of molecules. His name is now associated with the *Gibbs free energy*, a thermodynamic property of fundamental importance in phase and chemical equilibrium.

Chemical engineering thermodynamics is the subset that applies thermodynamics to processes of interest to chemical engineers. One important task is the calculation of energy requirements of a process and, more broadly, the analysis of energy utilization and efficiency in chemical processing. This general problem is discussed in the first part of the book, Chapters 2 through 7. Another important application of chemical engineering thermodynamics is in the design of separation units. The vapor-liquid separator in Figure 1-1 does more than just separate the liquid portion of the stream from the vapor. When a multicomponent liquid boils, the more volatile ("lighter") components collect preferentially in the vapor and the less volatile ("heavier") ones remain mostly in the liquid. This leads to partial separation of the initial mixture. By staging multiple such units together, one can accomplish separations of components with as high purity as desired. The determination of the equilibrium composition of two phases in contact with each other is an important goal of chemical engineering thermodynamics. This problem is treated in the second part of the book and the first part is devoted to the behavior of single-component fluids. Overall then, the chemical engineer uses thermodynamics to

1. Perform energy and material balances in unit operations with chemical reactions, separations, and fluid transformations (heating/cooling, compression/expansion),

2. Determine the various physical properties that are required for the calculation of these balances,

3. Determine the conditions of equilibrium (pressure, temperature, composition) in phase transformations and chemical reactions.

These tasks are important for the design of chemical processes and for their proper control and troubleshooting. The overall learning objective of this book is to provide the undergraduate student in chemical engineering with a solid background to perform these calculations with confidence.

1.1 Molecular Basis of Thermodynamics

All macroscopic behavior of matter is the result of phenomena that take place at the microscopic level and arise from force interactions among molecules. Molecules exert a variety of forces: direct electrostatic forces between ions or permanent dipoles;

induction forces between a permanent dipole and an induced dipole; forces of attraction between nonpolar molecules, known as van der Waals (or dispersion) forces; other specific chemical forces such as hydrogen bonding. The type of interaction (attraction or repulsion) and the strength of the force that develops between two molecules depends on the distance between them. At far distances the force is zero. When the distance is of the order of several Å, the force is generally attractive. At shorter distances, short enough for the electron clouds of the individual atoms to begin to overlap, the interaction becomes very strongly repulsive. It is this strong repulsion that prevents two atoms from occupying the same point in space and makes them appear as if they possess a solid core. It is also the reason that the density of solids and liquids is very nearly independent of pressure: molecules are so close to each other that adding pressure by any normal amounts (say 10s of atmospheres) is insufficient to overcome repulsion and cause atoms to pack much closer.

Intermolecular Potential The force between two molecules is a function of the distance between them. This force is quantified by intermolecular potential energy, $\Phi(r)$, or simply *intermolecular potential*, which is defined as the work required to bring two molecules from infinite distance to distance r. Figure 1-3 shows the approximate intermolecular potential for CO_2. Carbon dioxide is a linear molecule and its potential depends not only on the distance between the molecules but also on their relative orientation. This angular dependence has been averaged out for simplicity. To interpret Figure 1-3, we recall from mechanics that force is equal to the negative derivative of the potential with respect to distance:

$$F = -\frac{d\Phi(r)}{dr}. \tag{1.1}$$

That is, the magnitude of the force is equal to the slope of the potential with a negative sign that indicates that the force vector points in the direction in which the potential decreases. To visualize the force, we place one molecule at the origin and a test molecule at distance r. The magnitude of the force on the test molecule is equal to the derivative of the potential at that point (the force on the first molecule is equal in magnitude and opposite in direction). If the direction of force is towards the origin, the force is attractive, otherwise it is repulsive. The potential in Figure 1-3 has a minimum at separation distance $r_* = 4.47$ Å. In the region $r > r_*$ the slope is positive and the force is attractive. The attraction is weaker at longer distances and for r larger than about 9 Å the potential is practically flat and the force is zero. In the region $r < r_*$ the potential is repulsive and its steep slope indicates a very strong force that arises from the repulsive interaction of the electrons surrounding the molecules. Since the molecules cannot be pushed much closer than about $r \approx r_*$,

1.1 Molecular Basis of Thermodynamics

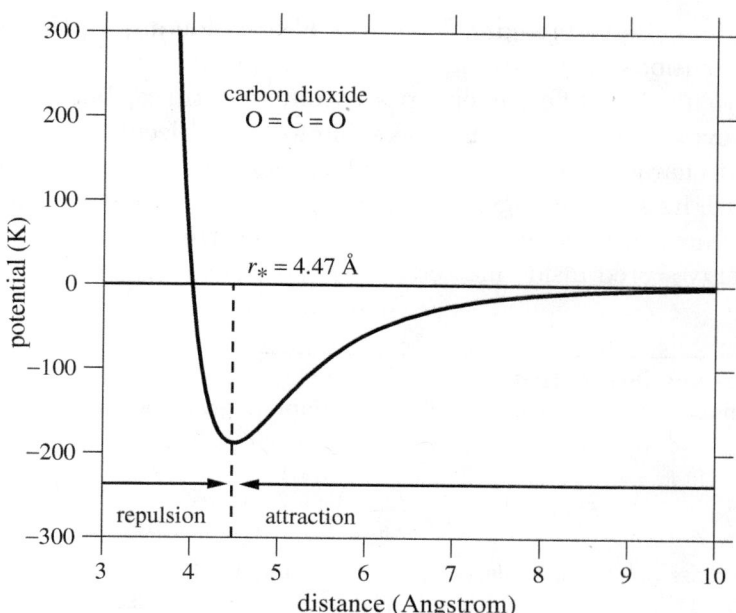

Figure 1-3: Approximate interaction potential between two CO_2 molecules as a function of their separation distance. The potential is given in kelvin; to convert to joule multiply by the Boltzmann constant, $k_B = 1.38 \times 10^{-23}$ J/K. The arrows show the direction of the force on the test molecule in the regions to the left and to the right of r_*.

we may regard the distance r_* to be the effective diameter of the molecule.[1] Of course, even simple molecules like argon are not solid spheres; therefore, the notion of a molecular diameter should not be taken literally.

The details of the potential vary among different molecules but the general features are always the same: Interaction is strongly repulsive at very short distance, weakly attractive at distance of the order of several Å, and zero at much larger distances. These features help to explain many aspects of the macroscopic behavior of matter.

Temperature and Pressure In the classical view of molecular phenomena, molecules are small material objects that move according to Newton's laws of motion, under the action of forces they exert on each other through the potential interaction. Molecules that collide with the container walls are reflected back, and the force of

---

1. The closest center-to-center distance we can bring two solid spheres is equal to the sum of their radii. For equal spheres, this distance is equal to their diameter.

this collision gives rise to pressure. Molecules also collide among themselves,[2] and during these collisions they exchange kinetic energy. In a thermally equilibrated system, a molecule has different energies at different times, but the distribution of energies is overall stationary and the same for all molecules. Temperature is a parameter that characterizes the distribution of energies inside a system that is in equilibrium with its surroundings. With increasing temperature, the energy content of matter increases. Temperature, therefore, can be treated as a measure of the amount of energy stored inside matter.

NOTE

Maxwell-Boltzmann Distribution
The distribution of molecular velocities in equilibrium is given by the Maxwell-Boltzmann equation:

$$f(v) = 4\pi v^2 \left(\frac{m}{2\pi k_B T}\right)^{3/2} e^{-mv^2/2k_B T}, \tag{1.2}$$

where m is the mass of the molecule, v is the magnitude of the velocity, T is absolute temperature, and k_B is the Boltzmann constant. The fraction of molecules with velocities between any two values v_1 to v_2 is equal to the area under the curve between the two velocities (the total area under the curve is 1). The velocity v_{max} that corresponds to the maximum of the distribution, the mean velocity $\bar{v}$, and the mean of the square of the velocity are all given in terms of temperature:

$$v_{max} = \left(\frac{2k_B T}{m}\right)^{1/2}, \quad \bar{v} = \left(\frac{8k_B T}{\pi m}\right)^{1/2}, \quad \overline{v^2} = \frac{3k_B T}{m}.$$

The Maxwell-Boltzmann distribution is a result of remarkable generality: it is independent of pressure and applies to any material, regardless of composition or phase. Figure 1-4 shows this distribution for water at three temperatures. At the triple point, the solid, liquid, and vapor, all have the same distribution of velocities.

Phase Transitions The minimum of the potential represents a stable equilibrium point. At this distance, the force between two molecules is zero and any small deviations to the left or to the right produce a force that points back to the minimum. A pair of molecules trapped at this distance r_* would form a stable pair if it were not for their kinetic energy, which allows them to move and eventually escape from the minimum. The lifetime of a trapped pair depends on temperature. At high temperature, energies are higher, and the probability that a pair will remain trapped is low. At low temperature a pair can survive long enough to trap additional molecules and form a small cluster of closely packed molecules. This cluster is a nucleus of

---

2. Molecular collisions do not require solid contact as macroscopic objects do. If two molecules come close enough in distance, the steepness of the potential produces a strong repulsive force that causes their trajectories to deflect.

1.1 Molecular Basis of Thermodynamics

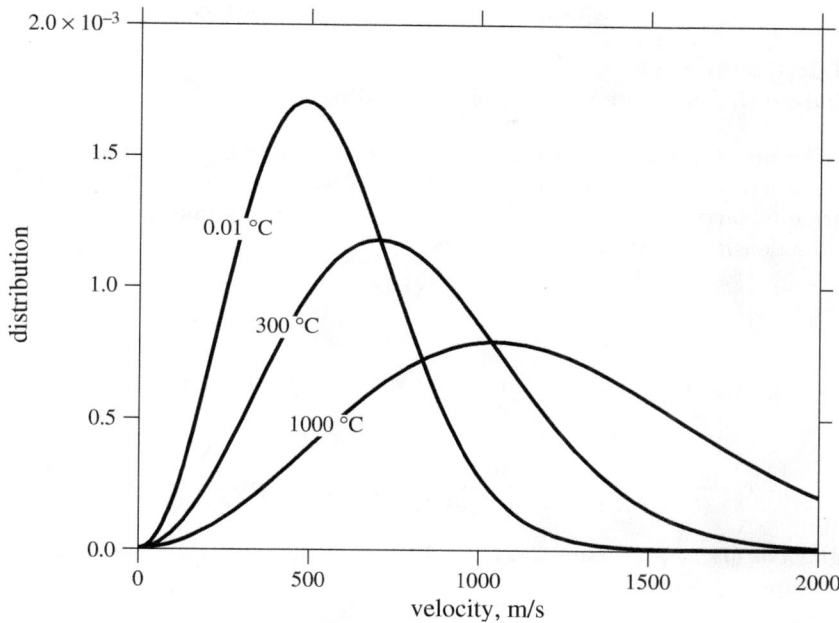

Figure 1-4: Maxwell-Boltzmann distribution of molecular velocities in water.

the liquid phase and can grow by further collection to form a macroscopic liquid phase. Thus we have a molecular view of vapor-liquid equilibrium. This picture highlights the fact that to observe a vapor-liquid transition, the molecular potential must exhibit a combination of strong repulsion at short distances with weak attraction at longer distances. Without strong repulsion, nothing would prevent matter from collapsing into a single point; without attraction, nothing would hold a liquid together in the presence of a vapor. We can also surmise that molecules that are characterized by a deeper minimum (stronger attraction) in their potential are easier to condense, whereas a shallower minimum requires lower temperature to produce a liquid. For this reason, water, which associates via hydrogen bonding (attraction) is much easier to condense than say, argon, which is fairly inert and interacts only through weak van der Waals attraction.

NOTE

Condensed Phases
The properties of liquids depend on both temperature and pressure, but the effect of pressure is generally weak. Molecules in a liquid (or in a solid) phase are fairly closely packed so that increasing pressure does little to change molecular distances by any appreciable amount. As a result, most properties of liquids are quite insensitive to pressure and can be approximately taken to be functions of temperature only.

Example 1.1: Density of Liquid CO_2
Estimate the density of liquid carbon dioxide based on Figure 1-3.

Solution The mean distance between molecules in the liquid is approximately equal to r_*, the distance where the potential has a minimum. If we imagine molecules to be arranged in a regular cubic lattice at distance r_* from each other, the volume of N_A molecules would be $N_A r_*^3$. The density of this arrangement is

$$\rho = \frac{M_m}{N_A r_*^3},$$

where M_m is the molar mass. Using $r_* = 4.47$ Å $= 4.47 \times 10^{-10}$ m, $M_m = 44.01 \times 10^{-3}$ kg/mol,

$$\rho = \frac{44.01 \times 10^{-3} \text{ kg/mol}}{(6.022 \times 10^{23} \text{ mol}^{-1})(4.47 \times 10^{-10} \text{ m})^3} = 818 \text{ kg/m}^3.$$

Perry's *Handbook* (7th ed., Table 2-242) lists the following densities of saturated liquid CO_2 at various temperatures:

T (K)	ρ (kg/m^3)
216.6	1130.7
270	947.0
300	680.3
304.2	466.2

According to this table, density varies with temperature from 1130.7 kg/m^3 at 216.6 K to 466.2 kg/m^3 at 304.2 K. The calculated value corresponds approximately to 285 K.

Comments The calculation based on the intermolecular potential is an estimation. It does not account for the effect of temperature (assumes that the mean distance between molecules is r_* regardless of temperature) and that molecules are arranged in a regular cubic lattice. Nonetheless, the final result is of the correct order of magnitude, a quite impressive result given the minimal information used in the calculation.

Ideal-Gas State Figure 1-3 shows that at distances larger than about 10 Å the potential of carbon dioxide is fairly flat and the molecular force nearly zero. If carbon dioxide is brought to a state such that the mean distance between molecules is more than 10 Å we expect that molecules would hardly register the presence of each other and would largely move independently of each other, except for brief close encounters. This state can be reproduced experimentally by decreasing pressure (increasing volume) while keeping temperature the same. This is called the *ideal-gas state*. It is a *state*—not a gas—and is reached by any gas when pressure is reduced sufficiently. In the ideal-gas state molecules move independently of each other and without the influence of the intermolecular potential. Certain properties in this state become universal for all gases regardless of the chemical identity of their molecules. The most important example is the ideal-gas law, which describes the pressure-volume-temperature relationship of any gas at low pressures.

1.2 Statistical versus Classical Thermodynamics

Historically, a large part of thermodynamics was developed before the emergence of atomic and molecular theories of matter. This part has come to be known as *classical thermodynamics* and makes no reference to molecular concepts. It is based on two basic principles ("laws") and produces a rigorous mathematical formalism that provides exact relationships between properties and forms the basis for numerical calculations. It is a credit to the ingenuity of the early developers of thermodynamics that they were capable of developing a correct theory without the benefit of molecular concepts to provide them with physical insight and guidance. The limitation is that classical thermodynamics cannot explain why a property has the value it does, nor can it provide a convincing physical explanation for the various mathematical relationships. This missing part is provided by statistical thermodynamics. The distinction between classical and statistical thermodynamics is partly artificial, partly pedagogical. Artificial, because thermodynamics makes physical sense only when we consider the molecular phenomena that produce the observed behaviors. From a pedagogical perspective, however, a proper statistical treatment requires more time to develop, which leaves less time to devote to important engineering applications. It is beyond the scope of this book to provide a bottom-up development of thermodynamics from the molecular level to the macroscopic. Instead, our goal is to develop the knowledge, skills, and confidence to perform thermodynamic calculations in chemical engineering settings. We will use molecular concepts throughout the book to shed light to new concepts but the overall development will remain under the general umbrella of classical thermodynamics. Those who wish to pursue the connection between the microscopic and the macroscopic in more detail, a subject that fascinated some of the greatest scientific minds, including Einstein, should plan to take an upper-level course in statistical mechanics from a chemical engineering, physics, or chemistry program.

The Laws of Classical Thermodynamics Thermodynamics is built on a small number of axiomatic statements, propositions that we hold to be true on the basis of our experience with the physical world. Statistical and classical thermodynamics make use of different axiomatic statements; the axioms of statistical thermodynamics have their basis on statistical concepts; those of classical thermodynamics are based on behavior that we observe macroscopically. There are two fundamental principles in classical thermodynamics, commonly known as the first and second law.[3] The first law expresses the principle that matter has the ability to store energy *within*. Within the context of classical thermodynamics, this is an axiomatic statement since its physical explanation is inherently molecular. The second law of thermodynamics expresses the principle that all systems, if left undisturbed, will move towards equilibrium –never away from it. This is taken as an axiomatic principle because we cannot prove it without appealing to other axiomatic statements. Nonetheless, contact with the physical world convinces us that this principle has the force a universal physical law.

Other laws of thermodynamics are often mentioned. The "zeroth" law states that, if two systems are in thermal equilibrium with a third system, they are in equilibrium with each other. The third law makes statements about the thermodynamic state at absolute zero temperature. For the purposes of our development, the first and second law are the only two principles needed in order to construct the entire mathematical theory of thermodynamics. Indeed, these are the only two equations that one must memorize in thermodynamics; all else is a matter of definitions and standard mathematical manipulations.

The "How" and the "Why" in Thermodynamics Engineers must be skilled in the art of *how* to perform the required calculations, but to build confidence in the use of theoretical tools it is also important to have a sense *why* our methods work. The "why" in thermodynamics comes from two sources. One is physical: the molecular picture that gives meaning to "invisible" quantities such as heat, temperature, entropy, equilibrium. The other is mathematical and is expressed through exact relationships that connect the various quantities. The typical development of thermodynamics goes like this:

(a) Use physical principles to establish fundamental relationships between key properties. These relationships are obtained by applying the first and second law to the problem at hand.

(b) Use calculus to convert the fundamental relationships from step (a) into useful expressions that can be used to compute the desired quantities.

Physical intuition is needed in order to justify the fundamental relationships in step (a). Once the physical problem is converted into a mathematical one (step [b]),

---

3. The term *law* comes to us from the early days of science, a time during which scientists began to recognize mathematical order behind what had seemed up until then to be a complicated physical world that defies prediction. Many of the early scientific findings were known as "laws," often associated with the name of the scientist who reported them, for example, Dalton's law, Ohm's law, Mendel's law, etc. This practice is no longer followed. For instance, no one refers to Einstein's famous result, $E = mc^2$, as Einstein's law.

physical intuition is no longer needed and the gear must shift to mastering the "how." At this point, a good handle of calculus becomes indispensable, in fact, a prerequisite for the successful completion of this material. Especially important is familiarity with functions of multiple variables, partial derivatives and path integrations.

1.3 Definitions

System

The *system* is the part of the physical world that is the object of a thermodynamic calculation. It may be a fixed amount of material inside a tank, a gas compressor with the associated inlet and outlet streams, or an entire chemical plant. Once the system is defined, anything that lies outside the system boundaries belongs to the *surroundings*. Together system and surroundings constitute the *universe*. A system can interact with its surroundings by exchanging mass, heat, and work. It is possible to construct the system in such way that some exchanges are allowed while others are not. If the system can exchange mass with the surroundings it called *open*, otherwise it is called *closed*. If it can exchange heat with the surroundings it is called *diathermal*, otherwise it is called *adiabatic*. A system that is prevented from exchanging either mass, heat, or work is called *isolated*. The universe is an isolated system.

A *simple system* is one that has no internal boundaries and thus allows all of its parts to be in contact with each other with respect to the exchange of mass, work, and heat. An example would be a mole of a substance inside a container. A *composite* system consists of simple systems separated by boundaries. An example would be a box divided into two parts by a firm wall. The construction of the wall would determine whether the two parts can exchange mass, heat, and work. For example, a permeable wall would allow mass transfer, a diathermal wall would allow heat transfer, and so on.

Example 1.2: Systems
Classify the systems in Figure 1-5.

Solution In (a) we have a tank that contains a liquid and a vapor. This system is inhomogeneous, because it consists of two phases; closed, because it cannot exchange mass with the surroundings; and simple, because it does not contain any internal walls. Although the liquid can exchange mass with the vapor, the exchange is internal to the system. There is no mention of insulation. We may assume, therefore, that the system is diathermal.

In (b) we have the same setup but the system is now defined to be just the liquid portion of the contents. This system is simple, open, and diathermal. Simple, because there are no internal walls; open, because the liquid can exchange mass with the vapor by evaporation or condensation; and diathermal, because it can exchange heat with the vapor. In this case, the insulation around the tank is not sufficient to render the system adiabatic because of the open interface between the liquid and vapor.

In (c) we have a condenser similar to those found in chemistry labs. Usually, hot vapor flows through the center of the condenser while cold water flows on the outside, causing the vapor to condense. This system is open because it allows mass flow through its boundaries. It is composite because of the wall that separates the two fluids. It is adiabatic because it is insulated from the surroundings. Even though heat is transferred between the inner and outer tube, this transfer is internal to the system (it does not cross the system bounds) and does not make the system diathermal.

Comments In the condenser of part (c), we determined the system to be open and adiabatic. Is it not possible for heat to enter through the flow streams, making the system diathermal? Streams carry energy with them and, as we will learn in Chapter 6, this is in the form of enthalpy. It is possible for heat to cross the boundary of the system inside the flow stream through conduction, due to different temperatures between the fluid stream just outside the system and the fluid just inside it. This heat flows slowly and represents a negligible amount compared to the energy carried by the flow. The main mode heat transfer is through the external surface of the system. If this is insulated, the system may be considered adiabatic.

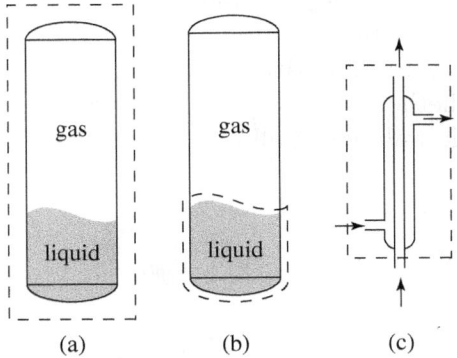

Figure 1-5: Examples of systems (see Example 1.2). The system is indicated by the dashed line. (a) Closed tank that contains some liquid and some gas. (b) The liquid portion in a closed, thermally insulated tank that also contains some gas. (c) Thermally insulated condenser of a laboratory-scale distillation unit.

Equilibrium

It is an empirical observation that a simple system left undisturbed, in isolation of its surroundings, must eventually reach an ultimate state that does not change with time. Suppose we take a rigid, insulated cylinder, fill half of it with liquid nitrogen at atmospheric pressure and the other half with hot, pressurized nitrogen, and place a wall between the two parts to keep them separate. Then, we rupture the wall between the two parts and allow the system to evolve without any disturbance from the outside. For some time the system will undergo changes as the two parts mix. During this time, pressure and temperature will vary, and so will the amounts of liquid and vapor. Ultimately, however, the system will reach a state in which no more changes are observed. This is the equilibrium state.

Equilibrium in a simple system requires the fulfillment of three separate conditions:

1. *Mechanical equilibrium*: demands uniformity of *pressure* throughout the system and ensures that there is no net work exchanged due to pressure differences.

2. *Thermal equilibrium*: demands uniformity of *temperature* and ensures no net transfer of heat between any two points of the system.

3. *Chemical equilibrium*: demands uniformity of the *chemical potential* and ensures that there is no net mass transfer from one phase to another, or net conversion of one chemical species into another by chemical reaction.

The chemical potential will be defined in Chapter 7.

Although equilibrium appears to be a static state of no change, at the molecular level it is a dynamic process. When a liquid is in equilibrium with a vapor, there is continuous transfer of molecules between the two phases. On an instantaneous basis the number of molecules in each phase fluctuates; overall, however, the molecular rates to and from each phase are equal so that, on average, there is no net transfer of mass from one phase to the other.

Constrained Equilibrium If we place two systems into contact with each other via a wall and isolate them from the rest of their surroundings, the overall system is isolated and composite. At equilibrium, each of the two parts is in mechanical, thermal, and chemical equilibrium at its own pressure and temperature. Whether the two parts establish equilibrium with *each other* will depend on the nature of the wall that separates them. A diathermal wall allows heat transfer and the equilibration of temperature. A movable wall (for example, a piston) allows the equilibration of pressure.

A selectively permeable wall allows the chemical equilibration of the species that are allowed to move between the two parts. If a wall allows certain exchanges but not others, equilibrium is established only with respect to those exchanges that are possible. For example, a fixed conducting wall allows equilibration of temperature but not of pressure. If the wall is fixed, adiabatic, and impermeable, there is no exchange of any kind. In this case, each part establishes its own equilibrium independently of the other.

Extensive and Intensive Properties

In thermodynamics we encounter various properties, for example, density, volume, heat capacity, and others that will be defined later. In general, *property* is any quantity that can be measured in a system at equilibrium. Certain properties depend on the actual amount of matter (size or extent of the system) that is used in the measurement. For example, the volume occupied by a substance, or the kinetic energy of a moving object, are directly proportional to the mass. Such properties will be called *extensive*. Extensive properties are additive: if an amount of a substance is divided into two parts, one of volume V_1 and one of volume V_2, the total volume is the sum of the parts, $V_1 + V_2$. In general, the total value of an extensive property in a system composed of several parts is the sum of its parts. If a property is independent of the size of the system, it will be called *intensive*. Some examples are pressure, temperature, density. Intensive properties are independent of the amount of matter and are *not* additive.

As a result of the proportionality that exists between extensive properties and amount of material, the ratio of an extensive property to the amount of material forms an intensive property. If the amount is expressed as mass (in kg or lb), this ratio will be called a *specific* property; if the amount is expressed in mole, it will be called a molar property. For example, if the volume of 2 kg of water at 25 °C, 1 bar, is measured to be 2002 cm³, the specific volume is

$$V = \frac{2002 \times 10^{-6} \text{ m}^3}{2 \text{ kg}} = 1.001 \times 10^{-3} \text{ m}^3/\text{kg} = 1.001 \text{ cm}^3/\text{g},$$

and the molar volume is

$$V = \frac{2002 \times 10^{-3} \text{ m}^3}{2 \text{ kg}} \times \frac{18 \times 10^{-3} \text{ kg}}{\text{mol}} = 1.8018 \times 10^{-5} \frac{\text{m}^3}{\text{mol}} = 18.018 \frac{\text{cm}^3}{\text{mol}}.$$

In general for any extensive property F we have a corresponding intensive (specific or molar) property:

$$\begin{aligned} F_{\text{extensive}} &= F_{\text{molar}} \times (\text{total number of moles}) \\ &= F_{\text{specific}} \times (\text{total mass}). \end{aligned} \quad (1.3)$$

1.3 Definitions

The relationship between specific and molar property is

$$F_{\text{molar}} = F_{\text{specific}} \cdot M_m, \tag{1.4}$$

where M_m is the molar mass (kg/mol).

NOTE

Nomenclature
We will refer to properties like volume as extensive, with the understanding that they have an intensive variant. The symbol V will be used for the intensive variant, whether molar or specific. The total volume occupied by n mole (or m mass) of material will be written as V^{tot}, nV, or mV. No separate notation will be used to distinguish molar from specific properties. This distinction will be made clear by the context of the calculation.

State of Pure Component

Experience teaches that if we fix temperature and pressure, all other intensive properties of a pure component (density, heat capacity, dielectric constant, etc.) are fixed. We express this by saying that the state of a pure substance is fully specified by temperature and pressure. For the molar volume V, for example, we write

$$V = V(T, P), \tag{1.5}$$

which reads "V is a function of T and P." The term *state function* will be used as a synonym for "thermodynamic property." If eq. (1.5) is solved for temperature, we obtain an equation of the form

$$T = T(P, V), \tag{1.6}$$

which reads "T is a function of P and V." It is possible then to define the state using pressure and molar volume as the defining variables, since knowing pressure and volume allows us to calculate temperature. Because all properties are related to pressure and temperature, the state may be defined by any combination of two intensive variables, not necessarily T and P. Temperature and pressure are the preferred choice, as both variables are easy to measure and control in the laboratory and in an industrial setting. Nonetheless, we will occasionally consider different sets of variables, if this proves convenient.

NOTE

Fixing the State
If two intensive properties are known, the state of single-phase pure fluid is fixed, i.e., all other intensive properties have fixed values and can be obtained either from tables or by calculation.

State of Multicomponent Mixture The state of a multicomponent mixture requires the specification of *composition* in addition to temperature and pressure. Mixtures will be introduced in Chapter 8. Until then the focus will be on single components.

Process and Path

The thermodynamic plane of pure substance is represented by two axes, T and P. A point on this plane represents a state, its coordinates corresponding to the temperature and pressure of the system. The typical problem in thermodynamics involves a system undergoing a change of state: a fixed amount of material at temperature T_A and pressure P_A is subjected to heating/cooling, compression/expansion, or other treatments to final state (T_C, P_C). A change of state is called a *process*. On the thermodynamic plane, a process is depicted by a *path*, namely, a line of successive states that connect the initial and final state (see Figure 1-6). Conversely, any line on this plane represents a process that can be realized experimentally. Two processes that are represented by simple paths on the TP plane are the constant-pressure (or isobaric) process, and the constant-temperature (or isothermal) process. The constant-pressure process is a straight line drawn at constant pressure (path AB in Figure 1-6); the constant-temperature process is drawn at constant temperature (path BC). Any two points on the TP plane can be connected using a sequence of isothermal and isobaric paths.

Processes such as the constant-pressure, constant-temperature, and constant-volume process are called *elementary*. These are represented by simple paths during which one state variable (pressure, temperature, volume) is held constant. They

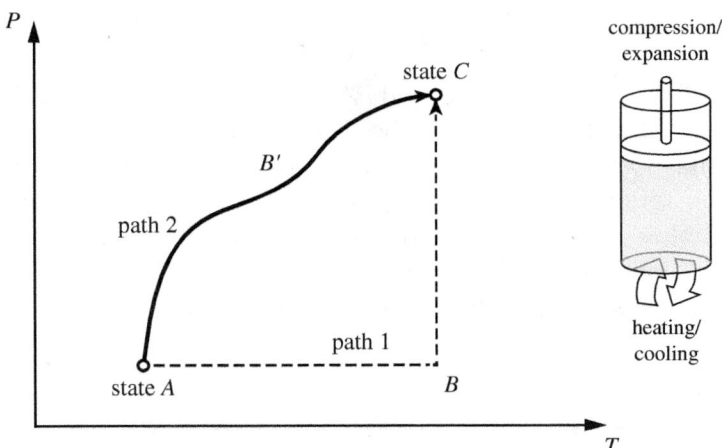

Figure 1-6: Illustration of two different paths (ABC, $AB'C$) between the same initial (A) and final (C) states. Paths can be visualized as processes (heating/cooling, compression/expansion) that take place inside a cylinder fitted with a piston.

1.3 Definitions

are also simple to conduct experimentally. One way to do this is using a cylinder fitted with a piston. By fitting the piston with enough weights we can exert any pressure on the contents of the cylinder, and by making the piston movable we allow changes of volume due to heating/cooling to take place while keeping the pressure inside the cylinder constant. To conduct an isothermal process we employ the notion of a *heat bath*, or heat reservoir. Normally, when a hot system is used to supply heat to a colder one, its temperature drops as a result of the transfer of heat. If we imagine the size of the hot system to approach infinity, any finite transfer of heat to (or from) another system represents an infinitesimal change for the large system and does not change its temperature by any appreciable amount. The ambient air is a practical example of a heat bath with respect to small exchanges of heat. A campfire, for example, though locally hot, has negligible effect on the temperature of the air above the campsite. The rising sun, on the other hand, changes the air temperature appreciably. Therefore, the notion of an "infinite" bath must be understood as relative to the amount of heat that is exchanged. A constant-temperature process may be conducted by placing the system into contact with a heat bath. Additionally, the process must be conducted in small steps to allow for continuous thermal equilibration. The constant-volume process requires that the volume occupied by the system remain constant. This can be easily accomplished by confining an amount of substance in a rigid vessel that is completely filled. Finally, the adiabatic process may be conducted by placing thermal insulation around the system to prevent the exchange of heat.

We will employ cylinder-and-piston arrangement primarily as a mental device that allows us to visualize the mathematical abstraction of a path as a physical process that we could conduct in the laboratory.

Quasi-Static Process

At equilibrium, pressure and temperature are uniform throughout the system. This ensures a well-defined state in which, the system is characterized by a single temperature and single pressure, and represented by a single point on the PT plane. If we subject the system to a process, for example, heating by placing it into contact with a hot source, the system will be temporarily moved away from equilibrium and will develop a temperature gradient that induces the necessary transfer of heat. If the process involves compression or expansion, a pressure gradient develops that moves the system and its boundaries in the desired direction. During a process the system is not in equilibrium and the presence of gradients implies that its state cannot be characterized by a single temperature and pressure. This introduces an inconsistency in our depiction of processes as paths on the TP plane, since points on this plane represent equilibrium states of well-defined pressure and temperature. We resolve this difficulty by requiring the process to take place in a special way, such that the displacement of the system from equilibrium is *infinitesimally* small. A process conducted in such manner is called *quasi static*. Suppose we want to increase the temperature of the system from T_1 to T_2. Rather than contacting the system with a bath at temperature T_2, we use a bath at temperature $T_1 + \delta T$, where δT is a small number, and let the system equilibrate with the bath. This ensures that the temperature of the system is nearly uniform (Fig. 1-7). Once the system is equilibrated to temperature $T_1 + \delta T$, we place it into thermal contact with another bath at temperature $T_1 + 2\delta T$, and repeat the process until the final desired temperature is reached. Changes in pressure are conducted in the same manner. In general, in a quasi-static process we apply small changes at a time and wait between changes for the system to equilibrate. The name derives from the Latin *quasi* ("almost") and implies that the process occurs as if the system remained at a stationary equilibrium state.

Quasi Static is Reversible A process that is conducted in quasi-static manner is essentially at equilibrium at every step along the way. This implies that the system can retrace its path if all inputs (temperature and pressure differences) reverse sign. For this reason, the quasi-static process is also a *reversible* process. If a process is conducted under large gradients of pressure and temperature, it is neither quasi static nor reversible. Here is an exaggerated example that demonstrates this fact. If an inflated balloon is punctured with a sharp needle, the air in the balloon will escape and expand to the conditions of the ambient air. This process is *not* quasi static because expansion occurs under a nonzero pressure difference between the air in the ballon and the air outside. It is not reversible either: we cannot bring the deflated balloon back to the inflated state by reversing the action that led to the expansion, i.e., by "de-puncturing" it. We can certainly restore the initial state by patching the balloon and blowing air into it, but this amounts to performing an entirely different process. The same is true in heat transfer. If two systems

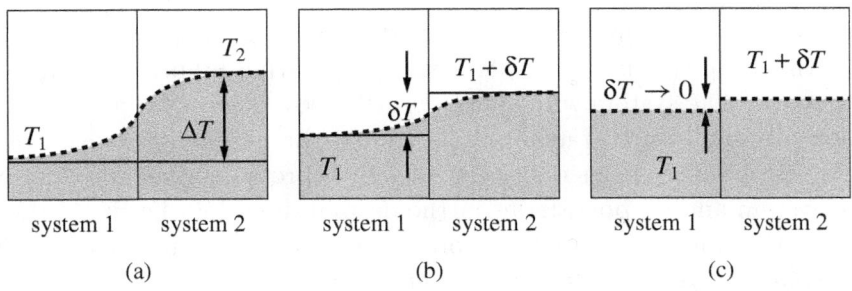

Figure 1-7: (a) Typical temperature gradient in heat transfer. (b) Heat transfer under small temperature difference. (c) Quasi-static idealization: temperatures in each system are nearly uniform and almost equal to each other.

1.3 Definitions

exchange heat under a finite (nonzero) temperature difference, as in Figure 1-7(a), reversing ΔT is not sufficient to cause heat to flow in the reverse direction because the temperature gradient inside system 1 continues to transfer heat in the original direction. For a certain period of time the left side of system 1 will continue to receive heat until the gradient adjusts to the new temperature of system 2. Only when a process is conducted reversibly is it possible to recover the initial state by *exactly retracing* the forward path in the reverse direction. The quasi-static way to expand the gas is to perform the process against an external pressure that resists the expansion and absorbs all of the work done by the expanding gas. To move in the forward direction, the external pressure would have to be slightly lower than that of the gas; to move in the reverse direction, it would have to be slightly higher. In this manner the process, whether expansion or compression, is reversible. The terms *quasi static* and *reversible* are equivalent but not synonymous. *Quasi static* refers to *how* the process is conducted (under infinitesimal gradients); *reversible* refers to the characteristic property that such process can retrace its path exactly. The two terms are equivalent in the sense that if we determine that a process is conducted in a quasi-static manner we may conclude that it is reversible, and vice versa. In practice, therefore, the two terms may be used interchangeably.

NOTE

About the Quasi-Static Process
The quasi-static process is an idealization that allows us to associate a path drawn on the thermodynamic plane with an actual process. It is a mental device that we use to draw connections between mathematical operations on the thermodynamic plane and real processes that can be conducted experimentally. Since this is a mental exercise, we are not concerned as to whether this is a practical way to run the process. In fact, this is a rather impractical way of doing things: Gradients are desirable because they increase the rate of a process and decrease the time it takes to perform the task. This does not mean that the quasi-static concept is irrelevant in real life. When mathematics calls for an infinitesimal change, nature is satisfied with a change that is "small enough." If an actual process is conducted in a way that does not upset the equilibrium state too much, it can then be treated as a quasi-static process.

Example 1.3: The Cost of Doing Things Fast
You want to carry a cup of water from the first floor to your room on the second floor, 15 feet up. How much work is required?

Solution A cup holds about 250 ml, or 0.25 kg of water. The potential energy difference between the first and second floor is

$$\Delta E_p = mgh = (0.25 \text{ kg})(9.81 \text{ m/s}^2)(15 \text{ ft})(0.3048 \text{ m/ft}) = 11.2 \text{ J}.$$

Therefore, the required amount of work is 11.2 J—*but only if the process is conducted in a quasi-static way!* Picture yourself hurrying up the stairs: The surface of the liquid is not level but forms ripples that oscillate. The liquid might even spill if you are a little careless, by jumping over the brim. This motion of the liquid takes energy that was not accounted for in the previous calculation. By the time you reach your room you will have consumed *more* than 11.2 J. Suppose you want to recover the work you just spent. You could do this by dropping the glass to the first floor onto some mechanical device with springs and other mechanical contraptions designed to capture this work. However, you will only be able to recover the 11.2 J of potential energy. The additional amount that went into producing the ripples will be unavailable. This energy is not lost; it is turned into internal energy, as we will see in Chapter 3. You can still extract it, but this will require more work than the amount you will recover, as we will learn in Chapter 4.

Here is the quasi-static way to conduct this process: Take small, careful steps, trying to avoid getting the liquid off its level position. That way the liquid will always stay level (in equilibrium) and the amount of work you will do will exactly match the calculation. Of course, this will take more time, but it will require no more than the theoretical work. The point is that, doing things in a nonquasi-static manner is fast but also carries a hidden cost in terms of extra work associated with the presence of gradients and nonequilibrium states. Doing things in a quasi-static manner is inconvenient, but more efficient, because it does not include any hidden costs.

1.4 Units

Throughout this book we will use primarily the SI system of units with occasional use of the American Engineering system. The main quantities of interests are pressure, temperature, and energy. These are briefly reviewed below. Various physical constants that are commonly used in thermodynamics are shown in Table 1-1.

Pressure Pressure is the ratio of the force acting normal to a surface, divided by the area of the surface. In thermodynamics, pressure generates the forces that give

Table 1-1: Thermodynamic constants

Avogadro's number	$N_A = 6.022 \times 10^{23}$ mol^{-1}
Boltzmann's constant	$k_B = 1.38 \times 10^{-23}$ J/K
Ideal-gas constant	$R = k_B N_A = 8.314$ J/mol K
Absolute zero	0 K = 0 R = -273.15 °C = -459.67 °F

1.4 Units

rise to mechanical work. The SI unit of pressure is the pascal, Pa, and is defined as the pressure generated by 1 N (newton) acting on a 1 m² area:

$$\text{Pa} = \frac{\text{N}}{\text{m}^2} = \frac{\text{J}}{\text{m}^3}. \tag{1.7}$$

The last equality in the far right is obtained by writing $\text{J} = \text{N} \cdot \text{m}$. The pascal is an impractically small unit of pressure because 1 N is a small force and 1 m² is a large area. A commonly used multiple of Pa is the bar:

$$1 \text{ bar} = 10^5 \text{ Pa} = 100 \text{ kPa} = 0.1 \text{ MPa}. \tag{1.8}$$

An older unit of pressure, still in use, is the Torr, or mm Hg, representing the hydrostatic pressure exerted by a column of mercury 1 mm high. In the American Engineering system of units, pressure is measured in pounds of force per square inch, or psi. The relationship between the various units can be expressed through their relationship to the standard atmospheric pressure:

$$1 \text{ atm} = \begin{cases} 1.013 \text{ bar} = 1.013 \times 10^5 \text{ Pa} \\ 760 \text{ Torr} = 760 \text{ mm Hg} \\ 14.696 \text{ psi} \end{cases} \tag{1.9}$$

Temperature Temperature is a fundamental property in thermodynamics. It is a measure of the kinetic energy of molecules and gives rise the sensation of "hot" and "cold." It is measured using a thermometer, a device that obtains temperature indirectly by measuring some property that is a sensitive function of temperature, for example, the volume of mercury inside a capillary (mercury thermometer), the electric current between two different metallic wires (thermocouple), etc. In the SI system, the absolute temperature is a fundamental quantity (dimension) and its unit is the kelvin (K). In the American Engineering system, absolute temperature is measured in rankine (R), whose relationship to the kelvin is,

$$1 \text{ K} = 1.8 \text{ R}. \tag{1.10}$$

Temperatures measured in absolute units are always positive. The absolute zero is a special temperature that cannot be reached except in a limiting sense.

In practice, temperature is usually measured in empirical scales that were originally developed before the precise notion of temperature was clear. The two most widely used are the Celsius scale and the Fahrenheit. They are related to each other and to the absolute scales as follows:[4]

$$\frac{T}{°F} = 1.8 \frac{T}{°C} + 32 \tag{1.11}$$

---

4. The notation $T/°C$ reads, "numerical value of temperature expressed in °C." For example, if $T = 25$ °C, then $T/°C$ is 25.

$$\frac{T}{\text{K}} = \frac{T}{^\circ\text{C}} + 273.15 \tag{1.12}$$

$$\frac{T}{\text{R}} = \frac{T}{^\circ\text{F}} + 459.67, \tag{1.13}$$

where the subscript in T indicates the corresponding units. The units of absolute temperature are indicated without the degree (°) symbol, for example, K or R; the units in the Celsius and Fahrenheit scales include the degree (°) symbol, for example, °C, °F. Although temperatures are almost always measured in the empirical scales Celsius or Fahrenheit, it is the absolute temperature that must be used in all thermodynamics equations.

Mole (mol, gmol, lb-mol) The mole[5] is a defined unit in the SI system such that 1 mol is an amount of matter that contains exactly N_A molecules, where $N_A = 6.022 \times 10^{23}$ mol^{-1} is Avogadro's number.[6] The mass of 1 mol is the *molar mass* and is numerically equal to the molecular weight multiplied by 10^{-3}kg. For example, the molar mass of water (molecular weight 18.015) is

$$M_m = 18.015 \times 10^{-3} \text{ kg/mol}. \tag{1.14}$$

The symbol M_m will be used to indicate the molar mass. The number of moles n that correspond to mass m is

$$n = \frac{m}{M_m}. \tag{1.15}$$

The pound-mol (lb-mol) is the analogous unit in the American Engineering system and represents an amount of matter equal to the molecular weight expressed in lbm. The relationship between the mol and lb-mol is

$$1 \text{ lb-mol} = 454 \text{ mol}. \tag{1.16}$$

Energy The SI unit of energy is the joule, defined as the work done by a force 1 N over a distance of 1 m, also equal to the kinetic energy of a mass 1 kg with velocity 1 m/s:

$$\text{J} = \text{N} \cdot \text{m} = \frac{\text{kg} \cdot \text{m}^2}{\text{s}^2}. \tag{1.17}$$

The kJ (1 kJ = 1000 J) is a commonly used multiple.

As a form of energy, heat does not require its own units. Nonetheless, units specific to heat remain in wide use today, even though they are redundant and

---

5. The symbol of the unit is "mol" but the name of the unit is "mole," much like the unit of SI temperature is the kelvin but the symbol is K.
6. The units for Avogadro's number are number of molecules/mol, and since the number of molecules is dimensionless, 1/mol.

1.4 Units

require additional conversions when the calculation involves both heat and work. These units are the cal (calorie) and the Btu (British thermal unit) and are related to the joule through the following relationships:

$$\text{cal} = 4.18 \text{ J} = 4.18 \times 10^{-3} \text{ kJ} \tag{1.18}$$

$$\text{Btu} = 1.055 \times 10^3 \text{ J} = 1.055 \text{ kJ}. \tag{1.19}$$

Some unit conversions encountered in thermodynamics are shown in Table 1-2.

Table 1-2: Common units and conversion factors

Magnitude	Definition	Units	Other Units and Multiples
Length	–	m	1 cm = 10^{-2} m
			1 ft = 0.3048 m
			1 in = 2.54×10^{-2} m
Mass	–	kg	1 g = 10^{-3} kg
			1 lb = 0.4536 kg
Time	–	s	1 min = 60 s
			1 hr = 3600 s
Volume	(length)^3	m^3	1 cm^3 = 10^{-6} m^3
			1 L = 10^{-3} m^3
Force	(mass)×(acceleration)	N = kg m s^{-2}	1 lbf = 4.4482 N
Pressure	(force)/(area)	Pa = N m^{-2}	1 Pa = 10^{-3} kJ/m^3
	= (energy)/(volume)		1 bar = 10^5 Pa
			1 psi = 0.06895 bar
Energy	(force)×(length)	J = kg m^2 s^{-2}	1 kJ = 10^3 J
	= (mass)×(velocity)^2		1 Btu = 1.055 kJ
Specific energy	(energy)/(mass)	J/kg	1 Btu/lbm = 2.3237 kJ/kg
Power	(energy)/(time)	W = J/s	1 Btu/s = 1.055 kW
			1 hp = 735.49 W

1.5 Summary

Thermodynamics arises from the physical interaction between molecules. This interaction gives rise to temperature as a state variable, which, along with pressure, fully specifies the thermodynamic state of a pure substance. Given pressure and temperature, all intensive properties of a pure substance are fixed. This means that we can measure them once and tabulate them as a function of pressure and temperature for future use. Such tabulations exist for many substances over a wide range of conditions. Nonetheless, for engineering calculations it is convenient to express properties as mathematical functions of pressure and temperature. This eliminates the need for new experimental measurements—and all the costs associated with raw materials and human resources—each time a property is needed at conditions that are not available from tables. One goal of chemical engineering thermodynamics is to provide rigorous methodologies for developing such equations.

Strictly speaking, thermodynamics applies to systems in equilibrium. When we refer to the pressure and temperature of a system we imply that the system is in equilibrium so that it is characterized by a single (uniform) value of pressure and temperature. Thermodynamics also applies rigorously to quasi-static processes, which allow the system to maintain a state of almost undisturbed equilibrium throughout the entire process.

1.6 Problems

Problem 1.1: The density of liquid ammonia (NH_3) at 0 °F, 31 psi, is 41.3 lb/ft^3.
a) Calculate the specific volume in ft^3/lb, cm^3/g and m^3/kg.
b) Calculate the molar volume in ft^3/lbmol, cm^3/mol and m^3/mol.

Problem 1.2: The equation below gives the boiling temperature of isopropanol as a function of pressure:

$$T = \frac{B}{A - \log_{10} P} - C,$$

where T is in kelvin, P is in bar, and the parameters A, B, and C are

$$A = 4.57795, \quad B = 1221.423, \quad C = -87.474.$$

1.6 Problems

Obtain an equation that gives the boiling temperature in °F, as a function of $\ln P$, with P in psi. *Hint:* The equation is of the form

$$T = \frac{B'}{A' - \ln P} - C'$$

but the constants A', B', and C' have different values from those given above.

Problem 1.3: a) At 0.01 °C, 611.73 Pa, water coexists in three phases, liquid, solid (ice), and vapor. Calculate the mean thermal velocity ($\bar{v}$) in each of the three phases in m/s, km/hr and miles per hour.
b) Calculate the mean translational kinetic energy contained in 1 kg of ice, 1 kg of liquid water, and 1 kg of water vapor at the triple point.
c) Calculate the mean translational kinetic energy of an oxygen molecule in air at 0.01 °C, 1 bar.

Problem 1.4: The intermolecular potential of methane is given by the following equation:

$$\Phi(r) = a\left[\left(\frac{\sigma}{r}\right)^{12} - \left(\frac{\sigma}{r}\right)^{6}\right]$$

with $a = 2.05482 \times 10^{-21}$ J, $\sigma = 3.786$ Å, and r is the distance between molecules (in Å).
a) Make a plot of this potential in the range $r = 3$ Å to 10 Å.
b) Calculate the distance r_* (in Å) where the potential has a minimum.
c) Estimate the density of liquid methane based on this potential.
Find the density of liquid methane in a handbook and compare your answer to the tabulated value.

Problem 1.5: a) Estimate the mean distance between molecules in liquid water. Assume for simplicity that molecules sit on a regular square lattice.
b) Repeat for steam at 1 bar, 200 °C (density 4.6×10^{-4} g/cm^3).
Report the results in Å.

Problem 1.6: In 1656, Otto von Guericke of Magdeburg presented his invention, a vacuum pump, through a demonstration that became a popular sensation. A metal sphere made of two hemispheres (now known as the Magdeburg hemispheres) was evacuated, so that a vacuum would hold the two pieces together. Von Guericke would then have several horses (by one account, 30 of them, in two teams of 15) pulling, unsuccessfully, to separate the hemispheres. The demonstration would end with the opening of a valve that removed the vacuum and allowed the hemispheres to separate. Suppose that the diameter of the sphere is 50 cm and the sphere is completely evacuated. The sphere is hung from the ceiling and you pull the other half with the force of your body weight. Will the hemispheres come apart? Support your answer with calculations.

Chapter 2
Phase Diagrams of Pure Fluids

The strength of molecular interactions is determined by the mean intermolecular distance and the property that most directly reflects this distance is molar density, or its reciprocal, molar volume. The approximate relationship between molar volume and mean intermolecular distance is given by (see Example 1.1),

$$r \sim \left(\frac{V}{N_{\text{Av}}}\right)^{1/3},$$

where N_A is Avogadro's number. The packing density of molecules in given volume reflects the strength of the potential interaction. In gases (large V), distances are large and interactions weak. In liquids, the opposite is true. The volume that is occupied by a fixed number of molecules depends both on temperature and pressure. The relationship between volume, pressure, and temperature is of fundamental importance and its mathematical form is known as *equation of state*. In this chapter we examine this relationship in graphical and mathematical form. The learning objectives of this chapter are to develop the following skills:

1. Using the PVT graph to identify the phase of a pure fluid.
2. Working with tabulated values of P, V, T (steam tables).
3. Identify the region of applicability of the ideal-gas law and the truncated virial equation.
4. Working with cubic equations of state.
5. Working with generalized correlations for the compressibility factor.
6. Representing processes on the PV graph.

2.1 The *PVT* Behavior of Pure Fluid

The molar volume, V, is the volume occupied by 1 mol of the substance, and is the inverse of the molar density, ρ:[1]

$$V_{\text{molar}} = \frac{1}{\rho}, \qquad (2.1)$$

---

[1]. We will use the same general symbol, ρ, for both the molar density and the mass density. We will annotate them differently only if they both appear in the same equation.

The *specific* volume is the volume occupied by 1 kg of substance. The specific volume, the mass density, and the molar densities are related to each other:

$$V_{\text{specific}} = \frac{1}{\rho'} = \frac{1}{M_m \rho} = \frac{V_{\text{molar}}}{M_m}, \tag{2.2}$$

where ρ' is the mass density (kg/m^3) and M_m is molar mass (kg/mol). The volume occupied by a given amount of matter depends on temperature and pressure: it decreases under compression and (for most substances) increases upon heating. The relationship between P, V, T is characteristic of a substance but the general features of this relationship are common to all pure fluids and will be examined below.

The relationship between volume, pressure, and temperature is represented graphically by a three-dimensional surface whose general shape is shown in Figure 2-1. This graph has been rotated to show pressure in the vertical axis, with the mesh lines on the surface representing lines of constant temperature (left to right), and lines of constant volume (front to back). A point on the surface gives the molar volume at the indicated pressure and temperature. The bell-shaped curve facing the pressure-volume plane is the vapor liquid boundary. To its left is the liquid region (steep part of the surface), to its right the vapor region. At sufficiently low temperatures a system exhibits a phase transition to a solid; this region is not shown on this graph.[2]

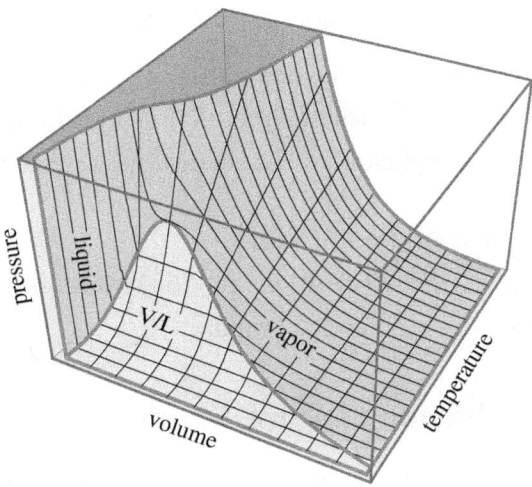

Figure 2-1: The *PVT* surface of a pure fluid.

---

2. The surface in Figure 2-1 was calculated using the Soave-Redlich-Kwong equation of state (see Section 2.6), which is appropriate for liquids and gases but not for solids.

2.1 The PVT Behavior of Pure Fluid

The three-dimensional representation is useful for the purposes of visualizing the PVT relationship but is impractical for routine use. We work, instead, with projections of the PVT surface on one of the three planes, PV, PT, or VT. A projection is a view of the three-dimensional surface from an angle perpendicular to the projection plane and reduces the graph into a two-dimensional plot. The most commonly used projections are those on the PV and the PT planes.

The PV Graph

The PV graph is the projection on the PV plane and is shown in Figure 2-2. The characteristic feature of this graph is the vapor-liquid region, represented by a bell-shaped curve that consists of two branches, saturated liquid (to the left) and saturated vapor (to the right). The two branches meet at the top and this point defines the *critical point* of the fluid. Temperature is indicated by contours of constant temperature (isotherms). These are lines with the general direction from the upper left corner to the lower right. The isotherm that passes through the critical point corresponds to the critical temperature, T_C. This isotherm has an inflection point at the critical point, namely, its first and second derivatives are both zero:

$$\left(\frac{\partial P}{\partial V}\right)_T = \left(\frac{\partial^2 P}{\partial V^2}\right)_T = 0, \quad \text{at } V = V_c, \; T = T_c. \tag{2.3}$$

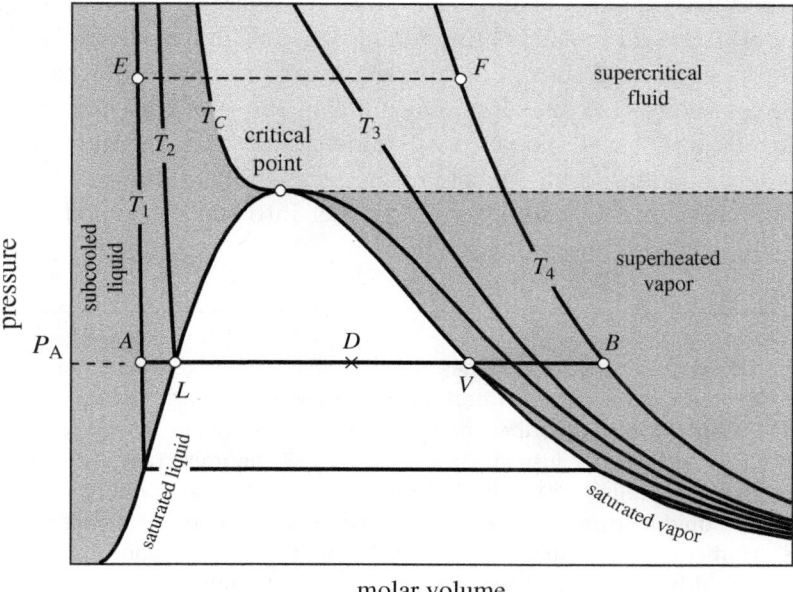

Figure 2-2: The PV graph of pure fluid (the solid phase is not shown).

From there on, it decreases smoothly into the vapor region. The region to the left of the saturated liquid is called *subcooled* liquid to indicate that its temperature is below the boiling point that corresponds to its pressure. For example, at A the temperature is T_1, lower than the saturation temperature (T_2) that corresponds to pressure P_A. Alternatively, it is called *compressed* liquid to indicate that pressure is higher than the saturation pressure that corresponds to its temperature (the terms subcooled and compressed liquid are used interchangeably). Vapor to the right of the saturated line is *superheated* because its temperature is higher than the boiling temperature that corresponds to its pressure. At B, for example, temperature (T_4) is higher than the saturation temperature (T_2) that corresponds to its pressure.

The organization of information on the PV graph can be better understood by conducting a heating or cooling process and following the path on the graph. Suppose we add heat under constant pressure starting with liquid at state A, which is at pressure P_A and temperature $T_A = T_1$. The process is depicted by the line AB, drawn at constant pressure P_A. Between states A and L, heating causes the volume to increase somewhat but the increase is relatively small because the thermal expansion of liquids is small. At point L the liquid is saturated and at the verge of boiling. Adding heat at this point causes liquid to evaporate and produce more vapor, moving the state along the line LV. During this process both pressure and temperature remain constant (line LV is both an isotherm and an isobar). At point V all the liquid has evaporated and the system is saturated vapor. Adding more heat causes temperature and molar volume to increase and moves the state along the line VB. If we start at state B and perform a constant-pressure cooling process, we will observe the reverse course of events. During BV, the vapor is cool. At point V, the state is saturated vapor at the verge of condensation. Removing heat at this point causes vapor to condense until the steam becomes 100% saturated liquid (state L). Upon further cooling, the system moves further into the subcooled region.

NOTE

Boiling in Open Air

If the heating/cooling process that is described here is conducted in an open container, for example, by heating water in an open flask at atmospheric pressure, the behavior will be somewhat different than the one described here. In an open container, water forms vapor at any temperature below boiling, not just at the boiling point. The important difference is that in an open container we are dealing with a *multicomponent* system that contains not only water, but also *air*. A vapor-liquid mixture with two or more components behaves differently from the pure components. Multicomponent phase equilibrium is treated in the second part of this book and until then, it should be understood that we are dealing with pure fluids. The process described by the path AB may be thought to take place inside a sealed cylinder fitted with a piston and initially filled with liquid containing no air at all.

2.1 The PVT Behavior of Pure Fluid

The Critical Point The critical point is an important state and its pressure and temperature have been tabulated for a large number of pure substances. In approaching the critical point from below, the distance between points L and V decreases, indicating that the molar volume of the saturated liquid and saturated vapor come closer together. At the critical point the two saturated phases coincide: vapor and liquid become indistinguishable and the phase boundary seizes to exist. The region of the phase diagram above P_c and T_c is referred to as supercritical fluid. No isotherms or isobars in this region intersect the vapor-liquid boundary. If point E is heated isobarically to final state F, one will observe a continuous transition from a dense, liquid-like state, to a dilute, gaslike state. In the supercritical region the notions of "liquid" and "vapor" are not helpful. These terms are meaningful when both phases can exist simultaneously and can be identified as distinct from each other. The term *supercritical fluid* avoids these ambiguities.

Properties near the critical point are quite different compared to states at lower temperatures and pressures. As the difference between vapor and liquid becomes less clear near the critical point, the liquid becomes substantially more compressible than typical liquids. This is indicated on the PV graph by the gentle slope of the isotherm as it approaches the critical point. Isotherms below but near the critical temperature (not shown in Figure 2-2) show similar behavior. The usual approximation that treats liquids as incompressible is acceptable only at temperatures well below the critical. In the supercritical region, the behavior of a fluid is somewhere between that of a liquid and a gas. The gentle slope of the isotherms indicates that the fluid is quite compressible, even at high, liquidlike densities (low molar volumes). Other properties, in particular, the solubility of various nonvolatile solutes, are often found to be quite enhanced compared to the subcritical region. As an example, the enhanced solubility of caffeine in supercritical carbon dioxide ($T_c = 304.1$ K, $P_c = 73.8$ bar) makes it possible to use carbon dioxide as a solvent to extract caffeine from coffee, thus avoiding the use of other solvents with potential toxic effects.

A Special Limit: The Ideal-Gas State If the molar volume is increased sufficiently, the effect of molecular interactions decreases, and in the limit that it becomes infinite, it vanishes completely. When this condition is met we say that the system is in the *ideal-gas state*. The mathematical specification of the ideal-gas state is :

$$V \to \infty, \quad \text{at constant } T, \tag{2.4}$$

or, equivalently,

$$P \to 0, \quad \text{at constant } T. \tag{2.5}$$

The stipulation of constant temperature is necessary. Without it, it would be possible to maintain the system in the liquid (or even solid) phase, even at very low pressures, thus never reaching a state where intermolecular distances are large. On the PV graph, the ideal-gas state is found near the lower-right corner.

In the ideal-gas state, the PVT relationship is universal for all gases, regardless of chemical composition and this relationship is given by the ideal-gas law:

$$\boxed{PV^{\text{ig}} = RT,} \qquad \text{ideal-gas state} \qquad (2.6)$$

where V^{ig} is *molar* volume, T is absolute temperature, and R is a universal constant (ideal-gas constant), whose value in the SI system is

$$R = 8.314 \text{ J/mol K}. \qquad (2.7)$$

The superscript "ig" will be used to indicate results that are valid *only* in the ideal-gas state. The ideal-gas law should be viewed as the limiting form of the equation of state of any real fluid when pressure is reduced under constant temperature. Even though the ideal-gas state represents an idealization (infinite distance between molecules), in practice eq. (2.6) provides satisfactory results if the actual state of a gas is sufficiently close to the ideal-gas state. More often than not, engineering problems require calculations at conditions where the ideal-gas law is not valid. It is important to be aware of the limitations of the ideal-gas law and never use it without proper justification.

NOTE

Directions to the Ideal-Gas State
The specification in eq. (2.5) offers simple directions to the ideal-gas state from any initial state: move along the current isotherm until the phase is vapor and pressure is sufficiently low (or the volume sufficiently large). The pressure does not have to be absolute vacuum as long as it is "sufficiently low." What constitutes a sufficiently low pressure will be answered in Section 2.4.

Two-Phase Region—The Lever Rule A point E in the two-phase region represents a two-phase system that contains portions of liquid and vapor. Both phases are at the same temperature and pressure but each has its own molar volume. Therefore, the state at E should be viewed as a mixture of states L (saturated liquid) and V (saturated vapor). Line LV that connects the two pure phases is a *tie line*. Points along the tie line have the same temperature and pressure but differ with respect to the amount of liquid and vapor they contain. A vapor-liquid mixture is also referred to as *wet vapor*. The amount of vapor as a fraction of the total mass is the *quality* of the vapor, x_V. It varies from 0% (point L, saturated liquid) to 100% (point V, saturated vapor). The mass fraction of the liquid is $x_L = 1 - x_V$.

Consider a two-phase system (state E in Figure 2-3) that contains n_L moles of liquid and n_V of vapor. The molar fractions of vapor and liquid are[3]

$$x_V = \frac{n_V}{n_L + n_V}, \quad x_L = \frac{n_L}{n_L + n_V}. \qquad (2.8)$$

---

3. For pure components, mass and molar fraction in a liquid/vapor mixture are the same.

2.1 The PVT Behavior of Pure Fluid

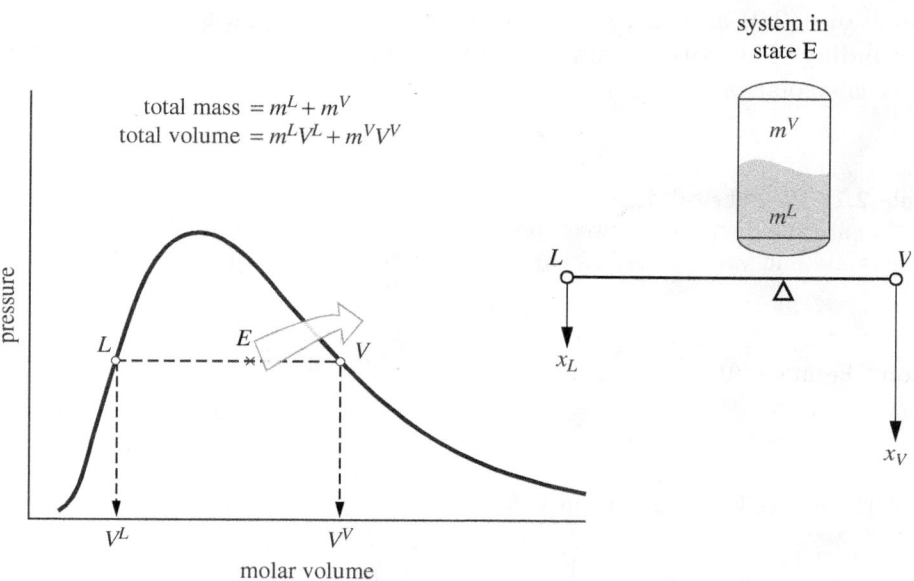

Figure 2-3: Setup for the application of the lever rule on a tie line.

where $n_V + n_L = n$ is the total mass. The total volume of the two-phase mixture is the sum of the liquid and vapor portions:

$$V_{\text{tot}} = n_L V_L + n_V V_V.$$

The molar volume is obtained through division by the total number of moles, $n_{\text{tot}} = n_L + n_V$:

$$V = \frac{V_{\text{tot}}}{n_{\text{tot}}} = \frac{n_L V_L}{n_L + n_V} + \frac{n_V V_V}{n_L + n_V} = x_L V_L + x_V V_V. \qquad (2.9)$$

This equation gives the molar volume of the two-phase system, if the fractions of vapor and liquid are known. Solving for the liquid and vapor fractions and using $x_L + x_V = 1$, we obtain

$$x_V = \frac{V - V_L}{V_V - V_L}, \quad x_L = \frac{V_V - V}{V_V - V_L}. \qquad (2.10)$$

These equations give the fraction of liquid and vapor, if the volume of the two-phase system is known. The two equations in (2.10) are known as the lever rule: if LV is viewed as a lever with force x_L acting on point L, force x_V on point V, and the pivot placed at E, the lever would be in mechanical equilibrium (see Example 2.1).

Equations (2.9) and (2.10) are applicable on a molar or specific basis with the understanding that both volume and phase fractions are to be expressed on the same basis (molar or specific).

Example 2.1: Why "Lever" Rule?
If EV, LE, are the distances between points (E, V) and (L, E) respectively (Fig. 2-3), show that the liquid and vapor fractions corresponding to point E satisfy the relationship

$$x_V(VE) = x_L(LE).$$

Solution From eq. (2.10), the ratio of x_V to x_L is

$$\frac{x_V}{x_L} = \frac{V - V_L}{V_V - V}.$$

The difference $V - V_L$ is the distance LE and the difference $V_V - V$ is the distance VE. Then we have,

$$\frac{x_V}{x_L} = \frac{(LE)}{(VE)} \quad \Rightarrow \quad x_V(VE) = x_L(LE).$$

This result states that if two forces with magnitude x_L and x_V act on points L and V, respectively, their torques with respect to point E are equal. This property gives the name "lever" rule to eq. (2.10). The lever analogy has be used as a mnemonic trick to memorize the equation. No memorization is required, however: eq. (2.10) can be derived easily when needed by applying a straightforward mass balance.

The *PT* Graph

The PT graph is the projection of the three-dimensional surface (Figure 2-1) on the pressure/temperature plane. This graph is shown in Figure 2-4. The vapor-liquid boundary is shown by the line FC and corresponds to the vapor/liquid dome in Figure 2-1 viewed from its side. Line FS marks the solid/liquid boundary. This line is nearly vertical because the melting temperature is not affected strongly by pressure. Unlike the vapor/liquid boundary (FC), which terminates at the critical point, there is no critical point on the solid/liquid boundary. The intersection between the solid/liquid and the liquid/vapor line is the triple point of the fluid. At this point—and only at this point—in the system, all three phases coexist in equilibrium. Most of our applications will be in the liquid and vapor regions, that is, to the right of line FS. Line FC marks the boundary between the vapor phase (to the right)

2.1 The PVT Behavior of Pure Fluid

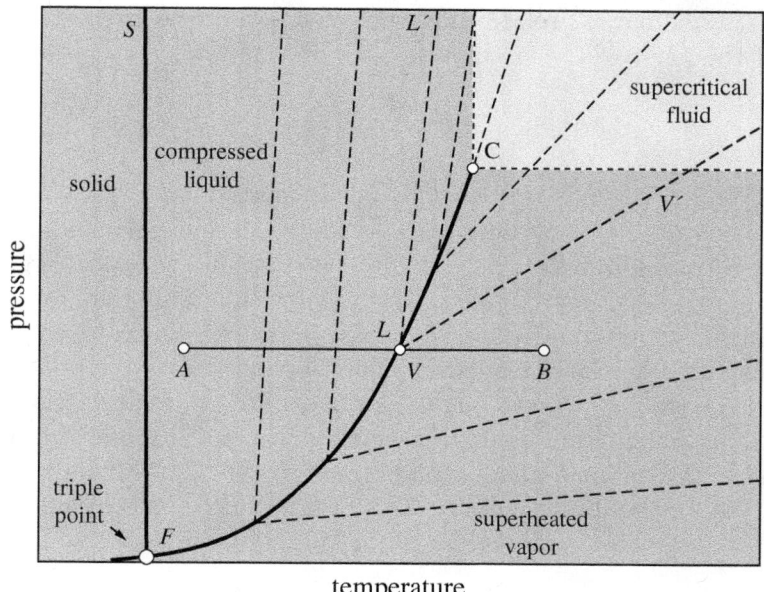

Figure 2-4: Pressure-temperature graph of pure fluid showing the solid, liquid, vapor and supercritical regions. Dotted lines are lines of constant molar volume.

and the liquid (to the left and above). The dotted lines are lines of constant molar volume, also known as isochores. Molar volume increases from left to right, that is, moving on a line of constant pressure from lower temperature to high we encounter increasingly higher molar volumes. Line AB represents a constant-pressure process and corresponds to the same states as in Figure 2-2. Points L and V coincide on this graph (all tie lines are viewed from their edge in this projection). Line LL' is a line at constant volume $V = V_L$, equal to the volume of saturated liquid. Line VV' is a line at constant volume $V = V_V$, (saturated vapor). Although the two lines meet on the saturation line, they correspond to different values of the molar volume, as we recall from the PV graph.

In addition to marking the phase boundary, line FC expresses the relationship between saturation pressure and temperature. The saturation pressure generally increases quickly with temperature up to the critical point. There is no vapor-liquid transition above the critical point; therefore, the relationship between saturation pressure and temperature exists only below the critical point. The saturation pressure of pure component is an important physical property and a required parameter in many calculations of phase equilibria. Several equations have been developed to describe the mathematical relationship between saturation pressure and temperature. One of the most widely used is the Antoine equation:

$$\ln P^{\text{sat}} = A - \frac{B}{C+T} \qquad (2.11)$$

where A, B, and C are numerical constants specific to the substance. These constants have been obtained by numerical fitting against experimental data and are available for a large number of pure fluids. The Antoine equation is generally very accurate, but this accuracy is guaranteed only within a limited range of temperatures. This range must be indicated in the literature source used to obtain the Antoine constants. A recommended resource for parameters of the Antoine equation is the *Properties of Gases and Liquids*, by Poling, Prausnitz and O'Connell (5th ed., 2007), which contains extensive tabulations. A more limited tabulation can be found in the online database maintained by NIST.[4] The boiling temperature at 1 atm is the *normal* boiling point. This is a useful property that is often found in tables.

NOTE

Units in the Antoine Equation
The numerical values of the Antoine parameters depend on the working units of P^{sat} and T, which can vary from source to source. It is important to establish, by checking with the source of the data, what are the proper units of pressure and temperature to be used with a given set of parameters. Also important is to establish whether the logarithm on the left-hand side should be natural or base 10, since parameters may be given for either form of the equation.

Example 2.2: Antoine Equation for Water
The NIST databases give the Antoine equation of water in the form

$$\log_{10} P^{\text{sat}} = A - \frac{B}{T+C}$$

with pressure in bar and temperature in K. The constants for water are

Temperature range (K)	A	B	C
344 - 373	5.08354	1663.13	−45.622
379 - 573	3.55959	643.748	−198.043

---

4. http://webbook.nist.gov.

2.1 The PVT Behavior of Pure Fluid

The first set is valid in the temperature range 344 K to 373 K and the second in the range 379 K to 573 K. Calculate (a) the saturation pressure at 100 °C and (b) the boiling temperature at 5 bar.

Solution (a) Notice that the equation is given in terms of the base-10 logarithm of pressure and that the units of pressure and temperature must be in bar and kelvin, respectively.

By numerical substitution into the Antoine equation using the first set of parameters and $T = 100 + 273.15 = 373.15$ K, we obtain

$$\log_{10} P^{\text{sat}} = 0.00572986 \quad \Rightarrow \quad P^{\text{sat}} = 1.01328 \text{ bar}$$

The result is in excellent agreement with the value 1.01325 bar for the standard atmosphere.

(b) To obtain the boiling temperature of water given the pressure, we solve the Antoine equation for T:

$$T = \frac{B}{A - \log_{10} P^{\text{sat}}} - C. \tag{2.12}$$

We use the second set of parameters since the boiling temperature of water at pressures higher than 1.013 bar is above 373 K. We find

$$T = \frac{643.748}{3.55959 - \log_{10}(5)} - (-198.043) = 423.021 \text{ K} = 149.9 \text{ °C}.$$

This is within the range of the parameters and agrees well with the tabulated value, 151.8 °C.

Example 2.3: Determining the Phase

Based only on the Antoine equation given in Example 2.2, determine the phase of water at (a) 2 bar, 115 °C, and (b) 20 bar, 300 °C.

Solution (a) To answer this question we must either calculate the saturation pressure at the temperature of the system (115 °C), or the saturation temperature at the pressure of the system (2 bar). We use the second set of parameters for these calculations since temperatures are higher than 373 K.

At temperature 115 °C = 388.15 K, the saturation pressure is calculated from the Antoine equation to be 1.49 bar. Since the system pressure (2 bar) is higher, the state is compressed liquid (i.e., the pressure would have to be reduced isothermally to 1.49 bar to cause the liquid to boil).

Alternatively, the saturation temperature calculated from the Antoine equation at pressure 2 bar is 395.6 K = 122.4 °C. Since the system temperature is lower, it is compressed liquid.

(i.e., at constant pressure, temperature would have to be raised to 122.4 °C for the liquid to boil). Both calculations give the same answer.

(b) This part will be answered the same way. Again, the second set of Antoine parameters is used here. The saturation pressure at $T = 300\ °C = 573.15$ K is calculated to be 69.7 bar. Since the pressure of the system (20 bar) is lower than the saturation pressure, we conclude that the phase is superheated vapor.

Alternatively, the saturation temperature at 20 bar is calculated to be 483 K = 210 °C, less than the temperature of the system. From this again we conclude that the phase is superheated vapor.

2.2 Tabulation of Properties

As a state property, the molar (or specific) volume can be determined once as a function of pressure and temperature, and tabulated for future use. Tabulations have been compiled for a large number of pure fluids. In very common use are the steam tables, which contain tabulations of the properties of water. Steam is a basic utility in chemical plants as a heat transfer fluid for cooling or heating, as well as for power generation (pressurized steam), and its properties are needed in many routine calculations. Thermodynamic tables for water are published by the American Society of Mechanical Engineers (ASME) and are available in various forms, printed and electronic. A copy is included in the appendix. We will use them not only because water is involved in many industrial processes but also as a demonstration of how to work with tabulated values in general.

NOTE

Interpolations
When working with tabulated values it is necessary to perform interpolations if the desired conditions lie between entries in the table. Suppose that a table contains the values of a function $f(x)$ at $x = x_1$ and $x = x_2$ and we wish to obtain the value of f at an intermediate point x such that $x_1 < x < x_2$, we assume a linear relationship between f and x and write

$$f = f_1 + \frac{f_2 - f_1}{x_2 - x_1}(x - x_1), \tag{2.13}$$

where f_1, f_2 are the values of the function at points x_1, x_2, respectively. The procedure is shown graphically in Figure 2-5a. If the value of x is outside the interval (x_1, x_2), the same formula may be used and this calculation is called an *extrapolation*. Extrapolations should be avoided because they can be subject to large error. They may be used if the desired value is beyond the last tabulated entry, however, one cannot be certain about the accuracy of the result.

2.2 Tabulation of Properties

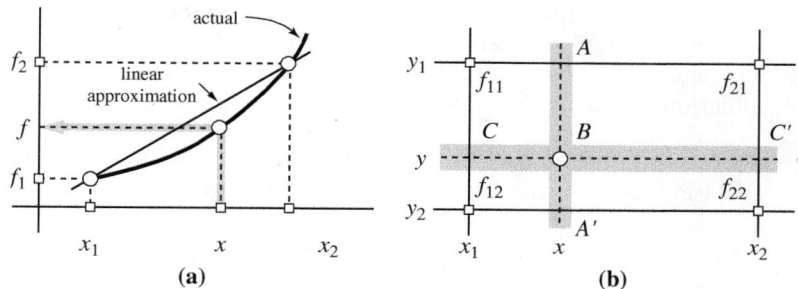

Figure 2-5: Linear interpolation: (a) simple interpolation; (b) double interpolation.

In thermodynamics we are usually dealing with functions of two variables, for example, $f(x, y)$. If point (x, y) is such that the value y is found in the table but the value of x falls between tabulated values, the above equation may be used to interpolate with respect to x. If both x and y fall between tabulated entries, a double interpolation is necessary. The procedure involves three simple interpolations, and can be outlined as follows (Figure 2-5b): first interpolate between the tabulated values at (x_1, y_1) and (x_1, y_2) to obtain the value of f at point C; do the same between points (x_2, y_1) and (x_2, y_2) to obtain f at point C'. Finally, interpolate between C and C' to obtain the value at the desired point B. Alternatively, interpolate to obtain points A and A' followed by interpolation between A and A' to obtain B—the result is the same. These steps can be combined into a single equation which takes the form:

$$f(x, y) = (1-a)(1-b)f_{11} + b(1-a)f_{12} + a(1-b)f_{21} + abf_{22} \qquad (2.14)$$

with

$$a = \frac{x - x_1}{x_2 - x_1}, \qquad b = \frac{y - y_1}{y_2 - y_1}. \qquad (2.15)$$

Here, the notation f_{ij} refers to $f(x_i, y_j)$. If both a and b are between 0 and 1, this calculation is indeed an interpolation and produces a result that is surrounded by the four tabulated values used in the calculation. This equation can be used for extrapolations outside these four values provided the distance is not large.

Example 2.4: Interpolation
Use the steam tables to determine the specific volume of water at 1.25 bar, 185 °C.

Solution From the superheated steam tables we find the following values for the specific volume (in m³/kg):

	$P_1 = 1$ bar	$P_2 = 1.5$ bar
$T_1 = 150$ °C	1.9367	1.2856
$T_2 = 200$ °C	2.1725	1.4445

In this problem, pressure (1.25 bar) and temperature (185 °C) both lie between the tabulated values. Therefore, a double interpolation is required. We do the calculation first by successive interpolations, then by applying equation (2.14) for double interpolations.

Method 1—successive interpolations: We first obtain the volume at $P = 1.25$ bar, $T_1 = 150$ °C by interpolation between the values listed in the first row of the above data:

$$V_1 = 1.9367 + \frac{1.2856 - 1.9367}{1.5 - 1}(1.25 - 1) = 1.6112 \text{ m}^3/\text{kg}.$$

This amounts to obtaining the volume at point C of Figure 2-5. Next, we obtain the volume at $P = 1.25$ bar, $T_1 = 200$ °C by interpolation between the values listed in the second row:

$$V_2 = 2.1725 + \frac{1.4445 - 2.1725}{1.5 - 1}(1.25 - 1) = 1.8085 \text{ m}^3/\text{kg}.$$

This amounts to calculating the volume at point C'. Finally, we interpolate between V_1 (point C) and V_2 (C') to obtain the volume at 1.25 bar, 185 °C:

$$V = 1.6112 + \frac{1.8085 - 1.6112}{200 - 150}(185 - 150) = 1.7493 \text{ m}^3/\text{kg}.$$

The result corresponds to point B in Figure 2-5.

Method II—double interpolation: To apply the double-interpolation formula, we take x to be temperature and y to be pressure, and f_{xy} to be the specific volume at temperature x and pressure y. The factors a and b are calculated from eq. (2.15):

$$a = \frac{185 - 150}{200 - 150} = 0.7, \quad b = \frac{1.25 - 1}{1.5 - 1} = 0.5.$$

The interpolated volume is

$$V = (1 - 0.7)(1 - 0.5)(1.9367) + (0.5)(1 - 0.7)(1.2856) +$$
$$(0.7)(1 - 0.5)(2.1725) + (0.7)(0.5)(1.4445) = 1.7493 \text{ m}^3/\text{kg}.$$

Both methods give the same answer, as they should.

Example 2.5: Locating the State in the Steam Tables
A 12 m³ pressurized vessel contains 200 kg of steam at 40 bar. What is the temperature?

Solution The specific volume of the steam in the tank is

$$V = \frac{12 \text{ m}^3}{200 \text{ kg}} = 0.06 \text{ m}^3/\text{kg}.$$

2.3 Compressibility Factor and the ZP Graph

We know two intensive properties, pressure and specific volume; the state, therefore, is fully specified. We must locate a point in the steam tables with $P = 40$ bar, $V = 0.06$ m³/kg. From the entries at 40 bar we obtain the following data:

$$T_1 = 300 \text{ °C} \quad V_1 = 0.0589 \text{ (m}^3\text{/kg)}$$
$$T_2 = 350 \text{ °C} \quad V_2 = 0.0665 \text{ (m}^3\text{/kg)}.$$

Interpolating for temperature at $V = 0.06$ m³/kg we have

$$T = 300 + \frac{350 - 300}{0.0665 - 0.0589}(0.06 - 0.0589) = 307.2 \text{ °C}.$$

Therefore, the temperature in the tank is 307.2 °C.

Example 2.6: Lever Rule
An additional 170 kg of steam is added to the tank of the previous example. If the final pressure is 50 bar, determine the temperature and phase of the contents of the tank.

Solution The new specific volume is

$$V = \frac{12 \text{ m}^3}{(200 + 170) \text{ kg}} = 0.03243 \text{ m}^3\text{/kg}.$$

This volume lies between the volume of the saturated liquid and saturated vapor at 50 bar:

$$P = 50 \text{ bar} \quad T = 263.94 \text{ °C}: \quad V_L = 0.0012864 \text{ m}^3\text{/kg}, \quad V_V = 0.039446 \text{ m}^3\text{/kg}.$$

The state, therefore, is wet steam at 50 bar, 263.94 °C. To determine the mass fractions of each phase we use the lever rule in eq. (2.10):

$$x_L = \frac{0.039446 - 0.03243}{0.039446 - 0.0012864} = 0.184 = 18.4\%,$$

$$x_V = 1 - x_L = 1 - 0.184 = 0.816 = 81.6\%.$$

Therefore, the quality of the steam in the tank is 81.6%.

2.3 Compressibility Factor and the ZP Graph

The compressibility factor, Z, is defined as the ratio

$$Z = \frac{PV}{RT}, \tag{2.16}$$

where V is the molar volume, P is pressure, and T is absolute temperature. It is a dimensionless quantity and a state function. In the ideal-gas state, $V^{\text{ig}} = RT/P$, and the compressibility factor is unity:

$$Z^{\text{ig}} = 1. \tag{2.17}$$

More precisely, this is the limiting value of the compressibility factor of any real gas when pressure is reduced to zero under constant temperature:

$$\lim_{P \to 0} Z = 1 \quad \text{(constant } T\text{)}. \tag{2.18}$$

This result states that the compressibility factor along a line of constant temperature goes to 1 as P is reduced to zero. In other words, on a graph of Z versus pressure, all isotherms at zero pressure must meet at $Z = 1$.

The compressibility factor represents an alternative way of presenting molar volume, since the molar volume can be obtained easily if the compressibility factor is known at a given pressure and temperature:

$$V = \frac{ZRT}{P}.$$

Mathematically, the equation of state can be represented as either a relationship between P, V, and T, or between Z, P, and T. The latter relationship is quite useful in presenting the volumetric behavior of fluids. Its graphical representation is given on the ZP graph, whose general form is shown in Figure 2-6. Here again we have the vapor-liquid boundary in the form of bell-shaped curve that is now seated on the vertical axis. The vapor region is at the top, the liquid at the bottom. Generally, compressibility factors in the liquid phase are smaller than those in the vapor phase because the molar volume of the liquid is small. All isotherms meet at $P = 0$, $Z = 1$, as anticipated on the basis of eq. (2.18). This point of convergence represents the ideal-gas state. The ZP graph illustrates the path to the ideal-gas state: it is approached by following an isotherm to zero pressure. Since all isotherms converge to this point, the ideal-gas state can be reached from any initial state. We can now see why there is no such a thing as an ideal-gas *state*: if such a gas existed, its ZP graph would consist of a single horizontal line at $Z = 1$, which would represent all isotherms. No substance exists that exhibits such behavior.

Although the mathematical definition places the ideal-gas state at a single point ($P = 0$, $Z = 1$), from a practical point of view we will consider a gas to be in the ideal-gas state if the compressibility factor is sufficiently close to 1. For calculations that do not require high accuracy we will assume a gas to be in the ideal-gas state if the compressibility factor is within $\pm 5\%$ of the theoretical value of 1. The pressure range over which this approximation is valid varies with temperature.

2.4 Corresponding States

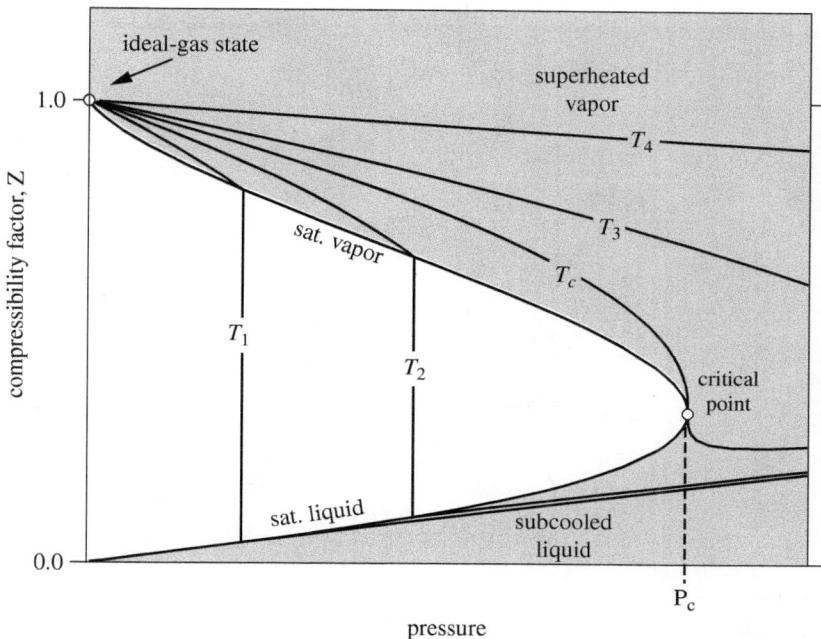

Figure 2-6: The ZP graph of pure fluid.

With reference to Figure 2-6, the isotherm at T_4 remains closer to 1 over a wider interval of pressures, compared to the isotherm at T_1, which has a larger negative slope and decreases faster. In general, to determine whether a gas at given pressure and temperature can be treated as ideal we must check with a ZP graph. We will return to this question in the next section.

2.4 Corresponding States

It is found experimentally that the ZP graphs of different fluids look very similar to each other as if they are scaled versions of a single, universal, graph. This underlying graph is revealed if pressure and temperature are rescaled by appropriate factors. We introduce a set of reduced (dimensionless) variables by scaling pressure and temperature with the corresponding values at the critical point:

$$P_r = \frac{P}{P_c}, \quad T_r = \frac{T}{T_c}. \tag{2.19}$$

Using the reduced coordinates it is possible to combine ZP data for several compounds on the same graph. Figure 2-7 shows the resulting graph for selected

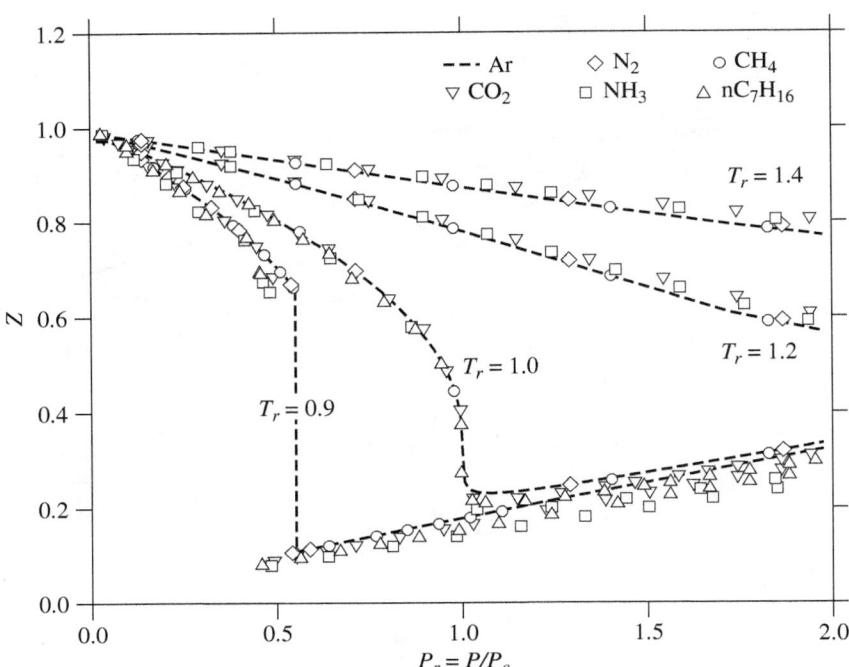

Figure 2-7: The compressibility factor of selected molecules as a function of reduced pressure and temperature. Data compiled from E. W. Lemmon, M. O. McLinden and D. G. Friend, "Thermophysical Properties of Fluid Systems." In *NIST Chemistry WebBook*, NIST Standard Reference Database No. 69, eds. P. J. Linstrom and W. G. Mallard, National Institute of Standards and Technology, Gaithersburg MD, 20899, http://webbook.nist.gov (retrieved December 11, 2010).

molecules. The remarkable feature of this graph is that the compressibility factor of the six compounds shown on this figure agree with each other quite well. Even though the dependence of the compressibility factor on temperature and pressure is different for each fluid, they all seem to arise from the same reduced function. This observation gives rise to the correlation of corresponding states, which is expressed as follows:

> *At the same reduced temperature and pressure, fluids have approximately the same compressibility factor.*

We express this mathematically by writing,

$$Z \approx Z(T_r, P_r). \tag{2.20}$$

2.4 Corresponding States

The practical implication is important: to the extent that eq. (2.20) is obeyed, the compressibility factor of any fluid can be described by a universal equation that is a function of reduced temperature and reduced pressure. Figure 2-7 is a graphical representation of this equation. Accordingly, the compressibility factor of a fluid can be determined at any pressure and temperature using just two parameters, the critical temperature and critical pressure.

The correlation of corresponding states is not an exact physical law and is not obeyed to the same degree of accuracy by all fluids. Although the agreement in Figure 2-7 between different fluids is impressive, it is not exact. The spread of the data points in Figure 2-7 is not due to experimental error (the values have been calculated from validated models) but reflect systematic deviations. These are more clearly seen near the saturation curve and in the liquid region. The correlation arises from similarities in the intermolecular potential. In general, agreement is very good in the gas phase, where interactions are unimportant. In the liquid region, interactions are important and the individual chemical character of molecules becomes more apparent. Even so, nonpolar molecules that are nearly spherical in shape (e.g., Ar, CH_4) agree remarkably well. Polar molecules or molecules with more complex structures (e.g., normal heptane) show the largest deviations because their interactions are more complex and dissimilar. Molecules like water, which is both polar and associates strongly via hydrogen bonding, exhibit even more deviations from this correlation. The principle of corresponding states should be treated as a working hypothesis that can provide useful but not always highly accurate estimates of the compressibility factor.

Acentric Factor and the Pitzer Method Eq. (2.20) is a two-parameter correlation because it requires two physical properties, critical temperature and critical pressure. To improve the predictive power of the principle of corresponding state while retaining its simplicity, a third parameter is introduced, the *acentric factor*. It is a dimensionless parameter that is defined according to the equation,

$$\omega = -1 - \log_{10} P_r^{\text{sat}}\big|_{T_r=0.7}, \qquad (2.21)$$

where P_r^{sat} is the reduced saturation pressure of the fluid at reduced temperature $T_r = 0.7$. It is a characteristic property of the fluid and is found in tables, usually along with the critical properties of the fluid. It was introduced by Pitzer as a measure of the sphericity of the molecule. More generally, it should be understood as a combined measure of the shape and polarity. Symmetric nonpolar molecules, such as Ar, have $\omega = 0$. These are called simple fluids and are found to obey the two-parameter correlation of corresponding states quite well. Nonspherical or polar molecules have a larger acentric factor. For most fluids the acentric factor is positive and in the range 0 to 0.4, although small negative values of ω are also possible.

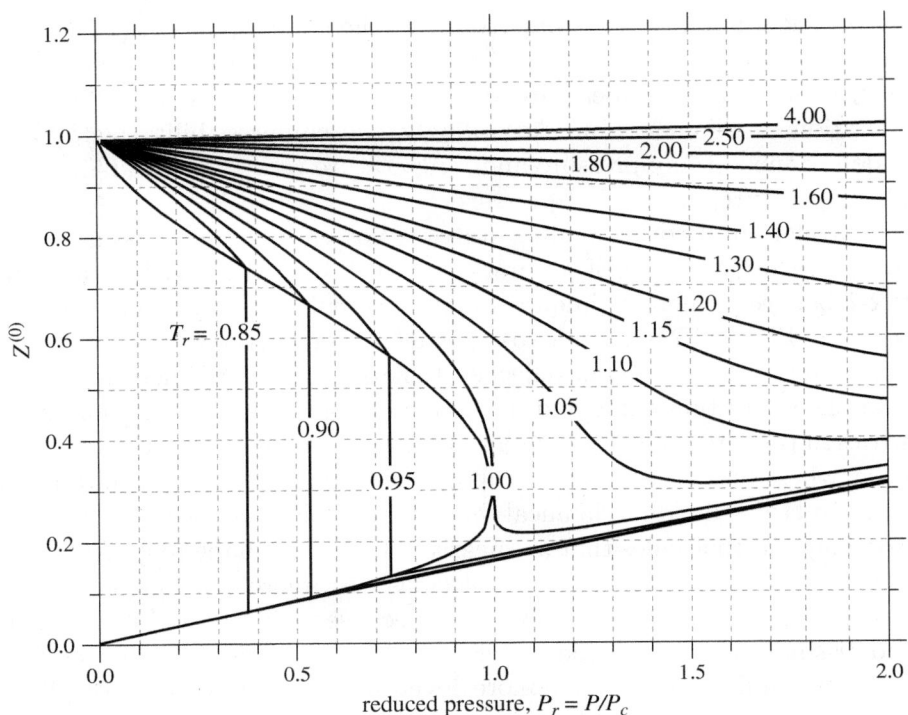

Figure 2-8: Generalized graph of $Z^{(0)}$ based on the Lee-Kesler method.

In the Pitzer method, the compressibility factor is expressed in the form:

$$Z = Z^{(0)} + \omega Z^{(1)} \qquad (2.22)$$

where $Z^{(0)}$ and $Z^{(1)}$ are universal functions that depend on T_r and P_r. Function $Z^{(0)}$ represents the compressibility factor of simple fluids ($\omega = 0$) and $Z^{(1)}$ represents a correction that is proportional to the acentric factor. The two functions $Z^{(0)}$ and $Z^{(1)}$ may be calculated once and tabulated against reduced pressure and temperature for future use. The resulting tables and charts are called *generalized* because they are not limited to a specific molecule.

Various methodologies have been developed for the calculation of the functions in eq. (2.22), but the most widely used is that of Lee and Kesler.[5] The Lee-Kesler result for $Z^{(0)}$ is plotted in Figure 2-8. Recall that this term represents the compressibility

---

5. B. I. Lee and M. G. Kesler. A generalized thermodynamic correlation based on three-parameter corresponding states. AIChE J., 21(3):510, 527 1975. doi: 10.1002/aic.690210313.

2.4 Corresponding States

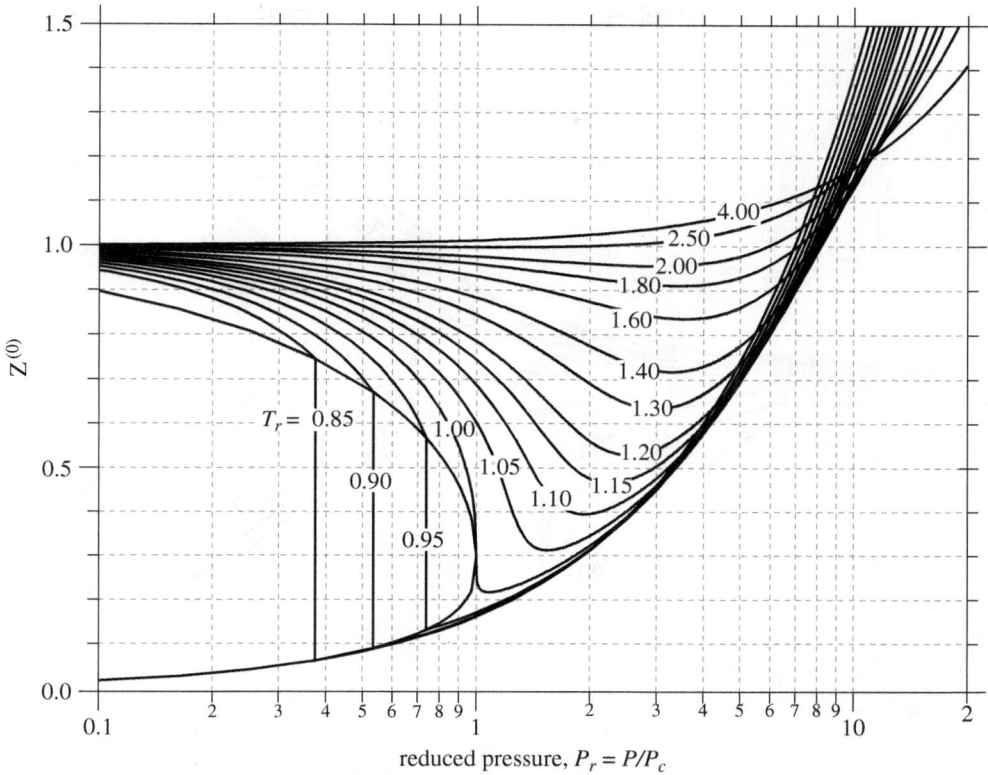

Figure 2-9: Generalized graph of $Z^{(0)}$ based on the Lee-Kesler calculation (extended range of pressures).

factor of a simple fluid, therefore, it has the familiar appearance of the ZP graph. Figure 2-9 shows the same graph in semi-logarithmic scales that cover an expanded range of pressures. The $Z^{(1)}$ is shown in Figure 2-10. The correction factor is zero in the vicinity of the ideal-gas state. It increases in absolute value with increasing pressure and it may take positive or negative values. The sharp changes in subcritical temperatures correspond to a shift from the vapor branch to the liquid branch of the isotherm. The vapor/liquid transitions in Figures 2-8, 2-9, and 2-10 apply *only* to simple fluids ($\omega = 0$). For nonsimple fluids the phase boundary also depends on the acentric factor ω and is generally shifted to lower reduced pressure compared to simple fluids.[6]

---

6. The determination of the precise location of the phase boundary is discussed in Chapter 7.

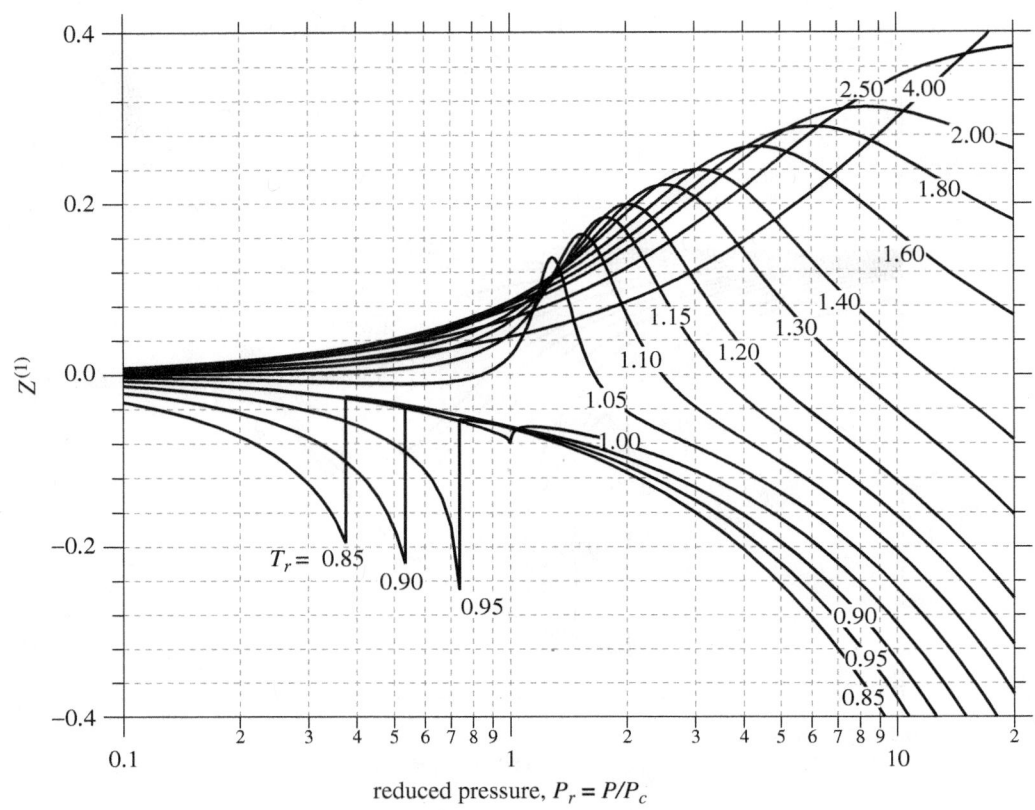

Figure 2-10: Generalized graph of $Z^{(1)}$ based on the Lee-Kesler calculation.

NOTE

Simple, Normal, and Polar Fluids

Noble gases such as Ar, Kr, Xe, and other spherical, nonpolar molecules such as CH_4, interact through similar potential that is spherically symmetric and which can be described through a combination of van der Waals attraction and hard-core repulsion. These fluids are called *simple*. Their acentric factor is zero or nearly zero, and they are described accurately by the two-parameter correlation of corresponding states. A notable exception is the group of *quantum* gases, He, Ne, and H_2, whose behavior at low temperatures is dominated by quantum effects.

Nonpolar molecules that deviate from spherical shape and small molecules of only moderate polarity are described quite well by the three-constant correlation that utilizes the acentric factor. The acentric factor, therefore, can be viewed as an empirical parameter that encompasses the effect of molecular shape, and to some extent polarity. Molecules in this category constitute the class of *normal* fluids, and include small molecules such as O_2, N_2, as well many lower hydrocarbons.

2.4 Corresponding States

Molecules that are strongly polar or associate strongly (e.g., via hydrogen bonding) are problematic when it comes to predicting their properties using correlations with a small number of parameters. The difficulty arises from the fact that the intermolecular potential of polar and associating molecules is fairly complex: it depends on the relative orientation of molecules and cannot be represented by a small number of parameters. Corresponding-states correlations are generally not accurate for such molecules except in the gas phase and away from the phase boundary.

As a rule, generalized methods, that is, equations or graphs and tables that depend on critical temperature, critical pressure and acentric factor, should be applied to *simple normal fluids*.

Example 2.7: Corresponding States
Use the Pitzer correlation with the Lee-Kesler graphs to estimate the density of carbon dioxide at 35 bar, 75 °C.

Solution First we collect the critical properties of CO_2:

$$P_c = 73.8 \text{ bar}, \quad T_c = 304.2 \text{ K}, \quad \omega = 0.225.$$

The reduced pressure and temperature are

$$P_r = \frac{35}{73.8} = 0.474, \quad T_r = \frac{75 + 273.15}{304.2} = 1.14.$$

From Figures 2-9 and 2-10 we find

$$Z^{(0)} = 0.886, \quad Z^{(1)} = 0.0145.$$

The compressibility factor is

$$Z = 0.886 + (0.225)(0.0145) = 0.889.$$

The molar volume is

$$V = \frac{ZRT}{P} = \frac{(0.889)(8.314 \text{ J/mol K})(348.15 \text{ K})}{35 \text{ bar} \cdot 10^5 \text{ Pa/bar}} = 6.957 \times 10^{-4} \text{ m}^3/\text{mol},$$

and the corresponding density is

$$\rho = \frac{M_m}{V} = \frac{(44 \text{ g/mol})}{6.957 \times 10^{-4} \text{ m}^3/\text{mol}} = 63.2 \text{ kg/m}^3.$$

The NIST web book reports a density of 59.72 kg/m^3 at these conditions.

Comments Reading the values of $Z^{(0)}$, $Z^{(1)}$ from the graphs cannot be done very accurately. Better results are obtained by interpolation from tabulated values. Tabulations can be found in many sources, including the original paper by Lee and Kesler, and in *Perry's Chemical Engineers' Handbook*. The numerical procedure for the calculation of these factors is given in Example 2.12.

Example 2.8: Using Steam as a Reference Fluid
Estimate the density of carbon dioxide at 35 bar, 75 °C using data only from the steam tables.

Solution According to the correlation of corresponding states, water and CO_2 have approximately the same compressibility factor at the same reduced state. The conditions given for CO_2 correspond to the reduced coordinates (see Example 2.7):

$$P_r = \frac{35}{73.8} = 0.474, \quad T_r = \frac{75 + 273.15}{304.2} = 1.14.$$

Using the critical constants of water, $P_c = 220.5$ bar, $T_c = 647.3$ K, the corresponding state for water is

$$P = (0.474)(220.5 \text{ bar}) = 104.573 \text{ bar}$$
$$T = (1.14)(647.3 \text{ K}) = 740.82 \text{ K} = 467.67 \text{ °C}.$$

The volume is obtained by interpolation from the steam tables. We find $V = 0.0303815$ m³/kg, or

$$V = (0.0303815 \text{ m}^3/\text{kg})(18.015 \times 10^{-3} \text{ kg/mol}) = 0.000547323 \text{ m}^3/\text{mol}.$$

The compressibility factor is

$$Z = \frac{PV}{RT} = \frac{(104.573 \text{ bar} \times 10^5 \text{ Pa/bar})(0.000547323 \text{ m}^3/\text{mol})}{(8.314)(740.82 \text{ K})} = 0.9293.$$

According to the corresponding-states correlation, this is also equal to the compressibility factor of carbon dioxide at 35 bar, 348.15 K. The molar volume of carbon dioxide is

$$V = \frac{ZRT}{P} = \frac{(0.9293)(8.314 \text{ J/mol K})(348.15 \text{ K})}{(35 \text{ bar})(10^5 \text{ Pa/bar})} = 7.685 \times 10^{-4} \text{ m}^3/\text{mol}$$

and the density is

$$\rho = \frac{M_m}{V} = \frac{44 \times 10^{-3} \text{ kg/mol}}{7.685 \times 10^{-4} \text{ m}^3/\text{mol}} = 57.3 \text{ kg/m}^3.$$

Comments The agreement is quite good with the reported value from the NIST tables [1], which is 59.72 kg/m³. Generally, however, water is not a good reference fluid because of its polarity and hydrogen-bonding ability. In this example the result was satisfactory because the state is vapor. In the gas phase, the chemical character of the molecule is not as important as in liquids because intermolecular distances are large and the effect of molecular interactions is small. Accordingly, very dissimilar molecules behave in very much the same way, as far as volumetric properties are concerned. If the state were liquid, the above calculation would have likely produced a rather poor approximation for the density of carbon dioxide.

2.5 Virial Equation

All isotherms on the ZP graph begin at $P = 0$, $Z = 1$, and grow outward with pressure. This behavior can be described mathematically as power series in P:

$$Z = 1 + b(T)P + c(T)P^2 + \cdots, \qquad (2.23)$$

where $b, c, \cdots$ are coefficients that are characteristic of the isotherm, in other words, for a given fluid they depend only on temperature. With $P = 0$, we obtain $Z = 1$, that is, this equation gives the proper result in the ideal-gas state. In the vicinity of the ideal-gas state, the most significant term in the series is the linear term. In this region isotherms are fairly linear in P. Upon further increasing pressure, first the quadratic and then other higher-order terms become increasingly more important and isotherms are no longer linear. If instead of P and T we select V and T as the state variables, a similar series expansion can be constructed in terms of volume:

$$Z = 1 + \frac{B(T)}{V} + \frac{C(T)}{V^2} + \cdots. \qquad (2.24)$$

Here the expansion is done in powers of inverse volume, which guarantees that in the ideal-gas state ($V = \infty$) the compressibility factor is $Z = 1$. This series expansion is known as the *virial* equation and its coefficients as the *virial* coefficients: B is the second virial coefficient, C is the third, and so on. These are characteristic of the fluid and depend on temperature only.

Eqs. (2.23) and (2.24) are different expansions (one is in P, the other in V) and their coefficients are different. They are, however, related. The relationship for the first two coefficients is given by the following equations

$$b = \frac{B}{RT}. \qquad (2.25)$$

$$c = \frac{C - B^2}{(RT)^2}. \qquad (2.26)$$

$$\vdots$$

As a matter of nomenclature, only eq. (2.24) is known as "virial," and its coefficients as "virial coefficients," whereas eq. (2.23) is referred to as the pressure expansion of the compressibility factor.

NOTE

Second Virial Coefficient and Intermolecular Potential
The second virial coefficient is related to the intermolecular potential according to the equation,

$$B = 2\pi \int_0^\infty \left(1 - e^{-\Phi(r)/kT}\right) r^2 dr,$$

where Φ(r) is the intermolecular potential, k is the Boltzmann factor, T is absolute temperature, and the integral is evaluated by letting the distance r between two molecules range from 0 to ∞. If the intermolecular potential is weak relative to kT, the ratio Φ(r)/kT is small, the exponential term is nearly equal to 1, and the second virial coefficient is nearly zero. Under these conditions, deviations from ideal-gas behavior are small. The magnitude of B increases when the potential is strong relative to temperature, a condition that leads to stronger deviations from ideal-gas behavior.

Truncated Virial Equation If all the coefficients of the expansion were known, the virial equation would provide complete description of the vapor isotherm up to the point it intersects the saturated vapor line. In practice, only the second virial coefficient is widely available and as a result, this equation is most often used in truncated form by dropping terms that include higher coefficients. Either series can be used, but eq. (2.23) is more convenient because it expresses the compressibility factor in terms of pressure and temperature. Dropping the quadratic and higher terms in eq. (2.23), the compressibility factor becomes

$$Z \approx 1 + \frac{BP}{RT}. \tag{2.27}$$

This equation is a linear approximation of the isotherm and is valid in the region near $P = 0$ that the actual isotherm is linear in pressure. Eq. (2.27) should be considered as a first-order correction to the ideal-gas law that provides us with the equation of state at pressures that are low but not sufficiently so to assume ideal-gas behavior. The range of pressures that satisfy this condition depends on temperature. Figure 2-8 may be used as a guide: Eq. (2.27) is to be used only within a pressure interval from 0 to pressure P such that the isotherm in this interval is a straight line. This pressure interval is narrow at lower temperatures but expands with increasing temperature.

As we can verify by checking with a ZP graph, there is always a neighborhood of pressures near $P = 0$ where linearity is observed. We also see that the linear range depends on temperature and the linear interval generally increases with increasing temperature. We refer to the linear range of an isotherm as a region of *moderate nonideality* because the second virial coefficient suffices to describe deviations from ideality. Writing $Z = PV/RT$ and with some rearrangement, eq. (2.27) becomes

$$P(V - B) \approx RT. \tag{2.28}$$

The truncated virial equation provides us with a simplified equation of state in the vicinity of the ideal-gas state where the ideal-gas law alone is not sufficiently accurate.

2.5 Virial Equation

Pitzer Method for the Second Virial Coefficient A useful empirical correlation for the second virial coefficient is based on the Pitzer method. In this method, the dimensionless ratio, BP_c/RT_c is expressed in the form

$$\frac{BP_c}{RT_c} = B^0 + \omega B^1 \tag{2.29}$$

where

$$B^0 = 0.083 - \frac{0.422}{T_r^{1.6}} \tag{2.30}$$

$$B^1 = 0.139 - \frac{0.172}{T_r^{4.2}}. \tag{2.31}$$

Since the second virial coefficient is a function of temperature only, the coefficients B^0 and B^1 are independent of P_r and functions of T_r, only. These equations can be used to estimate the second virial coefficient if other data are not available.

Example 2.9: Comparison of Methods

Calculate the molar volume of ethylene at 40 °C, 90 bar, using the (a) ideal-gas law, (b) the truncated virial equation, and (c) the Pitzer correlation with the Lee-Kesler values for $Z^{(0)}$, $Z^{(1)}$.

Solution The critical parameters of ethylene are

$$P_c = 50.41 \text{ bar}, T_c = 282.34 \text{ K}, \omega = 0.087.$$

The reduced coordinates are

$$P_r = \frac{90 \text{ bar}}{50.41 \text{ bar}} = 1.78536,$$

$$T_r = \frac{313.15 \text{ K}}{282.34 \text{ K}} = 1.10912.$$

(a) Ideal-gas law: The ideal-gas molar volume is

$$V^{ig} = \frac{RT}{P} = \frac{(8.314 \text{ J/mol K})(313.15 \text{ K})}{(90 \text{ bar})(10^5 \text{ J/m}^3 \text{ bar})} = 2.893 \times 10^{-4} \text{ m}^3/\text{mol}.$$

(b) Virial equation: We first calculate the second virial coefficient using eqs. (2.29), (2.30), and (2.31):

$$B^0 = 0.083 - \frac{0.422}{(1.10912)^{1.6}} = -0.274556$$

$$B^1 = 0.139 - \frac{0.172}{(1.10912)^{4.2}} = 0.0276699$$

$$B^0 + \omega B^1 = -0.274556 + (0.087)(0.0276699) = -0.272149$$

$$B = \frac{(8.314 \text{ J/mol K})(282.34 \text{ K})(-0.272149)}{(50.41 \times 10^5 \text{ Pa})} = -1.267 \times 10^{-4} \text{ m}^3/\text{mol}.$$

The compressibility factor is calculated form the truncated virial equation, eq. (2.27)

$$Z = 1 + \frac{(-1.267 \times 10^{-4} \text{ m}^3/\text{mol})(90 \times 10^5 \text{ Pa})}{(8.314 \text{ J/mol K})(313.15 \text{ K})} = 0.5619,$$

and the molar volume is

$$V = \frac{ZRT}{P} = \frac{(0.5619)(8.314 \text{ J/mol K})(313.15 \text{ K})}{(90 \times 10^5 \text{ Pa})} = 1.626 \times 10^{-4} \text{ m}^3/\text{mol}.$$

Lee-Kesler correlation The values of $Z^{(0)}$, $Z^{(1)}$, are

$$Z^{(0)} = 0.41822, \quad Z^{(1)} = 0.140794,$$

and the compressibility factor is

$$Z = Z^{(0)} + \omega Z^{(1)} = (0.41822) + (0.087)(0.140794) = 0.4305.$$

The molar volume is

$$V = \frac{ZRT}{P} = \frac{(0.4305)(8.314 \text{ J/mol K})(313.15 \text{ K})}{(90 \times 10^5 \text{ Pa})} = 1.245 \times 10^{-4} \text{ m}^3/\text{mol}.$$

The literature value from the NIST database [1] is $1.22 \times 10^{-4} \text{ m}^3/\text{mol}$.

Comments We summarize these results in the table below:

method	Z	V (10^{-4} m^3/mol)
ideal gas	1	2.893
virial	0.5619	1.626
Lee Kesler	0.4305	1.245
literature	0.4216	1.220

The graphical comparison of these results is shown on the ZP graph in Figure 2-11. The gas calculation assumes constant compressibility factor at all pressures and temperatures. It is not expected to be a correct approximation at the present state because the pressure is quite high and above the critical. The virial equation estimates Z by linear extrapolation based on the slope of the isotherm at $P = 0$. It gives good results as long as the isotherm is approximately linear, up to about $P_r = 0.5$, in this case. At higher pressures the isotherm is nonlinear and the truncated virial is not accurate any longer. Additional virial coefficients would be needed in this region. The Pitzer correlation with the Lee-Kesler values is remarkably close to the literature result. This is expected since ethylene is a nonpolar molecule. We also note that the state is in the vapor phase, a region where generalized equations exhibit higher accuracy.

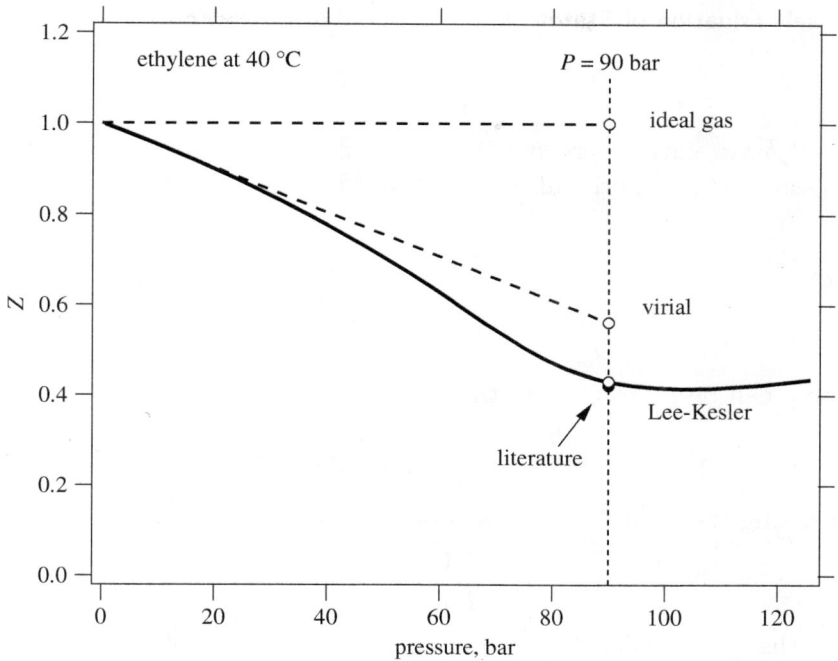

Figure 2-11: Calculated isotherms of ethylene at 40 °C using the ideal-gas law, the truncated virial equation, and the Pitzer method with the Lee-Kesler values of $Z^{(0)}$ and $Z^{(1)}$ (see Example 2.9).

2.6 Cubic Equations of State

For engineering calculations it is important to have equations of state that are accurate over a wide range of pressures and temperatures. The ideal gas law is very simple to use, but its validity is restricted to gases at low pressures. The truncated virial equation is applicable over a somewhat wider range of pressures, but only for gases. If the pressure is high or the phase liquid, neither of these equations can be used. Numerous empirical equations of state have appeared in the literature to overcome these difficulties. Such equations usually have some theoretical basis, but the primary consideration is sufficient accuracy for engineering applications. It is typical for these equations to contain parameters that are fine-tuned to improve accuracy. No single mathematical equation of state can describe all fluids. Nonetheless, it is convenient in having one equation whose mathematical form is the same for many fluids but with parameters that are specific to a particular fluid. Among the most important engineering equations of state is the family of cubic equations, which can be viewed as variants of the van der Waals equation of state.

Van der Waals Equation of State The van der Waals equation of state is given by

$$P = \frac{RT}{V-b} - \frac{a}{V^2}, \tag{2.32}$$

where a and b are parameters specific to the fluid and are given in terms of the critical pressure, P_c, and critical temperature, T_c, of the fluid:

$$a = \frac{27}{64} \frac{(RT_c)^2}{P_c} \tag{2.33}$$

$$b = \frac{1}{8} \frac{RT_c}{P_c}. \tag{2.34}$$

The equation can be rearranged in the form:

$$\frac{PV}{RT} = \frac{V}{V-b} - \frac{a}{RTV}, \tag{2.35}$$

and upon taking the limit $V \to \infty$, we obtain

$$\frac{PV}{RT} \to 1.$$

This proves that the van der Waals equation has the correct behavior in the ideal-gas limit.

The van der Waals equation can be rearranged to obtain the compressibility factor in terms of pressure and temperature. Starting with eq. (2.35) and using $V = ZRT/P$ to eliminate V, we obtain the following expression for Z:

$$Z = \frac{Z}{Z - bP/RT} - \frac{aP}{Z(RT)^2} = \frac{Z}{Z-B'} - \frac{A'}{Z},$$

where A' and B' are defined for convenience as

$$A' = \frac{aP}{(RT)^2}, \quad B' = \frac{bP}{RT}. \tag{2.36}$$

Rearranging to solve for Z, the previous result takes the form

$$Z^3 - (B'+1)Z^2 + A'Z - A'B' = 0. \tag{2.37}$$

The equation for the compressibility factor is a cubic polynomial in Z. This is the reason that the van der Waals equation is referred to as a cubic equation of state. The parameters A' and B' depend on pressure and temperature. If T and P are given, the compressibility factor can be calculated from eq. (2.37). As a cubic polynomial, this equation may have one or three real roots. Multiplicity of roots is a characteristic property of equations of state that are capable of describing phase transitions.

2.6 Cubic Equations of State

The predictions of the van der Waals equation are generally in good qualitative agreement with experimental data, but the accuracy is not sufficiently high for routine engineering calculations. Moreover, the equation makes use of only two physical parameters, critical pressure and critical temperature, and as we saw in Section 2.4, these are not enough to provide accurate estimates of the compressibility factor. This has motivated the development of variants of the van der Waals equation with increased accuracy. Two of the most commonly used equations in current practice are the Soave-Redlich-Kwong (SRK) and the Peng-Robinson (PR) equations of state.

NOTE

van der Waals and His Equation
The van der Waals equation was the first equation proposed that is capable of predicting the phase behavior of vapors *and* liquids. It was developed by van der Waals in his doctoral thesis, in 1873. This equation introduces in an approximate but effective way the effect of molecular repulsion and attraction, an element that is important in describing phase transformation, as discussed in Section 1.1. Repulsion is represented by the term $RT/(V-b)$, which ensures that at high pressures the molar volume approaches a constant value b. This effectively accounts for the strong repulsion at close molecular distances, an interaction that gives molecules the appearance of possessing a solid core. Attraction is represented by the term $-a/V^2$. This term is negative because the effect of attractive forces is to pull molecules together and reduce the pressure of the fluid. The factor V^2 in the denominator has its origin in the van der Waals force, which is the typical form of molecular attraction. The simultaneous presence of terms representing attraction and repulsion endows this equation with the ability to describe both the vapor *and* the liquid region of the phase diagram. In 1910, van der Waals received the Nobel Prize in Physics.

Example 2.10: van der Waals Equation at the Critical Point
The van der Waals equation has an isotherm with an inflection point that satisfies eq. (2.3). Here we show how this allows us to obtain the parameters a and b and the compressibility factor at the critical point.

Solution The first and second derivatives of P in eq. (2.32) with respect to V at constant T are

$$\left(\frac{\partial P}{\partial V}\right)_T = -\frac{RT}{(V-b)^2} + \frac{2a}{V^3},$$

$$\left(\frac{\partial^2 P}{\partial V^2}\right)_T = \frac{2RT}{(V-b)^3} - \frac{6a}{V^4}.$$

At the critical point ($T = T_c$, $V = V_c$) both derivatives are zero:

$$-\frac{RT_c}{(V_c - b)^2} + \frac{2a}{V_c^3} = 0,$$

$$\frac{2RT_c}{(V_c - b)^3} - \frac{6a}{V_c^4} = 0.$$

These may be solved for a and b:

$$a = \frac{9RT_c V_c}{8}, \quad b = \frac{V_c}{3}, \tag{2.38}$$

which gives the parameters a and b in terms of the critical volume and critical temperature. We substitute these values into the van der Waals equation at $V = V_c$, $T = T_c$, to obtain the pressure at the critical point:

$$P_c = \frac{RT_c}{V_c - b} - \frac{a}{V_c^2} = \frac{3RT_c}{8V_c}.$$

This result can also be written as

$$\frac{P_c V_c}{RT_c} = \frac{3}{8}, \tag{2.39}$$

and shows that the critical compressibility factor according the van der Waals equation is $Z_c = 3/8 = 0.375$. This value is rather high (most critical compressibility factors are in the range $0 - 0.3$) and a sign that the van der Waals equation will probably be not very accurate. We solve eq. (2.39) for the critical volume,

$$V_c = \frac{3RT_c}{8P_c}$$

and substitute this value into eq. (2.38):

$$a = \frac{27}{64} \frac{(RT_c)^2}{P_c}, \quad b = \frac{RT_c}{8P_c}.$$

These now give the parameters a and b in terms of critical pressure and critical temperature. With these values, the van der Waals equation will place the critical point exactly at the experimental critical temperature and critical pressure of the fluid; however, the critical volume from the van der Waals equation will not match the experimental critical volume unless the critical compressibility of the fluid happens to be 0.375. The main point, however, is that the van der Waals equation produces reasonable results at various pressures and temperatures using only two critical constants as a guide (see Example 2.11).

Soave-Redlich-Kwong (SRK) Equation of State The Soave-Redlich-Kwong equation of state is given by

$$P = \frac{RT}{V - b} - \frac{a}{V(V + b)}, \tag{2.40}$$

where a and b are parameters specific to the fluid and are given in terms of the critical pressure, critical temperature, and acentric factor:

$$a = 0.42748 \frac{R^2 T_c^2}{P_c} \left[1 + \Omega \left(1 - T_r^{1/2}\right)\right]^2, \tag{2.41}$$

$$b = 0.08664 \frac{R T_c}{P_c}, \tag{2.42}$$

$$\Omega = 0.480 + 1.574\omega - 0.176\omega^2. \tag{2.43}$$

Here, $T_r = T/T_c$ is the reduced temperature. The equation for the compressibility factor is

$$Z^3 - Z^2 + (A' - B' - B'^2)Z - A'B' = 0, \tag{2.44}$$

with the parameters A' and B' defined in eq. (2.36).

Peng-Robinson (PR) Equation of State The Peng-Robinson equation of state is given by

$$P = \frac{RT}{V - b} - \frac{a}{V^2 + 2bV - b^2}. \tag{2.45}$$

The parameters a and b are related to the critical pressure, critical temperature and acentric factor as follows:

$$a = 0.45724 \frac{R^2 T_c^2}{P_c} \left[1 + \Omega \left(1 - T_r^{1/2}\right)\right]^2, \tag{2.46}$$

$$b = 0.07780 \frac{R T_c}{P_c}, \tag{2.47}$$

$$\Omega = 0.37464 + 1.54226\omega - 0.26992\omega^2. \tag{2.48}$$

where $T_r = T/T_c$ is the reduced temperature. The equation for Z is

$$Z^3 + (B' - 1)Z^2 + Z(A' - 3B'^2 - 2B') - A'B' + B'^3 + B'^2 = 0, \tag{2.49}$$

with the A' and B' given by eq. (2.36).

2.7 PVT Behavior of Cubic Equations of State

To generate a pressure-volume isotherm, temperature is fixed and the pressure is calculated for various volumes. We demonstrate the behavior of isotherms in Figure 2-12 using the Soave-Redlich-Kwong equation for ethylene ($T_c = 282.35$ K, $P_c = 50.418$ bar, $\omega = 0.0866$). All volumes lie to the right of b, which for ethylene has the value 4.03395 m^3/mol. The isotherm at $T = 340$ K has the general shape of supercritical isotherms and decreases gently as volume is increased. The critical

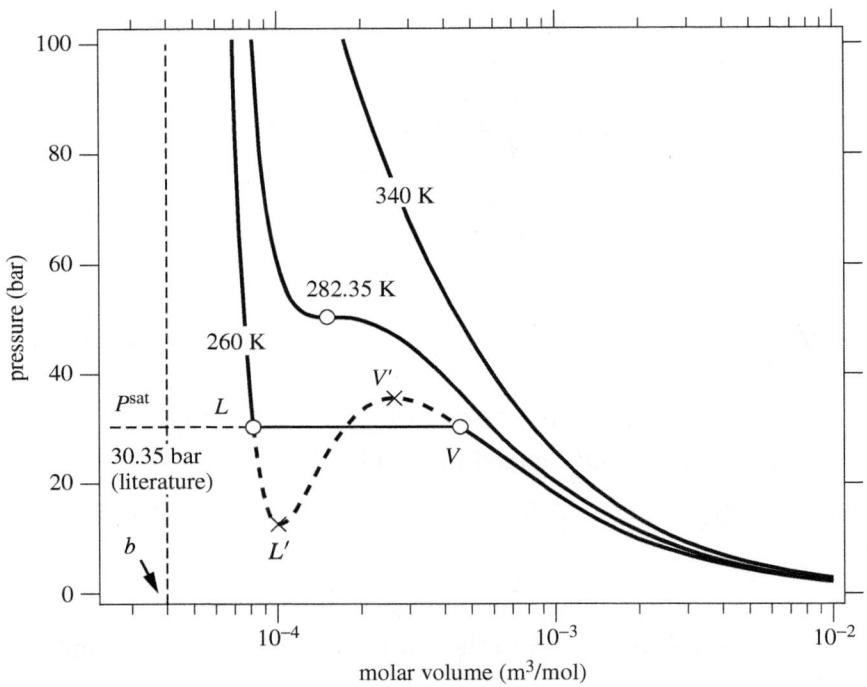

Figure 2-12: Isotherms of ethylene calculated by the Soave-Redlich-Kwong equation.

isotherm ($T_c = 282.35$ K) exhibits an inflection point, which occurs at the critical pressure. At this point the first and second derivative of pressure with respect to volume are both zero. The subcritical isotherm ($T = 260$ K) has unusual behavior: it decreases steeply in the liquid region, as expected, but then is goes through a minimum and a maximum before it emerges in the vapor region, where it resumes the expected behavior. The "hump," shown by the dashed line in Figure 2-12, represents metastable and unstable parts of the isotherm and must be removed. To do this, we draw a horizontal line at the saturation pressure at 260 K, which from tables is found to be 30.25 bar. The intersection of this line with the liquid branch of the isotherm defines the saturated liquid (point L) and its intersection with the vapor branch defines the saturated vapor (point V). The corrected isotherm is obtained by drawing the tie line between points L and V and removing the dashed part of the isotherm.

Unstable and Metastable Parts of Subcritical Isotherm Along the portion of the isotherm between points L' (local minimum of the isotherm) and V' (local maximum) the isotherm has a positive slope, which implies that volume increases with increasing pressure. This unphysical behavior makes this part of the isotherm

2.7 PVT Behavior of Cubic Equations of State

mechanically unstable. Suppose that the system is brought to a point between L' and V'. If we increase the external pressure by a small amount, the volume contracts and the state moves to the left. This causes the pressure of the fluid to decrease and the volume to contract even more, as the external pressure remains unchanged. This unstable behavior continues until the system reaches point L'. The branches LL' and $V'V$ are mechanically stable but thermodynamically *metastable*. In LL' the liquid is supersaturated, meaning that it is in the liquid phase, even though its pressure is lower than the saturation pressure. The stable phase is vapor, but the metastable liquid state will be maintained for some time, if care is taken to avoid contaminants and surface defects on the container walls, which can act as nucleation sites. Eventually, thermal fluctuations will cause a phase change to take place from the supersaturated phase to the one that is most stable. The situation is analogous along the metastable vapor part, VV'.

NOTE

Choosing Among Multiple Roots of Cubic Equations of State
When working with cubic equations of state it is important to remember that subcritical isotherms obtained by direct calculation of the equation of state include the metastable/unstable loop. This requires special attention when solving for V at temperatures below the critical. Suppose we seek the molar volume at specified pressure P and temperature T. The solution is graphically given on the PV graph by the intersection between the isotherm at T and the horizontal line drawn at pressure P. As we see in Figure 2-12, it is possible to have one, or three points of intersection, each of them corresponding to a molar volume that satisfies the equation of state at the given pressure and temperature. If there is only one intersection point, there is no ambiguity and this point defines the unknown volume. If there are three intersections only one (or two, if the system is saturated) is physically acceptable. To determine the correct volume we must determine the saturation pressure at the given temperature, T. This can be obtained from tables or using the Antoine equation. The proper volume is now selected as follows:

1. If pressure is *higher* than the saturation pressure, the phase is compressed liquid: select the *smallest* root.

2. If pressure is *lower* than the saturation pressure, the phase is superheated vapor: select the *largest* root.

3. If pressure is equal to the saturation pressure, then the system is saturated: select the smallest root to be that of the saturated liquid, and the largest root to be that of the saturated vapor.

The middle root is *never* selected. Cubic equations of state may also have spurious roots for the volume to the left of b. These roots are not physical and should be rejected. Acceptable roots must be positive and larger than b. In terms of the compressibility factor, acceptable roots must be positive and larger than Pb/RT.

2.8 Working with Cubic Equations

The typical problem involving cubic equations of state asks for the molar volume at given pressure and temperature. This volume can be calculated by solving the equation of state for V. Alternatively, it can be calculated by solving the cubic equation (see eq. [2.44]) for Z. The latter method is recommended for a number of reasons. The equation for Z is dimensionless; this reduces the chances of errors involving units. The values of Z generally range from 0 to about 1; by contrast, molar volume spans several orders of magnitude. Finally, as a cubic equation, its roots can be computed using standard formulas (see below). Alternatively, the roots can be obtained by iterative numerical procedures. Numerical methods require suitable starting guesses. For the gas-phase root, a starting guess is $Z = 1$; for the liquid-phase root, the starting guess is $Z = B = Pb/RT$. Nonetheless, there is no guarantee that these guesses will converge to the proper roots, therefore, the results of iterative numerical procedures must always be checked.

NOTE

Equations for the Roots of Cubic Equation
Here is the analytic procedure for the calculation of the real roots of a cubic polynomial. The equations shown here are taken from the book Numerical Recipes:[7] but they can be found in many other books, including *Perry's Chemical Engineers' Handbook*.

1. Given the cubic polynomial with real coefficients

$$x^3 + ax^2 + bx + c = 0$$

   first compute Q and R as shown below:

$$Q = \frac{a^2 - 3b}{9}, \quad \text{and} \quad R = \frac{2a^3 - 9ab + 27c}{54}.$$

2. If $R^2 < Q^3$, the polynomial has three *real* roots. Calculate them as follows:

$$x_1 = -2\sqrt{Q}\cos\frac{\theta}{3} - \frac{a}{3}$$

$$x_2 = -2\sqrt{Q}\cos\frac{\theta + 2\pi}{3} - \frac{a}{3}$$

$$x_3 = -2\sqrt{Q}\cos\frac{\theta - 2\pi}{3} - \frac{a}{3}$$

   where

$$\theta \equiv \arccos\frac{R}{\sqrt{Q^3}}.$$

   The roots as obtained here are not necessarily sorted.

---

7. W. H. Press, S. A. Teukolsky, W. T. Vetterling, and B. P. Flannery. *Numerical Recipes in Fortran*, 2nd ed. (New York: Cambridge University Press, 1992), p. 179.

2.8 Working with Cubic Equations

3. If $R^2 > Q^3$, the polynomial has only one real root. Calculate the parameters A and B as follows:

$$A = -\text{sign}(R)\left[|R| + \sqrt{R^2 - Q^3}\right]^{1/3}$$

and

$$B = \begin{cases} Q/A & \text{if } A \neq 0 \\ 0 & \text{if } A = 0 \end{cases}$$

where $\text{sign}(x)$ is the function that returns the sign of x. The real root is

$$x_1 = (A + B) - \frac{a}{3}.$$

Since the complex roots are of no interest to us, we do not provide the formulas for their calculation.

Notice that the various parameters defined here (for example, a, R, etc.) are not related to thermodynamic quantities with the same symbol defined elsewhere in the book.

Example 2.11: Solving Cubic Equations for the Volume
Use the van der Waals, the Soave-Redlich-Kwong, and the Peng-Robinson equations to determine the molar volumes of the saturated phases of ethylene at 260 K (the saturation pressure is 30.35 bar). Compare with the tabulated values,

$$V_L = 0.0713 \text{ l/mol}, \quad V_V = 0.456 \text{ l/mol}.$$

Solution First we collect the critical properties and acentric factor of ethylene:

$$T_c = 282.35 \text{ K}, \quad P_c = 50.418 \text{ bar}, \quad \omega = 0.0866.$$

van der Waals Equation The parameters a, b, of the van der Waals equation are calculated from eqs. (2.33) and (2.34):

$$a = 0.461099 \text{ J m}^3/\text{mol}^2, \quad b = 5.81999 \times 10^{-5} \text{ m}^3/\text{mol}.$$

Rather than solving the pressure form of the van der Waals equation we will work with the cubic equation in Z. The parameters A and B are calculated at $T = 260$ K, $P = 30.25$ bar:

$$A' = 0.299492, \quad B' = 0.0817142.$$

The cubic equation is calculated from eq. (2.37):

$$-0.0244728 + 0.299492 Z - 1.08171 Z^2 + Z^3 = 0.$$

Using the formulas for the roots of cubic equations, we find three real roots:

$$Z_1 = 0.16559, \quad Z_2 = 0.209004, \quad Z_3 = 0.707119.$$

Since the system is at saturation pressure, we assign the smallest root to the compressibility factor of the liquid and the largest root to the compressibility factor of the vapor (the middle root is not used):

$$Z_L = 0.16559, \quad Z_V = 0.707119.$$

The molar volumes are calculated using $V = ZRT/P$:

$$V_L = Z_L \frac{RT}{P} = (0.16559) \frac{(8.314 \text{ J/mol K})(260 \text{ K})}{(30.35 \text{ bar})(10^5 \text{ Pa/bar})} = 1.1794 \times 10^{-4} \text{ m}^3/\text{mol}$$

$$V_V = Z_V \frac{RT}{P} = (0.707119) \frac{(8.314 \text{ J/mol K})(260 \text{ K})}{(30.35 \text{ bar})(10^5 \text{ Pa/bar})} = 5.03637 \times 10^{-4} \text{ m}^3/\text{mol}.$$

Notice that temperature is fairly close to the critical (282.35 K), as a result, the vapor volume is only about five times bigger than the liquid volume.

Soave-Redlich-Kwong Equation The calculations for the SRK equation are done in a similar fashion. The results are summarized below:

$$a = 0.490727 \text{ J m}^3/\text{mol}^2 \qquad b = 4.03395 \times 10^{-5} \text{ m}^3/\text{mol}$$

$$A' = 0.318736 \qquad B' = 0.0566377$$

The cubic equation is

$$-0.0180525 + 0.258891 Z - Z^2 + Z^3 = 0,$$

with roots

$$Z_1 = 0.11477, \quad Z_2 = 0.246108, \quad Z_3 = 0.639122.$$

Selection of the proper roots is done as in the van der Waals case:

$$Z_L = 0.11477, \quad Z_V = 0.639122.$$

Finally, the molar volumes are

$$V_L = Z_L \frac{RT}{P} = 0.817434 \times 10^{-4} \text{ m}^3/\text{mol}$$

$$V_V = Z_V \frac{RT}{P} = 4.55207 \times 10^{-4} \text{ m}^3/\text{mol}.$$

Peng-Robinson Equation The results are summarized below:

$$a = 0.520397 \text{ J m}^3/\text{mol}^2 \qquad b = 3.6224 \times 10^{-5} \text{ m}^3/\text{mol}$$

$$A' = 0.338008 \qquad B' = 0.0508589$$

The cubic equation is

$$-0.0144725 + 0.22853Z - 0.949141Z^2 + Z^3,$$

with roots

$$Z_1 = 0.10164, \quad Z_2 = 0.230944, \quad Z_3 = 0.616557.$$

The compressibility factors of the saturation liquid and saturated vapor are

$$Z_L = 0.10164, \quad Z_V = 0.616557.$$

The corresponding molar volumes are

$$V_L = Z_L \frac{RT}{P} = 0.723918 \times 10^{-4} \text{ m}^3/\text{mol}$$

$$V_V = Z_V \frac{RT}{P} = 4.39135 \times 10^{-4} \text{ m}^3/\text{mol}.$$

Summary The results are compared in the table below:

	V_L (m^3/mol)	(% error)	V_V (m^3/mol)	(% error)
vdW	1.17940×10^{-4}	(+65)	5.03637×10^{-4}	(+10.5)
SRK	8.17434×10^{-5}	(+14)	4.55207×10^{-4}	(−0.14)
PR	7.23918×10^{-5}	(+1.5)	4.39135×10^{-4}	(−3.7)
literature	7.13×10^{-5}	—	4.56×10^{-4}	—

The van der Waals equation has the largest error, most notably in the liquid region. The Peng-Robinson equation gives the lowest error in the liquid, but the Soave-Redlich-Kwong equation is better in the vapor region.

2.9 Other Equations of State

Cubic equations of state are useful because they are capable of handling both liquids and gases, but they are not the only type of equations that have this feature. The mathematical requirement is that subcritical isotherms must exhibit the characteristic h-shaped portion, with an unstable part between the liquid and vapor branches. This behavior can be reproduced by other equations that are not necessarily cubic in Z. One that is worth mentioning is the Benedict-Webb-Rubin equation (BWR), which has the form,

$$P = \frac{RT}{V} + \frac{B_0 RT - A_0 - C_0/T^2}{V^2} + \frac{bRT - a}{V^3} + \frac{\alpha a}{V^6} + \frac{c}{T^2 V^3}\left(1 + \frac{\gamma}{V^2}\right)\exp\left(-\frac{\gamma}{V^2}\right). \tag{2.50}$$

It requires eight parameters (A_0, B_0, C_0, a, b, c, α, and γ) that are specific to the fluid. The Benedict-Webb-Rubin equation is modeled after the virial equation and expresses pressure as a finite sum of powers of $1/V$, up to the sixth power. The exponential term on the right is meant to account for the higher terms of the series that have been dropped. A modified form of this equation was used by Lee and Kesler[8] in the calculation of $Z^{(0)}$ and $Z^{(1)}$. This equation is not cubic but its subcritical isotherms have the same general behavior as those in Figure 2-12, namely, they exhibit an unstable part where the isotherm has a positive slope.

Example 2.12: Calculation of Lee-Kesler Factors, Z^0, $Z^{(1)}$

Lee and Kesler computed the values of $Z^{(0)}$ and $Z^{(1)}$ using a modification of the BWR equation of state. Written for the compressibility factor, the modified BWR equation is

$$Z = 1 + \frac{b}{v} + \frac{c}{v^2} + \frac{d}{v^5} + \frac{c_4}{T_r^3 v^2}\left(\beta + \frac{\gamma}{v^2}\right)\exp\left(-\frac{\gamma}{v^2}\right) \qquad [\text{A}]$$

with

$$b = b_1 - b_2/T_r - b_3/T_r^2 - b_4/T_r^3,$$
$$c = c_1 - c_2/T_r + c_3/T_r^3,$$
$$d = d_1 + d_2/T_r.$$

Here, $T_r = T/T_c$ is the reduced temperature and v is a dimensionless volume, defined as

$$v = \frac{VP_c}{RT_c}.$$

The term $Z^{(0)}$ at fixed pressure and temperature is calculated from eq. [A] using the set of parameters for simple fluid in Table 2-1. The term $Z^{(1)}$ is calculated from

$$Z^{(1)} = \frac{Z^{\text{ref}} - Z^{(0)}}{0.3978}, \qquad (2.51)$$

where Z^{ref} is the compressibility factor calculated from the BWR equation at the same pressure and temperature as $Z^{(0)}$, using the constants for the reference fluid. As an example, we will calculate the terms $Z^{(0)}$ and $Z^{(1)}$ at $T_r = 0.95$, $P_r = 0.6$.

---

8. B. I. Lee and M. G. Kesler. A generalized thermodynamic correlation based on three-parameter corresponding states, *AIChE J.*, 21(3):510, 527 1975, http://dx.doi.org/10.1002/aic.690210313.

2.9 Other Equations of State

Solution *Calculation of $Z^{(0)}$:* At $T_r = 0.95$, the BWR constants for simple fluid calculated from Table 2-1 are

$$b = -0.368474, \quad c = 0.00399187, \quad d = 0.0000812003,$$
$$c_4 = 0.00399187, \quad \beta = 0.65392, \quad \gamma = 0.060167.$$

The BWR equation gives the compressibility factor as an explicit function of temperature and volume. Since we are given pressure and temperature, we must first obtain the volume. To do this, we write the compressibility factor in terms of v,

$$Z = \frac{PV}{RT} = \frac{P_r v}{T_r},$$

and equate this result to the compressibility factor from eq. [A]:

$$\frac{P_r v}{T_r} = 1 + \frac{b}{v} + \frac{c}{v^2} + \frac{d}{v^5} + \frac{c_4}{T_r^3 v^2}\left(\beta + \frac{\gamma}{v^2}\right)\exp\left(-\frac{\gamma}{v^2}\right). \qquad [C]$$

The only unknown in this equation is v. The equation does not have analytic solution for the roots and must be solved numerically. At the given conditions, this equation has three roots for v, as we can confirm graphically. The three roots are

$$v_1 = 0.177567, v_2 = 0.2377, v_3 = 1.10314.$$

These correspond to the three intersections in Figure 2-13 between the isotherm of simple fluid and the horizontal line at $P_r = 0.6$. From Figure 2-8 we determine that at $T_r = 0.95$, $P_r = 0.6$ the phase is vapor, therefore, we choose the largest root: $v = 1.10314$. The value of $Z^{(0)}$ is

$$Z^{(0)} = \frac{P_r v}{T_r} = \frac{(0.6)(1.10314)}{(0.95)} = 0.6967.$$

Calculation of $Z^{(1)}$: The BWR constants for the reference fluid at $T_r = 0.95$ are

$$b = -0.414282, \quad c = -0.00196143, \quad d = 0.000056529, \quad c_4 = -0.00196143,$$
$$\beta = 1.226, \quad \gamma = 0.03754.$$

Eq. [C] is now solved for v with the new set of parameters, at the same reduced temperature ($T_r = 0.95$) and pressure ($P_r = 0.6$). Again, we find three roots for v:

$$v_1' = 0.145721, \quad v_2' = 0.270816, \quad v_3' = 1.03321.$$

These correspond to the three intersections in Figure 2-13 between the isotherm of the reference fluid and the horizontal line at $P_r = 0.6$. The proper root is v_3' because we determined the phase to be vapor. The compressibility factor is

$$Z^{\text{ref}} = \frac{P_r v}{T_r} = \frac{(0.6)(1.03321)}{0.95} = 0.6526.$$

Finally, the term $Z^{(1)}$ is

$$Z^{(1)} = \frac{Z^{\text{ref}} - Z^{(0)}}{0.3978} = \frac{0.6526 - 0.6967}{0.3978} = -0.1110.$$

Comments The calculation is straightforward, but some attention is required in solving the BWR equation for volume. Since we do not have the benefit of an analytical expression for the roots, this must be done by a trial-and-error method. The solution obtained in this manner depends on the initial guess. When the equation has three roots, one must be careful to identify the largest and the smallest (the middle root does not have to be calculated—this was done in this example for the sake of completeness). The most foolproof method is to plot the Z isotherm from eq. [A] against volume, draw a horizontal line at the given P_r, and inspect its intersection with the isotherm. To locate a particular root, begin the trial-and-error method with a guess as close to that root as possible.

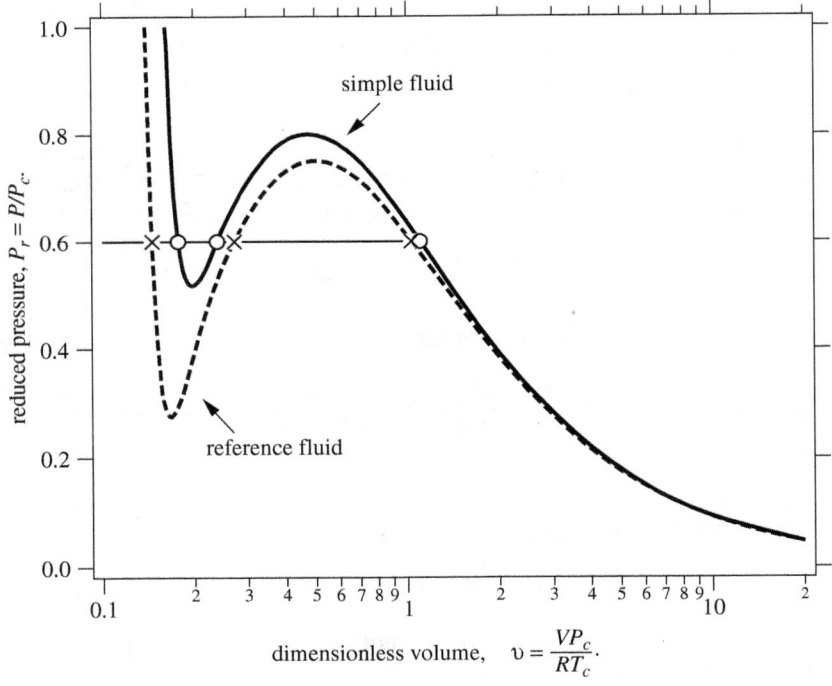

Figure 2-13: Isotherms of the modified BWR equation for the simple and reference fluid at $T_r = 0.95$ (see Example 2.12).

2.10 Thermal Expansion and Isothermal Compression

Table 2-1: Constants of the modified BWR used in the Lee-Kesler correlation (see Example 2.12).

Constant	Simple Fluid	Reference Fluid	Constant	Simple Fluid	Reference Fluid
b_1	0.118119	0.202658	c_3	0	0.016901
b_2	0.265728	0.331511	c_4	0.042724	0.041577
b_3	0.15479	0.027655	d_1	1.55488×10^{-5}	4.87360×10^{-5}
b_4	0.030323	0.203488	d_2	6.23689×10^{-5}	7.40336×10^{-6}
c_1	0.0236744	0.0313385	β	0.65392	1.226
c_2	0.0186984	0.0503618	γ	0.060167	0.03754

NOTE
Another Look at the Pitzer Correlation
Using eq. 2.51 in the Pitzer equation, the compressibility factor is expressed in the form

$$Z = Z^{(0)} + \omega \frac{Z^{\text{ref}} - Z^{(0)}}{\omega^{\text{ref}}},$$

where $\omega^{\text{ref}} = 0.3978$. With $\omega = 0$, this gives $Z^{(0)}$; that is, $Z^{(0)}$ is the compressibility factor of a simple fluid, as we have already seen. Setting $\omega = \omega^{\text{ref}}$ we obtain

$$Z = Z^{\text{ref}},$$

that is, Z^{ref} is the compressibility factor of a fluid with acentric factor $\omega = \omega^{\text{ref}}$. We may now interpret the Pitzer correlation as a linear interpolation between the compressibility factor of a simple fluid ($\omega = 0$) and that of the reference fluid (ω^{ref}). In the Lee-Kesler method, the reference fluid is octane ($\omega^{\text{ref}} = 0.3978$).

2.10 Thermal Expansion and Isothermal Compression

The effect of temperature and pressure on volume is quantified by two coefficients, the volumetric coefficient of thermal expansion, and the coefficient of isothermal compression. The volumetric coefficient of thermal expansion is defined as

$$\beta = \frac{1}{V}\left(\frac{\partial V}{\partial T}\right)_P, \qquad (2.52)$$

and gives the fractional change of volume per unit change of temperature under constant pressure. Its units in the SI system are K^{-1}. The coefficient of isothermal compressibility is defined as[9]

---
9. *Not* to be confused with the compressibility factor, Z.

$$\kappa = -\frac{1}{V}\left(\frac{\partial V}{\partial P}\right)_T, \tag{2.53}$$

and gives the fractional change of volume under pressure at constant temperature. Since volume decreases with increasing pressure, the negative sign is used to ensure that κ is a positive number. Using these definitions, the equation of state can be expressed in a alternative form. First, we write the differential of V in terms of P and T:

$$dV = \left(\frac{\partial V}{\partial T}\right)_P dT + \left(\frac{\partial V}{\partial P}\right)_T dP. \tag{2.54}$$

Using eqs. (2.52) and (2.53) to eliminate the partial derivatives that appear in eq. (2.54), this equation becomes

$$\frac{dV}{V} = \beta dT - \kappa dP. \tag{2.55}$$

If the coefficients β and κ are known, this equation can be integrated to calculate changes in V.

These coefficients are themselves functions of pressure and temperature, but for liquids, the effect of pressure is quite weak; we may take them to depend on temperature only. The coefficient of isothermal compressibility expresses the degree to which the volume of a liquid responds to pressure. Since liquids at temperatures well below the critical are nearly incompressible, the coefficient of isothermal compressibility is approximately zero. The coefficient of thermal expansion is usually small, as liquids expand much less than gases, but it is not zero. *Perry's Chemical Engineers' Handbook* provides data on the thermal expansion of selected liquids.

2.11 Empirical Equations for Density

Rackett Equation A useful and accurate method for the calculation of liquid molar volumes at saturation is the Rackett equation. In its modified form, it gives the molar volume of saturated liquid as

$$V_L = \frac{RT_c}{P_c} Z_R^{1+(1-T_r)^{2/7}} \tag{2.56}$$

where V_L is the molar volume at saturation, T_c is the critical temperature, P_c is the critical pressure, $T_r = T/T_c$ is the reduced temperature and Z_R is a parameter specific to the fluid. *Perry's Chemical Engineers' Handbook* [2] lists the values for

2.11 Empirical Equations for Density

some common fluids[10]. If Z_R is not available, it can be replaced by the compressibility factor at the critical point, $Z_c = P_c V_c/RT_c$, which is usually tabulated. If we use the critical compressibility factor in eq. (2.56), the Rackett equation is simplified and takes the form

$$V_L = V_c Z_c^{(1-T_r)^{2/7}} \tag{2.57}$$

where V_c is the critical volume. The Rackett equation is empirical but quite accurate, even in the above simplified form, which contains no adjustable parameters.

NOTE

Molar Volume of Compressed Liquid
At temperatures well below critical, liquids are practically incompressible and we may assume $\kappa \approx 0$. In this approximation, an isotherm on the PV graph is almost perpendicular to the volume axis. Accordingly, the molar volume of a compressed liquid is essentially the same as that of the saturated liquid at the same temperature. The volume of saturated liquid is often tabulated, or it can be calculated from empirical equations such as the Rackett equation. Therefore, the volume of compressed liquid can be estimated quite accurately from the volume of the saturated liquid at the same temperature. This approximation breaks down close to the critical point where the liquid phase becomes quite compressible.

Example 2.13: Thermal Expansion of Acetone
Perry's Chemical Engineers' Handbook provides the following equation for the thermal expansion of liquids (p. 2-131 in Ref. [2]):

$$V = V_0 \left(1 + a_1 t + a_2 t^2 + a_3 t^3\right)$$

where V_0 is the specific volume at 0 °C, V is the volume at temperature t, and t is in °C. For acetone, the values of the parameters a_1, a_2, a_3, are

$$a_1 = 1.3240 \times 10^{-3} \quad a_2 = 3.8090 \times 10^{-6} \quad a_3 = -0.87983 \times 10^{-8}.$$

Calculate the volumetric coefficient of thermal expansion of acetone at 20 °C and the percent increase in volume upon a temperature increase from 20 °C to 30 °C at constant pressure.

Solution We apply eq. (2.52) to the equation for V given in the problem statement, noting that the derivative must be taken in terms of absolute temperature whereas the equation is given with temperature in °C. Using $T = t + 273.15$, the required derivative is

$$\beta = \left(\frac{1}{V}\right)\left(\frac{dV}{dt}\right)\left(\frac{dt}{dT}\right) = \frac{3a_3 t^2 + 2a_2 t + a_1}{a_3 t^3 + a_2 t^2 + a_1 t + 1}. \tag{A}$$

---

10. Table 2-396 in Ref. [2].

In this equation, t is in °C, and the result is in K^{-1} or in °C^{-1} (with respect to temperature *differences*, as in dT, one degree in the Celsius scale is equal to one kelvin). Using the values of the parameters given in the problem statement at $t = 20$ °C, we find

$$\beta_{20 \,°C} = 1.426 \times 10^{-3} \text{ K}^{-1}.$$

The volume change upon increasing temperature from t_1 to t_2 is found from eq. (2.54). Keeping pressure constant ($dP = 0$), integration with respect to temperature gives

$$\ln \frac{V_2}{V_1} = \int_{t_1}^{t_2} \beta \, dt. \qquad \text{[B]}$$

The coefficient β is a function of temperature and is given by eq. [A]. A quick approximation is obtained by assuming β to be constant and equal to some average value between t_1 and t_2. The above integral then becomes

$$\ln \frac{V_2}{V_1} \approx \bar{\beta}(t_2 - t_1) \quad \Rightarrow \quad \frac{V_2}{V_1} = e^{\bar{\beta}(t_2 - t_1)}. \qquad \text{[C]}$$

For the average β we will use the simple average between the values at 20 °C and at 30 °C. The coefficient at 30 °C is calculated from eq. [A] and we find $\beta_{30 \,°C} = 1.466 \times 10^{-3}$ K^{-1}. The mean β is

$$\bar{\beta} = \frac{(1.426 + 1.466) \times 10^{-3}}{2} = 1.446 \times 10^{-3} \text{ K}^{-1}.$$

The volume change is calculated from eq. [C] using $t_2 - t_1 = 10$ °C $= 10$ K:

$$\frac{V_2}{V_1} = e^{\bar{\beta}(t_2 - t_1)} = e^{(1.446 \times 10^{-3})(10)} = 1.01456.$$

The volume increases by $V_2/V_1 - 1 = 1.45\%$. If instead of using an average β we perform the integration in eq. [B] using the full form of β from eq. [A], we find $V_2/V_1 = 1.01457$. The approximate calculation in this case is essentially the same as the exact result.

Example 2.14: Constant-Volume Heating of Liquid
A glass container is filled with acetone at 25 °C and sealed, leaving no air inside. Determine the pressure that develops in the container when it is heated to 35 °C. The isothermal compressibility of acetone is (see Table 2-188 in *Perry's Chemical Engineers' Handbook* [2]),

$$\kappa = 52 \times 10^{-6} \text{ bar}^{-1}$$

You may ignore the expansion of the container.

Solution If the expansion of the container is neglected, the volume of the acetone (total, specific, or molar) remains constant. Using $dV = 0$ in eq. (2.55) we obtain a relationship between temperature and pressure:

2.11 Empirical Equations for Density

$$0 = \beta dT - \kappa dP \quad \Rightarrow \quad \frac{dP}{dT} = \frac{\beta}{\kappa}.$$

Treating β and κ as constants, integration gives

$$\Delta P = \frac{\beta}{\kappa} \Delta T.$$

Using the average value of β between temperatures 20 °C and 30 °C, calculated in Example 2.13, and the value for κ given in the problem statement, the pressure increase is

$$\Delta P = \frac{(1.446 \times 10^{-3} \text{ K}^{-1})(10 \text{ K})}{(52 \times 10^{-6} \text{ bar}^{-1})} = 278 \text{ bar}.$$

The glass container will most likely break.

Comments This result may seem counterintuitive: if volume increases only by 1.45% when temperature increases by 10 °C, why is it that the pressure generated at fixed volume is so high? This behavior is a consequence of the steepness of the isotherms in the liquid region. Figure 2-14 shows a graph of these isotherms calculated from eq. (2.54) with the values of β and κ for acetone. These lines decrease very steeply, reflecting the fact that volume changes very little with pressure. During constant-volume heating, the state moves along a vertical line from the initial state at 1 bar, 20 °C, to final state at 30 °C. Because isotherms are very steep, pressure increases substantially along a constant-volume path. This behavior is common to all substances that are fairly incompressible. This is, the reason that frozen water pipes burst in winter and water bottles in the freezer break. The physical reason is that molecules in liquids and solids are at close distances from each other, and very close to the point that the intermolecular potential becomes very strongly repulsive (see Figure 1-3). In this region, very small changes to the position of molecules can create enormous repulsion. When we increase temperature, molecules have enough energy for such small changes in position. Normally, the resulting repulsion causes the system to expand but if the volume is constrained, the resulting pressure can be enormous.

Example 2.15: Rackett Equation
Use the simplified Rackett equation to estimate the density of saturated water at 100 °C.

Solution We collect the following data for water:

$$V_c = 56 \times 10^{-6} \text{ m}^3/\text{mol}, \quad Z_c = 0.229, \quad T_c = 647.3 \text{ K}.$$

At $T = 100$ °C $= 373.15$ we have $T_r = 0.5765$. Using the Rackett equation we find

$$V = 56 \times 10^{-6} \times 0.229^{0.4235^{0.2857}} = 17.67 \text{ cm}^3/\text{mol} = 0.9819 \text{ cm}^3/\text{g}$$

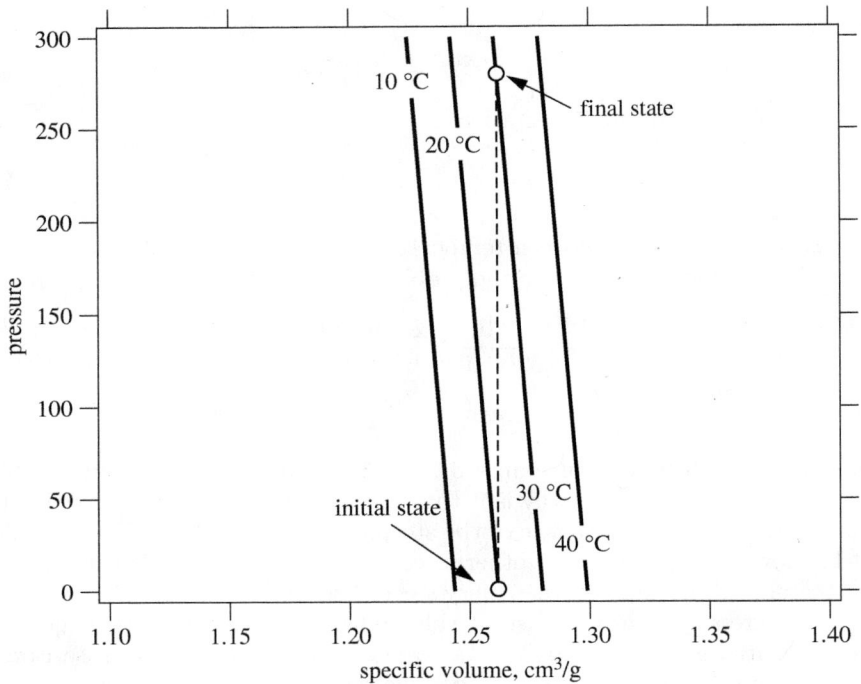

Figure 2-14: Pressure-volume isotherms of acetone (see Example 2.14).

from which we obtain the density

$$\rho = \frac{1}{0.9819} = 1.018 \text{ g/cm}^3.$$

The value from the steam tables is 0.001043 m^3/kg $= 1.043$ cm^3/g, a difference of -2.4%.

Example 2.16: Density of Compressed Liquid
Estimate the density of water at 100 °C, 5 bar.

Solution Under the given conditions water is compressed liquid. The pressure is low compared to the critical point; therefore, we may take the volume to be equal to that of saturated liquid at 100 °C. This value is 0.001043 m^3/kg.

2.12 Summary

The equation of state is a fundamental property of fluids that describes the relationship between molar volume, temperature, and pressure. This relationship is expressed mathematically as an equation between V, P, and T, or alternatively, between Z, P, and T. The PV graph is a graphical representation of this relationship and a convenient way to present the phase behavior of a pure component. It is a good idea to draw a qualitative PV graph when solving problems involving heating, cooling, compression, or expansion of pure fluids, as a way of becoming oriented in thermodynamic space. Simple processes on this graph is represented by simple paths and help visualize any phase transformations that may occur.

The accurate prediction of the molar volume, or, equivalently, of the compressibility factor, is a requirement in engineering calculations. Several methodologies were discussed in this chapter:

- *Tabulated values.* Tabulations are available for a large number of pure components. The steam tables is one such example. Usually, however, property tabulations are not as detailed or extensive as the steam tables. Tables are generally the most accurate source for properties. They are not convenient, however, for large-scale calculations.

- *Generalized graphs.* These graphs are correlations based on corresponding states. Using just three physical constants, critical pressure and temperature, and acentric factor, one can estimate the molar volume of a pure fluid over a very wide range of conditions. This is a very important advantage but it comes with certain limitations. The method is approximate and is based on graphs that have been tweaked to provide overall good accuracy for several different fluids. Necessarily, the agreement will be better for some and worse for others. The method should be used only for normal fluids that are not polar to a significant degree. This covers a large number of compounds but excludes many industrially important molecules, chiefly among them, water.

- *Equations of state.* An equation of state in mathematical form has the advantage that it can be used in repetitive calculations and is especially suited for computer-based calculations. Industrial software for chemical process design makes extensive use of such equations. The equations discussed in this chapter (van der Waals, Soave-Redlich-Kwong, and Peng-Robinson) incorporate the principle of corresponding states in their constants, which can be computed for any fluid given the critical pressure, critical temperature, and acentric factor. These equations must be used with the same caution as generalized graphs: they should be applied to nonpolar fluids only. The constants of these equations, specifically the part that involves the acentric factor, have been obtained from fitting data for several different fluids to obtain overall good agreement among them. It is possible to modify the constants of an equation of state to improve accuracy for a particular substance, however, the constants given here are specifically for normal fluids.

Of the many equations that were discussed here, two deserve a special note:

> *Ideal-gas law.* The ideal-gas law is the theoretical limit of any equation of state at low pressure. It is a universal limit but its range of applicability is limited to the region we call ideal-gas state. You should resist the temptation to apply the ideal-gas law without justification.
>
> *Virial equation.* The virial equation has rigorous basis on theory. However, its truncated form, it should be treated as an extrapolation to somewhat higher pressures that those in the ideal-gas state.

- *Empirical equations.* Empirical equations usually have no basis on theory but give nonetheless very accurate predictions. The Rackett equation is one such example. The limitation is that the range of applicability is typically narrow. The Rackett equation, for example, is only good along the line of saturated liquid.

The availability of several alternative methodologies is a toolbox that facilitates the job of the engineer. It is the responsibility of the engineer to choose the right tool from this toolbox for a given situation.

2.13 Problems

Problem 2.1: a) Use the steam tables to determine the phase of water (liquid or vapor) at following conditions: 25 °C, 1 bar; 80 °C, 10 bar; 120 °C; 50 bar.
b) The vapor pressure of bromobenzene at 40 °C is 10 mm Hg. Determine the phase of bromobenzene at 40 °C, 1 atm.
c) The normal boiling point of fluorobenzene is 84.7 °C. What is the phase of fluorobenzene at 25 °C, 1 atm?

2.13 Problems

Problem 2.2: Determine the temperature and phase of water from the following information. If the phase is a vapor-liquid mixture, report the fraction of vapor and liquid.
a) The specific volume of water is 100 cm^3/g and the pressure 40 bar.
b) The specific volume of water is 100 cm^3/g and the pressure 6 bar.

Problem 2.3: Use the steam tables to do the following:
a) A drum 3.5 m^3 in volume contains steam at 1 bar, 210 °C. Determine the mass of steam in the drum.
b) Wet steam with quality 15% vapor is to be stored under pressure at 20 bar in a thermally insulated vessel. What is the temperature?
c) If the total mass to be stored is 525 kg, what is the required volume of the vessel? The conditions are the same as in part (b).

Problem 2.4: A closed tank contains a vapor-liquid mixture of steam at 45 bar with liquid content 25% by mass.
a) Heat is added until the pressure becomes 80 bar. What is the temperature?
b) The pressure is changed until the contents become 100% saturated vapor. What is that pressure?
c) With pressure held constant at 45 bar, steam is added to or removed from the tank, as needed, until the contents are 25% liquid by volume. Calculate the mass of steam that must be added or removed and report it as a percentage of the total mass originally in the tank.

Problem 2.5: The following data are available for a gas:

T (°C)	P (bar)	V (m^3/kg)
25	0.5	1.1240
25	15	0.0341

To calculate the molar volume at 25 °C, 12 bar, our team came up with two different suggestions: (a) linear interpolation for V between the given pressure, or (b) linear interpolation for the density $\rho = 1/V$ between the two pressures. Which method do you recommend and why? Under what conditions is your recommendation accurate?

Problem 2.6: A 12 m^3 tank contains a liquid/vapor mixture of steam at 15 bar. The volume of the liquid in the tank is 0.5 m^3.
a) What is the temperature?
b) What is the total mass in the tank?

c) What is quality of the steam?
d) 87% of the mass in the tank is removed while keeping pressure constant at 15 bar. What is the final temperature in the tank?

Problem 2.7: An eight-liter pressure cooker contains a mixture of steam and liquid water at 2 bar. Through a level indicator we can see that the liquid occupies 25% of the volume inside the cooker.
a) What is the temperature inside the cooker?
b) What is the total mass of water (liquid plus vapor) in the cooker?
c) What is the mass fraction of the liquid?
d) The cooker, while it remains sealed, is placed under running water until its temperature cools to 25 °C. What is the pressure in the cooker?
e) What force does it take to unseal the cooker? The cover is circular with a radius of 20 cm.

Problem 2.8: a) A pressure cooker is filled to the brim with water at 80 °C and the lid is locked. The temperature is then changed until the contents become saturated liquid. What is the temperature and pressure at that point?
b) A closed pressure cooker contains 50% by volume liquid and 50% water vapor at 1 bar. The pressure is then changed until the point where the contents become a single phase. Is that phase saturated liquid or saturated vapor?
c) A closed rigid vessel that contains a pure fluid is cooled until the contents become saturated vapor. Determine whether the initial state is superheated vapor, compressed liquid, or vapor/liquid.
d) A closed rigid vessel that contains a pure fluid is heated until the contents become saturated liquid. Determine whether the initial state is superheated vapor, compressed liquid, or vapor/liquid.

Problem 2.9: A tank whose volume is 12 m^3 contains 6.2 kg of water at 1.4 bar.
a) What is the phase (liquid, vapor, liquid/vapor mixture)?
b) What is the temperature?
c) We add more steam to the tank while maintaining its temperature constant at the value calculated in part b. As a result, the pressure in the tank increases. Determine how much water (in kg) of steam must be added to bring the steam in the tank to the point of condensation.
d) Draw a qualitative PV graph and show the path of the process for part c.

Problem 2.10: a) Use data from the steam tables to construct the PV graph of water. Show the saturated liquid, the saturated vapor, the critical point. Include the isotherms at 100 °C, 200 °C, 300 °C and 400 °C. Make two plots, one using linear axes and one in which the pressure axis is linear but the volume axis is logarithmic.

2.13 Problems

b) Make a $Z-P$ plot of water using data from the steam tables showing the same information as the PV graph above.

Problem 2.11: Determine whether the truncated virial equation is valid for ethane at the following states:
a) 10 bar, 25 °C.
b) Saturated vapor at 10 bar.
c) 10 bar, −35 °C.
Additional data: The boiling point of ethane at 10 bar is −29 °C.

Problem 2.12: The R&D division of your company has released the following limited data on proprietary compound X-23:

T (°C)	P (bar)	ρ (kg/m^3)
25	0.01	0.0177
25	20	39.8

Saturation pressure at 25 °C: 64.3 bar
Critical temperature: below 35 °C

Using this incomplete information estimate as best as you can the following:
a) Phase of X-23 at 12 bar, 25 °C.
b) The molar mass of X-23.
c) The second virial coefficient at 25 °C.
d) The required volume of a tank that is needed to store 20 kg of X-23 at 12 bar, 25 C.
e) State clearly and justify as best as you can all your assumptions and the methods you use.

Problem 2.13: Methane is stored under pressure in a 1 m^3 tank. The pressure in the tank is 20 bar and the temperature is 25 °C.
a) Calculate the compressibility factor of methane in the tank from the virial equation truncated after the second term.
b) What is the amount (moles) of methane in the tank?
c) You want to store twice as much methane in the tank at the same temperature. What will be the pressure in the tank?
d) Is it appropriate to use the virial equation for this problem? Explain.
At 25 °C the second virial coefficient of methane is -4.22×10^{-5} m^3/mol.

Problem 2.14: Use the truncated virial equation to answer the questions below:
a) A 5 m^3 tank contains nitrogen at 110 K, 7 bar. How many kg of nitrogen are in the tank?
b) The tank is cooled until the contents become saturated vapor. What is the pressure and temperature in the tank?

c) Is the use of the truncated virial equation justified in this problem?
Additional data: The saturation pressure of nitrogen is given by the following empirical equation:
$$P^{\text{sat}}(T) = e^{14.9542 - \frac{588.72}{-6.6+T}}$$
with P^{sat} is in mm Hg and T is in kelvin.

Problem 2.15: a) Calculate the second virial coefficient of water at 200 °C using only data from the steam tables.
b) Use the truncated virial equation along with the second virial coefficient calculated above to estimate the volume of water at 200 °C, 14 bar. How does this value compare with the volume obtained from the steam tables? Discuss this comparison.

Problem 2.16: 1000 kg of methane is to be stored in a tank at 25 °C, 75 bar. What is the required volume of the tank? (*Hint:* Use the Pitzer equation with the Lee-Kesler values.)

Problem 2.17: 200 kg of carbon dioxide are stored in a tank at 25 °C and 70 bar.
a) Is carbon dioxide an ideal gas under the conditions in the tank?
b) What is the volume of the tank?
c) How much carbon dioxide must be removed for the pressure of the tank to fall to 1 bar?
The molecular weight of CO_2 is 44 g/mol.

Problem 2.18: A full cylinder of ethylene (C_2H_4) at 25°C contains 50 kg of gas at 80 bar.
a) Is ethylene an ideal gas under these conditions? Explain.
b) What is the volume of the cylinder?
c) What is the pressure in the cylinder after 90% of the ethylene has been removed, if temperature is 25°C?
The molecular weight of ethylene is 28 g/mol.

Problem 2.19: Use the Lee-Kesler method to answer the following:
2000 kg of krypton is to be stored under pressure in a tank at 110 bar, 20 °C. The tank is designed to withstand pressures up to 180 bar.
a) Determine the volume of the tank.
b) Is it safe to store 2500 kg in the tank at 25 °C?
c) Is the Lee-Kesler method appropriate?

Problem 2.20: A tank is divided by a rigid, thermally conducting partition into two equal parts, A and B, each 10 m³ in volume. Part A contains saturated liquid

2.13 Problems

n-butane at 20 °C, 2.07 bar; part B contains saturated n-butane vapor, also at 20 °C, 2.07 bar. Each part is equipped with a safety alarm that will go off if pressure exceeds 40 bar.
a) How many moles of n-butane is in part A of the tank?
b) How many moles of n-butane is in part B of the tank?
c) The tank is heated slowly in such a way that the temperature in both parts rises at the same rate. As soon as the alarm goes off, the heating stops. Which alarm goes off, that of part A or part B?
d) What is the temperature when the alarm sounds?
Additional data: The volume expansivity and the isothermal compressibility of liquid n-butane are given below and may be assumed to be constant:

$$\beta = 2.54 \times 10^{-3} \text{ K}^{-1}, \quad \kappa = 3.4 \times 10^{-4} \text{ bar}^{-1}.$$

Problem 2.21: a) A tank contains 10,000 kg of xenon at 132 °C, 82 bar. The plant supervisor asks you to remove xenon and fill the tank with 10,000 kg of steam at 200 °C. What is the pressure in the tank when it is filled with steam?
b) After the tank has been filled with steam, 5000 kg are withdrawn for use elsewhere in the plant. What is the pressure, if temperature remains at 200 °C?

Problem 2.22: The boiling point of o-xylene at 1 bar is 139 °C.
a) What is the state of o-xylene at 0.1 bar, 200 °C?
b) 100 moles of o-xylene are to be loaded in a tank at 0.1 bar, 200 °C. What is the required volume of the tank?
c) The tank can safely withstand pressures up to 44.9 bar. How much o-xylene can be stored in the tank under maximum pressure at 200 °C?

Problem 2.23: a) Determine the percent change in volume when olive oil is heated at constant pressure from 18 °C to 40 °C.
b) Olive oil is stored in a full container at 18 °C. Determine the pressure that will develop in the container if the temperature in the storage room rises to 40 °C.
Additional data: The volume expansion of olive oil is given by the empirical equation

$$V = V_0(1 + a_1 t + a_2 t^2 + a_3 t^3)$$

where t is temperature in °C, V_0 is the volume at 0 °C, and the coefficients in the equation are

$a_1 \times 10^3$	$a_2 \times 10^6$	$a_3 \times 10^8$
0.6821	1.1405	−0.539

The coefficient of isothermal compression is

$$\kappa = 52 \times 10^{-6} \text{ bar}^{-1}.$$

(Data from *Perry's Chemical Engineers' Handbook*, 7th ed., Tables 2-147 and 2-188.)

Problem 2.24: A 0.5 m³ tank will be used to store CO_2 at 20 °C. Using the SRK equation answer the following:
a) Determine the maximum amount (kg) of CO_2 that can be stored safely if the tank can withstand a maximum pressure of 70 bar.
b) Repeat if the maximum pressure is 60 bar.
c) Repeat at 50 bar.

Problem 2.25: Use the SRK equation to answer the following:
a) 5000 kg of isobutane is to be stored in a tank at 60 psi, 70 °F. What is the required tank volume?
b) Since the temperature in the summer can get as high as 95 °F, determine the pressure that the tank must withstand to avoid rupture.

Problem 2.26: Use the SRK equation to calculate the molar volume of isobutane at the following states:
a) 30 °C, 1 bar.
b) 30 °C, 10 bar.
c) Saturated liquid at 30 °C.
d) Saturated vapor at 30 °C.
For parts c and d, compare with the saturated molar volumes reported in the NIST Web Book.
Additional information: The saturation pressure at 30 °C is 4.05 bar.

Problem 2.27: Use the SRK equation to perform the following calculations for isobutane to make a graph that shows three isotherms, one at 30 °C, one at the critical temperature, and one at 150 °C. Make the axis logarithmic in volume and linear in pressure, and select the range in the two axes so that the graph is not crowded and its important features are seen clearly. Report volume in m³/mol, pressure in bar and temperature in °C. Annotate the graph and label the axes properly.

Problem 2.28: The parameters β and κ of a substance are reported to be functions of pressure and temperature and are given below:

$$\beta = 1/T, \qquad \kappa = 1/P.$$

a) Determine the equation of state. Assume that at pressure P_0 and temperature T_0 the volume is known to be V_0. Your final answer then should be in terms of P, V, T, P_0, T_0 and V_0.
b) Is this equation appropriate for liquids?

2.13 Problems

Problem 2.29: Calculate the coefficient of isothermal compressibility of isobutane as saturated liquid and saturated vapor at 30 °C, $P^{\text{sat}} = 4.05$ bar, using the SRK equation. Report the result in bar^{-1}.

Hint: Recall from calculus that

$$\left(\frac{\partial V}{\partial P}\right)_T = 1 \bigg/ \left(\frac{\partial P}{\partial V}\right)_T.$$

Problem 2.30: Use the steam tables to calculate the value of β of water at the following states:
a) 1 bar, 25 °C.
b) 20 bar, 25 °C.
c) 1 bar, 200 °C.

Problem 2.31: Use the Rackett equation to estimate the volume expansivity of liquid ethanol at 25 °C. State your assumptions clearly.

Chapter 3

Energy and the First Law

A classical mechanical system is characterized by a set of mechanical state variables: velocity, elevation in a gravitational field, electrical charges, and so forth. If these variables are given and the external fields are known, the system is fully specified and its behavior at any instant of time, past or future, can be calculated. Thermodynamic systems are characterized by an additional state variable: *temperature*. Unlike mechanical variables, which describe external interactions, temperature characterizes the *internal* state of macroscopic matter. Temperature is a measure of the energy stored inside matter in various forms, which we collectively call *internal energy*. Temperature gives rise to another type of energy exchange that is not encountered among purely mechanical systems, *heat*: when two systems, each at its own temperature, are put into thermal contact, energy in the form of heat flows from the higher temperature to the lower temperature. For the complete energy balance, heat and internal energy must both be accounted for.

The incorporation of heat effects into the energy balance constitutes one of the fundamental principles of thermodynamics known as the *first law*. This chapter is devoted to the mathematical formulation of the first law and in the definition of two thermodynamic properties, *internal energy* and *enthalpy*, both of which are important in the calculation of energy balances.

Instructional Objectives In this chapter we will formulate the mathematical statement of the first law for a closed system and will learn how to:

1. Do energy balances in closed systems.

2. Distinguish between path and state functions.

3. Use the steam tables to calculate internal energy and enthalpy.

4. Use heat capacities to calculate changes in internal energy and enthalpy.

5. Apply the energy balance to systems undergoing vaporization or condensation.

6. Perform calculations of internal energy and enthalpy in the ideal-gas state.

3.1 Energy and Mechanical Work

Heat, work and energy are measured in the same units but they represent different physical entities. We begin with work and energy, which are familiar concepts from mechanics. Energy is the ability of a system to produce work, namely, ability to cause the displacement of a force. Consider an object with mass m resting on the floor (Figure 3-1). To lift it to the top of the table at height Δz above ground, we must supply an amount of work equal to $W = mg\Delta z$. When the mass is resting on the desk, its potential energy has increased by $\Delta E_p = mg\Delta z$, and the work done to lift the object is now part of the energy of the body and is stored as potential energy. The energy added to the system will stay with it for as long it remains at rest on the table. This could be hours, days, or millennia: potential energy is preserved for as long as the *state* of the system, in this case elevation, is preserved. The more general conclusion to be drawn from this analysis is that energy is a *property of the state* and as such, a *storable* quantity. We can recover it by letting the mass drop to the floor and use its kinetic energy to, say, catapult a small projectile into the air, or to load a spring, or to push a nail into the floor.

Let us focus on work now. Work takes place when a force is displaced. If we lift the weight ourselves, this work is supplied through the action of our muscle. Once the object is placed on the table, there is no more displacement, thus no more work: work is exchanged *during* a *process*. It characterizes, not the state of the system, but the transition of the system between states (from the floor to the top of

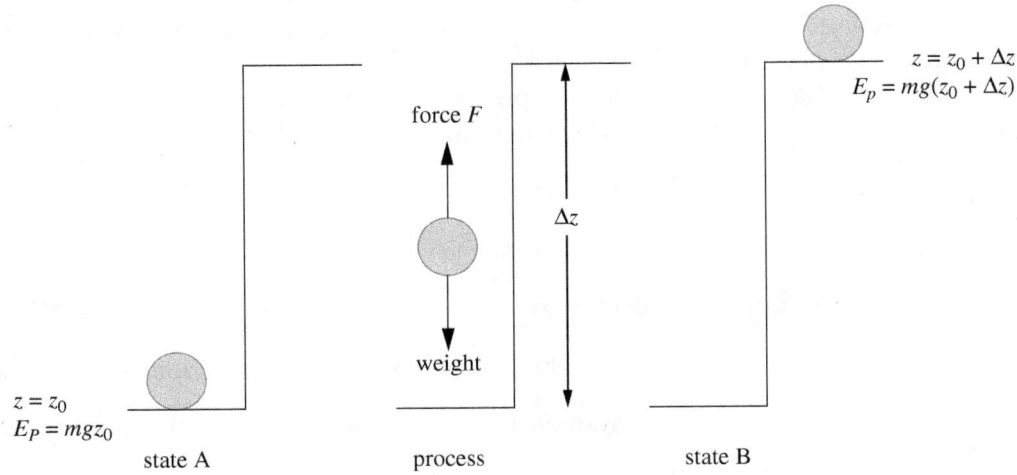

Figure 3-1: Potential energy and mechanical work. From state *A* to state *B* the potential energy increases by the amount of work needed to lift the weight. This work must be supplied from the surroundings, that is, from a force external to the system.

3.1 Energy and Mechanical Work

the desk). In thermodynamic terms, the picture illustrated in Figure 3-1 would be described as follows: *A system initially in equilibrium state A undergoes a process that brings it to final state B; during this process the system exchanges work with the surroundings so that the energy change of the system is equal to the amount of work exchanged:*

$$\Delta E_{AB} = W. \tag{3.1}$$

This equation reads, "as a result of the process, the energy of the system in state B has increased relative to A by the amount of work added." Also notice that work represents energy that passes from *one* system into *another*. In this example, our muscle (chemical energy) transfers work to the lifted object, where it is converted into potential energy. As a transfer quantity, work is characterized by a direction, "from" (the muscle) "to" (the system). Finally, when work passes into a new system, it is converted into some form of energy (in our example, potential), which now is property of the state. Let us summarize:

- Energy is *storable*; it remains in the system for as long as the state of the system is preserved. It is a *state* function.

- Work is energy in *transit*: it appears when energy is passed from one system into another.

- Work is *not* a storable quantity as such.[1] Work that enters a system must be stored in some form of energy.

- Work is associated with a direction "from" one system in which it originates, "to" another, where it is transferred to.

To indicate the direction of the transfer we adopt a sign convention:

Sign convention for work: Work is positive if it enters the system, negative if it exits.

When the surroundings do the work, the work is positive because it is absorbed by the system. When the work is negative, it is done by the system and absorbed by the surroundings. Clearly, the sign depends on the definition of the system since work that enters one system (the "system") exits the other (the "surroundings") and vice versa. To avoid ambiguity, the definition of the system must be made clear when we report positive or negative values of work.

---

1. By this we mean that it is incorrect to say that a system "stores work"; the proper statement is, the system "stores energy."

NOTE

Forms of Energy

As a matter of classification, energy comes in two basic forms, kinetic, and potential. Kinetic is energy stored in the motion of a moving object. Potential energy is a more general category and encompasses several forms that arise from various conservative forces, such as gravitational fields, electric charges, spring forces, and others. All forms of potential energy have the common characteristic that they are described by a potential function $\Phi(x, y, z)$ that depends on the space coordinates, x, y, z, and has the property that the force in a given direction is equal to the negative derivative of the potential with respect to the corresponding coordinate:

$$F_x = -\frac{\partial \Phi(x,y,z)}{\partial x}, \quad F_y = -\frac{\partial \Phi(x,y,z)}{\partial y}, \quad F_z = -\frac{\partial \Phi(x,y,z)}{\partial z}.$$

As an example, in the gravitational potential $\Phi(x, y, z) = mgz$, the force (weight) in the z direction is

$$F_z = -\frac{\partial (mgz)}{\partial z} = -mg,$$

with the negative sign indicating that the weight points downwards. The forces in the x and y direction are both zero. The intermolecular potential introduced in Section 1.1 is another example of potential energy.

3.2 Shaft Work and *PV* Work

A direct method by which to exchange work with a fluid is through the use of a mechanical device. When wind blows through a turbine, it produces work that is manifested in the rotational motion of the turbine. A centrifugal pump imparts work to a fluid through the rotation of an impeller. This type of work is called *shaft* work. It will generally be recognizable through the presence of a rotating or reciprocating shaft that absorbs work from a fluid or imparts work to it.

Another more subtle form of work is associated with movement of the system boundaries. Thermodynamic systems make mechanical contact through the pressure that is exerted on the boundaries that separate them. At equilibrium, this pressure is equal on both sides of the boundary. If a pressure imbalance arises between the system and its surroundings, the boundaries of the system must move in response to the mechanical force. Such imbalance may arise from the application of a mechanical force that acts to compress or expand the system, or through the application of heat, which causes the volume to expand or contract. The movement of boundaries involves the exchange of work, which we call PV work.

To obtain an equation for the PV work, consider the compression of a gas in a cylinder fitted with a piston whose area is a (Fig. 3-2b). Compression is done

3.2 Shaft Work and PV Work

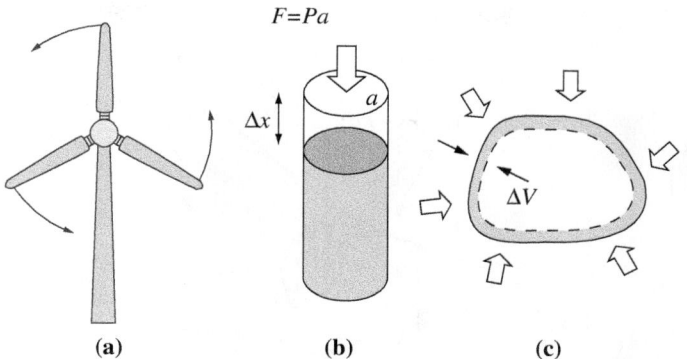

Figure 3-2: Examples of work: (a) shaft work (wind turbine); (b) *PV* work for compression/expansion of a gas in a cylinder; (c) *PV* work associated with volume changes.

through the application of an external pressure P_{ex}, which produces a force on the piston, $F = P_{\text{ex}} a$, and causes the piston to move by dx. The amount of work associated with this process is equal to the force on the piston, $P_{\text{ex}} a$, multiplied by the displacement dx. Noting that the product adx is equal to the change of volume, $-dV$, the amount of work is

$$dW = -(P_{\text{ex}} a)\, dx = -P_{\text{ex}}\, dV. \qquad (3.2)$$

The negative sign is consistent with the convention adopted for work: in compression, dV is negative and the work is positive (it is added to the system); in expansion, dV is positive and the work is negative (work is transferred from the system to the surroundings).

Equation (3.2) gives the work in terms of the external pressure that causes the change. It is more convenient, however, to relate the work to the pressure of the system. For compression, the external pressure must be higher than the pressure of the system, P, which means, $P_{\text{ex}} = P + \delta P$, where δP is a positive increment; for expansion, δP is negative. If the process is conducted in a quasi-static manner, then $\delta P \to 0$ and $P_{\text{ex}} \to P$. Such process is called *mechanically reversible* and in this case the PV work is

$$\boxed{dW_{\text{rev}} = -PdV} \qquad \text{mechanically reversible process} \qquad (3.3)$$

This is the differential amount of work along a small step of a process that causes a volume change. The total amount of work involved in taking the system from an initial state A to final state B along a reversible path is obtained by integration:

$$\boxed{W_{\text{rev}} = -\int_A^B PdV} \qquad \text{mechanically reversible process} \qquad (3.4)$$

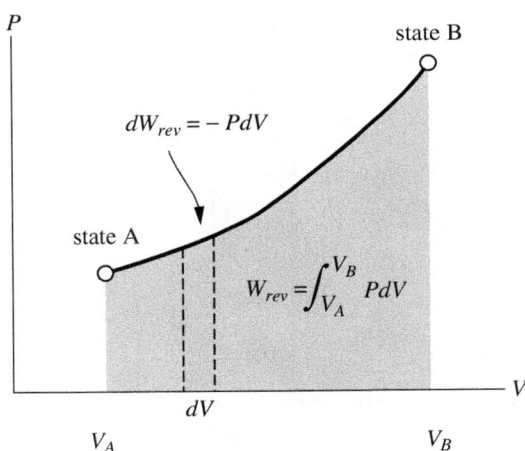

Figure 3-3: Graphical interpretation of reversible PV work.

This result is general and not limited to gases, nor to objects of cylindrical shape. It applies to any system, gas, liquid, or solid, whose volume changes by compression or expansion, under or against the opposing pressure of the surroundings. The only requirement is that the process must be mechanically reversible.

Reversible PV work has a simple graphical interpretation on the PV graph: it is equal in absolute value to the area under the path of the process when this is plotted on the PV graph (Figure 3-3). The sign indicates whether the work is added to or removed from the system. For a path that moves in the direction of increasing V, the area is by convention positive and the work is negative: the system expands and does work against the surroundings. For a process that moves in the direction of decreasing V (compression), the area is negative and the work is positive.

An important conclusion is that PV work depends on the *entire path* that connects the two states. Different paths between the same two states generally correspond to different areas, and thus different amounts of work. Therefore, the PV work is a *path* function whose value depends not only on the initial and final states, but on the entire path. This is in contrast to thermodynamic properties (state functions) whose value depends on the state alone.

Example 3.1: *PV Work in Expansion*
A cylinder fitted with a piston contains 1 liter of gas. The piston has a 1-in diameter and weighs 5 kg. How much work is needed to expand the gas reversibly to twice its volume against the pressure of the atmosphere (1.013 bar)?

3.2 Shaft Work and PV Work

Solution The gas expands against the combined pressure of the piston and the atmosphere. The weight of the piston is $Mg = (5 \text{ kg})(9.81 \text{ m/s}^2) = 49.05$ N and the area of the piston is $\pi D^2/4 = \pi (2.54 \times 10^{-2} \text{ m})^2 = 5.067 \times 10^{-4} \text{ m}^2$. The pressure exerted by the piston is

$$P' = \frac{49.05 \text{ N}}{5.067 \times 10^{-4} \text{ m}^2} = 0.968 \text{ bar}.$$

The total pressure in the cylinder is

$$P = P' + P_0 = 0.968 \text{ bar} + 1.013 \text{ bar} = 1.981 \text{ bar}.$$

The expansion is to take place against an external pressure of 1.981 bar, which remains constant during the process. If the volume doubles, the change in volume is $1 \text{ L} = 1000 \text{ cm}^3 = 10^{-3} \text{ m}^3$. The corresponding amount of work is

$$W = \int P dV = P \Delta V = (1.981 \text{ bar})(1 \times 10^{-3} \text{ m}^3) = -198.1 \text{ J}.$$

The negative sign indicates that this work is done *by* the system on to the surroundings.

Example 3.2: *PV* Work Using the Soave-Redlich-Kwong Equation of State
Ethylene is compressed reversibly in a closed system. The compression is conducted isothermally at 350 K, from initial pressure 20 bar to final pressure 55 bar. Calculate the work using the SRK equation of state.

Solution First, we write the SRK equation in the form,

$$P = \frac{RT}{V-b} + \frac{a}{b}\left(\frac{1}{V+b} - \frac{1}{V}\right).$$

To obtain the work, this pressure must be integrated at constant temperature as a function of V:

$$W = -\int_{V_1}^{V_2} P dV = -RT \ln \frac{V_2 - b}{V_1 - b} + \frac{a}{b}\left(\ln \frac{V_2 + b}{V_1 + b} - \ln \frac{V_2}{V_1}\right). \quad [A]$$

Numerical substitutions The critical parameters of ethylene and the acentric factor are:

$$T_c = 282.35 \text{ K}, \quad P_c = 50.418 \text{ bar}, \quad \omega = 0.0866.$$

Using these values, the constants a and b are:

$$a = 0.404344 \text{ J m}^3/\text{mol}^2, \quad b = 4.03395 \times 10^{-5} \text{ m}^3/\text{mol}.$$

The volume V_1 must be obtained by solving the SRK equation. As in Example 2.11, we solve for the compressibility factor first. At the initial state (20 bar, 350 K), the equation for Z is

$$Z^3 - Z^2 + 0.0670103 Z - 0.00264794 = 0.$$

This equation has one real root, $Z = 0.931084$, from which we obtain the molar volume of the initial state:

$$V_1 = \frac{ZRT}{P} = \frac{(0.931084)(8.314 \text{ J/mol K})(350 \text{ K})}{20 \times 10^5 \text{ J/m}^3} = 0.00135468 \text{ (m}^3/\text{mol}).$$

In the final state, the equation for the compressibility factor is

$$Z^3 - Z^2 + 0.180579 Z - 0.020025 = 0.$$

This equation has one real root, $Z = 0.806979$, and the corresponding molar volume is

$$V_2 = \frac{ZRT}{P} = \frac{(0.806979)(8.314 \text{ J/mol K})(350 \text{ K})}{55 \times 10^5 \text{ J/m}^3} = 0.00117411 \text{ (m}^3/\text{mol}).$$

Finally, by numerical substitution into eq. [A], the isothermal work is

$$W = 385.6 \text{ J/mol}.$$

The work is positive, as it should be, since volume decreases.

Example 3.3: Isothermal Compression of Steam
Steam undergoes reversible isothermal compression at 350 °C, from initial pressure 20 bar to 40 bar. Calculate the amount of work.

Solution This problem is very similar to the previous one (see Example 3.2). Here, however, the pressure-volume relationship is given in tabular form and the integration of the PV work must be done numerically. First, we collect values of the specific volume V at 350 °C at various pressures between the final and initial state:

P (bar)	V (m³/kg)	P (bar)	V (m³/kg)
20	0.1386	32	0.08454
22	0.1255	34	0.07923
24	0.1146	36	0.07451
26	0.1053	38	0.07028
28	0.0974	40	0.06647
30	0.09056		

One way to perform the integration is using the trapezoidal rule within each pressure interval. An alternative method, which is generally more accurate, is to fit an equation

3.2 Shaft Work and PV Work

to the data and perform the integration analytically using the fitted equation. We will illustrate the use of both methods.

Trapezoidal rule: In this method, the integral of n tabulated values $\{x_i, f_i\}$ of a function $f(x)$ is given by the following approximation:

$$\int_{x_1}^{x_n} f(x)\,dx \approx \frac{1}{2}\sum_{i=2}^{n}(y_i + y_{i-1})(x_i - x_{i-1}).$$

Using $x_i = V_i$, $y_i = P_i$, we construct the following table:

i	V_i	P_i	$\frac{1}{2}(P_i + P_{i-1})(V_i - V_{i-1})$
1	0.1386	20	
2	0.1255	22	-0.2751
3	0.1146	24	-0.2507
4	0.1053	26	-0.2325
5	0.0974	28	-0.2133
6	0.09056	30	-0.1984
7	0.08454	32	-0.1866
8	0.07923	34	-0.1752
9	0.07451	36	-0.1652
10	0.07028	38	-0.1565
11	0.06647	40	-0.1486
			$\sum = -2.0021$

From the sum of the last column we find,

$$\int_{V_1}^{V_2} P\,dV = -2.0021 \text{ bar m}^3/\text{kg} = -200.2 \text{ kJ/kg}.$$

The work is given by the negative of the above integral:

$$W = -\int_{V_1}^{V_2} P\,dV = 200.2 \text{ kJ/kg}.$$

Fitting the data: The data are fitted to a quadratic polynomial in V,

$$P = a_0 + a_1 V + a_2 V^2$$

with coefficients

$$a_0 = 81.4217, \quad a_1 = -801.366, \quad a_2 = 2598.63.$$

We use the quadratic equation to calculate the integral between V_1 and V_2:

$$W = -\int_{V_1}^{V_2} P\,dV = -a_0(V_2 - V_1) - \frac{a_1}{2}(V_2^2 - V_1^2) - \frac{a_2}{3}(V_2^3 - V_1^3).$$

Using the fitted parameters and $V_1 = 0.1386$ m^3/kg, $V_2 = 0.06647$ m^3/kg, we find

$$W = 1.998 \text{ bar m}^3/\text{kg} = 199.8 \text{ kJ/kg}.$$

Figure 3-4 shows the data and the fitted line. In this case both methods give nearly identical results because the tabulation includes closely spaced points.

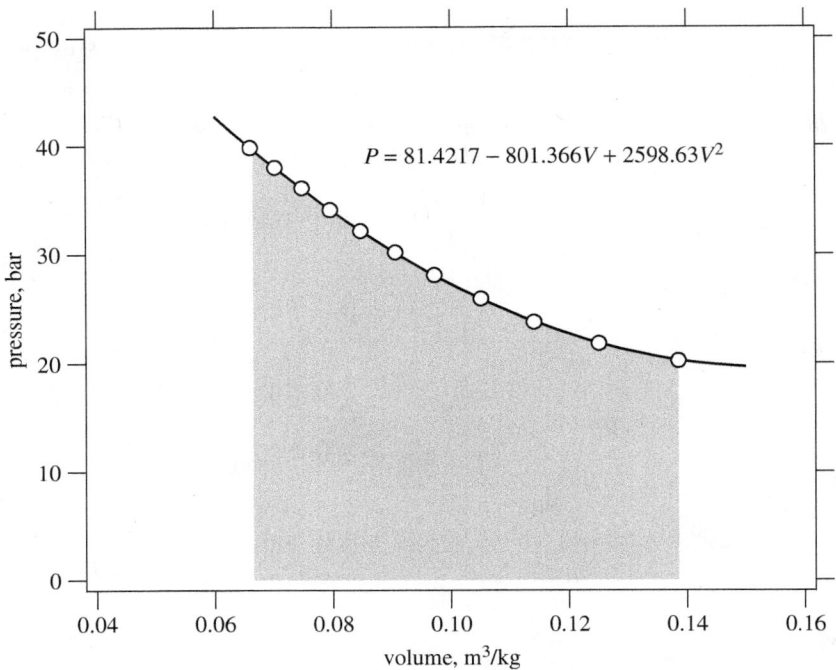

Figure 3-4: Calculation of isothermal PV work at 350 °C from steam tables. The solid line is a quadratic fit to the points from the tables (see Example 3.3).

3.3 Internal Energy and Heat

Molecules possess energy of various forms. All molecules possess kinetic energy due to the motion of the center of mass in space (translational kinetic energy). In addition to translational kinetic energy, polyatomic molecules possess rotational kinetic energy, manifested as motion relative to the center of mass, and vibrational energy, as motion of the constituent atoms relative to the equilibrium length of their chemical bonds. Molecules also possess potential energy as a result of interactions between different molecules as well as between atoms of the same molecule. These interactions can be loosely visualized as springs whose compression or extension requires energy.[2] For our purposes it is not necessary to consider these molecular modes of energy storage in any detail. It suffices to say that matter, regardless of chemical composition or phase, is capable of storing energy internally. We refer to these combined storage modes as *internal energy* and we will use the symbol U. Like all forms of energy, internal energy is a *state function* and for a pure substance, it is a function of pressure and temperature.

---

2. To apply the spring analogy to intermolecular interactions we imagine the spring constant to be variable with distance: the intermolecular spring stiffens when highly compressed but becomes very soft when it is overextended.

3.3 Internal Energy and Heat

The molecular nature of matter gives rise to a different type of energy transfer, *heat*. The mean kinetic energy of molecules in a hot substance is higher than that in a cold substance. The nature of molecular collisions is such, that when molecules with low kinetic energy are mixed with molecules of higher energy, energy is exchanged in a way that gives all molecules the same energy on average. The effect of collisions then is to spread the energy among molecules, much like the cue ball distributes its energy among all balls on a pool table. This process amounts to transferring energy from "hot" molecules to "cold" ones. Collisional transfer of energy does not require intimate mixing and it can take place when two solids are placed into contact, or when two fluids come into contact with a common solid wall. Molecules in the solid phase, though restricted in the range of their motion, move vigorously about their equilibrium position and collide with their neighbors, thus absorbing energy from higher-energy molecules and transferring it to those with lower energy. The rate of this transfer varies among materials, but this affects only how long it takes to reach equilibrium, not the final equilibrium itself. A thermal *insulator*, we may note here, is a material that offers a resistance to heat transfer so that the amount transferred over the duration of a typical application is negligibly small. Over time, however, heat will escape, even through an insulator, and thermal equilibrium will be established across all systems that are in contact with each other.

Heat shares some important characteristics with work:

1. It is a *transient* form of energy that is observed during a change of state (process); once the system is in equilibrium with its surroundings there is no net heat transfer because both system and surroundings are at the same temperature.[3]

2. As a transient form, heat is not a storable mode of energy: once it enters a system, it is stored as *internal energy*. It is incorrect to say that a system "contains heat," or to speak of energy that "is converted into heat."

3. It has a direction, from the system to the surroundings, or vice versa.

4. It is a *path* function whose value is determined by the entire path of the process. This property of heat is not obvious at the moment but will become so in the next section.

Sign convention for heat: Heat is positive if it is transferred to the system from the surroundings, negative if it is transferred from the system to the surroundings.

---

[3]. On an instantaneous basis, individual collisions may transfer energy in either direction, from the system to the bath or the other way around, even at equilibrium. Overall, however, there is no net transfer between system and surroundings if the conditions of thermal equilibrium are met.

3.4 First Law for a Closed System

All material systems possess internal energy. They may possess various other forms of energy as well. In systems of interest to chemical engineers the most relevant forms are kinetic energy due to bulk motion (for example, a moving fluid in a pipe) and potential energy, due to changes in elevation (for example, a piping network that spans various elevations). Consider a closed system, originally at state A, undergoing a change of state to final state B, as shown in Figure 3-5. Changes of state invariably require the exchange of energy between the system and the surroundings. Such exchanges must be in the form of work (PV, shaft, or both) and/or heat. Suppose that the net amount of work that is exchanged during this process is W, and the net amount of heat is Q. The work W is the combined amount of PV and shaft work involved in the process. As a result of the transfer of work and heat, the energy content of the system changes during a process. By energy conservation, the energy content in the final state (E_B) is increased relative to that at the initial state (E_A) by the amount of work and heat added to the system:

$$E_B = E_A + Q + W. \tag{3.5}$$

This we can write in the equivalent form,

$$\Delta E_{AB} = Q + W, \tag{3.6}$$

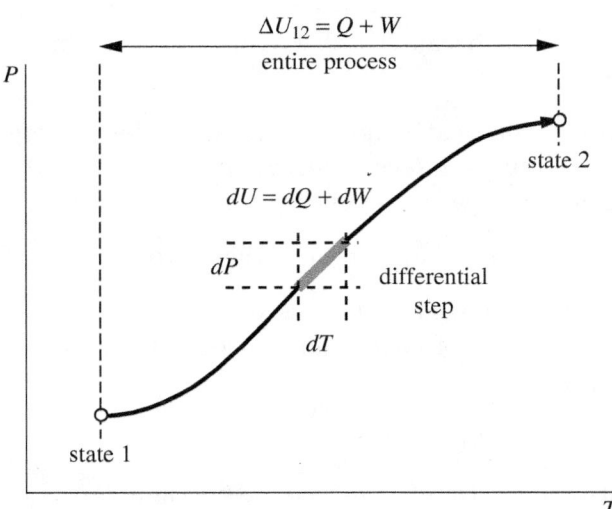

Figure 3-5: Schematic illustration of the differential and integral expressions of the first law.

3.4 First Law for a Closed System

where $\Delta E_{AB} = E_B - E_A$ is the difference in the energy between the initial and final states. The terms E_A and E_B refer to the total energy stored in the system in all possible forms. In general, this includes internal (U), kinetic (E_k), and potential (E_p) energy:[4]

$$E = U + E_k + E_p. \tag{3.7}$$

Equation (3.6) may now be written as

$$\Delta(U + E_k + E_p)_{AB} = Q + W. \tag{3.8}$$

It will often be the case that kinetic and potential energy are either not present or their contribution to the energy balance is negligible. Whenever these conditions are met, these energy terms may be dropped and the first law takes the simpler form,

$$\boxed{\Delta U = Q + W,} \tag{3.9}$$

with the subscript AB dropped from ΔU for simplicity. This equation is the mathematical statement of the first law for a closed system. It extends the familiar energy balance to include heat and internal energy, two terms that arise from molecular phenomena. The significance of the first law is that it relates internal energy, an "invisible" form of energy, to heat and work, both of which are directly measurable. If eq. (3.9) is applied to a differential step along the process for a small change of state from (P, T) to $(P + dP, T + dT)$, we obtain

$$\boxed{dU = dQ + dW.} \tag{3.10}$$

This equation is the differential of eq. (3.9). Conversely, eq. (3.9) is the integrated form of this differential between two states.

NOTE

The Operator $\Delta(\cdots)$

The operator Δ is introduced as shorthand notation for a change of a property between two states. Given a property X and a change of state from A to B, ΔX_{AB} is defined as

$$\Delta X_{AB} = X_B - X_A.$$

If no subscripts are given, the difference will be understood to be (final) − (initial):

$$\Delta X = X_{\text{final}} - X_{\text{initial}}.$$

Here, X stands for any state function. It is *incorrect* to write ΔW or ΔQ: neither work nor heat are state functions; therefore, they cannot be represented by a difference Δ. A differential

---
4. If other forms of energy are relevant, they should be included on the right-hand side as needed.

amount of work or heat is written as *dW*, or *dQ*; for the integrated quantity we will write *W* and *Q*, *without* a Δ in front. The notation δQ and δW will be used in numerical calculations to denote finite approximations of *dQ*, *dW*, respectively.

NOTE

A Brief History of the First Law

Throughout history humans have valued mechanical work more than heat. Work makes cars move, airplanes fly, electrons perform binary computations. However, sources of raw work in nature, while plentiful, are often in forms that are not immediately usable without considerable engineering ingenuity: waterfalls, wind, ocean currents, avalanches. Heat, on the other hand, is plentiful and easily available by burning wood, coal, or oil. By converting heat into work we obtain a valuable form of energy from a plentiful form of energy. The technological development that made such conversion possible on a grand scale was the steam engine: heat is used to produce high-pressure steam, which is then used to push a piston and produce shaft work. This engineering development attracted interest in the theoretical problems raised by these new machines, in particular, how to maximize the amount of work that can be extracted from a given amount of heat. The answer to this problem was made difficult by the confusion surrounding the nature of heat. A popular scientific theory that persisted during the 18th and 19th centuries was the "caloric" theory of heat, which believed heat to be an invisible fluid contained inside matter. Heat was given its own units (calorie, Btu), which survive to date, and which are based on the temperature rise that follows the addition of heat.[5] James Prescott Joule demonstrated that mechanical work raises temperature by a specific amount per unit of work, thus establishing the equivalence between heat and work. The microscopic explanation for heat would come with the development of statistical thermodynamics, through the works of William Thomson (Lord Kelvin), Boltzmann, Clausius, and Maxwell.

Between 1845 and 1850, Joule performed several of his now famous experiments that established the equivalence between heat and work. The highlight of the experiments was the measurement of the temperature rise in a system (water) resulting from the application of mechanical work. That work can cause temperature to increase was hardly a new discovery, as anyone who has rubbed their hands on a cold day knows. Joule's breakthrough was to demonstrate that the temperature increase is *proportional* to the amount of work, thus proving that the unit of heat (calorie) is proportional to the unit for work—now named after Joule himself. He performed his experiment in a paddle-wheel apparatus consisting of a stirrer in the form of paddles. To rotate the stirrer, Joule used the motion of falling weights via pulleys and strings attached to the shaft of the stirrer. This arrangement allowed Joule to calculate precisely the amount of work applied. By making the stirrer in the shape of a paddle instead of a more streamlined shape, the rotational kinetic energy of the stirrer is more efficiently transferred into the fluid via viscous force. From the proportionality between the applied work and the resulting

---

5. The calorie is the amount of heat that raises the temperature of 1 g of water by 1 °C. The Btu is the amount of heat that raises the temperature of 1 lb of water by 1 °F.

temperature rise, Joule was able to obtain a fairly accurate numerical value for the work equivalent of heat. He found that one degree Fahrenheit is equivalent to 890 pounds falling from the altitude of one foot.

Example 3.4: Analysis of Joule's Experiment

In this example we consider a variation of Joule's experiment: a thermally insulated vessel contains 10 kg of water. The liquid is stirred by an impeller driven by a 1 kW motor. If the motor runs for 1 min and all of the work it produces is transferred to the liquid, analyze the experiment on the basis of the first law and report the relevant amounts of heat, work, and internal energy.

Solution We define the system to be the insulated container with the water inside it. This is a closed system. Because of the insulation, no heat can be exchanged between the system and the surroundings, therefore, $Q = 0$. By first law then,

$$\Delta U = W.$$

The amount of work that enters the system, expressed on a per-mass basis, is

$$W = \frac{(1000 \text{ W})(60 \text{ s})}{10 \text{ kg}} = 6 \text{ kJ/kg}.$$

The change in the internal energy of the water is equal to work, 60 kJ. If we divide this value by the amount of water we obtain the change of the *specific* internal energy:

$$\Delta U = 6 \text{ kJ/kg}.$$

In this experiment, mechanical work is converted into internal energy (not into heat!). The energy of the impeller is first converted into kinetic energy of the moving liquid, and then into internal energy by the action of viscous forces. The increase of internal energy is manifested through the temperature rise.

3.5 Elementary Paths

Constant-Volume Heating

Consider the process in which a closed system is heated under constant volume. There is no PV work because the volume of the system is fixed. We assume there is no shaft work either. Then, the first law gives,

$$Q = \Delta U \quad \text{(constant-volume process)}. \tag{3.11}$$

Therefore, the amount of heat that is exchanged under constant volume is equal to the change of internal energy, provided that no shaft work is present. Recall that shaft work, if present, will be made obvious through the presence of mechanical devices such as impellers, pumps, compressors, turbines, and the like. If a problem makes no mention of any such devices, it will be assumed that no shaft work is involved, without explicitly stating this assumption each time.

Example 3.5: Constant-Volume Cooling
A sealed metal cylinder contains steam at 1 bar, 500 °C. How much heat must be removed at constant volume in order to produce saturated vapor?

Solution *Outline:* This is a constant-volume process; therefore, the heat is equal to the change of internal energy between the initial (A) and final (B) state:

$$Q = U_B - U_A.$$

The initial state is known. In the final state we know the specific volume (must be equal to the specific volume in the initial state) and that the state is saturated vapor. These two pieces of information are sufficient to fix the final state. Once the final state is known, the heat is calculated by the above equation.

Numerical substitutions: From the steam tables at the initial state we read:

$$V_A = 3.5656 \text{ m}^3/\text{kg}, \quad U_A = 3132.2 \text{ kJ/kg}.$$

At the final state water is saturated vapor with $V_B = V_A = 3.5656$ m^3/kg. The desired temperature is between 78 °C and 80 °C. By interpolation we find:

$$T_B = 78.8 \text{ °C}, \quad U_B = 2480.1 \text{ kJ/kg}.$$

The amount of heat that must be exchanged with the surroundings is

$$Q = U_B - U_A = (2480.1 - 3132.2) \text{ kJ/kg} = -652.1 \text{ kJ/kg}.$$

It is negative; therefore, it represents cooling.

Constant-Pressure Heating

A common way to exchange heat is under constant pressure. The expansion/contraction of volume that accompanies heating/cooling involves the exchange of PV work with the surroundings, and this must be accounted for in the energy balance. We assume that the PV work is exchanged reversibly. Experimentally, this means that heat must be added or removed at a gentle rate to allow temperature to

3.5 Elementary Paths

equilibrate and the expansion or contraction to take place in an controlled manner. Applying eq. (3.4) and noting that pressure is constant, the work is

$$W = -\int_{V_A}^{V_B} P dV = -P \int_{V_A}^{V_B} dV = -P(V_B - V_A),$$

with subscripts A and B referring to the initial and final state, respectively. Substituting this result in eq. (3.9) and solving for the heat, we find,

$$\Delta U_{AB} = Q - P(V_B - V_A) \quad \Rightarrow \quad Q = (U_B + PV_B) - (U_A + PV_A).$$

In the last result, the right-hand side is the difference of the term $U + PV$ evaluated at the final state B, and the same term evaluated at the initial state A (recall that $P_B = P_A = P$ for this process). This grouping of state functions is itself a state function. This motivates the definition of a new thermodynamic property, *enthalpy*:

$$\boxed{H = U + PV.} \tag{3.12}$$

Using this definition, the heat in constant-pressure process is

$$Q = \Delta H \quad \text{(constant-pressure process)}. \tag{3.13}$$

Therefore, in constant-pressure heating of a closed system, the amount of heat that is exchanged is equal to the change in enthalpy. Enthalpy is a state function and has the same units as internal energy (kJ/kg or J/mol). It is the relevant thermodynamic in the energy balance of open systems, as we will see in Chapter 6. As the large majority of chemical processes involves open systems, such as units with multiple inlet and outlet flow streams, enthalpy is used far more than internal energy and is tabulated more extensively. In fact, tabulations of internal energy are uncommon because it can be calculated from tabulations of enthalpy and volume.

Example 3.6: Enthalpy of Steam
Use the steam tables to obtain the enthalpy of steam at 1 bar, 300 °C, and compare the value calculated by application of the definition, eq. (3.12).

Solution From the steam tables at $P = 1$ bar, $T = 300$ °C, we find

$$H = 3074.5 \text{ kJ/kg}.$$

At the same pressure and temperature we also find

$$V = 2.6389 \text{ m}^3/\text{kg}, \quad U = 2810.7 \text{ kJ/kg}.$$

To apply eq. (3.12) we first calculate the product PV:

$$PV = \left(1 \text{ bar} \times 10^5 \frac{\text{J/m}^3}{\text{bar}} \times 10^{-3} \frac{\text{kJ}}{\text{J}}\right)(2.6389 \text{ m}^3/\text{kg}) = 263.89 \text{ kJ/kg}.$$

The enthalpy calculated from eq. (3.12) is

$$H = U + PV = (2810.7 \text{ kJ/kg}) + (263.89 \text{ kJ/kg}) = 3074.54 \text{ kJ/kg}.$$

The calculated value agrees with the tabulated value within the precision of the table. Small round-off errors should be expected due to truncation in the tables.

Example 3.7: Constant-Pressure Cooling of Steam
8.5 kg of steam at 600 °C, 15 bar, are cooled at constant pressure by removing 6200 kJ of heat. Determine the final temperature and the amount of PV work that is exchanged with the surroundings.

Solution To specify the final state we need two intensive variables. The final pressure is known since it is constant during the process and equal to the initial pressure. The missing piece of information is the enthalpy of the final state, which will be obtained by energy balance.

For constant-pressure heating, $Q = H_2 - H_1$, which solved for H_2 gives

$$H_2 = H_1 + Q.$$

The amount of heat, per kg of steam, is

$$Q = \frac{-6200 \text{ kJ}}{8.5 \text{ kg}} = -729.4 \text{ kJ/kg}.$$

From the steam tables at the initial state,

$$P_1 = 15 \text{ bar}, \quad T_1 = 600 \text{ °C}: \quad U_1 = 3294.5 \text{ kJ/kg}, \quad H_1 = 3694.6 \text{ kJ/kg}.$$

The final enthalpy is now calculated to be

$$H_2 = H_1 + Q = 3694.6 \text{ kJ/kg} + (-729.4) \text{ kJ/kg} = 2965.2 \text{ kJ/kg}.$$

In the final state we know pressure ($P_2 = P_1 = 15$ bar) and enthalpy ($H_2 = 2965.2$ kJ/kg). By inspection in the steam tables at $P = 15$ bar, the final temperature must be between 250 °C and 300 °C. The exact temperature is obtained by interpolation:

$$P = 15 \text{ bar} \quad T = 250 \text{ °C} \quad H = 2924.0 \text{ kJ/kg}$$
$$T' = 300 \text{ °C} \quad H' = 3038.3 \text{ kJ/kg}.$$

3.5 Elementary Paths

By interpolation at $H = H_2 = 2965.2$ kJ/kg,

$$T_2 = 250 + \frac{300 - 250}{3038.3 - 2924.0}(2965.2 - 2924.0) = 268\ °C.$$

To calculate the work, we use eq. (3.9):

$$\Delta U_{12} = Q + W \quad \Rightarrow \quad W = \Delta U_{12} - Q.$$

Since pressure and temperature are known at the final state, the internal energy is obtained directly from the table. This requires one additional interpolation:

$$P = 15\ \text{bar} \quad T = 250\ °C \quad U = 2696.0\ \text{kJ/kg}$$
$$T' = 300\ °C \quad U' = 2783.7\ \text{kJ/kg}.$$

$$U_2 = 2696.0 + \frac{2783.7 - 2696.0}{(300 - 250)}(268 - 250) = 2727.6\ \text{kJ/kg}.$$

Finally, the work is

$$W = 2727.6 - 3294.5 - (-729.4) = 162.5\ \text{kJ/kg}.$$

The work calculated here is per kg of steam. The total amount of work for the entire amount of steam is $W = (8.5\ \text{kg})(162.5\ \text{kJ/kg}) = 1381.4$ kJ. The work is positive, which means, it is done by the surroundings and is absorbed by the system.

Example 3.8: Work and Heat are Path Functions
Steam undergoes the following reversible process in a closed system: from initial conditions 10 bar, 400 °C, to 550 °C under constant pressure, then to 8 bar under constant volume. Determine the energy balances. How would the energy balances change if steam from the same initial state were first cooled at constant volume to 8 bar, then heated at constant pressure to the same final state?

Solution The process is shown by the path $1 \rightarrow 2 \rightarrow 3$ in Figure 3-6. The properties for states 1 and 2 are obtained directly form the steam tables; the state at 3 requires interpolation. The information is summarized in the table below:

	State 1	State 2	State 3
P (bar)	10	10	8
T (°C)	400	550	389
V (m^3/kg)	0.3066	0.3777	0.3777
U (kJ/kg)	2957.8	3210.4	2942.1
H (kJ/kg)	3264.4	3588.1	3244.3

The energy balances are performed along each branch separately.

$1 \to 2$: In this constant-pressure segment, heat is equal to the enthalpy change:

$$Q_{12} = H_2 - H_1 = (3588.1 - 3264.4) \text{ kJ/kg} = 323.7 \text{ kJ/kg}.$$

The amount of heat is positive, i.e., the system is being heated. The work is given by eq. (3.4), which for a constant-pressure process is

$$W_{12} = -P_1(V_2 - V_1) = -(10 \text{ bar})\left(100\frac{\text{kJ/m}^3}{\text{bar}}\right)(0.3777 - 0.3066) \text{ (m}^3/\text{kg}) = -71.1 \text{ kJ/kg}.$$

This work is produced by the system as a result of expanding against the surroundings.

$2 \to 3$: In this segment volume is constant. The heat is equal to the change of internal energy:

$$Q_{23} = U_3 - U_2 = (2942.1 - 3210.4) \text{ kJ/kg} = -268.3 \text{ kJ/kg}.$$

No PV work is exchanged:

$$W_{23} = 0 \text{ kJ/kg}.$$

Entire process: $1 \to 2 \to 3$: The results are summarized in the table below:

	Q (kJ/kg)	W (kJ/kg)
1→2	323.7	−71.1
2→3	−268.3	0
1→3	55.4	−71.1

Overall balance: The change in internal energy between the initial and final states is

$$\Delta U_{13} = U_3 - U_1 = (2942.1 - 2957.8) \text{ kJ/kg} = -15.7 \text{ kJ/kg}.$$

This is equal to the sum of heat and work:

$$Q_{123} + W_{123} = 55.4 + (-71.1) = -15.7 \text{ kJ/kg}.$$

Alternate Path When the process is conducted by first applying a constant-V path followed by a constant-pressure one, the combined path is the one marked as $1 \to 4 \to 3$ in Figure 3-6. The calculation can be done by following the same procedure as with the previous path. However, the calculation can be simplified by noting that several required quantities have computed already.

The change in internal is the same as in the previous case since the initial and final states are the same:

$$\Delta U_{13} = U_3 - U_1 = (2942.1 - 2957.8) \text{ kJ/kg} = -15.7 \text{ kJ/kg}.$$

The work corresponds to the area under the path and is calculated as

$$W_{143} = -P_3(V_3 - V_1) = -(8 \text{ bar})\left(100\frac{\text{kJ/m}^3}{\text{bar}}\right)(0.3777 - 0.3066) \text{ (m}^3/\text{kg}) = -56.9 \text{ kJ/kg}.$$

3.5 Elementary Paths

The heat is calculated by application of the first law:

$$Q_{143} = \Delta U_{13} - W_{143} = -15.7 \text{ kJ/kg} - (-56.9) \text{ kJ/kg} = 41.2 \text{ kJ/kg}.$$

The results for the two cases are summarized below:

Path	Q (kJ/kg)	W (kJ/kg)	ΔU (kJ/kg)
1→2→3	55.4	−71.1	−15.7
1→4→3	41.2	−56.9	−15.7

Comments

- Internal energy is a state function: Its difference between two states is determined by the states alone and not by the path that is used to connect these states.
- Work and heat are path functions: Their value depends on the entire path that connects two states.
- That work is a path function is demonstrated by the fact that work corresponds to the area under the path. Clearly, the path $1 \to 2 \to 3$ involves more work because it covers a larger area. Since the sum of work plus heat is independent of the path (it is equal to the change of the internal energy), we conclude that heat must be a path function as well.

Constant-Temperature Process

An isothermal process is one conducted at constant temperature and its path is represented by an isotherm. Experimentally, it is conducted by placing the system

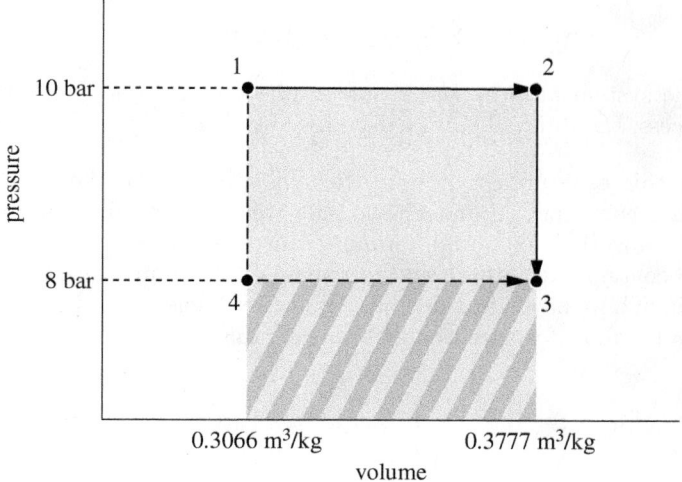

Figure 3-6: Energy balances along different paths (Example 3.8).

inside a heat bath. A typical example of an isothermal process would be compression or expansion. In general, this process involves the simultaneous exchange of both work and heat, as demonstrated with an example.

Example 3.9: Isothermal Compression of Steam
Steam is compressed isothermally at 350 °C in a closed system. Compression is conducted reversibly from initial pressure 20 bar to final pressure 40 bar. Determine the amounts of work and heat exchanged between the system and its surroundings.

Solution The governing equation is the first law:

$$\Delta U_{AB} = Q + W.$$

The internal energy is obtained from the steam tables:

	State 1	State 2
P (bar)	20	40
U (kJ/kg)	2860.5	2827.4

We find
$$\Delta U_{12} = -33.1 \text{ kJ/kg}.$$

The work for this process was calculated in Example 3.3, where we found

$$W = 199.81 \text{ kJ/kg}.$$

The heat is finally calculated from the first law:

$$Q = \Delta U_{12} - W = -232.91 \text{ kJ/kg}.$$

In summary, the system absorbs 199.81 kJ/kg of work and rejects 232.91 kJ/kg of heat. During the process, the internal energy decreases by 33.1 kJ/kg.

Comments In this example the system exchanges heat with the bath even though its temperature stays constant. To understand why this is so, recall that during compression the system absorbs work. This would normally cause the temperature to increase. However, the system is in contact with a bath and to remain at the temperature of that bath it must reject heat. Although we have identified heat as a form of energy that requires a temperature difference, we must allow for situations where such difference is infinitesimal.

3.6 Sensible Heat—Heat Capacities

In the absence of phase transitions, the exchange of heat is accompanied with a temperature change. Such heat is called *sensible* because it can be sensed with a thermometer. The amount of heat needed to produce a given temperature change varies from one substance to another and is quantified by the heat capacity, defined as the ratio of the amount of heat exchanged to the temperature change observed:

$$\text{heat capacity} = \left.\frac{dQ}{dT}\right|_{\text{path}}. \tag{3.14}$$

The amount of heat depends on the path of the heating process. We have encountered two simple paths, the constant-volume and the constant-pressure path. Each is characterized by its own heat capacity.

Constant-Volume Heat Capacity

In constant-volume heating, eq. (3.11) gives $dQ = dU$. Applying this result to eq. (3.14) we obtain the constant-volume heat capacity:

$$C_V = \left(\frac{\partial U}{\partial T}\right)_V. \tag{3.15}$$

Thus, even though the definition of the heat capacity was motivated by heat, the heat capacity is a partial derivative of a state function (internal energy), and itself a state function. The above can also be expressed in the differential form,

$$dU = C_V \, dT \quad \text{(constant } V\text{)}. \tag{3.16}$$

This equation gives the change in internal energy during a small step along a path of constant volume. It can be integrated between two states to give the change in internal energy between these states:

$$\boxed{\Delta U_{AB} = \int_{T_A}^{T_B} C_V \, dT.}_{\text{(const. } V\text{)}} \tag{3.17}$$

The condition "constant V" specifies the *integration path*, that is, the initial state, final state, and all states between are at the same volume.

NOTE

Derivatives and Integrals With Functions of Multiple Variables

In dealing with functions of multiple variables we encounter various partial derivatives. For a function of two variables, $F(x, y)$, the partial derivative

$$\frac{\partial F}{\partial x}$$

is the usual derivative of F with respect to x, while keeping y constant. In thermodynamics we are free to choose the set of variables that we take to be independent. Although we prefer to express properties as a function of P and T, we will often use other combinations of variables. To make clear which are the independent variables, a partial derivative must indicate all of them. This is done by adopting the notation

$$\left(\frac{\partial F}{\partial x}\right)_y$$

with the differentiation variable appearing in the denominator, as usual, and the remaining variables (in this case, y) appearing as subscripts outside the derivative. Notice that the following two derivatives below are *not* the same:

$$\left(\frac{\partial U}{\partial T}\right)_V \neq \left(\frac{\partial U}{\partial T}\right)_P$$

The left-hand side is a partial derivative of the function $U(T, V)$ whereas the right-hand side is a derivative of the function $U(T, P)$; these are two different mathematical functions.

A partial derivative can be broken up to form a differential, which can then be integrated:

$$\left(\frac{\partial F}{\partial x}\right)_y = A \quad \Rightarrow \quad dF = A\, dx \quad \text{(const. } y\text{)}.$$

Integration between two states A and B produces the change of F between the two states:

$$F_B - F_A = \int_{x_A}^{x_B} A\, dx.$$
$$\text{(const. } y\text{)}$$

Upon separating the derivative into a differential, the specification "const. y," which originated as a subscript in the partial derivative, must be explicitly written down. When the differential is integrated, the condition "const. y" becomes the specification of the *integration path*, namely, a line on the xy plane along which we evaluate the integral. Integration paths are necessary when working with functions of multiple variables: When we integrate with respect to one variable, the path specifies what to do with the remaining variables.

Proper notation is important for consistency but also serves the practical purpose of facilitating learning by organizing mathematical thought: It is not necessary to memorize the fact that eq. (3.17) applies only to constant volume processes, as long as one can recall the standard mathematical steps that connect this result to the basic definition in eq. (3.15).

3.6 Sensible Heat—Heat Capacities

Constant-Pressure Heat Capacity

For constant-pressure process, $dQ = dH$, from eq. (3.13). Applying this result to eq. (3.14) we obtain the constant-pressure-heat capacity:

$$C_P = \left(\frac{\partial H}{\partial T}\right)_P. \tag{3.18}$$

As an immediate consequence, the enthalpy change along a path of constant pressure is given by the integral,

$$\boxed{\Delta H_{AB} = \int_{T_A}^{T_B} C_P \, dT.} \tag{3.19}$$
$$\text{(const. } P\text{)}$$

Both C_P and C_V have dimensions of energy per mass per temperature. In the SI system, their values are usually reported in J/mol K, or kJ/kg K. The term *specific heat* is sometimes used when the heat capacity is reported on a per-mass basis. Equation (3.19) may be written in simpler terms by making use of the mean heat capacity, $\overline{C}_P$, which we define as

$$\overline{C}_P = \frac{\int_{T_1}^{T_2} C_P \, dT}{T_2 - T_1}. \tag{3.20}$$

Using this definition, eq. (3.19) becomes

$$\Delta H_{AB} = \overline{C}_P (T_B - T_A). \tag{3.21}$$

The use of the mean heat capacity does not eliminate the need for the integration because $\overline{C}_P$ depends on the end temperatures, T_A and T_B, and must be recalculated each time the interval changes. Written in the above form, however, eq. (3.21) suggests that if the heat capacity does not change much with temperature in the interval T_A to T_B, it may be taken to be constant and equal to its value at T_A or at T_B. This approximation should be made only for quick estimates, when high accuracy is not critical.

Why Are C_V and C_P Different? The mathematical answer is that they represent different partial derivatives. The thermodynamic answer is that each is associated with heating along different paths. We may also provide a third explanation based on physical arguments. Consider the following two experiments, shown schematically in Figure 3-7: in the first one, a closed system is heated under constant volume until its temperature increases by ΔT. In the second experiment, an identical system at the same initial state is heated under constant pressure until its temperature increases

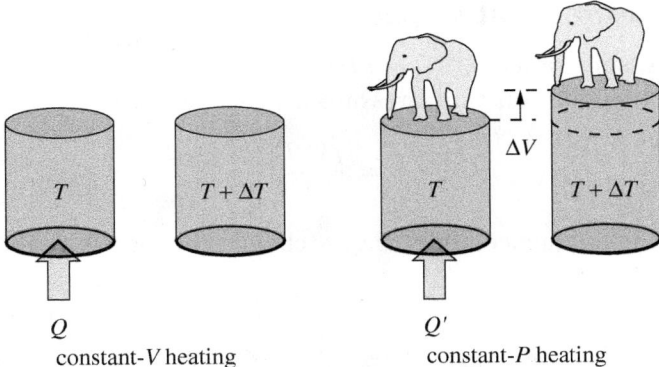

Figure 3-7: Constant-volume and constant-pressure heating: to achieve the same temperature increase, constant-pressure heating must provide more heat because of the work needed to expand the volume against the surroundings.

by the same amount ΔT. Let Q_V and Q_P be the amounts of heat for the constant-V and the constant-P experiments, respectively. How do these amounts compare? In constant-volume heating, the amount Q_V is used to increase the temperature by ΔT. In constant-pressure heating the amount Q_P is used to increase temperature by the same amount *and* to expand the volume against the surroundings. This argument suggests that $Q_P > Q_V$, which further implies that $C_P > C_V$. We also expect the difference between C_P and C_V to be larger for gases, because of the large volume expansion, and smaller for liquids and solids, whose volume expands much less. These qualitative conclusions are indeed true.

The practical utility of the heat capacities is twofold. First, they allow us to calculate heat in constant-volume and constant-pressure processes. This is useful in energy balances. Second, they allow us to calculate changes in internal energy and enthalpy. This allows us to calculate these properties using equations rather than tables, or to obtain their values in states that are not found in tables. There is a limitation, however. Equation (3.17) may be used only between two states of the same volume, and eq. (3.19) only between two states of the same pressure. The general calculation of properties between any two states will be discussed in Chapter 5.

Example 3.10: Heat Capacities and Heat
According to eq. (3.14), heat capacities are defined in terms of heat. Does this make C_V and C_P *path* functions?

Solution No. Equation (3.14) applies to a specific path (constant-volume for C_V, constant-pressure for C_P). Once the path is specified, the amount of heat that is exchanged (dQ) is uniquely defined and depends on the local state of the system. It is best to view eqs. (3.15) and (3.18) as the definitions of the heat capacities because they make clear that both C_V and C_P are state functions.

Effect of Pressure and Temperature on Heat Capacity

As state functions, C_V and C_P depend on pressure and temperature. This dependence is illustrated for C_P in Figure 3-8, which shows the constant-pressure heat capacity of water as a function of pressure at selected temperatures. In general, the heat capacity of the liquid is higher than that of the vapor; heat capacity is a strong function of pressure in the vapor phase but almost independent of pressure in the liquid. In both phases, C_P is fairly sensitive on temperature. Following an isotherm to zero pressure we reach the ideal-gas limit and the corresponding value is the ideal-gas heat capacity, C_P^{ig}. In the ideal-gas state the heat capacity is independent of pressure but it remains a function of temperature (notice that the isotherms in Figure 3-8 do not converge to a single value at $P = 0$).

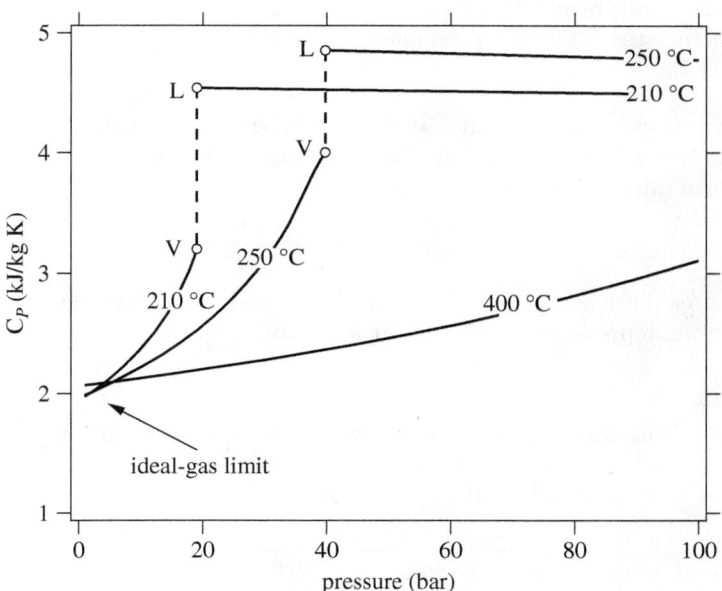

Figure 3-8: C_P of steam as a function of pressure and temperature. Points V, L, mark the saturated vapor and saturated liquid, respectively.

Ideal-Gas Heat Capacity Ideal-gas heat capacities have been compiled for a large number of pure components and the data are usually presented in the form of empirical correlations that give C_P^{ig} as a function of temperature. One common representation is in the form of a polynomial in temperature:

$$\frac{C_P^{ig}}{R} = a_0 + a_1 T + a_2 T^2 + a_3 T^3 + a_4 T^4, \tag{3.22}$$

where R is the ideal-gas constant, T is temperature, and a_i, $i = 1, \cdots, 4$, are parameters specific to the fluid. This equation gives the ratio of the ideal-gas heat capacity to the ideal-gas constant and the heat capacity is obtained by multiplying the right-hand side by R. One must be careful about the proper units of T. These units cannot be inferred from the equation and must be specified explicitly (this information should be obtained from the source that specifies the equation and the parameters). Parameters of this equation for a large number of substances are given in *The Properties of Gases and Liquids* (5th ed., 2007).

Liquid Heat Capacities The heat capacity in the liquid phase is nearly independent of pressure for the same reasons discussed for other liquid properties (see Section 1.1). The liquid heat capacity is expressed as a function of temperature by empirical equations, often polynomial in T.

Example 3.11: C_P of Steam
Use data from the steam tables to calculate the heat capacity of saturated water vapor at 250 °C.

Solution The saturation pressure at 250 °C is 39.76 bar. To calculate the C_P from tabulated data we will approximate the partial derivative in eq. (3.18) with a finite difference between two states at the same pressure

$$C_P = \left(\frac{\partial H}{\partial T}\right)_P \approx \frac{\Delta H}{\Delta T}.$$

The temperatures must be in the vicinity of 250 °C and the corresponding enthalpies must both be at the same pressure, 39.77 bar. At saturation, we read the following values of the enthalpy:

$$T = 250 \text{ °C}, \quad P^{\text{sat}} = 39.76 \text{ bar}, \quad H_V = 2801.0 \text{ kJ/kg}.$$

To calculate the heat capacity of the vapor we must obtain the enthalpy at a neighboring temperature at 39.76 bar. We choose $T_2 = 300$ °C and obtain the enthalpy by interpolation from the values at $P_1 = 38$ bar and $P_2 = 40$ bar:

$T = 300$ °C	
$P_1 = 38$ bar	$H_1 = 2968.4$ kJ/kg
$P_2 = 40$ bar	$H_2 = 2961.7$ kJ/kg

3.6 Sensible Heat—Heat Capacities

By interpolation at $P = 39.76$ bar,

$$H' = 2968.4 + \frac{2961.7 - 2968.4}{40 - 38}(39.76 - 38) = 2962.5 \text{ kJ/kg}.$$

The heat capacity of the vapor is

$$C_P \approx \frac{2962.5 - 2801.0}{300 - 250} = 3.23 \text{ kJ/kg K}.$$

Comments The heat capacity will normally be obtained from handbooks or from fitted equations as a function of temperature. The point of this exercise is to show that it is related to other tabulated properties and that it may be calculated from them, if needed.

Example 3.12: Constant-Pressure Cooling of Liquid
Saturated liquid acetone at 380 K, is cooled under constant pressure to a final temperature of 300 K. Determine the enthalpy, and internal energy at the final state, as well as the amount of heat and PV work exchanged with the surroundings. Perry's Handbook reports the following properties of saturated liquid at 380 K (see *Perry's Chemical Engineering Handbook*, 7th ed., Table 2-226, p. 2-206):

$$T = 380 \text{ K}, \quad P^{\text{sat}} = 4.52 \text{ bar}, \quad H_L = 127 \text{ kJ/kg}, \quad V_L = 0.001464 \text{ m}^3/\text{kg}.$$

The heat capacity of the liquid is

$$C_P = 135.6 - 0.177\,T + 0.0002837\,T^2 + 6.89 \times 10^{-7}\,T^3$$

with T in K and C_P in J/mol K.

Solution For constant-pressure process, the heat is

$$Q = \int_{T_1}^{T_2} C_P\,dT.$$

Integrating the given expression for the heat capacity between temperatures $T_1 = 380$ K and $T_2 = 300$ K, we obtain (the molar mass of acetone is 58.08×10^{-3} kg/mol)

$$Q = -10{,}865.8 \text{ J/mol} = -187.1 \text{ kJ/kg}.$$

The enthalpy at the final state is

$$H_2 = H_1 + Q = 127 + (-187.1) = -60.1 \text{ kJ/kg}.$$

The internal energy in the final state is

$$U_2 = H_2 - P_2 V_2, \qquad \text{[A]}$$

where $P = P^{\text{sat}} = 4.52$ bar is the constant pressure of this process, and V_2 is the specific volume in the final state. This volume is somewhat smaller due to thermal contraction and to calculate it we need the volumetric coefficient of thermal expansion. This will be done in the same way as in Example 2.13. Assuming a mean value for the volumetric coefficient of thermal expansion, the volume at the final state is

$$V_2 = V_1 e^{\bar{\beta}(T_2 - T_1)}.$$

Using the data in Example 2.13, the average coefficient between 380 K and 300 K is calculated to be $\bar{\beta} = 0.00648347$ K^{-1} and the volume V_2 in the final state is

$$V_2 = (0.001464 \text{ m}^3/\text{kg}) e^{(0.00648347 \text{ K}^{-1})(300-380) \text{ K}} = 0.0008715 \text{ m}^3/\text{kg}.$$

The internal energy is now obtained using eq. [A]:

$$U_2 = (-60.1 \text{ kJ/kg}) - (4.52 \text{ bar}) \left(\frac{10^2 \text{ kJ/m}^3}{\text{bar}} \right) (0.0008715 \text{ m}^3/\text{kg}) = -60.47 \text{ kJ/kg}.$$

The PV work is given by

$$W = -\int_{V_1}^{V_2} P dV = -P(V_2 - V_1), \qquad [\text{B}]$$

where $P = P^{\text{sat}} = 4.52$ bar is the constant pressure of this process, and V_1, V_2, are the specific volumes in the initial and final state. Finally, the PV work is calculated from eq. [A]:

$$W = -(4.52 \text{ bar}) \left(\frac{10^2 \text{ kJ/m}^3}{\text{bar}} \right) (0.0008715 - 0.001464) \text{ m}^3/\text{kg} = 0.2678 \text{ kJ/kg}.$$

This work is positive, that is, it is absorbed by the system.

There are several points worth noting about these results:

- This problem is very similar to that in Example 3.7. Indeed, the main difference is that in that example properties were calculated from tabulated values (steam tables) whereas here the calculations were done using mathematical relationships among various properties. Which method is to be preferred will generally depend on the available information.

- The enthalpy and the internal energy in the final state were found to be negative. There is nothing wrong with this result. Enthalpy and internal energy are measured relative to a reference state. A negative value simply means that the property has a lower value than that at the reference state. The choice of the reference state will be discussed in Chapter 5.

- The PV work in this example is much less than that calculated in Example 3.7. This is due to the relative incompressibility of liquids. Because the volume of the liquid

3.6 Sensible Heat—Heat Capacities

does not yield much under pressure, the displacement of the force exerted by the outside pressure is very small.

- The internal energy in the final state ($U_2 = -60.47$ kJ/kg) and the enthalpy ($H_2 = -60.1$ kJ/kg) are nearly the same. This is generally true for compressed liquids away from the critical point. The difference between H and U is equal to the product PV, and V for liquids is very small. Unless the pressure is very high, the approximation

$$U \approx H$$

is generally acceptable. This can be confirmed by comparing the enthalpy and internal of saturated liquid water in the steam tables. The approximation is very good for solids as well.

Example 3.13: Properties of Compressed Liquid
Use information from the steam tables and the heat capacity of liquid water to obtain the enthalpy and internal energy of water at 100 °C, 20 bar.
Additional information: The heat capacity of liquid water is

$$C_P = a_0 + a_1 T + a_2 T^2 + a_3 T^3 + a_4 T^4$$

with C_P in J/mol K, T in kelvin, and the constants a_i given by

$$a_4 = 9.37 \times 10^{-9}, \quad a_3 = -1.41 \times 10^{-5}, \quad a_2 = 8.13 \times 10^{-3}, \quad a_1 = -2.09, \quad a_0 = 2.76 \times 10^2.$$

Solution At 100 °C, 20 bar, water is compressed liquid. Assuming that enthalpy and internal energy are practically independent of pressure, we can relate the properties of compressed liquid to those of the saturated liquid in two different ways, along a path of constant temperature, or a path of constant pressure (Figure 3-9).

Constant-temperature path If enthalpy and internal energy are assumed to be functions of temperature only, the properties at state A can be set equal to those of the saturated liquid at the same temperature (state B):

$$H \approx H_L(100\ °C) = 419.10\ \text{kJ/kg}$$

$$U \approx U_L(100\ °C) = 418.99\ \text{kJ/kg}.$$

Constant-pressure path The constant-pressure path connects state A to the saturation line at the same isobar (state C) at $T_C = 212.38$ °C. We express the enthalpy at A in terms of the enthalpy at C plus the change from C to A:

$$H_A = H_C + \Delta H_{CA},$$

with $H_C = 908.62$ kJ/kg, from the steam tables. Along the path CA pressure is constant and ΔH_{CA} is calculated from eq. (3.21):

$$\Delta H_{CA} = \overline{C}_P(T_A - T_C).$$

The mean heat capacity of water between $T_A = 100$ °C (373.15 K) and $T_C = 212.38$ °C (485.53 K) is

$$\overline{C}_P = \frac{a_0(T_C - T_A) + \tfrac{1}{2}a_1(T_C^2 - T_A^2) + \tfrac{1}{3}a_2(T_C^3 - T_A^3) + \tfrac{1}{4}a_3(T_C^4 - T_A^4) + \tfrac{1}{5}a_4(T_C^5 - T_A^5)}{T_C - T_A},$$

and with numerical substitution,

$$\overline{C}_P = 80.16 \text{ J/mol K} = 4.453 \text{ kJ/kg K}.$$

The enthalpy at A is now calculated to be

$$H_A = H_C + \Delta H_{CA} = 908.62 \text{ kJ/kg} + (4.453 \text{ kJ/kg K})(373.15 - 485.53) \text{ K} = 408.2 \text{ kJ/kg}.$$

The two results are within 2.6% of each other. In this case, the constant-temperature path is simpler. Nonetheless, it is useful to be aware of alternative methods in case the state at B is not known.

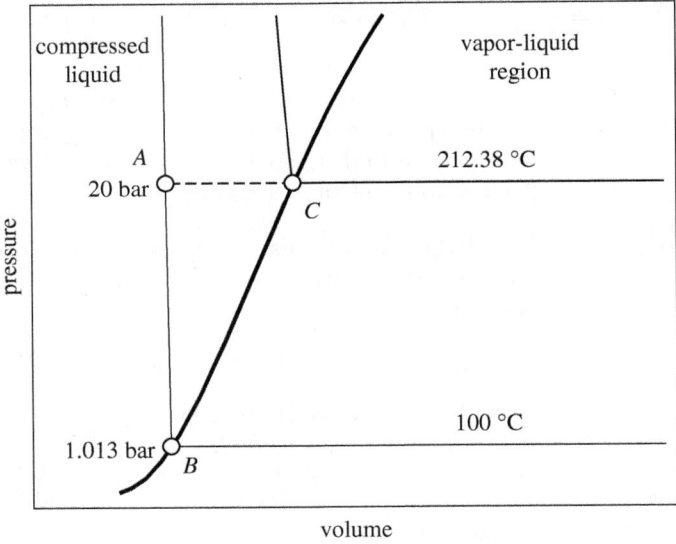

Figure 3-9: Calculation of the enthalpy of compressed liquid from values at saturation (see Example 3.13).

3.7 Heat of Vaporization

> **NOTE**
>
> **Properties of Compressed Liquid**
>
> At temperatures well below the critical, enthalpy (and internal energy) is approximately independent of pressure. Under this approximation, eq. (3.21) can be used to calculate the enthalpy difference between two states in the compressed liquid region, not necessarily at the same pressure. Alternatively, the enthalpy of compressed liquid may be approximated by the enthalpy of the saturated liquid at the same temperature.

3.7 Heat of Vaporization

To convert the saturated liquid into vapor of the same pressure and temperature, we must provide the energy required to free molecules from the attractive forces that hold the liquid together. This energy is called *heat of vaporization* and since this process is at constant pressure, it is equal to the difference of the enthalpy of the saturated liquid and saturated vapor:

$$\Delta H_{\text{vap}} = H_V - H_L. \tag{3.23}$$

This amount of energy is also called *latent* heat of vaporization, to indicate that it is not accompanied by any temperature change. Writing $H = U + PV$ for the enthalpy, the enthalpy of vaporization can also be expressed as

$$\Delta H_{\text{vap}} = U_V - U_L + P^{\text{sat}}(V_V - V_L) = \Delta U_{\text{vap}} + P^{\text{sat}} \Delta V_{\text{vap}}. \tag{3.24}$$

Here, ΔU_{vap} is the internal energy of vaporization, ΔV_{vap} is the change in volume from the saturated liquid to the saturated vapor, and P^{sat} is the saturation pressure. The heat of vaporization can be calculated from tabulated properties of the saturated liquid and saturated vapor.

A state that lies in the vapor-liquid region (point C in Figure 3-10) consists of a mixture of saturated vapor and saturated liquid. If the mass fraction of liquid is x_L and the mass fraction of vapor is x_V, the specific enthalpy of the two-phase system is

$$H = x_L H_L + x_V H_V. \tag{3.25}$$

Using $x_L + x_V = 1$, the above relationship can be solved for the mass fractions:

$$x_L = \frac{H_V - H}{H_V - H_L}, \quad x_V = \frac{H - H_L}{H_V - H_L}. \tag{3.26}$$

Equations (3.25) and (3.26) are expressions of the lever rule for enthalpy. It will be left as an exercise to show that internal energy also satisfies the lever rule. An

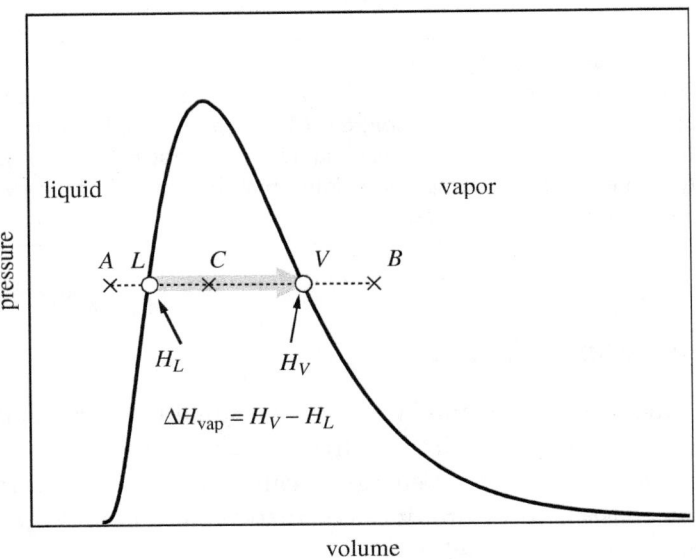

Figure 3-10: Liquid-vapor transition at constant pressure.

alternative form of the two expressions in eq. (3.26) is obtained by solving each expression for the enthalpy of the two-phase system. Noting that the denominator in both expressions is the enthalpy of vaporization, the result is

$$H = H_V - x_L \Delta H_{\text{vap}} = H_L + x_V \Delta H_{\text{vap}}. \tag{3.27}$$

This equation has a simple interpretation that can be useful in calculations: If we start with the saturated vapor and remove an amount of heat that is equal to fraction a of the heat of vaporization ($0 < a < 1$), the fraction of the vapor that condenses is $x_L = a$. Similarly, if we start with saturated liquid and add an amount of heat equal to fraction a of the heat of vaporization, the fraction of the liquid that vaporizes is $x_V = a$. This relationship is helpful when the amount of heat is not enough to carry the system entirely through the vapor-liquid region.

NOTE
Lever Rule
The lever rule applies to the molar or specific version of any extensive property. So far we have encountered three such properties: volume, enthalpy, internal energy.

3.7 Heat of Vaporization

Pitzer Correlation for ΔH_{vap}   The heat of vaporization can be obtained from various sources. It can often be found in tables directly, or calculated from tabulated values of the enthalpy of the saturated vapor and liquid. Several empirical equations have been developed for the heat of vaporization, and these are reviewed in *The Properties of Gases and Liquids* by Polling, Prausnitz, and O'Connell (5th ed., 2007). Here we mention one that is based on Pitzer's method:

$$\frac{\Delta H_{vap}}{RT_c} = 7.08(1 - T_r)^{0.354} + 10.95\omega(1 - T_r)^{0.456}, \tag{3.28}$$

where T_c is the critical temperature, T_r is the reduced temperature, and ω is the acentric factor. This should be treated as an approximation, to be used if more reliable data are not available. As with all correlations based on corresponding states, this equation should be applied to normal fluids.

Consider the constant-pressure path AB in Figure 3-10: The initial state is compressed liquid and the final state is superheated vapor. The enthalpy change between initial and final state is

$$\Delta H_{AB} = H_B - H_A. \tag{3.29}$$

If H_A and H_B are available from tables, this difference can be immediately determined. An alternative calculation is to break the path into three parts: AL, LV, and VB. The first part involves constant-pressure heating of a liquid (sensible heat only). The enthalpy change for this part is

$$\Delta H_{AL} = \int_{T_A}^{T^{sat}} C_P^{liq} dT,$$

where C_P^{liq} is the constant-pressure heat capacity of the liquid, and T^{sat} is the saturation pressure at the pressure of the process. The enthalpy change between L and V is the equal to the enthalpy of vaporization:

$$\Delta H_{LV} = \Delta H_{vap}.$$

In the last part, VB, the vapor is heated under constant pressure. The enthalpy change for this part is

$$\Delta H_{VB} = \int_{T^{sat}}^{T_B} C_P^{vap} dT.$$

The enthalpy change for the complete path $ALVB$ is the sum of the parts:

$$\Delta H_{AB} = \int_{T_A}^{T^{sat}} C_P^{liq} dT + \Delta H_{vap} + \int_{T^{sat}}^{T_B} C_P^{vap} dT. \tag{3.30}$$

In this form, the parts of the path that involve sensible heat are expressed as integrals of $C_P\, dT$, using the heat capacity of the appropriate phase. Equations (3.29) and (3.30) are equivalent and either one can be used. The choice between the two will depend on the type of information that is available.

Example 3.14: Energy Balances With Phase Change: Using Tables

Steam at 1.013 bar, 200 °C is cooled under constant pressure by removing an amount of heat equal to 1000 kJ/kg. Determine the final state.

Solution Since this is a constant-pressure process, $Q = \Delta H_{AB}$. In the initial state, $P_A = 1.013$ bar, $T_A = 200$ °C, and from the steam tables we find $H_A = 2875.3$ kJ/kg. The enthalpy at the final state is

$$H_B = H_A + Q = (2875.3 - 1000) \text{ kJ/kg} = 1875.3 \text{ kJ/kg}.$$

Since we know pressure ($P_B = P_A = 1.013$ bar) and enthalpy in the final state, the state is fully specified. Searching the steam tables at constant $P = 1.013$ bar we see that the enthalpy of B is between the enthalpies of the saturated liquid and saturated vapor:

$$H_V = 2676 \text{ kJ/kg}, \quad H_L = 419.1 \text{ kJ/kg}.$$

We conclude that the system is a vapor-liquid mixture at 1.013 bar, therefore, $T_B = T^{\text{sat}}(1.013 \text{ bar}) = 100$ °C. To calculate the fraction of the vapor we use the lever rule. Using eq. (3.26), the quality of the steam is

$$x_V = \frac{H_B - H_L}{H_V - H_L} = \frac{1875.3 - 419.1}{2676 - 419.1} = 0.645.$$

Therefore, the final state is wet vapor at 1.013 bar with 64.5% quality.

Example 3.15: Energy Balances With Phase Change: Using C_P and ΔH_{vap}

Toluene vapor at 1.013 bar, 150 °C, is cooled under constant pressure to a final temperature of 22 °C. (a) Determine the amount of heat that must be removed. (b) Repeat the calculation if the final state is a vapor-liquid mixture that contains 47% vapor by mass.

Solution We will perform the calculations assuming tabulated values are not available. Toluene at room temperature is liquid; therefore, the process involves condensation. We will need the heat capacities of the vapor and liquid. We will assume that the heat capacity of the vapor is sufficiently close to the ideal-gas heat capacity, which is available from the literature. Both heat capacities are given by a polynomial equation of the form,

$$C_P/R = a_0 + a_1 T + a_2 T^2 + a_3 T^3 + a_4 T^4$$

3.7 Heat of Vaporization

with T in kelvin and with the following parameters:

	a_0	a_1	a_2	a_3	a_4
liquid	1.68559×10^1	-1.83185×10^{-2}	8.35939×10^{-5}	0	0
vapor	3.866	3.558×10^{-3}	1.3356×10^{-4}	-1.8659×10^{-7}	7.69×10^{-11}

We will also need the heat of vaporization. For this, we will use the Pitzer equation, eq. (3.28),
$$\frac{\Delta H_{\text{vap}}}{RT_c} = 7.08(1-T_r)^{0.354} + 10.95\omega(1-T_r)^{0.456}.$$

The critical constants of toluene are
$$T_C = 591.75 \text{ K}, \quad P_C = 41.08 \text{ bar}, \quad \omega = 0.264.$$

The normal point of toluene at 1.013 bar is $T^{\text{sat}} = 383.79$ K and the corresponding reduced temperature is
$$T_r = \frac{383.79}{591.75} = 0.6486.$$

The heat of vaporization from the Pitzer equation is found to be
$$\Delta H_{\text{vap}} = 32883.4 \text{ J/mol}.$$

The heat is calculated by splitting the processes into three parts:

1. *Cooling from initial vapor state to saturated vapor.* The heat for this part is
$$Q_1 = \int_{423.15}^{383.79} C_P^{\text{vap}} dT,$$
where C_P^{vap} is the heat capacity of the vapor. Using the polynomial equation with the coefficients given above, we find
$$Q_1 = -5505.44 \text{ J/mol}.$$

2. *Condensation of saturated vapor to saturated liquid.* The heat for this part is the negative of the heat of vaporization:
$$Q_2 = -32883.4 \text{ J/mol}.$$

3. *Cooling of saturated liquid at constant pressure.* This heat is
$$Q_3 = \int_{383.79}^{295.15} C_P^{\text{liq}} dT,$$
where C_P^{liq} is the heat capacity of the liquid. Using the equation for the heat capacity given above, the result is
$$Q_3 = -14978.9 \text{ J/mol}.$$

The heat for the complete process is

$$Q = Q_1 + Q_2 + Q_3 = (-5505.44 - 32883.4 - -14978.9) \text{ J/mol} = -53367.7 \text{ J/mol}.$$

Notice that most of the heat is related to the condensation of the vapor.

(b) The final state in this case is a vapor-liquid mixture that contains a vapor fraction $x_V = 0.47$ and a liquid fraction $x_L = 0.53$. We still need to cool the vapor to the saturation point, a process that requires Q_1 amount of heat. To produce the vapor-liquid mixture, we need to condense a fraction x_L of the vapor; this takes $x_L \Delta H_{\text{vap}}$ amount of heat. Therefore,

$$Q' = Q_1 - x_L \Delta H_{\text{vap}} = -5505.44 - (0.53)(32883.4) = -22933.6 \text{ J/mol}.$$

3.8 Ideal-Gas State

The properties of the ideal-gas state will be discussed in detail in Chapter 5, but some important results will be introduced here on the basis of qualitative molecular arguments. In the ideal-gas state, intermolecular interactions are unimportant because distances between molecules are large, beyond the range of such interactions. If the volume of the system is increased at constant temperature, there should be no change in internal energy: since temperature remains constant, the kinetic energy of the molecules (including translational, rotational, vibrational, etc.) is the same before and after the expansion; and since there is no interaction between molecules, there is no other mode of energy storage available to the system. Therefore, internal energy remains constant as long as temperature is not changed. In other words, the internal energy in the ideal-gas state is a function of temperature only:

$$U^{\text{ig}} = U^{\text{ig}}(T). \tag{3.31}$$

The enthalpy of the ideal gas is $H^{\text{ig}} = U^{\text{ig}} + PV^{\text{ig}} = U^{\text{ig}} + RT$. The right-hand side is a function of temperature only. Therefore, enthalpy in the ideal-gas state is also independent of pressure and a function of temperature only:

$$H^{\text{ig}} = H^{\text{ig}}(T). \tag{3.32}$$

These results greatly simplify the calculation of internal energy and enthalpy in the ideal-gas state. Consider two states (P_1, T_1) and (P_2, T_2) on the PT plane (Figure 3-11). To calculate the enthalpy difference between these states we construct

3.8 Ideal-Gas State

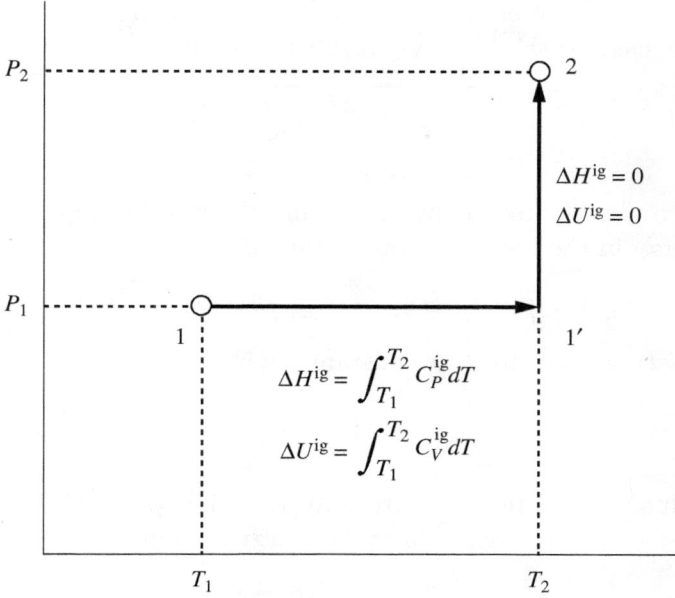

Figure 3-11: Path for the calculation of ΔH and ΔU between any two states in the ideal-gas state.

the composite path $11'2$, which consists of a constant-pressure segment $(11')$ and a constant-temperature segment $(1'2)$. Along the constant-pressure part, the enthalpy is given by eq. (3.19):

$$\Delta H_{11'}^{ig} = \int_{T_1}^{T_2} C_P^{ig} \, dT.$$

In the second segment, temperature is constant. Since enthalpy in the ideal-gas state depends only on temperature, there is no enthalpy change along this part:

$$\Delta H_{1'2}^{ig} = 0.$$

The enthalpy change for the combined path is the sum of the parts:

$$\boxed{\Delta H_{12}^{ig} = \int_{T_1}^{T_2} C_P^{ig} \, dT.} \qquad (3.33)$$

This result gives the enthalpy change between *any* two states in the ideal-gas limit. Since enthalpy is a state function, the result is the same for any integration path between the two states, therefore, eq. (3.33) has general validity. A similar analysis on the VT plane leads to the following result for the internal energy:

$$\Delta U_{12}^{ig} = \int_{T_1}^{T_2} C_V^{ig} dT. \qquad (3.34)$$

Another useful result is obtained by expressing the relationship between enthalpy and internal energy in the ideal-gas state. Starting with eq. (3.12) and using $V^{ig} = RT/P$, we have

$$H^{ig} = U^{ig} + PV^{ig} = U^{ig} + RT.$$

Differentiation with respect to T at constant P gives

$$\left(\frac{\partial H^{ig}}{\partial T}\right)_P = \left(\frac{\partial U^{ig}}{\partial T}\right)_P + R.$$

Since U^{ig} is a function of temperature only (i.e., independent of P and V), the following two partial derivatives of internal energy are equal

$$\left(\frac{\partial U^{ig}}{\partial T}\right)_P = \left(\frac{\partial U^{ig}}{\partial T}\right)_V.$$

The same is true for enthalpy.[6] Thus we conclude that

$$C_P^{ig} = C_V^{ig} + R. \qquad (3.35)$$

This relationship allows us to calculate the constant-volume heat capacity in the ideal-gas state from tabulations of the constant-pressure heat capacity. For this reason, C_V^{ig} is not tabulated separately.

Example 3.16: Isothermal Compression of Steam
Steam undergoes reversible isothermal compression at 350 °C, from 20 bar to 40 bar. Assuming steam to be in the ideal-gas state, calculate the heat and work, as well as the change in enthalpy and internal energy. Compare the results to those obtained in Example 3.9 using the steam tables. The ideal-gas heat capacity of steam is

$$C_P^{ig}/R = 4.395 - 4.186 \times 10^{-3} T + 0.00001405 T^2 - 1.564 \times 10^{-8} T^3 + 6.32 \times 10^{-12} T^4$$

with T is in K.

---

6. For enthalpy we have,
$$\left(\frac{\partial H^{ig}}{\partial T}\right)_P = \left(\frac{\partial H^{ig}}{\partial T}\right)_V.$$
Since H^{ig} is also a function of T only, it is independent of P or V.

3.8 Ideal-Gas State

Solution The reversible work is

$$W = -\int_{V_1}^{V_2} P\,dV.$$

In the ideal-gas state, $P = RT/V$, and with T constant, the integration gives,

$$W = -RT\int_{V_1}^{V_2} \frac{dV}{V} = -RT\ln\frac{V_2}{V_1} = -RT\ln\frac{P_1}{P_2}.$$

By numerical substitution,

$$W = -(8.314 \text{ J/molK})(623.15 \text{ K})\ln\frac{(20 \text{ bar})}{(40 \text{ bar})} = 3591.1 \text{ J/mol} = 199.34 \text{ kJ/kg}.$$

The temperature is the same in the initial and final states, and since U and H in the ideal-gas state depend on temperature only, both remain constant during the process:

$$\Delta U = 0, \Delta H = 0.$$

The heat is obtained by application of the first law:

$$Q = \Delta U - W = -199.34 \text{ kJ/kg}.$$

The comparison with the results from the steam tables, obtained in Example 3.9, are summarized below:

	Ideal gas	Steam tables	
W	199.34	199.81	(kJ/kg)
Q	−199.34	−232.91	(kJ/kg)
ΔU	0	−33.1	(kJ/kg)
ΔH	0	−44.3	(kJ/kg)

Comments The work between the two methods is in good agreement but the result for the heat is off by about 15%. This is due to the fact that the change in internal energy is not exactly zero, as predicted by the ideal-gas law. This change is still pretty small (33 kJ/kg), but because the overall amount of heat is not very large, the percent difference is appreciable. This is a warning that the ideal-gas law should be used only with justification. In this problem, the compressibility factor in the final state (which is the one further removed from ideality) is $Z_2 = 0.9245$, which is not within the acceptable range for the ideal-gas law.

Reversible Adiabatic Process This is a special process in which the system exchanges reversible work with the surroundings but no heat. Experimentally, this can be accomplished by thermally insulating the system and conducting the process in a quasi-static manner. Such system can either be compressed or expanded. The conditions "reversible" and "adiabatic" fully specify the path, as we will see below. The process is not specific to ideal-gases, but the calculation of this path for a general fluid will be delayed until Chapter 4.

For an adiabatic process, $dQ = 0$, and application the first law gives

$$W = \Delta U_{12} = \int_{T_1}^{T_2} C_V^{ig} dT. \tag{3.36}$$

To calculate this work we must know the initial and final temperatures. In typical problems, we usually know the initial state (pressure and temperature) and one property (pressure, temperature, or volume) in the final state. To fully specify the final state we must fix one more intensive property and this is done through the equation that describes the path. The path of the process is expressed as a relationship between any two variables, either P and T, P and V, or T and V. If the path is obtained in terms of one of these pairs, all the others can be obtained by using the ideal-gas law to eliminate one variable in terms of another. This procedure is developed below.

We begin by applying the differential form of the first law, eq. (3.10), along a small step of the process. For adiabatic process, $dQ = 0$, and for reversible work, $dW = -PdV$. Then, the first law gives

$$dU^{ig} = dW = -PdV^{ig}.$$

Writing $dU^{ig} = C_V^{ig} dT$ for the internal energy, this becomes

$$C_V^{ig} dT = -PdV^{ig} = -RT\frac{dV^{ig}}{V^{ig}}. \tag{3.37}$$

This result we can rewrite in the form

$$\frac{dV^{ig}}{V^{ig}} = -\frac{C_V^{ig}}{R}\frac{dT}{T}. \tag{3.38}$$

This equation gives the path in differential form by relating changes in volume to changes in temperature. Integration from the initial state (T_1, V_1^{ig}) to the final (T_2, V_2^{ig}) gives

$$\ln \frac{V_2^{ig}}{V_1^{ig}} = -\frac{1}{R}\int_{T_1}^{T_2} C_V^{ig} \frac{dT}{T}. \tag{3.39}$$

The heat capacity in the ideal-gas state is a function of temperature; therefore, it must be left inside the integral. This equation can be expressed in simpler form if we make use of the log-T mean heat capacity between temperatures T_1 and T_2. The general definition of the mean log-x value of a function is given in eq. (3.47). Applying this definition to C_V^{ig}, we obtain

$$\overline{C}_{V\log}^{ig} = \frac{\int_{T_1}^{T_2} C_V^{ig} \frac{dT}{T}}{\ln(T_2/T_1)}. \tag{3.40}$$

3.8 Ideal-Gas State

Using this definition, eq. (3.39) can be written in the simpler form:

$$\boxed{\frac{V_2^{ig}}{V_1^{ig}} = \left(\frac{T_2}{T_1}\right)^{-\overline{C}_{V\log}^{ig}/R}.} \qquad (3.41)$$

The equation expresses the relationship between any two states 1 and 2 along the reversible adiabatic path. To express the path as a relationship between pressure and temperature, we write $V^{ig} = RT/P$ and use this to eliminate volume from eq. (3.41). The result is

$$\frac{P_2}{P_1} = \left(\frac{T_2}{T_1}\right)^{1+\overline{C}_{V\log}^{ig}/R}. \qquad (3.42)$$

The exponent of the temperature term can be expressed in terms of the log-T mean C_P^{ig}:

$$1 + \frac{\overline{C}_{V\log}^{ig}}{R} = \frac{R + \overline{C}_{V\log}^{ig}}{R} = \frac{\overline{C}_{P\log}^{ig}}{R}. \qquad (3.43)$$

With this result, eq. (3.42) becomes

$$\boxed{\frac{P_2}{P_1} = \left(\frac{T_2}{T_1}\right)^{\overline{C}_{P\log}^{ig}/R}.} \qquad (3.44)$$

Finally, we express the path in terms of pressure and volume. To do this, we write $T = PV^{ig}/R$ and use this result to eliminate temperature from eq. (3.42):

$$\frac{P_2}{P_1} = \left(\frac{P_2 V_2^{ig}}{P_1 V_1^{ig}}\right)^{\overline{C}_{P\log}^{ig}/R} \Rightarrow \frac{P_2}{P_1} = \left(\frac{V_2^{ig}}{V_1^{ig}}\right)^{-\overline{C}_{P\log}^{ig}/\overline{C}_{V\log}^{ig}}. \qquad (3.45)$$

Using the symbol $\overline{\gamma}$ for the ratio of the heat capacities, $\overline{\gamma} = \overline{C}_{P\log}^{ig}/\overline{C}_{V\log}^{ig}$, the result can be written in the more common form,

$$\boxed{P_1 \left(V_1^{ig}\right)^{\overline{\gamma}} = P_2 \left(V_2^{ig}\right)^{\overline{\gamma}}.} \qquad (3.46)$$

Equations (3.41), (3.44), and (3.46) are alternative expressions for the reversible adiabatic path in the ideal-gas state.

If the heat capacity is constant (independent of temperature), the log-x is equal to the heat capacity $\overline{C}_{P\log}^{ig} = C_P^{ig}$. This suggests that if the heat capacity does not change much over the temperature interval of the problem it can be approximately taken to be constant, $\overline{C}_{P\log}^{ig} \approx C_P^{ig}$. Then, the reversible adiabatic path is simplified and becomes,

$$\frac{P_2}{P_1} \approx \left(\frac{T_2}{T_1}\right)^{C_P^{ig}/R},$$

$$P_1\left(V_1^{ig}\right)^\gamma \approx P_2\left(V_2^{ig}\right)^\gamma,$$

where $\gamma = C_P^{ig}/C_V^{ig}$. This eliminates the need for the calculation of the log-x heat capacity. This approximation is acceptable if the heat capacity is an insensitive function of temperature, or if the temperature interval is too small to make much of a difference in the heat capacity.

NOTE

Log-x Mean
Equation (3.47) and eq. (3.40) involve integrals of the form

$$\int_{x_1}^{x_2} f(x)\frac{dx}{x}.$$

This is a weighted integral of $f(x)$ with weights $1/x$. If function $f(x)$ is replaced by a constant value, $\bar{f}$, the integral becomes

$$\int_{x_1}^{x_2} \bar{f}\frac{dx}{x} = \bar{f}\ln\frac{x_2}{x_1}.$$

The two integrals are equal if $\bar{f}$ has the value

$$\boxed{\bar{f}_{\log} = \frac{\int_{x_1}^{x_2} f(x)\frac{dx}{x}}{\ln\frac{x_2}{x_1}}.} \qquad (3.47)$$

This defines the mean log-x value of f in the interval (x_1, x_2). This value is a weighted mean of f that gives the same result as f when integrated in the interval x_1 to x_2 with weights $1/x$. With the substitution $y = \ln x$, eq. (3.47) becomes

$$\bar{f}_{\log} = \frac{\int_{y_1}^{y_2} F(y)dy}{y_2 - y_1},$$

where $F(y)$ is f expressed as a function of $y = \ln x$. Therefore, the log-x mean is the usual arithmetic mean with respect to $\ln x$. The subscript $_{\log}$ will be used to distinguish this from the arithmetic mean of f. The value of the log-T mean depends on x_1 and x_2; therefore, the mean must be recalculated if the interval changes. The use of the log x in equations like eq. (3.44) does not eliminate the need for integrations, it just simplifies the notation. However, if $f(x)$ does not vary much between x_1 and x_2, it may be replaced by a constant value, which may be approximated as $\bar{f}_{ml} \approx f(x_1)$, or $\bar{f}_{ml} \approx f(x_2)$, or $\bar{f}_{ml} \approx (f(x_1) + f(x_2))/2$. Treating f as constant is acceptable if the range (x_1, x_2) is narrow or if f is not a sensitive function of x.

3.8 Ideal-Gas State

Example 3.17: Reversible Adiabatic Expansion of Nitrogen
Nitrogen expands reversibly in an insulated cylinder fitted with a piston from 500 K, 5 bar to a final pressure of 1 bar. Determine the final temperature as well as the amount of work assuming nitrogen to be in the ideal-gas state. The heat capacity of nitrogen is

$$C_P^{ig}/R = a_0 + a_1 T + a_2 T^2 + a_3 T^3 + a_4 T^4$$

with T in kelvin and

$a_0 = 3.539$, $a_1 = -0.000261$, $a_2 = 7.0 \times 10^{-8}$, $a_3 = 1.57 \times 10^{-9}$, $a_4 = -9.9 \times 10^{-13}$.

Solution The relationship between pressure and temperature in the initial and final states is given by eq. (3.44). Solving for T_2,

$$T_2 = T_1 \left(\frac{P_2}{P_1}\right)^{R/\overline{C}_{P\log}^{ig}}. \tag{A}$$

Since the logarithmic mean capacity depends on T_2, this equation must be solved by trial and error. We start by assuming $\overline{C}_{P\log}^{ig}$ to be equal to the C_P^{ig} at the initial temperature, $T_1 = 500$ K:

$$\frac{C_P^{ig}(500 \text{ K})}{R} = 3.560.$$

Using this value in eq. (A), the temperature is

$$T_2 = (500 \text{ K}) \left(\frac{1 \text{ bar}}{5 \text{ bar}}\right)^{1/3.560} = 318.164 \text{ K}.$$

We now calculate the logarithmic mean capacity between 500 K and 318.164 K by integration of the equation for C_P^{ig}. For the polynomial expression given here, the result is

$$\frac{\overline{C}_{P\log}^{ig}}{R} = \frac{a_0 \ln \frac{T_2}{T_1} + a_1(T_2 - T_1) + \frac{a_2}{2}(T_2^2 - T_1^2) + \frac{a_3}{3}(T_2^3 - T_1^3) + \frac{a_4}{4}(T_2^4 - T_1^4)}{\ln(T_2/T_1)} = 3.524.$$

Using this value we calculate a revised temperature from eq. [A]:

$$T_2 = (500 \text{ K}) \left(\frac{1 \text{ bar}}{5 \text{ bar}}\right)^{1/3.524} = 316.697 \text{ K}.$$

This procedure is repeated until the temperature does not change any more. These iterations are summarized below:

Iteration	C_P^{ig}/R	T_B (K)
1	3.560	318.164
2	3.524	316.697
3	3.524	316.688
4	3.524	316.688

Therefore, $T_B = 316.688$ K.

The work is calculated using eq. (3.36), which we write as

$$W = \int_{T_A}^{T_B} \left(C_P^{ig} - R\right) dT = \int_{T_A}^{T_B} C_P^{ig} dT - R(T_B - T_A).$$

The integral is evaluated using the polynomial expression for the heat capacity:

$$\int_{T_A}^{T_B} C_P^{ig} dT = R\left(a_0(T_2 - T_1) + \frac{a_1}{2}(T_2^2 - T_1^2) + \frac{a_2}{3}(T_2^3 - T_1^3)\right.$$
$$\left. + \frac{a_3}{4}(T_2^4 - T_1^4) + \frac{a_4}{5}(T_2^5 - T_1^5)\right)$$
$$= -5,374.21 \text{ J/mol}.$$

The work is

$$W = -5,374.21 \text{ J/mol} - (8.315 \text{ J/mol K})(316.688 \text{ K} - 500 \text{ K}) = -3,850.2 \text{ J/mol}.$$

The path of this process is shown in Figure 3-12 by the line AD.

Comments The heat capacity of nitrogen changes little between the initial and final temperature, and this is the reason that the final temperature is not much different from that obtained after just one iteration.

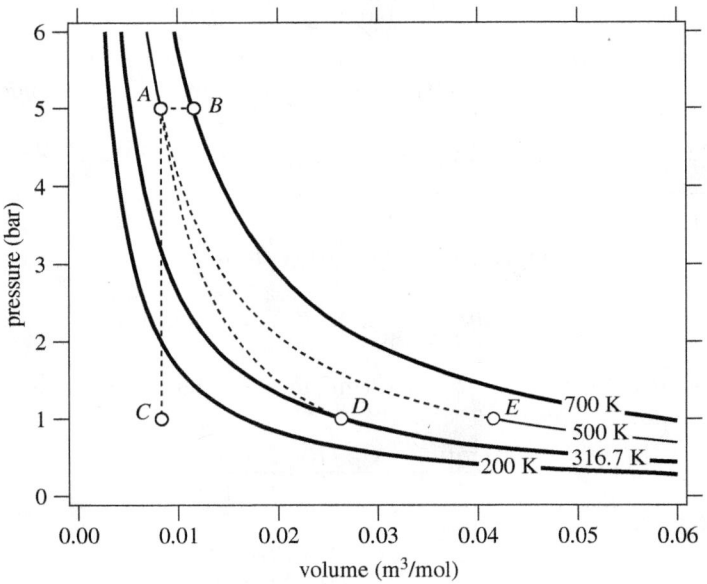

Figure 3-12: Elementary paths in the ideal-gas state (data for nitrogen): constant-pressure (AB), constant-volume (AC), reversible adiabatic (AD), isothermal (AE).

3.9 Energy Balances and Irreversible Processes

As a statement of energy conservation, the first law is applicable to any process, whether it is reversible or not. The change of internal energy, ΔU, between the initial and final state is fixed by the initial and final state and independent of the path used to connect them. This difference gives the *net* amount of energy that the system must exchange with its surroundings in order for this change to take place. The exchange is generally a combination of heat and work, but the individual amounts of heat and work depend on the specific path. For mechanically reversible paths, the work is calculated by eq. (3.3). If the process is irreversible and involves work, one must be careful because eq. (3.3) is not applicable. If the process does not involve work, the calculation is done in the usual way. Implicit in all of these calculations is the assumption that the system is internally uniform so that its state can be described by a uniform pressure and temperature.

Example 3.18: Heat Transfer
A 10 kg piece of hot copper at 450 °C is quenched in an open tub that contains 10 kg of water at 20 °C. Calculate the final temperature assuming no heat losses to the environment. The heat capacities of copper and water are $C_{Pc} = 0.38$ kJ/kg K, $C_{Pw} = 4.184$ kJ/kg K, and may be assumed independent of temperature.

Solution This is a constant-pressure process; therefore, the amount of heat is equal to the change in enthalpy. If we take the system to be copper and water together, no heat is exchanged with the surroundings:

$$Q = 0 = \Delta H^{\text{tot}} \quad \Rightarrow \quad \Delta H_w^{\text{tot}} + \Delta H_c^{\text{tot}} = 0.$$

For the enthalpy change of water and copper we use eq. (3.19), and with constant heat capacity, the result is:

$$\Delta H_w^{\text{tot}} = m_w C_{Pw}(T_f - T_w)$$

$$\Delta H_c^{\text{tot}} = m_c C_{Pc}(T_f - T_c).$$

where m_c, m_w, are the mass of copper and water, respectively, T_c, and T_w are the initial temperatures, and T_f is the common final temperature. Substituting these expressions into the energy balance we obtain an equation in which the only unknown is T_f. Solving for the final temperature we obtain

$$T_f = \frac{m_c C_{Pc} T_c + m_w C_{Pw} T_w}{m_c C_{Pc} + m_w C_{Pw}}.$$

By numerical substitution,

$$T_f = 74.8 \text{ °C}.$$

Comments This process is irreversible because heat is exchanged between systems at different temperatures.

Example 3.19: Adiabatic Mixing
A rigid insulated tank contains 5 kg of steam at 15 bar, 300 °C. A second tank, also insulated and rigid, contains 12 kg of steam at 8 bar, 500 °C. The two tanks are connected via a pipe and a valve that is initially closed. The valve is opened and the two compartments are allowed to exchange contents. Determine the temperature and pressure at the end state. Is this process reversible?

Solution We take the system to be the two tanks together. Initially, this is a composite system consisting of two simple systems separated by a closed valve. The process consists of the transient flow between the two tanks. Since the tanks are rigid and insulated, neither heat nor PV work are exchanged: $Q = W = 0$. The first law for this process reads

$$\Delta U^{\text{tot}} = U_f^{\text{tot}} - U_i^{\text{tot}} = 0.$$

Here, U^{tot} is the total energy of the system and the subscripts i and f and refer to the initial and final states, respectively. The initial internal energy is

$$U_i^{\text{tot}} = m_1 U_1 + m_2 U_2$$

where m_i is the mass and U_i is the specific internal energy in tank $i = 1, 2$. In the final state,

$$U_f^{\text{tot}} = (m_1 + m_2) U_f.$$

We combine these equations and solve for the internal in the final state:

$$U_f = \frac{m_1 U_1 + m_2 U_2}{m_1 + m_2}. \qquad [A]$$

Since the initial state in each tank are known, this result will allow us to calculate the specific internal energy of the final state. This gives us one intensive property. Since the total volume is constant for this process, we also have $V_i^{\text{tot}} = V_f^{\text{tot}}$:

$$m_1 V_1 + m_2 V_2 = (m_1 + m_2) V_f$$

which solved for V_f gives

$$V_f = \frac{m_1 V_1 + m_2 V_2}{m_1 + m_2}. \qquad [B]$$

This provides a second intensive property in the final state; therefore, this state is fully specified.

Calculations From the steam tables we collect the following information for the initial state in each tank:

	Tank 1	Tank 2
m (kg)	5	12
P (bar)	15	8
T (°C)	300	500
V (m³/kg)	0.1697	0.4433
U (kJ/kg)	2783.7	3126.5

3.9 Energy Balances and Irreversible Processes

The specific internal energy and volume in the final state are calculated from eqs. [A] and [B]:

$$U_f = \frac{(5 \text{ kg})(2783.7 \text{ kJ/kg}) + (12 \text{ kg})(3126.5) \text{ kJ/kg}}{(5+12) \text{ kg}} = 3025.7 \text{ kJ/kg},$$

$$V_f = \frac{(5 \text{ kg})(0.1697 \text{ m}^3/\text{kg}) + (12 \text{ kg})(0.4433 \text{ m}^3/\text{kg})}{(5+12) \text{ kg}} = 0.3628 \text{ m}^3/\text{kg}.$$

The above two intensive properties fully specify the final state. To do this, we must locate a state in the steam tables with these values of internal energy and volume. This task is somewhat tricky, however, because the steam tables are organized by temperature and pressure, not by internal energy and volume. To solve this problem, we need to locate four entries in the steam tables, not necessarily consecutive, such that the desired state is found near them. For best accuracy, one should look for four states that bracket the desired value most closely; however, this is not a strict requirement. The table below lists four such states:

P (bar)	6.5		6.5
T (°C)	200		450
V (m³/kg)	0.3241		0.5102
U (kJ/kg)	2637.4		3044.1
		$\vdots$	
P (bar)	6		6
T (°C)	200		450
V (m³/kg)	0.3521		0.5530
U (kJ/kg)	2639.4		3044.6

We must interpolate between these values for pressure and temperature, given the internal energy and volume. The reverse problem would be easier: If we knew pressure and temperature and were asked to calculate internal energy and volume, we would perform a double interpolation. Using x for temperature and y for pressure, the corresponding equations are

$$U_f = (1-a)(1-b)U_{11} + b(1-a)U_{12} + a(1-b)U_{21} + abU_{22}, \qquad [\text{C}]$$

$$V_f = (1-a)(1-b)V_{11} + b(1-a)V_{12} + a(1-b)V_{21} + abV_{22} \qquad [\text{D}]$$

with

$$a = \frac{T - x_1}{x_2 - x_1}, \quad b = \frac{P - y_1}{y_2 - y_1}. \qquad [\text{E}]$$

In our case we know U_f and V_f and want to find P and T. Noting that the only unknowns in eqs. [C] and [D] are the two parameters a and b, the solution should be now clear: solve eqs. [C] and [D] for a and b, then obtain pressure and temperature from the two equations in [E].

The numerical values of the required parameters are obtained from the tabulated values and are summarized below:

$$x_1 = 200\ °C \qquad U_{11} = 2639.4\ kJ/kg \qquad V_{11} = 0.3521\ m^3/kg$$
$$x_2 = 450\ °C \qquad U_{12} = 2637.4\ kJ/kg \qquad V_{12} = 0.3241\ m^3/kg$$
$$y_1 = 6\ bar \qquad U_{21} = 3044.6\ kJ/kg \qquad V_{21} = 0.553\ m^3/kg$$
$$y_2 = 6.5\ bar \qquad U_{22} = 3044.1\ kJ/kg \qquad V_{22} = 0.5102\ m^3/kg$$

The values of a and b are

$$a = 0.95927, \quad b = 4.3128,$$

and using eq. [E], the temperature and pressure of the final state are

$$T = 440\ °C, \quad P = 8.16\ bar.$$

We note that this state is found to lie somewhat farther away from the points used in the interpolation (the pressure in fact lies outside the pressure interval of the interpolation). If a more accurate answer is needed, one should use the above solution to guess that the answer must lie between 400 °C and 450 °C in temperature, and between 8 bar and 9 bar in pressure. A refinement could then be obtained by double interpolation between the four states. This is left as an exercise.

Is the process reversible? Clearly not. When the valve is opened, the two tanks interact under a finite difference of pressure and temperature. However, there is no work involved in this problem; therefore, the irreversible nature of the process poses no difficulties in the calculation.

Example 3.20: Irreversible Expansion Against Vacuum
A closed insulated cylinder is divided into parts by a piston that is held into place by latches. One compartment is evacuated, the other contains steam at 7.5 bar, 300 °C. The latches are removed and the gas expands to fill the entire volume of the cylinder. When the system reaches equilibrium, its pressure is 1 bar. Determine the final temperature.

Solution *Outline of solution method:* No heat is exchanged with the surroundings; therefore $Q = 0$. The gas expands against vacuum; therefore, no PV work is exchanged with the surroundings, that is $W = 0$. The first law reduces to

$$\Delta U_{12} = 0 \quad \Rightarrow \quad U_1 = U_2$$

where U_1, U_2, is the internal energy in the initial and final states, respectively. The initial state is specified; therefore, U_2 can be evaluated. We also know the pressure in the final state. We know two intensive properties, U_2 and P_2; therefore the state is fully specified.

3.9 Energy Balances and Irreversible Processes

Numerical calculations: The initial state is known and is obtained immediately from the steam tables:
$$P_1 = 7.5 \text{ bar}, \quad T_1 = 300 \text{ °C}, \quad U_1 = 2798.6 \text{ kJ/kg}.$$

In the final state we know the internal energy, $U_2 = U_1 = 2798.6$ kJ/kg, and the pressure, $P_2 = 1$ bar. By interpolation in the tables we find
$$T_2 = 292 \text{ °C}.$$

The temperature decreases by 8 °C.

Comments This is clearly an irreversible process because expansion takes place against a nonzero pressure difference between system and surroundings. Because expansion is done against vacuum, no PV work is involved. In fact, in this process the system reduces its pressure (and its temperature somewhat) without exchanging any energy with the surroundings.

Example 3.21: Irreversible Expansion of Ideal Gas
Repeat the previous example treating steam as an ideal gas. Is the ideal-gas assumption appropriate?

Solution The working equation is
$$\Delta U_{12} = 0.$$
In the ideal-gas state, internal energy is a function of temperature only. Internal energy always increases with temperature (why?), and the only way to satisfy the above equation is if $T_2 = T_1$. We conclude, therefore,
$$T_2 = 300 \text{ °C}.$$

Is the ideal-gas assumption appropriate? From the steam tables at the $P_1 = 7.5$ bar, $T_1 = 300$°C, we find $V_1 = 0.3462$ m³/kg. The compressibility factor is
$$Z_1 = \frac{(7.5 \text{ bar})(0.3462 \text{ m}^3/\text{kg})}{(8.314 \text{ J/mol K})(573.1 \text{ K})}\left(\frac{10^5 \text{ Pa}}{\text{bar}}\right) = 0.9808.$$
In the final state, pressure is lower and the compressibility factor is expected to be even closer to 1. In both states, therefore, the compressibility factor is sufficiently close to 1 to assume ideality. Nonetheless, the ideal-gas solution overestimates the final temperature by 8 °C.

Calculations based on the assumption of ideal-gas behavior must always be done very carefully. If a more accurate method is available (tables, equations, etc.), it should be preferred.

Example 3.22: Irreversible Expansion Against Pressure
An insulated cylinder fitted with a weightless latched piston contains steam at 5 bar, 300 °C. The piston is unlatched and the steam expands against ambient air at $P = 1$ bar. The temperature of steam in the final state is 142 °C. Determine the energy balance for this process.

Solution The process is adiabatic ($Q = 0$) and the first law reduces to

$$\Delta U_{AB} = W.$$

The initial and final states are both known; therefore, the work can be obtained from this equation.

Numerical calculations From the steam tables we obtain (the internal energy in the final is obtained by interpolation),

	Initial state	Final state
P (bar)	5	1
T (°C)	300	142
U (kJ/kg)	2803.3	2570.6

The work is
$$W = \Delta U = (2570.6 - 2803.3) \text{ kJ/kg} = -232.7 \text{ kJ/kg}.$$

The work is negative because it is done by the system on the surroundings.

Comments This process is irreversible because expansion is done under a nonzero pressure difference; therefore, eq. (3.4) cannot be used. In this problem we were able to calculate the work from the energy balance because ΔU and Q were both known. It is interesting to calculate the amount of reversible work that is needed to push the surroundings. This work is

$$W' = -P_{ex}\Delta V$$

where $P_{ex} = 1$ bar is the external pressure and ΔV is change in the specific volume of steam, equal to the volume of the atmosphere displaced by the piston during the expansion. The specific volumes in the initial and final states are 0.5226 m³/kg and 1.8982 m³/kg, respectively, therefore, the reversible work is

$$W' = -\left(1 \text{ bar} \frac{10^2 \text{ kJ/m}^3}{\text{bar}}\right)(1.8982 - 0.5226) \text{ (m}^3\text{/kg)} = -137.6 \text{ kJ/kg}.$$

We notice that the work expended by the system is more than what is required to push the surroundings. The difference represents a penalty that must be paid for conducting the process irreversibly. As a result of this extra work the temperature of the surroundings will increase somewhat, but this temperature change is imperceptible because we take the surroundings to be infinite in size.

3.10 Summary

The first law extends the energy balance to include *heat*, and in doing so it introduces *internal energy* as a state function. Internal energy represents the collective ways in which matter is capable of storing energy. During a process, changes in internal energy can be tracked through the exchange of heat and work that take place between the system and its surroundings. Enthalpy,

$$H = U + PV, \qquad [3.12]$$

is a state function with units of energy. Both internal energy and enthalpy are important in energy balances. As state functions, internal energy and enthalpy can be obtained from tables, if available, or they can be calculated using suitable methodologies. So far we have formulated equations for the calculation of internal energy and enthalpy in two special cases:

- *Internal energy in constant-volume path:*

$$\Delta U_{AB} = \int_{\substack{T_A \\ \text{const. } V}}^{T_B} C_V \, dT. \qquad [3.17]$$

- *Enthalpy in constant pressure path:*

$$\Delta H_{AB} = \int_{\substack{T_A \\ \text{const. } P}}^{T_B} C_P \, dT. \qquad [3.19]$$

In Chapter 5 we will learn how to calculate properties between any two states without the limitation of constant-volume or constant-pressure paths. A special case of practical importance is the *ideal-gas* state. In this special limit, internal energy and enthalpy are functions of temperature only and changes between *any* two states are given by

$$\Delta H_{12}^{\text{ig}} = \int_{T_1}^{T_2} C_P^{\text{ig}} \, dT \qquad [3.33]$$

$$\Delta U_{12}^{\text{ig}} = \int_{T_1}^{T_2} C_V^{\text{ig}} \, dT. \qquad [3.34]$$

It is a recurring theme in thermodynamics that calculations in the ideal-gas state are simpler than in any other state. However, *the ideal-gas equations should never be used without adequate justification, and even then, only as an approximation!*

As a statement of energy conservation, the first law is the starting point for all energy balances in a closed system. Problems of this type typically require the calculation of heat and work. The amount of heat that is exchanged can, under special conditions, be related to internal energy or enthalpy:

$$Q = \Delta U \quad \text{(constant-volume process)} \quad [3.11]$$

$$Q = \Delta H \quad \text{(constant-pressure process)}. \quad [3.11]$$

In the calculation of work we must distinguish between shaft work and PV work. There is no special formula for shaft work in a closed system; this type of work, if present, will be calculated from mechanical considerations of device that is used to produce or absorb this type of work. The PV work is of fundamental importance in thermodynamics and it arises whenever the boundaries of a system move in the presence of a forcing or opposing external pressure. In the special case that the exchange of work is conducted in mechanically reversible manner, the PV work is

$$W_{\text{rev}} = -\int_A^B PdV. \quad [3.4]$$

Care must be taken when applying energy balances to mechanically irreversible processes because in such cases the work cannot be calculated using eq. (3.4).

3.11 Problems

Problem 3.1: A gas is confined in a cylinder sealed by a piston 20 cm in diameter whose mass is 30 kg. The pressure in the room is 1 bar.
a) What is the pressure inside the cylinder?
b) Heat is added to the cylinder causing the gas to expand. The piston rises and finally rests at 50 cm above its initial level. How much work was done during this process?
c) What is the pressure inside the cylinder at the end of part (b)?

Problem 3.2: The PVT behavior of a certain gas is described by the equation $P(V-b) = RT$ where b is a constant.
a) Show that when $P \to 0$, this gas becomes ideal.
b) Calculate the work done by this gas when it expands reversibly under constant temperature from initial volume V_A to final volume V_B.
c) Obtain the second and the third virial coefficients.

Problem 3.3: a) At 0 °C, 1 atm, the density of liquid water is 1 g/cm^3 and that of ice is 0.917 g/cm^3. Calculate the amount of work (in joule) that is exchanged when 1 liter of liquid water freezes to produce ice at 0 °C and 1 atm. Use the proper sign convention!

3.11 Problems

b) If this work could be converted into kinetic energy of this quantity of water, what would be the speed? Give your answer in m/s and in mph.
c) If the work of part (a) were used to raise this quantity of water by a distance h, what would be that distance? Report the result in m and in ft.

Problem 3.4: a) Calculate the amount of work necessary for the reversible compression of 1 kg of steam from 3 bar to 7 bar. The compression is to take place in a cylinder fitted with a weightless piston at the constant temperature of 250 °C.
b) Calculate the amount of heat, if any, associated with this process.

Problem 3.5: a) Calculate the amount of work necessary for the reversible expansion of 1 kg of steam from 10 bar to 5 bar. The expansion is to take place in a cylinder fitted with a weightless piston at the constant temperature of 400 °C.
b) Calculate the amount of heat associated with this process.

Problem 3.6: A sealed tank contains saturated steam at 5 bar. The volume of the tank is 1 m^3.
a) Heat is added until the pressure in the tank doubles. What is the final temperature?
b) Calculate the necessary amount of heat.

Problem 3.7: A closed rigid tank contains saturated water vapor at 100 °C. The tank has a safety relief valve that will go off at 2.5 bar. How much heat can be added to the steam before the valve goes off?

Problem 3.8: A sealed pressure cooker contains steam at 5 bar and 200 °C. Subsequently, the cooker is cooled until the contents become saturated vapor.
a) What is the final temperature?
b) How much heat is removed during the cooling step?

Problem 3.9: A sealed rigid tank is filled with 1 kg of steam at 200 °C and 10 bar.
a) What is the volume of the tank?
b) 0.5 kg of steam are removed from the tank. If the temperature remains at 200 °C, what is the new pressure in the tank?
c) What is the state of the steam at the end of part b (e.g., liquid, vapor, saturated, superheated)?
d) How much heat must be added to bring the pressure back to 10 bar?
e) What is the temperature at the end of step (d)?

Problem 3.10: A closed and sealed rigid tank contains 11 kg of liquid water at equilibrium with 1.17 kg of vapor at a pressure of 2 bar. The tank is equipped with two alarms: a pressure alarm that will sound if the pressure exceeds 10 bar; and a temperature alarm that will sound if the temperature exceeds 200 °C.
a) What is the volume of the tank?
b) Calculate the internal energy of the liquid-vapor mixture in the tank.
c) If we add heat to the tank continuously, which alarm will sound first?
d) We stop the heating when the contents are saturated vapor. How much heat has been added?
e) Which alarm(s), if any, will be sounding at the end of step (d)?
f) Draw a qualitative PV graph and sketch the path of the processes involved. Mark all the relevant isotherms and pressures.

Problem 3.11: a) A closed tank contains steam with 20% moisture at 3 bar. Heat is added until one of the two phases disappears. Which phase disappears and what is the final pressure and temperature?
b) Another closed tank contains steam with 99.9 % moisture at 3 bar. Heat is added until one of the two phases disappears. Which phase disappears and what is the final pressure and temperature?
c) A third closed tank contains a vapor-liquid mixture of toluene at 1 atm. Ninety-eight percent of the toluene (by mole) is in the liquid phase. Heat is added until one of the two phases disappears. Which phase disappears? The normal boiling point of toluene is 384 K.

Problem 3.12: A sealed rigid tank of 1 m^3 capacity contains 1.2 kg of water at 1 bar and 99.63 °C. To this tank we add heat until the pressure increases to 3 bar.
a) What is the temperature at the end of the heating?
b) Calculate the amount of vapor before and after the heating.
c) Calculate the amount of heat added.

Problem 3.13: A sealed rigid tank contains 1 kg of wet steam at 300 °C. The fraction of the liquid in the tank is not known. Upon adding heat to the tank it is observed that the vapor condenses. The heating stops when the system becomes saturated liquid. At that point the pressure in the tank is 148 bar.
a) What fraction of the water is initially liquid?
b) How much heat was added?
c) What is the minimum amount of liquid that must be present in the tank initially, in order to observe condensation upon heating?
d) What would be observed upon heating, if the initial amount of liquid was less than the minimum value calculated in part (c)?
e) How do you explain the fact that heating can result in condensation of the vapor?

3.11 Problems

Problem 3.14: A 0.5 m^3 sealed rigid vessel contains 15 kg of water (phase unknown) at 120 °C. Heat is added or removed, as needed, until the steam becomes a single saturated phase.
a) Determine the state of water before heating.
b) Determine the pressure and temperature at the end of heating.
c) Determine the amount of heat.

Problem 3.15: A 3 m^3 tank is divided into two parts via a thermally insulated wall. One part has a volume of 2 m^3 and contains a vapor-liquid mixture of steam, 7% liquid, at 5 bar. The other part contains steam at 10 bar, 500 °C. A valve is opened and allows the two parts to thoroughly mix.
a) Determine the final pressure and temperature in the tank assuming no losses to the surroundings.
b) Determine the amount of heating or cooling that must be supplied in order to make the final pressure 5 bar.

Problem 3.16: A closed tank, 1 m^3 in volume is filled with water at 1 bar 80 °C. A second identical tank contains 20 kg of steam at 20 bar.
a) What is the mass of water in tank 1?
b) What is the temperature and phase in tank 2?
c) The contents of the two tanks are mixed together in a third closed tank whose volume is 2 m^3. Heat is added or removed, as needed, to bring the final temperature to 175 °C. What is the final pressure?

Problem 3.17: An insulated cylinder is divided into two parts: one part has a volume of 5 L and contains steam at 1 bar, 200 °C; the other part has a volume of 2 L and contains steam at 2 bar, 400 °C. The partition that divides the cylinder is removed and the contents are allowed to reach equilibrium without heat losses to the surroundings.
a) Determine the pressure and temperature in the final state assuming steam to be an ideal-gas. Is ideality an acceptable assumption?
b) Repeat using the steam tables.

Problem 3.18: Wet steam with quality 63.1% exchanges heat in a closed system under constant pressure $P = 6$ bar. Determine the final temperature and the amount of PV work in each of the following cases:
a) The system is cooled by removing 1500 kJ/kg of heat.
b) The system is heated by adding 1500 kJ/kg of heat.
Report the work using the proper sign convention.

Problem 3.19: a) 200 kJ of heat is added under constant pressure to 1 kg of steam initially at 30 bar, 400 °C. What is the final temperature?
b) What is the change in internal energy during this process?

Problem 3.20: Steam at 200 °C and 10 bar is cooled under constant pressure until it becomes saturated liquid.
a) What is the final temperature?
b) How much heat is removed from the steam?
c) Calculate the work involved in this process, if any.

Problem 3.21: a) Water at 1 bar, 20 °C is heated at const P to produce steam with quality 95%. What is the amount of heat?
b) What is the final state if the amount of heat is 2100 kJ/kg?

Problem 3.22: 1 mol of liquid octane is heated in a closed system at constant pressure of 11 bar from 8 °C until it becomes saturated liquid ($T^{\text{sat}} = 240$ °C). Calculate:
a) The amount of heat.
b) The amount of work.
c) The enthalpy at the initial state.
d) The internal energy at the initial state.
Additional data: The enthalpy of the saturated liquid at 1.1 bar is 38.592 kJ/mol. The heat capacity of the liquid is

$$C_P/R = 27.0423 - 0.0224477\,T + 0.000115337\,T^2$$

with T in kelvin. For the liquid volume, use the Rackett equation. You may assume that the liquid is incompressible.

Problem 3.23: Steam is heated in closed system from 20 bar, 300 °C to 450 °C. Determine the amount of heat and work in the following cases:
a) Heating is conducted at constant pressure.
b) Heating is conducted at constant volume. What is the final pressure?
c) Use these results to estimate the C_P and C_V of steam.
The process may be assumed to be mechanically reversible.

Problem 3.24: Wet steam at 200 °C with 80% moisture is cooled by removing 600 kJ/kg of heat.
a) Determine the final state (pressure and temperature and quality, if a two-phase system) if cooling is at constant pressure.
b) Determine the final state if cooling is at constant volume.

3.11 Problems

Problem 3.25: Use the steam tables to calculate the C_V and C_P of water at the following states:
a) 25 °C, 1 bar.
b) 50 bar, 400 °C.

Problem 3.26: a) A cylinder with a volume of 3 L contains 2.031 mol of butane at 50 °C. Determine the enthalpy and internal energy of the contents in the tank (report in J/mol).
b) How much heat must be added or removed to the system in part (a) under constant volume to turn the contents into saturated vapor? What is the temperature?
c) How much heat must be removed from the system in part (a) under constant pressure to bring the temperature to 0 °C?
d) State your assumptions clearly.
The following data are available:

Properties of saturated butane

T^{sat} (°C)	P (bar)	ρ_L (mol/L)	H_L (J/mol)	ρ_V (mol/L)	H_V (J/mol)
0	1.03	10.336	11624	0.047	34017
10	1.48	10.148	12984	0.067	34846
20	2.08	9.955	14374	0.091	35676
30	2.83	9.755	15795	0.123	36504
40	3.78	9.547	17252	0.162	37329
50	4.96	9.331	18745	0.211	38147
60	6.38	9.104	20279	0.270	38953
70	8.09	8.864	21858	0.343	39742
80	10.12	8.610	23485	0.432	40508
90	12.49	8.335	25167	0.542	41239
100	15.26	8.037	26912	0.677	41923

Heat capacity of liquid: $C_{P,\text{liq}} = 153$ J/mol K
Heat capacity of vapor: $C_{P,\text{vap}} = 118$ J/mol K

Problem 3.27: Ammonia is heated under constant pressure of 5 bar, from 0 °C to 200 °C. Determine the amount of heat. Additional information is given below:

T^{sat}(at 5 bar) :	4.14	°C
ΔH^{vap}(at 5 bar) :	21.242	kJ/mol
C_P^L :	79.022	J/mol K
C_P^V :	46.726	J/mol K

Problem 3.28: Ammonia vapor is cooled under a constant pressure of 5 bar. The initial temperature is 120 °C and the amount of heat that is removed is 10 kJ/mol. Determine the final temperature. If the system is a vapor-liquid mixture, determine the mass fractions in each phase. Use the data given in problem 3.27.

Problem 3.29: 1 mol of liquid hexane at 10 bar, 50 °C, is mixed with 2 mol of hexane vapor at 250 °C, 10 bar. The process takes place adiabatically in a closed system at constant pressure. Determine the final temperature and if the system is a vapor-liquid mixture report the mass fraction in each phase.
Additional data:

T^{sat} (at 10 bar)	= 165.7 °C
$H^{sat,L}$ (at 10 bar)	= 22.645 kJ/mol
$H^{sat,V}$ (at 10 bar)	= 43.320 kJ/mol
C_P^L	= 263.64 J/mol K
C_P^V	= 226.34 J/mol K

Problem 3.30: 1 mol of liquid hexane at 10 bar, 25 °C, is mixed adiabatically and under constant pressure with hexane vapor at 180 °C. Determine the moles of the vapor that must be mixed with the liquid in order to produce a final state that consists of a vapor-liquid mixture with 75% (by mass) liquid. You may use the data given in problem 3.29.

Problem 3.31: Methanol at 25 °C, 8 bar is heated under constant P in closed system.
a) Determine the amount of heat to produce 95% vapor
b) Determine the final state if $Q = 32000$ J/mol
c) Determine the final state if $Q = 52000$ J/mol
Additional data:

T^{sat} (8 bar)	= 128.09 °C
ΔH^{vap}	= 30135 J/mol
C_{PL}	= 112.40 J/mol/K
C_{PV}	= 130.0 J/mol/K

Problem 3.32: 1 kg of liquid water, initially at 40 °C, 2 bar, is heated under constant pressure. If the amount of heat added is 1200 kJ, determine the final temperature and the phase of the system in the final state (if vapor-liquid mixture, report the mass fraction of each phase).

Problem 3.33: a) Obtain the heat of vaporization of steam at 30 bar.
b) Saturated liquid water at 30 bar is heated until the quality is 75%. What is the amount of heat?
c) Saturated liquid water at 30 bar is heated by adding 750 kJ/kg of heat at constant pressure. What is the final state?
d) What is the final state if the amount of heat that is added to the saturated liquid is 1950 kJ/kg?

3.11 Problems

Problem 3.34: a) Water at 1 bar, 20 °C is heated at const P to produce steam with quality 95%. What is the amount of heat?
b) Steam at 1 bar 400 °C is cooled by removing 300 kJ/kg of heat at constant pressure. What is the final temperature? If the state is a vapor-liquid mixture, report the mass fraction in each phase.

Problem 3.35: Water at 20 °C 30 bar is heated in a closed system under constant pressure.
a) If the quality in the final state is 75%, what is the amount of heat?
b) If the amount of heat is 2000 kJ/kg, what is the final temperature?
c) If the amount of heat is 3000 kJ/kg, what is the final temperature?

Problem 3.36: An insulated tank is divided into two equal parts. Both halves contain the same ideal gas but one is at 300 K and 2 bar while the other one is at 400 K and 1 bar. The partition is suddenly removed and the system is allowed to reach equilibrium. What is the temperature and pressure when equilibrium has been established?
Additional data: The heat capacities of this ideal gas may be assumed to be constant in the temperature range of this problem.

Problem 3.37: At 120 °C the saturation pressure of ethanol is 4.3 bar.
a) What is the state of ethanol at 120 °C and 10 bar (i.e., compressed liquid, saturated liquid/vapor, superheated vapor)? At 140 °C and 1 bar?
b) Can we assume that ethanol vapor at 1800 °C and 20 bar behaves as an ideal gas?
c) A rigid vessel contains 1 mol of ethanol at 1800 °C and 20 bar. How much heat must be removed in order for the pressure to drop to 10 bar?
Additional data: The C_P of ethanol is 173 J/mol/K.

Problem 3.38: Oxygen in a closed system is compressed adiabatically from 25 °C and 1 bar to a final temperature of 450 °C. Then it is cooled under constant pressure until the temperature falls to 5 °C. Assuming oxygen to be an ideal-gas:
a) Draw a PV graph on the computer and show the path of the process.
b) Calculate the final pressure.
c) Calculate ΔU, ΔH, Q, and W for the entire process.
d) Is it correct to assume that oxygen is an ideal gas under the conditions of this problem? You may use $C_P^{ig} = 29.4$ J/molK.

Problem 3.39: Saturated steam at 1 bar is heated at constant pressure to a final temperature of 500 °C in a closed system. Treating steam as an ideal gas, determine the amount of heat and work, and the change in enthalpy and internal energy.

Compare with the steam tables and discuss. The ideal-gas heat capacity of steam is given by

$$C_P^{ig}/R = 4.395 - 0.004186\,T + 0.00001405\,T^2 - 1.564 \times 10^{-8}\,T^3 + 6.32 \times 10^{-12}\,T^4$$

with T in kelvin.

Problem 3.40: Steam undergoes reversible adiabatic expansion from 500 °C, 10 bar. Assuming ideal-gas state, answer the following:
a) What is the pressure if the final temperature is 350 °C?
b) What is the temperature if the final pressure is 7 bar?
Additional data: The ideal-gas heat capacity of steam is

$$\frac{C_P^{ig}}{R} = 4.395 - 4.186 \times 10^{-3}\,T + 1.405 \times 10^{-5}\,T^2 - 1.564 \times 10^{-8}\,T^3 + 6.32 \times 10^{-12}\,T^4$$

with T in kelvin.

Chapter 4

Entropy and the Second Law

The first law, discussed in the previous chapter, expresses a fundamental principle, energy conservation. All processes observed in nature are bound to satisfy this principle. To put this differently, a process is physically impossible if it violates the first law. The opposite, however, is not true: not every process that satisfies the first law is in fact possible. Consider this familiar situation: if we throw a piece of ice in warm water and leave the system undisturbed, the ice will melt and the water will become colder. This process satisfies the first law, as the amount of energy that is absorbed by the ice is equal to the energy (heat) rejected by the warm water. Now, consider the same process in reverse: a glass of water spontaneously produces the original piece of ice floating on warm water. Such a process does not violate the first law, if all the heat that leaves the ice goes to make the rest of the water warm. Yet, this process is impossible. We can of course, separate the liquid into two parts, freeze one, heat the other, then put them together to produce the original system. However, such a process is *not* spontaneous; to take place it requires our intervention. By contrast, the melting of ice in the original experiment requires no such intervention whatsoever. There are numerous similar examples of processes that take place spontaneously in one specific direction but never in the opposite one: oxygen and nitrogen placed into contact with each other will mix without further interference from the outside but air does not spontaneously unmix into oxygen and nitrogen; a bouncing ball must eventually come to rest by converting its kinetic energy into internal energy, but a ball at rest will not spontaneously begin to bounce by converting some of its internal energy into kinetic energy; and so on. These situations fall under the same general paradigm: a system brought to state A and left undisturbed reaches state B; but if the same system is brought to state B and is left undisturbed, it will not reach A. Since the process $A \to B$ is observed to take place, it obviously satisfies the first law. This means that the process $B \to A$ satisfies the first law as well; yet it does not take place. We conclude that there is another principle at work that determines whether a process that satisfies the first law is indeed possible or not. This principle is known as *the second law of thermodynamics* and in its heart lies entropy, the most fundamental concept in thermodynamics.

What You Will Learn in This Chapter. In this chapter we formulate the mathematical statement of the second law for a closed system and learn how to:

1. Calculate entropy changes of real fluids along special paths.
2. Determine, after the fact, whether a process was conducted reversibly or not.
3. Apply the first and second law to determine if a proposed process is thermodynamically feasible.
4. Apply the notion of a cycle to determine thermodynamic efficiency.

4.1 The Second Law in a Closed System

To fix ideas, consider the following experiment: an insulated cylinder is divided by a movable wall into two compartments, one that contains a hot pressurized gas, and one that contains a colder, low-pressure gas (state A). The separating wall, which is conducting, is held into place with latches. The latches are removed and the system is left undisturbed to reach its final equilibrium state (state B), which, experience teaches, is a state of uniform pressure and temperature across both compartments. This process is spontaneous and adiabatic. Our development is motivated by the tendency of isolated systems like this to reach equilibrium. This tendency defines a preferred direction: *towards* the equilibrium state, *never* away from it. We recognize that such directionality cannot be represented by an equality; it requires an *inequality* so that if it is satisfied in one direction it is violated in the opposite one. At equilibrium, this inequality must reduce to an *exact equality*, so that the direction of change is neither forward nor backward—without this special condition, a system would never reach equilibrium. Moreover, the quantity in the inequality that fixes this direction must be such that if its change is positive in one direction, it will be negative in the opposite direction, so that the process may proceed in one direction only. It must, therefore, be a *state* function. Finally, since the state of a nonequilibrium system is nonuniform but consists of various parts in their own local states, the property we are after must be *extensive*, so that all parts of the system contribute to the determination of the direction to equilibrium. With these ideas in mind we formulate the second law as follows:

Formulation of Second Law

A spontaneous adiabatic process in a closed system proceeds in the direction that satisfies the inequality

$$\left[dS\right]_{\text{adiabatic}} > 0, \qquad (4.1)$$

4.1 The Second Law in a Closed System

where S is a new extensive state function, which we call *entropy*. At equilibrium, the inequality reduces to an exact equality:

$$\left[dS\right]_{\text{equilibrium}} = 0. \tag{4.2}$$

By extension, the equality also applies to an adiabatic process that is conducted quasi-statically (i.e., to a *reversible adiabatic* process) because such process is infinitesimally away from equilibrium at all times. The two results in eqs. (4.1) and (4.2) can be combined into the single condition,

$$\left[dS\right]_{\text{adiabatic}} \geq 0, \tag{4.3}$$

with the understanding that the equal sign applies to reversible adiabatic process.

Since entropy is a state function, a small change dS is expressed as the difference $S_2 - S_1$ between two neighboring states; it follows that for a step in the reverse direction, the entropy change is $S_1 - S_2 = -dS$. In other words, if the entropy change is positive in one direction, it is negative in the opposite: the direction of the process at each step is unambiguously revealed by entropy.

Relating Entropy to Measurable Quantities

For entropy to be of any practical value, we must be able to relate it to quantities that can be measured experimentally. Here is how we develop this relationship. Since entropy is a state function, we can express it as a mathematical function of two intensive properties. We choose internal energy U, and volume V, and write $S = S(U, V)$. This unusual choice is perfectly permissible.[1] The differential of entropy in terms of these independent variables is

$$dS = \left(\frac{\partial S}{\partial U}\right)_V dU + \left(\frac{\partial S}{\partial V}\right)_U dV. \tag{4.4}$$

This is a general equation for the change of entropy following a differential change of state. We will apply it to a *mechanically reversible* process. We begin by expressing

---

1. This choice of state variables is not arbitrary but guided by the process in Figure 4-1. An isolated system exchanges neither heat nor PV (or other) work with its surroundings. Its internal energy and volume, therefore, remain constant. The conditions of constant internal energy and constant volume, which appear in the partial derivatives in eq. (4.4), represent the external constraints imposed to the process depicted in Figure 4-1.

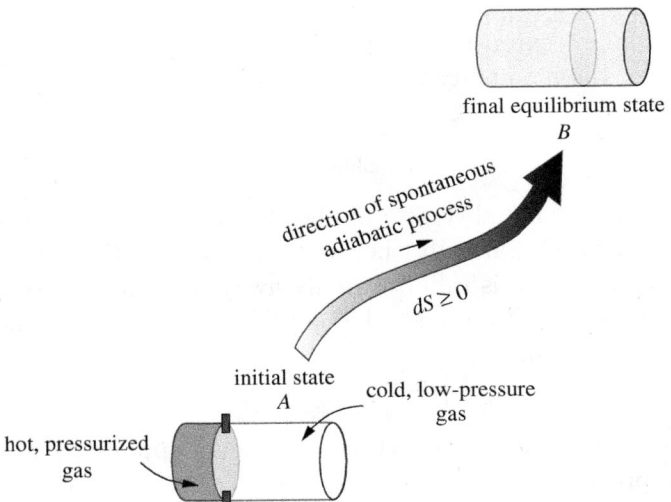

Figure 4-1: Spontaneous adiabatic process: equilibration between two parts initially held at different pressure and temperature.

dU from the first law, $dU = dQ_{\text{rev}} + dW_{\text{rev}}$, and use $dW_{\text{rev}} = -PdV$:

$$dU = dQ_{\text{rev}} - PdV,$$

We substitute this result in eq. (4.4), which now takes the form

$$dS = \left(\frac{\partial S}{\partial U}\right)_V dQ_{\text{rev}} + \underbrace{\left[\left(\frac{\partial S}{\partial V}\right)_U - P\left(\frac{\partial S}{\partial U}\right)_V\right]}_{0} dV. \qquad (4.5)$$

The quantity in square brackets is zero by virtue of eq. (4.2): in the special case of reversible adiabatic process ($dQ_{\text{rev}} = 0$) dS must be zero and the only way to satisfy this condition is if the bracketed quantity is identically equal to zero. As the last step we *define* the absolute temperature as the inverse of the partial derivative that is multiplied to dQ_{rev} in the above equation:

$$T \equiv \frac{1}{\left(\frac{\partial S}{\partial U}\right)_V}. \qquad (4.6)$$

Applying this definition to eq. (4.5) we obtain an important relationship between entropy, heat and temperature:

$$\boxed{dS = \frac{dQ_{\text{rev}}}{T}.} \qquad (4.7)$$

With this result we have related entropy to measurable quantities (heat, temperature). We now have a simple methodology to calculate entropy changes between

4.2 Calculation of Entropy

two states: connect the states with a mechanically reversible path, and compute the quantity dQ_{rev}/T along a differential step. The change of entropy between the two states is the integral of this differential over the entire path:

$$\boxed{\Delta S_{AB} = \int_A^B \frac{dQ_{\text{rev}}}{T}.} \tag{4.8}$$

Since entropy is a state function, *any* reversible path will yield the same result. Therefore, the choice of the path is not important and this allows us to choose according to convenience. The fact that entropy, a state function, is related to heat, a path function, should be no reason for concern: different paths will yield different amounts of heat (and different increments dQ_{rev} along the way) but the integral of dQ_{rev} between two points must yield the same value regardless of path.[2] Equation (4.7) also reveals the units of entropy: in extensive form it has units of energy divided by temperature (kJ/K); its intensive form is usually reported in kJ/kg K or J/mol K.

NOTE

Entropy and Temperature

In eq. (4.6) we identified the partial derivative $(\partial S/\partial U)_V$ with $1/T$. We rewrite this as

$$\boxed{T = \left(\frac{\partial U}{\partial S}\right)_V} \tag{4.9}$$

and adopt it as the formal definition of absolute temperature. It is not obvious why this partial derivative represents the familiar temperature, but then again, we have not provided, until now, any formal definition of temperature beyond the qualitative ("gives rise to the sensation of hot and cold"). We will establish this equivalency through demonstrations that show the above definition is fully consistent with our intuitive notions of temperature. For instance, in Example 4.15 we show that heat flows from higher to lower temperature (and not the other way around), and in Section (4.9) that temperature at equilibrium is indeed uniform throughout the system.

4.2 Calculation of Entropy

Constant-Pressure Path, No Phase Change

A rather simple problem is the calculation of ΔS between two states at the same pressure. To perform this calculation we devise a reversible constant-pressure path

---

2. If still unconvinced, recall that the equation for the reversible work can be written as $dV = -dW_{\text{rev}}/P$, thus relating volume (a state function) to reversible work (a path function).

between the two states. If no phase change occurs, the heat along this path is

$$dQ = dH = C_P \, dT.$$

With this result, eq. (4.8) becomes

$$\boxed{\Delta S_{12} = \int_{T_1}^{T_2} \frac{C_P}{T} dT.}_{(\text{const. } P)} \qquad (4.10)$$

This equation gives the entropy difference between two states at the same pressure but at different temperatures. Since entropy is a state function, it does not matter whether the actual process that brought the system from the initial state to the final was a constant-pressure process, some more complicated path, or even an irreversible process. The integration path used to calculate a state function between two states, and the actual path that took the system from the initial state to the final, are entirely independent.

Equation (4.10) may be expressed in simpler form by making use of the log-T mean heat capacity, introduced in eq. (3.47). Then, the entropy change is

$$\Delta S_{12} = \overline{C}_{P\log} \ln \frac{T_2}{T_1}. \qquad (4.11)$$

If the heat capacity does not vary much over the temperature interval from T_1 to T_2, then $\overline{C}_{P\log} \approx C_P$. In this case, entropy may be calculated from eq. (4.11) with the log-T mean heat capacity replaced by C_P.

Constant-Volume Path, No Phase Change

A similar equation can be written for a path of constant volume with no phase change. In this case the heat is

$$dQ = dU = C_V \, dT$$

and the entropy change is

$$\boxed{\Delta S_{12} = \int_{T_1}^{T_2} \frac{C_V}{T} dT.}_{(\text{const. } V)} \qquad (4.12)$$

This can also be written in the simpler form,

$$\Delta S_{12} = \overline{C}_{V\log} \ln \frac{T_2}{T_1}. \qquad (4.13)$$

where $\overline{C}_{V\log}$ is the log-T mean constant-volume heat capacity.

4.2 Calculation of Entropy

Example 4.1: Entropy Change at Constant Pressure
Toluene is heated at constant pressure from 12 °C, 1 bar, to 55 °C. Determine the entropy change of toluene.

Solution We will calculate the entropy from eq. (4.8) along a path of constant pressure. By first law, the amount of heat in a differential step along this path is

$$dQ = C_P \, dT.$$

With this result, eq. (4.8) becomes

$$\Delta S_{12} = \int_{T_1}^{T_2} \frac{C_P}{T} \, dT. \qquad [A]$$

The heat capacity of liquid toluene is

$$C_P/R = a_0 + a_1 T + a_2 T^2,$$

with T in kelvin and

$$a_0 = 16.8559 \quad a_1 = -0.0183185 \quad a_2 = 8.35939 \times 10^{-5}.$$

Applying this to eq. [A] we obtain

$$\Delta S_{12} = R \left[a_0 \ln \frac{T_2}{T_1} + a_1 (T_2 - T_1) + \frac{a_2}{2} (T_2^2 - T_1^2) \right].$$

Numerical substitution with $T_1 = 285.15$ K, $T_2 = 328.15$ K, gives $\overline{C}_{P\log} = 158.76$ J/mol K and the corresponding entropy change is

$$\Delta S_{12} = (158.76 \text{ J/mol K}) \ln \frac{328.15 \text{ K}}{285.15 \text{ K}} = 22.30 \text{ J/mol K}.$$

Example 4.2: Entropy of Solid
Calculate the entropy change for copper, from initial state at 120 °C, 1 bar, to 20 °C, 5 bar. The heat capacity of copper is $C_P = 0.38$ kJ/kg K.

Solution The two states are at different pressure. Since, however, pressure has negligible effect on the entropy of incompressible substances, eq. (4.10) may be used:

$$\Delta S = \int_{T_1}^{T_2} C_P \frac{dT}{T} = C_P \ln \frac{T_2}{T_1}.$$

By numerical substitution,

$$\Delta S = (0.38 \text{ kJ/kg K}) \ln \frac{(20 + 273.15) \text{ K}}{(120 + 273.15) \text{ K}} = -0.1115 \text{ kJ/kg K}.$$

Comments Temperature inside the logarithm must be in absolute units!

Example 4.3: Entropy of Compressed Liquid
Use the steam tables to estimate the entropy of water at 100 °C, 20 bar.

Solution Water at 100 °C, 20 bar is compressed liquid and its entropy is not listed in the steam tables that are provided in the appendix. It may be estimated, however, from the tabulated values at saturation. Since the entropy of liquids is largely independent of pressure and a function of temperature only, we may relate it to entropy on the saturation line in two different paths, one of constant temperature and one of constant pressure (see a very similar calculation of enthalpy in Example 3.13).

Constant-Pressure Path. Drawing a line of constant pressure from state $T_A = 100$ °C, $P_A = 20$ bar to the saturated liquid at the same pressure ($P_C = 20$ bar, $T_C = T^{\text{sat}} = 212.38$ °C), we write the following relationship for the entropies at states A and C (see also Figure 3-9):

$$S_A = S_C + \overline{C}_{P\log} \ln \frac{T_A}{T_C}.$$

where $S_C = 2.4470$ kJ/kg K, obtained from the system tables. The heat capacity of water as a function of temperature is given by the equation,

$$C_P = 15.354 - 0.116117\,T + 0.000451\,T^2 - 7.842 \times 10^{-7}\,T^3 + 5.206 \times 10^{-10}\,T^4,$$

with T in kelvin and C_P in kJ/kg K. Using this equation, the log-T mean heat capacity between $T_A = 100$ °C $= 373.15$ K, and $T_C = 212.38$ °C $= 485.53$ K, is calculated to be

$$\overline{C}_{P\log} = 4.341 \text{ kJ/kg K}.$$

Using these results the entropy at A is

$$S_A = S_C + \overline{C}_{P\log} \ln \frac{T_A}{T_C} = 2.4470 \text{ kJ/kg K} + (4.341 \text{ kJ/kg K}) \ln \frac{373.15 \text{ K}}{485.53 \text{ K}} = 1.304 \text{ kJ/kg K}.$$

Constant-Temperature Path. Assuming entropy to be a function of temperature only, an isotherm is also a line of constant entropy. Therefore, the entropy at 100 °C, 20 bar may be taken to be approximately the same as the entropy of saturated liquid at 100 °C:

$$S_A \approx S_L(100 \text{ °C}) = 1.307 \text{ kJ/kg K}.$$

This result agrees very well with the one obtained using the constant-pressure path.

4.2 Calculation of Entropy

TIP

Entropy of Solids and Liquids
The entropy of incompressible phases (solids, liquids far from the critical point) is nearly independent of pressure. Entropy in this case depends only on temperature and may be calculated using eq. (4.10), even if pressure is not the same in the two states.

Entropy and Phase Change

The entropy of vaporization is defined as the difference between the entropy of saturated vapor and saturated liquid:

$$\Delta S_{\text{vap}} = S_V - S_L. \tag{4.14}$$

The entropy of vaporization has a simple relationship to the enthalpy of vaporization. We begin with eq. (4.8), which we apply across the tie line from saturated liquid to saturated vapor. Along this path T is constant and $Q = \Delta H_{\text{vap}}$. Therefore,

$$\Delta S_{\text{vap}} = \frac{\Delta H_{\text{vap}}}{T^{\text{sat}}}. \tag{4.15}$$

Consider now a point inside the vapor-liquid region (point C in Figure 4-2). It represents a mixed-phase system with a fraction x_L of liquid and x_V of vapor. The entropy of this state is obtained by adding the contributions of the liquid and vapor fractions:

$$S = x_L S_L + x_V S_V. \tag{4.16}$$

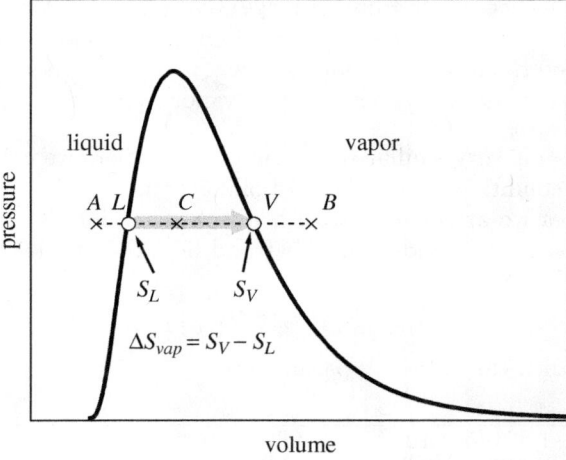

Figure 4-2: Entropy change during vapor-liquid transition.

Using $x_L + x_V = 1$ we solve this equation for x_L and x_V and find,
$$x_L = \frac{S_V - S}{S_V - S_L}, \quad x_V = \frac{S - S_L}{S_V - S_L}. \tag{4.17}$$
This is the familiar lever rule. It applies here because entropy is an extensive property.

Example 4.4: Entropy of Vaporization
Use the steam tables to obtain the entropy of vaporization of water at its normal boiling point.

Solution From the steam tables we obtain the entropy of saturated liquid and vapor water at 100 °C:
$$S_L = 1.3070 \text{ kJ/kg K}, \quad S_V = 7.3541 \text{ kJ/kg K}.$$
The entropy of vaporization is
$$\Delta S_{\text{vap}} = S_V - S_L = (7.3541 - 1.3070) \text{ kJ/kg K} = 6.0471 \text{ kJ/kg K}.$$
Alternatively, it can be calculated from the enthalpy of vaporization:
$$\Delta S_{\text{vap}} = \frac{H_V - H_L}{T} = \frac{(2675.6 - 419.10) \text{ kJ/kg}}{(100 + 273.15) \text{ K}} = 6.04717 \text{ kJ/kg K}.$$
The small difference (6.6×10^{-5} kJ/kg K) is due to round-off error.

Example 4.5: Energy Balances with Phase Change: Using C_P and ΔH_{vap}
Toluene vapor at 1.013 bar, 150 °C, is to be cooled under constant pressure to final temperature of 22 °C. (a) Determine the entropy change. (b) Repeat the calculation if the final state is a vapor-liquid mixture that contains 47% vapor by mass.

Solution This problem is very similar to Example 3.15, where we calculated the enthalpy changes for the same conditions. The normal boiling point of toluene is $T^{\text{sat}} = 383.79$ K; therefore, in both cases we are dealing with a phase transition. The vapor phase will be treated as an ideal gas. Both the ideal-gas heat and liquid-phase heat capacities are given by the polynomial function
$$C_P/R = a_0 + a_1 T + a_2 T^2 + a_3 T^3 + a_4 T^4,$$
with T in kelvin and with the following parameters:

	a_0	a_1	a_2	a_3	a_4
Liquid	16.8559	-1.83185×10^{-2}	8.35939×10^{-5}	0	0
Vapor	3.866	3.558×10^{-3}	1.3356×10^{-4}	-1.8659×10^{-7}	7.69×10^{-11}

4.2 Calculation of Entropy

The heat of vaporization of toluene was calculated in Example 3.15 and was found to be

$$\Delta H_{\text{vap}} = 32883.4 \text{ J/mol}.$$

A. The calculation is split into three parts:

1. *Cooling from initial vapor state to saturated vapor (BV).* The entropy change for this part is

$$\Delta S_{BV} = \int_{423.15}^{383.79} C_P^{\text{ig}} \frac{dT}{T} = -13.65 \text{ J/mol K}.$$

2. *Condensation of saturated vapor to saturated liquid (VL).* The entropy for this part is the negative of the entropy of vaporization:

$$\Delta S_{VL} = -\Delta S_{\text{vap}} = \frac{-\Delta H_{\text{vap}}}{T^{\text{sat}}} = -85.68 \text{ J/mol K}.$$

3. *Cooling of saturated liquid at constant pressure (LA).* The entropy change for this part is

$$\Delta S_{LA} = \int_{383.79}^{295.15} C_P^{\text{liq}} \frac{dT}{T} = -44.22 \text{ J/mol K}.$$

The total entropy change is

$$\Delta S = -13.65 - 85.68 - 44.22 = -143.55 \text{ J/mol K}.$$

B. The final state (C) in this case is a vapor-liquid mixture that contains a vapor fraction $x_V = 0.47$ and a liquid fraction $x_L = 0.53$. The entropy change from the initial state up to saturated vapor is the same as before. When the saturated vapor is converted in a vapor-liquid mixture with liquid fraction x_L, the entropy change is $-x_L \Delta S_{\text{vap}}$ (see also Example 3.15 for a very similar calculation of enthalpy). Therefore,

$$\Delta S = \Delta S_{BV} - x_L \Delta S_{\text{vap}} = -13.65 - (0.53 \times 85.68) = -59.05 \text{ J/mol K}.$$

Comments The calculation of entropy in paths that involve both sensible and latent heat is very similar to the calculation of enthalpy.

Entropy Change of Bath

A heat bath is a system so large that its temperature does not change in any appreciable amount when it absorbs or rejects heat. The entropy change is calculated using eq. (4.10). Since temperature is constant, the integral reduces to

$$\Delta S_{\text{bath}} = \frac{Q_{\text{bath}}}{T_{\text{bath}}}, \tag{4.18}$$

where Q_{bath} is defined with respect to the bath, that is, positive if added to the bath, negative otherwise. Usually, the bath is part of the surroundings, used as a heat source or sink for the system of interest. In situations like this, heat is calculated with respect to the system and has the opposite sign. The entropy change of the bath, expressed in terms of the heat of the system, is

$$\Delta S_{\text{bath}} = -\frac{Q}{T_{\text{bath}}}, \quad (4.19)$$

where $Q = -Q_{\text{bath}}$ is the heat defined with respect of the system. The example below illustrates this calculation.

TIP

Entropy of Bath
In writing eq. (4.19) we assume that the heat has been calculated with reference to the system. Another way to write the same equation is in the form

$$\Delta S_{\text{bath}} = \pm \frac{|Q|}{T_{\text{bath}}},$$

with the $+$ sign if the bath receives the heat from the system, and the $-$ sign if it transfers it to the system. It follows that when the bath absorbs heat its entropy increases, and when it rejects heat its entropy decreases. Use this result as a check whenever you perform a bath calculation.

Example 4.6: Entropy Change of Bath
A piece of copper ($C_P = 0.38$ kJ/kgK), initially at 250 °C, is left to cool in open air at 1 bar, 22 °C. Calculate the entropy change of the copper and of the air.

Solution The final temperature of the copper is equal to the temperature of the air. Initial and final states are at the same pressure, therefore, eq. (4.10) applies. The entropy change of the copper is

$$\Delta S_{\text{copper}} = C_P \ln \frac{T_B}{T_A} = (0.38 \text{ kJ/kg K}) \ln \frac{(22 + 273.15 \text{ K})}{(250 + 273.15) \text{ K}} = -0.2175 \text{ kJ/kg K}.$$

Air acts as a bath and its entropy is calculated using eq. (4.19). First, we calculate the heat with respect to the system (copper):

$$Q = \int_{T_A}^{T_B} C_P \, dT = C_P(T_B - T_A) = (0.380 \text{ kJ/kg K})(295.13 \text{ K} - 523.15 \text{ K}) = -86.64 \text{ kJ/kg}.$$

This heat is negative because it is rejected by the copper. The entropy of the bath is

$$\Delta S_{\text{bath}} = -\frac{Q}{T_{\text{bath}}} = \frac{-(-86.64 \text{ kJ/kg})}{295.15 \text{ K}} = 0.2935 \text{ kJ/kg K}.$$

4.2 Calculation of Entropy

> The entropy change is positive because the bath absorbs this amount of heat. Alternatively, the heat defined with respect to the bath is $Q_{\text{bat}} = 86.64$ kJ/kg, therefore, the entropy change of the bath may also be calculated as
>
> $$\Delta S_{\text{bath}} = \frac{Q_{\text{bath}}}{T_{\text{bath}}} = \frac{86.64 \text{ kJ/kg}}{295.15 \text{ K}} = 0.2935 \text{ kJ/kg K},$$
>
> which gives the same answer.

Entropy in the Ideal-Gas State

In the previous sections we demonstrated the calculation of entropy between two states at different temperature but the same pressure. The more general calculation between any two states will be presented in Chapter 5. Here, we show that such general calculation is quite simple in the special case that both states are in the ideal-gas state. Consider two states, 1 and 2, on the PT plane, as in Figure 4-3. We construct a two-step mechanically reversible path, first at constant pressure to final temperature T_2 (state $2'$), then at constant temperature to final pressure P_2. The calculation of entropy will be performed along each step separately using eq. (4.8) and the results will be added at the end.

Along branch $12'$, pressure is constant, therefore, eq. (4.10) applies:

$$\Delta S_{12'}^{\text{ig}} = \int_{T_1}^{T_2} \frac{C_P^{\text{ig}}}{T} dT. \tag{4.20}$$

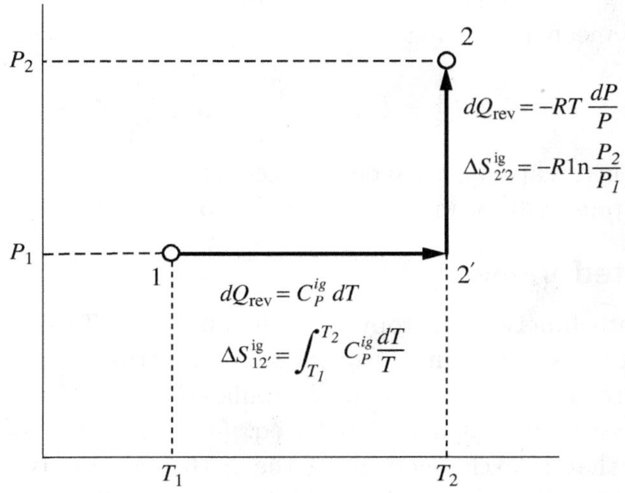

Figure 4-3: Path for the calculation of entropy changes in the ideal-gas state.

Along branch 2'2, the process is isothermal, as well as mechanically reversible. Application of the first law gives

$$dU^{ig} = dQ_{rev} - PdV^{ig}.$$

The change in internal energy in the ideal-gas state is given by eq. (3.34), and since temperature is constant, $dU^{ig} = 0$. Therefore, the amount of heat is

$$dQ_{rev} = PdV^{ig}.$$

We use the ideal-gas law to write $V^{ig} = RT/P$; differentiation at constant temperature gives

$$dV^{ig} = d\left(\frac{RT}{P}\right) = -RT\frac{dP}{P^2}.$$

The heat, therefore, is

$$dQ_{rev} = -RT\frac{dP}{P}.$$

The entropy change along this step is now calculated by applying eq. (4.8) with the above result for the heat:

$$\Delta S^{ig}_{2'2} = \int_{2'}^{2} \frac{dQ_{rev}}{T} = -R\int_{P_1}^{P_2} \frac{dP}{P} = -R\ln\frac{P_2}{P_1}.$$

The entropy change between states 1 and 2 is $\Delta S^{ig}_{12} = \Delta S^{ig}_{12'} + \Delta S^{ig}_{2'2}$, or,

$$\boxed{\Delta S^{ig}_{12} = \int_{T_1}^{T_2} \frac{C^{ig}_P}{T} dT - R\ln\frac{P_2}{P_1}.} \tag{4.21}$$

Using the log-T mean heat capacity (see eq. [3.40]), this result can also be expressed as

$$\Delta S^{ig}_{12} = \overline{C}^{ig}_{P\log} \ln\frac{T_2}{T_1} - R\ln\frac{P_2}{P_1}. \tag{4.22}$$

The log-T mean heat capacity may be replaced by C^{ig}_P if the heat capacity does not vary much with temperature in the interval T_1 to T_2.

Using Tabulated Values

Finally, as a state function, entropy can be tabulated. The steam tables in the appendix include the specific entropy of water in the saturated region (liquid, vapor) and in the superheated vapor region. If tabulated values are not available, then entropy changes may be calculated from eq. (4.8). The calculation requires the amount of heat that is exchanged along the path and this is generally obtained by application of the first law. This methodology is applied below to a number of special cases.

Example 4.7: Calculation of Entropy of Steam
Steam is heated under constant pressure from 15 bar, 300 °C to 500 °C. Determine the entropy change.

Solution Entropy is a tabulated property in the steam tables. We find:

	P (bar)	T (°C)	S (kJ/kg K)
State 1	15	300	6.9199
State 2	15	500	7.5716

The change in entropy is

$$\Delta S_{12} = S_2 - S_1 = (7.5716 - 6.9199) \text{ kJ/kg K} = 0.6517 \text{ kJ/kg K}.$$

Comments The path of the process that is given in the problem statement is not relevant. Entropy is a state function, therefore, the change ΔS_{12} depends solely on the entropy in the initial and final state. In this case, the values of the entropy in the two states were obtained directly from tables without the need to consider paths.

4.3 Energy Balances Using Entropy

Equation (4.7) relates entropy to temperature and heat. If this equation is solved for heat, we obtain,

$$dQ_{\text{rev}} = TdS. \tag{4.23}$$

This equation can be used to calculate the amount of heat, provided that the process is mechanically reversible. The full amount of heat for a process is given by the integral of this differential from initial to final state:

$$Q_{\text{rev}} = \int_A^B TdS. \tag{4.24}$$

This equation has a simple graphical interpretation on the TS graph: heat in a mechanically reversible process is equal to the area under the path (Figure 4-4). The relationship of heat to temperature and entropy is analogous to the relationship of work to pressure and volume, and further illustrates the fact that heat is a *path* function. Heat is positive if the path moves in the direction of increasing entropy (from left to right), and negative if the path is in the direction of decreasing entropy. If the path is closed, that is, the process is a *cycle* that returns the system to its initial state, the net amount of work is equal to the area enclosed inside the path. This is

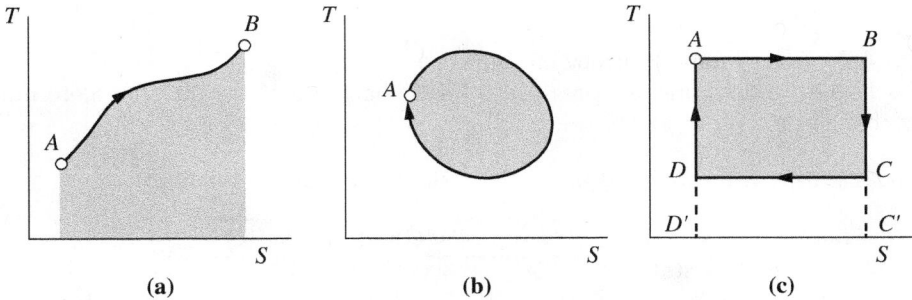

Figure 4-4: Paths on the TS graph: (a) open path—this process adds heat to the system; (b) closed path (cycle)—in clockwise direction this cycles absorbs a net amount of heat equal to the area inside the cycle and produces an equal amount of work; (c) Carnot cycle; it consists of two isothermal parts, AB, CD, and two isentropic segments, BC, DA.

shown easily by noting that the area $DCC'D'$ in Figure 4-4 is positive, during step AB, but negative during step CD, and thus is canceled. This result applies to any closed path, not just the rectangular paths because any arbitrary closed path can be represented by a series of interconnected thin (differential) rectangular cycles. A rectangular cycle on the TS graph, operating reversibly, is known as a *Carnot* cycle and is discussed in more detail in Section 4.5.

The TS plane offers yet another way of representing the thermodynamic state of a system. On this plane, isotherms are shown by horizontal lines, while vertical lines represent lines of constant entropy (*isentropic* lines). Along an isentropic line, $dS = 0$; according to eq. (4.7), this condition is met if a process is a mechanically reversible adiabatic process. Thus we conclude that a *reversible adiabatic process is also isentropic*.

Reversible Adiabatic Process in the Ideal-Gas State

The entropy of ideal gas is given in eq. (4.22). For an isentropic process, $\Delta S^{\mathrm{ig}} = 0$. Applying this condition to eq. (4.22) we obtain a relationship between pressure and temperature along the isentropic path:

$$0 = \overline{C}_{P\mathrm{log}}^{\mathrm{ig}} \ln \frac{T_2}{T_1} - R \ln \frac{P_2}{P_1} \quad \Rightarrow \quad \frac{P_2}{P_1} = \left(\frac{T_2}{T_1}\right)^{\overline{C}_{P\mathrm{log}}^{\mathrm{ig}}/R}. \qquad [3.44]$$

The last result is identical to the one obtained in Section 3.8 (see eq. [3.44]). Therefore, the path of the reversible adiabatic process can be derived either through application of the first law, as was done in Section 3.4 of Chapter 3, or by applying the isentropic condition, as was done here.

4.3 Energy Balances Using Entropy

Example 4.8: Reversible Adiabatic Compression of Steam

Steam is compressed in a closed system from 1 bar, 200 °C to 20 bar. Calculate the work if the compression is reversible and adiabatic. Repeat the calculation assuming steam to be an ideal gas.

Solution *(a) Using the steam tables.* From the steam tables we find $S_A = 7.8356$ kJ/kg K. The reversible adiabatic process is isentropic; therefore, $S_B = S_A = 7.8356$ kJ/kg K. In the final state we know pressure (20 bar) and entropy; therefore, the state is fully specified. The temperature and internal energy are obtained by interpolation at $P_B = 20$ bar, $S_B = 7.8356$ kJ/kg K and the results are summarized in the table below:

	A	B
P (bar)	1	20
T (°C)	200	652.3
U (kJ/kg)	2658.2	3385.03
S (kJ/kg K)	7.8356	7.8356

The amount of work is

$$W = \Delta U_{AB} = (3385.03 - 2658.2) \text{ (kJ/kg)} = 726.83 \text{ kJ/kg}.$$

(b) Treating steam as an ideal gas. Solving eq. 3.44 for the final temperature we obtain

$$T_B = T_A \left(\frac{P_B}{P_A}\right)^{R/\overline{C}_{P\log}^{ig}}. \qquad [A]$$

The ideal-gas heat capacity of steam is

$$C_P^{ig}/R = 4.4 - 4.19 \times 10^{-3} T + 1.41 \times 10^{-5} T^2 - 1.56 \times 10^{-8} T^3 + 6.32 \times 10^{-12} T^4,$$

with T in kelvin. Equation [A] must be solved numerically for the final temperature. One way to do this by iteration is as follows: as an initial guess, use $\overline{C}_{P\log}^{ig} = C_P^{ig}(T_A)$ and calculate the final temperature from eq. [A]; using this temperature calculate a new value of $\overline{C}_{P\log}^{ig}$ and calculate a new temperature from eq. [A]. This procedure is repeated until temperature does not change appreciably any more. These iterations are summarized below:

Iteration	$\overline{C}_{P\log}^{ig}/R$	T_B (K)
1	4.220	962.30
2	4.490	922.10
3	4.468	925.03
4	4.470	924.82
5	4.470	924.84
6	4.470	924.84

The procedure converges to final temperature $T_B = 924.84$ K $= 651.7$ °C. The amount of work is

$$W = \int_{T_A}^{T_B} C_V^{ig} dT = \int_{T_A}^{T_B} (C_P^{ig} - R) dT = \int_{T_A}^{T_B} C_P^{ig} dT - R(T_B - T_A).$$

The integral is calculated using the polynomial expression for the heat capacity and the result is 13145.8 J/mol, or

$$W = (13145.8 \text{ J/mol}) \frac{10^{-3} \text{ kJ/J}}{18.011 \times 10^{-3} \text{ kg/mol}} = 729.7 \text{ kJ/kg}.$$

Comments The agreement with the calculation from the steam tables is very good. This is because steam at the conditions of this problem is fairly close to the ideal-gas state.

Example 4.9: Using Entropy to Calculate Heat

Steam undergoes reversible, isothermal compression in a closed system from initial state $T_A = 400\ °C$, $P = 24$ bar to $T_B = 400\ °C$, $P_B = 40$ bar. Calculate the amount of work and heat exchanged.

Solution In reversible processes, the exact equality $dS = dQ/T$ holds (see eq. [4.7]) and this can be useful in obtaining the amount of heat by calculating the entropy change first. This is the approach we follow in this example.

By first law
$$\Delta U_{AB} = W + Q.$$

Applying eq. (4.7) and noting that temperature remains constant through the process, we obtain
$$dQ = TdS \quad \Rightarrow \quad Q = \int_A^B TdS = T\int_A^B dS = T(S_B - S_A).$$

From steam tables we find
$$U_A = 2940.9 \text{ kJ/kg} \quad S_A = 7.0375 \text{ kJ/kg K}$$
$$U_B = 2920.6 \text{ kJ/kg} \quad S_B = 6.7712 \text{ kJ/kg K}.$$

Therefore,
$$\Delta U_{AB} = -20.3 \text{ kJ/kg}$$
$$Q = (400 + 273.15)(6.7712 - 7.0375) = -179.26 \text{ kJ/kg}$$
$$W = \Delta U_{AB} - Q = -20.3 + 179.26 = 158.96 \text{ kJ/kg}.$$

There is another way to do this problem: since the process is reversible process, the PV work may be calculated as
$$W_{\text{rev}} = -\int_A^B PdV.$$

This integration may be done numerically, as was done in Example 3.3. Once the work is known, heat may be calculated from the energy balance. This procedure is more calculation-intensive however. The solution that utilizes entropy is much simpler in this case.

4.4 Entropy Generation

A system and its surroundings constitute the universe, a closed adiabatic system. For such system the inequality of the second law applies, and thus we have

$$\Delta S_{\text{sys}}^{\text{tot}} + \Delta S_{\text{sur}}^{\text{tot}} \geq 0.$$

According to this result, when a system undergoes a process, the net change of entropy in the system and its surroundings is positive, if the process is irreversible, or zero, in the special case of a reversible process. This nonnegative change represents a net generation of entropy,

$$S_{\text{gen}} = \Delta S_{\text{sys}}^{\text{tot}} + \Delta S_{\text{sur}}^{\text{tot}}. \tag{4.25}$$

The second law may then be expressed as

$$\boxed{S_{\text{gen}} \geq 0,} \tag{4.26}$$

which may be stated as follows:

> *All processes must result in nonzero entropy generation; a process that results in negative entropy generation is not feasible.*

The entropy generation for a process is calculated by combining the entropy changes of all parts of the universe that were affected by the process. Suppose that a system undergoes a differential process during which it exchanges the amount dW and dQ with the surroundings. The entropy generation is

$$dS_{\text{gen}} = dS_{\text{sys}} + dS_{\text{sur}} \geq 0.$$

Treating the surroundings as a bath at temperature T', its entropy change is

$$dS_{\text{sur}}^{\text{tot}} = -\frac{dQ}{T'}.$$

Its precise amount depends on the temperature T', and increases as the temperature difference between system (at temperature T) and surroundings (T') increases.[3]

---

3. If the heat is removed from the system ($dQ < 0$), the temperature of the surroundings must be lower than that of the system ($T' \leq T$) and we get

$$-\frac{dQ}{T'} \geq -\frac{dQ}{T}.$$

If the heat is added to the system ($dQ > 0$), the bath temperature must be higher ($T' > T$) and the same inequality holds. That is, the surroundings make the smallest possible contribution to entropy generation when the temperature is the same as that of the system.

The best case corresponds to $T' = T$, such as, when the bath has the same temperature as the system:

$$dS_{\text{gen}} = dS - \frac{dQ}{T}. \tag{4.27}$$

Since $dS_{\text{gen}} \geq 0$, the above results leads to the following inequality for the entropy change of the system:

$$\boxed{dS \geq \frac{dQ}{T} \quad \text{or} \quad \Delta S \geq \int \frac{dQ}{T}.} \tag{4.28}$$

To reach these inequalities we assumed that heat transfer between system and surroundings is reversible (i.e., $T = T'$). If it is not, dS is even larger (according to eq. [4.27]) and the inequality is even stronger. This inequality arises from mechanical irreversibilities within the system. If all processes inside the system are done reversibly, then the inequality reduces to an exact equality and the two expressions in eq. (4.28) revert to eqs. (4.7) and (4.8), as they should.

NOTE

Entropy of the Universe and Irreversibility

Entropy generation is the net change of entropy in the universe as a result of a process. This total entropy can never decrease; it stays constant, if the process is reversible. Otherwise it must increase. Entropy is a *state* function and the monotonic increase of entropy means that the universe as a whole *cannot* return to its previous state, following an irreversible process. This is precisely why we refer to such processes as *irreversible*. Irreversibility must always be understood with respect to the entire universe. Parts of the universe can be restored to their previous states, but this can only be done at the expense of other parts, whose states must change. When expressed in such stark terms, the relentless march of entropy to ever-higher values with no possibility of return sounds as a harsh principle that imposes a severe limitation to what is possible in nature. Such arguments eventually turn into philosophical inquiries on the relationship between entropy and the irreversible passage of time, and ultimately about human perception of physical reality. It is not our intention to follow that road, fascinating as it may be. When its various implications make the meaning of entropy fuzzy, it is useful to return the starting point and interpret entropy as the physical property that points evolving systems towards their state of equilibrium.

4.5 Carnot Cycle

The first law introduces the notion that matter is a medium for energy storage: energy can be added in the form of either heat or work, and is stored as internal energy. This energy can be later extracted in the form of a combination of heat and work. This is precisely what the steam engine does: water absorbs heat and in the process it becomes pressurized steam; work is extracted through the expansion of the

4.5 Carnot Cycle

steam. A practical question of engineering relevance is, can all of the heat that enters the system be recovered as work? If not, what is the maximum possible fraction that can be recovered? To address the question of "maximum" work, we must first establish a baseline. This baseline is provided by the notion of *thermodynamic cycle*. A cycle is any process that returns the system back to the same state where it was found initially. Why use a cyclic process as a baseline? Consider the analogous problem in accounting: Two people receive loans from the bank and invest the money in various projects. To determine who made more money, the loans must be repaid first. This amounts to restoring the system (the bank) to its initial state. Neglecting to do this would lead to meaningless comparisons since the person with the largest debt could claim to be the richest of the two. The notion of the cycle requires us to consider the costs of restoring the system to its original state before we can make pronouncements of efficiency.

A special cycle of fundamental theoretical importance is the Carnot cycle, depicted graphically in the TS graph in Figure 4-5. It is a *reversible* cycle that consists of two isothermal branches, AB and CD, and two isentropic branches, BC and DA. The isothermal branches are one at high temperature T_H, and one at lower temperature T_L ($T_H > T_L$).

The cycle may be implemented experimentally. We examine each portion of the path separately.

1. *Isothermal step AB.* This step represents reversible isothermal heating. The amount of heat in this part is given by eq. (4.24). Noting that temperature is constant, this heat is

$$Q_H = \int_A^B T_H dS = T_H(S_B - S_A) > 0. \qquad (4.29)$$

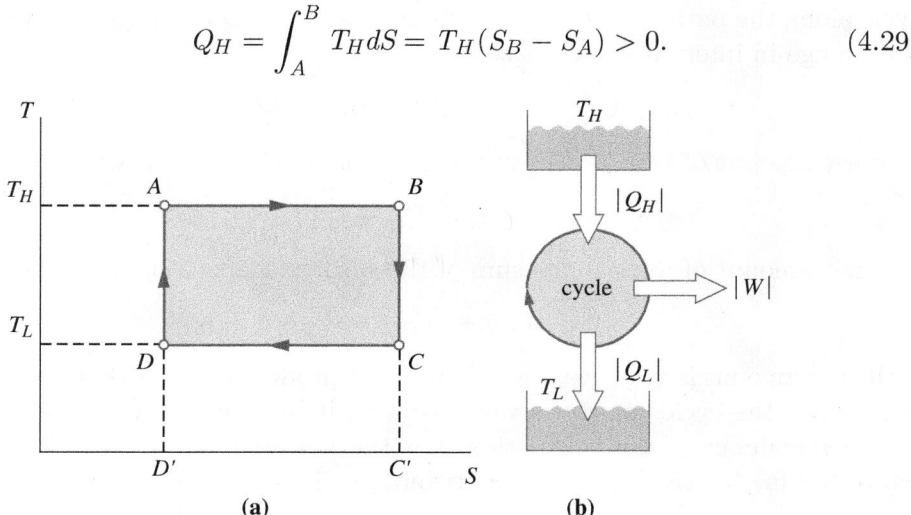

Figure 4-5: (a) Carnot cycle on the TS plane; (b) pictorial representation of the Carnot cycle.

Since $S_B > S_A$, this heat is positive and must be added to the system. This step also involves some amount of PV work (W_{AB}), as volume changes due to heating.

2. *Isentropic step BC.* This step represents reversible adiabatic *expansion*. During this step the system produces an amount of work, W_{BC}.

3. *Isothermal step CD.* This step is the reverse of that in AB and represents reversible isothermal cooling. The amount of heat is

$$Q_L = \int_C^D T_L dS = T_L(S_D - S_C) < 0. \qquad (4.30)$$

The system also exchanges PV work (W_{CD}) due to volume changes that occur with cooling.

4. *Isentropic step DA.* This is the reverse of step BC. It represents reversible adiabatic compression during which the system absorbs an amount of work, W_{DA}.

These heat and work exchanges are indicated by the arrows in Figure 4-5 that show the direction of the exchange.

First-Law Analysis of the Carnot Cycle The overall energy balance along the path $ABCDA$ is

$$\Delta U = Q + W, \qquad (4.31)$$

where Q and W are the net amounts of heat and work, respectively. Over a complete cycle along the path $ABCDA$, the system is returned to its original state; therefore, the change in internal energy is zero:

$$\Delta U = 0.$$

The net amount of heat is the sum of the amounts exchanged with the two baths:

$$Q = Q_H + Q_L.$$

The net amount of work is the sum of the work exchanged in each part:

$$W = W_{AB} + W_{BC} + W_{CD} + W_{DA}.$$

If the net amount is negative, then there is net production of work; if the net amount is positive, the cycle consumes work, that is, it requires work in order to operate. In the remainder of the calculation it is the net amount of work that we will be interested in, not the individual contributions in each step.

4.5 Carnot Cycle

Second-Law Analysis of the Carnot Cycle The entropy change of the system is zero because its state is restored at the end of one cycle:

$$\Delta S_{\text{sys}} = 0.$$

The entropy of the surroundings is affected in the two baths that exchange heat with the system. The entropy change in each of these baths is given by eq. (4.19):

$$\Delta S_H = \frac{-Q_H}{T_H}, \quad \Delta S_L = \frac{-Q_L}{T_L}.$$

Since the Carnot cycle is reversible, the entropy generation is zero:

$$S_{\text{gen}} = \Delta S_{\text{sys}} + \Delta S_{\text{sur}} = 0 \quad \Rightarrow \quad 0 - \frac{Q_H}{T_H} - \frac{Q_L}{T_L} = 0. \tag{4.32}$$

We solve this equation for Q_L, substitute the result into eq. (4.31) and solve the resulting equation for the work. We find

$$W = -Q_H \left(1 - \frac{T_L}{T_H}\right). \tag{4.33}$$

The heat Q_H is positive (see eq. [4.29]); so is the quantity in the parenthesis, because $T_H > T_L$. This makes the net work negative, in other words, work is *produced* by the cycle. This cycle is a heat engine: It receives heat from a source at high temperature T_H and produces a net amount of work.

Thermodynamic Efficiency In addition to producing a net amount of work, the cycle produces an amount of heat Q_L, which it rejects to the surroundings at the lower temperature T_L. The fraction of the input heat that is converted into work is from eq. (4.33),[4]

$$\frac{-W}{Q_H} = 1 - \frac{T_L}{T_H} < 1. \tag{4.34}$$

Since $T_L < T_H$, the right-hand side is less than 1. We conclude that the cycle cannot convert the entire amount of heat it receives into work. Moreover, the fraction of heat that is converted into work depends only on temperature, but it is independent of the physical properties of the fluid. The Carnot efficiency, η, is defined as

$$\eta_{\text{Carnot}} = 1 - \frac{T_L}{T_H}, \tag{4.35}$$

---

4. Since W is negative, $-W$ is the absolute value of the work produced.

and gives the fraction of the heat that enters the Carnot cycle at the temperature of the hot reservoir that is converted into work. The remaining fraction is rejected as heat at the lower temperature T_L. It is not possible to exceed the efficiency of the Carnot cycle: Given two heat reservoirs, one at T_H and one at T_L, the maximum fraction of the heat that enters the cycle is solely a function of the two temperatures and is given by eq. (4.35). The proof is straightforward. The entropy generation of any cycle that exchanges heat with the two reservoirs is

$$S_{\text{gen}} = -\frac{Q_H}{T_H} - \frac{Q_L}{T_L} \geq 0,$$

from which we obtain

$$Q_L \leq -Q_H \frac{T_L}{T_H}.$$

Applying the first law on this cycle and solving for Q_L we obtain

$$0 = Q_H + Q_L + W \quad \Rightarrow \quad Q_L = -Q_H - W.$$

Combining the two results and solving for the ratio $-W/Q_H$ we find

$$\frac{-W}{Q_H} \leq 1 - \frac{T_L}{T_H}. \tag{4.36}$$

On the left-hand side we have the efficiency of the unspecified cycle, while on the right-hand side we have the efficiency of the Carnot cycle. That is,

$$\left(\eta\right)_{\text{any cycle}} \leq \left(\eta\right)_{\text{Carnot}}, \tag{4.37}$$

with the understanding that all cycles are restricted to exchange heat with the same two reservoirs. The equal sign applies if the unspecified cycle operates reversibly. We recognize now that the Carnot efficiency is achieved when a cycle is operated reversibly between two heat reservoirs. Exceeding the Carnot efficiency is not possible because it would amount to negative entropy generation, which is equivalent to a spontaneous process that moves an isolated system away from equilibrium.

Real power plants operate differently from the Carnot cycle but the big picture is the same: An energy source at high temperature is used to supply heat to a process, whose net result is to produce an amount of work. In industrial production of power the heat source is either a fossil fuel such as coal or a nuclear fuel. In either case a chemical or nuclear reaction releases heat at an elevated temperature, which acts as the high-temperature reservoir. Invariably, part of this heat must be rejected to the surroundings but at a lower temperature. As a result, only a fraction of the heat released by the fuel is converted into work. The Carnot efficiency gives the maximum possible fraction of work that can be extracted, given the temperature

4.5 Carnot Cycle

of the heat source and the heat sink (usually the surroundings). The efficiency of real processes is lower than the theoretical efficiency because of irreversibilities that cannot be avoided in practice.

The technical reasons for the unavoidable rejection of heat at the lower-temperature reservoir will become more clear when we consider actual cycles in Chapter 6. The fundamental reason, however, is evident in Figure 4-5. According to eq. (4.30), the amount of heat Q_L is equal in absolute value to the area $DCC'D'$. Clearly, it is impossible to complete the cycle without rejecting this amount of heat at the lower reservoir. The area is zero only if T_L is equal to absolute zero, and in this case the Carnot efficiency becomes 1. For any other value of T_L, the amount of Q_L is nonzero and the efficiency of the cycle less than 1. In practice, T_L is fixed by the temperature of the ambient surroundings. It is possible of course to produce temperatures lower than ambient using refrigeration, but we will see in Chapter 6 that this requires work and cancels the benefit from the higher efficiency of the cycle.

Example 4.10: Internal Combustion Engine
Assuming the temperature obtained in the combustion of gasoline to be about 1000 °C, estimate the theoretical efficiency of the internal combustion engine.

Solution The combustion temperature represents T_H; it is the temperature at which the input heat is supplied. The rejected heat will go to the surroundings. For T_L we choose 300 K. The theoretical efficiency is

$$\eta = 1 - \frac{300}{1273.15} = 0.76 = 76\%.$$

Typical engines have thermal efficiencies in the range 45% to 50%.

Example 4.11: Carnot Cycle Using an Ideal Gas
A fluid undergoes the Carnot cycle in Figure 4-5 with $T_L = 300$ °C, $T_H = 500$ °C. Calculate the amounts of heat and work for each step, as well as the thermodynamic efficiency of the cycle (net amount of work produced over amount of heat absorbed) assuming the fluid is in the ideal-gas state.

Solution We apply the first law along each step of the cycle.

Isothermal Heating, AB. Along the isothermal part AB, $\Delta U_{AB} = 0$, and by first law, $Q = -W$. For a reversible isothermal process, the work is

$$-Q_{AB} = W_{AB} = -\int_A^B P dV = -\int_A^B \frac{RT_H}{V} dV = -RT_H \ln \frac{V_B}{V_A}.$$

Reversible Adiabatic Expansion, BC. This is a reversible adiabatic step during which no heat is exchanged. The first law gives,

$$\Delta U_{BC} = W_{BC} \Rightarrow W_{BC} = \overline{C}_{V\log}^{ig}(T_L - T_H).$$

Isothermal Step, CD. This step is analogous to *AB*:

$$-Q_{CD} = W_{CD} = -RT_L \ln \frac{V_D}{V_C}.$$

Reversible Adiabatic Compression, DA. This step is analogous to *BC*: There is no heat, and the amount of work is

$$W_{DA} = C_V(T_H - T_L).$$

A relationship can be written for the molar volume at the four corners of the cycle. Applying eq. (3.41) along the reversible isentropic path, *BC*, we have

$$\frac{V_C}{V_B} = \left(\frac{T_L}{T_H}\right)^{-\overline{C}_{V\log}^{ig}/R}.$$

Similarly, along the reversible adiabatic step, *DA*,

$$\frac{V_A}{V_D} = \left(\frac{T_H}{T_L}\right)^{-\overline{C}_{V\log}^{ig}/R}.$$

Combining these results we obtain the following relationship,

$$\frac{V_B}{V_A} = \frac{V_C}{V_D} = x,$$

where x is the common values of the two volume ratios. Using this result, the energy balances can be summarized in the concise form below:

Step	Q	W	ΔU
AB	$RT_H \ln x$	$-RT_H \ln x$	0
BC	0	$\overline{C}_{V\log}^{ig}(T_L - T_H)$	$\overline{C}_{V\log}^{ig}(T_L - T_H)$
CD	$-RT_L \ln x$	$RT_L \ln x$	0
DA	0	$\overline{C}_{V\log}^{ig}(T_H - T_L)$	$\overline{C}_{V\log}^{ig}(T_H - T_L)$
Total	$R \ln x (T_H - T_L)$	$-R \ln x (T_H - T_L)$	0

Thermodynamic Efficiency. The net amount of work is $-R \ln x (T_H - T_L)$. This is negative (why?), which means that it is produced by the cycle. The amount of heat that enters the cycle is $Q_{AB} = RT_H \ln x$. The thermodynamic efficiency is

$$\frac{-W}{Q_{AB}} = \frac{T_H - T_L}{T_H} = 1 - \frac{T_L}{T_H}.$$

Therefore, the thermodynamic efficiency of the cycle is exactly equal to that predicted by the second law.

4.5 Carnot Cycle

Numerical Substitutions. To calculate the amounts of heat and work we need the heat capacity of the fluid, which is not given. The efficiency, however, can be calculated since temperatures are known:

$$\eta = 1 - \frac{T_L}{T_H} = 1 - \frac{(300 + 273.15) \text{ K}}{(500 + 273.15) \text{ K}} = 0.259 = 25.9\%.$$

Comments The efficiency of the cycle does not depend on the properties of the gas. In general, the efficiency of the Carnot cycle is independent of the working fluid and a function of temperature only.

Example 4.12: Carnot Cycle Using Steam

Steam undergoes a Carnot cycle between temperatures 500 °C and 300 °C. During the isothermal heating step, the steam expands from pressure 50 bar to 30 bar. Determine the states in each of the four corners A, B, C, and D of the cycle (see Figure 4-5), calculate the thermodynamic energy balances and determine the thermodynamic efficiency of the cycle.

Solution The isothermal step is from 50 bar, 500 °C (state A) to 30 bar, 500 °C. State C is at 300 °C and the same entropy as B; this state is determined by interpolation in the tables. State D is at 300 °C and at the same entropy as A; this state is also determined by interpolation. The properties of the four states are summarized below. The specific volume and enthalpy are not needed in the calculations, but they are shown below for the sake of completeness.

	A	B	C	D
P (bar)	50	30	7.98	13.40
T (°C)	500	500	300	300
V (m³/kg)	0.0686	0.1162	0.3250	0.1910
U (kJ/kg)	3091.6	3108.5	2797.63	2786.99
H (kJ/kg)	3434.5	3457.0	3056.95	3042.61
S (kJ/kg K)	6.9778	7.2356	7.2356	6.9778

The energy balances are calculated using the first law,

$$\Delta U = Q + W.$$

In the isothermal parts, heat is calculated using eq. (4.24), and work is calculated from the first law, above. In the isentropic parts, $Q = 0$ and the work is obtained from the first law. These calculations are summarized below:

	Q (kJ/kg)	W (kJ/kg)	ΔU (kJ/kg)
AB	199.318	−182.418	16.9
BC	0	−310.865	−310.865
CD	−147.758	137.114	−10.6444
DA	0	304.61	304.61
Total	51.56	−51.56	0

The cycle absorbs the amount of heat $Q_H = 199.318$ kJ/kg, rejects the amount $Q_L = -147.758$ kJ/kg and produces a net amount of work $W = -51.56$ kJ/kg. The thermodynamic efficiency is

$$\frac{-W}{Q_H} = \frac{51.56 \text{ kJ/kg}}{199.318 \text{ kJ/kg}} = 0.259 = 25.9\%.$$

The theoretical efficiency of the Carnot cycle is

$$\eta = 1 - \frac{T_L}{T_H} = 1 - \frac{(300 + 2731.15) \text{ K}}{(500 + 273.15) \text{ K}} = 0.259 = 25.9\%.$$

In agreement with the second law, the efficiency of the Carnot cycle is independent of the properties of steam and a function of temperature only. *Any* fluid operating between the same two temperatures will have the same Carnot efficiency.

NOTE

Entropy, Reversibility, and Real Processes

In Chapter 1 we introduced the notion of the quasi-static process and explained that such process is also reversible. The second law provides us with an independent and unambiguous way to determine whether a process is reversible or not: If entropy generation is positive, the process is irreversible; if it is zero, the process is reversible. If the entropy generation is negative, the process cannot take place as described. The only information that is required are the initial and final states of the system and of the surroundings. It is not necessary to know the path, nor how the process was conducted (quasi-statically or not). In other words, the second law allows us to characterize a process as reversible or irreversible by examining the state of the universe before and after a process. Since "quasi-static" implies "reversible" and vice versa, when a process results in entropy generation one must be able to identify gradients that give rise to irreversibilities. Similarly, if gradients are identified, one must expect an entropy analysis to show positive generation. Why are such processes irreversible? One answer to this question was given in Chapter 1, where we explained that only a quasi-static process can be reversed and retrace its exact path in either direction. The second law provides us with a different way to view this question. Since entropy is a state function, generation of entropy means that the state of the universe after an irreversible process is different. And since restoring the state of the universe would require negative entropy generation, such process is not feasible. In other words, the initial state of the universe cannot be restored following an irreversible process.

The impossibility of restoring the effects of irreversible processes may appear as a dramatic statement to some, but there is nothing controversial about it. It simply restates in quantitative terms what we know by experience to be true, namely, that spontaneous processes are driven by gradients. More specifically, they are driven by the *dissipation* of gradients: A hot system and a cold system exchange heat and during the process their temperatures come closer together;

a pressurized gas expands against surroundings at a lower pressure and during this process the pressures come closer together; a blob of ink spreads in water and the composition of the solution becomes more uniform.[5] At the end of these processes, the universe as a whole has moved closer to equilibrium by erasing some gradients that existed before. It is possible, of course, to reintroduce gradients in a uniform system: through heating and cooling we can create temperature gradients in an equilibrated system; a gas can be divided into parts of which one may be compressed, the other expanded, to produce a pressure difference; ink and water can be separated and then remixed to produce a concentrated blob. All of these processes, however, require the exchange of energy, which may only come at the expense of other gradients. In other words, to restore those parts of the universe that were affected as a result of an irreversible process is possible only by altering the state of other parts.

A reversible process is a special process that allows us to take advantage of gradients without losing them to dissipation. Suppose that an amount of heat Q must be added to a system at temperature T. This heat must be available at some higher temperature T'. Normally, we would place the two systems into thermal contact until they exchange the required amount of heat. This process is irreversible and results in positive entropy generation, as we found in Example 4.15. Here is one way to achieve this heat transfer reversibly. Operate a Carnot cycle between temperatures T' and T such that the amount of heat that is rejected at T is equal to Q. This process accomplishes two things: It delivers the required amount of heat to the system, but also produces a certain amount of mechanical work. This work may be used to reverse the process (transfer heat from the system at T to the bath at higher temperature), or to produce other gradients elsewhere in the universe (e.g., compress a gas). From this perspective, an irreversible process represents a lost opportunity to produce work. It is this aspect of irreversibility that is of relevance in thermodynamics because it attaches a "work penalty" to irreversibility. In general, a process that produces work will produce less than the maximum possible amount, if conducted irreversibly. A process that consumes work will consume more than the minimum theoretical amount under irreversible conditions. The best case scenario (maximum amount of work produced, minimum amount of work consumed) corresponds to the minimization of entropy generation, that is, to the reversible condition, $S_{\text{gen}} = 0$.

4.6 Alternative Statements of the Second Law

The second law is expressed in many equivalent forms and different textbooks adopt different statements. We will discuss some of these in order to provide a connection to statements you may find in the thermodynamics literature and to show how these are consistent with the approach taken here.

---

5. As we will learn in Chapter 10, the spreading of the ink blob is driven by gradients in the chemical potential of the ink and the water. These gradients are being erased by the diffusion process.

Alternative statement 1: For any change of state, the following inequality holds true:
$$dS \geq \frac{dQ}{T}.$$
This becomes an exact equality if the process is reversible.

We recognize this as eq. (4.28), which we derived previously. In many textbooks this is the preferred statement of the second law because the relationship between entropy and heat is immediate and does not require additional derivations. The problem with approach is that this inequality must be accepted without any physical justification. Mathematically, however, this statement is completely equivalent to our definition: for an adiabatic process, eq. (4.28) indeed reduces to eq. (4.3) and from there on, we recover all of the other results obtained in this chapter.

Two statements of historical significance are given below. One is due to Rudolph Clausius, the other due to William Thomson (Lord Kelvin) and Max Planck:

Alternative statement 2 (Clausius): It is not possible to construct a device whose sole effect is to transfer heat from a colder body to a hotter body.

Alternative statement 3 (Kelvin-Planck): It is not possible to construct a device that operates in a cycle and whose sole effect is to produce work by transferring heat from a single body.

These statements express the second law in terms of observations that may be defended through our empirical knowledge of the physical world. The Clausius statement essentially states the impossibility of spontaneous heat flow from lower temperature to higher temperature. The key condition in this statement is that such transfer cannot be "sole effect" of the device. This leaves room for refrigerators, which require work to pump heat from lower temperature to higher temperature. Absent such energy input, heat cannot travel from low to high temperature. The Kelvin-Plank statement states that the efficiency of a power cycle cannot reach 1. It can be shown that starting with either one of these statements we obtain eq. (4.2). The example below outlines the equivalence between the Clausius statement and the fundamental inequality of the second law. If these derivations sound complicated, lengthy, and indirect it is because they follow the historical development of entropy. It is a credit to the intellect of Carnot, Clausius, Kelvin, and Planck that they were able to discover entropy without the benefit of our modern understanding of thermodynamics, based only on observations of irrefutable physical behavior.

4.6 Alternative Statements of the Second Law

Example 4.13: Clausius Statement
Prove that the Clausius statement is a necessary consequence of the second law, as formulated in this chapter.

Solution Suppose a device (cycle) transfers heat between two reservoirs at temperatures T_H and T_L, with $T_H > T_L$. Application of the first and second law gives

$$Q_H + Q_L + W = 0,$$
$$-\frac{Q_H}{T_H} - \frac{Q_L}{T_L} = S_{\text{gen}} \geq 0.$$

Here, Q_H is the amount of heat exchanged with the hot reservoir, Q_L is the heat exchanged with the cold reservoir, and W is the net amount of work. These exchanges are all defined with respect to the system that undergoes the cycle. We solve the first equation for Q_H and substitute into the second equation:

$$\frac{Q_L + W}{T_H} - \frac{Q_L}{T_L} = S_{\text{gen}} \quad \Rightarrow \quad \frac{W}{T_H} - \frac{Q_L}{T_H}\left(\frac{T_H}{T_L} - 1\right) = S_{\text{gen}}.$$

If the device transfers heat from the lower temperature to the higher temperature, Q_L must be positive, that is, the heat is added to the cycle from T_L, and the cycle eventually delivers Q_H to the reservoir at higher temperature T_H. Since entropy generation cannot be negative, the process is possible only if the device exchanges work. Solving for this work,

$$W \geq Q_L \frac{T_H - T_L}{T_L}.$$

According to this result, the work must be positive (both terms on the right-hand side are positive) and *at least* equal to

$$Q_L \frac{T_H - T_L}{T_L}.$$

The minimum required work corresponds to reversible operation ($S_{\text{gen}} = 0$). If the cycle is irreversible, it will require more work. Positive work means that we (the surroundings) must supply this amount to the device. This is precisely what a refrigerator does: It uses work to "pump" heat from lower to higher temperature (more about refrigeration in Chapter 6). Without supplying work from the surroundings, the process is impossible. This proves the Clausius statement.

Comments The Clausius statement refers to a "device," which we took it to mean "cycle." A noncyclic process would leave the state of the working fluid altered, but this contradicts the stipulation that the *sole* effect of such device should be to transfer heat between two reservoirs. Therefore, the Clausius device, if it exists, must be a cycle.

Example 4.14: The Clausius Statement as an Equivalent Statement of the Second Law
Starting with the Clausius statement, define entropy and prove the inequality in eq. (4.3).

Solution The point of this exercise is to show that the Clausius statement, which essentially states that heat cannot flow from lower to higher temperature, logically leads to the existence of entropy and to all of the other consequences of the second law we discussed in this chapter. This is optional reading for those who want to look at the second law from yet a different point. The proof is given as a broad outline, leaving the details as an exercise.

1. First, show that no cycle can exceed the efficiency of the Carnot cycle. This is done as follows. Suppose that a cycle between temperatures T_H and $T_L < T_H$ exceeds the efficiency of a Carnot cycle between the same two temperatures. We reverse the Carnot cycle to operate as a refrigerator and use the work produced by the super-efficient cycle to power this refrigerator. By energy balance, this arrangement is shown to transfer heat from the low-temperature bath to the high-temperature bath. Since this arrangement does not receive work from the surroundings (the work transfer from the super-efficient cycle to the Carnot refrigerator is an internal transfer when the two cycles are taken to the system), it violates the Clausius statement. Therefore, no cycle can exceed the efficiency of the Carnot.

2. Next, show that the efficiency of a Carnot cycle that operates using an ideal gas as the working fluid is
$$\eta_{\text{Carnot}} = 1 - \frac{T_L}{T_H}.$$
This calculation was done already in Example 4.11 by applying the first law alone, that is, without making any references to entropy, which in the Clausius approach has not been defined yet.

3. Show that the integral
$$\int \frac{dQ_{\text{rev}}}{T}$$
is independent of the integration path, that is, it represents a state function. To do this, we first point out that the equation for the Carnot efficiency can be expressed in the form
$$\frac{Q_H}{T_H} + \frac{Q_L}{T_L} = 0.$$
Next we consider two interconnected Carnot cycles, as in Figure 4-6. When both cycles operate in the same direction (both clockwise, or both counterclockwise), their common part is canceled and the combined cycle is represented by the outline that encloses the two cycles, a closed path. This cycle exchanges heat with four baths, two at elevated temperatures (T_{H1} and T_{H2}) and two at lower temperatures (T_{L1} and T_{L2}). For this combined cycle, the above relationship between amounts of heat and temperature becomes
$$\frac{Q_{H1}}{T_{H1}} + \frac{Q_{H2}}{T_{H2}} + \frac{Q_{L1}}{T_{L1}} + \frac{Q_{L2}}{T_{L2}} = 0.$$

4.6 Alternative Statements of the Second Law

This we write in the more compact form as

$$\sum \frac{Q_{\text{rev}}}{T} = 0,$$

where Q_{rev} is the amount of heat that is absorbed by the system divided by the temperature of the system (for reversible heat transfer, the temperature of the bath matches the temperature of the system), and the summation is over all heat baths with which the system exchanges heat. If the reversible adiabatic paths are drawn close enough, any arbitrary closed path can be represented by a series of infinitesimally small Carnot cycles, each of which exchanges an amount dQ_{rev} of heat with a bath at T. The previous result then is generalized to

$$\oint \frac{dQ_{\text{rev}}}{T} = 0,$$

which states that the integral of the quantity Q_{rev}/T over a closed path is zero. This result implies that this integral, calculated between two states A and B, is independent of the integration path: if the integral did depend on the path, then going from A to B and returning back to A through different paths would yield a different result each time. But as we showed already, the integral must be zero along any closed path. We conclude that the integral of dQ_{rev}/T is independent of the path, therefore, it must be a *state* function. This defines entropy:

$$dS = \frac{dQ_{\text{rev}}}{T}.$$

4. The efficiency of an irreversible cycle operating between temperatures T_H and T_L, is less than that of the Carnot cycle between the same two temperatures. This condition is equivalent to the inequality

$$\frac{Q_H}{T_H} + \frac{Q_L}{T_L} \leq 0,$$

which, carried through the previous derivation leads to

$$\oint \frac{dQ}{T} \leq 0.$$

This can also be written as

$$\oint \frac{dQ}{T} \leq \oint \frac{dQ_{\text{rev}}}{T}.$$

For this result to be true along *any* closed path, we must have

$$\frac{dQ}{T} \leq \frac{dQ_{\text{rev}}}{T}.$$

But the right-hand side is dS; therefore, this result reads

$$dS \geq \frac{dQ}{T}.$$

For an adiabatic process, this becomes

$$\left[dS\right]_{\text{adiabatic}} \geq 0.$$

Thus we have proved that if we start with the Clausius statement we obtain the mathematical inequality of the second law introduced at the beginning of this chapter.

NOTE

Carnot Cycle and the "Work Value" of Heat

From an end-user perspective we can understand that heat and work are not equivalent in the sense that work is more highly valued than heat. Suppose we are given an amount of heat Q at temperature T. How much work can we extract from that amount? If we have access to a reservoir at lower temperature T_0 (and the surroundings are always a convenient choice), then we can build a cycle between T and T_0 and feed the amount Q into it to produce work. In this sense, the temperature difference is equivalent to a voltage difference in an electric circuit, or elevation difference in a gravitational system (a river flowing to sea level from a mountain spring), and provides an opportunity to produce work. If the power cycle is reversible, the fraction of heat that is converted into work is the Carnot efficiency, $1 - T_0/T$. The balance will appear as heat that is rejected at T_0; this amount can no longer be converted into work unless a reservoir at

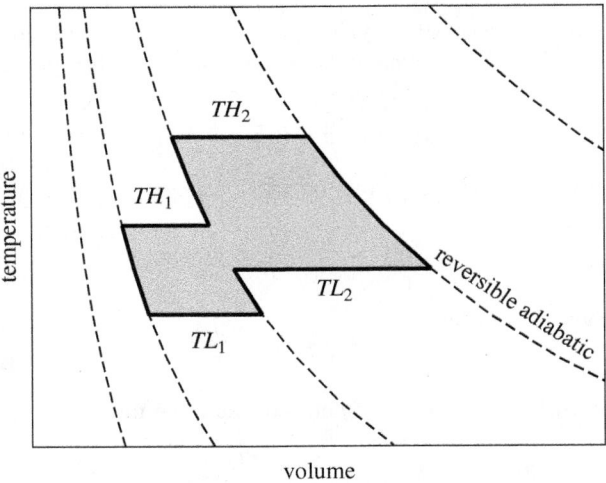

Figure 4-6: A cycle composed of two interconnected Carnot cycles. Using enough reversible adiabatic paths close to each other, any reversible closed path can be represented by a series of Carnot cycles. Along such closed path, the integral of dQ_{rev}/T is zero (see Example 4.14).

an even lower temperature becomes available. In this sense, the "work value" of the amount of heat Q is

$$\text{(Work value of heat } Q \text{ at temperature } T) = \left(1 - \frac{T_0}{T}\right) Q.$$

There are two important observations to make here: first, that the work value is always less that the amount of heat itself; and second, that the work value of heat depends on the temperature at which the heat is available and increases with increasing temperature. Sometimes we express this second point by saying that the quality of heat decreases as its temperature gets closer to that of the surroundings. The practical conclusion is that the usefulness of heat depends on its temperature. For example, if 100 kJ is available at 40 °C, we could use it to melt ice but not to boil water. That is, this heat is not as useful at 40 °C as it would be at, say 200 °C.

4.7 Ideal and Lost Work

We now want to answer the following question: How much work is necessary to change the state of a fixed amount of a substance from some initial state A to a final state B? Work is, of course, a path function, so the answer must depend on the specific path. Let's say that the change of state requires the consumption of work. Clearly, there is no upper limit to the amount of work that can be consumed because we can always devise arbitrarily long paths to reach the final state. But is there a *minimum* amount of work that is needed to produce this change? The answer must be yes, because, even if the change of state can be achieved entirely through heating with no direct input work, that heat has a corresponding price in terms of work.

To perform this calculation we adopt a path that takes the system from the initial to the final state using a single reservoir at T_0 for all heat exchanges. This reservoir can be used either as sink or as a source of heat. We can do that by setting up a cycle between T_0 and the system at T: If the heat transfer is in the direction of the lower temperature, the cycle will operate as an engine and will produce some work; if the transfer is in the direction of the higher temperature, then work will have to be consumed to move the heat against the temperature difference. Starting with the first law we have

$$\Delta U = Q + W,$$

where W is the net amount of work and includes the work in the cycle operating between the system and the reservoir. By second law, the entropy generation is

$$S_{\text{gen}} = \Delta S - \frac{Q}{T_0},$$

where ΔS is the entropy change of the system and $-Q/T_0$ is the entropy change of the surroundings. We solve the second law for Q, substitute into the first law and solve for the work:

$$W = \Delta U - T_0 \Delta S + T_0 S_{\text{gen}}. \tag{4.38}$$

Since entropy generation is always positive, the term $T_0 S_{\text{gen}}$ *adds* to the work that must be done. The minimum amount of work corresponds to $S_{\text{gen}} = 0$, or to a reversible process. This amount of work is called *ideal*:

$$\boxed{W_{\text{ideal}} = \Delta U - T_0 \Delta S,} \tag{4.39}$$

and depends on the initial and final states and also on the temperature of the reservoir—but not on the path of the process. The term $T_0 S_{\text{gen}}$, which represents extra work that must be done due to irreversibilities, is called *lost work*:

$$\boxed{W_{\text{lost}} = T_0 S_{\text{gen}} \geq 0.} \tag{4.40}$$

Returning to eq. (4.38) we may express the actual work W in terms of the ideal and lost work in the simple equation,

$$W = W_{\text{ideal}} + W_{\text{lost}}. \tag{4.41}$$

If the change of state requires work to take place (e.g., $W > 0$), the ideal work represents the *minimum amount* that must be consumed. If the change of state produces work (e.g., $W < 0$), the ideal work represents the *maximum amount* (in absolute value) that can be produced. In both cases, the lost work represents a penalty due to irreversibilities. If the process consumes work, the lost work is an extra amount that must be offered. If the process produces work, the lost work subtracts from the amount that could be produced.[6] In this sense, "lost" work refers to a lost *opportunity* to produce more work.

Example 4.15: Which Way Does Heat Flow?

Two systems, one at T_1 and the other at T_2, are placed in thermal contact with each other and in isolation of their surroundings. Use the second law to determine the direction of heat flow.

---

[6]. Recall that the lost work is always positive but work produced in negative. Then, the actual work produced can be written as

$$|W| = |W_{\text{ideal}}| - |W_{\text{lost}}| \quad \text{(process produces work)}.$$

The corresponding equation for a process that consumes work is

$$|W| = |W_{\text{ideal}}| + |W_{\text{lost}}| \quad \text{(process consumes work)}.$$

In this form it is clear that the lost work always represents a penalty.

4.7 Ideal and Lost Work

Solution Suppose that a small amount of heat, Q, defined with respect to system 1, is transferred between the two systems. For this process to be feasible the entropy change of the universe must be positive. This entropy is the sum of the entropy change in the two systems

$$S_{\text{gen}} = \Delta S_1^{\text{tot}} + \Delta S_2^{\text{tot}} \geq 0.$$

We are working with a small amount of heat so that the temperature of the two systems can be assumed to remain constant during the transfer. The entropy changes in the two systems are,

$$\Delta S_1^{\text{tot}} = \frac{Q}{T_1}, \quad \Delta S_2^{\text{tot}} = \frac{-Q}{T_2}.$$

The entropy generation is

$$S_{\text{gen}} = \frac{Q}{T_1} - \frac{Q}{T_2} = Q\left(\frac{T_2 - T_1}{T_1 T_2}\right) \geq 0.$$

If $Q > 0$, that is, system 1 receives this amount of heat, the inequality states that T_2 must be higher than T_1. If, on the other hand, $Q < 0$, meaning system 1 rejects this heat, the inequality requires that T_1 be larger than T_2. In both cases, heat flows from the higher to the lower temperature. The opposite direction is in violation of the second law.

Comment Because the second law is not intuitive, inevitably some will question its validity. It is true that the second law cannot be proven, in the same sense that it cannot be proven that all electrons carry the same charge—if by "proof" we mean that *all* electrons should be checked for compliance. We accept these statements as true on the basis of our experiential knowledge: there has been no reproducible observation to our knowledge that contradicts them. This example demonstrates that a violation of the second law would make it possible for heat to flow from a *lower* to a *higher* temperature, a result that we indisputably consider a physical impossibility. It is in fact possible to pump heat from lower temperature to higher temperature, as the kitchen refrigerator does, but this requires work from the surroundings. When two systems are placed into thermal contact in isolation from other surroundings, as in the above example, heat has only one direction to go: the direction that results in positive entropy generation.

Example 4.16: Feasibility of Process
Looking for ways to improve energy utilization in a chemical plant, a young intern proposes to design a process that will use steam in a closed system to produce work. The steam is available at 10 bar, 650 °C, and will be delivered at 1 bar, 150 °C. The intern expects to generate 1200 kJ/kg from this process. A heat source is available at 200 °C, should the process need any heat. Determine whether the company should go ahead with this plan.

Solution At a preliminary stage of design, the question is whether a process as described is feasible or not, that is, whether it conforms with the restrictions imposed by physical principles, in this case, the first and second law. If physical principles are violated, the process is clearly impossible. If it does not violate any principles, other considerations, such as a cost-benefit analysis, will have to be taken into account before a final decision is made. Our job is to determine whether the process is thermodynamically feasible or not. To make such determination, a detailed design is not necessary, as long as the broad features of the process are specified. In this case we know the effect on the system (initial and final state of steam), the amount of work that is exchanged, and the temperature of the surroundings that will supply heat, if necessary.

We collect the following information from the steam tables:

	A	B
T (°C)	650	150
P (bar)	10	1
U (kJ/kg)	3386.1	2582.9
S (kJ/kg K)	8.1557	7.6147

We begin by applying the first law, from which we calculate the amount of heat:

$$\Delta U_{AB} = Q + W \quad \Rightarrow \quad Q = \Delta U_{AB} - W,$$

and by numerical substitution,

$$Q = (2582.9 - 3386.1) \text{ kJ/kg} - (-1200) \text{ kJ/kg} = 396.8 \text{ kJ/kg}.$$

This amount of heat must be added to the system in order to satisfy the first law. Next, we apply the second law. The entropy change of the system is ΔS_{AB}; the entropy change of the surroundings is $-Q/T'$, where $T' = 473.15$ K is the temperature of the source that provides the heat. The entropy generation is

$$S_{\text{gen}} = \Delta S_{AB} - \frac{Q}{T'} = (7.6147 - 8.1557) \text{ kJ/kg K} - \frac{396.8 \text{ kJ/kg}}{473.15 \text{ K}} = -1.380 \text{ kJ/kg K}.$$

The entropy generation is *negative*, therefore, the process is not possible as described.

Example 4.17: Maximum Work
How should the specifications of the proposed design in Example 4.16 be modified to make it thermodynamically feasible?

4.7 Ideal and Lost Work

Solution A feasible design must satisfy both laws of thermodynamics. Assuming that the process exchanges heat with a single bath at temperature T', the two laws impose the following requirements:

$$\Delta U = Q + W,$$
$$\Delta S - \frac{Q}{T'} = S_{\text{gen}}.$$

Solving the first law for heat and substituting into the second law we obtain the condition,

$$S_{\text{gen}} = \Delta S + \frac{W - \Delta U}{T'} \geq 0.$$

The design is thermodynamically feasible as long as the entropy generation is positive. That is, ΔS, ΔU, W, and T' must be selected such that the inequality is satisfied. For a numerical example, we will assume that in the new design, the initial and final state of the steam will remain the same as in the original design. We may then either increase T', decrease the absolute value of work (recall that W is negative, therefore, decreasing its absolute value increases the entropy generation), or both. We will use $T' = 473.15$ K as in the original design. Then, the work must satisfy the condition,

$$-W \leq T' \Delta S - \Delta U.$$

Substituting the numerical values (recall that $-W$ is the absolute value of the work produced),

$$-W \leq (473.15 \text{ K}) \times (7.6147 - 8.1557) \text{ (kJ/kg K)} - (2582.9, -3386.1) \text{ kJ/kg} = 547.2 \text{ kJ/kg}.$$

The *maximum* amount of work that can be extracted by this process is 547.2 kJ/kg. In other words, if we conduct a reversible process to bring the steam from initial state 10 bar, 650 °C, to 1 bar, 150 °C, the amount of work that will be extracted is 547.2 kJ/kg. If the process is conducted irreversibly, the amount of work will be less. The original design attempted to extract more work than would be possible under reversible operation of the process.

Example 4.18: Irreversible Expansion in Vacuum
A closed insulated cylinder is divided into parts by a piston that is held into place by latches. One compartment is evacuated, the other contains steam at 7.5 bar, 300 °C. The latches are removed and the gas expands to fill the entire volume of the cylinder. When the system reaches equilibrium, its pressure is 1 bar. Determine the entropy generation.

Solution The final temperature was calculated in Example 3.20 where it was found to be 292 °C. The entropy of the final state is obtained by interpolation at 1 bar, 292 °C.

	Initial state	Final state
P (bar)	7.5	1
T (°C)	300	292
U (kJ/kg)	2798.6	2798.6
S (kJ/kg K)	7.2660	8.1884

The process is adiabatic, therefore, the entropy of the surroundings remains unchanged during this process ($\Delta S_{\text{sur}} = 0$). The entropy generation is

$$S_{\text{gen}} = \Delta S_{\text{sur}} + \Delta S_{\text{sys}} = 0 + (8.1884 - 7.2660) \text{ kJ/kg K} = 0.9224 \text{ kJ/kg K}.$$

The entropy generation is positive and we conclude the process is irreversible. In Example 3.20 we concluded that the process is irreversible because the expansion is not quasi-static.

Example 4.19: Heat Transfer
Calculate the entropy generation in Example 3.18.

Solution In Example 3.18, we considered the thermal equilibration of a hot piece of copper with a colder amount of water under constant pressure. The entropy change of the copper and of the water is

$$\Delta S_{\text{c}}^{\text{tot}} = m_{\text{c}} C_{P,\text{c}} \ln \frac{T_f}{T_{\text{c}}} = -4.507 \text{ kJ/kg K},$$

$$\Delta S_{\text{w}}^{\text{tot}} = m_{\text{w}} C_{P,\text{w}} \ln \frac{T_f}{T_{\text{w}}} = 7.229 \text{ kJ/kg K},$$

with the temperatures in kelvin. The entropy generation is

$$S_{\text{gen}}^{\text{tot}} = (-4.507 + 7.229) \text{ kJ/kg K} = 3.222 \text{ kJ/kg K}.$$

The piece of copper loses heat and its entropy decreases. The water gains heat and its entropy increases. The net change in entropy is positive because heat is exchanged between systems at two different temperatures.

4.8 Ambient Surroundings as a Default Bath—Exergy

The ideal work for a process that changes the state of the system depends on the temperature of the surroundings, which play the role of heat source and heat sink for all the heat requirements of the process. In most situations, heat will usually be available at an elevated temperature, for example, at the flame temperature produced when natural gas is burned. One must observe, however, that even natural gas must first be brought from ambient conditions to the conditions of the flame, and this requires some additional effort and the expenditure of energy. More generally we should view the surroundings as a baseline state to which all other states tend if left unattended. If we leave a cup of hot coffee or a glass of iced water on the kitchen table, both will equilibrate with the temperature of the surroundings. And a car tire, given enough time, will leak and equilibrate with the pressure of the ambient air. These processes are driven by the tendency of systems to reach equilibrium. It is possible, of course, to remove a state from ambient conditions, but this now requires energy: to make hot coffee we must burn a fuel, to make iced water we need to power a refrigerator, to blow air in a tire we need a compressor. How much work is needed to *remove* a unit mass of a substance from ambient conditions T_0, P_0, to final temperature T and pressure P? The answer is given by the ideal work for the state change $(T_0, P_0) \to (T, P)$:

$$W = (U - U_0) - T_0(S - S_0).$$

For an expanding system $(P > P_0)$, this is the work that can be extracted from the process. For a system under compression $(P < P_0)$, it is the work required for the compression. Part of that work, $-P_0(V - V_0)$, to be specific, is associated with PV work on the surroundings. In expansion, this is the work to push the surroundings by $(V - V_0)$; in compression, this is work done by the surroundings towards the compression. If we subtract this work from the ideal work above we are left with the work that is available to *us*, the human operators of the process. We call this work *exergy*, or *available work*, or simply *availability*:

$$\mathcal{E} = (U - U_0) + P_0(V - V_0) - T_0(S - S_0). \qquad (4.42)$$

The term *exergy* derives from "ex(ternal)" and *ergon* (Greek for "work") and implies the useful amount of work that can be extracted from the process, if the process produces work (by contrast, "energy," which derives from *en* (Greek for "in") and *ergon*, refers to the total amount of stored energy). The term *available work* refers to the same idea. Equation (4.42) summarizes in a concise way the "attractive pull" of the surroundings. A system at the state of the surroundings has zero exergy and indeed, it is in no position to produce any work. To produce a deviation from this default state we must consume work. Though not immediately obvious from this equation, the exergy is *always positive*, which means that work must be supplied *to* the system in order to remove it from the pull of the surroundings.

Example 4.20: Exergy
Calculate the available work (exergy) of 1 kg of saturated steam at 2 bar, and of 1 kg of water at the triple point. Take the surroundings to be at 25 °C, 1 bar.

Solution We collect the following data from the steam tables:

	P (bar)	T (K)	V (m^3/kg)	U (kJ/kg)	S (kJ/kg K)
Surroundings	1	298.15	0.001003	104.835	0.36725
State 1	2	393.36	0.8857	2529.1	7.1269
State 2	0.006117	273.16	0.001	0	0

The available work of saturated steam at 2 bar (state 1) is

$$\mathcal{E} = (2529.1 - 104.835) \text{ kJ/kg}$$
$$+ (10^5 \text{ Pa}) (0.8857 - 0.001003) \text{ (m}^3\text{/kg)}(10^{-3} \text{ kJ/J})$$
$$- (298.15 \text{ K})(7.1269 - 0.36725) \text{ kJ/kg K}$$
$$= 497.345 \text{ kJ/kg}$$

Similarly, for liquid water at the triple point we find

$$\mathcal{E} = (0 - 104.835) \text{ kJ/kg}$$
$$+ (10^5 \text{ Pa}) (0.001 - 0.001003) \text{ (m}^3\text{/kg)}(10^{-3} \text{ kJ/J})$$
$$- (298.15 \text{ K})(0 - 0.36725) \text{ kJ/kg K}$$
$$= 4.66 \text{ kJ/kg}$$

Comments In both cases the work is positive (must be supplied), even though state 1 is at conditions above ambient and state 2 below.

NOTE

Exergy, Entropy Generation, Lost Work

Exergy is the ideal work to bring the system from ambient conditions to the current state (excluding the PV work associated with the surroundings). It is also equal to the maximum amount of useful work that can be produced when the state reverts to ambient. More generally, if the state changes from (T_1, P_1) to (T_2, P_2), the corresponding reversible work is the change in exergy between the two states:

$$W_{\text{rev}} = \mathcal{E}_2 - \mathcal{E}_1.$$

If the process is done reversibly, then the above reversible work is also equal to the actual work. In practice, irreversibilities will cost a work penalty equal to $T_0 S_{\text{gen}}$ that adds to work that must be supplied, and subtracts from the work that can be produced. Exergy, therefore, offers a way to assess the magnitude of irreversibilities, and in that sense it is equivalent to an analysis based on entropy generation. The use of exergy is common in processes whose main focus is to produce mechanical or electrical work. In the study of chemical processes we will use entropy generation as a way to assess irreversibilities and we will report the results in terms of lost work, $T_0 S_{\text{gen}}$, because then entropy generation is translated into units of work, which makes it much easier to appreciate its magnitude.

4.9 Equilibrium and Stability

We started this chapter by postulating the existence of entropy based on the undisputable observation that isolated systems have the tendency to reach a state of equilibrium. We return to this idea to show that entropy, via the second law, specifies in precise terms what we mean by *equilibrium*. To draw this connection, we start with eq. (4.9) to express the bracketed expression in eq. (4.5) in the form,

$$\left(\frac{\partial S}{\partial V}\right)_U = P \left(\frac{\partial S}{\partial U}\right)_V = \frac{P}{T}.$$

With this and eq. (4.6), the differential of entropy in eq. (4.4) is written in the simpler form,

$$dS = \frac{dU}{T} + \frac{PdV}{T}. \tag{4.43}$$

Let us now revisit the process in Figure 4-1. The system consists of two compartments, I and II, that can exchange energy via the partition, which is conducting and movable. The energy and volumes of the two parts satisfy the conditions

$$U^{\text{I}} + U^{\text{II}} = U^{\text{I}+\text{II}} = \text{constant},$$
$$V^{\text{I}} + V^{\text{II}} = V^{\text{I}+\text{II}} = \text{constant},$$

which follows from the fact that the overall system is insulated and its volume is constant. We may view the equilibration process between the two compartments as an exchange of energy (i.e., by changing the temperature in each compartment) and volume (by moving the partition) under the conditions that the total energy and volume are constant. Suppose that the amount of energy δU and the amount of volume δV are transferred from side II to side I. The entropy change in each compartment and in the total system are obtained from eq. (4.43):

Compartment I: $\quad \delta S^{\mathrm{I}} = \dfrac{\delta U}{T^{\mathrm{I}}} + \dfrac{P^{\mathrm{I}} \delta V}{T^{\mathrm{I}}}$

Compartment II: $\quad \delta S^{\mathrm{II}} = \dfrac{(-\delta U)}{T^{\mathrm{II}}} + \dfrac{P^{\mathrm{II}}(-\delta V)}{T^{\mathrm{II}}}$

Entire system (I + II): $\quad \delta S^{\mathrm{I+II}} = \left(\dfrac{1}{T^{\mathrm{I}}} - \dfrac{1}{T^{\mathrm{II}}}\right)\delta U + \left(\dfrac{P^{\mathrm{I}}}{T^{\mathrm{I}}} - \dfrac{P^{\mathrm{II}}}{T^{\mathrm{II}}}\right)\delta V$

If this transfer of energy and volume causes the entropy of the entire box to increase (i.e., $\delta S^{\mathrm{I+II}} > 0$), it means that the system is still on its way to equilibrium. At equilibrium, $\delta S^{\mathrm{I+II}} = 0$. For $\delta S^{\mathrm{I+II}}$ to be zero for small but arbitrary transfers of energy and volume between the two compartments, the multipliers of δU and δV in the above equation must be zero:

$$T^{\mathrm{I}} = T^{\mathrm{II}},$$
$$\frac{P^{\mathrm{I}}}{T^{\mathrm{I}}} = \frac{P^{\mathrm{II}}}{T^{\mathrm{II}}},$$

from which it also follows that

$$P^{\mathrm{I}} = P^{\mathrm{II}}.$$

We conclude from these equations that when the system reaches equilibrium, temperature and pressure must be uniform throughout the system. That is, entropy, as defined above, leads to the correct equilibrium conditions of thermal and mechanical equilibrium.

Equilibrium Conditions at Constant Temperature and Pressure

In an isolated system (constant energy, constant volume, and constant number of moles), the equilibrium state is the state that maximizes entropy:

$$(\Delta S \geq 0)_{U,V,n}.$$

In many practical situations we are dealing, not with isolated systems, but with systems that interact in various ways with their surroundings. An important case is a closed system that interacts with a bath at temperature T under constant pressure

4.9 Equilibrium and Stability

P. To study the equilibrium conditions for such a system, we apply the second law to the system and its surroundings:

$$\Delta S - \frac{Q}{T} = S_{\text{gen}}.$$

Using $Q = \Delta H$ (recall that the process occurs under constant pressure), this becomes

$$T\Delta S - \Delta H = TS_{\text{gen}}.$$

We now introduce a new thermodynamic property, the *Gibbs free energy*, G, which we define as follows:

$$\boxed{G = H - TS.} \tag{4.44}$$

For an isothermal process, such as the one considered here, the change in the Gibbs free energy is

$$\Delta G = \Delta H - T\Delta S, \qquad (T = \text{const.})$$

Combining these results we obtain

$$\Delta G = -TS_{\text{gen}} \leq 0.$$

This result states that during the process the Gibbs free energy decreases (ΔG is negative) until equilibrium is reached, at which point G does not change any more (ΔG is zero). In other words, the Gibbs free energy at equilibrium is *at a minimum*. Mathematically,

$$\boxed{\left(\Delta G \leq 0\right)_{T,P,n}} \tag{4.45}$$

This result is the mathematical condition of *stable equilibrium*. It may not be immediately obvious how it is possible to minimize G while keeping T, P and n constant. After all as a state function, the Gibbs free energy is fixed once T, P and n are specified. Here are some problems where the minimization of the Gibbs free energy is relevant:

- When a pure system can exist in more than one phases, as when the equation of state has multiple roots, the stable phase is identified as the one with the lowest Gibbs free energy. This is discussed in more detail in Chapter 7. In this case, both phases are at the same pressure and temperature, but each has its own value of the Gibbs free energy. The stable phase is the one with the lower Gibbs free energy.

- When a *multicomponent* system consists of multiple coexisting phases, the compositions in each phase are such that the Gibbs energy is the lowest possible. This is the basis for the calculation of multicomponent phase equilibrium and is discussed in Chapter 10.

- When a mixture is brought to conditions that species can react to produce new species, the equilibrium composition is the one that minimizes the Gibbs free energy.[7] This application is discussed in Chapter 14.

The inequality in eq. (4.45) is an equivalent statement of the second law. In its original formulation, the second law specifies the equilibrium state of an isolated system (i.e., one of constant U^{tot}, V^{tot}, and n_i) as the state that maximizes entropy. If the system is by its temperature, pressure and moles, the equilibrium state corresponds to *minimum Gibbs free energy*. Other equivalent criteria can be developed depending on the variables that are used to fix the overall system. These statements are summarized below by the inequalities shown below, in which the superscript ^{tot} is dropped for simplicity:

$$(\Delta G \leq 0)_{T,P,n} \quad [4.45]$$

$$(\Delta A \leq 0)_{T,V,n} \quad (4.46)$$

$$(\Delta H \leq 0)_{S,P,n} \quad (4.47)$$

$$(\Delta U \leq 0)_{S,V,n}. \quad (4.48)$$

where, A is the Helmholtz free energy, defined as

$$A = U - TS. \quad (4.49)$$

Here is how to read these results: if we fix temperature (via a bath) and the total volume of a closed system, the equilibrium state minimizes the Helmholtz free energy (eq. [4.46]); if we fix entropy and pressure, the equilibrium state corresponds to minimum enthalpy (eq. [4.47]); if we fix entropy and volume, the equilibrium state corresponds to minimum internal energy (eq. [4.48]). Of these, eq. (4.45) is the most important inequality and the one that we will use repeatedly throughout the rest of this book.

NOTE

Thermodynamic Potentials

According to the above results, the equilibrium state minimizes a thermodynamic function, G, A, H, or U, depending on the variables that are used to specify the overall state of the system. These inequalities introduce a direct analogy with the potential energy of mechanical systems. In mechanical systems, equilibrium may be interpreted in terms of the potential energy function. If $\Phi(x)$ is the potential energy as a function of distance x, the force is

$$F_x = -\frac{\partial \Phi}{\partial x}.$$

---

7. In this case it is the total mass (or number of *atoms*), rather than the number of moles that remains constant.

4.10 Molecular View of Entropy

Mechanical equilibrium corresponds to a point of zero force, that is, to a point where the slope of the potential with respect to distance is zero. For stable equilibrium the shape of the potential must be convex, that is, a point of *minimum* in the potential function. Then, any deviation from that point produces a force that points in the direction of decreasing potential, and thus brings the system back to the minimum. The mathematical requirements for the minimum are

$$\frac{\partial \Phi}{\partial x} = 0, \quad \frac{\partial^2 \Phi}{\partial x^2} > 0.$$

The first condition ensures a point of extremum (minimum or maximum), and the second condition ensures that the extremum is a minimum. The Gibbs free energy plays the role of a *thermodynamic potential function* at fixed temperature and pressure. By analogy to the mechanical system, the mathematical conditions that define a stable equilibrium state at constant pressure and temperature are:

$$dG = 0, \tag{4.50}$$

$$d^2 G > 0. \tag{4.51}$$

Equation (4.50) expresses the equilibrium condition (zero "force"), and eq. (4.51) the condition for stability (equilibrium point is at a minimum).

4.10 Molecular View of Entropy

All of the previous developments of entropy and equilibrium arise from a macroscopic point of view, that is, from the perspective of someone who observes phenomena through their effect on measurable quantities such as pressure, temperature, and the like. It is a fair question to ask, what happens microscopically that gives rise to these macroscopic observations? How do the laws of physics allow a system to approach equilibrium but prevent it from moving away from it? For a molecular

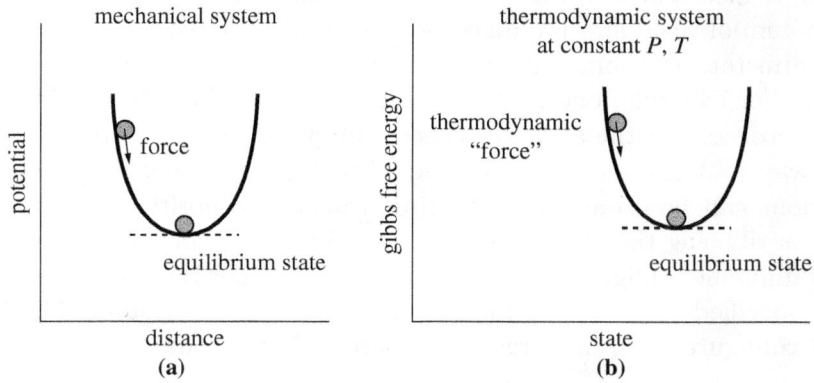

Figure 4-7: Gibbs free energy as a thermodynamic potential.

perspective we have to turn to statistical thermodynamics. As we mentioned before, a statistical treatment is beyond the scope of this book but this section is offered as a brief glimpse to the physical nature of entropy and the microscopic phenomena responsible for the effects of the second law.

Let us consider again the irreversible expansion experiment: a rigid, insulated tank is divided into two parts that are at different pressures and different temperatures. The partition is removed and the system is allowed to reach an equilibrium state in which pressure and temperature is uniform throughout. It is not very difficult to see the physical mechanism by which pressure becomes equalized. When there is a pressure differential, more molecules cross from the high-pressure part to the low pressure part than in the reverse direction, which cause a net transfer of mass from high pressure to low pressure and a corresponding decrease of the pressure difference. The equilibration of temperatures works in a similar manner: Molecules in hot regions have higher velocities and transfer some of their energy through collisions to slow moving molecules from colder regions of the system. Nature in other words works in ways that tend to erase gradients and ultimately produce equilibrium states characterized by uniform conditions. We view this state as one of *higher disorder* compared to what we had initially. Right after the partition is removed but before the system has had a chance to establish equilibrium, all the high-pressure molecules are on one side of the tank. This is a state of relatively high order because we know where to look for hot or cold molecules. At equilibrium, molecules are mixed and could be anywhere in the tank. This order/disorder picture may be easier to follow if we assume that the second half of the tank is evacuated before the partition was removed. We view the final state as one of higher disorder because we are less certain where a molecule is, since there is more space where the molecules can be found. The case can be made that spontaneous processes always lead to more disorder. In this view, entropy becomes a measure of disorder.

Statistical mechanics provides us with a precise way to define disorder. In an isolated system of fixed size (volume and mass) and energy, it is in principle possible to enumerate the different instantaneous configurations of molecules inside the system. For example, one configuration can be obtained by placing molecules at random positions in the available space and give each an arbitrary velocity but in such a way that the total energy is equal to the value specified initially. Other configurations can be obtained by shifting around the positions of the molecules and by redistributing the energy among them. Disorder can now be defined as the number of different configurations that are possible under the given conditions, i.e., under the specified size and total energy contained in the system. The larger the number of configurations, the larger the disorder, in the sense that our uncertainty

4.10 Molecular View of Entropy

as to which configuration is actually materialized at any point in time increases as the number of possibilities increases. Avoiding for the moment the question of how one could really count the apparently infinite number of such configurations, if that number is Ω, then the entropy of an isolated system is

$$S = k_B \ln \Omega, \qquad (4.52)$$

where k_B is Boltzmann's constant. It can be shown that eq. (4.52) is equivalent to the familiar entropy. Here, rather than a proof, we will provide a demonstration to connect this expression to more familiar results. Viewing entropy as a measure of disorder, there are two contributions that must be accounted for: one comes from changes in the number of ways that the total energy of the system can be divided among all molecules, and the other comes from changes in the number of ways to arrange these molecules in space. We will focus on an isothermal process in the ideal-gas state. Since energy in the ideal-gas state depends only on temperature, the only contribution we need to consider is that due to the number of arranging molecules in space. To enumerate the number of ways that we can arrange N molecules in a box of volume V^{tot} we divide the volume of the box into a grid of volume elements of size v_0, where v_0 is the volume of one molecule, as shown in Figure 4-8. This gives $M = V^{tot}/v_0$ number of grid elements. From combinatorial algebra we learn that the number of ways to place N identical molecules in M boxes is[8]

$$\Omega = \frac{M!}{N!(M-N)!}. \qquad (4.53)$$

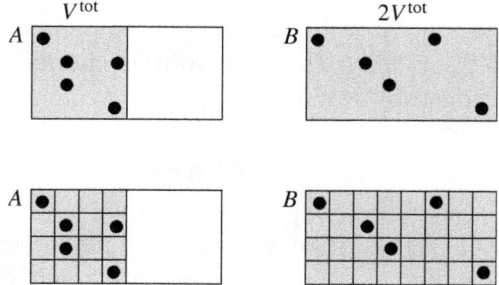

Figure 4-8: Molecular snapshots of a system of molecules at volume V^{tot}, temperature T (A) and volume $2V^{tot}$, temperature T (B). To enumerate the ways of arranging the molecules in the available volume we subdivide space into a grid of volume elements.

---

8. You can verify this result by explicitly counting the ways for some small N and M, say, $N = 2$, $M = 4$.

In typical thermodynamic systems N and M are very large numbers, of the order of 10^{23}, give or take a few quadrilion. For such large integers the logarithm of the factorial can be replaced by the following limiting expression,

$$\ln x! \to x \ln x - x. \tag{4.54}$$

Taking the logarithm of eq. (4.53), using the above limit we obtain

$$S^{\text{tot}}/k_B = \ln \Omega = M \ln M - M - N \ln N + N - (M-N) \ln(M-N) + M - N$$
$$= M \ln M - N \ln N - (M-N) \ln(M-N)$$
$$= M \ln \frac{M}{M-N} + N \ln \frac{M-N}{N}.$$

In the ideal-gas state molar density is very low, in other words, there are many more volume grid elements than there are molecules. This means that $M \gg N$, and leads to the conditions $\ln(M/(M-N)) \to \ln M/M = 0$ and $\ln((M-N)/N) \to \ln(M/N)$. Using these results, the entropy takes the form

$$S^{\text{tot}}/k_B = Nk \ln \frac{M}{N} = Nk \ln \frac{V^{\text{tot}}}{Nv_0}. \tag{4.55}$$

The number of moles, n, is $n = N/N_A$, where N_A is Avogadro's number. Recalling that $R = N_A k_B$, we obtain

$$S^{\text{tot}} = nR \ln \frac{V^{\text{tot}}}{nv_0 N_A} = nR \ln \frac{V}{c}$$

where $V = V^{\text{tot}}/n$ is the molar volume and $c = v_0 N_A$ is a constant. Let us now calculate the difference in entropy when the volume changes from V_1 to V_2. Using the above equation, the entropy change is

$$\Delta S_{12}^{\text{tot}} = nR \ln \frac{V_2}{V_1}.$$

If we further write $V_2/V_1 = P_1/P_2$ (recall that both states are at the same temperature), we obtain the familiar result for isothermal entropy change in the ideal-gas state:

$$\frac{\Delta S_{12}^{\text{tot}}}{n} = -R \ln \frac{P_2}{P_1}. \tag{4.56}$$

The point of this exercise is to show that eq. (4.52), whose basis is entirely molecular, produces the same final result for entropy as the classical macroscopic treatment.

Entropy and Probabilities Statistical mechanics provides a more general expression for entropy: if a system can exist in many microscopic states $i = 1, 2, \cdots, \Omega$, and if the probability to find the system in microstate i is p_i, then the entropy of the system is

$$S = -k_B \sum_{i=1}^{\Omega} p_i \ln p_i, \qquad (4.57)$$

where k_B is Boltzmann's constant and with the summation running through all microscopic states. By "microscopic state" we mean a molecular "snapshot" in time, namely, a specific arrangement of positions and velocities for all molecules. This equation embodies the probabilistic nature of entropy. Equation (4.52) is a special case of eq. (4.57) that is obtained if all microscopic states are equally probable (see Example 4.21). Equation (4.57) gives precise mathematical meaning to the qualitative notion of "disorder," which should be understood in a probabilistic sense: a system is in a state of higher disorder if the number of possible states where it can be found, weighted by the factor $-p_i \ln p_i$, increases.[9] Statistical mechanics also teaches how to evaluate the probability of microstates in different situations; this, however, is beyond the scope of our treatment.

Example 4.21: Equally Probable Microstates
Show that eq. (4.57) reduces to eq. (4.52) if all microscopic states are equally probable.

Solution There are Ω microscopic states in total. If all have the same probability, then this probability is

$$p_i = \frac{1}{\Omega}.$$

Inserting this result in eq. (4.57) we have,

$$S = -k_B \sum_{i=1}^{\Omega} \left(\frac{1}{\Omega}\right) \ln\left(\frac{1}{\Omega}\right) = -k_B \underbrace{\left[\frac{-\ln \Omega}{\Omega} + \frac{-\ln \Omega}{\Omega} + \cdots\right]}_{\Omega \text{ terms}} = k_B \ln \Omega.$$

According to statistical mechanics, in an isolated system all possible microscopic states are equally probable.

4.11 Summary

In this chapter we have dealt with two basic problems: One has to do with the various consequences of the second law; the second has to do with how to calculate entropy. We summarize them in reverse order.

---

9. According to eq. (4.57), entropy increases not just by the number of available microstates but by $-p_i \ln p_i$ per microstate. Microstates that are highly probable contribute more than those that have a lower probability.

Calculation of Entropy

- The basic formula for the calculation of entropy is eq. (4.7),

$$\Delta S_{AB} = \int_A^B \frac{dQ_{\text{rev}}}{T}, \qquad [4.7]$$

according to which, entropy is obtained by integrating the quantity dQ/T along a reversible path. The general procedure is to calculate dQ by applying the first law along a step of the process. In the special case that the path is one of constant pressure, the calculation is performed using the heat capacity according to eq. (4.10):

$$\Delta S_{12} = \int_{\substack{T_1 \\ \text{const. } P}}^{T_2} \frac{C_P}{T} dT. \qquad [4.10]$$

This equation applies provided that there is no phase change. If a phase change does occur, the entropy of vaporization must be included. Since entropy is an extensive property, the usual forms of the lever rule apply.

- Entropy calculations in the ideal-gas state are straightforward. The entropy change between any two states in the ideal gas is

$$\Delta S_{12}^{\text{ig}} = \int_{T_1}^{T_2} \frac{C_P^{\text{ig}}}{T} dT - R \ln \frac{P_2}{P_1}. \qquad [4.21]$$

These equations summarize the various ways by which we can calculate entropy. Of course, entropy may also be obtained from thermodynamic tables, if available.

Consequences of the Second Law The second law imposes an important limitation to all processes in nature: The entropy change in the universe may be zero or positive, but not negative. This requirement points a process always in the direction of equilibrium and has several important implications:

- Unless a process is conducted reversibly, it changes the universe in an irreversible way: the state of the universe cannot be fully restored following such process.

- A process that results in negative entropy generation is not feasible. This is a useful check when we design processes on paper.

- A thermal gradient has the potential to produce work. If an amount of heat is available at temperature T_H, a fraction of that heat can be converted into work. The maximum possible fraction that can be converted is given by the Carnot efficiency:

$$\eta_{\text{Carnot}} = 1 - \frac{T_L}{T_H}, \quad [4.35]$$

where T_L is a lower temperature. The Carnot efficiency refers specifically to a *cycle* operating between two temperatures, T_H and T_L.

4.12 Problems

Problem 4.1: Calculate the entropy change of steam between states $P_1 = 36$ bar, $T_1 = 250$ °C, and $P_2 = 22$ bar, $T_2 = 400$ °C by direct application of the definition of entropy. *Hint:* devise a path of constant pressure followed by a path of constant volume that connects the two states; use tabulated values of U and H to calculate the required heat capacities.

Problem 4.2: A 2 kg piece of copper at 200 °C is taken out of a furnace and is let stand to cool in air with ambient temperature 25 °C. Calculate the entropy generation as a result of this process. *Additional data:* The C_P of copper is 0.38 kJ/kg K.

Problem 4.3: A bucket that contains 1 kg of ice at -5 °C is placed inside a bath that is maintained at 40 °C and is allowed to reach thermal equilibrium. Determine the entropy change of the ice and of the bath. Additional information: $C_{P,\text{liq}} = 4.18$ kJ/kgK, $C_{P,\text{ice}} = 2.05$ kJ/kgK, $\Delta H_{\text{fusion}} = 334$ kJ/kg.

Problem 4.4: To convince yourself that the entropy of a bath changes even though its temperature remains unchanged, consider the following case: A tub that contains water at 40 °C receives 100 kJ of heat. Calculate the final temperature of the water and its entropy change if the mass of the water in the tub is:
a) 1 kg.
b) 10 kg.
c) 1000 kg.
d) Compare your results to the calculation $\Delta S = Q_{\text{bath}}/T_{\text{bath}}$. What do you conclude? (For water use $C_P = 4.18$ kJ/kg K.)
e) Prove that when the mass of the bath approaches infinity the entropy change of the bath is

$$\Delta S_{\text{bath}} = \frac{Q_{\text{bath}}}{T_{\text{bath}}}.$$

Hint: Recall that $\ln x \approx x - 1$ when $x \approx 1$.

Problem 4.5: Ethanol vapor at 0.5 bar and 78 °C is compressed isothermally to 2 bar in a reversible process in a closed system.
a) Draw a qualitative PV graph and show the path of the process. Mark all the relevant pressures and isotherms.
b) Calculate the enthalpy change of the ethanol.
c) Calculate the entropy change of the ethanol.
d) What amount of heat is exchanged between the ethanol and the surroundings during the compression?
Assume that ethanol vapor is an ideal gas with $C_P = 45$ J/K mol. The saturation pressure of ethanol at 78 °C is 1 bar.

Problem 4.6: One mol of methanol vapor is cooled under constant pressure in a closed system by removing 25000 J of heat. Initially, the system is at 2 bar, 200 °C.
a) Determine the final temperature.
b) Calculate the entropy change of methanol.
Additional data: boiling point of methanol at 2 bar: 83 °C; heat capacity of vapor: 60 J/mol K; heat capacity of liquid: 80 J/mol K; heat of vaporization at 2 bar: 36740 J/mol.

Problem 4.7: Calculate the entropy of toluene in the following states:
a) 10 bar, 300 °C
b) Vapor-liquid mixture at 10 bar with quality 25%.
c) 10 bar, 20 °C.
d) 15 bar, 20 °C.
The following data are available:

Saturation temperature at 10 bar:	216.80	°C
Enthalpy of saturated liquid at 10 bar:	21.757	kJ/mol
Entropy of saturated liquid at 10 bar:	49.582	J/mol K
Enthalpy of vaporization at 10 bar:	25.36	kJ/mol
C_P of liquid:	227.49	J/mol K
C_P of vapor:	189.37	J/mol K

Problem 4.8: Steam is compressed by reversible isothermal process from 15 bar, 250 °C, to a final state that consists of a vapor-liquid mixture with a quality of 32%.
a) Calculate the amount of work.
b) Calculate the amount of heat that is required to maintain isothermal conditions. Is this heat added or removed from the system?

Problem 4.9: Steam in a closed system is compressed by reversible isothermal process in a heat bath at 250 °C, starting from an initial pressure of 15 bar. During

4.12 Problems

the process, the steam transfers 2000 kJ/kg of heat to the bath. Determine the final state and the amount of work involved.

Problem 4.10: Steam in a closed system is compressed by reversible isothermal process in a heat bath at 250 °C, starting from an initial pressure of 15 bar. During the process, the steam receives 400 kJ/kg of work. Determine the final state and the amount of heat exchanged with the bath.

Problem 4.11: Steam at 1 bar, 150 °C, is compressed in a closed system by reversible isothermal process to final pressure of 20 bar. Determine the amount of work and the entropy change of the steam and of the heat bath that is used to maintain the process isothermal.

Problem 4.12: Steam at 1 bar, 150 °C is compressed in a closed system by reversible adiabatic process to final pressure of 20 bar. Determine the final temperature and the amount of work.

Problem 4.13: Nitrogen is compressed adiabatically in a closed system from initial pressure $P_1 = 1$ bar, $T_1 = 5$ °C to $P_2 = 5$ bar. Due to irreversibilities there is an entropy generation equal to 4.5 J/mol K.
a) Calculate the temperature at the end of the compression.
b) Calculate the amount of work. How does it compare to the reversible adiabatic work?
Assume nitrogen to be an ideal gas under the conditions of the problem. Is this a reasonable assumption?

Problem 4.14: Methanol at 2 bar and 65 °C expands isothermally to 0.2 bar through a reversible process.
a) Draw a qualitative PV graph and show the path of the process. Mark all the relevant pressures and isotherms. The process is conducted in a closed system.
b) Calculate the enthalpy change of the methanol.
c) Calculate the entropy change of the methanol.
d) What amount of heat is exchanged between the methanol and the surroundings during the compression?
Assume that the vapor phase is an ideal gas with $C_P = 45$ J/mol K. The boiling point of methanol at 1 bar is 65 °C.

Problem 4.15: Calculate ΔS for steam from 250 °C, 0.05 bar to 250 °C, 6 bar assuming steam to be an ideal gas. Compare to the value in the steam tables.

Problem 4.16: a) Calculate the amount of work necessary for the reversible isothermal expansion of 1 kg of steam from 10 bar to 5 bar at 400 °C.
b) Calculate the amount of heat associated with this process.

Problem 4.17: A rigid insulated tank is divided into two parts, one that contains 1 kg of steam at 10 bar, 200 °C, and one that contains 1 kg of steam at 20 bar, 800 °C. The partition that separates the two compartments is removed and the system is allowed to reach equilibrium. What is the entropy generation?

Problem 4.18: A young engineer notices in her plant that 1-kg blocks of brick are routinely removed from a 800 °C oven and are let stand to cool in air at 25 °C. Conscious about cost-cutting and efficiency, she wonders whether some work could be recovered from this process. Calculate the maximum amount of work that could be obtained. The C_P of brick is 0.9 kJ/kg K. Can you come up with devices that could extract this work?

Problem 4.19: A Carnot cycle operating between 600 °C and 25 °C absorbs 1000 kJ of heat from the high-temperature reservoir. The work produced is used to power another Carnot cycle which transfers 1000 kJ of heat from 600 °C to a reservoir at higher temperature T_H.
a) Calculate the amount of work exchanged between the two cycles.
b) Calculate the temperature T_H.
c) Calculate the amount of heat that is transferred to the reservoir at T_H.

Problem 4.20: A Carnot cycle operates in a closed system using steam: saturated liquid at 20 bar (state A) is heated isothermally until it becomes saturated vapor (state B), expanded by reversible adiabatic process to 10 bar (state C), partially condensed to state D, and finally compressed by reversible adiabatic process to initial state A.
a) Obtain the properties (T, P, V, U, H, and S) at the four states, A, B, C, and D, and summarize the results in a table.
b) Calculate the amount of heat and work involved in each of the four legs of the process.
c) Calculate the net amount of work that is produced.
d) Calculate the ratio of the net work over the amount of heat absorbed at the high temperature of the cycle. How does this value compare to the theoretical efficiency of the Carnot cycle that is calculated based on the two operating temperatures?

Problem 4.21: A Carnot cycle using steam in a closed system operates between 700 °C and 500 °C. During the isothermal step the steam expands from 20 bar to 10 bar. Perform the energy and entropy balances, calculate the efficiency, and compare to the theoretical value.

Chapter 5

Calculation of Properties

So far we have encountered several thermodynamic properties: molar volume, internal energy, enthalpy, entropy, heat capacities. They are all state functions, meaning that they are fixed once the state is fixed. An immediate consequence of practical importance is that thermodynamic properties can be tabulated once and for all. Tabulations have their limitations, however. Tables are impractical for mass-calculations such as those involved in large scale design, and of course it is impractical to tabulate properties of all conceivable substances under all conceivable states. Thus the need to develop methodologies for the numerical calculation of thermodynamic properties. In this chapter we have two main goals: (a) to develop relationships among the various thermodynamic properties and (b) to develop methodologies for the calculation of enthalpy and entropy as a function of pressure and temperature. Our focus continues to be on *pure fluids*. The extension to multicomponent systems is discussed in Chapter 9.

Instructional Objectives. In this chapter we will learn how to:

1. Manipulate the differential of thermodynamic properties.

2. Generate relationships between properties.

3. Calculate properties using the equation of state.

4. Properly apply reference states in the calculation of absolute properties.

This chapter makes extensive use of calculus. The most important concepts are reviewed here, but for more details you should consult a dedicated textbook.

5.1 Calculus of Thermodynamics

In dealing with pure substances we have two independent variables. Though we prefer to choose pressure and temperature, any two variables from the set $\{V, P, T, S, H, \cdots\}$ may be used for this purpose.[1] The same thermodynamic property may be

---

1. For example, in the PV graph we have chosen P and V as the independent variables. Isotherms and isobars on that graph are expressed functions of H and S.

written in various equivalent forms, depending on the pair that is chosen as the independent variables. This leads to a large number of relationships among the many properties. This is a good thing because it provides us with many alternative ways to perform a given calculation. It can also be a source of confusion if we lose track of the logic that governs these relationships and treat them instead as equations to be memorized. This logic is provided by the tools of multivariate calculus. In preparation for this discussion we first review some useful mathematical tools.

Thermodynamic properties are state functions. Mathematically,

$$F = F(X, Y),$$

which states that property F has a fixed value once the independent state variables X and Y are fixed. The differential of this function is given by

$$dF = \left(\frac{\partial F}{\partial X}\right)_Y dX + \left(\frac{\partial F}{\partial Y}\right)_X dY. \tag{5.1}$$

A useful mathematical property is the *triple-product* rule. We set $F = \text{const.}$, from which $dF = 0$. The differential then becomes

$$\left[0 = \left(\frac{\partial F}{\partial X}\right)_Y dX + \left(\frac{\partial F}{\partial Y}\right)_X dY\right]_{\text{const. } F} \Rightarrow \left[\left(\frac{\partial F}{\partial X}\right)_Y \frac{dX}{dY} = -\left(\frac{\partial F}{\partial Y}\right)_X\right]_{\text{const. } F}.$$

The last result is written more formally as

$$\left(\frac{\partial F}{\partial X}\right)_Y \left(\frac{\partial X}{\partial Y}\right)_F = -\left(\frac{\partial F}{\partial Y}\right)_X,$$

or, equivalently,

$$\left(\frac{\partial F}{\partial X}\right)_Y \left(\frac{\partial X}{\partial Y}\right)_F \left(\frac{\partial Y}{\partial F}\right)_X = -1. \tag{5.2}$$

This relationship exists between any three variables that are related by an equation. Since volume, pressure, and temperature are related via the equation of state, we obtain the following result as an immediate consequence of the triple-product rule:

$$\left(\frac{\partial P}{\partial V}\right)_T \left(\frac{\partial V}{\partial T}\right)_P \left(\frac{\partial T}{\partial P}\right)_V = -1. \tag{5.3}$$

The symmetry between F, X, and Y in the triple-product rule indicates that it does not matter which variable is taken to be the function and which are the independent variables since the equation $F = F(X, Y)$ can be solved to give X as a function of F and Y, or Y as a function of F and X.

5.1 Calculus of Thermodynamics

Example 5.1: Triple-Product Rule in the Ideal-Gas State
Calculate the three partial derivatives that appear in the triple-product rule between P, T, and V using the ideal-gas equation and confirm that their product is equal to -1.

Solution Starting with the ideal-gas law we have
$$P = \frac{RT}{V} \quad \Rightarrow \quad \left(\frac{\partial P}{\partial V}\right)_T = -\frac{RT}{V^2}.$$

We solve the ideal-gas equation for V and calculate its derivative with respect to T:
$$V = \frac{RT}{P} \quad \Rightarrow \quad \left(\frac{\partial V}{\partial T}\right)_P = \frac{R}{P}.$$

Finally, we solve for T and differentiate with respect to P:
$$T = \frac{PV}{R} \quad \Rightarrow \quad \left(\frac{\partial T}{\partial P}\right)_V = \frac{V}{R}.$$

The product of the three derivatives is
$$\left(\frac{\partial P}{\partial V}\right)_T \left(\frac{\partial V}{\partial T}\right)_P \left(\frac{\partial T}{\partial P}\right)_V = \left(-\frac{RT}{V^2}\right)\left(\frac{R}{P}\right)\left(\frac{V}{R}\right) = -\frac{RT}{PV} = -1.$$

Comments The example demonstrates that the triple-product rule is satisfied when P, V, and T are given by the ideal-gas equation. As a general equation, eq. (5.3) applies regardless of the form of the equation of state.

Example 5.2: Coefficient of Thermal Expansion Using the SRK Equation
Obtain an expression for the coefficient of thermal expansion of a fluid that is described by the Soave-Redlich-Kwong equation of state.

Solution The coefficient of thermal expansion is defined as
$$\beta = \frac{1}{V}\left(\frac{\partial V}{\partial T}\right)_P.$$

To calculate the derivative $(\partial V/\partial T)_P$, we have to solve the equation of state for V and perform the differentiation. However, the SRK equation is a cubic polynomial in V and cannot be expressed as an explicit function of P and T. We overcome this difficulty by using the triple-product rule to solve for the required derivative:
$$\left(\frac{\partial V}{\partial T}\right)_P = -\frac{1}{\left(\frac{\partial P}{\partial V}\right)_T \left(\frac{\partial T}{\partial P}\right)_V},$$

and this may be written as

$$\left(\frac{\partial V}{\partial T}\right)_P = -\frac{\left(\frac{\partial P}{\partial T}\right)_V}{\left(\frac{\partial P}{\partial V}\right)_T} = -\left(\frac{\partial P}{\partial T}\right)_V \bigg/ \left(\frac{\partial P}{\partial V}\right)_T.$$

In this form, the desired derivative involves derivatives of pressure with respect to temperature and volume. These can be calculated easily since the SRK equation gives pressure as an explicit function of these variables:

$$P = \frac{RT}{V-b} - \frac{a}{V(V+b)}.$$

The derivative with respect to volume is

$$\left(\frac{\partial P}{\partial V}\right)_T = -\frac{RT}{(V-b)^2} + a\frac{2V+b}{V^2(V+b)^2},$$

and the derivative with respect to temperature is

$$\left(\frac{\partial P}{\partial T}\right)_V = \frac{R}{V-b} - \frac{1}{V(V+b)}\frac{da}{dT},$$

where da/dT is the derivative of the parameter a with respect to temperature. This parameter is given by

$$a = 0.42748\frac{R^2 T_c^2}{P_c}\left(1 + \Omega\left(1 - \sqrt{T_r}\right)\right)^2,$$

where Ω stands for

$$\Omega = 0.48 + 1.574\omega - 0.176\omega^2.$$

Using $T_r = T/T_c$, the derivative with respect to temperature is

$$\frac{da}{dT} = -0.42748\frac{R^2 T_c}{P_c \sqrt{T_r}}\Omega\left(1 + \Omega\left(1 - \sqrt{T_r}\right)\right).$$

We now have all the elements for the calculation. The coefficient of thermal expansion is

$$\boxed{\beta = -\frac{1}{V}\left(\frac{\partial P}{\partial T}\right)_V \bigg/ \left(\frac{\partial P}{\partial V}\right)_T,} \tag{5.4}$$

with

$$\left(\frac{\partial P}{\partial V}\right)_T = -\frac{RT}{(V-b)^2} + \frac{(2V+b)\,a}{V^2(V+b)^2},$$

$$\left(\frac{\partial P}{\partial T}\right)_V = \frac{R}{V-b} - \frac{1}{V(V+b)}\frac{da}{dT},$$

5.1 Calculus of Thermodynamics

$$\frac{da}{dT} = -0.42748 \frac{R^2 T_c}{P_c \sqrt{T_r}} \Omega \left(1 + \Omega \left(1 - \sqrt{T_r}\right)\right),$$

$$\Omega = 0.48 + 1.574\omega - 0.176\omega^2.$$

Numerical substitutions The parameters of ethylene are

$$T_c = 282.35 \text{ K}, \quad P_c = 50.418 \text{ bar}, \quad \omega = 0.0866.$$

To continue with the calculation we need the molar volume at 400 K, 35 bar. This is obtained by solving the SRK equation for V and is found to be 8.82928×10^{-4} m^3/mol. The numerical results are summarized below:

$$T = 400 \text{ K}$$
$$P = 35 \text{ bar}$$
$$a = 0.364292 \text{ J m}^3/\text{mol}^2$$
$$b = 4.03 \times 10^{-5} \text{ m}^3/\text{mol}$$
$$da/dT = -7.55 \times 10^{-4} \text{ J m}^3/\text{mol}^2 \text{ K}$$
$$(\partial P/\partial V)_T = -3.69 \times 10^9 \text{ J mol/m}^6$$
$$(\partial P/\partial T)_V = 1.08 \times 10^4 \text{ J/m}^3 \text{ K}$$
$$V = 8.82928 \times 10^{-4} \text{ m}^3/\text{mol}$$
$$\beta = 3.31 \times 10^{-3} \text{ K}^{-1}.$$

Comments Equation (5.4) can be used with any cubic or other equation of state that gives pressure as an explicit function of volume and temperature. The partial derivatives of course will have to be recalculated for the corresponding equation of state.

Example 5.3: Differential and Integral Forms of a Function
Determine the differential of

$$f(x, y) = \frac{ax}{y - b},$$

confirm that it is exact, and integrate it from initial point (x_0, y_0), to final point (x_1, y_1).

Solution The partial derivatives are

$$\left(\frac{\partial f}{\partial x}\right)_y = \frac{a}{y - b}, \quad \left(\frac{\partial f}{\partial y}\right)_x = -\frac{ax}{(y - b)^2}.$$

Using these derivatives, the differential is

$$df = \frac{a}{y - b} dx - \frac{ax}{(y - b)^2} dy$$

The above derivatives satisfy the criterion of exactness:
$$\frac{\partial}{\partial y}\left(\frac{a}{y-b}\right) = -\frac{a}{(y-b)^2}, \quad \frac{\partial}{\partial x}\left(-\frac{ax}{(y-b)^2}\right) = -\frac{a}{(y-b)^2}.$$
This is to be expected: the differential of any function is an exact differential.

To integrate the differential we must define a path on the xy plane, namely, the relationship between x and y as we move from the initial point (x_0, y_0), to the final point (x_1, y_1). Though defining a path is necessary, *which* path we choose is unimportant because the result of the integration *must* be equal to $f(x_1, y_1) - f(x_0, y_0)$, a result that depends only the end points and not on the line that connects them. Here we choose the following two-step path: from (x_0, y_0) to (x_1, y_0) to (x_1, y_1). Conveniently, in the first segment of the path y_0 is held constant at $y = y_0$ and in the second segment x is constant at $x = x_1$.

From (x_0, y_0) to (x_1, y_0): Here $y = y_0$, which means $dy = 0$. Therefore, only the dx part is to be integrated while the value of y is set equal to y_0:
$$\text{(integral 1)} = \int_{x_0}^{x_1} \frac{a\, dx}{y_0 - b} = \frac{a(x_1 - x_0)}{y_0 - b}. \quad [\text{A}]$$

From (x_1, y_0) to (x_1, y_1): Here $x = x_1 = \text{const.}$, which means $dx = 0$. Only the dy part is to be integrated:
$$\text{(integral 2)} = -\int_{y_0}^{y_1} \frac{ax_1}{(y-b)^2}\, dy = ax_1\left(\frac{1}{y_1 - b} - \frac{1}{y_0 - b}\right). \quad [\text{B}]$$

Adding eqs. [A] and [B], the integrated differential is
$$\text{(integrated differential)} = \frac{ax_1}{y_1 - b} - \frac{ax_0}{y_0 - b}.$$
Notice that the right hand-side is indeed equal to the difference
$$f(x_1, y_1) - f(x_0, y_0).$$
Setting $x_1 = x$ and $y_1 = y$, the integrated differential is
$$\text{(integrated differential)} = \frac{ax}{y - b} - \underbrace{\frac{ax_0}{y_0 - b}}_{\text{const.}}.$$

Therefore, by integrating the differential we recover the original function to within a numerical constant.

Comments How is this relevant to thermodynamics? Set $a = R$, $x = T$, $y = V$, $f = P$, and the function of this example is revealed to be
$$P = \frac{RT}{V - b},$$

5.1 Calculus of Thermodynamics

which we recognize as a simplified version of the truncated virial equation (in the actual virial equation, b is a function of temperature, while in this example we treated b as a constant).

Exercise Perform the calculation with a different integration path (for example, constant x followed by constant y) and show that the final result is the same.

Exact Differential The differential of a function $F(X, Y)$ has the general form

$$dF = A\,dX + B\,dY, \tag{5.5}$$

where A and B are functions of X and Y and represent the partial derivatives of F:

$$A = \left(\frac{\partial F}{\partial X}\right)_Y, \quad B = \left(\frac{\partial F}{\partial Y}\right)_X. \tag{5.6}$$

We also have the following relationship between A and B

$$\boxed{\left(\frac{\partial A}{\partial Y}\right)_X = \left(\frac{\partial B}{\partial X}\right)_Y,} \tag{5.7}$$

which expresses the fact that the mixed derivative of F does not depend on the order of differentiation:

$$\left(\frac{\partial A}{\partial Y}\right)_X = \frac{\partial^2 F}{\partial Y\,\partial X} = \frac{\partial^2 F}{\partial X\,\partial Y} = \left(\frac{\partial B}{\partial X}\right)_Y. \tag{5.8}$$

While the differential of any function $F(X, Y)$ is always of the form in eq. (5.5), the opposite is not true: not every expression of the form in eq. (5.5) represents the differential of a function $F(X, Y)$, *unless* the two relationships in eq. (5.6) is also satisfied. If these relationships are indeed satisfied, then A and B are identified as the partial derivatives of $F(X, Y)$. As a further consequence, the integration of dF between two points depends only on the coordinates of the initial and final points and is independent of the integration path. Such differential is called *exact*. If eq. (5.7) is not satisfied, the differential is *inexact*. As a consequence, its integration produces different results depending in the path. This is another way of saying that if dF is an inexact differential, F *cannot* be represented by a function $F(X, Y)$, as such function integrated between (X_1, Y_1) and (X_2, Y_2) would produce $F(X_2, Y_2) - F(X_1, Y_1)$, *regardless* of the path.

The relevance to thermodynamics is this: thermodynamic properties are *state functions* and their differential with respect to any set of independent variables is *exact*. On the other hand, heat and work are *path* functions and their differentials are *inexact*. Our goal in this chapter is to develop differential expressions for enthalpy and entropy. These differentials will be of the form in eq. (5.5). Once we have the differential form of a property, we will be able to calculate changes between any two states by straightforward integration.

NOTE

Calculus and Thermodynamics
There is a correspondence between thermodynamic language and mathematics:

Thermodynamics		Calculus
state function	$\iff$	exact differential
path function	$\iff$	inexact differential
reversible process	$\iff$	integration path

The corresponding terms in this table are equivalent; in other words, they mean the same thing.

Example 5.4: Inexact Differential
The following differential expression has been proposed for the equation of state of a substance.

$$dP = \left(\frac{a}{V-b} + \frac{c}{V^2}\right) dT - \frac{aT}{(V-b)^2} dV,$$

where a, b, and c are non zero constants. Is such equation possible?

Solution The equation of state says that pressure is a function of temperature and volume. This means that the differential of P must be exact. To test this, we set

$$A = \frac{a}{V-b} + \frac{c}{V^2}, \quad B = -\frac{aT}{(V-b)^2},$$

and calculate the partial derivatives:

$$\left(\frac{\partial A}{\partial V}\right)_T = -\frac{a}{(V-b)^2} - \frac{2c}{V^3},$$

$$\left(\frac{\partial B}{\partial T}\right)_V = -\frac{a}{(V-b)^2}.$$

The right-hand sides are not the same, which means that the differential is not exact. We conclude that the differential given in the problem statement cannot represent an equation of state.

Exercise Integrate this differential between two states (T_0, V_0), (T_1, V_1), and show that the result is different depending on the path.

5.2 Integration of Differentials

A recurring problem in thermodynamics is the calculation of the change of a property F between two states (X_1, Y_1) and (X_2, Y_2). To perform this calculation, we devise a reversible path between the end states and consider a small step during which the state changes by (dX, dY). The corresponding change in F is given by its differential

$$dF = \left(\frac{\partial F}{\partial X}\right)_Y dX + \left(\frac{\partial F}{\partial Y}\right)_X dY. \tag{5.9}$$

To calculate the change ΔF_{AB} for a change of state from A to B we integrate this differential from the initial state to the final state:

$$\Delta F_{AB} = \int_{A\atop \text{path}}^{B} dF = \int_{A}^{B} \left[\left(\frac{\partial F}{\partial X}\right)_Y dX + \left(\frac{\partial F}{\partial Y}\right)_X dY\right]_{\text{path}}. \tag{5.1}$$

As we saw in Example 5.3, such integration requires the specification of a path, namely, an equation that relates the independent variables as they change from the initial to the final state. Nonetheless, the result of the integration must be equal to the difference between the value of F at the initial and final state,

$$\int_{A\atop \text{path}}^{B} dF = F_B - F_A.$$

Therefore, while a path is necessary to perform the integration, the result is *independent* of the path. This is true for all thermodynamic properties and has a very practical implication: to calculate the change of a thermodynamic property between two states we are free to choose the integration path any way we want. It makes sense then to pick a path that makes the calculation easier.

The geometric interpretation of this integration is shown in Figure 5-1. The independent variables are shown as axes on the horizontal plane and F is shown on the vertical axis. The state of the system is represented by a point on the XY plane. Function F is represented by a surface, which for simplicity is not shown. Integration is equivalent to calculating dF for small changes of state along the integration path and adding up all contributions. By the time the final state is reached, the integral must be equal to the difference $F_B - F_A$. Clearly, any path must yield the same value of the integral. Mathematically convenient paths are those that consist of segments in which one variable is held constant. Path $AB'B$ consists of a constant-X portion (AB') followed by a constant-Y part ($B'B$). In the constant-X part, $dX = 0$, thus only the Y-derivative is integrated. Similarly, along the constant-Y portion we have $dY = 0$, and the only derivative that is integrated is the one with respect to X. Such paths correspond to elementary processes, for example,

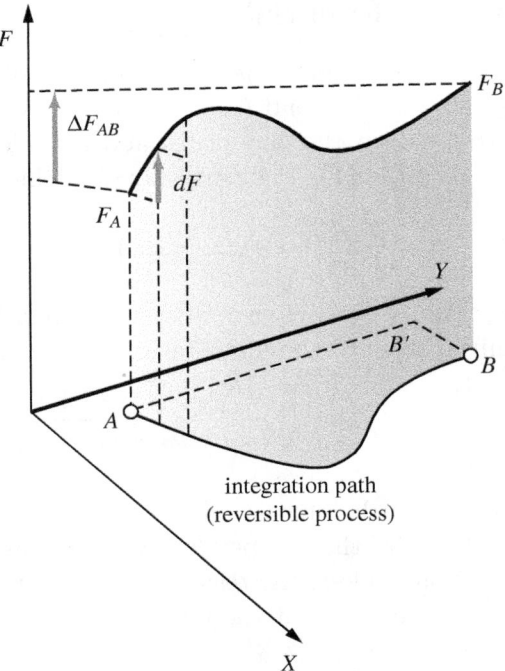

Figure 5-1: Geometric interpretation of the integration path. As the state traces the integration path on the XY plane, property F traces a line on the F(X, Y) surface.

isothermal isentropic, which correspond to constant temperature, constant entropy, and so on. For this reason, elementary processes are useful not only as experimental tools, but also as mathematical paths for the integration of thermodynamic differentials.

5.3 Fundamental Relationships

The Maxwell relationships are a set of relationships among various partial derivatives involving the following properties: T, S, P, and V. These, as well as additional relationships, can be obtained in a systematic and straightforward way. We begin by considering a closed system undergoing a reversible process without any shaft work. The system may exchange heat and PV work with the surroundings. We focus on a small step along the path. By first law,

$$dU = dQ + dW. \tag{5.10}$$

5.3 Fundamental Relationships

For a reversible process, the PV work is $-PdV$, and the second law gives $dQ = TdS$. We make these substitutions into the first law to obtain the first important equation:

$$\boxed{dU = TdS - PdV.} \tag{5.11}$$

This equation is of fundamental importance. Its thermodynamic interpretation is that it gives the change in internal energy as a result of a small change of state under reversible conditions. Its mathematical interpretation is that it expresses the differential of the internal energy, using entropy and volume as the independent variables. Internal energy is a state function, therefore, the above is an exact differential. As an immediate consequence we recognize the multipliers of dS and dV as partial derivatives of U (see eqs. [5.5] and [5.6]):

$$T = \left(\frac{\partial U}{\partial S}\right)_V, \quad P = -\left(\frac{\partial U}{\partial V}\right)_S. \tag{5.12}$$

A second result follows by applying the criterion of the exact differential, eq. (5.7), to these derivatives:

$$\left(\frac{\partial T}{\partial V}\right)_S = -\left(\frac{\partial P}{\partial S}\right)_V. \tag{5.13}$$

Analogous relationships can be obtained for enthalpy. Beginning with the definition $H = U + PV$, the differential of H is $dH = dU + PdV + VdP$. Substituting eq. (5.11) for dU we obtain

$$dH = TdS + VdP. \tag{5.14}$$

This is also of the form in eq. (5.5) and gives the differential of enthalpy with entropy and pressure as the independent variables. Now we identify the multipliers of dS and dP as the following partial derivatives:

$$T = \left(\frac{\partial H}{\partial S}\right)_P, \quad V = \left(\frac{\partial H}{\partial P}\right)_S. \tag{5.15}$$

An additional relationship is obtained by applying the criterion of exactness:

$$\left(\frac{\partial T}{\partial P}\right)_S = \left(\frac{\partial V}{\partial S}\right)_P. \tag{5.16}$$

We apply the same procedure to the Gibbs free energy, $G = H - TS$, whose differential is $dG = dH - TdS - SdT$. Using eq. (5.14), this becomes

$$dG = -SdT + VdP. \tag{5.17}$$

Following eq. (5.6) we make the identifications:

$$S = -\left(\frac{\partial G}{\partial T}\right)_P, \quad V = \left(\frac{\partial G}{\partial P}\right)_T. \tag{5.18}$$

Applying the criterion of exactness we obtain,

$$\left(\frac{\partial S}{\partial P}\right)_T = -\left(\frac{\partial V}{\partial T}\right)_P. \tag{5.19}$$

A final set of relationships is obtained for the Helmholtz free energy, $A = U - TS$. Its differential is $dA = dU - TdS - SdT$, and this, using eq. (5.11), becomes

$$dA = -SdT - PdV. \tag{5.20}$$

We identify the partial derivatives,

$$S = -\left(\frac{\partial A}{\partial T}\right)_V, \quad P = -\left(\frac{\partial A}{\partial V}\right)_T, \tag{5.21}$$

and obtain the following result by virtue of the exactness criterion:

$$\boxed{\left(\frac{\partial S}{\partial V}\right)_T = \left(\frac{\partial P}{\partial T}\right)_V.} \tag{5.22}$$

With this we have completed the derivations of this section.

All of these results are exact relationships among various properties. They are general and apply to any pure substance,[2] whether gas, liquid, or solid. All other relationships in thermodynamics may be considered as mathematical consequences of the results obtained here. Equations (5.13), (5.16), (5.19), and (5.22) are known as the Maxwell relationships. They relate various partial derivatives among the set of the four fundamental variables, P, V, T, and S, and they are very useful when we want to change from one set of independent variables to another. The complete results of these sections are summarized in Table 5-1.

Table 5-1: Summary of fundamental relationships

$dU = TdS - PdV$	$\left(\frac{\partial T}{\partial V}\right)_S = -\left(\frac{\partial P}{\partial S}\right)_V$	$T = \left(\frac{\partial U}{\partial S}\right)_V = \left(\frac{\partial H}{\partial S}\right)_P$
$dH = TdS + VdP$	$\left(\frac{\partial T}{\partial P}\right)_S = \left(\frac{\partial V}{\partial S}\right)_P$	$P = -\left(\frac{\partial U}{\partial V}\right)_S = -\left(\frac{\partial A}{\partial V}\right)_T$
$dG = -SdT + VdP$	$\left(\frac{\partial S}{\partial P}\right)_T = -\left(\frac{\partial V}{\partial T}\right)_P$	$V = \left(\frac{\partial H}{\partial P}\right)_S = \left(\frac{\partial G}{\partial P}\right)_T$
$dA = -SdT - PdV$	$\left(\frac{\partial S}{\partial V}\right)_T = \left(\frac{\partial P}{\partial T}\right)_V$	$S = -\left(\frac{\partial A}{\partial T}\right)_V = -\left(\frac{\partial G}{\partial T}\right)_P$

---

2. More specifically, these results apply to any closed system of *constant composition*, for example, any mixture undergoing changes of state as long as its composition does not change (through chemical reaction, for example).

5.4 Equations for Enthalpy and Entropy

There is a certain symmetry between these results: properties U, H, G, and A, all of which have units of energy, or energy per mass[3] appear as the dependent variables; properties P, V, T, and S, appear as independent variables in combinations that involve dP or dV, and dT or dS. Many more such relationships can be obtained. If the differential of dU is solved for dS, we obtain the differential of entropy in terms of U and V:

$$dS = \left(\frac{1}{T}\right) dU + \left(\frac{P}{T}\right) dV.$$

This too is an exact differential and can be used to produce additional set of relationships.[4] This procedure may be repeated with other properties as well. All these equations are a consequence of the fact that we have a wide choice for the pair of variables that we take to be independent. Mathematically, these derivations are an exercise in changing variables.

Example 5.5: Change of Variables
Express the partial derivative $(\partial H/\partial P)_T$ in terms of P, V, and T.

Solution To construct this partial derivative of H we start with eq. (5.14), which we divide by dP while keeping T constant:

$$\left[\frac{dH}{dP} = T\frac{dS}{dP} + V\right]_{\text{const. } T} \Rightarrow \left(\frac{\partial H}{\partial P}\right)_T = T\left(\frac{\partial S}{\partial P}\right)_T + V.$$

The remaining step is to express the entropy derivative in terms of P, V, T. This is provided by the Maxwell relationship in eq. (5.19). Thus finally we have

$$\left(\frac{\partial H}{\partial P}\right)_T = V - T\left(\frac{\partial V}{\partial T}\right)_P. \tag{5.23}$$

Comments This result is important and we will use it in the next section: it allows us to compute the derivative $(\partial H/\partial P)_T$ from the equation of state.

5.4 Equations for Enthalpy and Entropy

The stated goal for this chapter is to obtain equations for the enthalpy and entropy in terms of pressure and temperature.[5] We approach this as a calculus problem: if

---

[3]. We have written these equations in intensive form, such as for a unit of mass (1 mol or 1 kg) undergoing a change of state. All equations in Table 5-1 can be written for the extensive form of the properties involved, corresponding to mass m, or n number of moles.
[4]. Some of the results obtained in this manner will be redundant with the ones obtained already.
[5]. Equation (5.14) is not useful in this respect because the independent variables are entropy and pressure, rather than temperature and pressure.

enthalpy is taken to be a function of T and P, the general form of its differential must be,

$$dH = \left(\frac{\partial H}{\partial T}\right)_P dT + \left(\frac{\partial H}{\partial P}\right)_T dP.$$

We recognize the partial derivative with respect to temperature as the constant-pressure heat capacity. The derivative with respect to pressure was calculated in eq. (5.23). Combining these results,

$$\boxed{dH = C_P dT + \left[V - T\left(\frac{\partial V}{\partial T}\right)_P\right] dP.} \quad (5.24)$$

This is the desired equation and gives the differential of enthalpy with pressure and temperature as the independent variables. To obtain an analogous equation for entropy, we start with eq. (5.14), which we solve for dS:

$$dS = \frac{dH}{T} - \frac{V}{T} dP. \quad (5.25)$$

Using eq. (5.24) for dH we obtain

$$\boxed{dS = C_P \frac{dT}{T} - \left(\frac{\partial V}{\partial T}\right)_P dP.} \quad (5.26)$$

Equations (5.24) and (5.26) are important results: not only do they express enthalpy and entropy in terms of pressure and temperature, but their computation requires only the equation of state, which is needed to calculate the factor of the pressure term, and the heat capacity. Therefore, the calculation of enthalpy and entropy is reduced to the selection of a suitable equation of state. In the remainder of this chapter we will produce alternative forms of these equations, suitable for calculation with cubic equations of state. First, however, we point out that as an immediate consequence of these equations, we obtain simple expressions for the special case of constant-pressure process:

$$dH = C_p dT \quad \Rightarrow \quad \Delta H_{12} = \int_{T_1}^{T_2} C_P dT, \quad [3.19]$$
$$\text{(const. } P\text{)}$$

$$dS = C_P \frac{dT}{T} \quad \Rightarrow \quad \Delta S_{12} = \int_{T_1}^{T_2} C_P \frac{dT}{T}. \quad [4.10]$$
$$\text{(const. } P\text{)}$$

These results are not new, we encountered them previously in Chapters 3 and 4. That we recover them here serves a test of the validity of eqs. (5.24) and (5.26).

5.5 Ideal-Gas State

Table 5-2: Summary of ideal-gas equations

Equation of state	$PV^{ig} = RT$
Compressibility factor	$Z^{ig} = 1$
Heat capacities	$C_P^{ig} - C_V^{ig} = R$
Enthalpy	$dH^{ig} = C_P^{ig}\, dT, \qquad \Delta H_{12}^{ig} = \int_{T_1}^{T_2} C_P^{ig}\, dT$
Internal energy	$dU^{ig} = C_V^{ig}\, dT, \qquad \Delta U_{12}^{ig} = \int_{T_1}^{T_2} C_V^{ig}\, dT$
Entropy	$dS^{ig} = C_P^{ig}\dfrac{dT}{T} - R\dfrac{dP}{P}, \quad \Delta S_{12}^{ig} = \int_{T_1}^{T_2} C_P^{ig}\dfrac{dT}{T} - R\ln\dfrac{P_2}{P_1}$

5.5 Ideal-Gas State

The calculation of enthalpy and entropy according to eqs. (5.24) and (5.26) requires the equation of state. Our first application will be in the ideal-gas state. Starting with the ideal-gas law, we compute the derivative $(\partial V/\partial T)_P$:

$$V^{ig} = \frac{RT}{P} \quad \Rightarrow \quad \left(\frac{\partial V^{ig}}{\partial T}\right)_P = \frac{R}{P}.$$

For the coefficient of the dP term in the enthalpy equation we find:

$$V^{ig} - T\left(\frac{\partial V^{ig}}{\partial T}\right)_P = V^{ig} - \frac{RT}{P} = 0.$$

Applying these results to eqs. (5.24) and (5.26):

$$dH^{ig} = C_P^{ig}\, dT, \tag{5.27}$$

$$dS^{ig} = C_P^{ig}\frac{dT}{T} - R\frac{dP}{P}. \tag{5.28}$$

These results are not new, they are the differential forms of eqs. (3.33) and (4.21), obtained previously. For the internal energy, we use the definition of enthalpy to write $U = H - RT$, and take its differential:

$$dU^{ig} = dH^{ig} - R\, dT,$$

and use eq. (5.27) for dH^{ig}:

$$dU^{ig} = (C_P^{ig} - R)\, dT. \tag{3.34}$$

Here we identify the multiplier of dT as the constant-volume heat capacity, that is,

$$C_V^{ig} = C_P^{ig} - R. \qquad [3.35]$$

Thus we have recovered all the results obtained previously by independent methods. The ideal-gas properties are summarized in Table 5-2.

5.6 Incompressible Phases

The effect of pressure on enthalpy and entropy is described by the dP term in eqs. (5.24) and (5.26). The partial derivative that appears in this term can be expressed in terms of the coefficient of thermal expansion as follows,

$$\beta = \frac{1}{V}\left(\frac{\partial V}{\partial T}\right)_P \quad \Rightarrow \quad \left(\frac{\partial V}{\partial T}\right)_P = \beta V.$$

With this result, the equations for enthalpy and entropy become

$$dH = C_P\, dT + V(1 - \beta T)\, dP \qquad (5.29)$$

$$dS = C_P \frac{dT}{T} - \beta V dP. \qquad (5.30)$$

For condensed phases (solids, liquids away from the critical point), both β and V are small. Accordingly, the contribution of the terms $V(1 - \beta T)\, dP$ and $-\beta V dP$ to the enthalpy and entropy is generally negligible when compared to the contribution of the temperature term. This is the reason that we often take the enthalpy and entropy of compressed liquids to be independent of pressure. The accuracy of this approximation is tested in the example below.

Example 5.6: Effect of Pressure and Temperature on the Enthalpy and Entropy of Liquid
The volume-temperature relationship for liquid acetone is (see *Perry's Chemical Engineers' Handbook*, 7th ed., p. 2-131)

$$V = V_0\left(1 + a_1 t + a_2 t^2 + a_3 t^3\right),$$

where $V_0 = 1.228 \times 10^{-3}$ m^3/kg, V is the volume at temperature t (in °C), and the parameters a_1, a_2, a_3, are

$$a_1 = 1.3240 \times 10^{-3}, \quad a_2 = 3.8090 \times 10^{-6}, \quad a_3 = -0.87983 \times 10^{-8}.$$

Determine the sensitivity of enthalpy and entropy to pressure by calculating the partial derivatives,

$$\left(\frac{\partial H}{\partial P}\right)_T \quad \text{and} \quad \left(\frac{\partial S}{\partial P}\right)_T.$$

5.6 Incompressible Phases

Solution From eqs. (5.29) and (5.30), these derivatives are

$$\left(\frac{\partial H}{\partial P}\right)_T = V(1-\beta T), \quad \left(\frac{\partial S}{\partial P}\right)_T = -\beta V.$$

The values of these derivatives depend on temperature. For the purposes of illustrating the calculation we use $T = 300$ K. At $T = 300$ ($t = 26.85$ °C) the volume is

$$V = 1.275 \times 10^{-3} \text{ m}^3/\text{kg}.$$

The coefficient of thermal expansion is (see also Example 2.13)

$$\beta = \left(\frac{1}{V}\right)\left(\frac{dV}{dt}\right)\left(\frac{dt}{dT}\right) = \frac{3a_3 t^2 + 2a_2 t + a_1}{a_3 t^3 + a_2 t^2 + a_1 t + 1}.$$

At $t = 26.85$ °C,

$$\beta = 1.454 \times 10^{-3} \text{ K}^{-1}.$$

The pressure-derivative of enthalpy is

$$\left(\frac{\partial H}{\partial P}\right)_T = (1.275 \times 10^{-3} \text{ m}^3/\text{kg})\left(1 - (1.454 \times 10^{-3} \text{ K}^{-1})(300 \text{ K})\right)$$

$$= 7.189 \times 10^{-4} \frac{\text{m}^3}{\text{kg}} = 7.189 \times 10^{-4} \frac{\text{J/kg}}{\text{Pa}} = 0.07189 \frac{\text{kJ/kg}}{\text{bar}}.$$

For entropy, the result is

$$\left(\frac{\partial S}{\partial P}\right)_T = -(1.454 \times 10^{-3} \text{ K}^{-1})(1.275 \times 10^{-3} \text{ m}^3/\text{kg})$$

$$= -1.854 \times 10^{-6} \text{ m}^3/\text{kg K} = -1.854 \times 10^{-6} \frac{\text{(J/kg K)}}{\text{Pa}}$$

$$= -1.854 \times 10^{-4} \frac{\text{(kJ/kg K)}}{\text{bar}}.$$

Comments To put these results in perspective, let us compare the effect of temperature and pressure: the heat capacity of liquid acetone at 300 K is

$$C_P = 126.636 \text{ J/mol K} = 2.180 \text{ kJ/kg K},$$

that is, for every degree kelvin, enthalpy increases by 2.180 kJ/kg; by contrast, for every bar of pressure, it increases by only 0.07 kJ/kg.

For entropy, the sensitivity to temperature is given by the ratio C_P/T, which at 300 K is 0.422 kJ/kg K^2. Accordingly, the entropy changes by 0.422 kJ/kg K per degree kelvin, but only by -1.854×10^{-4} kJ/kg K for each bar of pressure.

The conclusion is that pressure has a very weak effect on the enthalpy and entropy of compressed liquid, and may be neglected, unless the pressure change is very large. The effect of temperature is much stronger. In fact, at the conditions of this example, one degree kelvin is equivalent to a few hundred bar!

5.7 Residual Properties

The calculation of properties in the ideal-gas state, summarized in Table 5-2, is straightforward: the equation of state is simple to manipulate, and the heat capacity is independent of pressure, which further simplifies the calculation. This simplicity is lost when we move away from the ideal-gas state. The difficulty is not so much that the equation of state is now more complex, the computer will take care of that, but rather that the heat capacity becomes a function of both temperature and pressure. It would be possible, with some effort, to tabulate the heat capacity in terms of temperature and pressure. Fortunately, this is not necessary. The difficulty is removed through the introduction of *residual* properties. In this approach, the calculation is done in two parts: first we obtain enthalpy and entropy as if the substance were in the ideal state, then we add an appropriate correction that makes the result exact. This correction is the *residual property*. Given a property $F(P, T)$, the corresponding residual property, $F^R(P, T)$ is defined by the equation,

$$F(P, T) = F^{ig}(P, T) + F^R(P, T), \tag{5.31}$$

where F^{ig} is the ideal-gas property at pressure P and temperature T, namely, the property obtained by applying the ideal-gas equations. Residual properties can be defined for any thermodynamic property that can be expressed as a function of pressure and temperature, that is, for all properties except pressure and temperature themselves. For a change of state, eq. (5.31) gives:

$$\Delta F_{12} = \Delta F_{12}^{ig} + F_2^R - F_1^R, \tag{5.32}$$

where ΔF_{12}^{ig} is calculated by the ideal-gas equations, shown in Table 5-2. For enthalpy and entropy, specifically, eq. (5.32) takes the form,

$$\boxed{\Delta H_{12} = \int_{T_1}^{T_2} C_P^{ig} dT + H_2^R - H_1^R,} \tag{5.33}$$

$$\boxed{\Delta S_{12} = \int_{T_1}^{T_2} C_P^{ig} \frac{dT}{T} - R \ln \frac{P_2}{P_1} + S_2^R - S_1^R.} \tag{5.34}$$

The usefulness of the residual properties should now be clear: the calculation of ΔH and ΔS reduces to the calculation of the residual enthalpy and entropy at the initial and final state. Notice that the only heat capacity that is needed in this calculation is the ideal-gas C_P^{ig}, which is available from tables. Of course, we still need expressions for these residuals. As it turns out, in deriving eqs. (5.24), (5.26), we have done most of the work already.

5.7 Residual Properties

NOTE

The Hypothetical Ideal-Gas State

In eqs. (5.33) and (5.34), the ideal-gas part represents a step of the calculation, a purely mathematical operation that computes the ideal-gas terms by the equations in Table 5-2. Sometimes we describe this part of the calculation by saying that the system is brought to the ideal-gas state at its own pressure P and temperature T. The "ideal-gas state at the system's own pressure and temperature" exists only as a mathematical operation, not as a physical state. If we insist on ascribing physical meaning to it, we would describe it as state in which molecular interactions are turned off, so that molecules act as point masses at the pressure and temperature of the system; this system would act as an ideal gas, no matter how high the pressure, and regardless of whether the actual phase is a gas, liquid, or solid. Of course, in the physical world we cannot turn molecular interactions on and off at will. In the virtual world of mathematics, we can. We call this the "hypothetical ideal-gas state," to indicate that we are not referring to the actual state of the system but to a mathematical recipe.

Residual Enthalpy

To obtain an equation for the residual enthalpy, we write $H^R = H - H^{ig}$ and take its differential at constant temperature:

$$dH^R = dH - dH^{ig} \quad \text{(const. } T\text{)}. \tag{5.35}$$

The term dH is given in eq. (5.24), the term dH^{ig} in eq. (5.27), and since temperature is constant, we set $dT = 0$ in both. This produces the following expression for the differential of the residual enthalpy:

$$dH^R = \left[V - T\left(\frac{\partial V}{\partial T}\right)_P\right] dP \quad \text{(const. } T\text{)}. \tag{5.36}$$

Decreasing pressure to zero at constant T, we reach the ideal-gas state. At this limit the residual enthalpy vanishes because the enthalpy of system is exactly equal to the ideal-gas contribution. Thus, at $P = 0$, $H^R = 0$, on all isotherms. We now integrate dH^R from $(P = 0, H^R = 0)$, to (P, H^R), along a line of constant temperature:

$$\boxed{H^R = \int_0^P \left[V - T\left(\frac{\partial V}{\partial T}\right)_P\right] dP.} \quad \text{(5.37)}$$
$$\text{(const. } T\text{)}$$

This is the desired result: it gives the residual enthalpy in state (P, T) in the form of an integral that may be evaluated if the equation of state is known.

Residual Entropy

An analogous equation for entropy is obtained in a similar manner. In analogy to eq. (5.35) we write

$$dS^R = dS - dS^{\text{ig}}, \quad (\text{const. } T) \tag{5.38}$$

We use eqs. (5.26) and (5.28) for dS and dS^{ig}, respectively, and set $dT = 0$. Substituting these results to the above equation we obtain

$$dS^R = \left[\frac{R}{P} - \left(\frac{\partial V}{\partial T}\right)_P\right]_{\text{const.} T} dP. \tag{5.39}$$

As with residual enthalpy, in the ideal-gas limit, $S^R = 0$. Integration from $P = 0$, $S^R = 0$ to P, S^R, at constant temperature leads to the following result for the residual entropy:

$$\boxed{S^R = \int_0^P \left[\frac{R}{P} - \left(\frac{\partial V}{\partial T}\right)_P\right] dP. \atop (\text{const. } T)} \tag{5.40}$$

Residual Volume

The residual volume is obtained by applying the definition of residual to volume. Using $V^{\text{ig}} = RT/P$ for the ideal-gas volume at pressure P and temperature T, the residual volume is

$$V^R = V - \frac{RT}{P}, \tag{5.41}$$

where V is the volume obtained from the equation of state. An alternative expression is obtained using the compressibility factor to write $V = ZRT/P$. Substituting this result into eq. (5.41) we obtain

$$V^R = \frac{RT}{P}(Z - 1), \tag{5.42}$$

which now gives the residual volume in terms of the compressibility factor.

Other Residual Properties

It is not necessary to derive equations for any other residual properties because these can all be related to V^R, H^R, and S^R. In general, all relationships among "regular" properties can also be written among the corresponding residual properties. For example, internal energy is related to enthalpy through the relationship,

$$U = H - PV.$$

In the ideal-gas state, this becomes

$$U^{\text{ig}} = H^{\text{ig}} - PV^{\text{ig}}.$$

5.7 Residual Properties

Taking the difference between the two, we obtain the relationship between the residual enthalpy and residual internal energy:

$$U^R = H^R - PV^R.$$

We can now go ahead and write the similar equations for other properties. For example, for Gibbs free energy and Helmholtz free energy we have,

$$G^R = H^R - TS^R,$$
$$A^R = U^R - TS^R.$$

As we see, U^R, G^R, and A^R can all be computed once H^R, S^R, and V^R are known.

Applications

Equations (5.37) and (5.40) are suitable for calculations with equations of state. The general procedure is this: express the volume in terms of pressure and temperature, compute the partial derivative $(\partial V/\partial T)_P$ and finally perform the integrations in eqs. (5.37) and (5.40). The simplest possible equation of state is the ideal-gas law; in this case, the residuals should vanish. The next simplest case is the truncated virial equation. These are examined in the examples that follow. Cubic equations of state require some additional work and are discussed in Section 5.8.

Example 5.7: Residual Properties in the Ideal-Gas State
Obtain the residual enthalpy, entropy, and volume in the ideal-gas state.

Solution Using $V = RT/P$, the partial derivative $(\partial V/\partial T)_P$ is

$$\left(\frac{\partial V}{\partial T}\right)_P = \frac{R}{P}.$$

Enthalpy: The quantity inside the integral of eq. (5.37) is

$$V - T\left(\frac{\partial V}{\partial T}\right)_P = V - \frac{RT}{P} = 0.$$

It follows then that $H^R = 0$.

Entropy: The quantity inside the integral of eq. (5.40) is

$$\frac{R}{P} - \left(\frac{\partial V}{\partial T}\right)_P = \frac{R}{P} - \frac{R}{P} = 0.$$

Again, we find $S^R = 0$.

Volume: In the ideal-gas state, $P \to 0$ and $Z \to 1$. These conditions create the indeterminacy $0 \div 0$ in eq. (5.42). This further implies that is possible for the residual volume to reach a constant value other than zero (see Example 5.8). This is not a problem, though. The ratio V^R/V is

$$\frac{V^R}{V} = \frac{RT(Z-1)}{PV} = \frac{Z-1}{Z},$$

which indeed goes to zero as the ideal gas is approached. In other words, even if V^R in the ideal-gas state is not zero, it represents an infinitesimal fraction of the molar volume and thus its contribution is negligible.

Example 5.8: Residual Properties from the Truncated Virial Equation
Obtain the residual enthalpy, entropy, and volume from the truncated virial equation.

Solution We use the truncated virial equation, eq. (2.28), to express volume in terms of temperature and pressure:

$$V = \frac{RT}{P} + B \quad \text{(moderate pressures)},$$

where B is the second virial coefficient, which we recall is a function of temperature.

Residual enthalpy. The partial derivative of volume with respect to temperature is

$$\left(\frac{\partial V}{\partial T}\right)_P = \frac{R}{P} + \frac{dB}{dT},$$

from which we obtain,

$$V - T\left(\frac{\partial V}{\partial T}\right)_P = \frac{RT}{P} + B - T\left(\frac{R}{P} + \frac{dB}{dT}\right) = B - T\frac{dB}{dT}.$$

To obtain the residual enthalpy, the above result is to be inserted in the integral in eq. (5.37). Noting that the term $B - TdB/dT$ is a function of temperature only, and that integration is at constant temperature, the result of the integration is

$$H^R = \left(B - T\frac{dB}{dT}\right) P. \tag{5.43}$$

Residual entropy. The procedure for entropy is similar. We evaluate the term that appears in the integral in eq. (5.40)

$$\frac{R}{P} - \left(\frac{\partial V}{\partial T}\right)_P = \frac{R}{P} - \left(\frac{R}{P} + \frac{dB}{dT}\right) = -\frac{dB}{dT}.$$

5.7 Residual Properties

This term depends only on temperature. Integration with respect to pressure at constant temperature gives,

$$S^R = -P\frac{dB}{dT}. \tag{5.44}$$

Residual volume. The residual volume is

$$V^R = \left(\frac{RT}{P} + B\right) - \frac{RT}{P} = B. \tag{5.45}$$

Notice that the residual volume does not go to zero in the ideal-gas state but to B. Nonetheless, the ratio V^R/V, which in this case can be expressed as

$$\frac{V^R}{V} = \frac{PB}{ZRT},$$

tends to zero as the ideal-gas state is approached ($P \to 0$, $Z \to 1$, $T = $ const.).

Comments Equations (5.43) and (5.44) may be used to calculate the residual terms if the second virial coefficient is known. The results are valid only in the range where the truncated virial equation applies, namely, at low enough pressures that the compressibility factor is approximately linear in P.

Example 5.9: Residual Properties of Methane Using the Virial Coefficient

The following data have been published for the second virial coefficient of methane (Thomas and van Steenwink, *Nature*, **187**, 229, [1960]):

T (K)	B (cm^3/mol)
108.45	-364.99
125.2	-267.97
149.1	-188.04
186.4	-126.20
223.6	-82.62
249.3	-68.53

Calculate the residual enthalpy and entropy at 220 K, 20 bar assuming the validity of the truncated virial equation. Is the assumption justified?

Solution The necessary equations were derived in Example (5.8) and require the value of the second virial coefficient and its derivative with respect to temperature. To obtain these values, it is convenient to fit the data to an empirical function of temperature:

$$B = -1.67676 \times 10^{-5} + \frac{5.85799 \times 10^{-3}}{T} - \frac{4.71233}{T^2},$$

where T in kelvin and B is in m^3/mol (the second virial coefficient is better fitted by a polynomial in inverse T rather than in T). The derivative with respect to temperature is obtained by differentiation of this equation:

$$\frac{dB}{dT} = \frac{9.42467}{T^3} - \frac{5.85799 \times 10^{-3}}{T^2}.$$

At $T = 220$ K the fitted equation gives $B = -8.75027 \times 10^{-5}$ m^3/mol and $dB/dT = 7.64079 \times 10^{-7}$ m^3/mol K.

The residual enthalpy is calculated using eq. (5.43):

$$H^R = \left(B - T\frac{dB}{dT}\right)P$$
$$= (-8.75027 \times 10^{-5} \text{ m}^3/\text{mol} - (220 \text{ K})(7.64079 \times 10^{-7} \text{ m}^3/\text{mol K}))(20 \times 10^5 \text{ Pa})$$
$$= -511.2 \text{ J/mol}.$$

For the residual entropy, eq. (5.44), the result is

$$S^R = -P\frac{dB}{dT} = -(20 \times 10^5 \text{ Pa})(7.64079 \times 10^{-7} \text{ m}^3/\text{mol K})$$
$$= -1.52816 \text{ J/mol K}.$$

Comments To determine whether these values are acceptable we must examine whether the truncated virial equation is valid at 220 K, 20 bar. The critical properties of methane are

$$T_c = 190.56 \text{ K}, \quad P_c = 45.99 \text{ bar}.$$

The reduced coordinates at 220 K, 20 bar are $T_r = 220/190.56 = 1.15$, $P_r = 20/45.99 = 0.44$. From Figure (2-8) we see the $T_r = 1.15$ isotherm is approximately linear at $P_r \approx 0.4$. Therefore, the truncated virial equation is valid in this region.

5.8 Pressure-Explicit Relations

The importance of eqs. (5.37) and (5.40) is that they relate residual enthalpy and entropy to the equation of state. In the above form, these equations are useful if the equation of state can be expressed as function of pressure and temperature. Cubic equations express pressure in terms of volume and temperature; to calculate residual properties from such equations, eqs. (5.37) and (5.40) must be converted so that the independent variables are V and T. The mathematical manipulations are shown on next page.

5.8 Pressure-Explicit Relations

Enthalpy. We separate the integral in eq. (5.37) into two and work with each term separately:

$$H^R = \int_0^P VdP - T \int_0^P \left(\frac{\partial V}{\partial T}\right)_P dP. \quad \text{(const. } T\text{)} \quad \text{(const. } T\text{)} \tag{5.46}$$

To transform the first term, we start with the differential $d(PV) = PdV + VdP$, which we solve for VdP:

$$VdP = d(PV) - PdV.$$

Next, we integrate both sides from $P = 0$ to P along a line of constant temperature, T. Noting that the lower limit is the ideal-gas state, we have:

$$\int_0^P VdP = PV - (PV)^{\text{ig}} - \int_0^P PdV$$
$$\text{(const. } T\text{)} \qquad\qquad \text{(const. } T\text{)}$$

$$= PV - RT - \int_0^P PdV. \tag{5.47}$$
$$\text{(const. } T\text{)}$$

Now we work on the second integral in eq. (5.46). We start with the triple-product rule in eq. (5.3), and split the derivative $(\partial P/\partial V)_T$ to obtain

$$\left(\frac{\partial V}{\partial T}\right)_P dP = -\left(\frac{\partial P}{\partial T}\right)_V dV, \quad \text{(const. } T\text{)}. \tag{5.48}$$

Using this result, the second integral in eq. (5.46) becomes

$$T\int_0^P \left(\frac{\partial V}{\partial T}\right)_P dP = -T\int_\infty^V \left(\frac{\partial P}{\partial T}\right)_V dV. \tag{5.49}$$
$$\text{(const. } T\text{)} \qquad\qquad \text{(const. } T\text{)}$$

In all integrals, the lower limit is the ideal-gas state. When integration is with respect to pressure, the lower limit is $P = 0$; when it is with respect to volume, the lower limit is $V = \infty$. Substituting eqs. (5.47) and (5.49) into (5.46), we obtain the desired result:

$$\boxed{H^R = PV - RT + \int_\infty^V \left[T\left(\frac{\partial P}{\partial T}\right)_V - P\right] dV.} \tag{5.50}$$
$$\text{(const. } T\text{)}$$

Entropy. A similar equation is obtained for entropy. We begin with eq. (5.40), which we write in the form,

$$S^R = R\int_0^P \frac{dP}{P} - \int_0^P \left(\frac{\partial V}{\partial T}\right)_P dP. \tag{5.51}$$
$$\text{(const. } T\text{)} \quad \text{(const. } T\text{)}$$

We begin with the relationship $P = ZRT/V$ and take the logarithm of both sides:

$$\ln P = \ln Z + \ln R + \ln T - \ln V.$$

We now take the differential of both terms under constant temperature:

$$\frac{dP}{P} = \frac{dZ}{Z} - \frac{dV}{V}, \quad \text{(const. } T\text{)}.$$

We integrate both sides under constant temperature from the ideal-gas ($P = 0$, $V = \infty$, $Z = 1$) state up to the current state (P, Z, V), and multiply both side by R, to obtain the following expression for the first integral in (5.51):

$$R \int_0^P \frac{dP}{P} = R \ln Z - R \int_\infty^V \frac{dV}{V}. \quad (5.52)$$
$$\text{(const. } T\text{)} \qquad\qquad \text{(const. } T\text{)}$$

Using eq. (5.48) obtained above, the second integral is

$$\int_0^P \left(\frac{\partial V}{\partial T}\right)_P dP = -\int_\infty^V \left(\frac{\partial P}{\partial T}\right)_V dV. \quad (5.53)$$
$$\text{(const. } T\text{)} \qquad\qquad \text{(const. } T\text{)}$$

Combining eqs. (5.52) and (5.53) into (5.51) we obtain the final result in the form

$$\boxed{S^R = R \ln \frac{PV}{RT} + \int_\infty^V \left[\left(\frac{\partial P}{\partial T}\right)_V - \frac{R}{V}\right] dV.} \quad (5.54)$$
$$\text{(const. } T\text{)}$$

Both eqs. (5.50) and (5.54) are expressed as integrals in V. These integrals are evaluated along an isotherm, starting at the ideal-gas state and ending at the state of interest.

5.9 Application to Cubic Equations

Equations (5.50) and (5.54) are now in suitable form for use with cubic equations of state. The general procedure is this: use the equation of state to calculate the partial derivative $(\partial P/\partial T)_V$, then perform the integrations in eqs. (5.50) and (5.54). We demonstrate the procedure using the Soave-Redlich-Kwong equation. The pressure in the SRK equation is given by

$$P = \frac{RT}{V-b} - \frac{a}{V(V+b)}. \quad [2.45]$$

5.9 Application to Cubic Equations

Noting that the parameter a is a function of temperature, the partial derivative of pressure with respect to T is,

$$\left(\frac{\partial P}{\partial T}\right)_V = \frac{R}{V-b} - \frac{da/dT}{V(b+V)},$$

from which we obtain,

$$T\left(\frac{\partial P}{\partial T}\right)_V - P = \frac{a - T(da/dT)}{V(b+V)}.$$

Inserting into eq. (5.50), the residual enthalpy is

$$H^R = PV - RT + (a - T(da/dT))\int_{\infty}^{V} \frac{dV}{V(V+b)}$$
$$\text{(const. } T\text{)}$$

$$= PV - RT + \frac{a - T(da/dT)}{b}\ln\frac{V}{V+b}.$$

The result may be expressed in the alternative form,

$$\boxed{H^R = RT(Z-1) + \frac{T(da/dT) - a}{b}\ln\frac{Z+B'}{Z},} \tag{5.55}$$

where we have used $V = ZRT/P$ to eliminate V in favor of Z, and $B' = Pb/RT$ is the dimensionless parameter previously defined in eq. (2.36). This streamlines the calculation of the residual enthalpy since working with the compressibility form of the equation of state is generally preferable. The residual entropy is obtained in the same manner, and the final result is

$$\boxed{S^R = R\ln(Z-B') + \frac{da/dT}{b}\ln\frac{Z+B'}{Z}.} \tag{5.56}$$

The derivative da/dT, which is needed for the calculation, is

$$\frac{da}{dT} = -0.42748\frac{R^2 T_c}{P_c}\frac{\left(1 + \Omega\left(1 - \sqrt{T_r}\right)\right)\Omega}{\sqrt{T_r}}. \tag{5.57}$$

These results are summarized in Table 5-3, which also includes results for the van der Waals and the Peng-Robinson equation. In all cases, B' is defined as $B' = Pb/RT$, where b is the corresponding parameter in the equation of state.

The important conclusion is that residual properties can be calculated from the equation of state. The general procedure is this: at given T and P, first solve for Z, then apply the equations for the residual enthalpy and entropy. If the cubic equation for Z has three positive roots, the proper root must be selected based on the phase of the system. The calculation is demonstrated in the examples that follow.

Table 5-3: Residual properties from cubic equations of state

van der Waals

$$H^R = RT(Z-1) - \frac{a}{V}$$

$$S^R = R\ln(Z - B')$$

Soave-Redlich-Kwong

$$H^R = RT(Z-1) + \frac{T(da/dT) - a}{b} \ln \frac{Z + B'}{Z}$$

$$S^R = R\ln(Z - B') + \frac{da/dT}{b} \ln \frac{Z + B'}{Z}$$

$$\frac{da}{dT} = -0.42748 \frac{R^2 T_c}{P_c} \frac{\left(1 + \Omega\left(1 - \sqrt{T_r}\right)\right)\Omega}{\sqrt{T_r}}$$

$$\Omega = 0.48 + 1.574\omega - 0.176\omega^2$$

Peng-Robinson

$$H^R = RT(Z-1) + \frac{T(da/dT) - a}{2\sqrt{2}\,b} \ln \frac{(1+\sqrt{2})B' + Z}{(1-\sqrt{2})B' + Z}$$

$$S^R = R\ln(Z - B') + \frac{(da/dT)}{2\sqrt{2}\,b} \ln \frac{(1+\sqrt{2})B' + Z}{(1-\sqrt{2})B' + Z}$$

$$\frac{da}{dT} = -0.45724 \frac{R^2 T_c}{P_c} \frac{\left(1 + \Omega\left(1 - \sqrt{T_r}\right)\right)\Omega}{\sqrt{T_r}}$$

$$\Omega = 0.37464 + 1.54226\omega - 0.26992\omega^2$$

Example 5.10: Residual Properties Using the SRK
Calculate the residual properties of ethylene at 250 K, 30 bar using the Soave-Redlich-Kwong equation of state.

Solution The critical constants of ethylene are

$$T_c = 282.35 \text{ K}, \quad P_c = 50.418 \text{ bar}, \quad \omega = 0.0866.$$

5.9 Application to Cubic Equations

From Perry's *Handbook* (p. 2–237 in *Perry's Chemical Engineers' Handbook*, 7th ed.) or other bibliographic source we find that the state is compressed liquid. Therefore, if the compressibility equation has three positive roots, the smallest will be selected. The calculation of the various parameters is summarized below:

$$
\begin{aligned}
a &= 0.501763 \quad \text{J m}^3/\text{mol}^2 \\
da/dT &= -0.00112077 \quad \text{J m}^3/\text{mol}^2 \text{ K} \\
b &= 0.0000403395 \quad \text{m}^3/\text{mol} \\
A &= 0.348434 \quad (-) \\
B &= 0.058224 \quad (-)
\end{aligned}
$$

The cubic equation for Z is

$$Z^3 - Z^2 + 0.28682 Z - 0.0202872 = 0.$$

and has three positive roots:

$$Z_1 = 0.105351, \quad Z_2 = 0.360538, \quad Z_3 = 0.534111.$$

We select the smallest root because the phase is liquid. With $Z = 0.105351$ we find

$$H^R = -10388.1 \text{ J/mol},$$
$$S^R = -37.6224 \text{ J/mol K}.$$

Comments Since residual properties represent corrections, their sign may be positive or negative. A negative value means that the actual property is less than what would be calculated by the ideal-gas equations.

Example 5.11: ΔH, ΔS Using the SRK
Use the Soave-Redlich-Kwong equation to calculate ΔH and ΔS for ethylene between $T_1 = 250$ K, $P_1 = 30$ bar, and saturated vapor at $T_2 = 170$ K, $P_2 = 1.0526$ bar.

Solution The calculation is done using eqs. (5.33) and (5.34):

$$\Delta H_{12} = \int_{T_1}^{T_2} C_P^{ig} dT + H_2^R - H_1^R, \tag{5.33}$$

$$\Delta S_{12} = \int_{T_1}^{T_2} C_P^{ig} \frac{dT}{T} - R \ln \frac{P_2}{P_1} + S_2^R - S_1^R. \tag{5.34}$$

The calculation requires the ideal-gas heat capacity and the residual properties at the two states. The ideal-gas heat capacity is

$$C_P^{ig}/R = 4.221 - 0.008782\, T + 0.00005795\, T^2 - 6.729 \times 10^{-8}\, T^3 + 2.511 \times 10^{-11}\, T^4,$$

with T in kelvin. The residual properties at the initial state were calculated in Example 5.10 and the calculation in the final state is done in the same manner. The results are summarized below:

	Initial State	Final State	
T	250	170	K
P	30	1.0526	bar
a	0.501763	0.604855	J m³/mol²
da/dT	-0.00112077	-0.00149223	J m³/mol² K
b	0.0000403395	0.0000403395	m³/mol
A	0.348434	0.0318711	$(-)$
B	0.058224	0.00300424	$(-)$
Z	0.105351	0.970362	$(-)$
H^R	-10388.1	-107.679	J/mol
S^R	-37.6224	-0.390262	J/mol K

Enthalpy. The ideal-gas enthalpy is calculated by evaluating the integral of C_P^{ig} between the temperatures of the two states:

$$\Delta H_{12}^{ig} = \int_{T_1}^{T_2} C_P^{ig} dT = -2906.5 \text{ J/mol}.$$

The enthalpy change between the two states is

$$\Delta H_{12} = (-2906.5) + (-107.679) - (-10388.1) = 7373.94 \text{ J/mol}.$$

Entropy. The ideal-gas entropy change is

$$\Delta S_{12}^{ig} = \int_{T_1}^{T_2} C_P^{ig} \frac{dT}{T} - R\ln\frac{P_2}{P_1} = (-13.95) - (-27.8514) = 13.9014 \text{ J/mol K}.$$

The entropy change between the two states is

$$\Delta S_{12} = (13.9014) + (-0.390262) - (-37.6224) = 51.1335 \text{ J/mol K}.$$

Comments Perry's *Handbook* reports the following values:

T (K)	P (bar)	H (kJ/kg)	S (kJ/kg K)
250	30	50.5	0.201
170	1.0526	318.04	2.0375

Using the molecular mass of ethylene, 28.054×10^{-3} kg/mol, the enthalpy and entropy difference between the two states are

$$\Delta H = (318.04 - 50.5) \text{ kJ/kg} \quad = 267.54 \text{ kJ/kg} \quad = 7505.57 \text{ J/mol}$$
$$\Delta S = (2.0375 - 0.201) \text{ kJ/kg K} \quad = 1.8365 \text{ kJ/kg K} \quad = 51.5212 \text{ J/mol K}.$$

These are very close to the values calculated from the Soave-Redlich-Kwong equation. In fact, the SRK error is about 1.8% for enthalpy, and 0.8% for entropy.

5.10 Generalized Correlations

In the previous section we discussed the calculation of residual properties from cubic equations of state. The calculations are straightforward, though somewhat time consuming. A quicker alternative is to use generalized graphs. In Chapter 2 we discussed the Pitzer method for calculating the compressibility factor in terms of reduced temperature, reduced pressure, and acentric factor. Analogous equations can be obtained for the residual enthalpy and entropy. In this approach, the residual enthalpy, made dimensionless by the product RT_c, is computed as

$$\frac{H^R}{RT_c} = h^{(0)} + \omega\, h^{(1)}, \qquad (5.58)$$

and the residual entropy, made dimensionless by R, is given by

$$\frac{S^R}{R} = s^{(0)} + \omega\, s^{(1)}. \qquad (5.59)$$

Here, $h^{(0)}, h^{(1)}, s^{(0)}, s^{(1)}$, are dimensionless functions of the reduced temperature and reduced pressure, and common for all fluids. Lee and Kesler have computed these factors and the results are shown in Figures 5-2 and 5-3. These graphs may be used for quick estimations of the residual properties. The same cautionary comments made in Section 2.4 apply here as well, namely, that these charts may be used with normal fluids but not with fluids that are highly polar or strongly associating.

Example 5.12: Residual Properties from Corresponding States
Obtain the residual enthalpy and entropy of ethylene at 250 K, 30 bar using the Lee-Kesler charts.

Solution The critical parameters of ethylene are

$$T_c = 282.35 \text{ K}, \quad P_c = 50.418 \text{ bar}, \quad \omega = 0.0866.$$

The reduced state is at

$$T_r = \frac{250 \text{ K}}{282.35 \text{ K}} = 0.885426, \quad P_r = \frac{30 \text{ bar}}{50.418 \text{ bar}} = 0.595026.$$

From the charts in Figures 5-2 and 5-3 we read (see Comment, below)

	(0)	(1)
h	−4.15072	−4.39868
s	−4.23095	−4.43083

The reduced residual enthalpy is

$$\frac{H^R}{RT_c} = (-4.15072) + (0.0866)(-4.39868) = -4.53164,$$

from which we obtain

$$H^R = (RT_c)(-4.53164) = (8.314 \text{ J/mol K})(282.35 \text{ K})(-4.53164) = -10637.8 \text{ J/mol}.$$

The residual entropy is

$$\frac{S^R}{R} = (-4.23095) + (0.0866)(-4.43083) = -4.61466,$$

and finally, the entropy is

$$S^R = -4.61466 R = (-4.61466)(8.314 \text{ J/mol K}) = -38.3663 \text{ J/mol K}.$$

Comments These values are in very good agreement with those calculated by the SRK equation in Example 5.10. This should not be surprising since generalized graphs also employ an equation of state (the modified Benedict-Webb-Rubin equation). Of course, using the generalized graphs is simpler than performing the numerical calculations in the SRK. However, the charts cannot be read with great accuracy, especially when the desired state does not lie on one of the plotted isotherms. The values of $h^{(0)}$, $h^{(1)}$, $s^{(0)}$, and $s^{(1)}$, used in the solution above, were actually computed with the equations used to produce the charts. The Lee Kesler charts should be viewed as a graphical shortcut to the calculation of residuals when high accuracy is not of great importance. Otherwise, a more detailed numerical calculation should be used.

5.11 Reference States

The equations we have developed up to this point can be used to calculate differences in enthalpy and entropy between states. To calculate absolute values we must also know the actual value of enthalpy and entropy at some state. This actual value, however, is not important if our ultimate interest is in calculating differences. Indeed, in all problems of practical interest, this will be the case.[6] And yet, it is convenient to calculate properties on absolute scale, as the steam tables demonstrate, because then differences can be calculated simply as algebraic differences between the values of a property at two states. Absolute properties are really differences from a state

---

[6]. In fact, both the first and the second law were formulated in terms of differences, rather than absolute values.

5.11 Reference States

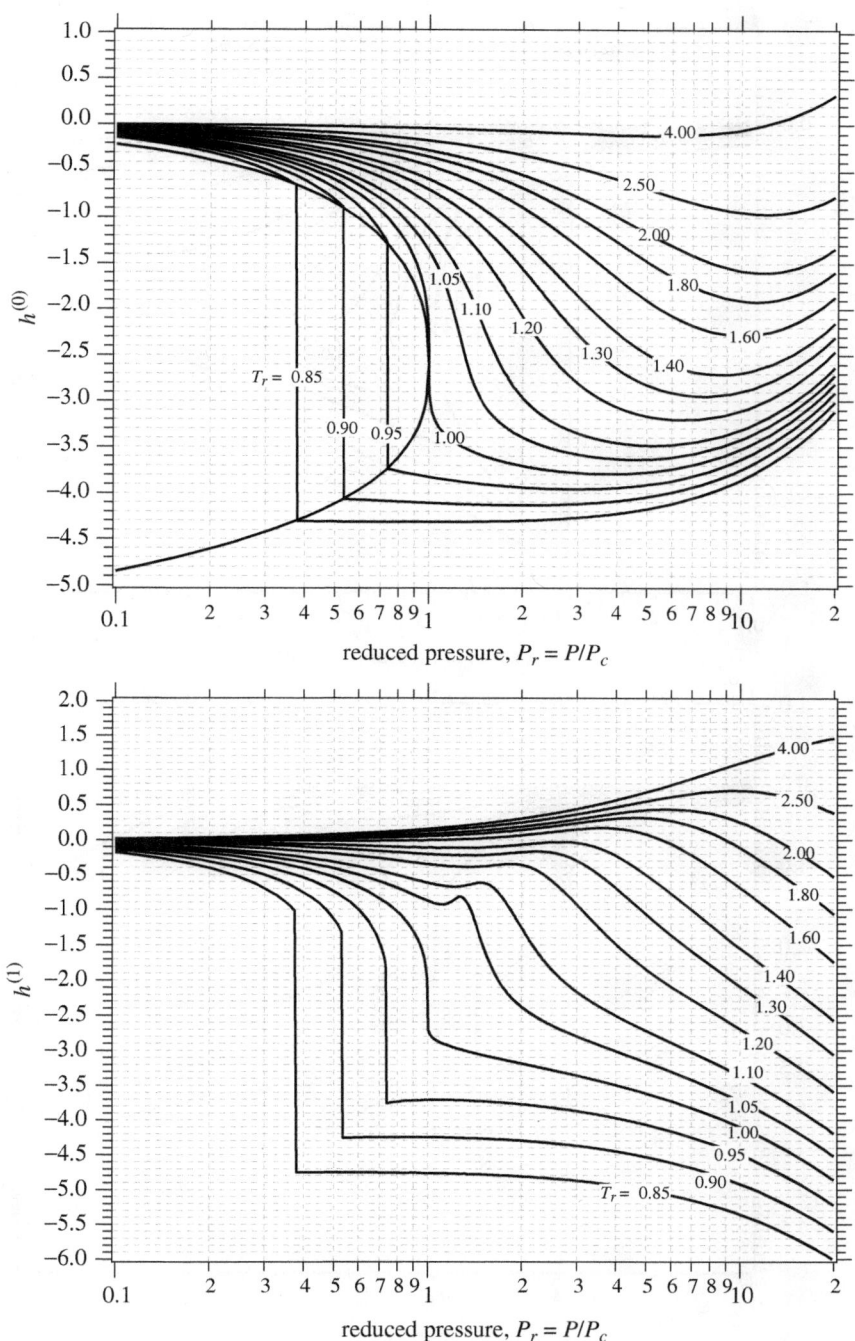

Figure 5-2: Lee-Kesler generalized graphs for the residual enthalpy.

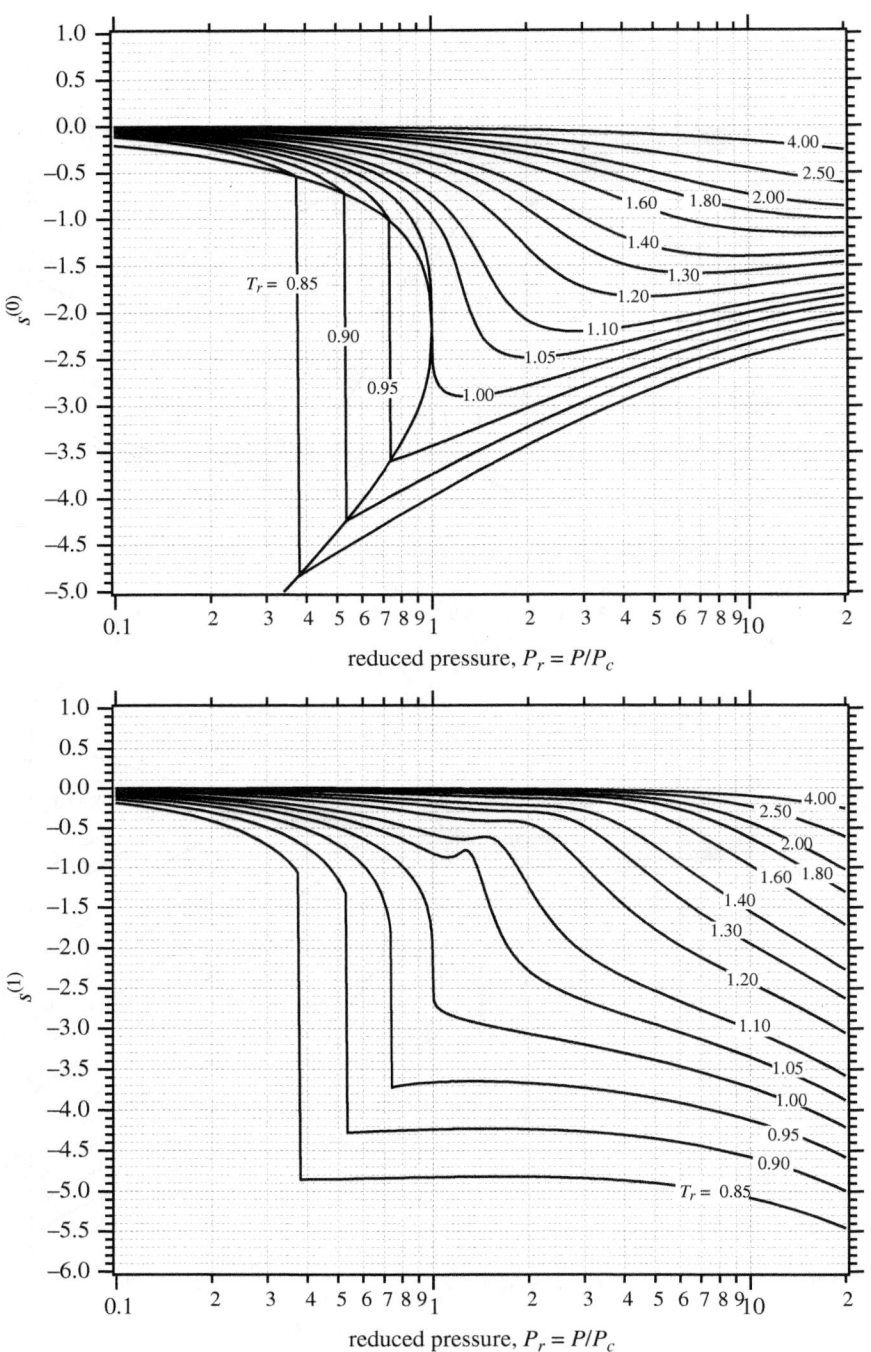

Figure 5-3: Lee-Kesler generalized graphs for the residual entropy.

5.11 Reference States

that we accept as a reference. This is common practice for many physical quantities, including potential energy, elevation, even kinetic energy.[7] To the extent that the choice of the reference does not affect differences, reference states may be chosen arbitrarily. With respect to enthalpy and entropy, the choice is usually made so as to simplify the overall calculation.

To develop an equation for the absolute properties, we start with eq. (5.31) and apply it to the enthalpy of the system at the state of interest, P, T, and at a reference state P_0, T_0, where the value H_0 of the enthalpy is presumed known:

$$H = H^{\text{ig}} + H^R \tag{5.60}$$

$$H_0 = H_0^{\text{ig}} + H_0^R, \tag{5.61}$$

where the subscript 0 is for the reference state, and unsubscripted properties are at the state of interest. We take the difference and solve for H:

$$\boxed{H = \Delta H_{\text{ref}}^{\text{ig}} + H^R - H_0^R + H_0,} \tag{5.62}$$

where $\Delta H_{\text{ref}}^{\text{ig}}$ is the ideal-gas enthalpy difference between the actual state and the reference state,

$$\Delta H_{\text{ref}}^{\text{ig}} = \int_{T_0}^{T} C_P^{\text{ig}} dT. \tag{5.63}$$

The corresponding equation for entropy is

$$\boxed{S = \Delta S_{\text{ref}}^{\text{ig}} + S^R - S_0^R + S_0,} \tag{5.64}$$

with

$$\Delta S_{\text{ref}}^{\text{ig}} = \int_{T_0}^{T} C_P^{\text{ig}} \frac{dT}{T} - R \ln \frac{P}{P_0}. \tag{5.65}$$

To complete the calculation, we must specify the numerical values of all properties at the reference point, namely, H_0, H_0^R, S_0, and S_0^R. Two common conventions are the following:

---

7. When we say that a car is traveling at a speed of 100 km per hour, we mean its speed with respect to the Earth, which itself is rotating around its own axis, moves around the sun, and travels with the solar system through the cosmos. These speeds are enormous. A car parked somewhere on the equator is actually rotating around the Earth's axis at about 1700 km per hour. Yet, it does not get a speeding ticket because with respect to the road it *is* at rest.

- *Actual enthalpy and entropy at (P_0, T_0) are set to zero.* By this convention we set $H_0 = 0$ and $S_0 = 0$.[8] The absolute enthalpy and absolute entropy are obtained from eqs. (5.62) and (5.64), which now become

$$H = \Delta H^{ig}_{ref} + H^R - H^R_0, \qquad (5.66)$$

$$S = \Delta S^{ig}_{ref} + S^R - S^R_0. \qquad (5.67)$$

This calculation requires the residual enthalpy and entropy at the refence state. These are constant and only need to be evaluated once. If the system at the reference pressure and temperature exists in multiple phases, the reference phase must be specified as well. For the steam tables, for example, the reference state is the saturated liquid at the triple point.

- *Ideal-gas enthalpy and entropy at (P_0, T_0) are set to zero.* By this convention we set the ideal-gas properties to zero: $H^{ig}_0 = 0$ and $S^{ig}_0 = 0$. From eq. (5.61) then, we obtain $H_0 = H^R_0$, and for the entropy, $S_0 = S^R_0$. With these results in eqs. (5.62) and (5.64), the absolute enthalpy and absolute entropy become

$$H = \Delta H^{ig}_{ref} + H^R, \qquad (5.68)$$

$$S = \Delta S^{ig}_{ref} + S^R. \qquad (5.69)$$

In this convention, therefore, we fix properties of the hypothetical ideal-gas state and refer to it as the *hypothetical ideal-gas reference state* at the specified pressure and temperature.

Mathematically, the only difference between the two conventions is in the additive terms H^R_0 and S^R_0. Both conventions, therefore, produce identical differences. The convention by which the actual properties at the reference state are set to zero is straightforward and makes physical sense: all enthalpies and entropies are measured relative to their values at the reference state. The hypothetical ideal-gas reference state is a bit simpler numerically, as it does not require the calculation of the residuals H^R_0 and S^R_0, but is less straightforward to explain physically because it involves a hypothetical state. In either case, the reference state should be understood *as a numerical recipe* for the calculation of the absolute enthalpy and entropy.

Example 5.13: Using Reference States
Calculate the enthalpy and entropy of ethylene at 250 K, 30 bar using as a reference state the saturated vapor at 170 K, 1.0526 bar. Repeat using the hypothetical ideal-gas state at 170 K,

---

8. The value of H_0, S_0, may be set to any constant, not necessarily zero. However, unless explicitly stated otherwise, we will take these constants to be zero.

5.11 Reference States

1.0526 bar as the reference state. For the residual properties, use the Soave-Redlich-Kwong equation.

Solution First we collect all necessary pieces for the calculation. The residual properties at the two states were calculated previously in Example 5.11:

	State of interest	Reference state	
T	250	170	K
P	30	1.0526	bar
Z	0.105351	0.970362	–
H^R	−10388.1	−107.679	J/mol
S^R	−37.6224	−0.390262	J/mol K

Using the ideal-gas heat capacity from Example 5.11, the ideal-gas enthalpy is

$$\Delta H^{\text{ig}}_{\text{ref}} = \int_{170 \text{ K}}^{250 \text{ K}} C_P^{\text{ig}} dT = 2906.5 \text{ J/mol},$$

and the ideal-gas entropy is

$$\Delta S^{\text{ig}}_{\text{ref}} = \int_{170 \text{ K}}^{250 \text{ K}} C_P^{\text{ig}} \frac{dT}{T} - R \ln \frac{P}{P_0} = -13.9014 \text{ J/mol K}.$$

Reference state is the saturated vapor at T_0, P_0. The enthalpy and entropy are calculated using eqs. (5.66) and (5.67):

$$H = (2906.5) + (-10388.1) - (-107.679) = -7373.94 \text{ J/mol},$$
$$S = (-13.9014) + (-37.6224) - (-0.390262) = -51.1335 \text{ J/mol K}.$$

Comparing these results with those in Example 5.11 we confirm that the absolute enthalpy (and entropy) at the state of interest is nothing but the enthalpy *change* from the reference state to the state of interest.

Reference state is the hypothetical ideal-gas state at T_0, P_0. This calculation is done using eqs. (5.68) and (5.69):

$$H = (2906.5) + (-10388.1) = -7481.62 \text{ J/mol},$$
$$S = (-13.9014) + (-37.6224) = -51.5238 \text{ J/mol K}.$$

The new results differ by H_0^R for enthalpy, and S_0^R for entropy.

Example 5.14: Other Absolute Properties
Calculate the internal energy of ethylene at 250 K, 30 bar using as reference state the saturated vapor at 170 K, 1.0526 bar.

Solution

Once we have the enthalpy, internal energy is calculated from its relationship to H:

$$U = H - PV = H - ZRT.$$

All that is needed is the compressibility factor at 250 K, 30 bar. This was calculated in Example 5.11 and was found to be $Z = 0.105351$. The internal energy is

$$U = -7373.94 \text{ J/mol} - (0.105351)(8.314 \text{ J/mol K})(250 \text{ K}) = -7592.91 \text{ J/mol}.$$

Comments Once enthalpy and entropy are known, all other properties can be calculated from their relationship to these two. Notice that we are allowed to fix only the enthalpy and entropy at the reference state. Internal energy, Gibbs free energy, and Helmholtz free energy may not be fixed arbitrarily, but they should be calculated through their relationship to H_0 and S_0. For example, the internal energy at the refence state is

$$U_0 = H_0 - Z_0 R T_0.$$

At the reference state, $H_0 = 0$, and $Z_0 = 0.970362$ (from the previous example). Then we find,

$$U_0 = -(0.970362)(8.314 \text{ J/mol K})(170 \text{ K}) = -1371.49 \text{ J/mol}.$$

As we see, the internal energy at the reference state is *not* zero.

5.12 Thermodynamic Charts

A useful representation of the properties of pure fluids is in the form of graphs. Thermodynamic charts, although not convenient for large-scale calculations, such as those required in the design of chemical processes, are very useful in visualizing a process and in obtaining the solution to small-scale problems graphically. A thermodynamic chart consists of two primary properties that are chosen as the axes and contains additional information in the form of phase boundaries and contours of various properties. The PV and ZP diagrams are examples of such charts that we encountered already. There is no limitation as to the properties that can be chosen to represent the axes and this freedom leads to various possible combinations. Three charts that find widespread use the pressure-enthalpy chart, the temperature-entropy chart, and the enthalpy-entropy chart, also known as Mollier chart. These are explained below.

Pressure-Enthalpy Chart In this graph, shown in Figure 5-4, the vertical axis is pressure, usually in logarithmic scale, and the horizontal axis is enthalpy. The vapor-liquid boundary sits on the horizontal axis with the liquid region on the left and

5.12 Thermodynamic Charts

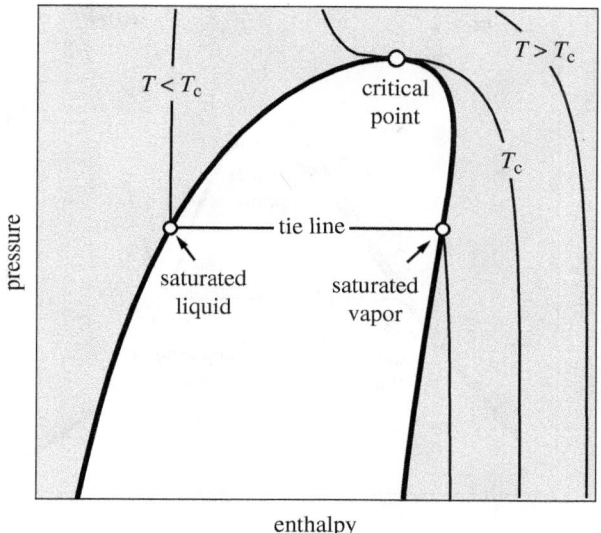

Figure 5-4: Pressure-enthalpy chart.

the vapor region on the right. The critical point is located at the very top of the vapor-liquid boundary. A horizontal line represents a constant-pressure process and a vertical line represents a constant-enthalpy (isenthalpic) process. Isotherms are lines with the general direction from the upper left to lower right corner. The shape of an isotherm depends on whether it is above or below the critical point. Starting from high pressure, a subcritical isotherm decreases sharply, intersects the phase boundary on the liquid side, moves horizontally to the saturated vapor, and resumes a sharp drop once in the superheated vapor region. The critical isotherm is tangent to the phase boundary at the critical point and bends noticeably in the region around it. Supercritical isotherms are shifted further up and do not intersect with the phase boundary. Isotherms are nearly vertical in the compressed liquid region (at high pressures to the left of the saturated liquid) and in the ideal-gas region (low pressures to the right of the saturated vapor). Both behaviors reflect the fact that enthalpy in the compressed liquid state and in the ideal-gas state is independent of pressure.

Temperature-Entropy Chart In this chart temperature is plotted on the vertical axis and entropy on the horizontal axis, as shown in Figure 5-5. The vapor-liquid boundary sits on the horizontal axis with the liquid region to the left, vapor region to the right, and the critical point at the top. Isotherms on this graph are represented by horizontal lines. Isobars are lines of constant pressure and move in the general

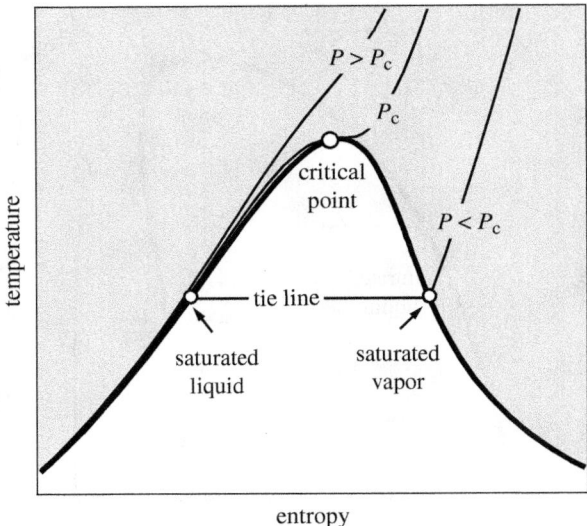

Figure 5-5: Temperature-entropy chart.

direction from the lower left corner to the upper right. Subcritical isobars intersect the saturated liquid, move horizontally until the saturated vapor, and move upwards once in the superheated vapor region. The critical isobar is tangent to the critical point, while supercritical isobars are shifted higher up and do not intersect the phase boundary. Isobars on the compressed liquid side are located very close to the saturated liquid, especially at temperatures below the critical. This is graphical manifestation of the fact that the entropy of the compressed liquid is very nearly equal to the entropy of the saturated liquid at the same temperature.

Enthalpy-Entropy Chart (Mollier Chart) In this chart the vertical axis is enthalpy and the horizontal axis is entropy (Figure 5-6). This chart is commonly known as the Mollier diagram, after Richard Mollier, the German engineer and professor who pioneered the use of this chart for steam. The phase boundary sits on the horizontal axis but has some unusual features compared to the PH and TS graphs. The liquid region is to the left of the phase boundary and the vapor region to the right, but the critical point is not at the top of the curve but rather shifted towards the left. Tie lines are straight but not horizontal and generally move in a diagonal direction. A tie line is a line of constant temperature *and* constant pressure, that is, an isotherm inside the vapor-liquid region coincides with the isobar that corresponds to the saturation pressure at that temperature. Outside the vapor-liquid region the paths of the isotherm and the isobar separate: The isotherm turns to the right and becomes nearly horizontal, while the isobar (dashed line in Figure 5-6) continues to

5.13 Summary

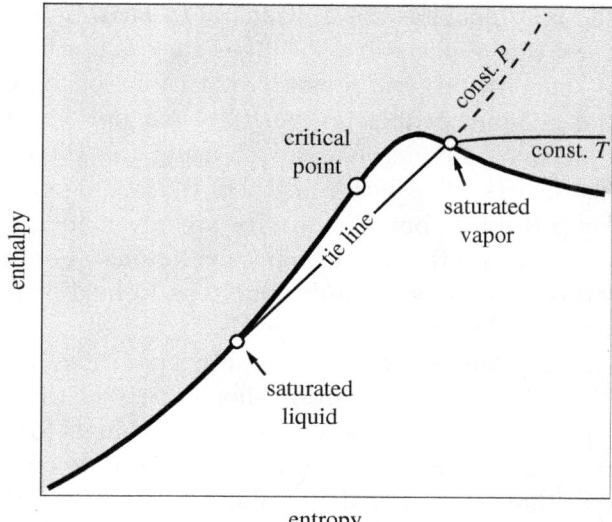

Figure 5-6: Enthalpy-entropy (Mollier) chart.

move upward. Isotherms and isobars also extend into the compressed liquid region (to the left of the saturated liquid) but these segments are not shown in Figure 5-6. The Mollier chart of steam is commonly used in calculations of steam turbines and compressors. Detailed Mollier charts found in the literature usually cover a smaller region than that shown in Figure 5-6 and focus mostly on the saturated line and the superheated vapor to the right of the critical point.

5.13 Summary

In this chapter we established various mathematical results, including relationships among various partial derivatives, and expressions for the differentials of properties with units of energy. The ultimate purpose of these derivations, and the most practical results of this chapter, are eqs. (5.33) and (5.34), which allow us to calculate enthalpy and entropy differences between *any* two states. This calculation requires two properties the ideal-gas heat capacity, and the equation of state. This is a result of great practical importance: it leads us to conclude that these two properties carry the entire thermodynamic DNA, so to speak, of any physical system.

The calculation of properties requires the residual enthalpy and entropy. These are intermediate properties and may be understood as corrections to the ideal-gas equations due to intermolecular interactions. If interactions could be turned off, the residual terms would drop off and all that is left then is the ideal-gas contribution.

In contrast with the real ideal-gas state, which is reached by reducing pressure at constant temperature, this hypothetical ideal-gas state is reached, hypothetically of course, at constant temperature *and* pressure by turning off interactions.

To calculate the residual properties we have two options: using an equation of state, such as the Soave-Redlich-Kwong, or using the Pitzer correlation along with the Lee-Kesler charts. The latter method is also based on an equation of state (Benedict-Webb-Rubin), but the results are given in graphical form, thus eliminating the need for lengthy numerical calculations. The need for graphical solutions is largely obsolete today, as desktop and hand-held computers are capable of very sophisticated calculations at high speeds.

A final comment on reference states. This subject is often a source of confusion to students, but there is no reason why it should be this way. A reference state represents a point from where we choose to measure things. We measure altitudes from sea level, but we might as well measure them from downtown State College, PA. If we were to do that, all mountains would become about 1500 ft shorter, but they would all decrease by the same amount, so when we calculate the vertical rise between any two locations on Earth using the new reference state, we would still get the same answer. Ultimately, a reference state adds a constant number to whatever we measure. Because this constant drops out when we calculate differences, it doesn't matter how we pick the reference point. We could measure altitudes from the center of the Earth and the fact that such choice sounds "unreasonable" does not mean we cannot use it.[9] If we specify a state at P_0, T_0, to be the reference, we mean that all enthalpies and entropies are reported as differences from that state. When we specify the ideal-gas state at P_0, T_0, we mean to say enthalpy and entropy are to be calculated as differences from the corresponding H and S that the substance would have at P_0, T_0, if we used ideal-gas equations to calculate them. But it is easier to simply say that the ideal-gas reference state means, "use eqs. (5.68), (5.69) to calculate enthalpy and entropy."

5.14 Problems

Problem 5.1: a) Show that the variation of the C_P with pressure at constant temperature is given by
$$\left(\frac{\partial C_P}{\partial P}\right)_T = -T\left(\frac{\partial^2 V}{\partial T^2}\right)_P$$
b) Use this result to show that the C_P and C_V in the ideal-gas state are independent of pressure.

---

9. What is so reasonable about sea level? It is not even level!

5.14 Problems

c) If the expansion coefficient β of a liquid is assumed to be approximately independent of pressure, what do you conclude for the relationship between C_P and pressure in the compressed liquid state?

Problem 5.2: A gas is described by the state equation $P(V-b) = RT$ where b is constant.
a) Obtain the residual entropy, enthalpy and internal energy as a function of pressure and temperature
b) Carbon dioxide undergoes reversible isothermal expansion in a turbine from $P_A = 20$ bar to $P_B = 1$ bar at a constant temperature of 300 K. Assuming that CO_2 is described by the above state equation with $b = 44.1 \times 10^{-6}$ m^3/mol, calculate ΔH, ΔS, the work produced and the amount of heat rejected or absorbed by the gas during the expansion.

Problem 5.3: Following a procedure analogous to that in Example 5.2, obtain an expression for the coefficient of thermal expansion based on the van der Waals equation of state. Use the result to calculate the value of β for oxygen at 20 °C, 30 bar.

Problem 5.4: Using the volume-temperature relationship for liquid acetone in Example 5.6 calculate the amount of heat and work involved when acetone is compressed isothermally and reversibly from 1 bar, 20 °C to 10 bar.

Problem 5.5: Estimate the constant-pressure heat capacity of oxygen at -50 °C, 38 bar, using the truncated virial equation. *Hint:* Use the truncated virial to calculate the enthalpy at two temperatures near -50 °C, 38 bar, then obtain the heat capacity by numerical differentiation.

Problem 5.6: Estimate the C_P of liquid hexane at 1 bar, 20 °C using the SRK equation of state. For a hint, see problem 5.5.

Problem 5.7: Perry's *Handbook* (*Perry's Chemical Engineers' Handbook*, 7th ed.) gives the ideal-gas heat capacity in the form,

$$C_P = +c_1 + c_2 \left(\frac{c_3}{T \sinh(c_3/T)} \right)^2 + c_4 \left(\frac{c_5}{T \cosh(c_5/T)} \right)^2,$$

(see Table 2-198, p. 2-178 in the 7th ed. of *Perry's Chemical Engineers' Handbook*).
a) Calculate the ideal-gas capacity of hydrogen cyanide at 300 K and report the value in J/mol K and in Btu/lbmol °F.
b) Calculate $\overline{C}_P^{ig}$ and $\overline{C}_{P\log}^{ig}$ in the temperature range 0 °C to 1000 °C (in J/mol K and in Btu/lbmol °F).

c) Calculate ΔH^{ig} and ΔS^{ig} of HCN for a change of state from 1 bar, 0 °C to 15000 mmHg, 1000 °C. Report the results for enthalpy in J/mol and in Btu/lbmol; report entropy in J/mol K and Btu/lbmol °F.
Additional data: The following identities may be helpful:

$$\int \left(\frac{a/x}{\sinh(a/x)}\right)^2 dx = a\,\coth(a/x)$$

$$\int \left(\frac{a/x}{\cosh(a/x)}\right)^2 dx = -a\,\tanh(a/x)$$

$$\int \left(\frac{a/x}{\sinh(a/x)}\right)^2 \frac{dx}{x} = \frac{a\,\coth(a/x)}{x} - \ln\left(\sinh(a/x)\right)$$

$$\int \left(\frac{a/x}{\cosh(a/x)}\right)^2 \frac{dx}{x} = \frac{-a\,\tanh(a/x)}{x} + \ln\left(\cosh(a/x)\right)$$

Problem 5.8: a) What is the residual volume of saturated liquid water at 1 bar?
b) Estimate the residual volume of ethanol vapor at 1 bar and 100 °C.
c) Use the generalized graphs to calculate the entropy change of 1 mole of ethanol undergoing isothermal compression from 1 bar to 100 bar along the critical isotherm.

Problem 5.9: n-Octane is compressed reversibly at constant temperature along the critical isotherm until the critical point is reached. The initial pressure is 1 bar. The pressure takes place in a closed system. Use the Lee-Kesler method to calculate the following:
a) What is the entropy change of n-octane?
b) Calculate the heat that must be supplied to the system to maintain isothermal conditions.
c) Calculate the necessary amount of work.

Problem 5.10: Propane is isothermally compressed from 0.01 bar, −51.4 °C to 17 bar. The process takes place reversibly in a closed system.
a) Draw the PV graph of the process.
b) Calculate the entropy change of propane.
c) How much heat is exchanged between the system and its surroundings? Is this heat added to or removed from the system?
Additional data: Use the Lee-Kesler graphs for enthalpy and entropy. The saturation pressure of propane at −51.4 °C is 0.66 bar. Take the ideal-gas C_P of propane at these conditions to be constant and equal to 67 J/mole K.

Problem 5.11: Isobutane is heated in a heat exchanger from 1 bar, 220 K to 300 K. Use the SRK equation to calculate the following:

5.14 Problems

a) The amount of heat.
b) The entropy generation if the heat is provided by a bath at 300 K.
Additional information: The ideal-gas heat capacity of isobutane is $C_P^{ig} = 96.5$ J/mol K.

Problem 5.12: Oxygen is compressed by reversible adiabatic process in a closed system, from 1 bar, 20 °C to 10 bar. Assuming oxygen to follow the SRK equation of state, calculate the amount of required work. *Hint:* Tabulate the entropy of oxygen at 10 bar at various temperatures and locate the temperature where the entropy is equal to that at 1 bar, 20 °C.

Problem 5.13: Using as reference state the hypothetical ideal-gas state at 70 K, 7.83 bar, and with the data given below, calculate the following.
a) The enthalpy of nitrogen at 70 K, 7.83 bar.
b) The enthalpy of saturated liquid nitrogen at 7.83 bar.
c) The enthalpy of vaporization of nitrogen at 7.83 bar.
Additional data:

Data for nitrogen

$P = 7.83$ bar	$T = 70$ K	$H^R = -5938$ J/mol
$P = 7.83$ bar	$T^{sat} = 100$ K	$H_L^R = -4974$ J/mol
		$H_V^R = -401$ J/mol
	$C_P^{ig} = 29$ J/mol K	

Problem 5.14: Use the SRK equation to calculate the following properties of isobutane:
a) Residual enthalpy and entropy at 10 bar, 300 K.
b) Enthalpy and entropy at 10 bar, 300 K using as reference state the saturated liquid at 1 bar ($T^{sat} = 266$ K).
c) Enthalpy and entropy at 10 bar, 300 K using as reference state the hypothetical ideal-gas state at 1 bar 266 K.
d) Enthalpy and entropy of vaporization at 1 bar.
The ideal-gas constant-pressure heat capacity may be taken to be constant and equal to 96.5 J/mol K.

Problem 5.15: Calculate the enthalpy and entropy of isobutane at 1 bar, 300 K, and 1 bar, 250 K using as reference state:
a) The actual state at $P = 20$ bar, $T = 400$ K.
b) The ideal-gas state at $P = 20$ bar, $T = 400$ K.

In each case calculate the amount of heat for constant pressure cooling of isobutane from 1 bar, 300 K to 250 K, and the entropy generation if cooling takes place inside a bath at $T_{\text{bath}} = 240$ K. The following data are available:

T (K)	P (bar)	H^R (J/mol)	S^R (J/mol K)
400	20	−2562.44	−4.63942
300	1	−163.626	−0.352764
250	1	−23224.5	−87.0989

$$C_P^{\text{ig}} = 96.5 \text{ J/mol K}$$

Problem 5.16: Use the Lee-Kesler method to do the following:
a) Calculate the entropy of propane at its critical point. The reference state is the ideal-gas state at the critical point.
b) Calculate ΔS of propane for an isothermal process that takes the substance from its critical point to pressure 1 Pa.
c) A member of our engineering team objects that the reference state is not valid because a substance is not ideal at the critical point. What is your response?

Problem 5.17: The steam tables are calculated with reference state the saturated liquid at the triple point ($P_{\text{triple}} = 0.006117$ bar, $T_{\text{triple}} = 0.01$ °C). Suppose we want to retabulate the properties of steam using a different reference state. Show how this can be done by calculating V, U, H, and S at 1 bar, 200 °C using the following reference states:
a) the saturated vapor at 10 bar;
b) the saturated vapor in the hypothetical ideal-gas state at 10 bar.
Additional data: The ideal-gas heat capacity of water is given by the following equation with T in kelvin.

$$C_P^{\text{ig}}/R = 4.395 - 0.004186\,T + 0.00001405\,T^2 - 1.564 \times 10^{-8}\,T^3 + 6.32 \times 10^{-12}\,T^4.$$

Chapter 6

Balances in Open Systems

In Chapters 3 (Section 3.4) and 4 we applied the first and second law to closed systems. Most chemical processes involve open systems that exchange mass with the surroundings, usually in the form of flow streams. These streams carry with them not only mass, but also energy and entropy. The application of the first and second law to open systems must, therefore, account for energy and entropy exchanges that take place through flow streams. A flow process may operate under steady-state or unsteady conditions. Under steady state, all properties at any given point are constant and do not vary with time (though they may vary from one point of the process to another). Steady state is the preferred mode of operation because it leads to continuous production and is simpler to maintain control. Nonetheless, unsteady-state conditions are also encountered, either during start-up and shut-down of a continuous process, or in batch processes that are designed to operate in discrete cycles. In this chapter we apply the first and second law to open systems that involve flow streams. The most general form of these balances is the unsteady-state form, from which steady state is obtained as a special condition. We then apply the energy and entropy balance to various common units encountered in chemical processes, as well as to larger processes involving multiple units. The learning objectives for this chapter are to:

1. Develop the balance equation for energy and entropy in open systems.

2. Obtain the steady-state equation as a special case of the general unsteady state balance.

3. Perform thermodynamic analysis on individual process units such as heat exchanger, turbines, pumps, compressors, and throttling valves.

4. Apply the first and second law to solve the material and energy balances in flow sheets that involve multiple units, and to analyze the operation of the process in terms of irreversible features and overall efficiency of energy utilization.

5. Perform thermodynamic analysis and preliminary design of power cycles and refrigeration units.

6. Learn to use various thermodynamic charts such as the entropy-temperature, enthalpy-entropy and enthalpy-pressure charts to perform graphical calculations of various units and processes.

6.1 Flow Streams

An open system exchanges mass with its surroundings. Most commonly this exchange takes place through flow streams, usually pipes, that carry material into and out of the system. Mass exchange does not have to occur through pipes. For example, the free surface of a liquid (or of an ocean) may exchange mass with the gas above it through evaporation, condensation, or absorption. However, in their majority, the situations we will encounter involve flow through pipes. Figure 6-1 gives a partial list of some common flow units encountered in chemical processes, which will be discussed in more detail in this chapter.

The mass flow rate in a stream will be denoted as $\dot{m}$ (kg/s), with the dot indicating "rate per unit time." Given the mass flow rate, the mass (in kg) that flows in time dt

$$dm = \dot{m}\ dt. \tag{6.1}$$

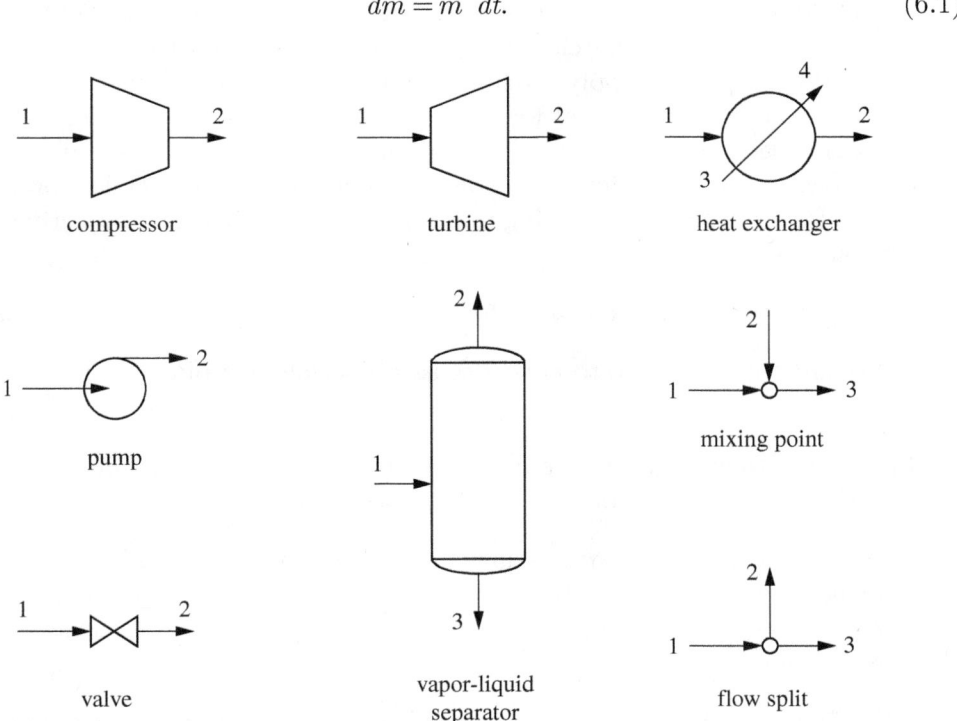

Figure 6-1: Schematic representation of some common flow equipment.

6.2 Mass Balance

When mass flows through a pipe, the mass flow rate can be expressed in terms of the mean velocity[1] of the fluid in the pipe:

$$\dot{m} = dva = \frac{va}{V}, \tag{6.2}$$

where v is the average velocity (m/s), a is the cross-sectional area of the pipe (m^2), d is the mass density (kg/m^3), and V is the specific volume (m^3/kg). Suppose a stream with mass flow rate $\dot{m}$ exchanges the amount of heat, Q, or shaft work, W_s, per kg of fluid (i.e., the units of Q and W_s are J/kg). The *rate* of the energy exchange is

$$\dot{Q} = \dot{m}\ Q, \tag{6.3}$$

$$\dot{W}_s = \dot{m}\ W_s, \tag{6.4}$$

with units of J/s, or W (watt). The amount of heat or work that is transferred within time dt is,

$$\dot{Q}\ dt = Q\ dm, \tag{6.5}$$

$$\dot{W}_s\ dt = W_s\ dm. \tag{6.6}$$

In the general case, the exchange rates $\dot{m}$, $\dot{Q}$, and $\dot{W}_s$ will vary with time. At steady state, however, all rates constant and do not vary with time.

6.2 Mass Balance

Mass is a conserved quantity,[2] therefore, all the mass that crosses the boundaries of a system must be accounted for. All conserved quantities satisfy the general balance equation,

$$(\text{in}) - (\text{out}) = (\text{accumulation in the system}). \tag{6.7}$$

The term *accumulation* is to be understood to represent either built up, if positive, or depletion, if negative. This balance equation applies at all times, including a small time interval dt. Consider the process shown in Figure 6-2. The system is defined by the dashed line and includes several interconnected units. For the purposes of computing the overall balance in the system, the internal details of the process are unimportant. What is important to know are the exchanges that cross the

---

1. The flow inside a pipe is not uniform because the fluid at the center of the pipe moves faster than the fluid near the pipe walls. The velocity profile in pipes is a subject of fluid mechanics. Here, we avoid such complicating factors by considering the average velocity of the fluid over the entire cross section of the pipe.
2. We are excluding the possibility of mass-to-energy conversions by nuclear reaction.

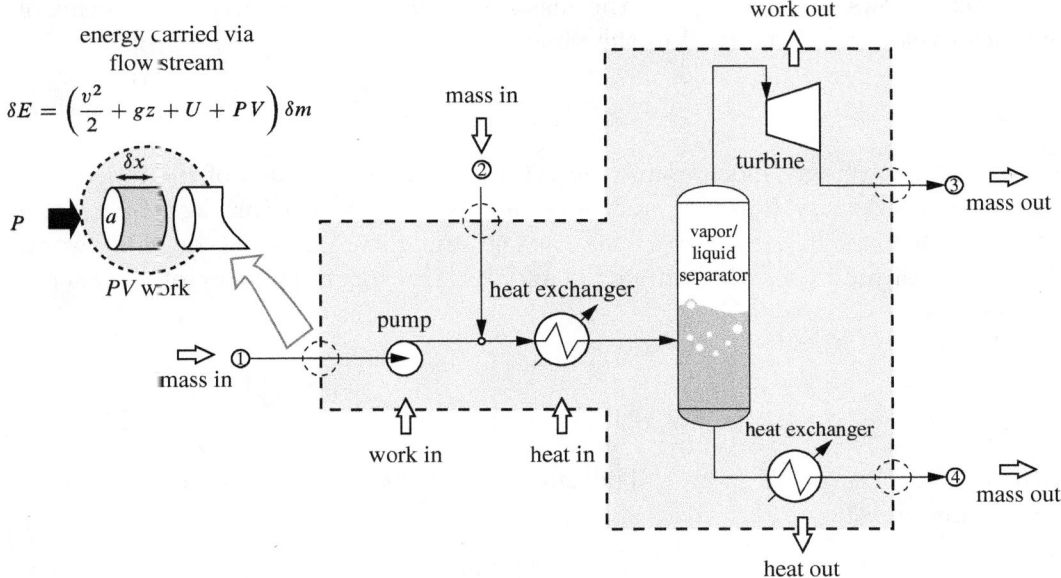

Figure 6-2: Setup for the energy balance in open system. The system is defined by the dashed line around it

system boundaries. In the case of mass, such exchange takes place only through flow streams: mass enters through streams 1 and 2, and exits through streams 3 and 4. Applying eq. (6.7) to the system over a small time interval dt, we have

$$dm_1 + dm_2 - dm_3 - dm_4 = dM^{\text{tot}}$$

where dM^{tot} refers to the change in the total mass contained within the system. We divide by dt and note that the terms on the left-hand side give the flow rates of the corresponding streams:

$$\dot{m}_1 + \dot{m}_2 - \dot{m}_3 - \dot{m}_4 = \frac{dM^{\text{tot}}}{dt}.$$

We rewrite this result in the general and more compact form,

$$\boxed{\frac{dM^{\text{tot}}}{dt} + \Delta \dot{m} = 0} \qquad (6.8)$$

where $\Delta \dot{m}$ is shorthand for

$$\Delta \dot{m} = \sum_{(\text{out})} \dot{m}_i - \sum_{(\text{in})} \dot{m}_i \qquad (6.9)$$

and the summations are understood to go over all inlet and outlet flow streams. Equation (6.8) expresses the unsteady-state mass balance in terms of rates, and states that the net rate at which mass enters the system is equal to the rate at which mass accumulates inside the system. At steady state, the accumulation term is by definition zero. The steady-state form of the material balance then is,

$$\Delta \dot{m} = 0 \tag{6.10}$$

or

$$\sum_{\text{(out)}} \dot{m}_i = \sum_{\text{(in)}} \dot{m}_i \tag{6.11}$$

This result indicates that at steady state the amount of mass that enters the system is exactly equal to the amount that exits.

NOTE

The Operator $\Delta(\cdots)$
In Chapter 3 we introduced the operator Δ to mean "final minus initial." Here, we encounter a more general interpretation of this operator, to mean "the sum of all that goes out (via flow streams) minus the sum of all that comes in." If we imagine the observer to move with the fluid, inlets represent the state "before" and outlets the state "after." This interpretation of the operator applies to all extensive properties, including energy and entropy.

6.3 Energy Balance in Open System

As a conserved quantity, energy obeys the general balance equation in eq. (6.7). To construct this equation we must first identify all the exchanges of energy between system and surroundings. We begin by noting that a flow system may exchange heat and shaft work with the surroundings. Heat is exchanged via heat exchangers; shaft work can be exchanged in a number of units, for example, pumps, compressors, turbines, and the like. These units will be considered in more detail later. Here, and for the purposes of constructing the overall energy balance of the process, we will take the total (net) amount of heat that is exchanged to be $\dot{Q}$, and the total (net) amount of shaft work to be $\dot{W}_s$. Exchanges are defined with the usual sign convention: they are positive if they enter the system, negative if they exit. Based on this convention, the heat and work must be counted as part of the energy that goes "in." Over a time interval dt, these amounts are

$$\text{heat and shaft work in}: \dot{Q}\,dt + \dot{W}_s\,dt \tag{6.12}$$

Since mass carries with it energy in various forms, energy is also exchanged via flow streams. In general, mass carries kinetic, potential and internal energy. Kinetic energy is due to the velocity of the fluid; potential energy due to gravity enters through the elevation of individual streams (not all pipes need to be at the same height); internal energy is of course an inseparable part of matter. The combined kinetic, potential, and internal energy that is carried by a mass dm, over a time interval dt, is

$$\frac{v^2}{2}\,dm + gz\,dm + U\,dm \tag{6.13}$$

where v is the mean velocity in the pipe z is the elevation with respect to a reference plane, and U is the specific internal energy of the fluid at the pressure and temperature of the particular location. There is an additional energy contribution that is not immediately obvious: to introduce this element of mass into the system, we must push from the outside against the pressure P of the fluid at this point and this requires work. This requires an amount of work that must be included in the balance. To calculate this work, we assume for simplicity that the pipe has a circular cross section with area a. The mass dm then has the shape of a cylindrical plug with length dx. The force that opposes this volume element is equal to the product of the pressure times the cross section of the element, Pa, and this force must be displaced by a distance dx. The required amount of work is

$$dW' = (Pa)\,dx = P(a\,dx).$$

The product $a\,dx$ is the volume of the plug and this may be expressed in terms of the specific volume, V:

$$a\,dx = V\,dm.$$

Combining these results, the amount of work needed to push the mass element, dm, into the system is

$$dW' = PV\,dm.$$

We recognize this as a form of PV work and note that it depends on the state (pressure, specific volume) at the point of entry. For inlet streams, this represents work that is done by the surroundings; for outlet streams, it is work that is done by the system. To summarize these calculations, the energy that is carried by a mass dm via flow steams is

$$\begin{aligned}
\text{(energy via flow stream)} &= \frac{v^2}{2}\,dm + gz\,dm + U\,dm + PV\,dm \\
&= \left(\frac{v^2}{2} + gz + U + PV\right)dm \\
&= \left(\frac{v^2}{2} + gz + H\right)dm. \tag{6.14}
\end{aligned}$$

6.3 Energy Balance in Open System

In the last equation we used the definition of enthalpy to combine U and PV into a single term. The net amount of energy that goes into the system is obtained by combining eqs. (6.12) and (6.14):

$$\text{(in)} - \text{(out)} = \sum_{\text{(in)}} \left\{ \left(\frac{v^2}{2} + gz + H \right) dm \right\} - \sum_{\text{(out)}} \left\{ \left(\frac{v^2}{2} + gz + H \right) dm \right\}$$
$$+ \dot{Q}\, dt + \dot{W}_s\, dt$$
$$= -\Delta \left\{ \left(\frac{v^2}{2} + gz + H \right) dm \right\} + \dot{Q}\, dt + \dot{W}_s\, dt. \qquad (6.15)$$

This net energy must be accounted for by the accumulation of energy inside the system. The accumulation is represented by the change of the total amount of internal energy in the system:[3]

$$\text{(accumulation)} = dU^{\text{tot}}. \qquad (6.16)$$

We equate the results in eqs. (6.15) and (6.16), divide both sides by dt, replace the ratio dm/dt by flow rate of the stream, $\dot{m}$, and write the final result in the form:

$$\boxed{\frac{dU^{\text{tot}}}{dt} + \Delta \left\{ \left(\frac{v^2}{2} + gz + H \right) \dot{m} \right\} = \dot{Q} + \dot{W}_s.} \qquad (6.17)$$

This equation gives the general unsteady-state energy balance equation in an open system. It is often referred to as the form of the first law for open systems. Below we consider a number of special cases.

Special Case: Closed System The closed system may be considered as a special case of an open system in which all flow streams have zero flow. This means that if the mass flows are turned off, eq. (6.17) must revert to the familiar expression for a closed system. Let us confirm this result. With all mass flow rates set to zero, eq. (6.17) becomes

$$\frac{dU^{\text{tot}}}{dt} = \dot{Q} + \dot{W}_s.$$

Multiplying both sides by dt, this becomes,

$$dU^{\text{tot}} = dQ^{\text{tot}} + dW_s^{\text{tot}}.$$

---

3. In principle, a system may also accumulate kinetic, potential, or any other type of energy. Accumulation of kinetic or potential energy means that the velocity or the vertical elevation of the system changes, as in a roller coaster, or in a rocket. These modes of energy accumulation are not relevant to *stationary* processes such as those considered here. This leaves internal energy as the only accumulation term.

This is precisely the form of the first law for a closed system of constant volume that exchanges heat and shaft work with the surroundings (see eq. [3.10]). The reason that the PV work does not appear here is that the form of the energy balance in eq. (6.17) implicitly assumes that the system has constant volume, i.e., its boundaries are rigid and thus the system is prevented from exchanging any PV work. For an open system with movable boundaries (a rubber balloon, for example), eq. (6.17) must be amended to include PV work in addition to any shaft work.

Special Case: Steady State For a process at steady state, the accumulation term is zero and the energy balance simplifies to

$$\Delta\left\{\left(\frac{v^2}{2} + gz + H\right)\dot{m}\right\} = \dot{Q} + \dot{W}_s. \tag{6.18}$$

It will often be the case that the contribution of kinetic and potential energy terms is much smaller than the contribution of the other terms. Whenever this is true, the steady-state energy balance is further reduced to

$$\Delta\left(\dot{m}\,H\right) = \dot{Q} + \dot{W}_s, \tag{6.19}$$

where the operator $\Delta\left(H\dot{m}\right)$ is understood to mean the sum of the quantity $H\dot{m}$ over all outlet streams minus the sum of the same quantity over all inlet streams. If the process has a single inlet and a single outlet stream, their mass flow rates at steady state must be equal (see eq. [6.10]). Equation (6.19) then becomes

$$\dot{m}\left(H_2 - H_1\right) = \dot{Q} + \dot{W}_s, \tag{6.20}$$

where $\dot{m}$ is the common value of the flow rate. Dividing both sides by $\dot{m}$ and using eqs. (6.3) and (6.4), this result becomes

$$\Delta H = Q + W_s, \tag{6.21}$$

where $\Delta H = H_2 - H_1$. This simplified form of the energy balance applies to steady-state processes with two streams (one inlet, one outlet), provided that kinetic and potential energy contributions are negligible.

6.4 Entropy Balance

Entropy is *not* a conserved quantity. The entropy of the universe increases as a result of irreversible processing, and the balance equation must be amended to include a generation term. We must also keep in mind that such equation must be applied, not to the system alone, but to the universe. Accordingly, the entropy balance must

6.4 Entropy Balance

be written for the system *and* the surroundings. With these considerations in mind we return to the system shown in Figure 6-2. The starting equation is the second law, according to which entropy generation is the sum of the entropy change of the system and the surroundings:

$$dS_{\text{gen}}^{\text{tot}} = dS_{\text{sys}}^{\text{tot}} + dS_{\text{sur}}^{\text{tot}}. \tag{6.22}$$

We now consider all changes in entropy over a short time dt. For the entropy change of the system we simply write,

$$dS_{\text{sys}}^{\text{tot}} = dS^{\text{tot}}, \tag{6.23}$$

where dS^{tot} (we will drop the subscript "sys") is the accumulation of entropy within the system. The entropy change of the surroundings has two contributions. One is through flow streams: streams 1 and 2 exit the surroundings taking entropy with them; streams 3 and 4 enter the surroundings, adding to their entropy. The entropy change due to these flows is

$$-S_1(\dot{m}_1\, dt) - S_2(\dot{m}_2\, dt) + S_3(\dot{m}_3\, dt) + S_4(\dot{m}_4\, dt) = -\sum_{(\text{in})}(\dot{m}\, S\, dt) + \sum_{(\text{out})}(\dot{m}\, S\, dt),$$

where S_i is the specific entropy of stream i and $\dot{m}_i\, dt$ is the corresponding amount of mass that flows through that stream in time dt. The second contribution to the entropy change of the surroundings comes from interactions with baths. Suppose for simplicity that all the heat that is exchanged between the system and the surroundings comes from a single bath at temperature T_{bath}. The amount of heat exchanged in time dt is $\dot{Q}\, dt$ (with the sign of $\dot{Q}$ based on the system), and the entropy change is

$$-\frac{\dot{Q}\, dt}{T_{\text{bath}}}.$$

The entropy change of the surroundings is the sum of the above contributions:

$$dS_{\text{sur}} = -\sum_{(\text{in})}(\dot{m}\, S\, dt) + \sum_{(\text{out})}(\dot{m}\, S\, dt) - \frac{\dot{Q}}{T_{\text{bath}}}\, dt,$$

which we write in the more compact form as

$$dS_{\text{sur}} = \left[\Delta(\dot{m}\, S) - \frac{\dot{Q}}{T_{\text{bath}}}\right] dt. \tag{6.24}$$

The entropy generation is obtained by adding the contributions of the system and the surroundings:

$$dS_{\text{gen}} = dS^{\text{tot}} + \left[\Delta(\dot{m}\, S) - \frac{\dot{Q}}{T_{\text{bath}}}\right] dt. \tag{6.25}$$

Dividing both sides by dt, this becomes

$$\boxed{\frac{dS^{\text{tot}}}{dt} + \Delta(\dot{m}\,S) = \frac{\dot{Q}}{T_{\text{bath}}} + \dot{S}_{\text{gen}}.}\qquad(6.26)$$

where $\dot{S}_{\text{gen}}$ stands for the rate of entropy generation, dS_{gen}/dt. Equation (6.26) is the general unsteady-state entropy balance equation for a flow process that exchanges heat with a bath. If the system exchanges heat with several baths, then an additional term of the form $\dot{Q}_i/T_i$ must be included for each bath. The general form of this equation is similar to the energy balance, eq. (6.17), with the notable difference[4] that the entropy balance contains a generation term. This term represents production of entropy and obeys the inequality of the second law,

$$\dot{S}_{\text{gen}} \geq 0. \qquad(6.27)$$

In the special case of reversible process the generation term is zero and the above relationship becomes an exact equality. Two special forms of the entropy balance are examined below

Steady-State Process At steady state the accumulation term dS^{tot}/dt is zero and the entropy equation can be solved for the generation term:

$$\dot{S}_{\text{gen}} = \Delta(\dot{m}\,S) - \frac{\dot{Q}}{T_{\text{bath}}}. \qquad(6.28)$$

The first term on the right-hand side is the entropy change of the fluid during a pass trough the process; the second term is the entropy change of the bath during the same time. At steady state, the sum of the two constitutes the entropy change of the universe.

Reversible Adiabatic Process For a reversible adiabatic process, $\dot{Q} = 0$ and $\dot{S}_{\text{gen}} = 0$. The entropy balance then reduces to

$$\frac{dS^{\text{tot}}}{dt} + \Delta(\dot{m}\,S) = 0. \qquad(6.29)$$

In this case, the entropy of the system is a conserved quantity (compare this equation with eq. [6.8] for the mass) since a reversible adiabatic process is also isentropic. In the special case of steady state, this result further simplifies to

$$\Delta(\dot{m}\,S) = 0, \qquad(6.30)$$

which states that the total entropy of the inlet streams is equal to the entropy of the outlet streams.

---

4. Also notice that shaft work does not appear explicitly in this equation. It is taken into account indirectly, through its effect on heat and the entropy of the streams.

6.4 Entropy Balance

In typical calculations involving irreversible processes, the unknown quantity is the entropy generation. This may be obtained from the entropy balance once the material and energy balances have been performed and the states of all streams are known. In the special case of *reversible* process, the rate of entropy generation is known ($\dot{S}_{\text{gen}} = 0$) and the entropy balance may be used as an additional equation in the calculation of the unknown quantities of the process.

NOTE

Steady State versus Equilibrium

It is important to distinguish between steady state and equilibrium. In steady state, all accumulation terms (mass, energy, entropy) are equal to zero:

Steady State: $\quad \dfrac{dM^{\text{tot}}}{dt} = 0, \quad \dfrac{dU^{\text{tot}}}{dt} = 0, \quad \dfrac{dS^{\text{tot}}}{dt} = 0.$

In strict equilibrium, the entropy generation and all exchanges between system and surroundings are zero:

Equilibrium: $\quad \dot{S}_{\text{gen}} = 0, \quad \dot{m} = 0, \quad \dot{Q} = 0, \quad \dot{W}_s = 0.$

The steady-state condition requires the rate of all exchanges ($\dot{m}$, $\dot{Q}$, etc.) to be constant. Accordingly, the rate of entropy production at steady state, $\dot{S}_{\text{gen}}$, is also constant but not necessarily zero, unless the process is reversible. A system in strict *equilibrium* produces no entropy and undergoes no changes in time with respect to any of its properties. For such system, the entropy generation, all accumulation terms, and the rate of all exchanges ($\dot{m}$, $\dot{Q}$, etc) are zero. A reversible process produces no entropy ($\dot{S}_{\text{gen}} = 0$) and may operate either at steady state (zero accumulation) or at unsteady state (nonzero accumulation):

Reversible Process: $\quad \dot{S}_{\text{gen}} = 0.$

By *strict equilibrium* we mean a state of permanent equilibrium in which there is no exchange of mass or energy between system and surroundings. This is to be contrasted with the *reversible (quasi-static) process*, which is allowed to exchange mass and energy with its surroundings, albeit under slow, quasi-equilibrium conditions, such that the entropy generation is zero.

Example 6.1: Adiabatic Production of Work

Steam produces 2 kW of mechanical work in an adiabatic steady-state flow process. The steam enters at 30 bar, 500 °C and exits at 7.5 bar, 300 °C. Determine the entropy generation.

Solution The entropy balance of adiabatic steady-state process is given by eq. (6.28) with $\dot{Q} = 0$. The mass flow rate of the inlet and outlet streams are equal (by virtue of the steady-state condition on the mass balance equation); therefore, the entropy balance becomes

$$\dot{S}_{\text{gen}} = \dot{m}\,(S_2 - S_1).$$

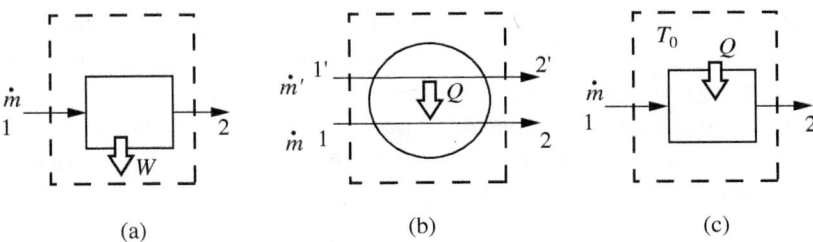

Figure 6-3: Processes with entropy generation (see examples in text): (a) work production in open system; (b) heat exchange between streams; (c); heat exchange between a stream and a heat bath whose temperature is T_0.

The mass flow rate will be computed from the energy balance, which for steady-state conditions reads,

$$\dot{m}(H_2 - H_1) = \dot{W}_s \quad \Rightarrow \quad \dot{m} = \frac{\dot{W}_s}{H_2 - H_1}.$$

Numerical substitutions. From steam tables we collect the following information:

	State 1	State 2
T (°C)	500	300
P (bar)	30	7.5
H (kJ/kg)	3457.0	3058.2
S (kJ/kgK)	7.2356	7.266

The mass flow rate is (recall that work produced by the system is negative)

$$\dot{m} = \frac{-2 \text{ kW}}{(3058.2 - 3457.0) \text{ kJ/kg}} = 5.0 \times 10^{-3} \text{ kg/s}.$$

The entropy generation is

$$\dot{S}_{\text{gen}} = (5.0 \times 10^{-3} \text{ kg/s})(7.266 - 7.2356) \text{ (kJ/kg K)} = 0.000152 \text{ kW/K} = \underline{0.152 \text{ W/K}}$$

Example 6.2: Heat Exchange Between Streams
Two streams of steam are brought into thermal contact as shown in Figure 6-3b. The cooler stream enters at 150 °C and exits at 300 °C and has a flow rate of 1 kg/s. The hot stream enters at 500 °C and exits at 400 °C. All streams are at constant pressure of 1 bar. Determine the unknown flow rate, the amount of heat exchanged between the streams, the rate of entropy generation, and the lost work.

6.4 Entropy Balance

Solution We collect the following data from the steam tables:

	1	2	1'	2'
T (°C)	150	300	500	400
P (bar)	1	1	1	1
H (kJ/kg)	2776.6	3074.5	3488.7	3278.5
S (kJ/kgK)	7.6147	8.2171	8.8361	8.5451

The system defined by the dashed line in Figure 6-3b is open and adiabatic. The energy balance gives

$$\dot{m}(H_2 - H_1) + \dot{m}'(H_{2'} - H_{1'}) = 0,$$

which we solve for the unknown flow rate:

$$\dot{m}' = -\frac{\dot{m}(H_2 - H_1)}{H_{2'} - H_{1'}}.$$

By numerical substitution we find

$$\dot{m}' = \frac{-(1 \text{ kg/s})(3074.5 - 2776.6)(\text{kJ/kg})}{(3278.5 - 3488.7)(\text{kJ/kg})} = 1.417 \text{ kg/s}$$

The amount of heat exchanged between the two streams is

$$|\dot{Q}| = \dot{m}(H_2 - H_1) = -\dot{m}'(H_{2'} - H_{1'}) = 297.9 \text{ kW}.$$

This heat is removed from stream $1' - 2'$ and is added to stream $1 - 2$.

Entropy generation. This is a steady-state adiabatic process and the entropy balance is given by eq. (6.28) with $\dot{Q} = 0$:

$$\dot{S}_{\text{gen}} = \dot{m}(S_2 - S_1) + \dot{m}'(S_{2'} - S_{1'}).$$

By numerical substitution,

$$\dot{S}_{\text{gen}} = (1.0 \text{ kg/s})(8.2171 - 7.6147) \text{ (kJ/kg K)} + (1.417 \text{ kg/s})(8.5451 - 8.8361) \text{ (kJ/kg K)}$$
$$= \underline{0.190 \text{ kW/K}}.$$

The lost work is

$$\dot{W}_{\text{lost}} = T_0 \dot{S}_{\text{gen}} = (300 \text{ K})(0.190 \text{ kW/K}) = \underline{57 \text{ kW}}.$$

Comments The amount of heat that is exchanged between the two streams does *not* enter in the calculation of the entropy generation because it is *internal* to the system. The contribution of this transfer is taken into account indirectly, through its effects on the affected streams. The heat must be accounted in the entropy balance only when it involves a bath, because in this case the bath is a closed system and its entropy change cannot be calculated as an entropy change of flow streams. The next two examples explore variations of this problem.

Example 6.3: Stream Acting as a Bath
In the process of Example 6.2, the flow rate of streams 1' and 2' is made very high, while the state of streams 1, 2, and 1' are the same as in the previous example. Repeat the calculation of entropy generation for these new conditions.

Solution Since stream 1 undergoes the same change of state as in the previous example, the amount of heat that is exchanged between the two streams is the same. The temperature of stream 2', however, is now different because a larger flow rate provides the same amount of heat. The general procedure for the calculation of this unknown temperature is based on the first law. In the previous example we obtained the following equation:

$$|\dot{Q}| = -\dot{m}'(H_{2'} - H_{1'}).$$

Solving this for the enthalpy of the exit stream we find

$$H_{2'} = H_{1'} - \frac{|\dot{Q}|}{\dot{m}'},$$

and the temperature of stream 2' is obtained from the known enthalpy and pressure of the stream. However, when the flow rate of stream 1'2' is very large ($\dot{m}' \to \infty$), we have, $H_{2'} \to H_{1'}$, and also $T_{2'} \to T_1$. That is, the flow rate is so high that stream 2' exits practically at the same temperature as the inlet stream. It further follows that the entropy of the exit stream is $S_{2'} \to S_{1'}$. This poses a numerical difficulty because the quantity $\dot{m}(S_{2'} - S_{1'})$, needed in the calculation of entropy generation, is indeterminate:

$$\dot{m}'(S_{2'} - S_{1'}) \to (\infty) \times (0).$$

We overcome this problem by treating the large flow rate as heat bath. This is acceptable because this stream exchanges heat at practically constant temperature:

$$\dot{m}'(S_{2'} - S_{1'}) \to -\frac{|\dot{Q}|}{T_{1'}},$$

so that the entropy generation is

$$\dot{S}_{\text{gen}} = \dot{m}(S_2 - S_1) - \frac{|\dot{Q}|}{T_{1'}}.$$

To avoid confusion with the sign of heat, we use absolute values. Since the hot stream loses heat, its entropy must decrease. By numerical substitution:

$$\dot{S}_{\text{gen}} = (1.0 \text{ kg/s})(8.2171 - 7.6147) \text{ (kJ/kg K)} - \frac{297.9 \text{ kJ/s}}{773.15 \text{ K}} = \underline{0.217 \text{ kW/K}}.$$

Comments The entropy generation is higher compared to the previous example, implying that there are more irreversible features when the flow rate is increased. How so? The net result of the increased flow rate is that the temperature of the hot stream remains constant

6.4 Entropy Balance

at its high value, 500 °C. As a result, the temperature difference between the hot and cold streams is higher than it was in the previous example, where the temperature of the hot stream was allowed to drop and get closer to that of the colder stream. By increasing the flow rate, the process was further removed from the assumptions of quasi-static operation and thus became more irreversible.

Question: It is suggested by a member of the engineering team that if the heat bath were operated with some suitable fluid, other than steam, it might be possible to decrease the generation of entropy. How should we respond to this suggestion?

Answer: We respond by pointing out that the entropy change of the bath does not depend on the properties of the hot stream. It depends only on the amount of heat that is exchanged, and on the temperature of the bath. Any substance under the same conditions would make exactly the same contribution to entropy generation. The suggestion has no technical merit.

Example 6.4: Heat Bath
Steam is to be heated in a flow process at constant pressure of 1 bar, from 150 °C, to 300 °C. Calculate the entropy generation.

Solution We are not given information about the stream that accomplishes this heating; therefore, we must make some assumptions about its conditions. Clearly, the heating fluid must enter at a temperature that is *at least* 300 °C, otherwise the steam will not be able to reach the desired final temperature. In theory, any temperature above 300 °C will do; in practice, we must have a sufficient temperature difference in order to accomplish the heat transfer within reasonable time. Let us assume that the inlet temperature of the heating fluid is 305 °C. We will further assume that the flow rate of the heating fluid is very high so that the temperature of the unknown stream remains practically constant through the exchanger. With these two assumptions the problem reverts to the exchange of heat between a system (steam) and a bath (the unknown heating fluid), as in the previous example:

$$S_{gen} = S_2 - S_1 - \frac{Q}{T_{bath}}.$$

The heat was found to be $Q = 298$ kJ/kg. With $T_{bath} = 305 + 273.15 = 578.15$ K, the entropy generation is

$$S_{gen} = (8.2171 - 7.6147) \text{ kJ/kg K} - \frac{298 \text{ kJ/kg K}}{578.15 \text{ K}} = 0.0871 \text{ kJ/kg K}.$$

Comments The entropy balance is applied to the system *plus* its surroundings, and for this reason, an exact calculation requires the detailed knowledge of processes in the surroundings. In this example we filled in the missing information by making *assumptions*. In particular, we assumed that the unknown stream is a *heat bath* with a temperature determined by

> the minimum ΔT required for practical heat exchange (~ 5 °C). The entropy generation that is obtained with these assumptions is a best-case scenario: if T_{bath} is higher than 305 °C, the entropy generation will be higher. If the stream is not a heat bath but undergoes a temperature change, then again the entropy generation is higher, as we found in example 6.2 (190 W/K versus 87 W/K in the present case). Therefore, by treating unknown streams as heat baths with minimum realistic ΔT we obtain an estimate for the lowest possible entropy generation. The absolutely lowest entropy generation would be obtained if we took the bath at 300 °C (the closest possible temperature to the system while still allowing for desired heating effect). However, this would require an extremely long time of thermal contact between the two streams.

6.5 Ideal and Lost Work

In Chapter 4 we introduced the ideal and lost work for a closed system. Now we extend these concepts to open systems. We return to the generic process in Figure 6-2, which we assume to run at steady state. The process exchanges a net amount of heat $\dot{Q}$ and a net amount of shaft work, $\dot{W}_s$, with its surroundings. As we did in Chapter 4, we will use a single reservoir at temperature T_0 as the source and sink for all heat transfers between the system and the surroundings. The steady-state energy and entropy balance equations are

$$\Delta(\dot{m}\,H) = \dot{Q} + \dot{W}_s, \tag{6.31}$$

$$\dot{S}_{\text{gen}} = \Delta(\dot{m}\,S) - \frac{\dot{Q}}{T_0}. \tag{6.32}$$

Eliminating the heat term between the two equations we obtain the following expression for the work:

$$\dot{W}_s = \Delta(\dot{m}\,H) - T_0\Delta(\dot{m}\,S) + T_0\dot{S}_{\text{gen}}. \tag{6.33}$$

This equation gives the amount of work involved in converting the input streams into output streams. The term $T_0\dot{S}_{\text{gen}}$ is positive and represents the lost work:

$$\boxed{\dot{W}_{\text{lost}} = T_0\dot{S}_{\text{gen}}.} \tag{6.34}$$

Just as in the case of closed system, it represents a penalty: it increases the work that must be consumed, if the process requires work, and reduces the amount of work that is extracted, if the process produces work. This penalty disappears only if the process is conducted reversibly. The reversible work is obtained from eq. (6.33) with $\dot{S}_{\text{gen}} = 0$:

$$\boxed{\dot{W}_{\text{ideal}} = \Delta(\dot{m}\,H) - T_0\Delta(\dot{m}\,S).} \tag{6.35}$$

6.5 Ideal and Lost Work

It represents the maximum amount of work that can be extracted, or the minimum that must be consumed, in order to convert the input streams into output.

Example 6.5: Entropy Generation in Power Plant
A 100 MW power plant produces entropy at the rate of 0.7 MW/K. How efficient is this plant?

Solution Using $T_0 = 300$ K, the lost work is

$$\dot{W}_{\text{lost}} = (0.7 \text{ MW/K})(300 \text{ K}) = 210 \text{ MW}.$$

This amount represents work that is *not* produced because of various irreversibilities. If the power plant operated fully reversibly, it would generate a total of

$$100 \text{ MW} + 210 \text{ MW} = 310 \text{ MW}.$$

The actual production is 100 MW, or $100/310 = 32\%$ of the maximum possible power. If all irreversibilities were eliminated, the plant would basically triple its production. Even if the entropy generation were reduced only by half, the power produced would double.

Comments This example shows how the concept of lost work places entropy generation into practical perspective.

Example 6.6: Ideal Work
A stream of steam at 20 bar, 220 C, is to be delivered at 15 bar, 500 C. How much work must be exchanged to produce this change of state?

Solution We collect the following data from the steam tables:

	P (bar)	T (°C)	H (kJ/kg)	S (kJ/kg K)
State 1	20	220	2821.9	6.3868
State 2	15	500	3473.6	7.5716

The ideal work is

$$W_{\text{ideal}} = \Delta H - T_0 \Delta S = (3473.6 - 2821.9) \text{ (kJ/kg)} - (300 \text{ K})(7.5716 - 6.3868) \text{ (kJ/kg K)}$$
$$= \underline{296.3} \text{ kJ/kg}.$$

This work is positive, that is, it must be added to the system. It represents the *minimum* amount of work that must be *consumed* in order to produce the desired change of state.

Comments Normally, a pressurized gas has the potential to produce work by expansion. In this case, however, the system must also be heated to a temperature above that of the surroundings. Adding heat is equivalent to consuming a certain amount of work, via a Carnot refrigerator, for example, in order to pump heat from the lower temperature of the surroundings to the system. In this example, the amount of work associated with the heating is more than the work associated with the expansion so that the process overall requires work.

Exercise Repeat the calculation if the final state is at 1 bar, 500 C.

Example 6.7: Ideal Work for Heat Transfer
Determine the ideal work that is required in order to add 1 kJ of heat to a system at temperature $T = 100\,°C$. Repeat for $T = -100\,°C$.

Solution Since the problem does not specify the states before and after the process, we cannot use eq. (6.35). However, we may still answer this problem by noting that ideal work corresponds to a reversible process that accomplishes the given task while using the surroundings as the only heat bath. We will do this by operating a Carnot cycle between the surroundings at T_0, and the system at T. Application of the first and second law to this Carnot cycle gives:

$$Q_0 + Q + W = 0,$$

$$-\frac{Q_0}{T_0} - \frac{Q}{T} = 0.$$

Eliminating Q_0 and solving for W gives

$$W = -Q\left(1 - \frac{T_0}{T}\right).$$

Case 1: $T = 100\,°C = 373.15\,K$. In this case heat is transferred from lower to higher temperature (from the surroundings to the system), so the Carnot cycle operates as a refrigerator (consumes work). For the numerical calculation we must set $Q = -1$ kJ because the energy balance as written above is with respect to the cycle (the amount Q that enters the system actually exits the cycle):

$$W = -(-1\text{ kJ})\left(1 - \frac{300\text{ K}}{373.15\text{ K}}\right) = 0.196\text{ kJ}.$$

According to this result, for every kJ of heat delivered at $100\,°C$ we must consume 0.196 kJ.

Case 2: $T = -100\,°C = 173.15\,K$. In this case heat is transferred from higher to lower temperature, that is, the Carnot operates as an engine (produces work). As before, $Q = -1$ kJ:

$$W = -(-1\text{ kJ})\left(1 - \frac{300\text{ K}}{173.15\text{ K}}\right) = -0.733\text{ kJ}.$$

6.5 Ideal and Lost Work

For every kJ of heat that is transferred to $-100\ °C$ we may produce 0.733 kJ of work.

Comments Transferring heat to temperatures above ambient carries a cost in terms of work. If heat is transferred to lower temperature, it carries a work bonus. In both cases, the equivalent amount of work (cost or bonus) is

$$W = |Q|\left(1 - \frac{T_0}{T}\right),$$

where $|Q|$ is the amount of heat that is transferred to the system. If $T > T_0$, the work is positive and represents amount to be consumed; if $T < T_0$, the work is negative and represents amount that can be produced.

Example 6.8: Choosing the Cooling Source
Water at 50 °C is to be cooled to in a flow process to 30 °C. This can be done either by mixing it with ice at 0 °C, or with water at 25 °C. Perform a second-law analysis of this process to determine which method is thermodynamically more efficient.

Additional data: The latent heat of ice melting is 332 kJ/kg. Assume the surroundings to be at 25 °C.

Solution The most efficient process is the one with the lowest entropy generation. Therefore, we need to calculate S_{gen} for each scenario. We will perform the calculation on the basis of 1 kg/s in the stream at 50 °C (stream 1). Using 2 for the cold stream (ice or cold water), and 3 for the stream that exits the process, the entropy generation is

$$\dot{S}_{\text{gen}} = (\dot{m}_1 + \dot{m}_2)S_3 - \dot{m}_1 S_1 + \dot{m}_2 S_2,$$

which we rewrite as

$$\dot{S}_{\text{gen}} = \dot{m}_1(S_3 - S_1) + \dot{m}_2(S_3 - S_2). \tag{a}$$

In this last equation, the first term on the right-hand side is the entropy change of stream 1 from its initial temperature to the final temperature of stream 3, and the second term is the entropy change of the cold stream from its initial temperature to the exit temperature in stream 3. The entropy terms can be calculated since all temperatures are known. What is still missing is the mass flow rate of stream 2. This is obtained from the energy balance:

$$\dot{m}_3 H_3 - \dot{m}_1 H_1 - \dot{m}_2 H_2 = 0,$$

Solving for $\dot{m}_2$ we obtain

$$\dot{m}_2 = -\frac{\dot{m}_1(H_3 - H_1)}{H_3 - H_2} = -\frac{\dot{Q}}{H_3 - H_2}, \tag{b}$$

where $\dot{Q} = \dot{m}_1(H_3 - H_1)$ is the heat with respect to stream 1. The numerical calculations are shown below for each of the two cases.

Using ice: The amount of heat $\dot{Q}$ is

$$\dot{Q} = \dot{m}_1 C_P (T_3 - T_1),$$

where $C_P = 4.18$ kJ/kg K is the heat capacity of the liquid (alternatively, the enthalpy change can be obtained from the steam tables; this is left as an exercise). We find

$$\dot{Q} = (1 \text{ kg/s})(4.18 \text{ kJ/kg K})(303.15 - 323.15) \text{ K} = -83.6 \text{ kW}.$$

The difference $H_3 - H_2$ for ice includes the latent heat of fusion plus the enthalpy change of the liquid from 0 °C to 30 °C:

$$H_3 - H_2 = \Delta H_{\text{fus}} + C_P(T_3 - T_2) = (332 \text{ kJ/kg}) + (4.18 \text{ kJ/kg K})(303.15 - 273.15)$$
$$\text{K} = 457.4 \text{ kJ/kg}.$$

The mass flow rate of stream 2 is

$$\dot{m}_2 = -\frac{-83.6 \text{ kW}}{457.4 \text{ kJ/kg}} = 0.183 \text{ kg/s}.$$

The entropy change of the feed stream is

$$\dot{m}_1(S_3 - S_1) = \dot{m}_1 C_P \ln \frac{T_3}{T_1} = (1 \text{ kg/s})(4.18 \text{ kJ/kg K}) \ln \frac{303.15}{323.15} = -0.267 \text{ kW/K},$$

and the entropy change of stream 2 is

$$\dot{m}_2(S_3 - S_2) = \dot{m}_2 \left(\frac{\Delta H_{\text{fus}}}{T_2} + C_P \ln \frac{T_3}{T_2} \right) =$$
$$(0.183 \text{ kg/s}) \left(\frac{332 \text{ kJ/kg}}{273.15 \text{ K}} + (4.18 \text{ kJ/kg K}) \ln \frac{303.15}{273.15} \right) = 0.302 \text{ kW/K}.$$

Just as in the calculation of the enthalpy, the entropy change of this stream includes the entropy of fusion ($\Delta S_{\text{fus}} = \Delta H_{\text{fus}}/T_{\text{fus}}$) plus the entropy change due to heating from T_{fus} to T_3.

Finally, the entropy generation is

$$\dot{S}_{\text{gen}} = (-0.267 \text{ kW/K}) + (0.302 \text{ kW/K}) = +0.0347 \text{ kW/K}.$$

Using cooling water: In the case of cooling water the enthalpy difference $H_3 - H_2$ is

$$H_3 - H_2 = C_P(T_3 - T_2) = (4.18 \text{ kJ/kg K})(303.15 - 298.15) \text{ (K)} = 20.9 \text{ kJ/kg}.$$

6.5 Ideal and Lost Work

The corresponding flow rate is

$$\dot{m}_2 = -\frac{-83.6 \text{ kW}}{20.9 \text{ kJ/kg}} = 4.0 \text{ kg/s}.$$

The entropy changes of the streams are:

$$\dot{m}_1(S_3 - S_1) = -0.267 \text{ kW/K}$$

$$\dot{m}_2(S_3 - S_2) = (4.0 \text{ kg/s})(4.18 \text{ kJ/kg K}) \ln \frac{303.15}{298.15} = 0.278 \text{ kW/K}.$$

The entropy generation is

$$\dot{S}_{\text{gen}} = (-0.267) + (0.278) = 0.0110 \text{ kW/K}.$$

These results are summarized in the table below. The table also shows the lost work, which is calculated as

$$\dot{W}_{\text{lost}} = T_0 \dot{S}_{\text{gen}}$$

with $T_0 = 300$ K.

	$\dot{Q}$	$\dot{m}_2$	$\dot{m}_1(S_3 - S_1)$	$\dot{m}_2(S_3 - S_2)$	$\dot{S}_{\text{gen}}$	$\dot{W}_{\text{lost}}$
Using ice	−83.6	0.183	−0.267	0.302	0.0347	10.4
Using water	−83.6	4.0	−0.267	0.278	0.0110	3.3
	(kW)	(kg/s)	(kW/K)	(kW/K)	(kW/K)	(kW)

As we see, the entropy generation in the case of ice is more than three times higher than that in the case of cooling water. The lost work for ice is more than triple that for cooling water. Here is what this means: If your coffee is too hot, cooling it with an ice cube is an unnecessary waste of resources because hot coffee can be cooled using room-temperature water, thus conserving the ice (which is more expensive due to the energy needed to produce it), for cooling something that would require subambient temperatures. This intuitive argument is expressed quantitatively by the calculation of entropy generation. The thermodynamic reason for this result has to do with irreversibilities in heat transfer. As we have discussed already, heat transfer between two different temperatures is an irreversible process. It becomes even more irreversible as the two temperatures become further apart. With cooling water, this temperature difference is smaller than it is when ice is used, resulting in a process that is less irreversible. When we have a choice as to how to exchange heat between various streams, we should opt for the arrangement that minimizes the temperature differences between the streams. These conclusions can be reached based either on entropy generation or on lost work. The advantage of lost work is that it translates entropy generation into work, which is much easier to appreciate. In this problem, if we choose cooling water instead of ice we will be saving $10.4 - 3.3 = 7.1$ kW of work. These savings, however, come at a price: using cooling water requires a higher mass flow rate and a bigger mixing tank. The final decision must take into consideration both capital costs and operating expenses. The main point, however, is that the second law (and the lost work) provides a systematic way to evaluate competing designs and search for an optimum solution.

6.6 Thermodynamics of Steady-State Processes

Chemical processes consist of interconnected unit operations such as heat exchangers, pumps and compressors, separation units, chemical reactors. Generally these are flow systems with several streams carrying mass into and out of a unit. To perform mass and energy balances in such units we will apply the principles developed in the previous chapter. The units we will consider in this chapter are: heat exchangers, compression devices (pumps, gas compressors) and expansion devices (turbines and throttling valves). We will not consider chemical reactors or separation units. These are treated with the same general tools developed in the previous chapter but they require the calculation of mixture properties, which is covered in Part II. Fairly complex processes can be designed by putting together simple units; we will discuss in more detail the preliminary design of power plants, refrigerators and liquefaction processes. By "preliminary" design we refer to the calculation of the mass and energy balances of the process. As we will see, this calculation is straightforward and requires very little information about the inner workings of the device under consideration.

Flow through Pipe

The most basic process is flow of a fluid through a pipe. When a fluid flows in a pipe its motion is resisted by the viscosity of the fluid, which represents molecular "friction" between adjacent layers of fluid that move at different velocities, and also resistance from the walls, which do not move. Viscosity and wall friction represent irreversibilities that dissipate the energy of the fluid into internal energy. To maintain steady flow through a pipe, we must apply mechanical work, usually via a pump. Here we are concerned with the macroscopic energy balance that accounts for changes in the various energy terms between the inlet and outlet of the pipe. Viscosity represents a microscopic mechanism which we will not consider explicitly except as a source of entropy generation. The pipe need not be cylindrical or have constant cross section. It may also span different elevations between inlet and outlet. We will assume that the velocity of the fluid is uniform at all points on a cross section. In reality, fluid velocity near the center of a pipe is faster than near the walls. By ignoring this variation we are dealing the average velocity of the fluid through the cross section. This average velocity is defined in terms of the mass flow rate (see also eq. [6.2]), as

$$v = \frac{V\dot{m}}{a} = \frac{\text{(volumetric flow rate)}}{\text{(cross sectional area)}}, \tag{6.36}$$

where V is the specific volume of the fluid, $\dot{m}$ is the mass flow rate, and a is the cross-sectional area of the pipe. Noting that the product $V\dot{m}$ is equal to the volumetric

6.6 Thermodynamics of Steady-State Processes

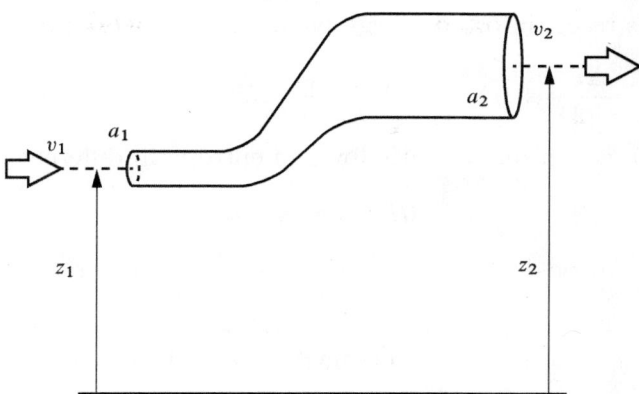

Figure 6-4: Schematic for the calculation of energy losses in flow through a pipe.

flow rate (m³/s), the mean velocity is simply equal to this rate divided by the cross-sectional area.

We consider now steady-state flow through a pipe, as shown schematically in Figure 6-4. The system may exchange work if a pump is present. To maintain the isothermal condition we must allow for the exchange of heat. Without such exchange, the dissipation of energy into internal energy via viscosity and wall friction would result in a temperature rise. To maintain the system at constant temperature T we will assume that heat is exchange with a bath with the same temperature as the fluid, $T_{\text{bath}} = T$. Since both the velocity and elevation change (the velocity changes due to variations of the cross-sectional area), kinetic and potential energy will be included. The steady-state energy balance is

$$\dot{m}\left(\frac{\Delta v_{12}^2}{2} + g\Delta z_{12} + \Delta H_{12}\right) = \dot{Q} + \dot{W}_s.$$

Here, $\Delta(\cdots)_{12}$ signifies the difference between inlet and outlet of the quantity in parenthesis. Since we are dealing with one stream at a steady state, the inlet and outlet have the same mas flow rate, $\dot{m}$. Dividing both sides by $\dot{m}$ we obtain,

$$\frac{\Delta v_{12}^2}{2} + g\Delta z_{12} + \Delta H_{12} = Q + W_s. \tag{6.37}$$

The corresponding entropy balance at steady state is

$$\dot{m}\Delta S_{12} = \frac{\dot{Q}}{T} + \dot{S}_{\text{gen}}$$

where we used the fact that the heat bath is at the temperature of the system. We divide both sides by $\dot{m}$ and solve for Q:

$$Q = T\Delta S_{12} - TS_{\text{gen}}$$

Substituting this result into the energy balance, eq. (6.37), gives

$$\frac{\Delta v_{12}^2}{2} + g\Delta z_{12} + \Delta H_{12} - T\Delta S_{12} + TS_{\text{gen}} = W_s. \tag{6.38}$$

Recall eq. (5.14), which relates enthalpy and entropy in differential form:

$$dH = TdS + VdP. \tag{5.14}$$

We integrate the above differential from inlet to outlet under constant temperature,

$$\Delta H_{12} = T\Delta S_{12} + \int_{1 \atop (\text{const } T)}^{2} VdP,$$

and substitute the result into (6.38):

$$\frac{\Delta v_{12}^2}{2} + g\Delta z_{12} + \int_{1 \atop (\text{const } T)}^{2} VdP + TS_{\text{gen}} = W_s. \tag{6.39}$$

This equation now expresses the macroscopic energy balance for isothermal flow in a pipe. It may further be simplified in the special case of an incompressible fluid. In this case V is constant and the energy balance equation becomes

$$\frac{\Delta v_{12}^2}{2} + g\Delta z_{12} + V(P_2 - P_1) + TS_{\text{gen}} = W_s. \tag{6.40}$$

This form of the mechanical balance is known as the Bernoulli equation. It applies to incompressible fluids and approximately to compressible gases as long as the pressure does not vary much. The Bernoulli equation gives the mechanical work that is required to produce the desired change in velocity, elevation, and pressure between the inlet and outlet of the pipe. The term TS_{gen} represents losses due to irreversibilities that arise from viscosity and wall friction. Its dependence on the viscosity of the fluid, the type of flow (laminar, turbulent), and the roughness of the walls is the subject of fluid mechanics and we will not consider them here in any detail. The main point we wish to make here is that these microscopic mechanisms result in dissipation of the mechanical energy of the fluid into internal energy. To see this, consider a horizontal segment of pipe ($\Delta z_{12} = 0$) with constant cross section ($\Delta v_{12}^2 = 0$) that does not include any pumps ($W_s = 0$). Since S_{gen} is positive, the exit pressure must be lower than the inlet pressure. This pressure drop arises from the resistance to the flow due to viscosity and wall friction. Therefore, to maintain flow at sufficient pressure, it is necessary to supply work via a pump.

6.6 Thermodynamics of Steady-State Processes

Example 6.9: Pressure Drop in Pipes
Steam flows in an insulated 100 m long straight pipe with a diameter of 2.5 cm. The steam enters the pipe at 30 bar, 400 °C, with mass flow rate 0.15 kg/s. Perry's *Handbook* gives the pressure drop per unit length of pipe as given by the empirical Blasius equation,

$$f = 0.079 \left(\frac{\mu}{\rho v d}\right)^{0.25} \left(\frac{2\rho v^2}{d}\right),$$

where L is the length of the pipe, D is its diameter, v is the average fluid velocity, ρ is the fluid density, and μ is the fluid viscosity. Calculate the pressure drop and the entropy generation. The viscosity of steam at 30 bar, 400 °C is 0.000023 Pa s.

Solution If the pressure drop is not very high, the steam may be taken to be incompressible and the Bernoulli equation may be used. Even though the system is adiabatic, rather than isothermal, as the Bernoulli equation assumes, if frictional losses are small it is reasonable to assume that any temperature rise due to them would be rather small. These assumptions will be rechecked at the end of the calculation.

From steam tables we find the specific volume of steam at the inlet conditions to be $V = 0.1162$ m³/kg corresponding to a density $\rho = 1/V = 8.60585$ kg/m³. The cross-sectional area of the pipe is

$$a = \frac{\pi D^2}{4} = \frac{\pi (0.025 \text{ m})^2}{4} = 0.000490874 \text{ m}^2.$$

From this we calculate the mean velocity:

$$v = \frac{V\dot{m}}{a} = \frac{(0.1162 \text{ m}^3/\text{kg})(0.15 \text{ kg/s})}{0.000490874 \text{ m}^2} = 35.5081 \text{ m/s}.$$

The pressure drop per unit length is calculated from the Blasius equation given above:

$$\frac{\Delta P}{L} = 0.079 \left(\frac{(0.000023 \text{ Pa s})}{(8.60585 \text{ kg/m}^3)(35.5081 \text{ m/s})(0.025)}\right)^{0.25} = 2856.49 \text{ Pa/m},$$

or,

$$\frac{\Delta P}{L} = (2856.49 \text{ Pa/m})(10^{-5} \text{ bar/Pa}) = 0.0285649 \text{ bar/m}.$$

Finally, the total drop over a length $L = 100$ m is

$$\Delta P = (100 \text{ m})(0.0285649 \text{ bar/m}) = 2.85 \text{ bar}.$$

The entropy generation is calculated from the Bernoulli equation. We set $\Delta z_{12} = 0$ (the pipe is horizontal) and $W_s = 0$ (there is no pump in this segment of the pipe). Ignoring the slight change of specific volume due to the pressure drop, we will also set $\Delta v_{12}^2 = 0$. Then, the entropy generation is

$$S_{\text{gen}} = -\frac{V(P_2 - P_1)}{T} = -\frac{(0.1162 \text{ m}^3/\text{kg})(-2.85 \times 10^5 \text{ Pa})}{(400 + 273.15) \text{ K}} = 4.930 \text{ J/kg K}.$$

This gives the entropy generation per kg of fluid; to obtain the rate of entropy generation per unit time we multiply by the mass flow rate:

$$\dot{S}_{\text{gen}} = \dot{m}\, S_{\text{gen}} = (0.15 \text{ kg/s})(4.930 \text{ J/kg K}) = 0.7396 \text{ W/K}.$$

Comments The pressure drop is indeed small, 0.0285 bar per m of pipe length. Of course, given enough length, the pressure drop will be significant. Pressure drop also increases with increasing flow rate, as we can confirm by repeating the calculation with a higher value of $\dot{m}$. Other factors that contribute to pressure drop are elbows, valves, and other common elements of piping networks that offer resistance to flow. In general, however, the pressure drop due to flow is rather small, from the point of view that its contribution to the energy balance is small compared to that of heat exchangers, compressors, and the like. In this example, the work equivalent of the frictional losses is

$$TS_{\text{gen}} = (673.15 \text{ K})(4.930 \text{ J/kg K}) = 3.319 \text{ kJ/kg}.$$

By contrast, the amounts of work and heat that are exchanged in typical process are of the order of 10^2 or 10^3 kJ/kg. As a measure of comparison, for example, the heat of vaporization of water at 1 bar is 2256 kJ/kg. From this point of view, it is acceptable to ignore the contributions of pressure losses to the overall energy balance in preliminary design of the process. This is not to say that pressure losses are unimportant; a real process will fail to operate if frictional loses are not taken into consideration in the final design.

The fact that frictional losses represent a small contribution to the overall energy balance also justifies the initial assumptions of the calculation. Since pressure drop is small, steam may indeed be treated as an incompressible fluid. The amount of work that is dissipated is correspondingly small (3.319 kJ/kg); therefore, the temperature rise that would be expected (recall that the pipe is insulated) must also be small. This temperature rise may be estimated by assuming that the enthalpy of steam at the exit has increased by the amount of frictional losses. Interpolation in the steam tables shows that the expected temperature rise over the entire pipe length is about 4 °C.

Adiabatic Mixing

The simplest way to combine streams is by bringing two or more pipes together. This is indicated as a mixing point on the process flow diagram (see Figure 6-5). For practical reasons the pressure of all streams around the mixing point is the same to avoid back flow from a high pressure stream into one at lower pressure. In some cases, mixing takes place inside a tank with or without stirring. In adiabatic mixing, no heat is exchanged between the streams and the surroundings. As a result, the temperature of the outlet stream must change to accommodate the energy of the

6.6 Thermodynamics of Steady-State Processes

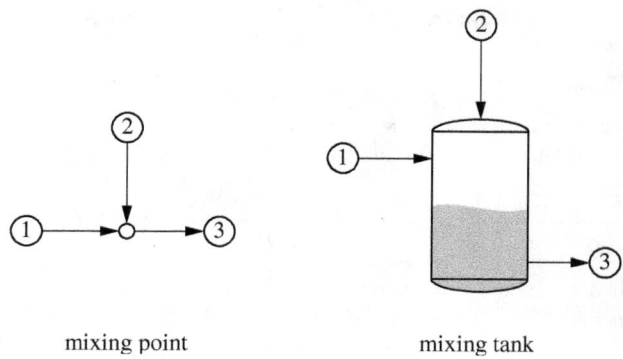

Figure 6-5: Adiabatic mixing of streams.

inlet streams. The steady-state balance equation for adiabatic mixing with no shaft work is

$$\Delta(\dot{m}H) = 0, \qquad (6.41)$$

and the entropy balance is

$$\Delta(\dot{m}S) = \dot{S}_{\text{gen}}. \qquad (6.42)$$

Mixing is an irreversible process, as the streams that are combined will generally have different temperatures. In the special case that the temperature of all inlet streams is the same (recall that all streams have the same pressure), then all streams, including the exit stream, have the same intensive properties and the entropy generation is zero.

Example 6.10: Adiabatic Mixing
Water (stream 1) at 20 °C, 2 bar is mixed with steam (stream 2) at 2 bar, 200 °C. The flow rate of the liquid stream is 2.0 kg/s and of the steam 5.0 kg/s. Mixing is adiabatic and the outlet stream is at 2 bar. Determine the temperature of the outlet stream and the entropy generation.

Solution First, by mass balance we have

$$\dot{m}_3 = \dot{m}_1 + \dot{m}_2 = (2) + (5) = 7 \text{ kg/s}.$$

Stream 1 is compressed liquid; its properties are taken to be those of saturated liquid at 25 °C:

$$H_1 = 83.92 \text{ kJ/kg}, \quad S_1 = 0.2965 \text{ kJ/kg K}.$$

The properties of stream 2 are

$$H_2 = 2870.8 \text{ kJ/kg}, \quad S_2 = 7.5081 \text{ kJ/kg K}.$$

The steady-state energy balance is

$$\dot{m}_3 H_3 - \dot{m}_1 H_1 - \dot{m}_2 H_2 = 0.$$

In this equation the only unknown is the enthalpy of stream 3. Solving for H_3,

$$H_3 = \frac{\dot{m}_1 H_1 + \dot{m}_2 H_2}{\dot{m}_3} = \frac{(2)(83.92) + (5)(2870.8)}{(7)} = 2074.55 \text{ kJ/kg}.$$

The state of stream 3 is fixed by its pressure (2 bar) and enthalpy (2074.55 kJ/kg). From the steam tables we find that this value of enthalpy lies in the vapor liquid region ($T^{\text{sat}} = 120.21$ °C):

	sat. L	sat. V
H (kJ/kg)	504.68	2706.2
S (kJ/kg K)	1.5301	7.1269

By lever rule, the vapor fraction is

$$x_V = \frac{H_3 - H_L}{H_V - H_L} = \frac{2074.55 - 504.68}{2706.2 - 504.68} = 0.713,$$

and the entropy is

$$S_3 = x_L S_L + x_V S_V = (1 - 0.713)(1.5301) + (0.713)(7.1269) = 5.5210 \text{ kJ/kg K}.$$

These results are summarized in the table below:

	1	2	3
$\dot{m}$ (kg/s)	2	5	7
P (bar)	2	2	2
T (°C)	20	200	120.21
H (kJ/kg)	83.92	2870.8	2074.55
S (kJ/kg K)	0.2965	7.5081	5.5211

Finally, the entropy generation is

$$\dot{S}_{\text{gen}} = \dot{m}_3 S_3 - \dot{m}_1 S_1 - \dot{m}_2 S_2 = (7)(5.5211) - (2)(0.2965) - (5)(7.5081) = 0.5141 \text{ kW/K}.$$

Heat Exchanger

A heat exchanger is a device used to transfer heat from one fluid to another. The basic idea is simple: two fluids come into thermal contact via a conducting wall that allows heat to pass from the hotter to the cooler fluid. In this arrangement, the fluids do not intimately mix. Typically, one fluid flows inside a pipe and the other in a shell

6.6 Thermodynamics of Steady-State Processes

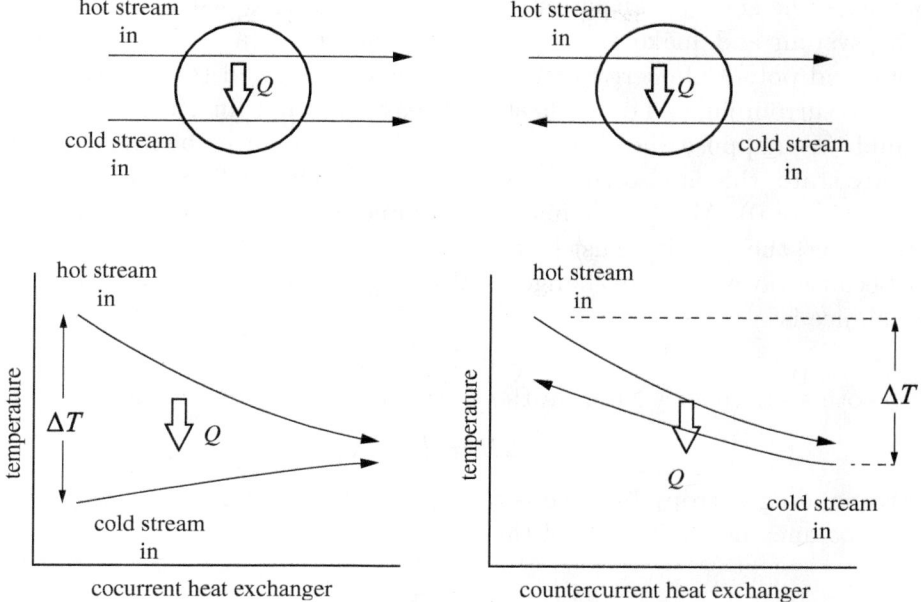

Figure 6-6: Cocurrent and countercurrent heat exchanger and their qualitative temperature profiles.

around that pipe. This design is called a shell-and-tube heat exchanger. Usually a large number of tubes are bundled together to increase the contact area and the rate of heat transfer. In terms of direction of flow, heat exchangers are classified as co- or countercurrent (Figure 6-6). In cocurrent arrangement both fluids flow in the same direction. The temperature gradient is highest at the inlet and smallest at the outlet. Given a large enough exchanger, the fluids would exit at a common temperature, corresponding to thermal equilibrium. This is not the case in practice because it would require a prohibitively large heat exchanger. In countercurrent arrangement the fluids flow in opposite directions and in this case the temperature difference between the fluids along the length of the heat exchanger is more uniform. A large variety of heat exchanger designs are covered in courses on heat transfer. For the purposes of overall energy and entropy balance, however, these design details are not important, as we will see below. Regardless of design, the driving force for the transfer of heat is the temperature difference between the two fluids at the inlet. A minimum temperature difference is required in order to accomplish the transfer at a practical rate. For our calculations we will assume as a rule of thumb a temperature difference of at least 5 °C between the temperature of the hot and cold streams.

To analyze the energy transfer in the heat exchanger we take one of the streams to be the system and make the following assumptions: (a) flow is steady state; (b) kinetic and potential energy terms can be neglected; and (c) there are no heat losses to the surroundings, i.e., all heat that leaves the hot fluid is absorbed by the colder fluid. We suppose the flow rate of the inlet stream to be $\dot{m}$, and since we have steady state, this is also the same at the outlet. We will neglect variations in elevation ($\Delta z_{12} = 0$). Moreover, unless the diameter of the pipe changes between inlet and outlet, the velocity must be the same ($\Delta v_{12}^2 = 0$). Finally, we set W_s equal to zero because no work is exchanged. With these assumptions, the steady-state energy balance is

$$\dot{m}\Delta H = \dot{Q}. \tag{6.43}$$

Dividing both sides by $\dot{m}$ we obtain the energy balance on a per-mass basis:

$$\Delta H = Q. \tag{6.44}$$

The corresponding entropy balance is obtained from eq. (6.28). Dividing by $\dot{m}$ we obtain the balance per unit mass of the fluid:

$$S_{\text{gen}} = \Delta S - \frac{Q}{T_{\text{bath}}}, \tag{6.45}$$

where T_{bath} is the bath with which the system exchanges heat.

Example 6.11: Heat Exchanger (1)
Steam at 1 bar is cooled from 300 °C to 150 °C. Determine the heat load and estimate the entropy generation.

Solution We collect the following data from the steam tables:

	T (°C)	P (bar)	H (kJ/kg)	S (kJ/kg K)
1	300	1.0	3074.5	8.2171
2	150	1.0	2776.6	7.6147

The heat is
$$Q = \Delta H_{12} = (2776.6 - 3074.5) \text{ kJ/kg} = -297.9 \text{ kJ/kg}.$$

To estimate the entropy generation we must make some assumptions about the stream that absorbs this heat. This stream must be no hotter than 150 °C; to allow for practical rate of heat transfer, we will take its temperature to be at 145 °C, and will treat it as a bath:

$$S_{\text{gen}} = S_2 - S_1 - \frac{Q}{T_{\text{bath}}} = (7.6147 - 8.2171) \text{ kJ/kg K} - \frac{-297.9 \text{ kJ/kg}}{(145 + 273.15) \text{ K}} = \underline{0.110 \text{ kJ/kg K}}.$$

The entropy of the stream decreases (it is being cooled) but this decrease is more than compensated by the entropy increase of the bath.

6.6 Thermodynamics of Steady-State Processes

Example 6.12: Heat Exchanger (2)
The steam in the previous example is cooled using water at 1 bar, 20 °C. The cooling water emerges at 40 °C at the exit of the heat exchanger. Determine the heat load and the entropy generation. Compare the result with the previous example.

Solution This example differs only in that the stream that absorbs the heat from the steam is now specified. We collect the properties of cold water from the steam tables approximating the enthalpy and entropy of the compressed liquid with those of the saturated liquid at the same temperature:

	T (°C)	P (bar)	H (kJ/kg)	S (kJ/kg K)
1'	20	1.0	83.92	0.2965
2'	40	1.0	167.54	0.5724

The heat load is the same as before:

$$Q = -297.9 \text{ kJ/kg}.$$

The entropy generation now involves a two streams and no bath:

$$\dot{S}_{\text{gen}} = \dot{m}(S_2 - S_1) + \dot{m}'(S_2' - S_1')$$

where $\dot{m}'$ is the mass flow rate of cold water. We obtain $\dot{m}'$ in terms of $\dot{m}$ through the energy balance. Noting that the sign of the heat must be reversed, the energy balance for the cold-water stream reads

$$-\dot{m}Q = \dot{m}'(H_2' - H_1') \quad \Rightarrow \quad \dot{m}' = \dot{m}\frac{-Q}{H_2' - H_1'}$$

and by numerical substitution,

$$\dot{m}' = \dot{m}\frac{+297.9 \text{ kJ/kg}}{(167.54 - 83.92) \text{ kJ/kg}} = 3.56254\dot{m}.$$

Substitution into the entropy balance gives the entropy generation:

$$\dot{S}_{\text{gen}}/\dot{m} = (7.6147 - 8.2171) \text{ kJ/kg K} + 3.56254\,(0.5724 - 0.2965) \text{ kJ/kg K}$$

$$= \underline{0.381 \text{ kJ/kg K}}.$$

Comments The entropy generation in this case is higher. This is because overall the temperature gradients between the steam and the cold water are higher compared to the bath, whose temperature is constant at 145 °C. The main point, however, is this: if the system that exchanges heat with the stream of interest is specified, the entropy generation is equal to the total change of entropy of all streams involved. If the system that exchanges heat with the system of interest is not specified, then one must make a suitable assumption as to the source or sink of that heat by invoking a heat bath.

Steam Turbine

A turbine is a device that extracts shaft work from a fluid. In its most basic form it is a fan that rotates under the action of the fluid, such as in water mills and wind mills, which have been in use for many centuries. Industrial gas turbines operate with a hot, pressurized gas, such as steam. Pressurized liquids can also be expanded to produce work and in this case the device is called an expander. The technical design details are different but the thermodynamic analysis of gas turbines and liquid expanders is done in the same way. The principle of operation is straightforward: the pressurized gas is forced to flow through converging nozzles that cause the velocity to increase. The fast-moving gas impinges on blades fixed on a rotor, which extracts shaft work in the form of rotational motion. Instead of actual nozzles, modern gas turbines utilize a set of stationary blades (stator) positioned right before the moving blades (rotor). These act as nozzles to guide the fluid onto the rotating blades. To maximize the amount of work, often multiple stages of nozzles and blades are incorporated into the same unit. Because the specific volume of an expanding gas increases, the cross section of the turbine increases in the direction of the flow.

The design and internal details of the turbine are not important when considering the overall energy balance between inlet and outlet. The operation is assumed to be adiabatic and at steady state. Variations in potential energy are unimportant ($\delta z_{12} = 0$). Even though the velocity of the gas varies through the turbine, the change in velocity between inlet and outlet can be neglected because it represents a small contribution to the overall balance. With these assumptions, the steady-state energy balance simplifies to

$$\boxed{\dot{m}\Delta H = \dot{W}_s} \quad \text{or} \quad \boxed{\Delta H = W_s}, \tag{6.46}$$

where ΔH is the enthalpy change between inlet and outlet. The entropy generation is obtained from eq. (6.28) with $\dot{Q} = 0$:

$$\dot{S}_{\text{gen}} = \dot{m}\Delta S \quad \text{or} \quad S_{\text{gen}} = \Delta S, \tag{6.47}$$

where ΔS is the entropy change between inlet and outlet.

If both inlet and outlet states are known, the work is easily computed from eq. (6.46). In process design we normally know the inlet state (pressure and temperature), which is usually determined by upstream processes, and the outlet pressure, which represents a design decision. The engineer is then called to determine both the exit temperature and the work produced in the turbine. This calculation requires the *efficiency* of the turbine, which is defined as

$$\eta = \frac{W_s}{W_s^{\text{rev}}}, \tag{6.48}$$

6.6 Thermodynamics of Steady-State Processes

where W_s^{rev} is the work for reversible expansion from the known inlet state to the known outlet *pressure*. This process is isentropic, since it is reversible and adiabatic. Under reversible operation the turbine produces the maximum possible work. Irreversibilities decrease the amount of work and thus the efficiency of the actual turbine is less than 100%. The efficiency of a turbine is characteristic of the device and usually available from the manufacturer. If the efficiency is known, the calculation of the turbine is done in two steps: first, we calculate the work and exit temperature for reversible operation, then we calculate the work and exit temperature for the actual operation.

Reversible Operation Let P_1 and T_1 refer to the inlet state and P_2 to the outlet pressure. Under isentropic conditions the gas would exit at a temperature T_2' such that

$$S(P_2, T_2') - S(P_1, T_1) = 0. \tag{6.49}$$

This equation fixes the unknown temperature T_2'. The corresponding reversible work is given by eq. (6.46):

$$W_s^{\text{rev}} = H(P_2, T_2') - H(P_1, T_1). \tag{6.50}$$

Actual Operation The actual work is calculated from the reversible work and the known efficiency:

$$W_s = \eta W_s^{\text{rev}}. \tag{6.51}$$

The exit temperature is determined from the energy balance, eq. (6.46), which takes the form

$$H(P_2, T_2) - H(P_1, T_1) = W_s. \tag{6.52}$$

In this equation the only unknown is T_2.

The procedure is demonstrated graphically in the Mollier chart in Figure 6-7. The reversible calculation is represented by the isentropic path 12', which on this graph is a vertical line starting at the known initial state and terminating at the known exit pressure. The actual exit state is at the same exit pressure but to the right of point 2', corresponding to positive ΔS_{12} (and higher temperature than state 2'). Accordingly, the actual work, $H_2 - H_1$, is smaller than the reversible amount. The numerical details of the calculation depend on the method used for the calculation of enthalpy and entropy. This can be done using an equation of state, tabulated values, or charts. The approach will be demonstrated with examples below.

Example 6.13: Steam Turbine using Steam Tables
A steam turbine expands steam from 500 °C, 40 bar to 1 bar. The efficiency of the turbine is 80%. Determine the amount of work produced and the conditions at the exit.

Solution The inlet conditions are known and from steam tables we find:

$$H_1 = 3445.8 \text{ kJ/kg}, \quad S_1 = 7.0919 \text{ kJ/kg K}.$$

Reversible work. The reversible work corresponds to isentropic operation between $T_1 = 500$ C, $P_1 = 40$ bar, and $P_2 = 1$ bar. The outlet state is defined by the conditions, $P_{2'} = P_2 = 1$ bar, $S_{2'} = S_1 = 7.0919$ kJ/kg K. This state is in the vapor liquid region. From the steam tables at 1 bar, 99.61 C, we find

	H (kJ/kg)	S (kJ/kg/K)
L	417.44	1.3026
V	2674.9	7.3588

The vapor fraction is

$$x_V = \frac{S_{2'} - S_L}{S_V - S_L} = \frac{7.0919 - 1.3026}{7.3588 - 1.3026} = 0.9559.$$

The enthalpy of the outlet stream is calculated using the lever rule,

$$H_{2'} = x_V H_V + (1 - x_V) H_L = (0.9559)(2674.9) + (1 - 0.9559)(417.44)$$
$$= 2575.41 \text{ kJ/kg},$$

and the reversible work is

$$W_s^{\text{rev}} = H_{2'} - H_1 = 2575.41 - 3445.8 = -870.387 \text{ kJ/kg}.$$

Actual operation. The actual work is

$$W_s = \eta W_s^{\text{rev}} = (0.8)(-870.387) = -696.31 \text{ kJ/kg}.$$

With this we now calculate the enthalpy in the actual exit stream:

$$H_2 = H_1 + W_s = 3445.8 + (-696.31) = 2749.49 \text{ kJ/kg}.$$

The exit state is now fully defined by its pressure (1 bar) and enthalpy (2749.49 kJ/kg). Its temperature and entropy are obtained from the table by interpolation. The exit state as well as the inlet and reversible outlet states are summarized below:

	Inlet	Outlet	Outlet (rev.)
P (bar)	40	1	1
T (C)	500	136.553	99.61
H (kJ/kg)	3445.8	2749.49	2575.41
S (kJ/kg K)	7.0919	7.54647	7.0919

6.6 Thermodynamics of Steady-State Processes

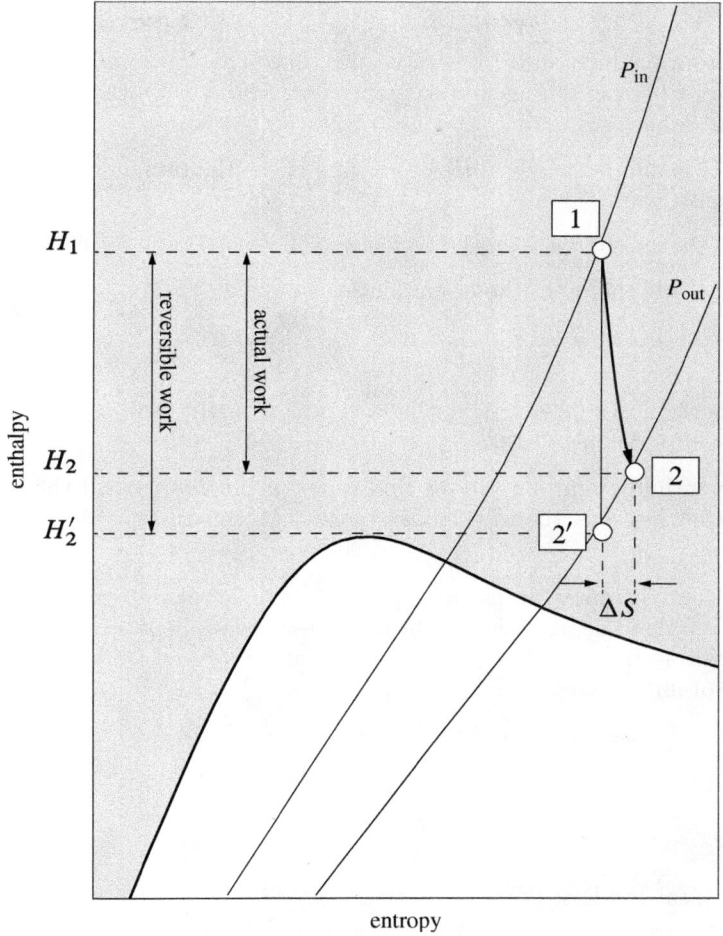

Figure 6-7: Graphical solution of turbine on Mollier graph.

Example 6.14: Steam Turbine using the Mollier Graph
Perform a graphical solution of the previous example on the Mollier chart.

Solution We outline the procedure using Figure 6-7 as a guide and leave the actual calculation as an exercise.

1. Determine the inlet state at the intersection of the isotherm and isobar that correspond to the inlet conditions (state 1).

2. Starting from the inlet state, draw a vertical line. This gives the path of the reversible (isentropic) operation. The intersection of this path with the isobar at the outlet conditions defines state $2'$.

3. Calculate the difference $H_{2'} - H_1$ by reading the enthalpies on the left axis; this gives the reversible work.

4. Calculate the actual work using the known efficiency.

5. Calculate the enthalpy of the actual outlet from
$$H_2 = H_1 + W_s.$$

6. Draw a horizontal line at H_2; its intersection with the isobar corresponding to the outlet pressure defines state 2.

7. Read the enthalpy and entropy of this state on the chart; read the temperature by locating the closest isotherm that passes through this state.

Comments Each of the above steps corresponds to the mathematical calculations given in the previous example. The graphical solution is very convenient. However, charts cannot be read with high accuracy, therefore, the graphical solution should be used only if great accuracy is not of importance.

Example 6.15: Ideal and Lost Work in Turbine
Calculate the entropy generation, and the ideal and lost work in Example 6.13.

Solution The process is adiabatic and at steady state. Accordingly, the entropy generation is equal to the entropy change between inlet and outlet streams (see eq. [6.28]):
$$S_{\text{gen}} = S_2 - S_1 = 7.54647 - 7.0919 = 0.4547 \text{ kJ/kg K}.$$

The lost work is
$$W_{\text{lost}} = T_0 S_{\text{gen}} = (300 \text{ K})(0.4547 \text{ kJ/kg K}) = 136.37 \text{ kJ/kg}.$$

The ideal work is
$$W_{\text{ideal}} = W_{\text{actual}} - W_{\text{lost}} = -832.68 \text{ kJ/kg}.$$

Comments Notice that the ideal work (-832.68 kJ/kg) is not equal to the reversible work (-870.387 kJ/kg) calculated in Example 6.13. This is because the two concepts are not

6.6 Thermodynamics of Steady-State Processes

> the same. Although both "ideal" and "reversible" work refer to a reversible path, these are two different paths. The reversible work calculated in Example 6.13 is the work for reversible operation between inlet state (P_1, T_1), and outlet pressure (P_2). The ideal work calculated here is the work for reversible operation between inlet state (P_1, T_1), and outlet *state* (P_2, T_2). To reach the outlet temperature reversibly, the system cannot follow an isentropic path. Instead, it follows some different path that exchanges heat with a reservoir at T_0.

Throttling

It is often necessary to reduce the pressure of a stream before leading it to the next step of the processes. This is most easily done through a partially open valve that offers sufficient resistance to flow as to maintain the pressure difference between inlet and outlet. Clearly, there is no shaft work involved. What is the outlet temperature? To analyze this process we make a number of simplifying assumptions. First we assume there are no heat losses to the surroundings. We further assume that the velocity of the fluid that emerges at the low-pressure end of the valve is not significantly lower than the inlet velocity. Normally, a fluid moving under a large pressure difference increases its velocity (see eq. [6.39]); however, if the resistance to flow is high, the kinetic energy is dissipated into internal energy. A process that satisfies these conditions is called *throttling*, or *Joule-Thomson* expansion. The process is encountered when a pressurized fluid expands via a path of high resistance to the flow, as when a fluid passes through a partially open valve, through a porous plug, through a small crack in the walls, and the like.

To analyze this process, we apply the steady-state energy balance to a system with one inlet and one outlet, that exchanges no work or heat, and with negligible contributions from kinetic or potential energy. Under these conditions, the energy balance simplifies to

$$\boxed{\Delta H_{12} = 0.} \tag{6.53}$$

Accordingly, during throttling the enthalpy of the fluid remains constant. In typical problems we know the inlet state (pressure and temperature) and the outlet pressure. Since the outlet enthalpy is also known, pressure and enthalpy fully specify the state of the outlet.

Throttling may be viewed as expansion in a turbine that is so irreversible that no work is produced. Why should one waste the potential of a pressurized gas by throttling it rather than using a turbine to expand it? Because using a valve is much cheaper to install and maintain (no moving parts) than a turbine. Replacing a throttling valve with a turbine will increase the efficiency of a process, but the decision to replace it would have to be made on a case-by-case basis, based on cost-benefit analysis.

Example 6.16: Throttling of Ideal Gas

Air is throttled from 5 bar, 80 °C to 1 bar. Calculate the temperature at the exit of the throttling valve.

Solution It is reasonable to treat air an ideal gas under the conditions of this problem. Equation (6.53) in this case becomes

Applying eq. (6.53) we obtain

$$\Delta H_{12} = 0 = \int_{T_1}^{T_2} C_P^{ig} \, dT = 0.$$

The only way that the integral on the right-hand side can be zero is if the two temperatures are the same. Otherwise, the heat capacity would have to take negative values in the interval (T_1, T_2), which is not possible (why?). Thus we conclude that the temperature at the exit will be the same as in the inlet.

Comments This result applies to the ideal-gas state in general and is a consequence of the fact that enthalpy in the ideal-gas state is a function of temperature only. Accordingly, a path of constant temperature is a path of constant enthalpy, and vice versa.

Example 6.17: Throttling of Compressed Liquid

Water is throttled from 1 bar, 60 °C, to 0.5 bar, and to 0.1 bar. In each case calculate the exit temperature and the entropy generation.

Solution *Throttling to 0.5 bar.* At inlet conditions water is a compressed liquid. Recall that for compressed liquid away from the critical point, an isotherm is also a line of constant volume, constant enthalpy, and constant entropy. The enthalpy at state 1 is then approximately equal to that of the saturated liquid at 60 °C. Using steam tables we find $H_1 = H^L(60\ °C) = 251.1$ kJ/kg. At the exit, water is still compressed liquid because it is above the saturation pressure, which at 60 °C is 0.1992 bar. Since a line of constant enthalpy in the compressed region is also a line of constant temperature, $T_2 = T_1 = 60\ °C$.

The entropy generation is equal to the entropy change of the stream:

$$S_{\text{gen}} = S_2 - S_1.$$

6.6 Thermodynamics of Steady-State Processes

Using the same approximation for the entropy as for the enthalpy, $S_1 \approx S_L = 0.8311$ kJ/kg K and $S_2 \approx S_L = 0.8311$ kJ/kg K, therefore,

$$S_{\text{gen}} \approx 0.$$

This is *not* to say that the process is isentropic but rather that the change in entropy is fairly small and below the accuracy of the approximation used to perform the calculations.

Throttling to 0.1 bar. At 0.1 bar the pressure is below the saturation pressure at 60 °C. The final state is in the vapor-liquid region and the temperature is equal to the saturation temperature at 0.1 bar (45.8°C). To determine the fraction of liquid and vapor, we first obtain the properties of the saturated phases at 0.1 bar:

	L	V
H (kJ/kg)	191.81	2583.9
S (kJ/kg K)	0.6492	8.1489

The liquid fraction is obtained from the known enthalpy using the lever rule:

$$x_L = \frac{H_V - H_1}{H_V - H_L} = \frac{2583.9 - 251.1}{2583.9 - 191.81} = 0.975.$$

The entropy of the exit stream is

$$S_2 = x_L S_L + (1 - x_L) S_V = (0.975)(0.6492) + (1 - 0.975)(8.1489) = 0.8357 \text{ kJ/kg K}.$$

Finally, the entropy generation is

$$S_{\text{gen}} = S_2 - S_1 = 0.8357 - 0.8311 = 0.00461 \text{ kJ/kg K}.$$

Comments To understand why there is no temperature difference when water is throttled to 0.5 bar, but there is a 14.2 °C drop when the final pressure is 0.1 bar, we examine the process on the PV graph in Figure 6-8. The initial state is compressed liquid (state A) and the isotherm AC is to a very good apprxoximation a line of constant enthalpy. As long as the final pressure is above the saturation pressure of the inlet (in this example, 0.1992 bar), the liquid will emerge from throttling at the same temperature. If the exit pressure is below the saturation pressure of the inlet, the system enters the vapor-liquid region. This means that some of the liquid evaporates, a process that requires energy. This energy comes at the expense of the fluid and as a result of it, its temperature drops. The temperature drop could be significant and throttling may be used as a way of cooling the fluid. As we will see later in this chapter, throttling is a basic component in refrigeration.

Exercise If we start with water at 60 °C, 1 bar, at what pressure should it be throttled to produce a temperature of 2 °C?

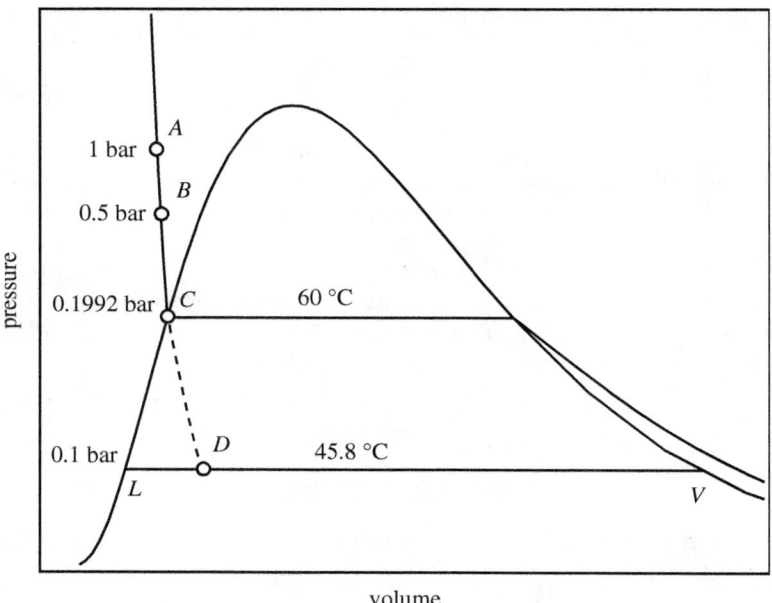

Figure 6-8: Throttling of compressed liquid (see Example 6.17).

Gas Compression

A compressor performs the opposite task of a turbine: it receives work and delivers the fluid at higher pressure. When the fluid is a liquid the device is called pump. The analysis of compressor and pumps is done on the same principles, but we will discuss each separately. There are many different designs of compressors. The axial-flow compressor is essentially a turbine working in reverse: the low pressure fluid is first accelerated by a rotating set of blades and then passes through a set of stationary blades that act as diverging nozzles and slow the flow. Multiple stages may be employed, as with turbines. Because the density of the fluid increases with pressure, the cross-sectional area of the compressor decreases in the direction of flow. Other designs use centrifugal motion, moving pistons, or helical screws to produce the compression. For the purposes of overall thermodynamic analysis, the design details are unimportant and what matters are the states of the inlet and outlet streams. We assume steady state, no heat losses to the surroundings, and negligible contribution of kinetic and potential energy terms. Under these conditions the energy balance reduces to

$$\boxed{\dot{m}\Delta H = \dot{W}_s} \quad \text{or} \quad \boxed{\Delta H = W_s} \tag{6.54}$$

6.6 Thermodynamics of Steady-State Processes

where ΔH is the enthalpy change between inlet and outlet. Similarly, the entropy generation is obtained from eq. (6.28) with $\dot{Q} = 0$:

$$\dot{S}_{\text{gen}} = \dot{m}\Delta S \quad \text{or} \quad S_{\text{gen}} = \Delta S, \tag{6.55}$$

where ΔS is the entropy change between inlet and outlet. The compressor consumes the minimum possible work under reversible operation. The actual work is higher due to irreversibilities. The efficiency of the compressor is defined as

$$\eta = \frac{W_s^{\text{rev}}}{W_s}. \tag{6.56}$$

This has the inverse form of the turbine efficiency in order to produce a number less than 100%. In the typical problem we know the inlet conditions (pressure and temperature), the outlet pressure, and the compressor efficiency. The calculation of the required work and of the exit temperature is done in complete analogy to the turbine calculation: first we compute the reversible work under the condition of isentropic operation, then we use the known efficiency to calculate the work. Finally, we calculate the enthalpy of the exit stream from the energy balance; this enthalpy and the known exit pressure are used to obtain all other properties at the outlet. These steps are demonstrated graphically on the Mollier chart in Figure 6-9.

Example 6.18: Compression of Steam
Saturated steam at 1 bar is compressed to 20 bar in a compressor with efficiency 75%. Determine the required work, exit temperature, and entropy generation.

Solution We begin by collecting the properties of steam at the inlet:

$$T_1 = 99.6 \text{ C}, \quad H_1 = 2675.4 \text{ kJ/kg}, \quad S_1 = 7.3598 \text{ kJ/kg K}.$$

Reversible operation. For reversible operation the exit state is at $P_2 = 20$ bar, $S_{2'} = S_1 = 7.3598$ kJ/kg K. By interpolation in the steam tables we find

$$T_{2'} = 474.626 \text{ K}, \quad H_{2'} = 3412.28 \text{ kJ/kg}.$$

The reversible work is

$$W_S^{\text{rev}} = H_{2'} - H_1 = (3412.28) - (2675.4) = 737.4 \text{ kJ/kg K}.$$

Actual operation. The actual work is calculated from the known efficiency:

$$W_s = \frac{W_s^{\text{rev}}}{\eta} = \frac{737.4 \text{ kJ/kg}}{0.75} = 983.2 \text{ kJ/kg}.$$

The exit enthalpy is calculated from the energy balance:

$$H_2 = H_1 + W_s = (2675.4) + (983.2) = 3658.07 \text{ kJ/kg}.$$

Finally, the state at the outlet is obtained by interpolation in the steam tables at $P_2 = 20$ bar, $H_2 = 3658.07$ kJ/kg:

$$T_2 = 585.6\ °C, \quad S_2 = 7.6679 \text{ kJ/kg K}.$$

The entropy generation is

$$S_{\text{gen}} = S_2 - S_1 = (7.6679) - (7.3598) = 0.3091 \text{ kJ/kg K}.$$

Comments The calculation can be performed graphically on the Mollier chart, as shown in Figure 6-9. The procedure is very similar to that for turbines and the calculation is left as an exercise.

Expansion and Compression of Liquids

Liquids can be expanded or compressed just as gases can. For technical reasons, the design of expansion and compression units for liquids is different from that of gases, but the thermodynamics analysis is the same. A turbine for liquids is usually called an expander, and a compressor for liquids, a pump. The analysis is based on the same equations as for gases, but we treat the subject separately because the relative incompressibility of liquids allows us to use short-cut approximations for the enthalpy and entropy changes of the fluid. The starting point is eqs. (5.29) and (5.30),

$$dH = C_P dT + V(1 - \beta T) dP, \qquad [5.29]$$

$$dS = C_P \frac{dT}{T} - \beta V dP, \qquad [5.30]$$

which give the enthalpy and entropy in terms of the coefficient of thermal expansion, β. For both expanders and pumps, the reversible work corresponds to isentropic operation between inlet state (P_1, T_1) and outlet pressure (P_2):

$$W_s^{\text{rev}} = \Delta H \Big|_{\text{isentropic}}. \qquad (6.57)$$

The isentropic enthalpy change is obtained from eqs. (5.29) and (5.30): we set $dS = 0$ in the entropy equation, solve for the term $C_P dT$ and substitute into the enthalpy equation. The result is

$$dH \Big|_{\text{isentropic}} = V dP.$$

6.6 Thermodynamics of Steady-State Processes

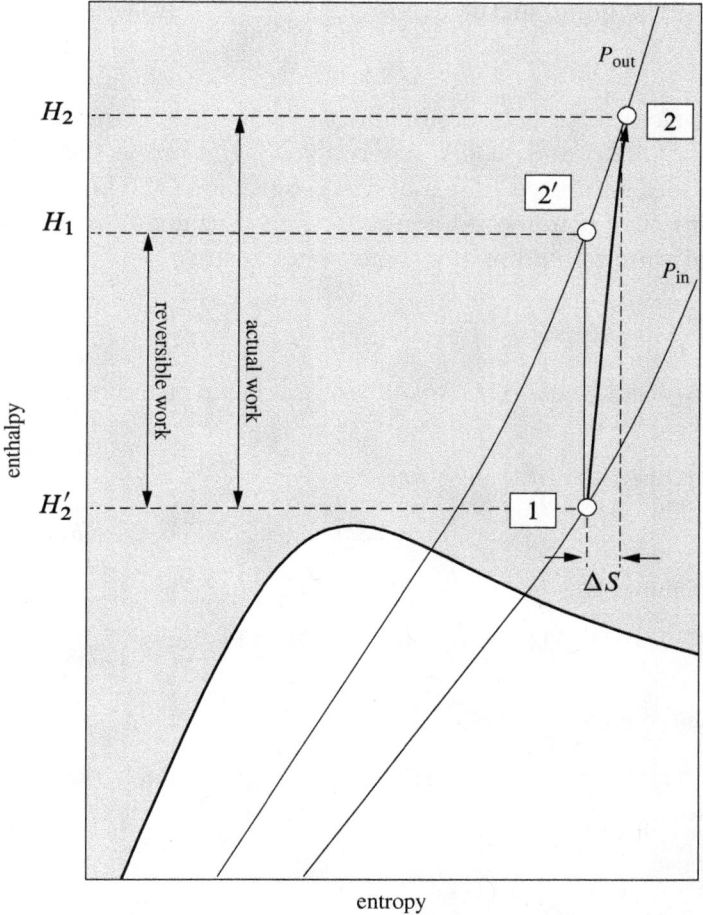

Figure 6-9: Graphical solution of compressor on Mollier graph.

We integrate between inlet and outlet treating V as constant (pressure has negligible effect on the volume of liquids), to obtain

$$W_s^{\text{rev}} = \Delta H \Big|_{\text{isentropic}} = V(P_2 - P_1) = \frac{P_2 - P_1}{\rho}, \qquad (6.58)$$

where $\rho = 1/V$ is the density of the fluid. The actual work is calculated from the known efficiency of the device:

$$W_s = \begin{cases} \eta_{\text{exp}} V(P_2 - P_1) & \text{(expander)} \\ \dfrac{V(P_2 - P_1)}{\eta_{\text{pump}}} & \text{(pump)}. \end{cases} \qquad (6.59)$$

The enthalpy of the liquid at the exit is

$$H_2 = H_1 + W_s,$$

and this value, along with pressure, define the exit state. If the value of β is known, we may use eqs. (5.29) and (5.30). Alternatively, the properties at the outlet may be calculated from saturated tables by interpolation at the known enthalpy (recall that the properties of compressed liquid are to a very good approximation equal to those of the saturated liquid at the same temperature).

Example 6.19: Calculation of Pump
Water is pumped from 1 bar, 20°C, to 20 bar. The pump efficiency is 75%. Calculate the required amount of work.

Solution The properties at the inlet state are

$$V_1 = 0.001002 \text{ m}^3/\text{kg}, \quad H_1 = 83.92 \text{ kJ/kg}, \quad S_1 = 0.2965 \text{ kJ/kg K}.$$

The work in the pump is

$$W_s = \frac{(0.001002 \text{ m}^3/\text{kg})(20-1) \times 10^2 \text{ kPa/bar}}{0.75} = 2.538 \text{ kJ/kg}.$$

The enthalpy at the pump exit is

$$H_2 = H_1 + W_s = (83.92) + (2.53) = 86.458 \text{ kJ/kg}.$$

The state at the exit of the pump is calculated by interpolation in the saturated tables at $H = 86.45$ kJ/kg:

$$T_2 = 20.6 \text{ °C}, \quad H_2 = 86.458 \text{ kJ/kg}, \quad S_2 = 0.3051 \text{ kJ/kg K}.$$

The entropy generation is

$$S_{\text{gen}} = 0.3051 - 0.2965 = 0.0086 \text{ kJ/kg K}.$$

Comments Compared to compression of steam over the same pressure difference (Example 6.18), the compression of liquid water involves much less work, and is accompanied by very small changes in the properties of the fluid except pressure. In general, the compression of liquid requires less work than the compression of gases because liquids are nearly incompressible. Consider compression of a fluid by a piston: Under an applied force F the piston moves by δx and the corresponding work is $F\delta x$. A compressible fluid yields under pressure and the displacement of the piston is substantial. For an incompressible fluid the displacement of the force is very small, and the work is correspondingly small.

6.7 Power Generation

Most industrial power generation is based on the ability of a compressed gas, often steam, to produce work through expansion in a turbine. Water from ambient conditions (a river or lake) is compressed and boiled to produce high-pressure steam, which expands in a turbine. The exhausted steam is recycled to the process to form a continuous closed loop. The basic process diagram of the power plant is shown in Figure 6-10. Here, in addition to the pump, boiler and turbine, we also encounter a condenser, which cools and condenses the exhausted steam before it is pumped again into the boiler. To see why the condenser is necessary, we examine the operation of the plant on the TS graph in Figure 6-11 (the compressed liquid region is exaggerated on this graph; in reality, lines of constant pressure lie very close to the line of saturated liquid). We trace the path of the process by starting at the pump, which compresses the liquid (state 4) to higher pressure (state 1), For isentropic compression the path on the TS graph is a vertical line moving up; if the pump efficiency is less than 100%, the actual path veers to the right, as it results in positive entropy change. The pressurized liquid is heated in the boiler to produce superheated steam (state 2), which then passes through the turbine. The exhausted stream is superheated vapor at lower pressure (condensation is to be avoided in the turbine). Since the efficiency of real turbines is less than 100%, the expansion follows a path to the

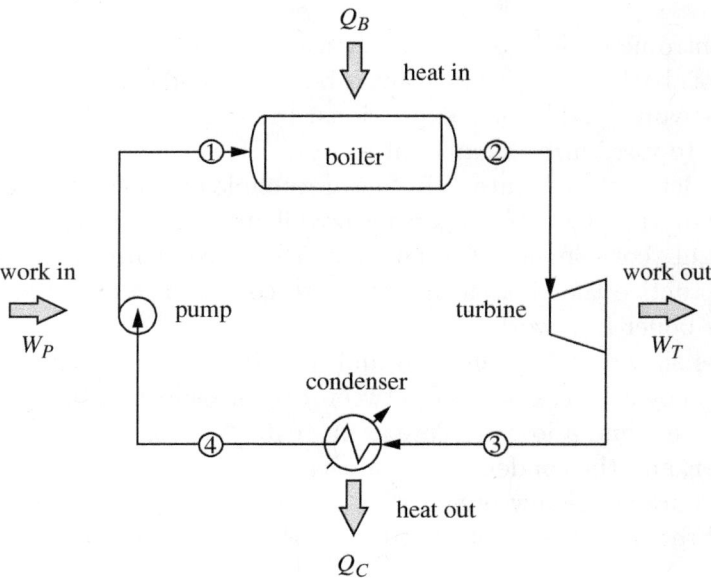

Figure 6-10: Rankine power plant.

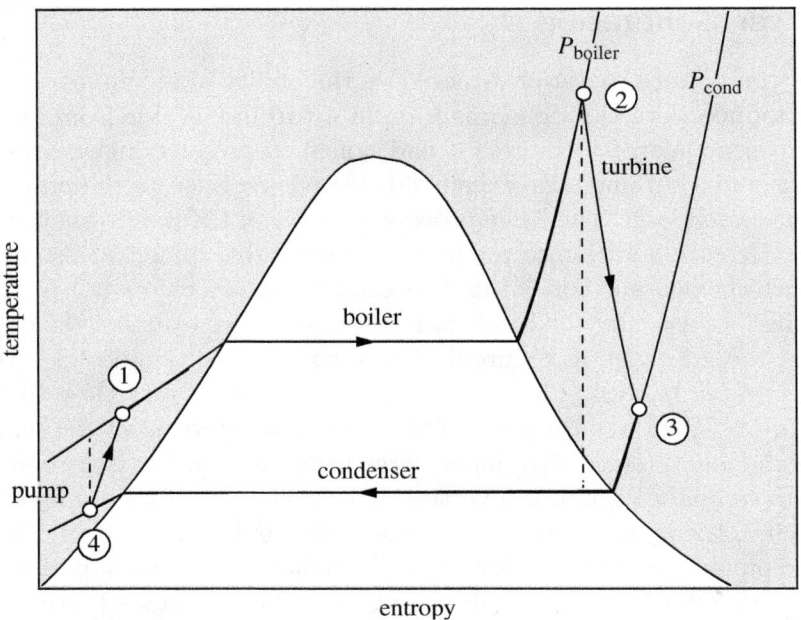

Figure 6-11: Rankine power plant on the *TS* graph. Dashed lines correspond to isentropic operation in the pump and turbine.

right of the isentropic path, shown by a vertical dashed line. To return this stream, which is still hot, back to the pump, it must be cooled and condensed. Why not cool the exhausted steam by additional expansions? In fact, steam is usually expanded in multiple stages to maximize the amount of work. However, expansion cannot move the state to the left, toward state 4. In fact, reversible expansion can only move the state vertically down (isentropic expansion), while irreversibilities cause the state to move to the right (corresponding to positive ΔS). A condenser is indeed necessary and shows in practice that it is not possible to convert the entire amount of heat that enters the boiler into work.

The process shown in Figures 6-10 and 6-11 is called a Rankine plant and in its basic form, shown here, operates between two pressures, a high pressure at the inlet of the turbine, and a low pressure at the exit. Ignoring pressure drop through pipes, the boiler, and the condenser, streams 1 and 2 are at the high pressure, while streams 3 and 4 are at the low pressure. In terms of energy inputs and outputs, the cycle absorbs heat in the boiler to produce work in the turbine. Part of the input

6.7 Power Generation

heat is rejected in the condenser. A small amount of work is also consumed in the pump. The overall balance is[5]

$$Q_B + Q_C + W_T + W_P = 0, \qquad (6.60)$$

where we understand Q_B and W_P to be positive, and Q_C and W_T to be negative. The net amount of work produced is

$$W = W_T + W_P. \qquad (6.61)$$

Since the amount of pumping work is much smaller than the work received in the turbine (see Example 6.19), there is a net production of work. The efficiency of the plant is defined as the ratio of the net work produced to the amount of heat in the boiler:

$$\eta = \frac{|W|}{|Q_B|}. \qquad (6.62)$$

This efficiency is limited by the second law: if the heat source that provides heat to the boiler is at T_B and the condenser rejects heat to a reservoir at T_C, the steady-state entropy balance gives[6]

$$S_{\text{gen}} = -\frac{Q_B}{T_B} - \frac{Q_C}{T_C}. \qquad (6.63)$$

Eliminating Q_C between eqs. (6.60) and (6.63) and solving for the ratio $-W/Q_B$, which is equal to the plant efficiency, we obtain

$$\eta = 1 - \frac{T_C}{T_B} - \frac{T_C S_{\text{gen}}}{Q_B}. \qquad (6.64)$$

Since both S_{gen} and Q_B are positive terms we must have,

$$\eta \leq 1 - \frac{T_C}{T_B}, \qquad (6.65)$$

which states that the efficiency of the plant is limited by the maximum Carnot efficiency that corresponds to the temperatures of the two heat baths. Temperature T_B corresponds to the temperature of the heat sources (for example, the flame

---

[5]. The energy terms in this balance are expressed per unit mass of the fluid. Alternatively, they can be expressed per unit time by multiplying the above values by the mass flow rate.
[6]. Start with the steady-state entropy balance, eq. (6.28), set $\Delta(\dot{m}S)$ to zero because the fluid operates in a cycle, and note that the plant exchanges heat with two baths.

temperature of coal combustion, if this is a coal-fired plant) and must be at least equal to the temperature of stream 2, which is the hottest stream in the process. Temperature T_C may be taken to be the same as T_4 (the coldest temperature in the plant), or equal to the temperature of the surroundings, since the cooling medium is usually ambient water or air.[7]

Example 6.20: Rankine Plant
A Rankine power plant using steam operates between pressures 60 bar and 1.013 bar. The turbine expands steam from 500 °C, 60 bar, to 1.013 bar. The liquid at the exit of the condenser is saturated. The efficiency of the turbine and of the pump is 75%. Calculate the energy balances, entropy generation, and thermodynamic efficiency. What are the sources of irreversibility?

Solution The key to performing calculations of this type is to determine the states of all streams. This means, mass flow rate, pressure, temperature, enthalpy, entropy, and phase; if the phase is a vapor-liquid mixture, we must also report the vapor or liquid fraction. With simple flow sheets involving a common flow rate for all streams, the flow rate does not have to be included and all energy terms are reported per kg of fluid. The solution can be organized by forming a table with the properties of all streams and update its entries as new quantities are computed. We will use the numbering of the streams shown in Figure 6-10. Notice that we know the pressure in all streams ($P_1 = P_2 = 60$ bar, $P_3 = P_4 = 1$ bar). We know the temperature of stream 2 (500 °C); therefore, all properties of this stream are known. We also know that stream 4 is saturated liquid at 1 bar; this information fixes the properties of this stream. The unknown streams are number 1 and 3. These will be calculated by energy balance in the pump and in the turbine. Therefore, the calculation of this plant reduces to one calculation around the pump and one around the turbine. In both cases we know the inlet streams and the efficiency of the unit; therefore, the calculation will proceed similar to Examples 6.19 and 6.13.

Calculation of turbine (Outline). The turbine exit for reversible operation is found at $P_3 = 1$ bar, $S_3 = S_2 = 6.8824$ kJ/kg K. This state is found to be in the vapor liquid region. By lever rule we find $H_{3'} = 2497.32$ kJ/kg from which we obtain the reversible work:

$$W^{\text{rev}} = -925.579 \text{ kJ/kg}.$$

The actual enthalpy of stream 3 is

$$H_3 = H_2 + \eta W^{\text{rev}} = 2728.72 \text{ kJ/kg}.$$

The temperature and entropy of stream 3 is obtained by interpolation at $P = 1$ bar, $H = 2728.72$ kJ/kg. We find

$$T_3 = 126.2 \text{ °C}, \quad S_3 = 7.4942 \text{ kJ/kg K}.$$

---

7. In cooling towers, often seen in large power plants, the cooling is provided by the atmosphere.

6.7 Power Generation

Calculation of pump (Outline). The specific volume of saturated water at 1 bar is $V = 0.00104$ m^3/kg. The work in the pump is

$$W_{\text{pump}} = \frac{V(P_1 - P_4)}{\eta} = 8.18133 \text{ kJ/kg}.$$

The enthalpy of stream 1 is

$$H_1 = H_4 + W_{\text{pump}} = 425.621 \text{ kJ/kg}.$$

The temperature and entropy of stream 1 is obtained by interpolation in the saturated liquid tables at $H_L = H_1 = 425.621$ kJ/kg. We find

$$T_1 = 101.5 \text{ °C}, \quad S_1 = 1.3245 \text{ kJ/kg K}.$$

Summary of results. The results of these calculations are summarized in the table below:

	1	2	3	4
P (bar)	60	60	1	1
T (°C)	101.5	500	126.2	99.61
H (kJ/kg)	425.621	3422.9	2728.72	417.44
S (kJ/kg K)	1.3245	6.8824	7.4942	1.3026
phase	comp. liq.	super'd vap.	superh'd vap.	sat. liq

The energy balances are:

$$
\begin{aligned}
Q_B &= H_2 - H_1 &&= 2997.28 \text{ kJ/kg} \\
Q_C &= H_4 - H_3 &&= -2311.28 \text{ kJ/kg} \\
W_{\text{turb}} &= H_3 - H_2 &&= -694.18 \text{ kJ/kg} \\
W_{\text{pump}} &= H_1 - H_4 &&= 8.18 \text{ kJ/kg} \\
W &= W_{\text{turb}} + W_{\text{pump}} &&= -686.00 \text{ kJ/kg} \\
\text{plant efficiency} &= |W|/|Q_B| &&= 22.9 \%
\end{aligned}
$$

To see how the plant efficiency compares to the theoretical maximum, we take T_B to be the temperature of stream 2 (T_B must be at least as high as T_2) and T_C to be T_4 (T_C should be at most as high as T_4):

$$\eta_{\text{max}} = 1 - \frac{(99.61 + 273.15) \text{ K}}{(500 + 273.15) \text{ K}} = 52\%.$$

The actual efficiency is less than half of the theoretical. If we take into account that the theoretical value is in fact a *low* estimate (T_B is actually higher than T_2 and T_C is lower than T_4), the actual operation fares even worse in comparison to the maximum possible. To assess the sources of this inefficiency, we perform an entropy analysis for each unit of the process.

Entropy Analysis The entropy generation in the boiler is

$$S_{\text{gen}}\bigg|_{\text{boiler}} = S_2 - S_1 - \frac{Q_B}{T_B} = 1.6812 \text{ kJ/kg K}.$$

Similarly for the condenser:

$$S_{\text{gen}}\Big|_{\text{cond}} = S_4 - S_3 - \frac{Q_C}{T_C} = 0.0089 \text{ kJ/kg K}.$$

The turbine operates adiabatically, therefore, the entropy generation is

$$S_{\text{gen}}\Big|_{\text{turb}} = S_3 - S_2 = 0.6118 \text{ kJ/kg K},$$

and similarly for the pump:

$$S_{\text{gen}}\Big|_{\text{pump}} = S_1 - S_4 = 0.0219 \text{ kJ/kg K},$$

These results are summarized below, along with the corresponding value for the lost work ($T_0 = 300$ K):

	S_{gen} (kJ/kg K)	$T_0 S_{\text{gen}}$ (kJ/kg)
Boiler	1.6812	504.4
Condenser	0.0089	2.7
Turbine	0.6118	183.5
Pump	0.0219	6.6
Total	2.3237	697.1

The largest contribution to entropy generation comes from the boiler. This is due to the large temperature difference between the temperature of the heat source and the stream that is being heated. The second most important contribution comes from the turbine, whose efficiency is 75%. The pump contributes very little because the work in the pump is pretty small. The condenser generates very little entropy because the temperature of the fluid changes very little between inlet and outlet, and since T_C has been assumed to be equal to the exit temperature, the overall gradient between the heat sink and the stream being cooled is small. In reality, T_C is closer to room temperature, as the condenser probably operates using ambient cooling water. In such case, the losses in the condenser would be larger. The ideal work, calculated using $T_0 = 300$ K, translates the entropy generation into a more tangible quantity, work. This is an estimate of the work that would be produced if the corresponding entropy generation term were driven to zero. The total lost work is about the same as the work produced, which is another way of saying that the plant operates at about half of its maximum efficiency. If we should commit the resources to improve the process, most effort should focus on the boiler and the turbine, since these are the major contributors to lost work. Improving the turbine amounts to replacing it with a unit with better efficiency. With respect to the boiler, one must reconsider the entire design of the plant in order to achieve a more efficient operation.

6.8 Refrigeration

To achieve temperatures below ambient we must use some type of refrigeration. In general, this requires work in order to maintain a temperature gradient between the desired low temperature and that of the surroundings. A common design is based on a familiar principle: if we are wet and sit by a breeze, it feels cool. This is because evaporation requires heat and this is drawn from our body, causing the temperature to drop. The breeze accelerates the evaporation rate by removing vapors from the immediate vicinity of the liquid, thus intensifying the cool feeling of coolness. However, it is the evaporation, not the breeze, that is responsible for the main effect. This situation can be replicated experimentally by expanding a liquid into the vapor-liquid region. As we saw in Example 6.17, if a compressed liquid is throttled (or expanded through a turbine) into the two-phase region, the temperature drops to the saturation temperature that corresponds to the pressure of the expanded liquid. Accordingly, the temperature is controlled by the pressure of the expanded liquid. This phenomenon forms the principle of operation of most refrigeration processes, from household refrigerators and to air-conditioning units, to industrial cryogenic processes. The basic vapor compression cycle is shown in Figure 6-12. Compressed liquid (stream 1) is throttled to lower pressure such that

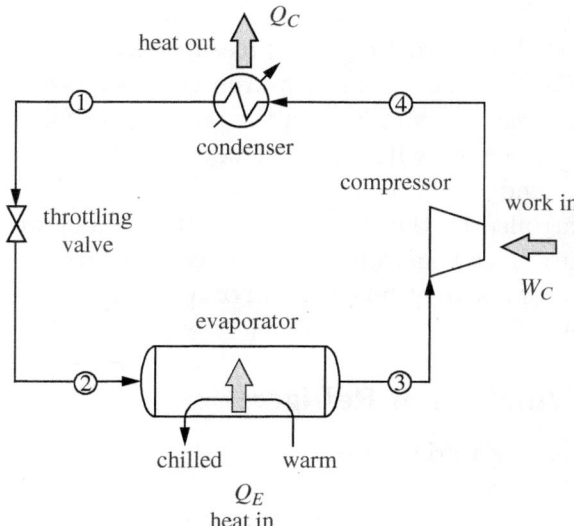

Figure 6-12: Vapor-compression refrigeration cycle.

the state at the exit of the valve is a vapor-liquid mixture (state 2). This cold stream passes through a heat exchanger in which the vapor-liquid mixture evaporates by drawing heat from the warm stream that is to be cooled. The refrigerant exits the evaporator as saturated vapor (stream 3). To return to throttling, the vapor must be compressed. The state at the exit of the compressor is superheated vapor (state 4). To regenerate the compressed liquid, the vapor is condensed in a heat exchanger (condenser), which completes the cycle. In terms of energy exchanges, the cycle absorbs heat in the evaporator, consumes work in the compressor, and rejects heat in the condenser. The net effect is to "pump" heat from the low temperature of the evaporator to the higher temperature in the condenser through the input of work. This cycle is essentially the reverse of the Rankine power plant. The throttling valve may be replaced by an expander; this would accomplish the expansion while also producing some work, which could be used towards the compression. Most units utilize a throttling valve instead because it is much simpler to operate and maintain, but also because the amount of work that can be extracted from the expansion of a liquid is rather small.

Refrigeration cycles are often represented on the PH chart, which may also be used to perform a graphical solution of the energy balances. Figure 6-13 shows the path of the vapor compression cycle on this chart. Ignoring pressure drops in pipes and heat exchangers, the cycle operates between two pressures, a high pressure that is achieved by the compressor, and a lower pressure that is accomplished by throttling. On the PH chart throttling is represented by a vertical line at constant enthalpy. Heat transfer in the evaporator and in the condenser takes place at constant pressure. For a reversible compressor, the path would follow a line of constant entropy; for a real compressor with an efficiency less than 100%, the compressed vapor emerges hotter and to the right of the isentropic path. The temperature of the evaporator is controlled by the throttling pressure. The lower this pressure, the colder the temperature. It is generally desirable, however, to maintain the pressure in the evaporator near or above atmospheric in order to avoid the expense of vacuum equipment.

Thermodynamic Analysis of Refrigeration

The energy balance on the refrigeration cycle is

$$Q_E + Q_C + W_C = 0, \qquad (6.66)$$

where Q_E is the amount of heat removed from the evaporator ($Q_E > 0$), Q_C is the amount of heat rejected in the condenser ($Q_C < 0$), and W_C is the work for compression ($W_C > 0$). The entropy balance is obtained by noting that the refrigerant operates as a cycle and thus does not contribute to the generation of entropy.

6.8 Refrigeration

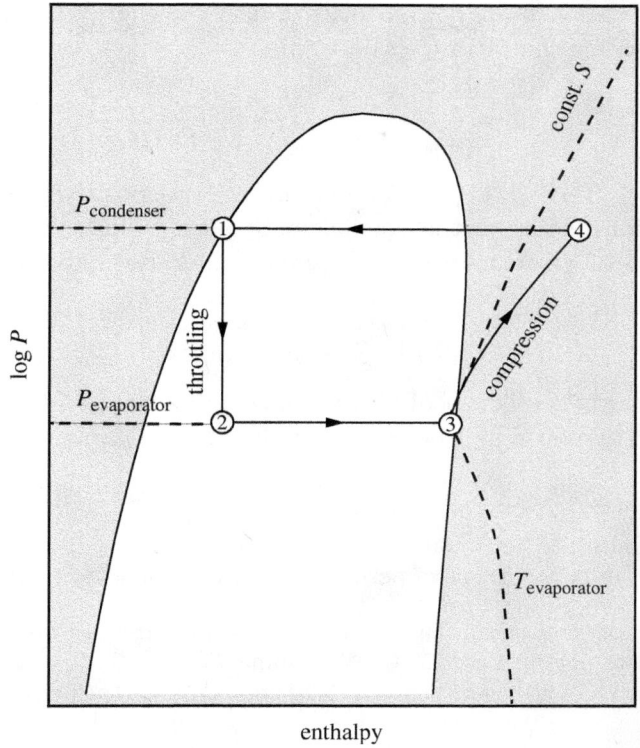

Figure 6-13: Refrigeration cycle on the *PH* graph.

Treating the evaporator and the condenser as baths at temperatures T_E and T_C, respectively, the entropy generation is

$$S_{\text{gen}} = -\frac{Q_C}{T_C} - \frac{Q_E}{T_E}. \tag{6.67}$$

The performance of refrigeration cycles is assessed via the coefficient of performance (cop), which is defined as the amount of heat extracted in the evaporator over the amount of work in the compressor:

$$\text{cop} = \frac{|Q_E|}{|W_C|} = \frac{Q_E}{W_C}. \tag{6.68}$$

This ratio represents the refrigeration capacity over the refrigeration "cost" in terms of work. Unlike the efficiency of the power plant, the coefficient of performance is not restricted to be less than one. Its maximum value is, however, limited by the second law. To determine the limiting value of the coefficient of performance, we solve the entropy balance for Q_C, substitute into the energy balance and solve the resulting equation for the ratio Q_E/W; we find:

$$\text{cop} = \frac{T_E}{T_C - T_E + (S_{\text{gen}} T_C T_E)/Q_E}. \tag{6.69}$$

The quantity $(S_{\text{gen}} T_C T_E)/Q_E$ in the denominator is positive, which implies that entropy generation decreases the coefficient of performance. The maximum value of the coefficient of performance is obtained under reversible operation. Setting $S_{\text{gen}} = 0$ we find

$$\text{cop}\Big|_{\text{max}} = \frac{T_E}{T_C - T_E}. \tag{6.70}$$

This is the maximum coefficient of performance and corresponds to a Carnot refrigerator that operates reversibly between temperatures T_E and T_C.

Example 6.21: Household Refrigerator
Estimate the maximum coefficient of performance of a household refrigerator.

Solution The lowest temperature in a refrigerator is in the freezer, which is typically maintained so that it does not exceed $-15\,°\text{C}$. We assume $T_E = -15\,°\text{C} = 258$ K. The condenser is air cooled and we will assume $T_C = 300$ K. The maximum coefficient of performance is

$$\text{cop}\Big|_{\text{max}} = \frac{258}{300 - 258} = 6.14.$$

That is, under reversible operation, each kJ consumed in the compressor pumps 6.14 kJ of heat in the evaporator. Actual appliances have a coefficient of performance closer to 4.

Example 6.22: Ammonia Refrigerator
Design a vapor compression cycle using ammonia as the refrigerant that will be used to maintain the temperature of a stream at $-20\,°\text{C}$. The condenser is air cooled and the ambient temperature may be assumed to be $25\,°\text{C}$. Assume a compressor efficiency of 80% and allow for $2\,°\text{C}$ of temperature difference between streams in the evaporator and condenser for practical heat transfer.

Solution We use the NIST WebBook for the properties of ammonia. We will set the evaporator at $-22\,°\text{C}$, $2\,°\text{C}$ below the required temperature. The saturation pressure at this temperature is 1.7379 bar. The condenser uses air at $26\,°\text{C}$ as the cooling medium. Allowing for a $2\,°\text{C}$ temperature difference, the temperature at the exit of condenser must be at least $28\,°\text{C}$. This corresponds to saturation pressure 10.993 bar and sets the compression

6.8 Refrigeration

pressure of the cycle. Thus we have determined the two pressures of the cycle. We set the state at the exit of the condenser to be saturated liquid, and at the exit of the evaporator to be saturated vapor. With reference to Figure 6-12, the states of streams 1 and 3 are fixed (saturated phases at known pressure) and their properties are shown below (also notice that the pressure of all streams are known).

	1	2	3	4
T (°C)	28	−22	−22	
P (bar)	10.993	1.7379	1.7379	10.993
H (kJ/kg)	475.25		1578.1	
S (kJ/kg K)	1.9281		6.4067	
Phase	sat L		sat V	

The remaining unknown streams are calculated by performing standard energy balances around the compressor and the throttling valve.

Compressor. It will be convenient to use the NIST WebBook to tabulate enthalpy and entropy at 10.993 bar over small increments of temperature. This will facilitate the interpolations needed in the calculation of the compressor. The following data were obtained at 20 °C increments:

	10.993 bar		
T (°C)	110	130	150
H (kJ/kg)	1848.3	1897.6	1946.9
S (kJ/kg K)	6.4061	6.5316	6.6509

For isentropic operation, the exit state is obtained by interpolation at $P_4 = 10.993$ bar, $S_{4'} = S_3 = 6.4067$ kJ/kg K. We find,

$$T_{4'} = 110.096 \text{ C}, \quad P_4 = 10.993 \text{ bar}, \quad H_{4'} = 1848.54 \text{ kJ/kg}, \quad S_{4'} = 6.4067 \text{ kJ/kg K}.$$

The reversible work is

$$W^{\text{rev}} = 1848.54 - 1578.1 = 270.436 \text{ kJ/kg},$$

and the enthalpy of stream 4 is found to be

$$H_4 = 1578.1 + \frac{270.436}{0.8} = 1916.1 \text{ kJ/kg}.$$

The remaining properties of stream 4 are obtained by interpolation at $P_4 = 10.993$ bar, $H_4 = 1916.1$ kJ/kg. We find

$$T_4 = 137.5 \text{ °C}, \quad S_4 = 6.5764 \text{ kJ/kg K}.$$

Throttling valve. Stream 2 is a vapor-liquid mixture with $H_2 = H_1 = 475.25$ kJ/kg K. The enthalpy and entropy of the saturated phases is

$$H_L = 242.67 \quad H_V = 1578.1 \quad (\text{kJ/kg})$$
$$S_L = 1.0896 \quad S_V = 6.4067 \quad (\text{kJ/kg K}).$$

The liquid fraction is obtained by the lever rule:

$$x_L = \frac{1578.1 - 475.25}{1578.1 - 242.67} = 0.826,$$

and the entropy of stream 2 is

$$S_2 = (0.826)(1.0896) + (1 - 0.826)(6.4067) = 2.0148 \text{ kJ/kg K}.$$

Summary of streams. The complete table of streams is shown below:

	1	2	3	4
T (°C)	28	−22	−22	137.5
P (bar)	10.993	1.7379	1.7379	10.993
H (kJ/kg)	475.25	475.03	1578.1	1916.1
S (kJ/kg K)	1.9281	2.0148	6.4067	6.5764
Phase	sat L	($x_L = 0.826$)	sat V	s/h V

Having evaluated all streams, we proceed to calculate the energy and entropy balances.

Energy balances. The energy balances in the evaporator, condenser, and compressor are:

$$\begin{aligned} Q_E &= H_3 - H_2 = 1103.07 \text{ kJ/kg} \\ Q_C &= H_1 - H_4 = -1440.85 \text{ kJ/kg} \\ W_C &= H_4 - H_3 = 338 \text{ kJ/kg} \\ \text{cop} &= Q_E/W_C = 3.26 \end{aligned}$$

The maximum coefficient of performance is estimated using $T_E = -20$ C $= 253.15$ K, and $T_C = 25$ °C $= 299.15$ K, corresponding to the temperatures of the baths that exchange heat with the evaporator and condenser, respectively. We find,

$$\text{cop}\Big|_\text{max} = \frac{253.15}{299.15 - 253.15} = 5.50$$

Entropy balances. The entropy balances are:

Evaporator	S_gen	$= S_3 - S_2 - Q_E/T_E =$	0.034523	kJ/kg K
Condenser	S_gen	$= S_1 - S_4 - Q_C/T_C =$	0.16818	kJ/kg K
Compressor	S_gen	$= S_4 - S_3 =$	0.1697	kJ/kg K
Throttling	S_gen	$= S_2 - S_1 =$	0.0867	kJ/kg K
Total			0.459103	

The main contributions come from the compressor and the condenser. The throttling valve contributes relatively little, corresponding to lost work of (0.0867 kJ/kg K)(300 K) = 26 kJ/kg, or about 7% of the work in the compressor. The contribution from the evaporator is also small because of the small temperature differential between the refrigerant, whose temperature stays constant during evaporation, and the chilled stream, which has been assumed to be at −20 °C. By contrast, the condenser operates under a larger temperature difference, air at 25 °C and superheated ammonia vapor at 137.5 °C.

Refrigerants

Refrigerants are classified by a code of the form RXYZ, where XYZ is a unique number assigned to the fluid. These codes are assigned by the American Society of Heating, Refrigerating and Air-Conditioning Engineers (ASHRAE). Water, for example is given the refrigerant code R718. Some common refrigerants are given in Table 6-1.[8] In principle any fluid can be used as refrigerant in a vapor compression cycle, however, some physical and practical limitations apply. For example, water freezes at 0 °C, therefore it cannot be used in subfreezing temperatures. Other limitations are imposed by practical operating constraints. It is generally desirable to operate at pressures above ambient. Lower pressures require vacuum equipment, which is more expensive. A second consideration is with respect to the operation of the condenser. The temperature at the compressor exit must be at least 20 °C or so to allow the use of ambient water or air as the coolant. This temperature determines the saturation pressure in the condenser and the minimum pressure that must be achieved by compression. The saturation pressures in the evaporator and condenser set the pressures of the cycle and the demand of the compressor. For this reason, the saturation pressure is the most important property of the fluid with respect to refrigeration.

Figure 6-14 shows the saturation pressure of some common refrigerants. If we are designing a refrigeration cycle with an evaporator at −20 °C, we could select ammonia, R12 or R134, which are very similar to each other. This would place the evaporator pressure at about 1 bar and the condenser at about 10 bar, requiring a compression ratio of 10:1. Ethane, on the other hand, would not be appropriate because it would require operation between ∼ 10 and 60 bar, a considerably higher pressure. If the evaporator requires much lower temperatures, say, −100 °C, ethane

Table 6-1: Some common refrigerants and their ASHRAE designation.

ASHRAE #	Name	Formula
R10	Carbon tetrachloride	CCl_4
R12	Dichlorodifluoromethane	CCl_2F_2
R114	1,2-Dichlorotetrafluoroethane	$C_2F_4Cl_2$
R134a	1,1,1,2-Tetrafluoroethane	$C_2H_2F_4$
R290	Propane	C_3H_8
R717	Ammonia	NH_3
R718	Water	H_2O
R744	Carbon dioxide	CO_2

---

8. For an explanation of the naming algorithm see www.epa.gov/ozone/geninfo/numbers.html.

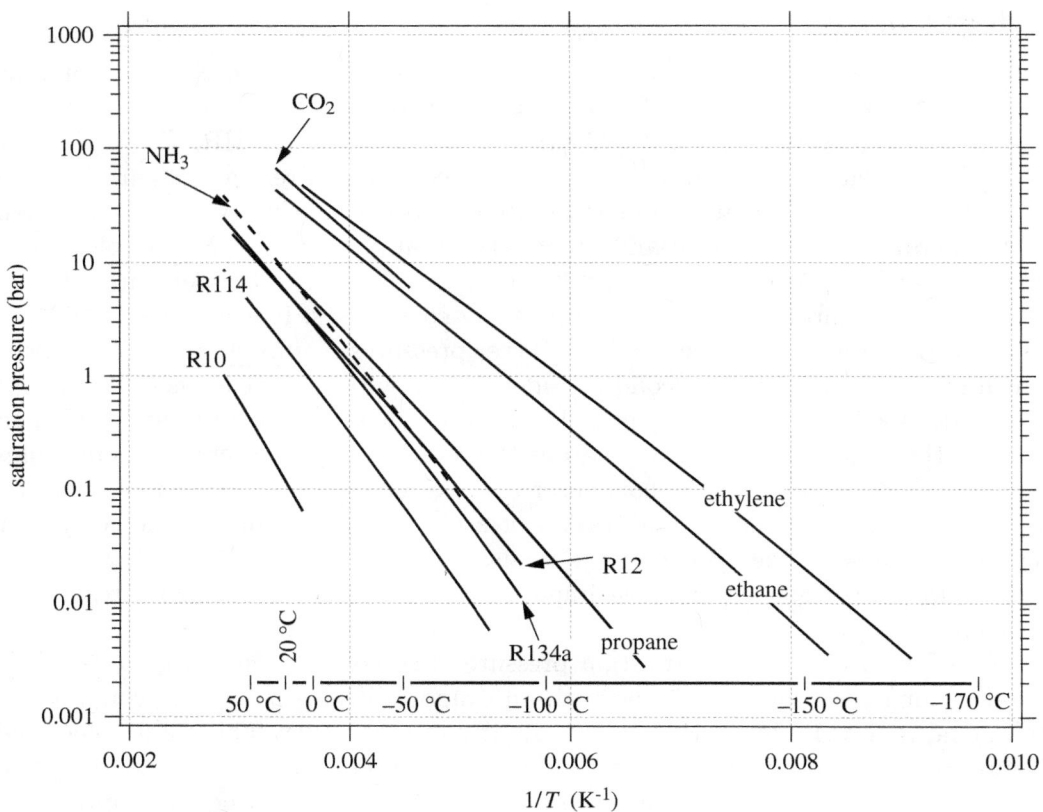

Figure 6-14: Saturation pressure of some common refrigerants.

(normal boiling point about −89 °C) would be a good candidate. This cycle would require a pressure of about 0.5 bar in the evaporator but at least 40 bar in the condenser. If the condenser is operated at lower pressure, condensation will require refrigeration. This is often more desirable than operating a single stage up to a very high pressure. This cascade design then, uses two refrigerators, one to produce the required low temperature, and another to condense the compressed vapor of the low-temperature refrigerant.

NOTE

Refrigerants and the Environment
An important consideration in choosing the refrigerant is chemical toxicity. Chlorofluorocarbons, or CFCs, have the general chemical formula $C_nF_mCl_{2n+2-m}$ and are remarkably inert. These fluids were developed after World War II and found widespread use as refrigerants (known

under the trade name Freon), but also as fire retardants and as propellants in household spray cans. An unintended consequence of their inert nature is that they do not decompose easily and have long lifetimes in the atmosphere, where they participate in reactions that deplete the ozone in the stratosphere. Since the late 70s their use is being phased out. Their replacements are molecules that are not fully halogenated but contain some hydrogen. One such molecule is tetrafluoroethane, also known as HFC-134a. The ozone impact of gases is reported as its ozone depletion potential (ODP), which measures the chemical effect of the molecule relative to that of trichlorofluoromethane (R11). Another property is the global warming potential (GWP), which characterizes the impact of the gas on global warming relative to that of carbon dioxide. For a list of ODP and GWP factors of various refrigerants, see www.epa.gov/ozone/science/ods/index.html.

6.9 Liquefaction

One application of refrigeration is in the liquefaction of air, nitrogen, and other gases that under ambient conditions are above their critical temperature. There are several industrial applications for liquefied gases. The separation of air into its components by distillation is the basis for the industrial production of inert gases such as nitrogen, argon, and helium. Liquid oxygen is used as a rocket fuel; natural gas (mostly methane) is liquefied for the purposes of increasing its density for transportation and storage purposes; liquid nitrogen provides a convenient heat bath for low temperature experiments. To appreciate the magnitude of the refrigeration problem, consider the liquefaction of nitrogen. Figure 6-15 shows the qualitative PH graph of nitrogen. At ambient conditions (state A) nitrogen is a gas above its critical temperature (-144 °C). To produce liquid nitrogen at 1 bar, the temperature must be lowered to -196 °C, its normal boiling point. The system can be brought into the two-phase region by throttling from a high pressure (state D). To produce such state, the nitrogen must be compressed (state B) and then cooled under constant pressure. Part of the required cooling can be done with water, but this will only cool the hot vapor to about 25 °C (state C). Further cooling to state D requires refrigeration. This may be provided by the cold vapor that is produced at the exit of the throttling valve.

A process that accomplishes these steps is shown in Figure 6-16. Liquid nitrogen is produced by throttling the compressed stream 5 to 1 bar (stream 6). The liquid fraction (stream 7) is collected as the product, and the vapor (stream 8) is used to chill the compressed nitrogen vapor, after it has been cooled to room temperature by a precooler using ambient water. This vapor is then mixed with the fresh nitrogen stream and is recycled through the process. This is known as the Linde liquefaction process. There is no external refrigeration, instead, nitrogen acts as its

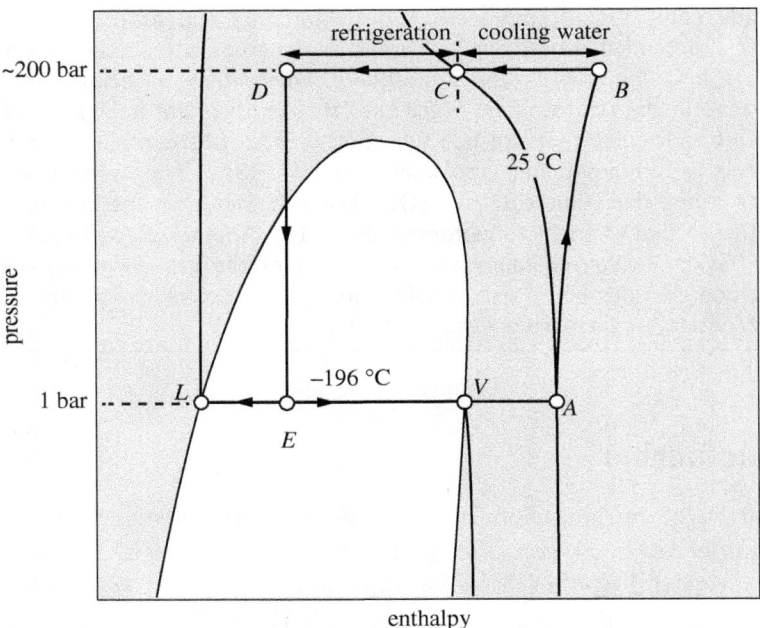

Figure 6-15: Illustration of the liquefaction of nitrogen on the *PH* graph (not to scale).

own refrigerant. For this process to work, the compressor pressure must be fairly high so that the precooler can accomplish a significant part of the cooling. If not, as more of the heat load is shifted from the precooler to the cooler, the process reaches a point where it does not produce enough cold vapor to sustain the required temperature of stream 5.

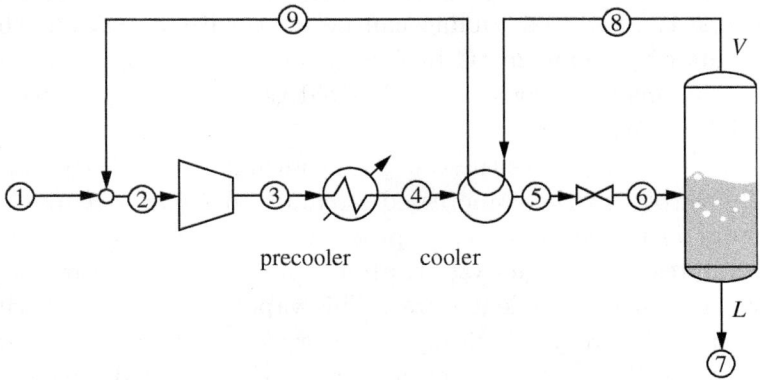

Figure 6-16: Linde process for gas liquefaction.

6.9 Liquefaction

Example 6.23: Liquefaction of Nitrogen

Nitrogen is liquefied in the process shown in Figure 6-16. Nitrogen gas is fed at 25 °C, 1 bar (stream 1). The gas is compressed to 180 bar in a compressor with an efficiency of 85%. The precooler uses ambient water as the cooling medium and cools the nitrogen to 30 °C. The cooler uses cold vapor from the separator as the cooling medium; this vapor leaves the cooler at 25 °C. Perform the energy and entropy balances on the basis of 1 kg/s of liquid nitrogen produced.

Solution We will use the NIST WebBook to calculate the properties of nitrogen. The pressures of all streams are known: streams 1, 2, 6, 7, 8, and 9 are at 1 bar; streams 3, 4, and 5 are at 180 bar. Streams 6, 7, and 8 is the boiling temperature of nitrogen at 1 bar (-195.91 °C). We notice now that the state of streams 1, 7, 8, and 9 is fully defined and their properties may be collected from the Webbook (see summary table below). In addition to the usual properties (P, T, H, S, and phase) we must also include in the table of streams the flow rates, since these are not the same among all streams. We set the basis to be 1 kg of liquid nitrogen in stream 7, and the unknown flow rate of vapor stream 8 to be x. By straightforward material balance, the flow rate of stream 1 is 1 kg/s, the flow rate of streams 8 and 9 are x, and all other flow rates are $x+1$. The known properties of the streams are summarized below:

	1	2	3	4	5	6	7	8	9
$\dot{m}$	1	$x+1$	$x+1$	$x+1$	$x+1$	$x+1$	1	x	x
P	1	1	180	180	180	1	1	1	1
T	25			30		196	196	196	25
H	309.27			285.15			-122.25	77.073	309.27
S	6.8392			5.2159			2.8312	5.411	6.8392

Here, x is the mass flow rate in stream 8 on the basis of 1 kg of liquid nitrogen, P is in bar, T is in °C, H is kJ/kg, and S is kJ/kg K.

First we perform an energy balance around a subsystem defined by the cooler, throttling valve, and separator. This part of the process exchanges no heat of work with the surroundings and involves three streams, stream 4 going in and streams 7 and 9 coming out.

Cooler/throttling/separator. The unknown flow rate x may be determined by application of the energy balance:

$$xH_9 + (1)H_7 = (1+x)H_4 \quad \Rightarrow \quad x = \frac{H_7 - H_4}{H_4 - H_9} = 16.89 \text{ kg/s}.$$

With this, all mass flow rates are now known. Since all vapor in stream 6 exits as stream 8 and all liquid exits as stream 7, the liquid and vapor fractions in stream 6 are

$$x_L = \frac{m_7}{m_7 + m_8} = 0.0559, \quad x_V = 1 - x_L = 0.9441.$$

The properties of stream 6 are now calculated by the lever rule:

$$H_6 = x_L H_L + x_V H_V = (0.0559)(-122.25) + (0.9441)(77.073) = 65.9314 \text{ kJ/kg}$$

$$S_6 = x_L S_L + x_V S_V = (0.0559)(2.8312) + (0.9441)(5.4116) = 5.26736 \text{ kJ/kg K}.$$

The enthalpy of stream 5 is

$$H_5 = H_6 = 65.9314 \text{ kJ/kg}.$$

The remaining properties are obtained by interpolation at P = 180 bar, H = 65.9314 kJ/kg. We find,

$$T_5 = -107.9 \text{ °C}, \quad S_5 = 4.2246 \text{ kJ/kg K}.$$

Mixing point. Stream 2 is calculated by energy balance around the mixing point. In this particular case the calculation is trivial because streams 1 and 9 that are mixed are in the same state; therefore, the resulting stream 2 is also at the same state.

Compressor. Stream 3 is obtained by energy balance around the compressor. This is done in the usual two-step method by first calculating the exit state for reversible operation, then for actual operation. The results of these calculations are summarized in table of streams below in which state 3' refers to reversible compression:

	2	(3')	3
P (bar)	1	180	180
T (°C)	25	949.1	1099.57
H (kJ/kg)	309.27	1355.7	1540.37
S (kJ/kg K)	6.8392	6.8392	6.98161

The results of these calculations are summarized in the table below:

	1	2	3	4	5	6	7	8	9
$\dot{m}$	1	17.89	17.89	17.89	17.89	17.89	1	16.89	16.89
P	1	1	180	180	180	1	1	1	1
T	25	25	1099.57	30	-107.856	-196	-196	-196	25
H	309.27	309.27	1540.37	285.15	65.93	65.93	-122.25	77.073	309.27
S	6.8392	6.8392	6.9816	5.2159	4.2246	5.2674	2.8312	5.4110	6.8392
Phase	V	V	s/c	s/c	s/c	$x_L = 5.59\%$	sat L	sat V	V

V = vapor; L = liquid, s/c = supercritical.

Energy and entropy balances. With all the streams known, the energy and entropy balances are calculated in the usual way. The energy balances are calculated by applying the steady-state energy balance,

$$\Delta(H\dot{m}) = \dot{Q} + \dot{W}.$$

This equation is applied to each unit noting that the compressor involves only work, the precooler involves only heat, and the rest of the units involve neither. Notice that the cooler

6.9 Liquefaction

exchanges heat between two streams of the process, so the net heat is reported as zero. The heat *load* in this heat exchange is

$$Q_{\text{cooler}} = |m_9 H_9 - m_8 H_8| = |m_5 H_5 - m_4 H_4| = 3921.81 \text{ kW}.$$

The entropy balances are calculated using the steady-state entropy balance equation:

$$S_{\text{gen}} = \Delta(S\dot{m}) - \frac{\dot{Q}}{T_{\text{bath}}},$$

where $\dot{Q}$ is the amount of heat exchanged with an external heat bath and T is the temperature of that bath. Only the precooler exchanges heat with the surroundings (cooling water). Taking the temperature of cooling water to be 25 °C we set $T_{\text{bath}} = 298.15$ K.

The results are summarized in the table below:

Unit	$\Delta(H\dot{m})$ (kW)	$\dot{Q}$ (kW)	$\dot{W}$ (kW)	$\dot{S}_{\text{gen}}$ (kW K)	$T_0 \dot{S}_{\text{gen}}$ (kW)
Mixing	0	—	—	0	0
Compressor	22,024.3	—	22,024.3	2.5478	764
Precooler	−22,455.8	−22,455.8	—	5,051.8	1,515,525
Cooler	0	—	—	6.3774	1,913
Throttling	0	—	—	18.66	5,597
Separator	0	—	—	0	0
Total	−431.5	−22455.8	22024.3	5079.33	1,523,800

(All values per 1 kg/s of liquid nitrogen produced)

Comments

- For every kg of liquid nitrogen produced, 16.89 kg of nitrogen must circulate through the system in a closed loop. This flow acts as the refrigerant: it is compressed, cooled, and throttled, as in usual refrigeration cycles. The only difference is that instead of using a standard evaporator to cool the incoming stream, it is directly mixed with it.

- The large recirculation flow also means the liquid fraction in stream 6 is small, only 5.6%. In this respect, the situation is somewhat different from that in Figure 6-15 in that state E (corresponding to stream 6) is much closer to state V that this figure shows. In other words, the process barely brings the state of the throttled fluid into the two-phase region. Producing more liquid would require further cooling in the cooler, and higher pressure in the compressor.

- The compression ratio (180:1) is large and the temperature at the exit of the compressor high (1100 °C). The compression should be done in stages so that each compression involves a smaller compression ratio with intercooling between stages.

- In terms of entropy production, by far the main contribution comes from the precooler. This amount was calculated assuming the cooling water to be at 25 °C. If this stream is specified in terms of flow rate, its contribution to the entropy generation would be calculated by replacing the bath term with $\dot{m}_{\text{cw}}\Delta S_{\text{cw}}$, where $\dot{m}_{\text{cw}}$ is the flow rate of this stream and ΔS_{cw} is its entropy change through the heat exchanger.

• Streams are always mixed at the same pressure to avoid backflows. This process is generally irreversible and results in entropy generation as the mixed streams are usually at different temperatures. In the above process, it is a lucky coincidence that streams 1 and 9 are mixed at same temperature. In this, special case mixing is a reversible process and results in no entropy production.

• It is generally a very good practice to include in the summary table the calculation of $\Delta(H\dot{m})$ for every unit and confirm that the energy balance is satisfied for all units. For example, the energy balance in the precooler was not used in the calculation because it is redundant with the other balances used. One must confirm, however, that the balance is indeed satisfied. This acts as a check of the calculation. As a further check, one should confirm the overall energy balance:

$$\Delta(H\dot{m}) = -431.52 \text{ kW},$$

$$\dot{Q} + \dot{W} = (-22455.8) + (22024.3) = -431.5 \text{ kW}.$$

These are important checks. It is the job of the engineer to certify the correctness of the calculation. It is important, therefore, to check the results in as many ways as possible to ensure the absence of errors.

Exercise Repeat the analysis by replacing the single-stage compressor with a series of compressors, each with a compression ratio of 4:1, and with intercooling down to 40 °C after each compression.

NOTE

On the Thermodynamic Analysis of Processes
The overall lesson from Example 6.23 is that flow sheets of moderate complexity, with several units and recycle streams, can be solved by hand in a systematic way, as long as the properties of the fluid are known, either from tables or from software applications. The calculation is always reduced to the energy balance around an individual unit, such as a compressor, a throttling valve, and the like. For relatively simple flow sheets, one will normally use the energy balance to calculate one unknown stream, then move serially to the next unknown stream and continue until all streams are known. With more complex flow sheets, such as the one here, several units may have to be solved simultaneously to obtain the unknown properties. For example, the separator, throttling valve, and precooler involve common unknown streams and cannot be solved as individual units. By combining such units to write an overall balance for the whole group it is

6.10 Unsteady-State Balances

sometimes possible to eliminate the intermediate unknowns. Otherwise, multiple units will have to be solved simultaneously. Once the streams are known, the energy and entropy balances are computed easily. Assumptions about baths need to be made only if a process stream exchanges heat with an unspecified stream. Though the main interest is in the energy balances, the entropy balance also provides very useful information for the sources of irreversibility. An efficient process should minimize entropy production because this leads to savings in terms of better energy utilization. Process design is beyond the scope of this book, but what should be clear is that the entropy analysis of a process is the starting point for optimizing the overall efficiency of the process.

6.10 Unsteady-State Balances

While the majority of processes encountered in plants are open processes running at steady state, there are also situations that involve unsteady-state operation. The analysis proceeds by the same general methodology as with steady-state, except that the accumulation terms are retained. However, unsteady state processes require special attention because some of the results obtained previously under steady state cannot be transferred automatically to unsteady-state processes. We examine below some typical examples of unsteady-state analysis.

Pressurizing a Tank

We consider the problem of filling a tank from a pressurized line, as shown in Figure 6-17. A main line carries the fluid at fixed pressure and temperature. A stream is drawn from this line and passes through a valve into a tank. The tank is

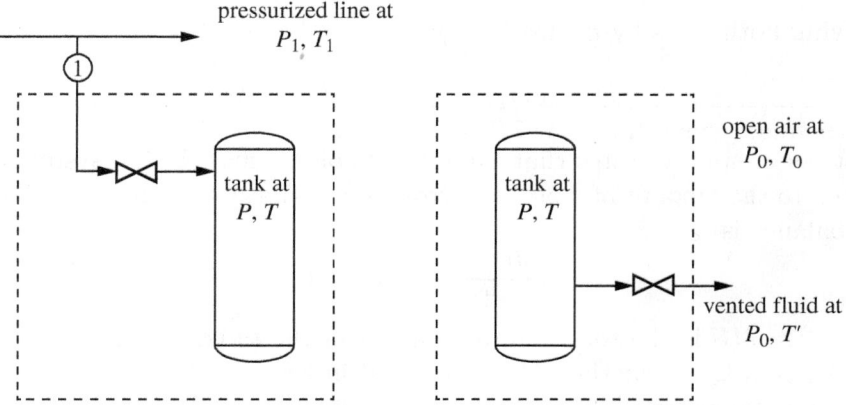

Figure 6-17: Pressurizing and venting of tank.

initially at a pressure lower than that of stream 1 and it may be fully evacuated, or partially filled. As fluid fills the tank, the pressure in the tank increases until it reaches the pressure of the supply line. The valve can be closed at any point, to produce a partially filled tank. We want to find the amount (mass) of fluid and its temperature at any point during filling. We will neglect kinetic and potential energy contributions. We will also assume that the process is adiabatic. If not, additional information must be given about possible heat exchange with the surroundings. We note, however, that heat transfer is generally much slower than mass flow, therefore, even if the tank is not insulated, we may still assume that any temperature change resulting from the expansion of the gas through the valve will resist equilibration with the surroundings over the duration of the process. In this respect, the process may be modeled as approximately adiabatic. To perform balances we define the system to be the tank and the valve, shown by the dashed line in Figure 6-17. We neglect the mass of the fluid in the line and in the valve and take the entire mass M of the system to be the mass of the gas inside the tank. We will further assume that pressure and temperature inside the tank is uniform. This may be a rather crude approximation, as there is definitely flow between the point where the pipe connects to the tank and the bulk of the tank. However, such details cannot be accounted for in our formulation of the energy balance, which is *macroscopic*. For this reason, we will have to accept this approximation. Initially the tank contains M_0 mass of the fluid at T_0, P_0. The main line is assumed to come from a large supply tank so that the conditions of stream 1 do not change as the tank is being filled.

We begin with the mass balance. With one inlet flow ($\dot{m}_1$) and no outlet, eq. (6.8) becomes

$$\frac{dM}{dt} - \dot{m}_1 = 0.$$

Multiplying both sides by dt this becomes

$$dM - dm_1 = 0 \quad \Rightarrow \quad dM = dm_1. \tag{6.71}$$

The last result simply states that the increase of the mass in the system over time dt is equal to the amount of mass that crosses into the system during this time. The energy balance is

$$\frac{dU^{\text{tot}}}{dt} - H_1 \dot{m}_1 = 0,$$

where $U^{\text{tot}} = MU$ is the total internal energy in the tank, and U is the specific or molar (depending on whether M is expressed in kg or mol) internal energy at the pressure and temperature of the tank. Multiplying by dt both sides of the energy balance we obtain

$$dU^{\text{tot}} - H_1 dm_1 = 0 \quad \Rightarrow \quad dU^{\text{tot}} = H_1 dm_1.$$

6.10 Unsteady-State Balances

The last result states that all the enthalpy that is carried into the system is accounted for as internal energy. This equation can be integrated from initial conditions ($U^{tot} = M_0 U_0$, $m_1 = 0$) up to the point that a total mass m has been introduced to the tank ($U^{tot} = MU$, $m_1 = m$). Noting that the mass in the tank at the final state is $M_0 + m$ and that H_1 is constant, the result of this integration is

$$(M_0 + m) U - M_0 U_0 = H_1 m,$$

which can further be written as

$$\boxed{U = H_1 \frac{m}{M_0 + m} + U_0 \frac{M_0}{M_0 + m}.} \tag{6.72}$$

This equation specifies an intensive property in the tank (specific or molar internal energy) at the point where it has admitted an amount of mass m. An additional property must be specified to fully define the state in the tank. This will depend on the specifics of the problem. For example, if the volume V^{tank} of the tank is known, then the specific volume is $V = V^{tank}/(M_0 + m)$, and this provides the additional intensive variable for the complete specification of the state.

Example 6.24: Filling a Tank with Steam

A 1 m³ rigid tank, initially evacuated, is connected to a steam main that supplies steam at 20 bar, 250 °C. The valve is opened and steam is allowed to flow into the tank. Calculate the amount of steam in the tank when the tank pressure is 10 bar.

Solution Since the tank is initially evacuated, $M_0 = 0$, and eq. (6.72) simplifies to

$$U = H_1. \qquad [A]$$

This results states that the specific internal energy in the tank at all times is equal to the specific enthalpy in the steam main. From the steam tables at the conditions of the supply line,

$$T_1 = 250\ °C, \quad P_1 = 20\ \text{bar}, \quad H_1 = 2903.2\ \text{kJ/kg}.$$

The state in the tank when the pressure is 10 bar is obtained by interpolation in the tables at $P = 10$ bar, $U = 2903.2$ kJ/kg. We find

$$T = 366.748\ °C, \quad V = 0.290572\ \text{m}^3/\text{kg}.$$

The mass in the tank is

$$m = \frac{V^{tank}}{V} = \frac{1\ \text{m}^3}{0.290572\ \text{m}^3/\text{kg}} = 3.44\ \text{kg}.$$

Repeating the calculation at several pressures we construct the following table that shows how pressure and temperature in the tank change during filling.

P (bar)	T (°C)	V (m^3/kg)	m (kg)
1	359.1	2.913160	0.34
5	362.5	0.581953	1.72
10	366.8	0.290572	3.44
15	370.9	0.193459	5.17
20	375.0	0.144911	6.90

Comments Throttling steam into the tank causes its temperature to rise. This throttling process, however, is not isenthalpic as in steady state. The difference is that eq. [A] is written, not between the inlet and outlet of the valve, as in eq. (6.53), but between the inlet of the valve and the *entire contents of the tank*. When a small amount of the gas passes through the valve, it mixes upon exit with the contents of the tank. We have assumed this mixing to be instantaneous, and it is this assumption that prevents us from analyzing in more detail the immediate exit point of the valve.

Example 6.25: Filling of Tank with Ideal Gas
Repeat the previous problem assuming steam to be an ideal gas with $C_P^{ig} = 35.62$ J/mol K.

Solution The working equation is
$$U = H_1, \qquad [A]$$
as in the previous example, but the calculation of enthalpy and entropy must now be done using the ideal-gas equations. Since the equations for the ideal-gas enthalpy and internal energy gives these properties as differences rather than as absolute values, we need to introduce a reference state for these calculations. We pick the reference state to be the state in the steam main and set the enthalpy. Next, we write an equation for the internal energy in the tank (at pressure P, temperature T) based on the reference state. To do this, we first express internal energy in terms of enthalpy at P, T, then we calculate this enthalpy relative to the reference state:

$$U(P, T) = H(P, T) - PV = H_1 + C_P^{ig}(T - T_1) - PV = H_1 + C_P^{ig}(T - T_1) - RT.$$

Equation [A] now becomes

$$H_1 + C_P^{ig}(T - T_1) + RT = H_1.$$

Solving for the temperature in the tank we find:

$$T = \frac{C_P}{C_P - R} T_1.$$

6.10 Unsteady-State Balances

By numerical substitution,

$$T = \frac{35.62 \text{ J/molK}}{(35.62 - 8.314) \text{ J/molK}} (250 + 273.15) \text{ K} = 682.4 \text{ K} = 409.3 \text{ °C}.$$

This temperature is independent of the pressure in the tank, that is, it is the same at all times. The molar volume of the steam in the tank when the pressure in the tank is 20 bar, is

$$V = \frac{RT}{P} = \frac{8.314 \text{ J/molK} \times 682.4 \text{ K}}{20 \cdot 10^5 \text{ Pa}} = 0.00283689 \text{ m}^3/\text{mol}.$$

The amount of steam in the tank is

$$n = \frac{V^t}{V} = \frac{1 \text{ m}^3}{0.00283689 \text{ m}^3/\text{mol}} = 352.5 \text{ mol} = 6.34 \text{ kg}.$$

Comments The main point of this exercise is to show how the same problem can be solved using either an equation of state (this example) or tabulated values (previous example). The above calculation also demonstrates one important aspect of reference states: they can be chosen *arbitrarily*. Notice that we never specified the value of the enthalpy in the reference state because it dropped out during the calculation. We often set this value to zero and make it disappear from the calculation precisely because it will eventually drop out anyway.

Also notice that once we have fixed enthalpy at the reference state we are *not* free to fix the internal energy as well. Internal energy must be calculated from the known values of enthalpy, pressure, and volume at the reference state using the definition $U = H - PV$.

Exercise Repeat the calculation using the Soave-Redlich-Kwong equation of state. Even though the SRK equation is not appropriate for water, the point of this challenge, should you choose to accept it, is to demonstrate that the solution to the problem is still given by eq. [A] and that the method by which we calculate the terms of this equation is merely a computational issue.

Venting a Tank

We consider now the opposite problem, venting a tank into the ambient surroundings. The tank contents are at pressure P and temperature T and the surroundings at P_0, T_0. Venting is done through a valve. Over a small time, dt, an amount of fluid, dm, leaves the tank and emerges in the surroundings. The pressure of this element at the venting point is the same as the pressure of the surroundings. Its temperature, however, is not that of the surrounding air but is determined by the process. Eventually, of course, this mass will equilibrate with the surroundings. Thermal equilibration, however, is slow and the process may be treated as adiabatics.[9] We will also assume the contents of the tank to be at a uniform state, effectively ignoring the pressure gradients that cause the fluid to flow from the bulk of the tank to the entrance of the valve. Under these assumptions, the gas is throttled from the conditions in the tank (pressure P, temperature T) to pressure P_0 and temperature T' at the exit of the vent. Starting with the mass balance equation, we have

$$\frac{dM}{dt} + \dot{m} = 0 \quad \Rightarrow \quad dM = -dm,$$

which says that the mass in the tank decreases by the amount withdrawn. Ignoring kinetic and potential energy effects, and setting $\dot{Q} = \dot{W}_s = 0$, the energy balance is

$$\frac{dU^{\text{tot}}}{dt} + \dot{m} H' = 0,$$

where $U^{\text{tot}} = MU$ is the total internal energy in the tank, U is the specific internal energy in the tank, and $H' = H(P_0, T')$ is the enthalpy at the venting point. We multiply the energy balance by dt and use $dm = -dM$ to obtain

$$\frac{d(MU)}{dM} = H'. \tag{6.73}$$

As the tank vents, its pressure and temperature change, and so does the temperature at the exit of the vent. Accordingly, H' varies with time and eq. (6.73) cannot be integrated yet until we determine how this variation is related to the state inside the tank. To establish this relationship we imagine the process to take place as follows: first an amount dm is removed from the tank; it is then throttled from pressure P, and temperature T, to the conditions at the exit of the vent. We assume now that the removal of the amount dm from the tank is done reversibly. This amounts to ignoring any pressure gradients and viscous effects in the flow of that mass from the bulk of the tank to the entrance of the valve. By removing the mass dm from the tank, the state in the tank will change. The corresponding entropy balance following this change is

$$\frac{d(MS)}{dt} + S\dot{m} = \dot{S}_{\text{gen}}\Big|_{\text{tank}} = 0 \quad \text{or} \quad d(MS) + S dm = 0,$$

where S is the specific entropy in the tank, also equal to the specific entropy carried away by the mass dm, and $\dot{S}_{\text{gen}}\big|_{\text{tank}}$ is the entropy generation for the removal of mass dm, which we have assumed to be zero. Expanding the differential in the right-hand side of the above equation we have

$$M dS + S dM + S dm = 0,$$

---

9. This is analogous to blowing air from the mouth: it comes out warm, even in a cold day, despite the fact that it is out in the open.

6.10 Unsteady-State Balances

and using $dm = -dM$, this becomes

$$MdS = 0 \quad \Rightarrow \quad S = \text{const.} \tag{6.74}$$

The result states that the specific entropy in the tank remains constant at all times; it is, therefore, equal to its value at the beginning of the venting process. This provides the additional equation for the integration of the energy balance equation. We return to eq. (6.73) and work on the derivative $d(MU)/dM$. First, we note that this derivative expresses changes in the tank that occur at *constant specific entropy*. Using $M = V^{\text{tank}}/V$, where V^{tank} is the tank volume and V is the specific volume in the tank, we obtain a very simple result that derivative:

$$\left(\frac{d(MU)}{dM}\right)_S = \left(\frac{\partial(U/V)}{\partial(1/V)}\right)_S = U - V\underbrace{\left(\frac{\partial U}{\partial V}\right)_S}_{-P} = U + PV = H,$$

where we used eq. (5.12) to recognize the partial derivative $(\partial U/\partial V)_S$ as the pressure in the tank. Finally, combining this result with eq. (6.73), we obtain the desired relationship between H' and the state in the tank:

$$H = H', \tag{6.75}$$

which states that the enthalpy of the vented fluid is equal to the enthalpy in the tank at the same moment. This, along with the isentropic condition in eq. (6.74), form two equations that fully define the process:

$$S(P, T) = S(P_i, T_i) \qquad [6.74]$$

$$H(P, T) = H(P_0, T') \qquad [6.75]$$

where (P, T) is the state in the tank and (P_0, T') is the state at the exit of the vent. One may wonder why we took such a circuitous path to arrive at eq. (6.75), which, after all, expresses the familiar result that throttling is isenthalpic. One reason is that, as we saw in the problem of pressurizing a tank, under unsteady-state conditions we do not always obtain the familiar results of steady-state analysis. Moreover, the isenthalpic condition is not enough to solve this problem, we need the isentropic condition as well. Our approach shows that the assumption of reversibility in the tank leads to *both* conditions. And by laying out the assumptions of the calculation, we know exactly under what circumstances we may expect the solution to apply in a real situation.

NOTE

Pressurizing and Venting
Although venting is the reverse of pressurization, we have obtained quite different results for the two cases. The reason is that these processes are *irreversible*: the path for venting is *not* the same as the reverse of the pressurization path.

Example 6.26: Venting Steam

Steam is vented from a pressurized tank to air at pressure 1 bar. The volume of the tank is 2.5 m³ and is initially at 350 °C and 20 bar. Determine the amount of steam in the tank when the pressure is 12 bar, and the temperature of the steam coming out from the vent at this point.

Solution The initial state in the tank is

$$P_i = 20 \text{ bar} \qquad T_i = 350 \text{ °C} \qquad V_i = 0.1386 \text{ m}^3/\text{kg}$$
$$U_i = 2860.5 \text{ kJ/kg} \qquad H_i = 3137.6 \text{ kJ/kg} \qquad S_i = 6.9582 \text{ kJ/kg K}.$$

The amount of steam in the tank initially is

$$M_i = \frac{V^{\text{tank}}}{V_i} = \frac{2.5 \text{ m}^3}{0.1386 \text{ m}^3/\text{kg}} = 18.04 \text{ kg}.$$

Pressure in the tank is 12 bar. The state in the tank is defined by the pressure ($P = 12$ bar) and entropy ($S = S_i = 6.9582$ kJ/kg K). By interpolation in the steam tables the properties in the tank are,

$$P = 12 \text{ bar} \qquad T = 281.4 \text{ °C} \qquad V = 0.2059 \text{ m}^3/\text{kg}$$
$$U = 2758.1 \text{ kJ/kg} \qquad H = 3005.1 \text{ kJ/kg} \qquad S = 6.9582 \text{ kJ/kg K}$$

The mass of steam in the tank is

$$M = \frac{V^{\text{tank}}}{V} = \frac{2.5 \text{ m}^3}{0.2059 \text{ m}^3/\text{kg}} = 12.14 \text{ kg}.$$

The state at the exit of the vent is determined by the conditions $H' = H = 3005.12$ kJ/kg and $P = 1$ bar. By interpolation in the tables, the temperature in the vent is

$$T' = 265.3 \text{ °C}.$$

In summary, when the pressure in the tank is 12 bar, the temperature in the tank is 281.4 °C, and the temperature at the vent is 265.3 °C.

Example 6.27: Venting an Ideal Gas

Repeat the previous problem assuming steam to be an ideal gas with $C_P^{\text{ig}} = 35.62$ J/mol K.

Solution The temperature in the tank when pressure is 12 bar is given by the isentropic condition,

$$T = T_i \left(\frac{P}{P_i}\right)^{R/C_P} = (623.15 \text{ K}) \left(\frac{12 \text{ bar}}{20 \text{ bar}}\right)^{8.314/35.62} = 554.9 \text{ K} = 281.8 \text{ °C}.$$

> The molar volume of the steam in the tank is
>
> $$V = \frac{RT}{P} = \frac{(8.314 \text{ J/mol K})(554.9 \text{ K})}{12 \times 10^5 \text{ Pa}} = 3.84463 \times 10^{-3} \text{ m}^3/\text{mol},$$
>
> and the corresponding amount in the tank is
>
> $$\frac{V^{\text{tank}}}{V} = \frac{2.5 \text{ m}^3}{3.84463 \times 10^{-3} \text{ m}^3/\text{mol}} = 650.258 \text{ mol} = 11.7 \text{ kg}.$$
>
> The temperature at the exit of the vent is given by the isenthalpic condition. In the ideal-gas state, an isenthalpic process is also isothermal, therefore,
>
> $$T' = T = 281.8 \text{ °C}.$$
>
> **Comments** The temperature and amount of steam in the tank are fairly close to the value obtained using the steam tables. However, the ideal-gas calculation predicts no temperature drop during throttling and thus overestimates the exit temperature by about 16 °C.

6.11 Summary

The mathematical form of the first and second law is different for open and for closed systems. In a closed system these equations are:

First law, closed system: $\Delta \left(U + \frac{v^2}{2} + gz \right)_{AB} = Q + W,$ [3.8]

Second law, closed system: $\Delta S_{\text{sys}} + \Delta S_{\text{sur}} = S_{\text{gen}},$ [4.26]

where W is the PV and shaft work combined, and Δ refers to the difference between before and after. In an open system with flow streams the corresponding equations are:

First law, open system: $\dfrac{dU^{\text{tot}}}{dt} + \Delta \left\{ \left(H + \frac{v^2}{2} + gz \right) \dot{m} \right\} = \dot{Q} + \dot{W}_s,$ [6.17]

Second law, open system: $\dfrac{dS^{\text{tot}}}{dt} + \Delta(\dot{m} S) = \dfrac{\dot{Q}}{T_{\text{bath}}} + \dot{S}_{\text{gen}}.$ [6.26]

Here, W_s is the shaft work only and Δ refers to the sum of all outlet streams minus the sum of all inlets. The most significant difference is that the thermodynamic property that appears inside the operator Δ is *internal energy*, if the system is closed, but *enthalpy*, if the system is open. Because the equations are different, it is important to correctly identify the system as open or closed before applying the energy and entropy balances.

With respect to open systems, we make the additional distinction between steady-state, and unsteady-state operation. At steady state the accumulation terms, dU^{tot}/dt, dS^{tot}/dt and dM^{tot}/dt, are zero. This reduces the energy balance from a differential equation into an algebraic equation. For this reason, steady-state processes are much simpler to calculate compared to unsteady-state processes.

Even the most complex process consists of simple units interconnected through streams. The simple units we encountered in this chapter encountered heat exchangers, pumps, compressors, turbines, throttling valves, vapor/liquid separators, and mixing points. Other common units in chemical plants are distillation columns, absorption towers, and chemical reactors. These are also analyzed by the general tools of the thermodynamics; however, they involve mixtures of multiple components, and we will postpone their discussion until we develop methods for the calculation of mixture properties.

The general strategy in working with flow sheets of low to moderate complexity is to break them down into smaller parts and apply the energy balance to one simple unit at a time with the goal to calculate the properties of all streams. Remember that if we know any two intensive properties of a stream, all other intensive properties are fixed. It is generally the practice to neglect pressure drops due to flow, at least at the preliminary stage of design. This means that the pressure in consecutive streams is taken to be the same unless the fluid passes through pumps, compressors, turbines, expanders, or throttling valves, units that are meant to change pressure. The energy and entropy balances are computed very easily once the streams are known. The purpose of the energy balance is, of course, to determine the shaft work and heat load requirements of the process. The entropy balance serves as a diagnostic tool to assess the overall efficiency of the process. By multiplying the entropy generation by a characteristic temperature T_0, which we take to be that of the surroundings, we translate the irreversible features of a process into units of work. This represents the amount of work that would be gained if the underlying irreversibility were eliminated. It is not possible, nor desirable, to eliminate all irreversible features. A certain magnitude of gradients is necessary to produce practical rates in the operation of a process. However, entropy analysis provides us with a way to compare alternative designs, uncover design flaws, and optimize the overall efficiency of the process so that it makes better utilization of the available resources.

6.12 Problems

Problem 6.1: A flow process produces work using water as the working medium. The process is as follows: steam of quality 50% at 1 bar is compressed adiabatically

6.12 Problems

to 20 bar; is heated isothermally by absorbing 3500 kJ/kg of heat; is expanded adiabatically to an unspecified final pressure; and is finally cooled isothermally until the steam returns to its initial state (steam of quality 50% at 1 bar). It is further reported that 53% of the heat added during the heating step is rejected during the cooling step. What is your evaluation of this process?

Problem 6.2: The power generation unit in your plant uses a hot exhaust gas from another process to produce work. The gas enters at 10 bar and 350 °C and exits at 1 bar and 40 °C. The process produces a net amount of work equal to 4,500 J/mol and it exchanges an unknown amount of heat with the surroundings.
a) Determine the amount of heat exchanged with the surroundings. Is this heat absorbed or rejected by the system?
b) Calculate the entropy change of the exhaust gas.
c) As a young and ambitious engineer you seek ways to improve the process. What is the maximum amount of work that you could extract from this system? Assume that the inlet and outlet conditions of the exhaust gas remain the same.
Additional data: Assume the surroundings to be at the constant temperature of 25 °C and the exhaust gas to be ideal with $C_P = 29.3$ J/mol K.

Problem 6.3: You are responsible for a high-temperature process that produces two exhaust streams, stream A at 600 °C, 10 bar, with molar flow rate 100 mole/s; and stream B at 200 °C, 20 bar, with molar flow rate 50 mole/s. A young engineer in your team comes to you with an idea to utilize these streams to produce heat and work which will save costs in your process. She quickly puts together a sketch for a process that produces 220 kW of heat, 1700 kW of work, and delivers the exhaust gases as a single stream at 1 bar. The heat will be used to produce saturated steam at 1 atm by boiling the saturated liquid.
a) What is the temperature of the exit stream?
b) Should you take this idea to your superiors for further discussions?
c) Could you use the heat from this process to boil water at 10 bar?
Additional information: Assume the exhaust gases can be treated as an ideal gas with $C_P = 30$ J/mole K.

Problem 6.4: You are in charge of a unit that produces two streams, A at 600 °C, 10 bar, with molar flow rate 100 mole/s; and B at 200 °C, 20 bar, with molar flow rate 50 mole/s. Those streams are currently throttled to 1 bar. An engineer in your team comes to you with an idea to utilize these streams to produce heat and work. She believes she can extract 1700 kW of work and 500 kW of heat that will be used to boil water at 1 bar. Her process, whose details are yet to be determined, will receive streams A and B as inlet and will have a single outlet stream that will be delivered at 1 bar.

a) What is the temperature of the exit stream?
b) Should you take this idea to your manager for further discussions?
c) Could you use the heat from this process to boil water at 10 bar?
Additional information: Assume the exhaust gases can be treated as an ideal gas with $C_P = 30$ J/mole K.

Problem 6.5: Liquid propane boils in a constant pressure boiler to produce propane vapor at 30 bar and 245 °C. At the inlet of the boiler the propane is saturated liquid.
a) Draw a PV graph and show this process. Show all the relevant temperatures and pressures as well as the critical isotherm and the critical pressure.
b) Calculate the entropy change of propane between the inlet and outlet of the boiler. Explain your calculations and procedure clearly.
Additional data
The saturation temperature at 30 bar is 78 °C.
The heat of vaporization at 78 °C is 8780 J/mol.
The ideal-gas C_P of propane vapor is 112 J/mol K.

Problem 6.6: Steam flows in a long, noninsulated pipe under a constant pressure of 2 bar. At the inlet of the pipe the temperature is 300 °C. Due to losses to the surroundings, the temperature at the exit of the pipe is 60 °C. Assume the surroundings to be at 25 °C.
a) Draw a PV graph for the steam. Show the path of the process and all the relevant pressures and isotherms.
b) Calculate the amount of heat removed from the steam.
c) Calculate the entropy change of the steam.
d) Calculate the entropy change of the surroundings. What is the entropy generation?

Problem 6.7: Steam at 5 bar, 300 °C is condensed to saturated liquid at 5 bar in a heat exchanger at a mass flow rate of 1 kg/s. Cooling is provided by water, which enters the heat exchanger at 20 °C 1 bar and exits at 60 °C. Neglecting pressure drop in pipes and heat losses to the surroundings, determine the flow rate of the cooling water, the amount of heat that is exchanged, and the rate of entropy generation.

Problem 6.8: Steam enters a cooling tower at 200 °C and exits at 60 °C. Cooling is supplied by the surroundings which are assumed to be at 30 °C. The process takes place under atmospheric pressure (1 bar).
a) Calculate the amount of heat removed from the steam.
b) Calculate the entropy change of the steam.
c) What is the entropy generation?
d) Draw a PV graph and show the path of this process.

6.12 Problems

Problem 6.9: Water at 1 bar, 20 °C is heated in a heat exchanger to produce saturated steam. The heat is provided by a stream of saturated steam at 10 bar that exits the heat exchanger as saturated liquid.
a) Calculate the amount of heat that is exchanged between the two streams.
b) Calculate the mass flow rate of the 10 bar stream.
c) Calculate the entropy generation.
Report the results per kg of liquid water entering the heat exchanger.

Problem 6.10: Stream 1 enters a tank carrying steam at 5 bar, 300 °C at the rate of 1 kg/s. The stream is cooled by adiabatic mixing with water at 5 bar, 40 °C (stream 2). If the resulting stream is a vapor-liquid mixture with quality 18%, calculate the mass flow rate of stream 2, the entropy generation, and the lost work.

Problem 6.11: A steam turbine generates power by expanding steam from 30 bar, 450 °C to 8 bar.
a) Determine the amount of work and the exit temperature if operation is reversible.
b) Repeat the calculation assuming an efficiency of 80%. How much is the lost work due to the reduced efficiency?

Problem 6.12: A thermally insulated turbine generates 750 kW of power by expanding steam from 500 C and 3.5 MPa to 200 C and 0.3 MPa.
a) What is the mass flow rate of steam through the turbine?
b) Because of failure in the insulation there is a loss of 60 kJ per kg of steam. What is the power produced by the turbine if all inlet and outlet conditions remain the same?
Assume negligible kinetic and potential energy changes.

Problem 6.13: Steam at 500 C and 40 bar passes through two turbines arranged in parallel, and their exit streams are combined into one stream. Turbine 1 has 100% efficiency, while the efficiency of turbine 2 is 75%. Both turbines exhaust at 1 bar. The stream that is formed by combining the exhausts of the two turbines is saturated vapor.
a) Determine the fraction of the total flow rate that passes through each turbine.
b) Determine the rate of entropy generation. What are the irreversible feature of this process?

Problem 6.14: A steam turbine generates 800 kW of work by expanding steam from 50 bar, 400 °C to 1 bar. At the exit of the turbine the steam contains 2% moisture.
a) Determine the flow rate of steam through the turbine.
b) Determine the efficiency.
c) Determine the rate of entropy generation in kJ/K s.

Problem 6.15: Methane passes through a compressor and is subsequently cooled in a heat exchanger. Methane enters the compressor at 1 bar, 20 °C with volumetric flow rate 30 ft³/min; it exits the heat exchanger at 10 bar, 20 °C. The compressor efficiency is 80%. Assuming methane to be in the ideal-gas state with $C_P^{ig} = 36.3$ J/mol K, do the following:
a) Calculate the work in the compressor (in J/s)
b) Calculate the required heat in the heat exchanger.
c) Calculate the entropy generation.

Problem 6.16: Air is compressed in a steady-state compressor. The air enters at 1 bar, 25 °C and exits at 5 bar, 200 °C.
a) Determine the amount of work.
b) Calculate the entropy generation.
c) What is the amount of work for reversible compression of air from 1 bar, 25 °C to 5 bar?
d) What would be the temperature at the exit of the compressor in part (c)? Assume air to be an ideal gas with $C_p = 30$ J/molK.

Problem 6.17: A new compressor has just been delivered to your plant. In order to check its performance you order your staff to make a test run using air. Your staff reports to you the following results: air at a flow rate of 50 mol/min was compressed from 1 bar and 25 °C to 10 bar. The temperature at the exit of the compressor was measured and found to be 450 °C.
a) What is the power (in kW)?
b) What is the efficiency? Assume air to be an ideal gas with $C_P = 3.5R$.

Problem 6.18: During a test of a gas turbine, the following data were recorded: air at a flow rate of 50 mol/min was expanded from 10 bar and 700 °C to 2 bar. The temperature at the exit of the turbine was measured and found to be 450 °C.
a) What is the power (in kW)?
b) What is the efficiency? Assume air to be an ideal gas with $C_P = 3.5R$.

Problem 6.19: Steam is compressed from 1 bar, 200 °C to 12 bar in a compressor operating at steady state.
a) Calculate the amount of work and the final temperature if compression is reversible.
b) Repeat the calculation if the efficiency of the compressor is 75% and report the lost work.

6.12 Problems

Problem 6.20: a) Water is throttled from 20 bar, 20 °C to 1 bar. What is the final temperature?
b) Repeat if the final pressure is 0.00706 bar.

Problem 6.21: Water at 15 bar and 20 °C is throttled to final pressure such that the temperature is 5 °C. Determine the pressure and the entropy generation. How much work could be produced if instead of throttling the final pressure is reached via expansion in a reversible expander?

Problem 6.22: a) An ideal gas ($C_P = 30$ J/mol K) is throttled from 20 bar, 25 °C to 1 bar. What is the temperature at the exit?
b) Water is throttled from 20 bar, 25 °C to 1 bar. What is the temperature at the exit? If a vapor-liquid mixture, report the liquid fraction.
c) A vapor-liquid mixture of water containing 50% liquid is throttled from 60 °C, to 0.1 bar. What is the temperature at the exit? If a vapor-liquid mixture, report the liquid fraction.

Problem 6.23: Water is pumped at a flow rate of 2.25 kg/min from 1 bar 20 °C to 15 bar. Determine the amount of work and the temperature of water at the exit if the efficiency of the pump is 78%.

Problem 6.24: A 10 kW pump is used to pump liquid ammonia from 5 bar and 0 °C to 25 bar. The efficiency of the pump is 62%.
a) What is the mass flow rate (kg/s) delivered by the pump?
b) What is the temperature at the exit of the pump?
c) Calculate the rate of entropy generation (in kW/K).
Additional data for liquid ammonia: density $= 0.64$ g/cm^3; $C_P = 1.1$ kJ/kg K; $\beta = 2.095 \times 10^{-3}$ K^{-1}.

Problem 6.25: A flow process utilizes steam as the working fluid. The steam, initially at 30 bar and 700 °C (state A), is cooled under constant pressure to a temperature of 380 °C (state B) and it subsequently expands adiabatically to a final pressure of 1 bar (state C) through a turbine whose efficiency is 75%.
a) Plot the path of a Mollier diagram.
b) Determine the energy balances.

Problem 6.26: A flow process that operates with steam has two inlet streams (A and B) and one outlet stream (C). Stream A is at 20 bar, 25 °C, and has a mass flow rate of 10 kg/s; stream B is saturated liquid at 20 bar and has a flow rate of 7 kg/s; stream C is at 20 bar. The process receives heat at the rate of 14,100 kW from a heat source at 250 °C.
a) What is the temperature of stream C?
b) Determine the entropy generation.

Problem 6.27: A Rankine steam power plant produces 0.5 MW of mechanical power by expanding steam from 60 bar, 700 °C, to 3 bar. The efficiency of the turbine and of the pump is 80%. Calculate the energy balances, determine the flow rate of steam, and determine the entropy generation in each unit.

Problem 6.28: The boiler of a Rankine power cycle absorbs 1000 kW of heat from a furnace at 700 °C. The power plant rejects 850 kW of heat in a river at 25 °C.
a) What is the efficiency of the cycle?
b) What is the rate of entropy generation?
c) What is the maximum possible power that could be produced by a plant given the temperatures of the furnace and the river?

Problem 6.29: The steam power plant shown in Figure 6-18 uses a two-stage turbine and one feedwater heater that regenerates the partly exhausted steam from the first turbine stage. The steam enters the turbine at 60 bar and 700 °C, and exhausts at 1 bar. Steam for the feedwater heater is extracted from the turbine at 5 bar and exits the heater as saturated liquid (stream 7), while stream 1 exits at a temperature that is 6 °C below that of stream 7. The condenser produces saturated liquid. Calculate the energy balances, the thermodynamic efficiency, and perform an entropy analysis around each unit of the process. What are the advantages of this design compared to the simple Rankine plant? *Additional data:* The efficiency of both stages in the turbine and of the pump is 80%.

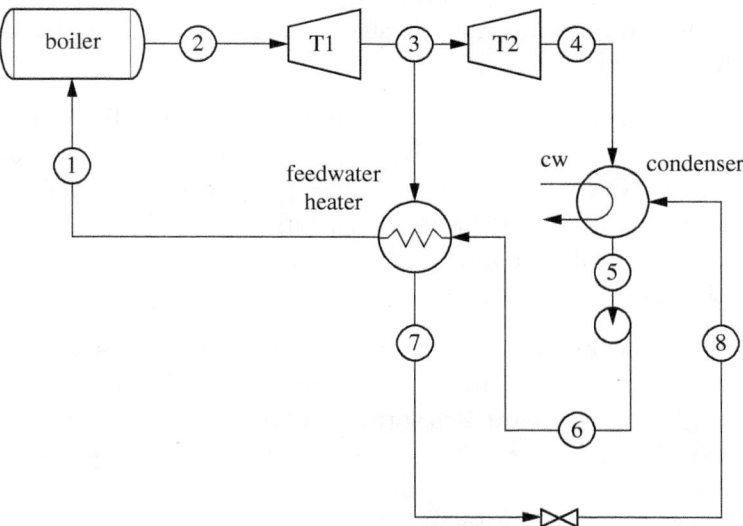

Figure 6-18: Process flow diagram for problem 6.29.

6.12 Problems

Problem 6.30: Design a vapor compression refrigeration unit using R290 as the refrigerant. The unit must be capable of pumping 1.5 refrigeration tons of heat at $-25°C$. The condenser will be operated using water at $25°C$. Determine the flow rate of the refrigerant and the coefficient of performance assuming the compressor to operate with 80% efficiency. *Additional information:* One standard refrigeration ton is 12,000 BTU/h (3,517 W).

Problem 6.31: The process shown in Figure 6-19 is part of a chiller unit that uses water as the refrigerant. The following data are available: stream 1 is a vapor-liquid mixture at 0.5 bar; stream 5 is saturated vapor at 5 °C; stream 6 is at 500 °C; stream 7 is at 100 °C. Calculate the following on the basis of 1 kg/s of water in stream 1:
a) Calculate the energy balances.
b) Determine the rate of entropy generation and identify the irreversible units in order of significance.
c) Show the process on a qualitative *PH* graph.

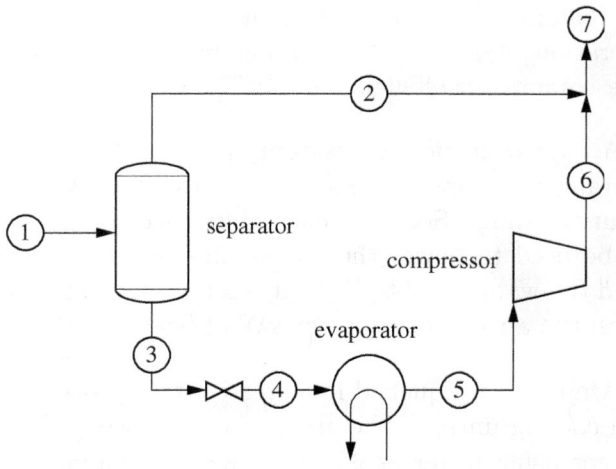

Figure 6-19: Process flow diagram for problem 6.31.

Problem 6.32: Methyl chloride at 2 bar and 65 °C is to be liquefied under constant pressure to produce saturated liquid.
a) Calculate the ideal work for this process assuming the surroundings to be at 25 °C.
b) The liquefaction will be carried out using a refrigeration cycle with a coefficient of performance $\omega = 3.5$. Calculate the work needed to operate this cycle.
c) What is the lost work for the cycle in part (b)?

The following information for methyl chloride is available: Saturation temperature at 2 bar: $-7\,°C$. ΔH_{vap} at 2 bar: 21.5 kJ/mol. C_P of vapor: 38 J/mol K. Assume CH_3Cl vapor may be assumed to be ideal. Is this a good assumption?

Problem 6.33: A standard vapor compression cycle uses tetrafluoroethane as the refrigerant. The refrigerant at the exit of the condenser is saturated liquid at 90 °F; at the exit of the evaporator it is saturated vapor at 0 °F. The cycle is powered by a compressor rated at 5 hp. The coefficient of performance of this cycle is 2.5.
a) How much heat is removed in the evaporator?
b) What is the flow rate of the refrigerant?
c) What is the entropy generation ?
d) What is the efficiency of the compressor?
NOTE: 1 hp = 2545 Btu/h.

Problem 6.34: You are designing a refrigeration cycle that will absorb 100 W of heat at -10 °F. The cycle will use ammonia as the refrigerant in a standard vapor compression cycle. Its condenser will be cooled by air, assumed to be at 80 °F. Calculate the energy balances, determine the flow rate of the refrigerant, and calculate the entropy generation. Assume a 10 °F temperature between streams in the heat exchangers and a compressor efficiency of 80%.

Problem 6.35: An air-conditioning system, powered by solar energy, is being designed to maintain the temperature inside a house at 20 °C against a temperature of 30 °C of the surroundings. Solar panels will be used to heat water to 200 °C and this energy will be used to power the air conditioner. Considering the house, the surroundings, and the water tank to be heat reservoirs, what is the power that must be supplied to run the air conditioner per kW of heat pumped out of the house?

Problem 6.36: Ammonia is liquefied in a steady-state process by first compressing the gas and then cooling until all the ammonia condenses. The cooling takes place in a heat exchanger using water at 90 °F. The ammonia enters the compressor at 80 psia and 80 °F. The efficiency of the compressor is 80%.
a) What is the minimum pressure that we must have at the exit of the compressor in order to achieve liquefaction?
b) Calculate the amount of work in the compressor.
c) Calculate the amount of heat that must be removed by the cooling water.

Problem 6.37: Figure 6-20 shows a two-stage refrigeration cycle with two evaporating temperatures, flash gas removal, and intercooling. The cycle operates using carbon dioxide as the refrigerant. The condenser (HE1) uses cooling water at 25 °C to produce saturated liquid (stream 2). One third of the mass flow rate at 2 is

6.12 Problems

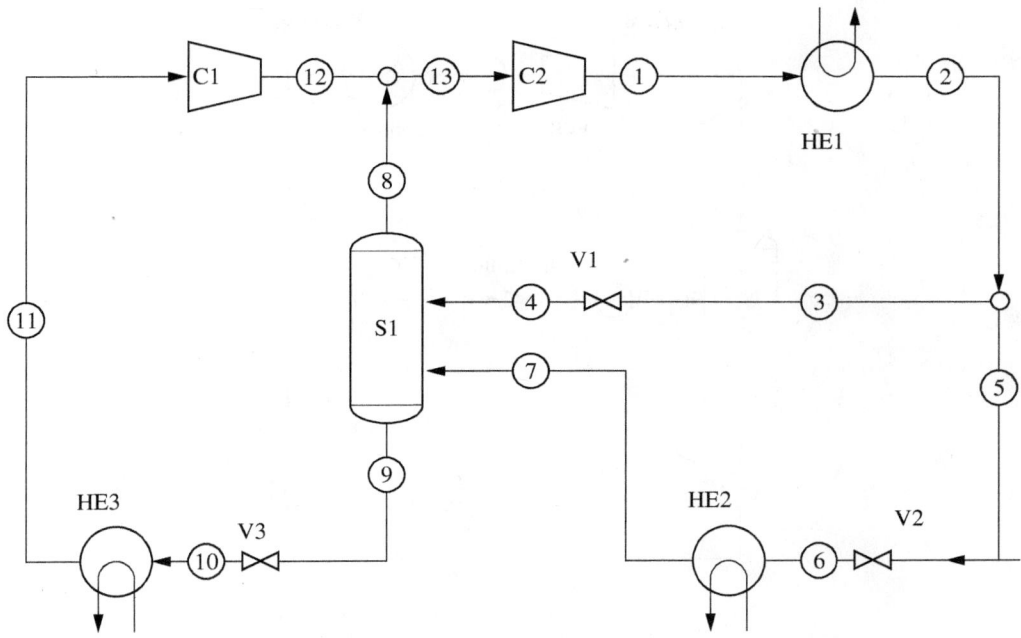

Figure 6-20: Process flow diagram for problem 6.37.

diverted to stream 3 with the remainder going into stream 5. Both evaporators (HE2 and HE3) produce saturated vapor. The flash gas separator (S1) removes all the vapor at the top (stream 8) and the liquid at the bottom (stream 9). The second stage HE3 removes heat at the rate of 100 W at $-40\ °C$. Both compressors increase the pressure of their respective incoming streams by the same factor and they both operate with an efficiency of 80%. This system must be capable of removing 100 kW of heat in evaporator HE3. Determine the flow rate of the refrigerant, the amount of work in the compressors, and the heat loads in all heat exchangers. Perform a second-law analysis and identify the unit that contributes most to the irreversible operation of this cycle. Finally, sketch a PH graph and show the states of all streams.

Problem 6.38: The refrigeration cycle in Figure 6-21 shows a vapor compression cycle with recirculation. The vapor-liquid separator after the throttling valve recycles the vapor to the compressor and passes the liquid to the evaporator. This process uses the refrigerant R-134a with the evaporator temperature set at $-20\ °C$, the condenser at $30\ °C$, and with compressor efficiency of 75%. Determine the energy and entropy balances and compare the efficiency of this process to the simple vapor-compression cycle without recirculation.

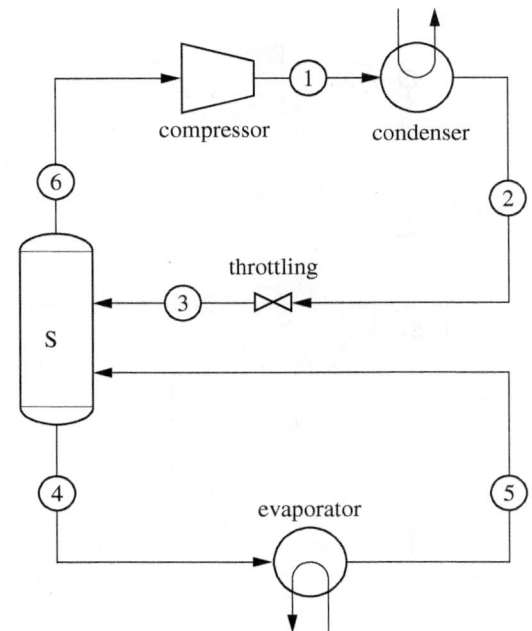

Figure 6-21: Process flow diagram for problem 6.38.

Problem 6.39: Natural gas, which you may take to be pure methane, is liquefied in the Claude process shown in Figure 6-22. The process operates between a high pressure of 30 bar and a low pressure of 1 bar. The inlet stream is at 1 bar, 20 °C, compression is to 30 bar, stream 13 is at 3 °C, the precooler uses water to cool the

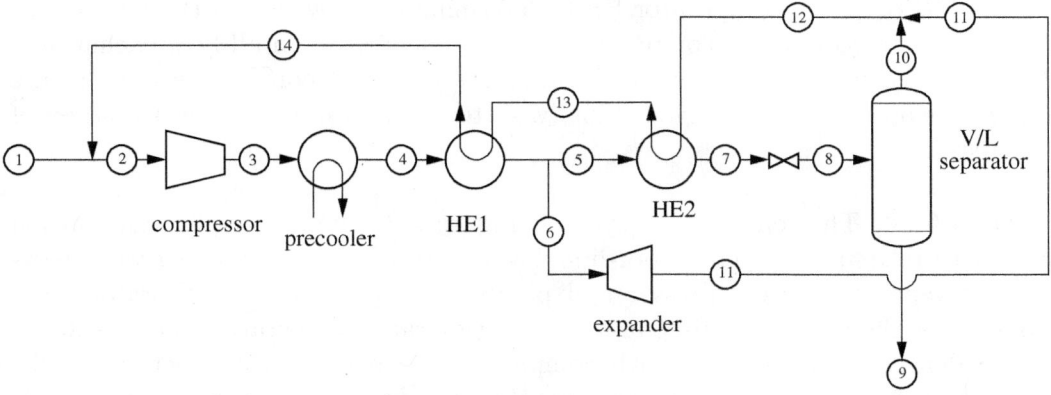

Figure 6-22: Process flow diagram for problem 6.39.

compressed gas to 30 °C, streams 5 and 6 are at 8 °C, and the mass flow rate in stream 2 is 20 times larger than $\dot{m}_1$. The efficiency of the compressor and the turbine is 85%. Determine the mass, energy, and entropy balances on the basis of 1 kg of liquified natural gas, and report the actual and ideal work. Discuss the features of this process compared to the Linde process.

Problem 6.40: Design a Linde process to liquefy oxygen. Pure oxygen is supplied at 1 bar, 25 °C and liquid oxygen is received at 1 bar. The fluid is compressed to 150 bar in a series of compressors whose efficiency is 80%. Precooling is done by water at 20 °C. Compression in done in stages with intercooling, to maximum temperature of 200 °C. Perform the energy balance and a second-law analysis of the process.

Problem 6.41: Oxygen is compressed from 1 bar, 25 °C, to 150 bar. To avoid overheating, compression is done in stages followed by intercooling with water at 30 °C so that the temperature exiting each stage of the compressor does not exceed 200 °C. Using data from the NIST WebBook, determine the number of stages, the work in each stage, and the amount of cooling, if the efficiency of all stages is 80%.

Problem 6.42: A 3 m^3 insulated tank contains steam at 1 bar, 150 °C. The tank is connected to a steam line that is maintained at 10 bar, 300 °C until the pressure in the tank is 5 bar. How much steam (kg) was transferred into the tank and what is the final temperature in the tank?

Problem 6.43: A 5 m^3 insulated tank contains steam at 10 bar, 200 °C. Three kilograms of steam is removed by venting to the atmosphere (1 bar, 20 °C). What is the pressure and temperature in the tank after venting?

Chapter 7

VLE of Pure Fluid

At the vapor-liquid boundary, a single-phase system splits into two phases,[1] each with its own properties (molar volume, enthalpy, entropy, etc.). The precise conditions under which phase splitting occurs is an important problem in thermodynamics. Up to this point we have relied on tabulated values and empirical equations, such as the Antoine equation, to establish the relationship between saturation temperature and pressure. In this chapter we develop a connection between the conditions at saturation and the equation of state. The key thermodynamic property that makes this connection possible is the Gibbs energy.

The learning objectives for this chapter are to:

1. Calculate the properties of two-phase systems.

2. Use the equation of state to construct the complete phase diagram of a pure fluid.

7.1 Two-Phase Systems

A system at saturation consists of a mixture of two phases. The extensive properties of the mixture are calculated by the lever rule. For the generic property F,

$$F = x_L F_L + x_V F_V, \tag{7.1}$$

where F_L and F_V are the properties (specific or molar) of the saturated phases and x_L, x_V, are the fractions (mass or mole) in each phase ($x_L + x_V = 1$). The properties of the saturated phases are calculated by the methods discussed in Chapter 5. The enthalpy of the saturated phases may be calculated from eq. (5.62),

$$H_V = \Delta H_{\text{ref}}^{\text{ig}} + H_V^R - H_0^R + H_0,$$
$$H_L = \Delta H_{\text{ref}}^{\text{ig}} + H_L^R - H_0^R + H_0,$$

where H_V^R and H_L^R are the residual properties of the saturated phases. These may be calculated from an equation of state, using the liquid root of the compressibility

---
1. A pure substance can form up to three coexisting phases.

equation for the saturated liquid, and the vapor root for the saturated vapor. The difference between H_V and H_L is equal to the enthalpy of vaporization. Noting that all other terms on the right-hand side except for H_V^R and H_L^R are common to both equations, the enthalpy of vaporization is simply equal to the difference of the residual terms:

$$\boxed{\Delta H_{\text{vap}} = H_V^R - H_L^R.} \tag{7.2}$$

The same arguments apply to the entropy of vaporization, for which we find

$$\boxed{\Delta S_{\text{vap}} = S_V^R - S_L^R.} \tag{7.3}$$

These results are of practical importance because they allow us to calculate the enthalpy and entropy of vaporization from the equation of state.

Example 7.1: Heat of Vaporization Using SRK
Calculate the enthalpy and entropy of vaporization of ethylene at 170 K using the SRK equation. The saturation pressure is 1.0526 bar.

Solution The critical constants of ethylene are

$$T_c = 282.35 \text{ K}, \quad P_c = 50.418 \text{ bar}, \quad \omega = 0.0866.$$

The residual properties of the saturated vapor were calculated in Example 5.10. The calculation for the saturated liquid is done in the same manner and the results are summarized below:

	Liquid	Vapor	
H^R	−13751.3	−107.679	J/mol
S^R	−80.5552	−0.390262	J/mol K

The heat of vaporization is

$$\Delta H_{\text{vap}} = (-107.679) - (-13751.3) = 13643.6 \text{ J/mol},$$

and the entropy of vaporization is

$$\Delta S_{\text{vap}} = (-0.390262) - (-80.5552) = 80.1649 \text{ J/mol K}.$$

Comments The literature value for the heat of vaporization (Perry's *Handbook*) is

$$\Delta H_{\text{vap}} = 481370 \text{ kJ/kg} = 13504.4 \text{ J/mol}.$$

The error in the SRK value is 1.03%.

Exercise Show that the computed values of the enthalpy and entropy of vaporization satisfy the fundamental relationship

$$\frac{\Delta H_{\text{vap}}}{T} = \Delta S_{\text{vap}}.$$

7.1 Two-Phase Systems

Example 7.2: Vapor-Liquid Mixture
Calculate the enthalpy and entropy of a vapor-liquid mixture of ethylene at 250 K, 23.307 bar, that contains 32% vapor by mass. The reference state is the saturated vapor at 170 K and residual properties are to be calculated by the Soave-Redlich-Kwong equation.

Solution We will first calculate the absolute properties of the saturated phases at 250 K, 23.207 bar, using eqs. (5.66), then we will obtain the enthalpy and entropy of the mixture using the lever rule. We begin by collecting the residual properties of the saturated phases and of the reference state.

	Sat. vapor	Sat. liquid	Ref. state	
T	250.	250.	170.	K
P	23.307	23.307	1.0526	bar
phase	vapor	liquid	vapor	—
a	0.501763	0.501763	0.604855	J m^3/mol^2
da/dT	-0.00112077	-0.00112077	-0.00149223	J m^3/mol^2 K
b	0.0000403395	0.0000403395	0.0000403395	m^3/mol
A	0.270698	0.270698	0.0318711	—
B	0.0452342	0.0452342	0.00300424	—
Z	0.709388	0.0832356	0.970362	—
H^R	-1802.27	-10318.7	-107.679	J/mol
S^R	-5.11985	-39.2464	-0.390262	J/molK

Using the ideal-gas heat capacity,

$$C_P^{ig}/R = 4.221 - 0.008782\,T + 0.00005795\,T^2 - 6.729 \times 10^{-8}\,T^3 + 2.511 \times 10^{-11}\,T^4,$$

the ideal-gas enthalpy at 250 K, 23.307 bar is

$$\Delta H_{\text{ref}}^{ig} = \int_{170\ \text{K}}^{250\ \text{K}} C_P^{ig}\,dT = 2906.5 \text{ J/mol}.$$

The enthalpy of the saturated vapor at 250 K is calculated according to eq. (5.66):

$$H_V = (2906.5) + (-1802.27) - (-107.679) = 1211.91 \text{ J/mol}.$$

Similarly, the enthalpy of the saturated liquid at 250 K is

$$H_L = (2906.5) + (-10318.7) - (-107.679) = -7304.48 \text{ J/mol}.$$

Finally, the enthalpy of the two-phase system is

$$H = (0.32)(1211.91) + (1 - 0.32)(-7304.48) = -4579.24 \text{ J/mol}.$$

The negative sign indicates that the enthalpy at the state of interest is lower than that at the reference state.

Comments Entropy is calculated in a similar manner. All the required information is given above and the calculation is left as an exercise.

Example 7.3: Condensing a Vapor-Liquid Mixture
Calculate the heat that must be removed from the vapor-liquid mixture of the previous example to produce the saturated liquid at 250 K.

Solution The required heat is equal to the enthalpy change between the initial state (enthalpy H_A) and the saturated liquid (H_L):

$$Q = \Delta H = H_L - H_A$$

where $H_L = -7304.48$ J/mol and $H_A = -4579.24$ J/mol, both calculated in the previous example. Therefore,

$$Q = (-7304.48) - (-4579.24) = -2725.24 \text{ J/mol}.$$

The heat is negative, that is, it must be removed.

Comments Once properties have been calculated relative to a reference state, energy calculations become a matter of algebraic differences. This convenience is the reason we have invested so much effort in formulating equations for the absolute enthalpy and entropy.

7.2 Vapor-Liquid Equilibrium

Away from the saturation point a fluid exists as a single phase, either liquid or vapor. The saturation point is a special state where the fluid can form both phases, which coexist in equilibrium with each other, both at the same temperature and pressure. We want to determine the precise conditions that define the saturation point. For this we turn to the Gibbs free energy and in particular to eq. (4.45), which states that the equilibrium state of a system at constant T, P, and n is such that the Gibbs free energy is at a minimum. We imagine the following experiment: a fixed number of moles of a fluid is placed in a closed system at such pressure and temperature as to form two phases. The pressure and temperature are then maintained constant (for example, using a movable piston and a heat bath). Suppose that we take a

7.2 Vapor-Liquid Equilibrium

small number of δn moles and transfer them from the liquid to the vapor. The new state is also an equilibrium two-phase system and since the Gibbs free energy is at a minimum, this transfer should cause no change in the total Gibbs free energy of the system. However, the Gibbs free energy of the individual phases changes because G is an extensive property and depends on the number of moles. Specifically, the Gibbs free energy of the liquid changes by $-G_L \delta n$ and that of the vapor by $+G_V \delta n$. For the total Gibbs free energy to remain constant regardless of the number of moles that are transferred, we must have

$$\boxed{G_L = G_V.} \qquad (7.4)$$

This important result states that, while the various molar properties (V, H, S, etc.) are different in the vapor and the liquid, the molar Gibbs free energy is the same in both phases.

NOTE

Chemical Potential

Equation (7.4) is one of the most important results in equilibrium and places the molar Gibbs energy alongside with pressure and temperature. We may now express the conditions of equilibrium in a single component by stating that pressure, temperature, *and* molar Gibbs free energy must be uniform throughout the entire system, regardless of the number of phases present.[2] For this reason, the molar Gibbs free energy of pure component is also known as *chemical potential* (symbol μ) because it is the thermodynamic potential of the chemical species at constant T, and P: if the chemical potential in one phase is higher than in another phase, the species will migrate into the phase with the lowest chemical potential; if the chemical potential is the same in two or more phases, then the species can exist with equal probability in any of these phases.

For a pure species, "chemical potential" and "molar Gibbs free energy" are synonymous. For mixtures, as we will find out in Chapter 8, there is a distinction as each component in the mixture has its own chemical potential. With reference to pure species we will continue to use the name "Gibbs free energy" and the symbol G. With reference to a species in a mixture we will use the term chemical potential and the symbol μ_i.

Clausius-Clapeyron Equation

One consequence of practical significance is that eq. (7.4) provides a relationship between saturation pressure and temperature. To extract this relationship, we return

---

2. Although in the derivation we assumed a vapor-liquid system, the same arguments apply to equilibrium in general, regardless of the number of phases that are present. This is because eq. (4.45) is a general statement of equilibrium and is not restricted with respect to the number or type of phases.

to eq. (5.17), which gives the differential of the Gibbs free energy. We write this equation for the saturated vapor and liquid:

$$dG_V = -S_V dT + V_V dP^{\text{sat}},$$
$$dG_L = -S_L dT + V_L dP^{\text{sat}}.$$

Because the difference between G_V and G_L is zero, we must have $dG_V - dG_L = 0$. Taking the difference between the two equations we have

$$0 = -\Delta S_{\text{vap}} dT + (V_V - V_L) dP^{\text{sat}}$$

and using $\Delta S_{\text{vap}} = \Delta H_{\text{vap}}/T$ the result becomes

$$\frac{dP^{\text{sat}}}{dT} = \frac{\Delta H_{\text{vap}}}{T(V_V - V_L)}.$$

This is an exact relationship between temperature and pressure at saturation. A practical result is obtained if we make the approximation that the liquid molar volume is negligible compared to the vapor volume (i.e., $V_V - V_L \approx V_V$), and that the vapor volume can be calculated by the ideal-gas law (i.e., $V_V \approx RT/P^{\text{sat}}$). Both approximations are acceptable at pressure well below the critical, and both become poor close to the critical point. Using these approximations, the relationship between temperature and saturation pressure becomes

$$\frac{dP^{\text{sat}}}{P^{\text{sat}}} \approx \frac{\Delta H_{\text{vap}}}{RT^2} dT.$$

The final approximation is to treat the enthalpy of vaporization as constant. Integrating from $(T', P^{\text{sat}'})$ to (T, P^{sat}), we obtain the Clausius-Clapeyron equation:

$$\boxed{\ln \frac{P^{\text{sat}}}{P^{\text{sat}'}} \approx -\frac{\Delta H_{\text{vap}}}{R}\left(\frac{1}{T} - \frac{1}{T'}\right).} \quad (7.5)$$

The enthalpy of vaporization is not constant, of course, it decreases with temperature and vanishes at the critical point. Therefore, the above result should be treated as an approximation over short temperature intervals. Despite its approximate nature, the Clausius-Clapeyron equation is useful in that it allows us to calculate the saturation pressure at a temperature, if its value is known at another. According to this equation, the logarithm of the saturation pressure is a linear function of inverse temperature with slope $-\Delta H_{\text{vap}}/R$. Saturation pressure is often plotted in semi-log axes against $1/T$. In this form the resulting graph is approximately linear.

NOTE

Antoine Equation

The Clausius-Clapeyron equation can be written as

$$\ln P^{\text{sat}} = -\frac{A}{T} + B,$$

where A and B are constants. This form suggests that the saturation pressure could be empirically fitted to such linear form. This linearity is, however, subject to the same assumptions used in the derivation, and its validity is limited. To obtain a better empirical fit, a third constant is introduced to write

$$\ln P^{\text{sat}} = -\frac{A}{T+C} + B.$$

This is the form of the *Antoine* equation. It is an empirical equation used to fit experimental data. Its mathematical form, however, is not arbitrary but represents an empirical modification of the Clausius-Clapeyron equation, which turns out to be fairly accurate over an extended range, compared to the equation it is based on. Even so, the Antoine equation cannot accurately fit the entire subcritical region of a fluid.

7.3 Fugacity

The starting point of phase equilibrium is eq. 7.4, which establishes the equality of the molar Gibbs energy of the phases that are present. In this Section we will obtain alternative forms of that equation that are suitable for calculations. First, we express the Gibbs free energy in terms of its residual

$$G = G^{\text{ig}} + G^R.$$

This equation can be written for the liquid and for the vapor. The ideal-gas term is the same in both phases because it depends only on temperature and pressure, which are the same in both phases. We conclude then that the residual Gibbs free energies of the two phases are also equal:

$$\boxed{G_L^R = G_V^R.} \tag{7.6}$$

This equation is equivalent to eq. (7.4) but has the advantage that it involves residual properties, whose calculation does not require a reference state. To further streamline the calculation of phase equilibrium we introduce a new property, *fugacity*, using the following definition:

$$\boxed{f = P e^{G^R/RT} = \phi P,} \tag{7.7}$$

where ϕ is the *fugacity coefficient* and is given by

$$\boxed{\ln \phi = \frac{G^R}{RT}.} \tag{7.8}$$

Since the residual Gibbs energy is the same in both phases, the fugacity at saturation satisfies the conditions:

$$f^V = f^L, \tag{7.9}$$

$$\phi^V = \phi^L. \tag{7.10}$$

The equality of fugacities is an alternative statement of the necessary and sufficient condition for phase equilibrium and the basis for all such calculations, whether we are dealing with pure substances or with mixtures. By contrast, the equality of the fugacity coefficients is a special result and applies to pure substances only.

Fugacity in the Ideal-Gas State

The limiting values of these properties in the ideal-gas limit follow easily by noting that $G^R = 0$ in this limit:

$$f^{ig} = P, \tag{7.11}$$

$$\phi^{ig} = 1. \tag{7.12}$$

The fugacity in the ideal-gas limit is equal to pressure, and the fugacity coefficient is equal to unity.

Relationship to the Gibbs Free Energy

The fugacity, an auxiliary property, is related to the Gibbs free energy, the fundamental thermodynamic property at equilibrium. To establish this relationship, first we write the Gibbs energy in the form,

$$G = G^{ig} + G^R = G^{ig} + RT \ln \frac{f}{P}.$$

We now take the difference between the Gibbs free energy of two states on the same isotherm:

$$G - G_0 = \underbrace{G^{ig} - G_0^{ig}}_{=RT \ln \frac{P}{P_0}} + RT \ln \frac{f/P}{f_0/P_0},$$

7.4 Calculation of Fugacity

where G, G^{ig} and f refer to state at T, P, while G_0, G_0^{ig} and f_0 refer to state T, P_0. Using $RT \ln P/P_0$ for the isothermal change of the ideal-gas Gibbs energy between pressures P and P_0, the final result is

$$G = G_0 + RT \ln \frac{f}{f_0}. \tag{7.13}$$

This establishes the relationship between Gibbs free energy and fugacity, and shows that if we know one of these properties, the other one can be easily obtained. Taking the differential of the above equation at constant temperature, and using eq. (5.17) for dG, the differential of fugacity is

$$d \ln f = \frac{V}{RT} dP, \quad (\text{const. } T). \tag{7.14}$$

We will use this equation to obtain the variation of fugacity along an isotherm.

NOTE
Why Fugacity

Fugacity is equivalent to the Gibbs free energy with respect to defining the conditions of phase equilibrium. It offers two practical conveniences over the Gibbs free energy: (a) it does *not* require a reference state and (b) it has a well-defined (and simple) value in the ideal-gas state. By contrast, the Gibbs free energy requires a reference state, and its value in the ideal-gas state ($P \to 0$) approaches $-\infty$. The name fugacity comes from the Latin *fugere* ("to flee") and refers to the tendency of species to "escape" to the more stable phase. It is best to think of fugacity as a defined quantity associated with the Gibbs free energy.

7.4 Calculation of Fugacity

Fugacity is a central property in phase equilibrium. Below we discuss several alternative methodologies that may be used to obtain the fugacity of pure component.

Compressed Liquids—Poynting Equation

In the compressed liquid region the molar volume is essentially independent of pressure. Equation (7.14) can then be integrated easily to obtain a relationship between the fugacity of a liquid at two different pressures on the same isotherm. With V and T constant, the result is

$$\ln \frac{f}{f_0} = \frac{V}{RT} \int_{P_0}^{P} dP = \frac{V(P - P_0)}{RT}. \tag{7.15}$$

The term in the right-hand side is known as the Poynting factor. This equation is valid to incompressible phases in general (liquids away from the critical point, and solids in general). A practical result is obtained if we choose the initial state to be the saturated liquid at the given temperature:

$$f(P, T) = f_L \exp\left(\frac{P - P^{\text{sat}}}{RT} V_L\right),$$

where P^{sat} is the saturation pressure, V_L, is the molar volume of the saturated liquid, and f_L is its fugacity. Using $f_L = \phi^{\text{sat}} P^{\text{sat}}$, the final result is

$$\boxed{f = \phi^{\text{sat}} P^{\text{sat}} \exp\left(\frac{P - P^{\text{sat}}}{RT} V_L\right).} \quad (7.16)$$

This equation allows us to obtain the fugacity in the compressed liquid region from tabulated values at saturation.

Example 7.4: Saturated Liquid
Calculate the fugacity and fugacity coefficient of saturated liquid water at 25 °C.

Solution The fugacity of the saturated liquid is equal to the fugacity of the saturated vapor

$$f_L = f_V = \phi^{\text{sat}} P^{\text{sat}}$$

The saturation pressure at 25 °C is 0.03166 bar and at such low pressure the vapor phase is ideal, that is, $\phi^{\text{sat}} \approx 1$. Therefore, the fugacity is equal to the saturation pressure,

$$f_L = f_V \approx 0.03166 \text{ bar},$$

and the fugacity coefficient is 1.

Example 7.5: Poynting Factor
Calculate the fugacity and fugacity coefficient of water at 25 °C, 100 bar, using data from the steam tables.

Solution Under these conditions water is a compressed liquid; therefore, the Poynting equation will be used. The liquid volume at 25 °C is $V_L = 1.003 \text{ cm}^3/\text{g} = 18 \times 10^{-6} \text{ m}^3/\text{mol}$. The Poynting factor is

$$\exp\left[(18 \times 10^{-6}) \frac{(100 - 0.03166) \times 10^5}{(8.314)(298)}\right] = 1.075.$$

7.4 Calculation of Fugacity

From this we calculate the fugacity, assuming the fugacity coefficient at saturation to be 1 because pressure is very low:

$$f = \phi^{\text{sat}} P^{\text{sat}} \times (1.075) = 0.034 \text{ bar}.$$

The fugacity coefficient is

$$\phi = \frac{0.034 \text{ bar}}{100 \text{ bar}} = 3.4 \times 10^{-4}.$$

Using Tabulated Properties

The fugacity is related to the Gibbs energy, which in turn is related to the enthalpy and entropy. It is possible, therefore, to calculate it from tabulated values of H and S. The calculation is demonstrated with the following example.

Example 7.6: Fugacity from Steam Tables
Use the steam tables to calculate the fugacity of water at 70 bar and 300 °C.

Solution We will calculate fugacity from eq. (7.13). The Gibbs free energy will be calculated from the steam tables based on its definitions:

$$G = H - TS.$$

To use eq. (7.13) we must find a state on the same isotherm where the Gibbs energy is known. We pick $P_0 = 0.05$ bar, a pressure sufficiently low that we may assume $f_0 \approx P_0$. Then, the fugacity is calculated as

$$f = P_0 \exp\left(\frac{G - G_0}{RT}\right),$$

and the fugacity coefficient as

$$\phi = \frac{f}{P}.$$

The results are summarized in the table below:

P (bar)	0.05	70
T (°C)	300	300
H (kJ/kg)	3076.9	2839.8
S (kJ/kg K)	9.603	5.9335
G (kJ/kg)	−2427.06	−560.986

The fugacity at 70 bar, 300 °C is

$$f = (0.05 \text{ bar}) \exp\left(\frac{(-560.986) - (-2427.06)}{(8.314 \text{ J/mol K})(573.15 \text{ K})} \frac{\text{kJ}}{\text{kg}} \frac{18.015 \times 10^{-3} \text{ kg}}{\text{mol}}\right) = 57.92 \text{ bar}.$$

The fugacity coefficient is
$$\phi = \frac{57.92 \text{ bar}}{70 \text{ bar}} = 0.8274.$$

Using the Compressibility Factor

Equation (7.14) gives the fugacity in differential form in terms of volume. It is useful to express this equation in terms of the compressibility factor. Subtracting the term $d\ln P = dP/P$ from both sides of that equation we obtain

$$d\ln \frac{f}{P} = \left(\frac{V}{RT} - \frac{1}{P}\right) dP, \quad \text{(const. } T\text{)},$$

which we rewrite as

$$d\ln \phi = (Z-1)\frac{dP}{P}, \quad \text{(const. } T\text{)}.$$

By integration from the ideal-gas state ($Z=1$, $\phi=1$) along an isotherm we obtain

$$\ln \phi = \int_0^P (Z-1)\frac{dP}{P}, \quad \text{(const. } T\text{)}. \tag{7.17}$$

According to this result, the fugacity coefficient can be obtained from an integral that involves the compressibility factor. The compressibility factor may be calculated from an equation of state, from tables, or it may be obtained experimentally.

A useful result is obtained if we restrict our attention to the part of the isotherm that is described by the truncated virial equation:

$$Z \approx 1 + \frac{BP}{RT}.$$

Since the second virial coefficient depends only on temperature and integration along an isotherm, the result is

$$\ln \phi \approx \frac{BP}{RT}. \tag{7.18}$$

Using $BP/RT = Z - 1$, this may be written in a simpler form as

$$\boxed{\ln \phi \approx Z - 1.} \tag{7.19}$$

This very simple approximation gives the correct answer in the ideal-gas state and is valid as an extrapolation to pressures sufficiently low that the compressibility factor may be approximated as a linear function of pressure.

7.4 Calculation of Fugacity

Example 7.7: Fugacity of Steam
Estimate the fugacity of steam at 300 °C 70 bar if the only information available is the density of steam at this state, 33.898 kg/m^3.

Solution We assume that the truncated virial is valid at this state (we will return to this assumption). The molar volume of steam at these conditions is

$$V = \frac{1}{33.898} = 0.0295 \text{ m}^3/\text{kg} = 5.314 \times 10^{-4} \text{ m}^3/\text{mol}.$$

The compressibility factor is

$$Z = \frac{(70 \times 10^5 \text{ Pa})(5.314 \times 10^{-4} \text{ m}^3/\text{mol})}{(8.314 \text{ J/mol K})(573.15 \text{ K})} = 0.780685.$$

The fugacity coefficient is

$$\ln \phi = 0.78685 - 1 = -0.219315 \quad \Rightarrow \quad \phi = 0.803$$

and the fugacity is

$$f = (0.803)(70 \text{ bar}) = 56.2 \text{ bar}.$$

Comments To assess the validity of the truncated virial we check with a generalized compressibility chart. The reduced coordinates of water at this state are

$$T_r = \frac{573.15 \text{ K}}{647.096 \text{ K}} = 0.886, \quad P_r = \frac{70 \text{ bar}}{220.64 \text{ bar}} = 0.317.$$

From Figure 2-8 we see that the reduced pressure is fairly low and the isotherm does not exhibit significant nonlinearity. Therefore, the assumption that the isotherm is linear is acceptable. This should be confirmed by an independent calculation of the fugacity from a different method. In Example 7.6 we calculated the fugacity using eq. (7.13) and found $f = 57.92$ bar. The present calculation is within 3% of that value.

Fugacity from Generalized Graphs

The residual Gibbs energy can be calculated from the residual enthalpy and entropy. Accordingly, all the methods developed for the residual enthalpy and entropy may be used to calculate G^R, and from eq. (7.8) the fugacity coefficient. A quick estimate can be obtained from generalized correlations. Instead of calculating H^R and S^R individually from Pitzer/Lee/Kesler graphs, separate graphs have been prepared that give the fugacity coefficient by the Pitzer expression,

$$\ln \phi = (\ln \phi)^{(0)} + \omega (\ln \phi)^{(1)}, \tag{7.20}$$

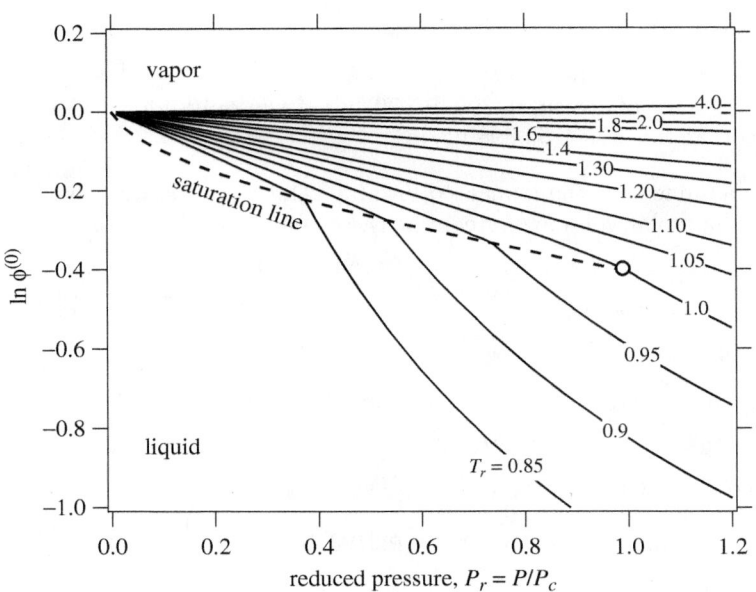

Figure 7-1: Generalized graph for $(\ln \phi)^{(0)}$. The dashed line is the saturation line between vapor and liquid.

where $(\ln \phi)^{(0)}$ and $(\ln \phi)^{(1)}$ are universal functions of reduced temperature and pressure. Figure 7-1 shows the term $(\ln \phi)^{(0)}$ calculated by the Lee-Kesler method, as a function of reduced pressure and temperature. This term represents the fugacity coefficient of a simple fluid ($\omega = 0$) and qualitatively resembles the fugacity graph of any pure fluid. All isotherms radiate outward from the ideal-gas state ($P_r = 0$, $\phi = 1$). Isotherms in this region are approximately linear in pressure, as expected on the basis of eq. (7.18). Subcritical isotherms intersect the saturation line and change direction as they pass from one phase into the other. Because $\phi_L = \phi_V$ at saturation, there is no jump in the value of the fugacity coefficient, however, the slope of the isotherm changes abruptly at this point. The saturation line terminates at the critical point ($P_r = T_r = 1$). Supercritical isotherms do not intersect with the saturation line. Figure 7-2 shows graphs of $(\ln \phi)^{(0)}$ and $(\ln \phi)^{(1)}$ over an extended range of pressures. Notice that the correction term, $(\ln \phi)^{(1)}$, has a jump across the phase boundary. Since this term is only a correction to a fugacity coefficient, not a fugacity coefficient in itself, the shape of this graph does not convey any physical meaning.

7.4 Calculation of Fugacity

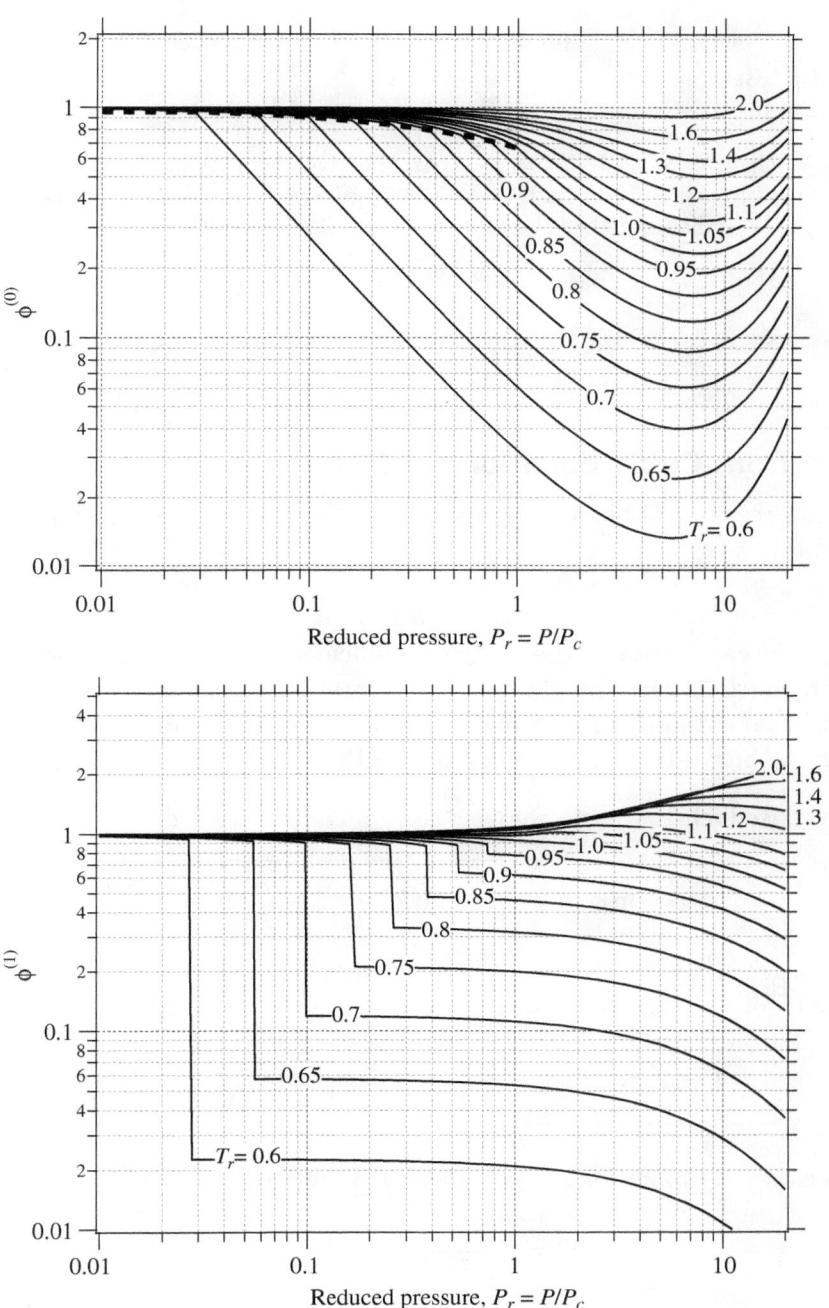

Figure 7-2: Generalized graph for $(\ln \phi)^{(0)}$ and $(\ln \phi)^{(1)}$ (extended range of pressures).

Example 7.8: Generalized Graphs
Calculate the fugacity of benzene at $T = 64\,°C$, $P = 34$ bar using the Lee-Kesler method.

Solution With $T_c = 562.1$ K, $P_c = 48.9$ bar, the given conditions correspond to $T_r = 0.6$, $P_r = 0.7$. From Figure 7-2, we find, $\phi^0 = 0.0433$, $\phi^1 = 0.0215$. Notice that these values are printed in italics indicating that the state is liquid. The fugacity coefficient is

$$\phi = (0.0433)(0.0215)^{0.210} = 0.0193$$

and the fugacity is $f = \phi P = (0.0193)(34) = 0.657$ bar.

Fugacity from Cubic Equations of State

Writing the residual Gibbs energy as $G^R = H^R - TS^R$, the fugacity coefficient is expressed as

$$\ln \phi = \frac{H^R}{RT} - \frac{S^R}{R}.$$

Therefore, the calculation of the fugacity coefficient by a cubic equation of state does not require much in terms of additional computations, since the required expressions for the residual enthalpy and entropy have been obtained already. In particular, the result for the Soave-Redlich-Kwong and the Peng-Robinson equation are

- Soave-Redlich-Kwong

$$\ln \phi = Z - 1 - \ln(Z - B') - \frac{A'}{B'} \ln \frac{Z + B'}{Z}. \qquad (7.21)$$

- Peng-Robinson

$$\ln \phi = Z - 1 - \ln(Z - B') - \frac{A'}{2\sqrt{2}B'} \ln \frac{Z + (1 + \sqrt{2})B'}{Z + (1 - \sqrt{2})B'}. \qquad (7.22)$$

In both cases the dimensionless parameters A' and B' are given by:

$$A' = \frac{aP}{(RT)^2}, \qquad B' = \frac{bP}{RT}. \qquad [2.36]$$

This calculation requires the compressibility factor at the pressure and temperature of interest. If the polynomial equation for Z has three real roots, the proper root must be selected based on the phase of the system.

Example 7.9: Fugacity from the SRK
Calculate the fugacity of CO_2 vapor at 4.5 °C, 15 bar, using the SRK equation.

Solution The critical constants and acentric factor of carbon dioxide are $T_c = 304.2$ K, $P_c = 73.8$ bar, and $\omega = 0.225$. The parameters of the SRK equation are

$$a = 0.3983 \text{ J m}^3/\text{mol}^2 \quad A = 0.112141,$$
$$b = 2.971 \times 10^{-5} \text{ m}^3/\text{mol} \quad B = 0.0193049.$$

The cubic equation for Z is

$$-0.00216488 + 0.0924639 Z - Z^2 + Z^3 = 0,$$

and has three real roots:

$$Z_1 = 0.0401352, \quad Z_2 = 0.0599377, \quad Z_3 = 0.899927.$$

Since the phase is vapor we select the largest root, $Z = 0.899927$. The fugacity coefficient is

$$\phi = 0.908246,$$

and the fugacity is

$$f = (0.908246)(15 \text{ bar}) = 13.62 \text{ bar}.$$

7.5 Saturation Pressure from Equations of State

We encountered the following equalities between properties of saturated phases of pure component:

$$G_L = G_V \quad [7.4]$$
$$G_L^R = G_V^R \quad [7.6]$$
$$f^V = f^L \quad [7.9]$$
$$\phi^V = \phi^L. \quad [7.10]$$

These are equivalent statements of phase equilibrium in a single-component system and any one of them can be used to study equilibrium. Mathematically, the equilibrium criterion is an equation that establishes the relationship between pressure and temperature at saturation:

$$\phi^V(T, P^{\text{sat}}) = \phi^L(T, P^{\text{sat}}).$$

If temperature is fixed, the saturation pressure may be obtained by solving this equation. This allows us to calculate the saturation pressure via the equation of state of the fluid. Subcritical isotherms from cubic and other analytic equations of state exhibit a metastable/unstable loop (Figure 7-3). The liquid part of the isotherm is represented by the steep branch, labeled AB on this figure, and the vapor part by the gentler branch CD. Points B and C mark the minimum and the maximum of the isotherm. The unstable part of the isotherm between points B and C has been removed. The problem now becomes to locate points L and V, which mark the saturated phases. Numerically, we seek a pressure such that the fugacity coefficients at L and V are the same. This may be done by trial and error. We guess a pressure and solve for the compressibility factor. In this region, there are always three real roots. We use the smallest root to calculate the fugacity coefficient of the liquid and the largest root to calculate the fugacity coefficient of the vapor. If the two fugacity coefficients are not equal, we pick another pressure and repeat until ϕ^L and ϕ^V agree to within an acceptable tolerance.

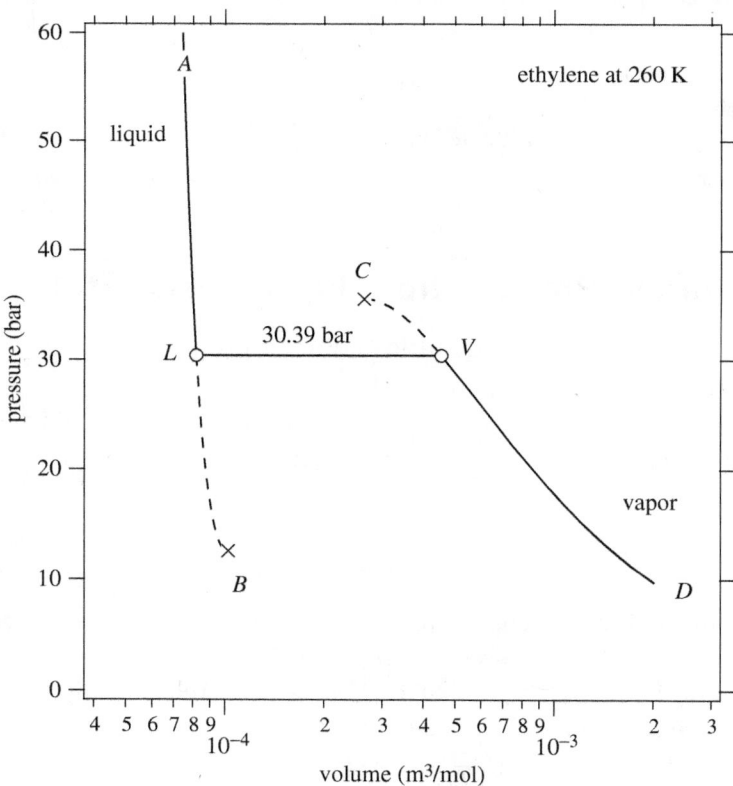

Figure 7-3: Determination of saturation pressure of ethylene using the SRK equation of state.

7.5 Saturation Pressure from Equations of State

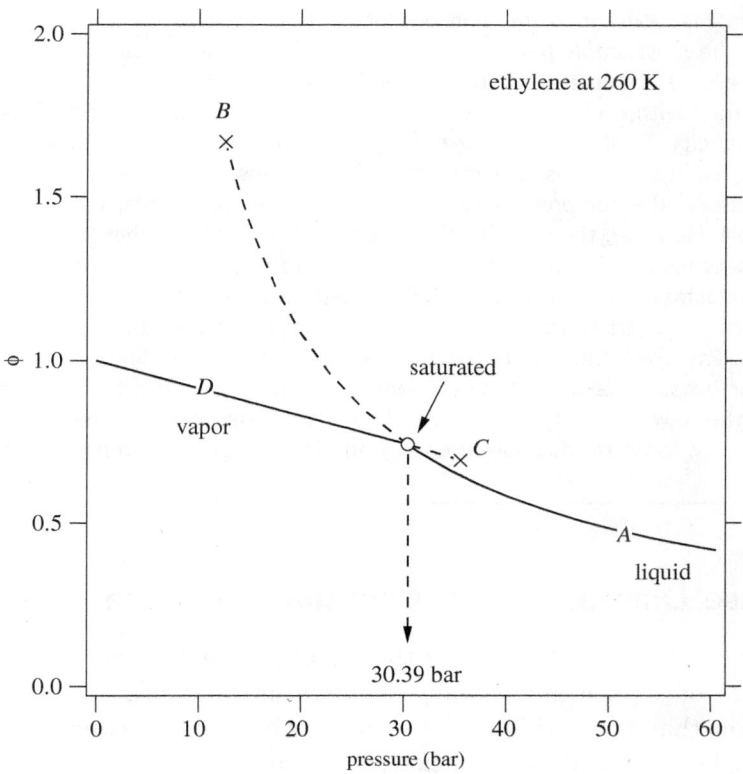

Figure 7-4: Fugacity coefficient of ethylene based on the SRK. Branches *AB* and *CD* correspond to the same ones in Figure 7-3.

The fugacity coefficients calculated by this procedure are shown in Figure 7-4. Branches AB and CD correspond to the same ones on the PV isotherm. The liquid fugacity coefficient at high pressures (point A) starts from small values, and increases with decreasing pressure. The fugacity coefficient of the vapor starts at 1 at low pressures and decreases approximately linearly with increasing pressure. At some point the two branches intersect. This defines the saturation point. Figure 7-4 represents a graphical determination of the saturation pressure.

NOTE
Stability
As discussed in Chapter 2 (see Section 2.7) the system is mechanically stable over the entire liquid and vapor branches, *AB* and *CD*. This means that over the range of pressures bracketed by the pressure of points *B* and *C* the system may exist as either vapor or liquid. In regions

where a system may exist in more than one phase that satisfies the criteria for stability, the system adopts the *most stable* phase, rendering the other phases *metastable*. When two phases are equally stable we obtain the saturation point, namely, a state where two (or more) phases exist in equilibrium with each other. We may identify the stable and metastable phases by comparing their fugacity. Suppose we follow the ϕ isotherm starting from point D. Initially, the compressibility equation has a single root and the only possible phase is the vapor. Beginning with some pressure, the compressibility equation has three real roots, indicating that a liquid phase is possible. However, the fugacity coefficient of this liquid is higher than that of the vapor: the liquid phase is *metastable*, meaning less stable than the vapor at the same pressure and temperature. A metastable phase may materialize briefly, but thermal fluctuations will eventually cause the system to revert to the more stable phase. At saturation, both phases have the same fugacity coefficient. Past this point, the situation is reversed: the liquid phase is more stable than the vapor because its fugacity coefficient is lower. In all cases the most stable phase is the one with the lower fugacity coefficient. This also means that the more stable phase has lower Gibbs energy, lower residual Gibbs energy and lower fugacity, compared to the metastable phase.[3]

7.6 Phase Diagrams from Equations of State

At this point we have reached one of the main goals we set out to achieve: obtain the properties of a pure fluid at any temperature and pressure. The two pieces of information that make this calculation possible are the equation of state and the ideal-gas heat capacity as a function of temperature. The equation of state allows the calculation of all residual properties and the determination of the phase boundary, namely of the saturation pressure and the properties of the saturated phases. Thus we have the tools to compute the entire phase diagram of the pure fluid. This calculation is demonstrated with the example below.

Example 7.10: Phase Diagram of Ethylene
Calculate the phase diagram of ethylene using the SRK equation. The reference state is saturated liquid at 110 K.

Solution We will demonstrate the calculation by obtaining all properties along an isotherm, from the compressed liquid region to the superheated vapor. The calculation can then be repeated with other temperatures to obtain the complete phase diagram. We pick $T = 260$ K.

---

3. To reach this conclusion we write $G = G^{ig} + G^R$. Since both phases are the same pressure and temperature, the term G^{ig} is common in both phases. Then, the phase with the lower fugacity coefficient has the lower residual Gibbs free energy and also the lower molar Gibbs free energy.

7.6 Phase Diagrams from Equations of State

Saturation pressure From Figure 7-3 we see that the saturation pressure at 260 K must be between the pressures of points B (~ 15 bar) and C (~ 35 bar). Using $P = 29$ bar as the trial pressure, we find the following roots for the compressibility factor:

$$Z_1 = 0.1104, \quad Z_2 = 0.2244, \quad Z_3 = 0.6651.$$

Using $Z_L = Z_1$ in eq. (7.21) we find $\phi_L = 0.7733$; with $Z_V = Z_3$ the same equation gives $\phi_V = 0.7539$. These values are not sufficiently close. Therefore we repeat the calculation with different pressures in small increments and construct the table below:

P (bar)	Z_L	Z_V	ϕ_L	ϕ_V	ϕ_L/ϕ_V
29	0.1104	0.6651	0.7733	0.7539	1.0256
30	0.1136	0.6461	0.7503	0.7452	1.0069
31	0.1169	0.6255	0.7289	0.7363	0.9898
32	0.1201	0.6026	0.7088	0.7274	0.9744

The saturation pressure is between 30 and 31 bar. By refining the pressure step (or by simple interpolation between the values in the table), we find

$$P^{\text{sat}} = 30.39 \text{ bar}.$$

The value reported in Perry's *Handbook* is 30.046 bar.

Properties. The enthalpy and entropy are calculated using eqs. (5.62) and (5.64):

$$H = \Delta H^{\text{ig}}_{\text{ref}} + H^R - H_0^R + H_0, \qquad [5.62]$$

$$S = \Delta S^{\text{ig}}_{\text{ref}} + S^R - S_0^R + S_0, \qquad [5.64]$$

with $H_0 = 0$, $S_0 = 0$. Since the reference state is the saturated liquid at $T_0 = 110$ K, we must first calculate the saturation pressure at this temperature. The calculation is done as in the previous step. We find,

$$P_0 = 0.00333 \text{ bar}, \quad H_0^R = -15565.8 \text{ J/mol}, \quad S_0^R = -141.505 \text{ J/mol K}$$

For subcritical isotherms the procedure for the calculation of the enthalpy and entropy is as follows:

1. Fix temperature.
2. Calculate the saturation pressure.
3. Fix pressure.
4. Determine the phase of the system at the given pressure: if $P < P^{\text{sat}}$, the phase is vapor; if $P > P^{\text{sat}}$ the phase is compressed liquid; if $P = P^{\text{sat}}$, the system is saturated.
5. Solve for the compressibility factor. If three real roots, pick according to the phase.

6. Calculate the residual properties as usual.

7. Calculate the ideal-gas parts of enthalpy and entropy.

8. Combine the ideal gas and residual parts to obtain the absolute properties.

9. Change the pressure and return to step 4 to repeat the calculation.

The calculation for supercritical isotherms is simpler. In this case there is no saturation pressure and the compressibility equation has one real root only. Therefore, steps 2 and 4 are skipped, otherwise the calculation is identical.

The table below shows results at three pressures, 40 bar (compressed liquid), 30.3929 bar (saturated vapor-liquid) and 20 bar (superheated vapor).

	Phase			
	Comp. liq.	Sat. liq.	Sat. vap.	S/H vap.
T (K)	260	260	260	260
P (bar)	40	30.3929	30.3929	20
Z	0.145477	0.114908	0.63825	0.797499
H (J/mol)	11056.1	11238.9	18428.9	19528.4
S (J/mol K)	60.04	61.04	88.70	95.46

Other properties may be computed from their relationship to the properties shown here. For example, the internal energy is calculated as $U = H - ZRT$, the Gibbs free energy as $G = H - TS$, and so on.

Figure 7-5 shows the pressure-enthalpy diagram of ethylene calculated by this method. The graph shows the phase boundary and representative isotherms.

7.7 Summary

In this chapter we developed the theoretical tools for the study of saturated phases, namely, phases that coexist in equilibrium with each other. The fundamental property is the Gibbs free energy, which has the same value in both phases. This fundamental equality is the basis of all the results obtained in this chapter. Fugacity and the fugacity coefficient are auxiliary variables introduced for convenience. They are both related to the Gibbs free energy but are more convenient to work with because they do not require a reference state. In addition, they reduce to very simple expressions in the ideal-gas state.

An important conclusion in this chapter is that all properties of a pure fluid at saturation can be calculated from the equation of state. The equation of state,

7.7 Summary

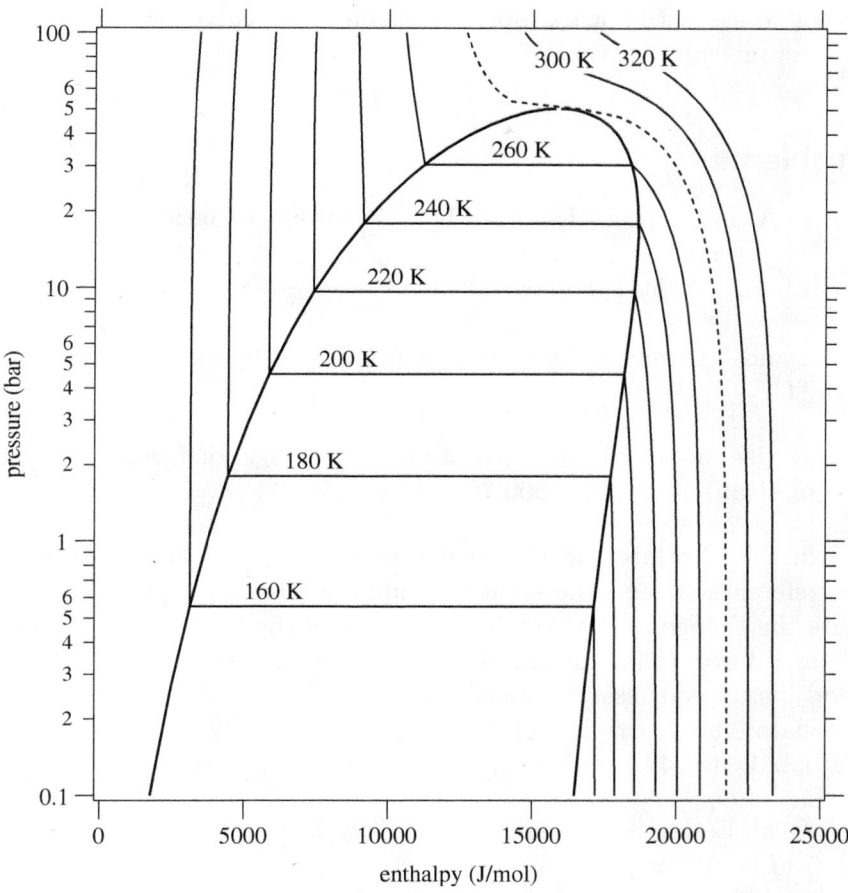

Figure 7-5: Pressure enthalpy graph of ethylene calculated from the Soave-Redlich-Kwong equation (see Example 7.10).

therefore, along with the ideal-gas heat capacity, provide a complete description of the thermodynamics of a pure fluid. Between the tools developed in the last two chapters we can calculate properties as a function of pressure and temperature at any point of the phase diagram. This benefit comes with one caveat: an accurate equation of state must be available for the fluid of interest. The equations of state discussed here are relatively simple and require a small number of specific properties of the components, namely, critical pressure, critical temperature, and acentric factor. It is quite impressive that such equations provide fairly accurate estimates of various properties. However, the use of these equations should be restricted to normal fluids. Even then, before a serious calculation is undertaken, the equation of state must be rigorously tested against known properties of the fluid. Only then may it be trusted to provide accurate results.

7.8 Problems

Problem 7.1: A pure fluid is described by the Antoine equation

$$\log_{10} P^{\text{sat}} = 4.00266 - \frac{1171.53}{-48.784 + T}$$

with P in bar and T in kelvin. How much heat is required to evaporate 1 mol of that liquid at 25 °C?

Problem 7.2: Use the steam tables to obtain the fugacity coefficient and ideal Gibbs free energy of steam at 200 bar, 500 °C.

Problem 7.3: a) Calculate the chemical potential of solid acetylene at its triple point. The reference state is the saturated liquid at the triple point.
b) Calculate the change in the chemical potential of the vapor when the pressure is reduced from 1.3 atm to 0.1 atm at the constant temperature of −84 °C.
c) State and justify your assumptions.
Additional data at the triple point: $T_{\text{triple}} = -84$ °C, $P_{\text{triple}} = 1.3$ atm, $V^{\text{solid}} = 34$ cm^3/mol, $V^{\text{liq}} = 42.7$ cm^3/mol, $V^{\text{vap}} = 12020$ cm^3/mol.

Problem 7.4: a) Estimate the fugacity of CO_2 ice (dry ice) at its triple point (216.55 K, 5.17 bar).
b) Calculate the fugacity of CO_2 ice at 216.55 K, 70 bar.
Additional information: density of dry ice: 97.5189 lb/ft^3.

Problem 7.5: Use the Lee-Kesler tables to calculate the chemical potential of benzene at the following states:
a) critical point.
b) saturated liquid at 200 °C.
c) 200 °C, 45 bar.
Additional data: The reference state is the ideal gas at 562.1 K, 48.9 bar. The saturation pressure at 200 °C is 14.3 bar.

Problem 7.6: a) A 45-L tank contains 13.5 mol of an unknown gas at 200 °C, 10 bar. Estimate as best as you can the fugacity of the gas.
b) Half of the gas is removed from the tank while maintaining the temperature at 200 °C. Determine the fugacity of the gas that remains in the tank.

7.8 Problems

Problem 7.7: Use the Pitzer correlation and the Lee-Kesler ϕ tables to calculate the fugacity of ethane at the following states:
a) 1 bar, 40 °C.
b) 50 bar, −70 °C.

Problem 7.8: A fluid that is gas under ambient conditions is stored in a 2 m³ tank. The amount of the gas in the tank is 150 mol. Estimate the fugacity of the gas. State your assumptions clearly.

Problem 7.9: Over a limited range of pressures, an unspecified gas obeys this equation of state,

$$\left(P - \frac{a}{V^2}\right) V = RT,$$

where a is constant.
a) Derive an expression for the fugacity coefficient.
b) At 30 °C, 1 bar, the compressibility factor of this gas is 0.88. Estimate its fugacity.
c) Calculate the residual Gibbs energy of the gas at 30 °C 1 bar.

Problem 7.10: The attached table from Perry's *Handbook* gives some properties of argon. Using data from this table estimate the following:
a) The fugacity at 300 K, 1 atm (state A).
b) The fugacity of the saturated liquid at 90 K (state B).
c) The ideal-gas Gibbs energy, G^{ig}, of the saturated liquid at 90 K.
d) The fugacity at 50 atm, 90 K (state C).
e) Draw a qualitative PV graph and show the states A, B, and C of the previous parts.

T (K)	P (atm)	ρ_L (mol/L)	ρ_V (mol/L)	H_L (J/mol)	H_V (J/mol)	S_L (J/mol K)	S_V (J/mol K)
90	1.321	34.47	0.1864	3099	9481	56.02	126.93

At atmospheric pressure and 300 K, the properties of argon are $\rho = 0.0406$ mol/L, $H = 13919$ J/mol, $S = 154.82$ J/mol K. (see Perry and Chilton, *Chemical Engineers' Handbook* 5th ed. (New York: McGraw-Hill, 1973), p. 3-159.)

Problem 7.11: Use the tabulated properties of Br in the table below to do the following:
a) Fugacity at 1 bar, 300 K.
b) Fugacity coefficient of saturated liquid bromine at 1 bar.
c) Fugacity and fugacity coefficient at 150 bar, 300 K.
Note: You may not use generalized correlations for this problem. Use only data given here and make any assumptions you deem appropriate or necessary.

Data for Br$_2$ from Perry's *Handbook*, 5th ed., p. 3-159, p. 2-222, Table 2-38

Temperature	Pressure	Volume (cm^3/g)		Enthalpy (J/g)		Entropy (J/g K)	
T (K)	P (bar)	Liquid	Vapor	Liquid	Vapor	Liquid	Vapor
260	0.042	0.3106	3195	−147.2	51.8	0.903	1.669
280	0.124	0.3168	1169	−138.9	56.2	0.933	1.629
300	0.310	0.3232	500.2	−131.6	60.6	0.956	1.597
320	0.680	0.3311	242.5	−124.2	64.8	0.978	1.570
340	1.330	0.3385	130.9	−112.3	71.1	1.004	1.539
360	2.384	0.3464	76.71	−108.6	73.1	1.026	1.531
380	4.010	0.3550	47.66	−100.6	76.9	1.048	1.515
400	6.390	0.3647	31.07	−93.4	80.6	1.063	1.501
420	9.730	0.3752	21.09	−85.8	84.0	1.084	1.488
440	14.25	0.3885	14.80	−77.7	87.1	1.103	1.477
460	20.17	0.4023	10.67	−69.0	89.9	1.122	1.467
480	27.75	0.4179	7.858	−59.7	92.2	1.142	1.457
500	37.21	0.4378	5.885	−49.3	94.0	1.161	1.448
520	48.81	0.4623	4.451	−37.7	95.0	1.183	1.438
540	62.80	0.4938	3.370	−24.0	94.8	1.207	1.428
560	79.41	0.5368	2.506	−7.1	92.5	1.237	1.414
580	98.90	0.6250	1.666	18.8	82.5	1.280	1.390
584.2^c	103.4	0.8475	0.848	64.8	64.8	1.356	1.356

^{c}critical point

Problem 7.12: Your company faxed you the following table with data for methane at 150 K. A few entries are missing (denoted by "···") because the fax machine is running out of ink.

a) Is it acceptable to assume ideal-gas state at 1 bar, 150 K? Justify your answer.
b) Calculate the fugacity of saturated liquid at 150 K.
c) Calculate the residual Gibbs energy at 150 K, 60 bar.
d) Calculate the ideal-gas Gibbs energy at 150 K, 60 bar.

Methane at 150 K

	1 bar	5 bar	10.41 bar	20 bar	40 bar
V (m^3/kg)	0.7661	···	0.061*	0.00279†	0.00277	0.00274
H (kJ/kg)	879.0	865.0	853.9*	···	429.8	430.8
S (kJ/kg K)	10.152	9.256	8.849*	···	6.003	5.973

*: saturated vapor; †: saturated liquid

7.8 Problems

Problem 7.13: The table below gives the compressibility factor of ethane at 500 K.

P (bar)	Z
1.013	0.995845
5	0.99122
10	0.985085
20	0.972817
40	0.949723
60	0.93096
80	0.912196
100	0.894876
150	0.871422

Use this information to obtain the fugacity of ethane at 100 bar, 500 K.

Problem 7.14: The table below shows data for the proprietary molecule X. Use these data to calculate the fugacity of this compound at the following conditions:
a) Saturated vapor at 160 °F.
b) At 200 psi, 160 °F.

P (psi)	T (F)	V (cu ft/lb)	
10	160	11.291	
14.696	160	7.641	
20	160	5.578	
30	160	3.669	
40	160	2.715	
50	160	2.14	
60	160	1.7575	
80	160	1.2781	
100	160	0.9888	
125	160	0.7558	
150	160	0.5983	
161.1	160	0.5437	(sat. vapor)
161.1	160	0.03285	(sat. liquid)

Additional data. The critical pressure of this compound is above 500 psi.

Problem 7.15: Some properties of saturated isobutane at 5 bar are given below:

$$T^{\text{sat}}(5 \text{ bar}) = 310.5 \text{ K}, \quad V^L = 0.00201585 \text{ m}^3/\text{kg}, \quad V^L = 0.0783803 \text{ m}^3/\text{kg}$$

Use this information to calculate the following:
a) The fugacity and fugacity coefficient at 0.1 bar, 310.5 K.
b) The fugacity and fugacity coefficient of saturated liquid at 5 bar.
c) The fugacity and fugacity coefficient at 310.5 K, 40 bar.
d) List and justify all of your assumptions.

Problem 7.16: a) Use the SRK equation of state to calculate the saturation pressure of ethane at -58.5 °C.
b) Use the SRK equation to construct the PV graph of ethane. Show isotherms at $T = -58.5$ °C and $T = T_c$.

Problem 7.17: Your R&D department works on a proprietary fluid X for which only limited data are disclosed. Among the available information is that the SRK equation at $T = 383.3$ K, $P = 8$ bar gives three real roots for Z. These roots and the corresponding residual properties calculated at each value of the three values of Z are listed below:

	1	2	3
Z	0.0332418	0.342355	0.624404
H^R (J/mol)	-34046.0	-6020.41	-3384.35
S^R (J/mol K)	-71.4816	-13.4281	-6.42023

Based on this information alone, how big of a tank would be needed to store 120 mol of this fluid at $T = 383.3$ K, $P = 8$ bar?

Problem 7.18: Use the SRK equation to obtain the saturation pressure of propane at 350 K. Perform the calculation as follows:
a) Determine the highest pressure at which the vapor phase is stable (P_1).
b) Determine the lowest pressure at which the liquid is stable (P_2).
c) Plot the fugacity coefficient of the vapor starting from a pressure close to the ideal-gas state up to the highest pressure where the vapor is stable. On the same graph make a plot of the fugacity coefficient of the liquid starting from the lowest pressure where the liquid is stable up to some pressure into the compressed-liquid region.
d) Identify the saturation pressure as the pressure where the liquid and vapor fugacity lines cross. How does this result compare with the literature value for the saturation pressure of propane at 350 K?

Problem 7.19: Write a numerical procedure to systematically calculate the saturation pressure of a fluid using the SRK equation of state at a given temperature T.

7.8 Problems

This can be done as follows: First, determine the highest pressure at which the vapor is stable (P') and the lowest pressure at which the liquid is stable (P'') at temperature T. The saturation pressure is located between these two values. Start with a pressure P between P' and P''. Obtain the three roots of the compressibility equation at P, T, and use the smallest root to calculate the fugacity of the liquid phase, and the largest root to calculate the fugacity of the vapor phase. If the fugacity of the liquid is lower than the fugacity of the vapor, choose a lower pressure (why?); otherwise choose a higher pressure. Repeat the calculation with the new pressure and continue until the fugacity of the vapor is nearly the same as that of the liquid. *Note:* At low temperatures P'' may be negative. In this case the saturation pressure is bracketed between $P = P'$ and $P = 0$.

a) Use your program to calculate V_L, V_V, ΔH_{vap}, and ΔS_{vap} of propane at 350 K.
b) Use your program to make a plot of the vapor-liquid boundary on the PV graph.

Problem 7.20: Use the program you wrote in Problem 7.19 to construct a PH graph for propane using as reference state the saturated liquid at 350 K. You may do this as follows: First, obtain the residual enthalpy of the saturated liquid at 350 K. Next, calculate P^{sat}, and the residual enthalpy and entropy of the saturated liquid and saturated vapor at various temperatures between T_c and 250 K. Using these residuals and the specified reference state, calculate the absolute enthalpy of the saturated phases. Plot P^{sat} versus H_L and on the same graph plot P^{sat} versus H_V. This will give the vapor-liquid boundary. To add isotherms, first tabulate H at various pressures. Start at a pressure close to the ideal-gas state and calculate the enthalpy. If the compressibility equation has three real roots, use the largest one, since the calculation is done for the vapor. Increase the pressure in small steps and repeat the calculation until you reach P^{sat}. Past $P = P^{\text{sat}}$, use the smallest root, since the calculation is now done for the liquid. The isotherm is obtained by plotting the results of this calculation as P versus H.

Problem 7.21: The SRK equation gives the following results for toluene at $T = 383.3$ K, $P = 8$ bar:

Z	0.0332418	0.342355	0.624404
H^R (J/mol)	-34046.0	-6020.41	-3384.35
S^R (J/mol K)	-71.4816	-13.4281	-6.42023

In this table, the first row gives the three roots of the cubic equation for Z, the second row gives the corresponding values of H^R, and the third row gives the corresponding values of S^R.

a) Report the enthalpy, entropy, and Gibbs energy of toluene at $T = 383.3$ K, $P = 8$ bar using as reference state the ideal-gas state at 383.3 K, 40 bar.
b) Calculate the fugacity and the fugacity coefficient at $T = 383.3$ K, $P = 8$ bar.
c) Calculate the fugacity at 383.3 K, 40 bar.

Additional data for toluene: Mw = 92.141 g/mol, $\omega = 0.262$, $T_c = 591.8$ K, $P_c = 41.06$ bar; boiling point at 1 atm: 383.8 K.

Part II

Mixtures

Chapter 8

Phase Behavior of Mixtures

The presence of multiple components adds a new dimension to the phase behavior of mixtures. In the pure component, molecules are always surrounded by similar species; in a mixture, they are surrounded by both like and unlike species. This gives rise to self-interactions between like molecules, and cross-interaction between unlike molecules. These interactions are much more pronounced in the liquid phase, where molecules are closely packed together. The balance of self- and cross-interactions creates phase behavior that is not seen in pure fluids. If cross-interactions are favorable, components form strong mixtures that are more difficult to separate. If cross interactions are unfavorable, the mixture is weaker and separation is easier. If they are strongly unfavorable, then components may exhibit partial miscibility. Additional variety of phase behaviors comes from the number of phases that can coexist simultaneously. With mixtures we encounter problems of vapor-liquid equilibrium (VLE), but also liquid-liquid (LLE) and liquid-liquid-vapor (LLVE) equilibrium. If a solid component is added, other combinations of equilibria are observed for example, solid-liquid, solid-liquid-vapor, etc. This enormous variety is made possible by the presence of additional components.

The phase behavior of mixtures forms the basis of industrial separations. What makes such separation possible is the fact that when a mixture is brought into a region of multiple coexisting phases, each phase has *its own composition*. Understanding the phase behavior of multicomponent systems is very important in the calculation of separation processes. In this chapter we review graphical representations of the phase behavior of binary and ternary systems. Since we are dealing with several independent variables, pressure, temperature, and composition, special conventions are used in order to represent information in two-dimensional graphs.

The learning objectives for this chapters are to:

1. Interpret the *Txy* and *Pxy* graph of binary mixtures.

2. Use phase diagrams to calculate material balances.

3. Work with ternary-phase diagrams.

8.1 The *Txy* Graph

A pure substance under constant pressure boils at constant temperature; when heat is added to the boiling liquid, it induces more evaporation, but no temperature change is observed until after all the liquid has evaporated. By contrast, a liquid *mixture* boils over a *range* of temperatures. As an example we consider a mixture of n-heptane and n-decane. The normal boiling points of the pure components are 98.1 °C for heptane, and 174.2 °C for decane. We form a solution that contains 50% by mole n-heptane (component 1) and observe the phase changes as the solution is heated, starting at room temperature, under constant pressure at 1 atm. To visualize such an experiment, imagine that the solution is placed in a cylinder fitted with a movable piston so that the pressure remains constant. The cylinder is sealed so no air is present. This is a closed system whose overall composition remains constant at 50% n-heptane by mol, at all times. The initial state, $T = 25$ °C, $P = 1$ atm, $x_1 = 0.5$ is shown by point A in Figure 8-1. At A the temperature is well below the

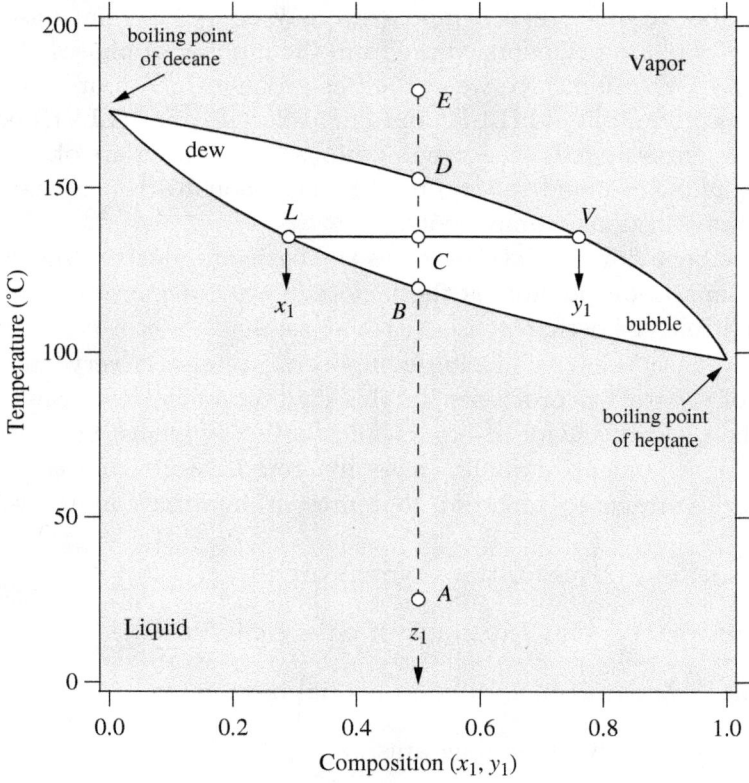

Figure 8-1: *Txy* graph of heptane (1) decane (2) at 1 atm.

boiling point of either compound, and the system is in the liquid phase. Heating the system at constant pressure is equivalent to moving up along the line AE, corresponding to a line of constant overall composition. When we reach 98.1 °C, the boiling point of pure heptane, the solution remains liquid and does not boil. Indeed, it does not begin to boil until about 120 °C. The point where the liquid begins to boil is called the *bubble temperature*, or bubble point of the solution (point B on the graph). As the liquid continues to boil, the temperature increases until a point is reached where all of the liquid has evaporated (point D). Further heating produces superheated vapor and the state moves higher up towards point E. These steps can be repeated in reverse by cooling the vapor. Starting with point E, cooling causes the temperature of the vapor to drop until condensation occurs at point D. This is the *dew* point and is identical to the point of complete evaporation of the liquid in the heating experiment. As more liquid condenses the temperature drops and this continues until the bubble point is reached (point B) where all the vapor has condensed. Further cooling produces a compressed liquid and moves the state towards point A.

If this procedure is repeated with different compositions in the range $x_1 = 0$ to 1 we obtain a series of bubble and dew points which define the bubble line and the dew line, respectively. These are shown as lines in Figure 8-1. Above the dew line the system is vapor; below the bubble line the system is liquid. Between the two lines the system consists of a mixture of two phases, each with its own composition. Consider a point C that lies between the bubble and the dew lines. This point represents a system with overall composition $z_1 = 0.5$, which we read off the horizontal axis. In reality, point C consists of a mixture of two phases, a liquid with composition $x_1 \approx 0.3$ (point L) and a vapor with $y_1 \approx 0.8$ (point V). Notice that the vapor and liquid compositions are different. In fact, the liquid contains less heptane (30% by mol) than the original liquid (50%), while the vapor contains more (80%). Since the liquid corresponding to C contains more decane that the liquid at B, it boils at a *higher* temperature, closer to the boiling point of decane. Thus, the reason that the temperature increases during boiling is that liquid progressively becomes more concentrated in the less volatile (heavier) component. The difference between the composition of the liquid and the composition of the vapor is a very important characteristic of multicomponent equilibrium and constitutes the basis for separating the components of a mixture.

The phase diagram in Figure 8-1 is a *Txy* graph. In this graph pressure is held constant, the vertical axis represents temperature, and the horizontal axis represents composition. Whether this is the composition of the liquid (x_1), of the vapor (y_1), or the overall composition (z_1) of a two-phase mixture depends on where we are on the phase diagram. From anywhere in the liquid, including the bubble line, the horizontal axis gives x_1; from anywhere in the vapor, including the dew line, the horizontal axis gives y_1. Between the dew and bubble lines the system consists of two phases. Directly below a point in this region we obtain the overall composition z_1.

The Lever Rule Let us return to point C inside the two-phase region. The line LV that connects the compositions of the vapor and the compositions of the liquid is a *tie* line. This is a horizontal line and passes through C since all three points, L, C, and V are at the same temperature. Let us assume that the liquid contains a fraction L of the total number of moles and the vapor contains the remaining fraction $V = 1 - L$. The concentration of heptane in the liquid is x_1, in the vapor y_1, and let's say that the overall concentration of heptane is z_1 (that would be the concentration of heptane in point C, 50% in our example). The mole balance on heptane and the total mole balance give the following two equations:

$$Lx_1 + Vy_1 = z_1 \tag{8.1}$$

$$L + V = 1. \tag{8.2}$$

Solving for the liquid and vapor fractions we obtain

$$L = \frac{y_1 - z_1}{y_1 - x_1}, \quad V = \frac{z_1 - x_1}{y_1 - x_1}. \tag{8.3}$$

This has the familiar form of the lever rule encountered in Chapter 2.

NOTE

Convention for x, y, z

We adopt the following convention for distinguishing the various compositions that appear when a mixture forms two phases: we use x_i for the mol fractions of the liquid phase, y_i for the mol fractions of the vapor phase, and z_i for the overall composition of a two-phase mixture. When a two phase mixture is fully condensed or fully evaporated, the composition of the single phase formed is z_i.

Example 8.1: Using the Lever Rule
A 50% by mole solution of heptane, decane is heated to 135 °C at 1 atm. Determine the state of the system (vapor, liquid, or vapor, liquid mixture). If this is a two-phase system, calculate the amount and composition of two phases.

Solution The conditions given in this example correspond to point C in Figure 8-1 which lies in the vapor-liquid range. The compositions of the two phases are read directly from the graph (normally a more accurate graph will be needed for this type of calculations).

We find $x_1 = 0.3$, $y_1 = 0.8$. The fractions of the liquid and vapor are calculated from the lever rule:
$$L = \frac{0.8 - 0.5}{0.8 - 0.3} = 0.6,$$
$$V = 1 - L = 0.4.$$

Therefore, 60% is in the liquid and the rest is vapor. Since point C is near the middle of the line VL but closer to the liquid side, we should have guessed that the liquid fraction is somewhat larger than 50%.

Example 8.2: Working with Phase Diagrams
What is the maximum concentration of heptane in the vapor that we can obtain by heating the solution of the previous example at 1 atm?

Solution The maximum concentration in the vapor is when the system is at the bubble point, point B in Figure 8-2. Then the vapor has the concentration of point V. From the graph we read $y_1 \approx 0.9$. The composition of the liquid at this point is $x_1 = 0.5$, the same as the initial solution. The lever rule then gives $L = 1$ and $V = 0$. Although the purity of the vapor is highest at that point, the *amount* collected is zero.

Example 8.3: Working with Phase Diagrams–2
A vapor mixture of heptane(1)-decane(2) with $y_1 = 0.5$ is condensed by cooling under constant pressure of 1 atm. What is the composition of the first drop to appear? What is the composition of the last bubble to condense?

Solution When the first drop condenses the system is at the dew point, D. If we draw the tie line at D, the composition of the first drop is read at the intersection of the tie line with the bubble line. We find $x_1 \approx 0.13$. When the last bubble condenses the system is at the bubble point, B. The corresponding composition is $y_1 \approx 0.9$.

8.2 The *Pxy* Graph

In the *Txy* graph, we plot the phase behavior of a binary system as a function of temperature at constant pressure. We may also make a plot as a function of pressure at constant temperature. The resulting phase diagram is a *Pxy* graph.

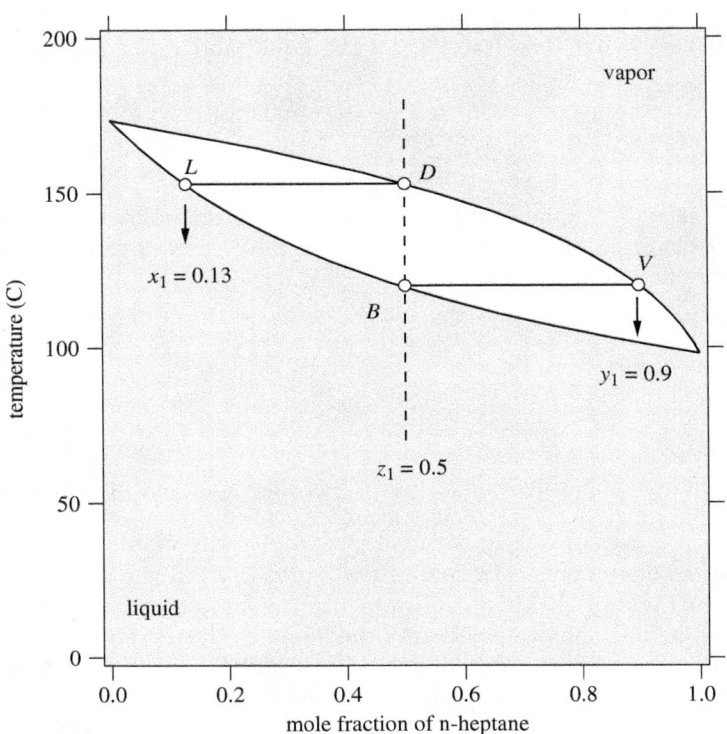

Figure 8-2: Compositions at the bubble and dew point.

The Pxy graph for the system heptane-decane at 120 °C is shown in Figure 8-3, and it resembles a Txy graph turned upside down. The liquid phase is now at the top of the graph (high pressure) and the vapor is at the bottom (low pressure). As with the Txy graph, the bubble line marks the boundary of the liquid, the dew line marks the boundary of the vapor, and the two lines meet at the saturation pressure of the pure components. In this case, heptane (component 1) is more volatile than decane; its saturation pressure is higher. As a result, the VLE envelope ascends in the direction of increasing x_1. The line $ABCDE$ corresponds to the same transitions as those discussed in Figure 8-1 but these are now caused by changing the pressure (expansion or compression) at constant temperature. At A the system is liquid; by decreasing pressure it reaches the bubble point at B; at C it consists of a mixture of liquid and vapor; at D it reaches the dew point; and by further decreasing pressure it evaporates completely. The transformations are reversible and by isothermally compressing the vapor in state E we eventually return to the initial state A.

To read the Pxy graph we follow the same principles as with Txy graphs. The horizontal axis reads composition (mol fraction of component 1). Points inside the

8.2 The Pxy Graph

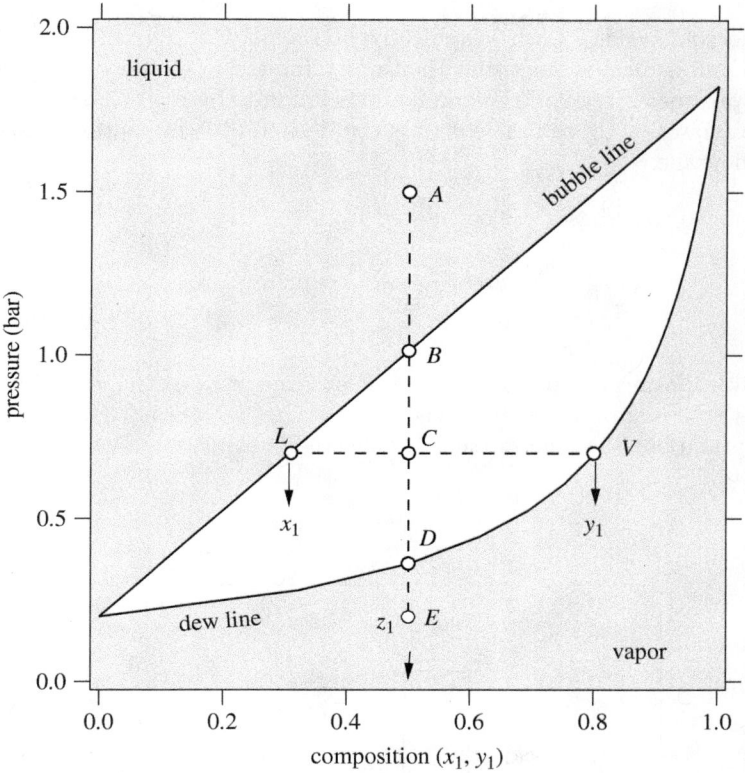

Figure 8-3: Pxy graph of heptane (1) decane (2) at 120 °C.

VLE envelope represent a two-phase system, and the horizontal axis gives the overall composition. The tie line is horizontal and points to the composition of the vapor (on the dew line) and the composition of the liquid (on the bubble line). The lever rule applies and can be used to find the liquid and vapor fractions.

Example 8.4: Flash Separator
When the pressure of a compressed liquid is brought between the bubble point and the dew point, the liquid "flashes," that is, it partially evaporates, creating a vapor that is enriched in the more volatile components and a liquid that is enriched in the heavier components. This simple process provides partial separation of the components.

A mixture of heptane/decane at 1.8 bar, 120 °C contains 80% by mol heptane. At what pressure should it be flashed to obtain a stream that is 90% rich in heptane? What are the composition of the liquid and the amounts of the two phases?

Solution The solution is shown graphically in Figure 8-4. The initial state is A. The final state is B and is located such that the tie line intersects the dew line at $y_1 = 0.9$. The corresponding pressure is read off the graph and is found to be $P_B = 1$ bar. The final system consists of two phases, a vapor that contains the desired 90% in heptane and a liquid that contains 50% heptane.

The amounts of the two phases are calculated from the lever rule:
$$L = \frac{0.9 - 0.8}{0.9 - 0.5} = 0.25,$$
$$V = 1 - L = 0.75.$$

Comments By choosing the pressure in the flash drum we can achieve different levels of separation. If the pressure is closer to the bubble line, we obtain a vapor that is highly enriched in heptane but the amount that is collected is small.

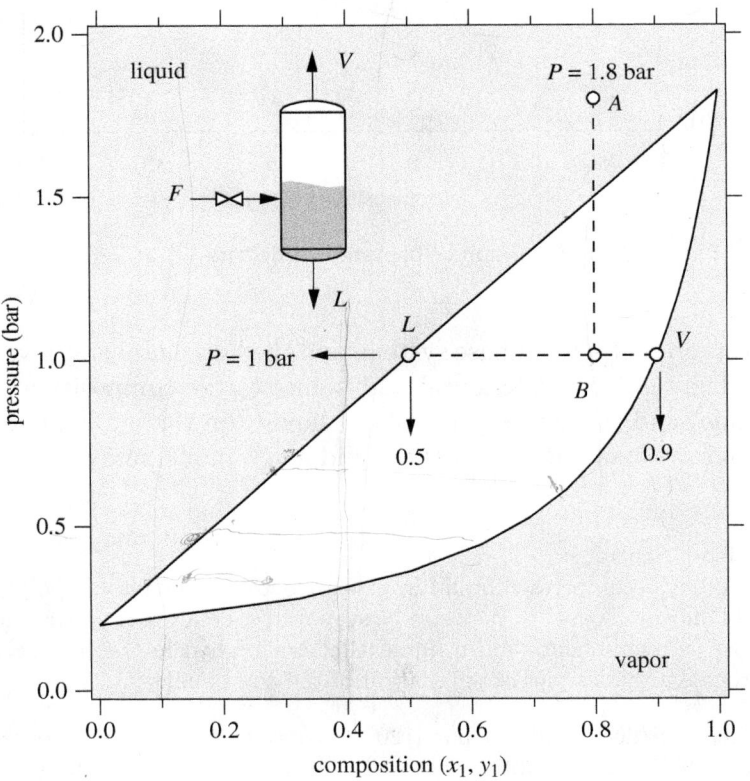

Figure 8-4: Flash separation (Example 8.4).

8.2 The Pxy Graph

Example 8.5: Multiple Flash Units
A solution of heptane (1)/decane (2) containing 80% by mol heptane at 1 bar is heated under constant pressure until half of the system is vapor. The liquid is separated from the vapor and is heated again until half of it vaporizes. This procedure, that is, creation of a 50-50 liquid/vapor mixture and separation of the liquid, is repeated for a total of five times. Calculate the amount, composition, and temperature of the remaining liquid.

Additional data Txy data for this system at 1 bar are shown below:

x_1	y_1	T (°C)	x_1	y_1	T (°C)
0.00	0.000	173.6	0.55	0.921	116.5
0.05	0.251	164.5	0.60	0.935	113.8
0.10	0.428	156.7	0.65	0.948	111.4
0.15	0.556	149.8	0.70	0.959	109.1
0.20	0.650	143.9	0.75	0.969	106.9
0.25	0.721	138.7	0.80	0.977	104.9
0.30	0.775	133.9	0.85	0.984	103.0
0.35	0.818	129.8	0.90	0.990	101.3
0.40	0.852	126.0	0.95	0.996	99.6
0.45	0.879	122.5	1.00	1.000	98.0
0.50	0.902	119.4			

Graphical Solution The solution is easy to construct graphically. Since the process is at constant pressure, the relevant phase diagram is the Txy graph. With the data given we construct the graph shown in Figure 8-5.

The feed is at $x_1 = 0.8$ and is brought to the two-phase region (point F_1) such that half of the liquid boils, that is, $L = V = 1/2$. Graphically, point F_1 is found by drawing a tie line so that the feed composition lies exactly in the middle. This requires some trial and error, but it can be done. The composition of the liquid fraction (L_1) is read from the graph, and it is about 0.6. This liquid becomes the new feed for the next step. We must now locate a point F_2 with the same composition as L_1 that also lies at equal distance from the dew and bubble lines (since $V = L = 1/2$ throughout this process). By repeating the process five times, we obtain point L_5. The composition and temperature are read off the graph. We find $x_1 \approx 0.02$ and $T \approx 170$ °C. Since each time we collect 50% of the original sample, after 5 steps we are left with $L_5 = 0.5^5 = 0.0313$ or 3.13% of the original feed. While this amount is small, it is almost pure decane ($x_2 = 1 - x_1 = 0.98$).

Numerical Solution The graphical solution is quite simple and fast, but reading numbers off the graph can only be done approximately. For more accurate answers, we must perform a numerical solution. This is done more easily once the graphical solution has been obtained. First, application of the lever rule gives

$$z = Lx + Vy \quad \Rightarrow \quad Lx + (1-L)y = z, \qquad [\text{A}]$$

where z is the composition of the feed (we drop the subscript "1" from the mole fractions with the understanding that all mole fractions are for heptane), x is the composition of the liquid fraction, (point L_1), and y that of the vapor (point V_1) and $L = 0.5$ is the liquid fraction. Point F_1 is at a temperature T_1 such that the bubble and dew compositions satisfy eq. [A]. Temperatures and mole fractions are given in tabular form, so we just have to find the right temperature by looking up the table. To facilitate the solution and avoid interpolations, we fit the tabular data to a polynomial function using a least-squares procedure:

$$x = 14.425 - 0.31757 T + 2.7789 \times 10^{-3} T^2 - 1.1277 \times 10^{-5} T^3 + 1.757 \times 10^{-8} T^4$$
$$y = 0.90716 + 5.5548 \times 10^{-3} T - 9.1222 \times 10^{-5} T^2 + 8.1846 \times 10^{-7} T^3$$
$$- 3.7472 \times 10^{-9} T^4.$$

We now repeat the steps of the graphical solution. In the fist step we solve for a temperature such that $Lx + (1 - L)y = 0.8$. By trial and error, we find $T = 111.32$ °C, $x = 0.651$ (this is the composition at L_1), and $y = 0.949$ (this is the composition at V_1). In the next step, the liquid composition $x = 0.651$ becomes the new feed composition. The same procedure is repeated with $z = 0.651$. The results of this calculation are summarized in the table below:

Step	T	z	x	y
1	111.32	0.8000	0.6513	0.9487
2	123.69	0.6513	0.4321	0.8704
3	143.32	0.4321	0.2062	0.6581
4	160.79	0.2062	0.0726	0.3397
5	169.41	0.0726	0.0221	0.1231

The final composition is given by x_1 at the fifth step. This solution contains 2.2% heptane and 97.8% decane.

NOTE
Distillation
Example 8.5 demonstrates the basic idea behind distillation. By partially boiling the solution, we enrich the vapor in the more volatile component and the liquid in the heavier component. By repeating this process as many times as necessary, we can achieve as high of a purity as desired. This is equivalent to a series of flash separators, as shown in Figure 8-6. The liquid from each separator is partially boiled at higher temperature and the vapor from each stage is partially condensed at lower temperature. In this manner, the liquid is continuously enriched in the less-volatile component, and the vapor in the more volatile one. In reality, distillation columns accomplish this task through a series of perforated trays, stacked vertically: the liquid drips down to hotter stages and the vapor bubbles up to cooler stages. The entire column is heated at the bottom. In packed columns, instead of discrete trays the column is filled with a packing material such as beads, rings, or other small objects. As the liquid trickles down and the vapor rises, the packing material forces the two phases into contact and helps them reach equilibrium with each other.

8.2 The Pxy Graph

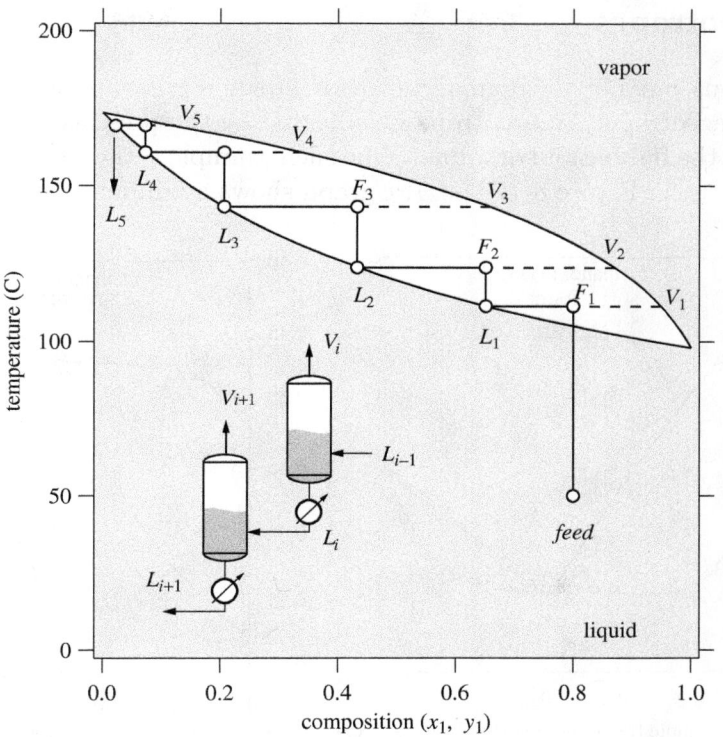

Figure 8-5: Txy graph for Example 8.5.

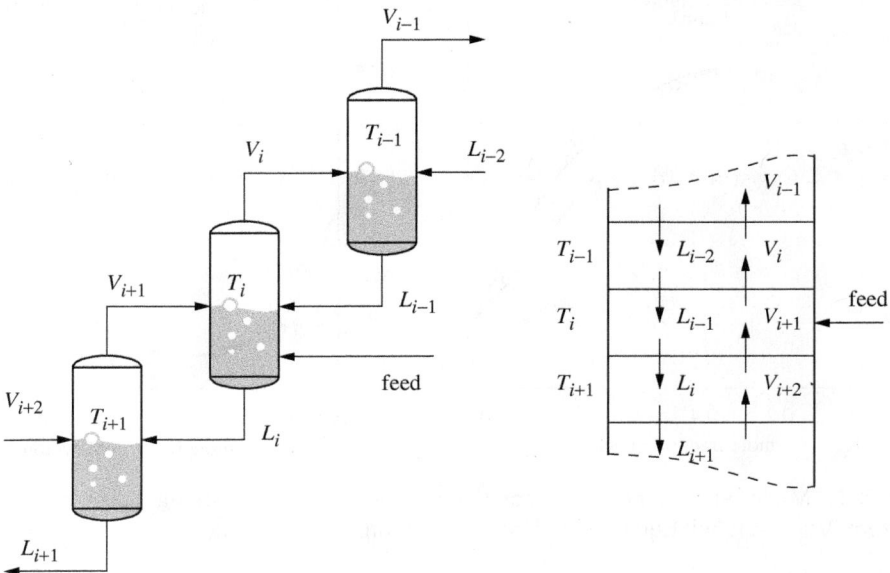

Figure 8-6: Staged separation (distillation). The column is heated at the bottom and cooled at the top ($T_i > T_{i-1}$).

8.3 Azeotropes

An anomalous but not uncommon behavior in some binary systems is the occurrence of an azeotrope. An azeotrope is indicated by the presence of a maximum or minimum in the bubble and dew lines. One such example is the system methanol(1)/CCl_4(2), shown in Figure 8-7. The Txy graph shows a minimum at which point the

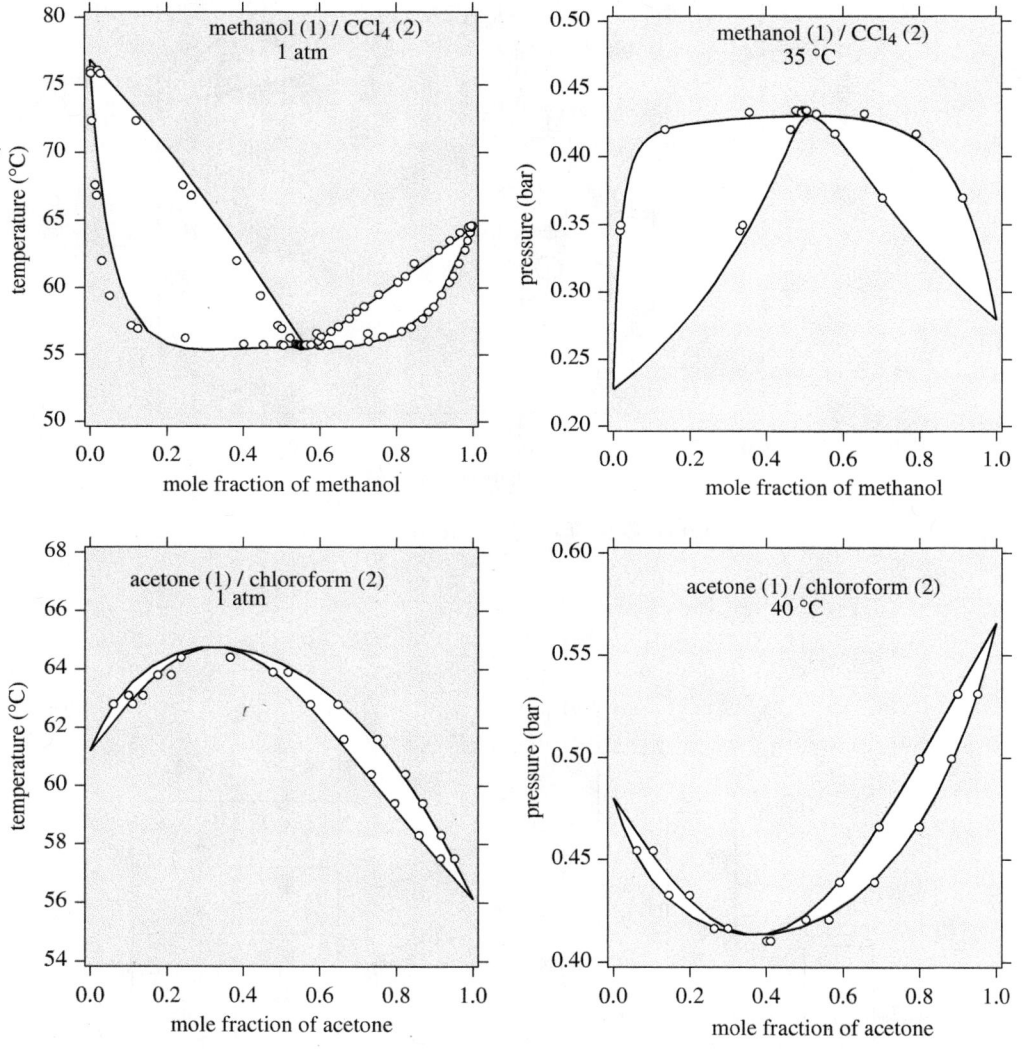

Figure 8-7: Minimum boiling azeotrope (top), and maximum boiling azeotrope (bottom). Data from Vapor-Liquid Equilibrium Data Collection, J. Gmehling, U. Onken, and W. Arlt, DECHEMA.

bubble and dew lines meet. There is a range of compositions for which the bubble temperature is lower than the boiling point of either pure compound, indicating that these solutions boil more easily than the pure components. In the Pxy graph the minimum boiling azeotrope is characterized by a maximum in the total pressure, that is, the solution is more volatile than the pure components. Maximum boiling azeotropic behavior is also encountered, though not as often. The Txy graph exhibits a maximum, which lies above the boiling point of either component. This solution is more difficult to boil compared to the pure components. The corresponding Pxy graph exhibits a minimum in the dew and bubble pressure.

The azeotropic point corresponds to the maximum or minimum on the phase diagram. At this point the bubble and dew lines meet and the composition of the vapor is identical to that of the liquid. A liquid solution that has exactly the azeotropic composition boils as a pure component: the composition of the vapor and the liquid phases are identical, while temperature remains constant throughout boiling. Similarly, a vapor with this concentration condenses as a pure component. The name *azeotrope* derives from the Greek and translates "not altered by boiling." Azeotropes are problematic in separations. Suppose we want to separate a solution of methanol in CCl_4 containing 10% methanol by mole. This composition lies to the left of the azeotrope. Subjecting this solution to distillation at 1 atm will produce a liquid that approaches pure carbon tetrachloride, and a vapor whose composition approaches the azeotropic point. This means that the vapor fraction cannot exceed the azeotropic concentration, preventing us from obtaining a stream of purified methanol. It is possible to alter the composition of the azeotrope and even remove the azeotrope altogether by changing the pressure and temperature. Alternatively, the addition of a suitable third component could have the same effect. In some cases, the formation of azeotropes is advantageous. Water and hydrochloric acid form an azeotrope; through boiling it is possible to collect a solution that has the azeotropic composition. Because its composition is precisely known (it depends only on the pressure during distillation), this solution may be used as a standard for acid/base titrations.

8.4 The xy Graph

Another graphical representation of binary vapor, liquid data, useful in material balances, is the *xy* graph, as shown in Figure 8-8. This graph gives the composition of the tie line by plotting the vapor mole fraction against the liquid mol fraction. In this graph, both axes run from 0 to 1 and the diagonal represents points where the vapor composition is equal to that of the liquid. By convention, the more volatile component is chosen as component 1; by this convention the *xy* graph generally lies

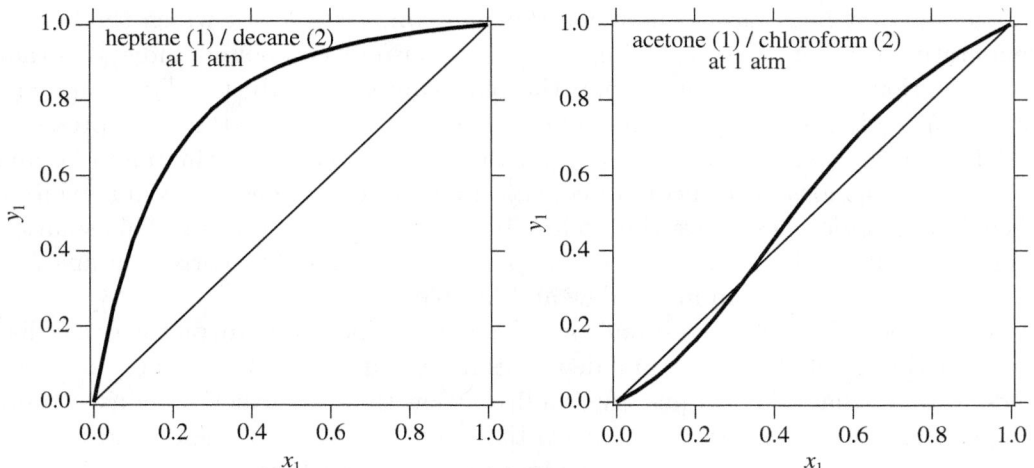

Figure 8-8: *xy* graphs for two binary systems.

above the diagonal. Azeotropes are easily identified on this graph because the *xy* line crosses the diagonal.

The distance of the *xy* line from the diagonal is a measure of the difference in the composition of the two phases, and thus a measure of the ease of separation. One way to quantify the difference between the compositions of the two phases is through the known as distribution coefficient, K_i, or K-factor for simplicity, which is defined as

$$K_i = \frac{y_i}{x_i}. \tag{8.4}$$

Good separation requires values substantially different from 1. A value larger than unity indicates that component i is preferentially found in the gas phase; a value less than 1 indicates preferential enrichment of the liquid phase. A related property is the relative volatility, α_{ij}, which is defined as

$$\alpha_{ij} = \frac{K_i}{K_j}.$$

In a binary mixture, this takes the form,

$$\alpha_{12} = \frac{y_1(1-x_1)}{x_1(1-y_1)}.$$

The K_1 factor of binary mixture can be obtained graphically from the *xy* graph: if we connect a point of the *xy* graph to the origin, the slope of this line is K_1. At $x_i = 1$ the value of K_i is unity for any component i; at $x_i = 0$ it reaches a limiting value that corresponds to the infinite dilution limit, namely, to the limit that the concentration of the component is reduced to zero by dilution in the other component.[1] That is, even though both y_i and x_i go to zero in this limit, the K_i factor is finite and equal to the slope of the xy line at $x_i = 0$.

If the K factors of a binary system are known as a function of pressure and temperature, the entire phase diagram can be constructed. As we will see in Chapter 10, thermodynamics provides methodologies for the estimation of these factors.

8.5 VLE at Elevated Pressures and Temperatures

The phase diagrams discussed so far extend over the entire range of the compositional axis, from $x_1 = 0$ (pure saturated component 2) to $x_1 = 1$ (pure saturated component 1). This behavior is observed when both components are below their respective critical points. If one or both components are above their critical point, the VLE region shrinks and does not cover the entire range of compositions. This situation is illustrated in Figure 8-9, which shows the Pxy graph of the system ethanol/water at several temperatures. At 200 °C the phase diagram has the usual form and extends over the entire compositional range. This system also exhibits a maximum boiling azeotrope. Upon increasing temperature, the phase diagram moves upwards but remains qualitatively the same, as long as the temperature remains below the critical of both components. At 275 °C ethanol above its critical temperature (240.77 °C) but water is below its own (374 °C). The phase diagram now does not reach all the way to the ethanol axis. Instead, the dew and bubble lines meet at some intermediate point. The phase diagram is read in the usual way: We draw a tie line, identify the liquid phase on the left (above the bubble line) and the vapor phase on the right (below the dew line). Where the bubble and dew lines meet we have a critical point, which is characterized by its own temperature, pressure, and composition. It is identified as the point where the tie line is tangent to the VLE curve. If temperature is increased, the phase diagram moves up and shrinks further. This behavior continues until the critical temperature of water is reached (not shown on this graph). The dashed line in pane (a) of Figure 8-9 tracks the critical point of the mixture as temperature increases and connects the critical pressures of the pure components (the critical temperature of water is outside the range of this graph). This behavior can also be studied on the xy graph, shown in Figure 8-9(b). When both components are below the critical points, the xy graph extends from $x_1 = 0$ to $x_1 = 1$.[2] Once temperature exceeds the critical temperature

---
1. The limit of infinite dilution should be visualized as a single molecule of component i completely surrounded by molecules of component j.
2. This line intersects the diagonal at the azeotropic point, but this detail cannot be seen clearly at the magnification of this graph.

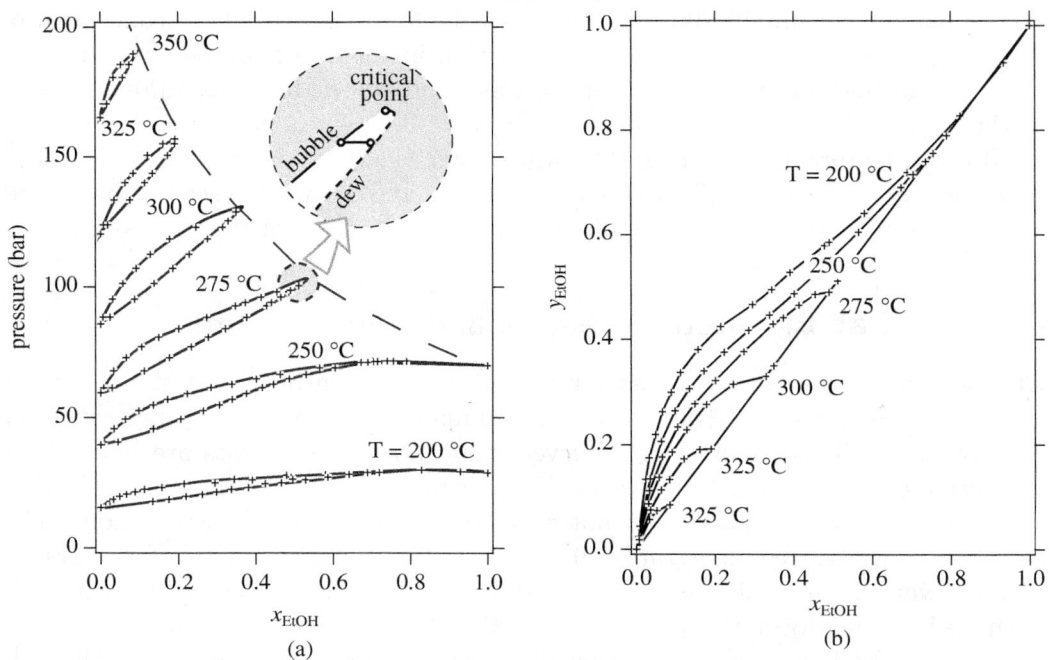

Figure 8-9: *Pxy* and *xy* graph of ethanol/water at elevated temperatures (data from Frank Barr-David and B. F. Dodge, *Journal of Chemical & Engineering Data*, 4 (2):107-121, 1959).

of compartment 1, the line intersects the diagonal and stops there. This marks the critical point of the mixture, where the composition of the two phases is identical.

Such incomplete phase diagrams occur commonly in systems involving typical gases (e.g., oxygen, nitrogen) in equilibrium with liquids (e.g., water) at room temperature. In this case the liquid is below its critical temperature but the gas above its own. Although such problems fall squarely in the scope of phase equilibrium they are often categorized as solubility problems. The thermodynamics of these systems are examined in Chapter 13.

8.6 Partially Miscible Liquids

Some liquids are only partially miscible in each other. This situation arises when the constituent molecules contain groups that have low affinity for each other. As an example we consider the system n-hexane/ethanol. Both molecules contain CH_2 and CH_3 groups which are very similar. Ethanol also contains a polar hydroxyl group, $-OH$, which has little affinity for the alkyl groups. A limited amount of hexane

8.6 Partially Miscible Liquids

can be accommodated in ethanol and, similarly, small amounts of ethanol can be dissolved in hexane. At certain compositions, however, the system splits into two separate liquid phases due to the lack of affinity between hydroxyls and hydrocarbons. If such a system is brought to boiling it becomes a three-phase system: two liquids and a vapor.

Figure 8-10 shows the Txy of the system hexane(1)/ethanol(2). There are three distinct phases: the vapor, one hexane-rich liquid phase to the right (marked as L_1), and one ethanol-rich liquid phase to the left (L_2). The rest of the graph represents coexistence between these phases. To find the phases present inside the two-phase regions, we draw a tie line until the phase boundaries are reached. To illustrate the behavior of the system in the two-phase region, consider the following experiment: Starting with pure ethanol at 60 °C, 1.96 bar, we add hexane in small amounts. Initially, all the hexane becomes incorporated into ethanol to form a homogeneous solution. Once the solubility limit of hexane in ethanol is reached, any additional of hexane forms a separate phase. Both phases contain both components, but at

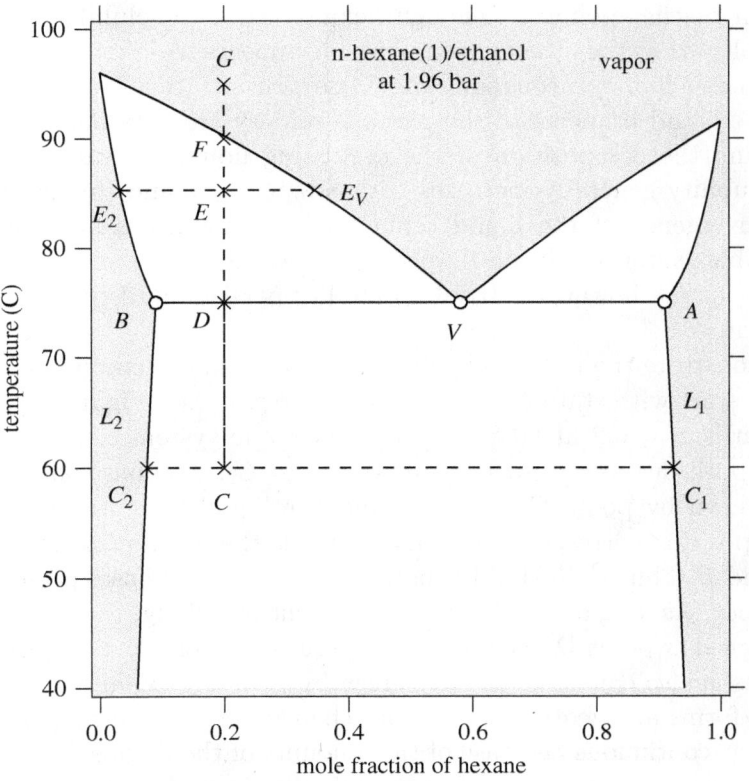

Figure 8-10: Txy graph of partially miscible liquids: hexane(1)/ethanol(2) at 1.96 bar.

different concentrations. The ethanol-rich phase (the phase which originated as pure ethanol) has the composition of point C_2, and the hexane-rich phase the composition of point C_1. The two-phase system is represented by a single point, C, which lies inside the two-phase region at the overall composition of the two-phase system. This composition is calculated by the lever rule:

$$z_i = L_1 x_i^{(1)} + L_2 x_i^{(2)},$$

where L_1 is the fraction of mass (or moles) in the hexane-rich phase, L_2 is the fraction of the ethanol-rich phase, and $x_i^{(1)}$, $x_i^{(2)}$, are the compositions (mass or mol fractions) of component i in the two phases. These compositions correspond to points C_1 and C_2 and are read off the horizontal axis of the graph. If we continue to add hexane, more and more of the ethanol in the ethanol-rich phase is incorporated into the hexane-rich phase. During this part of the process the system consists of two liquid phase. The compositions of these phases remain fixed at those of points C_1 and C_2, but the relative amounts of each liquid change, with the hexane-rich phase increasing at the expense of the ethanol-rich phase. When the overall composition reaches point C_1, enough hexane exists to solubilize all of the available ethanol so that the ethanol-rich phase disappears. Adding more hexane at this point produces a homogeneous solution whose concentration approaches that of pure hexane. The liquid branches of the phase boundary are determined experimentally by measuring the composition of the coexisting liquids at various temperatures. Mutual solubility generally increases with temperature and this is reflected in the (slight) convergence of the liquid branches as temperature increases. Some partially miscible systems become fully miscible at higher temperatures. In the case of hexane/ethanol, however, partial miscibility persists until the liquid reaches the boiling point.

To demonstrate the phase behavior of a two-liquid system in the boiling region, suppose we start with state C in the two-phase region (see Figure 8-10). The overall composition is $z_1 = 0.2$ at 60 °C, 1.96 bar and the system consists of two liquids whose compositions are given by points C_1 and C_2. The onset of boiling (bubble point) is shown by point D. At this point, the liquid boils and produces a vapor whose composition corresponds to point V, while the two liquids are represented by points A and B. Thus at the bubble point there are three phases present, two liquids and the vapor. As long as two liquids are present in boiling, the state of the system remains pegged at point D: the boiling temperature is constant, and the composition of all phases is also constant, and are given by points A, V, and B. In other words, the system forms an azeotrope. The only change observed during this stage of the process is the continuous decrease of the amounts of the liquids, and the increase in the amount of the vapor. Once any of the liquid phases is completely depleted, the state moves up. Which phase is depleted first depends on the overall composition

8.6 Partially Miscible Liquids

of the starting mixture. For example, state E consists of vapor in equilibrium with the ethanol-rich phase, as indicated by the phases at the end points of the tie line that passes through E; the hexane-rich phase has completely evaporated. At point F the system reaches its dew point. Further heating moves the state into the region of superheated vapor. The line CG on the Txy graph represents a path of heating under constant pressure and constant overall composition.

The corresponding Pxy graph is shown in Figure 8-11. Qualitatively, it resembles a Txy graph turned upside down. Since pressure has little effect on the mutual solubility of liquids, the boundaries between the two liquid phases, lines AA' and BB', are essentially vertical. To read this graph, we follow the same principles as with the Txy graph. First we label the single phases (hexane-rich liquid, ethanol-rich liquid, vapor). All other regions are areas where two or three phases are present. These phases are identified by drawing tie lines until they intersect a phase boundary.

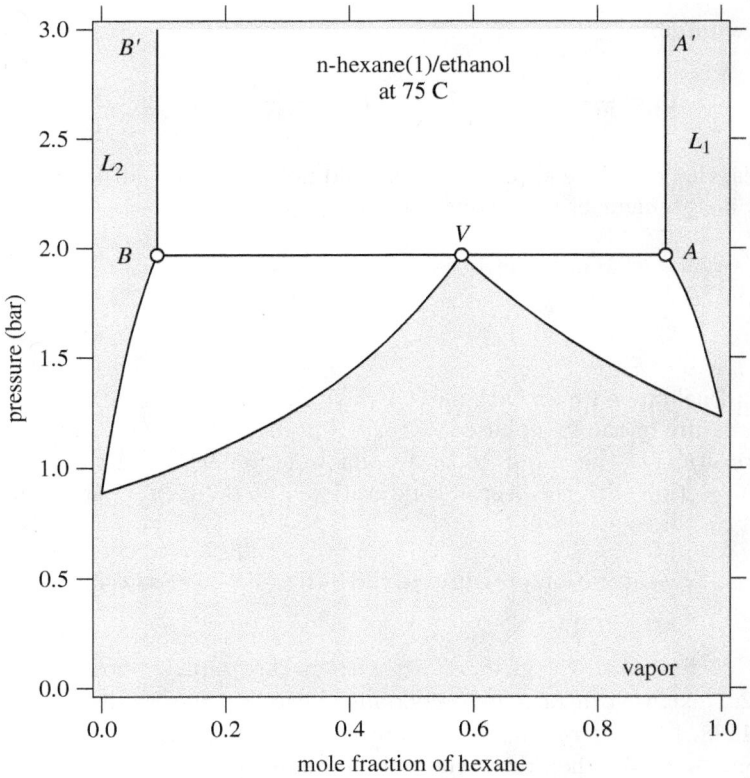

Figure 8-11: *Pxy* graph of partially miscible liquids – hexane(1)/ethanol(2).

Example 8.6: Partially Miscible Liquids and the Lever Rule
The system isobutane(1)/furfural(2) at 40 °C, 5 bar forms two liquid phases with the following compositions: $x_1^{(1)} = 0.9284$, $x_1^{(2)} = 0.1128$, where the superscripts (1) and (2) indicate the two liquid phases. How many phases are present and what is their composition when 0.7 mol of isobutane are mixed with 0.3 mol of furfural at 40 °C 10 bar?

Solution Pressure has negligible effect on solubility. Therefore, we can use the data, which refer to 5 bar, to answer the problem, which is given at 10 bar.

The overall composition, $z_1 = 0.7$, is in between the two equilibrium concentrations. This system is in the two-phase region: one phase contains a fraction L_1 by mole with composition $x_1^{(1)} = 0.9284$, while the other phase contains a fraction L_2 with composition $x_1^{(2)} = 0.1128$. By mole balance

$$1 = L_1 + L_2,$$
$$z_1 = x_1^{(1)} L_1 + x_1^{(2)} L_2,$$

from which we obtain

$$L_1 = \frac{z_1 - x_1^{(2)}}{x_1^{(1)} - x_1^{(2)}} = 72\%, \quad L_2 = \frac{x_1^{(1)} - z_1}{x_1^{(1)} - x_1^{(2)}} = 28\%.$$

In other words, the lever rule applies. This should not come as a surprise because the lever rule is simply a statement of mass conservation.

Example 8.7: Qualitative-Phase Diagram
The bubble pressure of the two-phase system isobutane(1)/furfural(2) at 40 °C is 4.66 bar and the composition of the vapor at that point is $y_1 = 0.99$. (a) Draw a qualitative Pxy graph; (b) if a solution with the overall composition $z_1 = 0.7$ is brought to boiling at 40 °C, which of the two phases will boil off first?

Additional data: The saturation pressures of the pure components at 40 °C are $P_1^{\text{sat}} = 4.956$ bar, $P_2^{\text{sat}} = 5$ mbar.

Solution (a) We first place the given information on the graph as shown by the open circles in Figure 8-12. Next, we draw the phase boundary between the two liquids at $x_1^{(1)} = 0.9284$ and $x_1^{(2)} = 0.1128$. In the Pxy graph these are vertical lines, AA', BB', extending upwards from the bubble line. We then draw the bubble line to go from P_1^{sat} to A, to B, to P_2^{sat}. Finally the dew line is drawn from P_1^{sat}, to V, to P_2^{sat}. Based on what we know so far, these

8.6 Partially Miscible Liquids

lines can only be drawn qualitatively. In Chapter 12 we will learn how we can draw them more accurately.

Notice that in this graph the composition of the vapor lies to the right of $x_1^{(1)}$ and the two lobes of the vapor-liquid region lie at opposite sides of the tie line ABV. The vapor-liquid region of phase (1) (isobutane-rich) is quite small. The vapor region also is very small because of the steepness of the dew line.

(b) From the graph we see that below the bubble pressure, a solution with the overall composition $z_1 = 0.7$ consists of a vapor in equilibrium with phase (2) (furfural-rich). That is, the isobutane-rich phase boils off first.

It is left as an exercise to draw the Txy graph of this system based on the information given above.

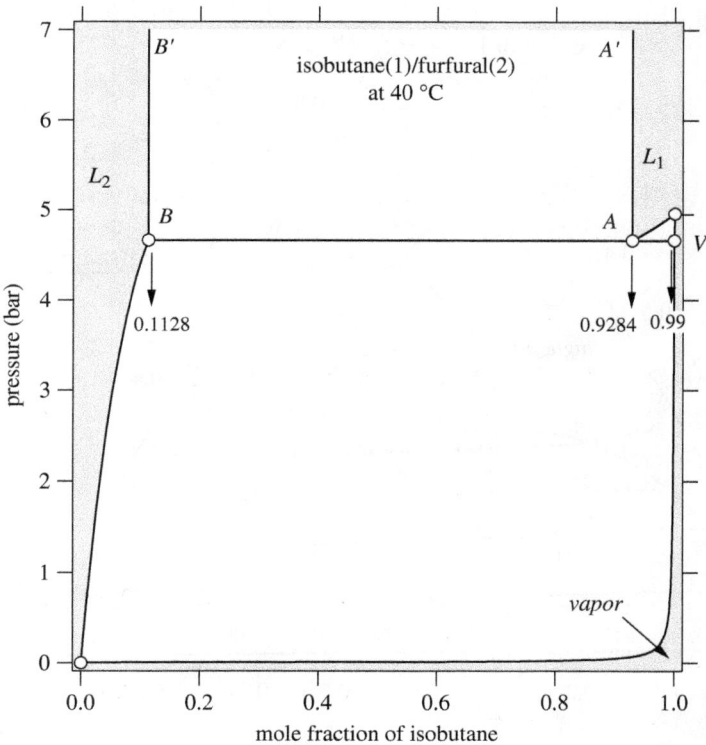

Figure 8-12: *Pxy* graph for the system isobutane(1)/furfural(2) at 40 °C. The composition of the two liquids and the vapor at the point of liquid-liquid-vapor equilibrium are shown on the graph.

8.7 Ternary Systems

When a soluble third component is added to two partially miscible liquids, we obtain a ternary system in which the third component (solute) is partitioned between the two liquid phases (solvents). The phase behavior of such systems is important in liquid-liquid extraction, a process that takes advantage of the differences in solubility to transfer a solute from one solvent into another. Since two mole fractions are required to represent composition in ternary systems, it is not possible to present temperature-composition or pressure-composition graphs in a two-dimensional plot. Instead, we map out the composition of the phases at constant pressure and temperature. This is done in triangular diagrams such as the one shown in Figure 8-13. In this graph the three components are placed at the vertices of an equilateral triangle.

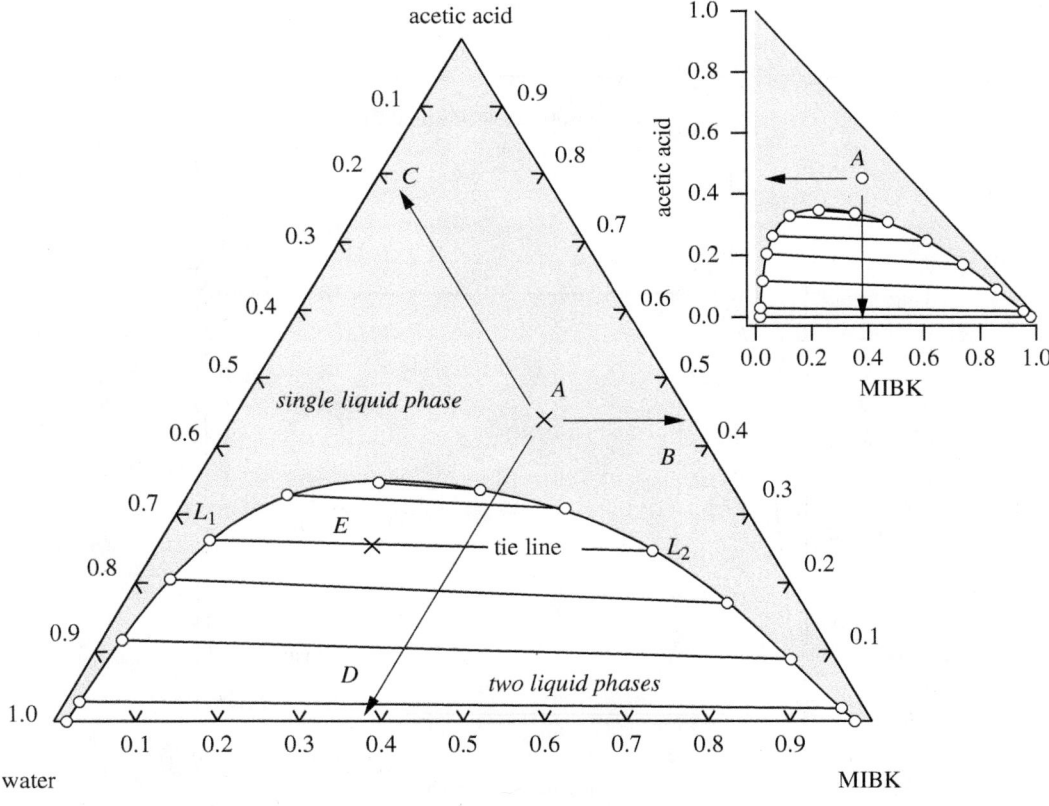

Figure 8-13: Phase diagram of acetic acid/water/methylisobutyl ketone at 25 °C. Concentrations are in weight percent.

8.7 Ternary Systems

Points inside the triangle represent possible concentrations of the three components. These concentrations are expressed on a percent basis, either by mass or by mole (the compositions in Figure 8-13 are given on a mass basis). The concentration of a component at any point inside the triangle is measured by the fractional distance of that point from the side opposite to the vertex of the component. The vertex corresponds to concentration 100% of that component, while points on the side opposite to the vertex correspond to 0% of that component. For example, point A lies at 44% of the maximum distance above the horizontal axis, which makes the concentration of acetic acid 44% by weight. Because of geometric similarity, fractional distances from a side do not have to be measured along vertical lines, thus for convenience we use grid lines parallel to one of the other sides. For example, the concentration of acetic acid at point A can be read by following the grid line AB to the right side of the triangle which is marked 0 at the bottom and 100 at the top. The concentration of water is 20%; this is found by following the grid line AC. The concentration of the third component can be determined the same way, or from the balance of the other two.

As a phase diagram, Figure 8-13 provides information similar to that obtained from Pxy and Txy graphs.[3] The shaded area represents a single liquid phase, and the white area represents separation into two liquid phases. Tie lines are drawn to indicate the composition of the two phases. For example, point E lies inside the two-liquid range, therefore, it phase separates into two liquid phases, L_1, which contains mostly water, and L_2, which contains mostly methylisobutyl ketone. The lever rule applies and can be used to obtain the relative amounts of the two liquid phases. Notice that the tie lines are not generally horizontal, and unless they are shown on the graph, we cannot tell the compositions at equilibrium.

The phase diagram of the system acetic acid/water/MIBK is typical of many ternary systems. The left side of the triangle represents mixtures of water and acetic acid (the concentration of methylisobutyl ketone along this side is 0). This side is entirely in the single-phase region, thus we conclude that acetic acid and water are fully miscible. Similarly, from the right side of the triangle we conclude that methylisobutyl ketone is fully miscible with acetic acid. By contrast, water and methylisobutyl ketone are only partially miscible because the bottom side, which represents mixtures of water and ketone (since the concentration of acetic acid along this side is 0), is in contact with the two phase region. This type of phase diagram in which a solute (here acetic acid) is fully miscible into two solvents (water and methylisobutyl ketone) but the two solvents are only partially miscible with each other is the most common type of ternary-phase diagram and is classified as type I.

---

3. However, on triangular plots both temperature and pressure are fixed.

More complex phase diagrams are also observed. In some, the two-phase region touches two axes, in some it touches no axis at all, and some exhibit two or more regions where phase separation occurs.

Because plotting data on equilateral triangles has certain inconveniences, sometimes ternary data are presented on a right triangle, as shown by the smaller inset in Figure 8-13. In this graph, the vertical axis represents one component, the horizontal axis the second component and the third component is obtained by mass balance from the two known fractions.

Example 8.8: Ternary Systems
5.76 kilograms of water are mixed with 3.072 kg acetic acid and 3.168 kg methylisobutyl ketone (MIBK) at 25 °C. Determine the phase of the mixture. If a two-phase system, report the amount and composition of each phase.

Solution The mixture has the overall composition (by weight):

$$z_\text{w} = 0.48, \quad z_\text{aa} = 0.256, \quad z_\text{mbik} = 0.264.$$

This places the state at point E in Figure 8-13, which lies in the two-phase region. We locate the tie line on which this point lies and read the composition of the equilibrium compositions of the two phases:

	Water	Acetic acid	MIBK
Phase 1	0.678	0.262	0.060
Phase 2	0.145	0.246	0.609

By mass balance on component i we have

$$z_i = L_1 x_i^{(1)} + L_2 x_i^{(2)},$$

where L_1, L_2, represent the mass fractions of the two phases and

$$L_1 + L_2 = 1.$$

Substituting into the mass balance equation and solving for L_1 we obtain

$$L_1 = \frac{z_i - x_i^{(2)}}{x_i^{(1)} - x_i^{(2)}}, \quad L_2 = 1 - L_1.$$

This is the familiar lever rule and applies to all three components:

$$\text{water} \qquad L_1 = \frac{0.48 - 0.678}{0.145 - 0.678} = 0.371,$$

$$\text{acetic acid} \qquad L_1 = \frac{0.256 - 0.262}{0.246 - 0.262} = 0.375,$$

$$\text{MIBK} \qquad L_1 = \frac{0.264 - 0.06}{0.609 - 0.06} = 0.372.$$

The system contains 37% (wt) of liquid phase 1 and 73% of liquid phase 2.

Comments The lever rule gives the same answer regardless of which component is used. The small discrepancy between these values arises from inaccuracies in reading the the composition of the equilibrium calculation of the tie line that passes through the given state. Often, ternary graphs will show selected tie lines, but if the desired state does not lie on one of them, we must obtain one through interpolation between the existing tie lines.

8.8 Summary

The most important feature of phase equilibrium in multicomponent systems is that the composition of phases is not the same.[4] If a mixture is brought into the two-phase region, we obtain phases that are preferentially enriched in one or more components. Separation methods such as distillation, absorption, and extraction operate based on this principle.

The phase diagram is a graphical representation of the phase of the system at a given pressure, temperature, and composition. Phase diagrams of binary systems are presented as Pxy or Txy graphs. A phase diagram consists of areas where a single phase exists, and areas where multiple phases (two or more) exist at equilibrium. The bubble line marks the boundary of the liquid phase and the dew line that of the vapor phase. Between the bubble and dew lines, the system consists of two phases at equilibrium. Tie lines connect the composition of the phases. The lever rule is a simple relationship between the composition of phases at equilibrium, the overall composition, and the relative amounts of each phase. It is nothing but a statement of mass conservation. The phase diagram of ternary systems is usually presented on an equilateral triangle. This graph shows the phase boundaries as a function of composition, at a given temperature and pressure. A series of such graphs is needed to study the phase behavior of a ternary system at various temperatures or pressures.

The solution of problems in phase equilibrium is facilitated enormously if the phase diagram is known. Several systems have been studied experimentally and the results have been collected in extensive databases. However, given the inexhaustible variety of components, compositions, temperatures, and pressures, the chemical engineer will invariably be faced with components, or conditions, for which such data are not available. A main goal in much of the rest of this book is the development of methods to predict phase behavior where data are not available.

---

4. There are some exceptions, azeotropes, for example, and critical points.

8.9 Problems

Problem 8.1: The data below give the dew and bubble temperature of methanol/ethanol mixtures at 1 bar.

mol fraction MeOH	T_{dew} (C)	T_{bubble} (C)	mol fraction MeOH	T_{dew} (C)	T_{bubble} (C)
0	78.00	78.00	0.6	70.82	68.99
0.1	76.95	76.32	0.7	69.35	67.71
0.2	75.85	74.72	0.8	67.76	66.49
0.3	74.70	73.19	0.9	66.05	65.31
0.4	73.49	71.73	1	64.20	64.20
0.5	72.20	70.33			

a) A solution containing 30% methanol (by mol) is flashed to 1 bar, 70.82 °C. Determine the phase of the system; if two phases, report the compositions and relative amounts of each phase.
b) What is the maximum mol fraction of methanol that can be achieved when a solution with 30% methanol is flashed to 1 bar? The maximum mol fraction of ethanol?
c) Make a Txy plot for this system and annotate it properly.
Note: Use linear interpolations and report mol fractions to the third decimal point.

Problem 8.2: Use the data for the system methanol-ethanol (from the previous problem) to design a separation process that takes a mixture with 20% methanol and produces a stream that contains 90% methanol by continuously flashing the vapor stream until the required purity is reached (the liquid streams are not recycled). The process is to be operated at 1 bar with a ratio $V/L = 0.5$ in all flash separators.
a) How many flash separators are needed?
b) Calculate the percentage of methanol recovered in the 90% stream relative to the methanol in the feed.
c) Change the vapor liquid ratio to $V/L = 0.75$ and repeat parts a and b.
d) Your boss has asked you to recommend a value for the vapor, liquid ratio V/L. What is your recommendation?

Problem 8.3: The data below are for the system 1,4 dioxane (1)/methanol(2) at 308.5 K.
a) What is the saturation pressure of methanol at 308.5 K?
b) What is the phase of a mixture that contains 60% dioxane at 308.5 K, 0.1 bar?
c) Determine the bubble and dew pressures of a mixture that contains 60% dioxane at 308.5 K.
These questions can be answered more easily if you plot the Pxy graph based on these data.

8.9 Problems

P/bar	x_1	y_1	P/bar	x_1	y_1
0.280	0.000	0.000	0.217	0.614	0.263
0.277	0.056	0.044	0.202	0.677	0.297
0.274	0.106	0.073	0.181	0.771	0.367
0.270	0.167	0.102	0.157	0.841	0.446
0.264	0.232	0.130	0.141	0.886	0.512
0.250	0.378	0.175	0.130	0.907	0.573
0.240	0.459	0.204	0.119	0.930	0.63
0.228	0.553	0.234	0.081	1.000	1.000

Problem 8.4: Use the data below for the system ethyl propyl ether (1)-chloroform (2) to answer the following questions:
a) What is the boiling point of chloroform at 0.5 bar?
b) Is this a maximum boiling or minimum boiling azeotrope?
c) What is the composition at the azeotropic point?
d) A mixture of the two components contains 80% by mol ethyl propyl ether. What is the phase of this mixture at 48.3 °C, 0.5 bar? If a two-phase system, report the composition of the two phases and their relative amounts.
e) One mol of a solution of these two components, whose bubble point at 0.5 bar is 48.3 °C, is mixed with chloroform until the final mixture contains 50% chloroform (by mol). How much chloroform is needed?

Ethyl Propyl Ether (1) - Chloroform (2) at 0.5 bar

T (°C)	x_1	y_1	T (°C)	x_1	y_1
42.9	0.000	0.000	49.0	0.470	0.455
43.0	0.020	0.010	49.1	0.520	0.520
43.9	0.065	0.029	48.9	0.567	0.592
45.4	0.156	0.089	48.3	0.652	0.720
46.4	0.215	0.142	47.6	0.745	0.815
47.6	0.296	0.223	46.7	0.822	0.872
48.3	0.362	0.302	45.7	0.907	0.937
48.7	0.410	0.375	44.6	1.000	1.000

Problem 8.5: With reference to Figure 8-7, consider the following experiment: We begin with 1 mol of carbon tetrachloride at 35 °C, 0.35 bar and add methanol dropwise at constant temperature and pressure until the the mixture contains 99% by mol methanol. Describe all phase changes observed along this path and report the amount of methanol (in moles) that has been added up to the point that a phase change is observed.

Problem 8.6: A mixture that contains 40% by mole n-heptane in n-decane is to be separated in a series of flash separators until a stream is obtained that contains at least 95% n-heptane. Determine the number of separators needed, their temperature, and the recovery of n-heptane if all separators are at 1.013 bar and $V/L = 3$ in all separators. Txy data are given below:

x_1	y_1	T (C)	x_1	y_1	T (C)
0.00	0	173.63	0.55	0.9206	116.53
0.05	0.25146	164.52	0.60	0.93525	113.84
0.10	0.42826	156.67	0.65	0.94825	111.37
0.15	0.55558	149.84	0.70	0.9593	109.08
0.20	0.6497	143.89	0.75	0.96874	106.93
0.25	0.72101	138.65	0.80	0.97689	104.93
0.30	0.77493	133.94	0.85	0.98395	103.04
0.35	0.81765	129.75	0.90	0.99011	101.26
0.40	0.85169	125.97	0.95	0.99551	99.58
0.45	0.87922	122.54	1.00	1.00028	97.99
0.50	0.90181	119.4			

Problem 8.7: A mixture of normal heptane (40% by mol) in normal decane is to be separated in two flash separators. The feed stream is led to separator 1; the vapor stream of separator 1 is fed to separator 2 while the liquid stream from separator 2 is recycled into separator 1. Both separators operate at 1 atm. Heat exchangers are used to ensure that all streams that enter a separator are at the same temperature as the separator. Determine if the problem is fully specified and if not, make any additional specifications and solve the material balances. Report the purity of heptane and decane in the product streams and the % recovery of each component. Additional data: Use the fitted equations for x and y as functions of T that are given in Example 8.5.

Problem 8.8: Two flash separators in series operate at 1 atmosphere total pressure. The feed (F_1) into the first drum is a binary mixture of methanol/water that is 55 mol % methanol with flow rate 10,000 kg moles/h. The first flash drum operates at 75 °C and its liquid stream is fed into the second drum. The second flash drum operates at temperature, T_2 and the liquid product composition (x_2) is 15 mol % methanol.
a) What is the fraction vaporized in the first flash drum and the total fraction of the initial feed vaporized?
b) What are y_1, y_2, x_1, and T_2?
In this problem, the subscripts 1 and 2 refer to separator 1 and 2.
Additional information: The bubble and dew temperature (in °C) as a function of the mol fraction of methanol (x) are given by the following equations:

$$T_{\text{bubble}} = 143.77x^4 - 356.52x^3 + 317.44x^2 - 138.94x + 99.037$$
$$T_{\text{dew}} = 36.74x^4 - 65.732x^3 + 24.357x^2 - 30.966x + 100.06$$

8.9 Problems

Problem 8.9: Carbon tetrachloride (1) and water (2) are essentially immiscible liquids. A stream contains 80% CCl_4 by mole and the remaining is water. The temperature is 100 °C.
a) If the pressure is 4 bar, what is the phase of the stream, liquid, vapor, or vapor, liquid mixture?
b) If the pressure is 2 bar, what is the phase?
c) A liquid with overall composition $x_1 = 0.8$ is brought to its bubble pressure at 100 °C. Which liquid boils off first?
d) What is the vapor fraction when the first liquid boils off? Explain your answers.
Additional data: The saturation pressure of CCl_4 at 100 °C is 1.95 bar.

Problem 8.10: The system 2-butanol(1)/water(2) at 1.013 bar exhibits partial miscibility. Data for this system are given below. Answer the following question neglecting the variability of mutual solubility with temperature.
a) Make a Txy graph and annotate single- and two-phase regions.
b) You are given 1 mol of a mixture that contains 2-butanol and water at 1 atm, 75°C. The system consists of a single phase but its composition is not known. You add water dropwise at 1 atm, 75°C, and you notice that a tiny amount of a second liquid phase appears when you have added 0.46 mol of water. What was the composition of the solution you started with?
c) Twenty mol of 2-butanol is mixed with 80 mol of water and the system is brought to boil until 75 mol is in the vapor phase. How many phases are present at this point and what is their composition?
d) A mixture that contains 82% 2-butanol by mol is flashed to 1 atm. If the desired mol fraction of 2-butanol at the exit of the flash separator is 90%, determine the temperature of the separator and the amount of 2-butanol that is recovered in the 2-butanol rich stream.

x_1 or y_1	T_{bub} (°C)	T_{dew} (°C)	x_1 or y_1	T_{bub} (°C)	T_{dew} (°C)
1.00	99.44	99.44	0.30	87.54	90.33
0.90	95.02	97.62	0.20	87.54	93.72
0.80	91.89	95.71	0.10	87.54	96.80
0.70	89.75	93.72	0.08	87.54	97.39
0.60	88.39	91.66	0.06	87.54	97.96
0.48	87.54	89.23	0.04	87.54	98.53
0.46	87.54	88.85	0.02	89.81	99.08
0.40	87.54	87.54	0.00	99.63	99.63

Problem 8.11: Acrylonitrile and water are partially miscible liquids. At 1 atm, a two-phase mixtures boils at 70.6 °C and the compositions of the three phases are:

	Vapor	Upper liquid phase	Lower liquid phase
Acrylonitrile	85.7	96.8	7.3
Water	14.3	3.2	92.7

The normal boiling point of acrylonitrile is 78 °C.
a) Seven mol of water is mixed with 5 mol of acrylonitrile at 1 bar, 25 °C. Assuming that the mutual solubility at 25 °C is the same as at 70.6 °C, how many phases are formed and what are their compositions and amounts?
b) How much water should be added to the mixture of part (a) to form a single-phase system?
c) Draw a qualitative Txy graph and mark the states involved in parts a and b.

Problem 8.12: Water and 1-butanol are partially miscible in each other. At 1 atm, the two-phase system boils at 93.0 °C and the composition of the three phases are (in mol %):

	Vapor	Upper liquid phase	Lower liquid phase
1-butanol	55.5	79.9	7.7
water	44.5	20.1	92.3

The boiling point of pure butanol at 1 atm is 117.7 °C.
a) A solution is formed by mixing 5 mol of water with 2 mol of 1-butanol at 50 °C. Assuming the mutual solubility of the two components at 50 °C is the same as at 93.0 °C, how many phases are formed? If more than one, indicate the composition and amount of each phase.
b) If the mixture of part (a) is brought to boiling, which liquid phase will disappear first, the butanol-rich or the water-rich phase?
c) It is desired to convert the mixture of part (a) into a single-phase solution. This can be done either by adding more water or by adding more butanol. Calculate the minimum amounts of water and butanol that are needed to produce a single-phase system.

Problem 8.13: Butyraldehyde(1) and water (2) are partially miscible liquids. At 1 bar, the two-phase system boils at 68°C, and the composition of the three phases are: $y_1 = 90.3\%$, $x_1^A = 96.8$ and $x_1^B = 7.1\%$ (all in mol percent, with superscripts A and B indicating the butyraldehyde-rich and water-rich phases, respectively). The normal boiling point of butyraldehyde is 75.7 °C.

8.9 Problems

a) Forty mol of butyraldehyde are mixed with 10 mol water at 1 bar and the system is boiled at constant pressure. Which phase boils off first?

b) With the overall composition of the previous part, how many phases are present and in what amounts (moles) when half of the total moles are in the vapor phase?

c) How many phases are present and in what amounts (moles) when half of the total moles are in the butyraldehyde-rich liquid?

Problem 8.14: Nitroethane (component 1) and octane (component 2) are only partially miscible at 31 °C. At this temperature the triple point is at 6.8 kPa, the mol fraction of nitroethane in the vapor is 0.63, and in the two liquids 0.25 and 0.85, respectively. The saturation pressure of the pure components at 31 °C are, $P_1^{sat} = 4.7$ kPa and $P_2^{sat} = 3.2$ kPa. All questions below refer to 31 °C.

a) Draw a qualitative P_{xy} graph of this system and mark all the phases.

b) 4.5 mol of nitroethane is mixed with 1.5 mol of octane at 1 bar. What phases are present? Determine the fraction of the total moles in each phase.

c) The mixture of part (b) is titrated by adding octane until the system forms a single phase. How many moles of octane were added?

d) The mixture of part (b) is brought to a boil by reducing pressure. Which liquid phase disappears first?

e) In part (d), what is the vapor fraction at the point that the first liquid phase completely boils off?

Problem 8.15: Glycerol(1) and acetophenone(2) are partially miscible. The bubble point of the two-liquid system at 140 °C is 0.15 bar. The mole fraction of acetophenone in the one liquid phase is 10%, in the other liquid phase 85%, and in the vapor 95%.

a) Draw a qualitative P_{xy} graph of this system. Place on the horizontal axis the mole fraction of the more volatile component. Annotate the graph properly, place all the available information on the graph, and identify the various phases.

b) Twenty mole of acetophenone are mixed with 60 mole of glycerol at 140 °C, 0.2 bar. How many phases are present? Show the state on the P_{xy} graph.

c) The solution of part b is brought into boiling by reducing the pressure while keeping the temperature at 140 °C. Which phase boils off first? How much vapor is present at the point that the first liquid boils off?

d) How many phases are present when 8% of the original system is in the vapor and what is their composition?

Additional data: Saturation pressures at 140 °C: P_1^{sat} : 0.00313 bar; P_2^{sat} : 0.17 bar.

Problem 8.16: The system 3-methyl-1-butanol (1)/ethanol (2)/water (3) exhibits partial miscibility. The data below give the equilibrium composition of the two phases at 20 °C (Kadir et al., *J. Chem. Eng. Data* 2008, **53**, 910–912):

	Phase 1			Phase 2	
x_1	x_2	x_3	x_1	x_2	x_3
0.662	0.000	0.338	0.006	0.000	0.994
0.606	0.038	0.356	0.006	0.014	0.980
0.534	0.079	0.387	0.006	0.027	0.968
0.477	0.121	0.403	0.006	0.040	0.954
0.393	0.145	0.463	0.006	0.056	0.938
0.339	0.175	0.486	0.007	0.071	0.921
0.274	0.207	0.520	0.008	0.082	0.910
0.199	0.215	0.587	0.015	0.105	0.881

a) Draw a phase diagram for this system.

b) Of the three binary systems that can be formed with these three components, which are fully miscible and which are not?

c) One mol of 3-methyl-1-butanol (1) is mixed with 0.241 mol of ethanol and 4.15 mol of water at 20 C. Determine the number of phases, their composition, and the number of moles in each phase.

Chapter 9

Properties of Mixtures

In the first part of this book we developed the principles (laws) of the thermodynamics and applied them to pure fluids. We now we want to extend their application to multicomponent systems. The thermodynamics laws themselves are general and apply to all systems, whether pure or multicomponent. Their mathematical expressions were developed in Chapter 6 and are summarized below:

$$\text{First law:} \quad \frac{dU^{\text{tot}}}{dt} + \Delta(\dot{m}H) = \dot{Q} + \dot{W}_s, \quad [6.17]$$

$$\text{Second law:} \quad \frac{dS^{\text{tot}}}{dt} + \Delta(\dot{m}S) = \frac{\dot{Q}}{T_{\text{bath}}} + \dot{S}_{\text{gen}}. \quad [6.26]$$

These equations require the enthalpy and entropy of the system and its various streams. The problem, then is how to calculate the properties of mixture. The properties of pure substance are functions of pressure and temperature. For mixtures, in addition to pressure and temperature we must consider *composition*. The goal in this chapter is to formulate equations for the properties of mixtures as a function of pressure, temperature and composition. Essentially, the task will be to incorporate composition as a new independent variable. We will express composition in terms of moles of each component,[1] therefore, when we refer to composition as an independent variable we will understand not a single variable but a set of variables. First we will develop formal mathematical expressions for properties, their differentials and their partial derivatives in terms of the independent variables, temperature, pressure and moles of component i. Next we will show how we can use equations of state to calculate the volume, enthalpy and entropy of mixtures. The mathematical formulation is based on the material developed in Chapter 5. A review of that chapter is recommended, as we will make frequent references to results obtained there.

The learning objectives in this chapter are to:

1. Formulate the differential of mixture properties with temperature, pressure and composition as the independent variables,

---

1. We will also use other measures of composition, for example, mass or mass fractions, various forms of concentration (molarity, molality), but for the most part we will express composition in terms of moles and mole fractions.

2. Apply equations of state to mixtures,

3. Calculate the properties of mixtures from a cubic equation of state.

9.1 Composition

Let us form a mixture of N components that contains n_i mol of component i, $i = 1, \cdots, N$. The total number of moles is

$$n = n_1 + n_2 + \cdots = \sum_{i=1}^{N} n_i.$$

We understand summations like this to run over all components and for simplicity the limits will be omitted. The mol fraction of component i is defined

$$x_i = \frac{n_i}{n_1 + n_2 \cdots} = \frac{n_i}{n}.$$

The mol fractions in a mixture satisfy the normalization condition,

$$\sum x_i = 1. \tag{9.1}$$

Mathematically, this means that of the N mol fractions only $N-1$ are independent. All extensive properties of mixtures are mathematical functions of pressure, temperature, and the number of moles of all components. Using the enthalpy as an example, we write

$$H^{\text{tot}} = H^{\text{tot}}(T, P, n_1, n_2, \cdots). \tag{9.2}$$

This makes H^{tot} a function of $N+2$ independent variables. The molar form of an extensive property is obtained by dividing the extensive property by the number of moles, for example,

$$H = H^{\text{tot}}/n.$$

The molar properties of a mixture are mathematical functions of pressure, temperature and mole fractions:

$$H = H(P, T, x_1, x_2, \cdots). \tag{9.3}$$

Since mol fractions are not independent but must satisfy the normalization condition in eq. (9.1), the above is a function of $N+1$ independent variables. To avoid mathematical inconveniences from the fact that the mole fractions are not all independent, we will formulate the basic theory for the *extensive* form of a property. Once we have expressions for the extensive property, we can obtain the molar property through division by the total number of moles.

9.1 Composition

The mass of the mixture is the sum of the masses of its constituents:

$$m = \sum_i m_i,$$

where m is the mass of the mixture and m_i is the mass of component i. The mass (or weight) fraction, w_i, of component i is

$$w_i = \frac{m_i}{m}.$$

The mass fractions also satisfy the normalization condition:

$$\sum w_i = 1.$$

The mass and number of moles are related as follows

$$m = n\overline{M}_m, \tag{9.4}$$

where $\overline{M}_m$ is the mean molar mass of the mixture and is weighted average of the molar masses M_{mi} of all components:

$$\overline{M}_m = \frac{\sum_i n_i M_{mi}}{\sum_i n_i} = \frac{\sum_i n_i M_{mi}}{n}. \tag{9.5}$$

The mean molar mass of mixture plays the same role as the molar mass of pure component and may be used to convert the mass of mixture into moles, and vice versa.

NOTE

Mixture Properties and Homogeneity
Intensive properties (for example, molar volume, molar enthalpy, etc.) are functions of *intensive variables*: temperature, pressure, mole fractions. Two mixtures with the same mole fractions have identical intensive properties at fixed pressure and temperature. The extensive property of the mixture is obtained by multiplying the molar property by the total moles:

$$H^{\text{tot}} = nH.$$

Expressing properties in terms of their corresponding variables, this equation is written as

$$H^{\text{tot}}(T, P, nx_1, nx_2, \cdots) = n H(T, P, x_1, x_2, \cdots) \tag{9.6}$$

where we used $n_i = nx_i$ to express the moles of component i in terms of mole fractions and total moles. The superscript ^{tot} is optional here and can be omitted as it is clear that the enthalpy on the left-hand side refers to n moles of mixture while the enthalpy on the right refers to 1 mol of mixture. Comparing the list of variables for the extensive (left-hand side) and the intensive property (right-hand side), it is as if the total number of moles can be "factored" out of all the terms that contain it. This expresses mathematically the fact that the properties of mixture depend on the *relative* amounts (i.e., mole fractions) of each component. The actual number of moles n makes a simple multiplicative contribution. In math language, the "factorization" of n means that extensive properties are *homogeneous* functions of number of moles n_i of component i. The homogeneity of thermodynamic properties leads to an important theoretical results, the Gibbs-Duhem equation, which discussed later on page 406.

9.2 Mathematical Treatment of Mixtures

Now we consider the mathematical problem of how to express properties as functions of pressure, temperature, *and* composition. We will develop equations for the generic extensive property F^{tot}, which will stand in for enthalpy, entropy, volume, and the like. This is a mathematical function of T, P, and the moles of all components:

$$F^{\text{tot}} = F^{\text{tot}}(T, P, n_1, n_2, \cdots).$$

By standard calculus the differential of this expression is

$$dF^{\text{tot}} = \left(\frac{\partial F^{\text{tot}}}{\partial T}\right)_{P,n_i} dT + \left(\frac{\partial F^{\text{tot}}}{\partial P}\right)_{T,n_i} dP + \sum_i \left(\frac{\partial F^{\text{tot}}}{\partial n_i}\right)_{P,T,n_j} dn_i.$$

This expression contains the partial derivatives with respect to every variable and each derivative is multiplied by the differential of the corresponding variable. In these derivatives, all variables except one are held constant. In the derivatives with respect to T or P, all n_i are held constant. This is indicated by the subscript n_i. In derivatives with respect to n_i, pressure, temperature, and all other moles ($n_j \neq n_i$) are held constant. This is indicated by the subscript n_j that appears in the derivative. We define the *partial molar property*, $\bar{F}_i$, as

$$\bar{F}_i = \left(\frac{\partial F^{\text{tot}}}{\partial n_i}\right)_{P,T,n_j}. \tag{9.7}$$

With this definition, the differential of an extensive property is written in the more compact form,

$$dF^{\text{tot}} = \left(\frac{\partial F^{\text{tot}}}{\partial T}\right)_{P,n_i} dT + \left(\frac{\partial F^{\text{tot}}}{\partial P}\right)_{T,n_i} dP + \sum \bar{F}_i dn_i. \tag{9.8}$$

This differential expresses the change in property F^{tot} upon a change of state by $(dT, dP, dn_1, \cdots)$.

9.2 Mathematical Treatment of Mixtures

The Partial Molar Property

Partial molar properties are defined for any property that has an extensive form for example, volume, enthalpy, etc. They are *intensive* properties and as such, they are functions of pressure, temperature, and mol fractions. To see how partial molar properties can be useful, consider the following thought experiment: a vessel A that contains a mixture (for example, a solution of several components) is poured into another vessel B. We will calculate the enthalpy in vessel B as it builds up during this process. The differential of H^{tot} is given by eq. (9.8); the process obviously takes place under constant pressure and constant temperature, therefore, $dT = 0$ and $dP = 0$. This simplifies the differential to the form,

$$dH^{\text{tot}} = \sum_i \bar{H}_i dn_i, \quad \text{constant } P, T.$$

During transfer from one vessel to another, pressure, temperature, and mole fractions are constant. Accordingly, all molar and partial molar properties also remain constant. Under these conditions, integration of the differential dH^{tot} with respect to n_i is trivial and the result is

$$H^{\text{tot}} = \sum_i n_i \bar{H}_i. \tag{9.9}$$

This result gives the enthalpy of the mixture in vessel B as a function of the amounts n_i of all components. This enthalpy is of course the same as in vessel A. In other words, eq. (9.9) is a general relationship between an extensive property and the corresponding partial molar properties. It states that

> *a partial molar property is the contribution (per mole) of a component to the total property of the mixture*

This remarkably simple relationship applies to any property; therefore, we can generalize the result by writing

$$F^{\text{tot}} = \sum_i n_i \bar{F}_i, \tag{9.10}$$

$$F = \sum_i x_i \bar{F}_i. \tag{9.11}$$

Partial molar properties obey the same equations as regular properties. For example, the partial molar internal energy, enthalpy, and volume are related through the equation,

$$\bar{H}_i = \bar{U}_i + P\bar{V}_i,$$

in complete analogy to the relationship $H = U + PV$.

NOTE

Euler's Theorem

Equation (9.10) is a special case of *Euler's theorem*. If a mathematical function satisfies the condition

$$f(ax, ay) = a^\nu f(x, y),$$

for any value of a, where ν is constant, it is called homogeneous in x and y of degree ν.[2] In the special case $\nu = 1$ we obtain,

$$f(ax, ay) = a f(x, y).$$

We take the derivative of both sides with respect to a, using the chain rule for the derivative in the left-hand side

$$\left(\frac{\partial f(ax, ay)}{\partial ax}\right)_{ay} (ax)' + \left(\frac{\partial f(ax, ay)}{\partial ay}\right)_{ax} (ay)' = f,$$

which we write as,[3]

$$f = x \left(\frac{\partial f}{\partial x}\right)_y + y \left(\frac{\partial f}{\partial y}\right)_x.$$

This is known as Euler's theorem. According to eq. (9.6), enthalpy, as well as all extensive properties of mixtures are homogeneous functions of the number of moles of each component with degree 1. Applying Euler's theorem with $x = n_1$, $y = n_2$, etc., we obtain

$$H^{\text{tot}} = n_1 \left(\frac{\partial H^{\text{tot}}}{\partial n_1}\right)_{P,T,n_{j\neq 1}} + n_2 \left(\frac{\partial H^{\text{tot}}}{\partial n_2}\right)_{P,T,n_{j\neq 2}} + \cdots,$$

which is the same as eq. (9.9).

Gibbs-Duhem Equation

Equations (9.10) and (9.11) express a simple but important relationship between partial molar properties and the corresponding property, total or molar, of the mixture. Another relationship can be derived for the differential of the partial molar property. We begin by taking the differential of eq. (9.10) keeping in mind that pressure and temperature are held constant:

$$dF^{\text{tot}} = \sum \bar{F}_i \, dn_i + \sum n_i \, d\bar{F}_i, \quad (\text{constant } P, T).$$

---

2. For example, $f = (x + 2y)xy$ is homogeneous of degree 3.
3. Here we have used

$$\left(\frac{\partial f(ax, ay)}{\partial ax}\right)_{ay} = \left(\frac{\partial f(x, y)}{\partial x}\right)_y = \left(\frac{\partial f}{\partial x}\right)_y,$$

since ax and ay may be renamed into two new variables, x and y.

9.2 Mathematical Treatment of Mixtures

The same differential is now calculated from eq. (9.8) by holding P and T constant:

$$dF^{\text{tot}} = \sum \bar{F}_i \, dn_i, \quad (\text{constant } P, T).$$

Comparing the last two results we conclude that

$$\sum n_i \, d\bar{F}_i = 0, \quad (\text{constant } P, T).$$

After dividing both sides by the total number of moles, this result is expressed in terms of mol fractions as

$$\boxed{\sum x_i \, d\bar{F}_i = 0,} \quad (\text{constant } T, P). \tag{9.12}$$

This result states that variations of the partial molar properties with composition at constant temperature and pressure are not independent but satisfy a zero-sum condition: if all variations are multiplied by the corresponding mole fraction, their sum must be zero.[4] This is known as the *Gibbs-Duhem* equation and applies to any molar property. In the special case of a two-component mixture, eq. (9.12) gives

$$x_1 d\bar{F}_1 + x_2 d\bar{F}_2 = 0.$$

Dividing both sides by dx_1 this can be written in the form

$$x_1 \frac{d\bar{F}_1}{dx_1} + x_2 \frac{d\bar{F}_2}{dx_1} = 0. \tag{9.13}$$

Here, the variations of the partial molar property are seen explicitly to depend on variations of the mol fractions. If the partial molar property of one component is known as a function of the mole fraction, the partial molar property of the other component can be evaluated by integration of this equation. Alternatively, if both partial molar properties are known (experimentally, for example), eq. (9.13) represents a test of thermodynamic consistency: if the measured values do not satisfy this condition, we would have reason to suspect that the data are not reliable.

Special Case: Systems of Constant Composition

The properties of mixtures depend on temperature, pressure and composition but some processes leave the composition unchanged while having an effect only on pressure or temperature. Typical examples are compression, expansion, heating or

---

4. We are referring to small variations that can be treated as differentials.

cooling, as long as there is only one phase present.[5] For such process, $dn_i = 0$ for all components and the differential of extensive property F^{tot} simplifies to

$$dF^{\text{tot}} = \left(\frac{\partial F^{\text{tot}}}{\partial T}\right)_{P,n_i} dT + \left(\frac{\partial F^{\text{tot}}}{\partial P}\right)_{T,n_i} dP, \quad \text{(const. composition)}.$$

The corresponding expression for the differential of the molar property:[6]

$$dF = \left(\frac{\partial F}{\partial T}\right)_{P,x} dT + \left(\frac{\partial F}{\partial P}\right)_{T,x} dP. \tag{9.14}$$

Here, the subscript x in the partial derivatives indicates that all *mole fractions* are held constant, since molar properties depend on the mol fractions, not the number of moles. This equation is identical to that for pure component. In fact, a mixture of constant composition behaves just as a pure component, except when phase change is involved.[7] This conclusion has important practical implications: It means that all of the relationships developed in Chapter 5 for pure components also apply to mixtures of constant composition. For example, by analogy to eq. (5.17), the differential of the molar Gibbs energy of mixture is

$$dG = -SdT + VdP, \quad \text{(constant composition)}, \tag{9.15}$$

where S and V is the molar entropy and molar volume of the mixture. Since n is constant, we also have

$$dG^{\text{tot}} = -S^{\text{tot}} dT + V^{\text{tot}} dP, \quad \text{(constant composition)}, \tag{9.16}$$

which is the same relationship between the total properties of the mixture.

NOTE

Residual, Partial Molar, and Residual Partial Molar
Analogous relationships to eq. (9.15) can be written for the *residual*, the *partial molar*, and the *residual partial molar* properties:

$$dG = -SdT + VdP,$$

$$dG^R = -S^R dT + V^R dP,$$

---

[5]. As we saw in Chapter 8, when a mixture is brought into the two-phase region there is an effect on composition since each phase has its own composition.
[6]. To obtain this result we write $dF^{\text{tot}}/n = d(F^{\text{tot}}/n)$, and treat n as constant, since the moles of all components are constant.
[7]. When phase change occurs, the system splits into phases of different compositions. This requires special treatment and will be discussed in Chapter 10.

9.3 Properties of Mixing

$$d\bar{G}_i = -\bar{S}_i dT + \bar{V}_i dP,$$
$$d\bar{G}_i^R = -\bar{S}_i^R dT + \bar{V}_i^R dP,$$

all of which apply under the condition of constant composition. The "residual" and the "partial molar" should be understood as mathematical operators: the residual operator removes the ideal-gas contribution from a property at constant P and T; the partial molar operator applies differentiation with respect to n_i at constant P, T, and $n_{j\neq i}$. The *residual-partial molar* operator does both. These operators are linear. For example, the residual of the sum of two properties is the sum of the residuals; the same is true for the partial molar operator (both operators leave temperature and pressure unaffected):

$$(A+B)^R = A^R + B^R,$$
$$\overline{(A+B)} = \bar{A} + \bar{B}.$$

The residual-partial molar operator gives the same result regardless of the order in which the individual operators are applied:

$$(\bar{A})^R = \overline{(A^R)}.$$

These mathematical properties have a simple practical result: they ensure that the same relationships can be written among all forms of a thermodynamic function, including molar, total, residual, partial molar, or residual partial molar.

9.3 Properties of Mixing

It is generally desirable to relate the properties of a mixture to those of the pure components, since pure-component properties are easier to calculate. To study this relationship, we form 1 mol of mixture by mixing its components under constant temperature and pressure, as shown in Figure 9-1. If the desired mol fractions are x_i, then we must mix x_i mol of component i. Before mixing, the total volume occupied by the pure components is

$$\sum_i x_i V_i,$$

where V_i is the molar volume of component i; after mixing, the mixture occupies volume V, the molar volume of the mixture. The change in volume is

$$\Delta V_{\text{mix}} = V - \sum_i x_i V_i. \tag{9.17}$$

This is called volume of mixing and it may be positive (volume expansion upon mixing), negative (contraction) or zero (no change). Similar expressions can be

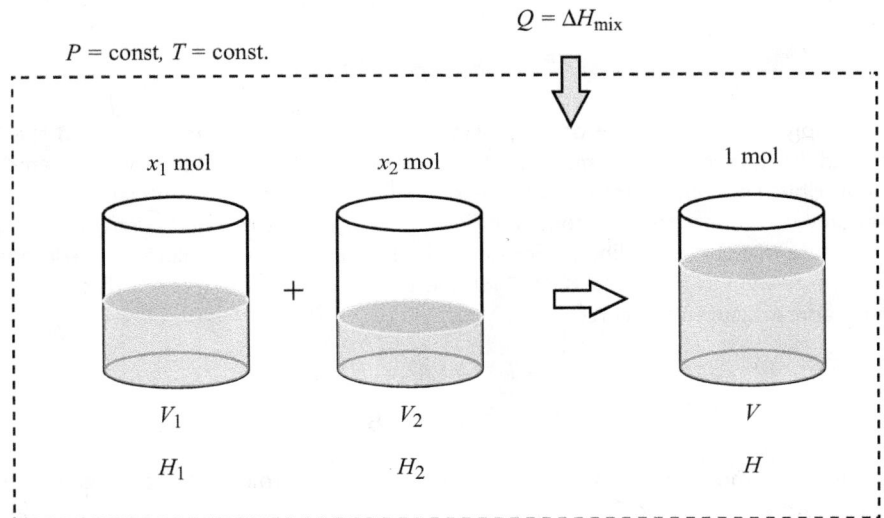

Figure 9-1: Formation of mixture from the pure components at constant pressure and temperature.

written for all extensive properties.[8] An important property is the enthalpy of mixing,

$$\Delta H_{\text{mix}} = H - \sum_i x_i H_i. \tag{9.18}$$

Since the mixing process is at constant pressure, the enthalpy of mixing represents the amount of heat that must exchanged in order to conduct the process isothermally. If $\Delta H_{\text{mix}} < 0$, the enthalpy of the solution is lower that of the pure components and the difference appears as heat transferred to the surroundings. Such process is called *exothermic*. If $\Delta H_{\text{mix}} > 0$, the mixture has higher enthalpy than the pure components. In this case the difference must be supplied from the surroundings and the process is called *endothermic*. If $\Delta H_{\text{mix}} = 0$, no heat is exchanged and the process is called *athermal*. Experimentally, the enthalpy of mixing may be determined from the heat that must be exchanged in order to mix the components isothermally, or from the resulting temperature change, if components are mixed adiabatically.

---

8. For any extensive property, its combined value over all pure components before mixing is $\sum x_i F_i$; after mixing it is F (on the basis of 1 mol of mixture). The change, therefore, is

$$\Delta F_{\text{mix}} = F - \sum_i x_i F_i.$$

Other properties of mixing are defined in the same way. The entropy of mixing, in particular, is
$$\Delta S_{\text{mix}} = S - \sum_i x_i S_i. \tag{9.19}$$
Similar expressions can be written for the internal energy, Gibbs free energy and Helmholtz free energy. The various properties of mixing are related among themselves through relationships analogous to those of the regular properties. For example,
$$\Delta H_{\text{mix}} = \Delta U_{\text{mix}} + P\Delta V_{\text{mix}} \tag{9.20}$$
$$\Delta G_{\text{mix}} = \Delta H_{\text{mix}} - T\Delta S_{\text{mix}} \tag{9.21}$$
and so on. Because of these relationships, if the primary properties ΔV_{mix}, ΔH_{mix} and ΔS_{mix} are known, either by measurement or by calculation, all others can be obtained from their relationship to them.

NOTE

On Subscripts
In dealing with mixtures we adopt the following notational convention: A subscript refers to the property of pure component whereas a property without a subscript refers to the mixture. A partial molar property (subscript and bar over the symbol) is a special case that refers to a property that characterizes a *component* in the *mixture*. In general, if an equation contains both mixture and component properties, it is understood that they are all at the same pressure and temperature, unless indicated otherwise.

9.4 Mixing and Separation

Mixing is a spontaneous process: if pure components are brought into contact, for example, oxygen and nitrogen in a closed vessel, sugar and water in a coffee pot, components mix to form a mixture. Even in systems that are partially or even sparingly miscible, such as water and oil, some mixing does take place until the solubility limit is reached. This process occurs through molecular motion, and while its rate is enhanced through the application of work (e.g., stirring), such interference is not necessary for the process to take place. As a spontaneous process, the mixing of pure components under constant temperature and pressure produces a mixture whose Gibbs free energy is lower than that of the pure components:
$$\Delta G_{\text{mix}} \leq 0. \tag{9.22}$$
From eq. (9.21) we see that there are two contributions to the Gibbs free energy of mixing: one comes from the entropy of mixing ΔS_{mix}, and the other from the

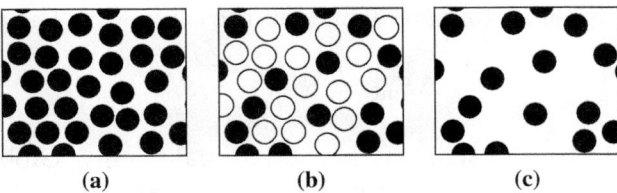

Figure 9-2: Molecular view of mixing: (a) pure component; (b) component in mixture; (c) component in mixture with the other components made invisible. In the mixed state a component has undergone isothermal "expansion" relative to the pure state. This gives rise to the entropy of mixing.

enthalpy of mixing ΔH_{mix}. Molecules in a mixture can assume more microscopic states compared to the pure components because they can be surrounded by various combinations of like and unlike species (see Figure 9-2). In molecular language, this means that the entropy of the mixture is higher than the entropy of the pure components combined, that is, $\Delta S_{\text{mix}} > 0$. The enthalpy of mixing may be positive, negative, or zero. If it is negative (mixing is exothermic), it contributes to the decrease of ΔG_{mix} and acts as a further driving force for mixing. If it is positive (mixing is endothermic), it increases the value of ΔG_{mix} and opposes mixing. If it is zero it makes no contribution to the Gibbs free energy of mixing and the process is driven entirely by entropy. Exothermic mixing indicates favorable interaction between unlike components, which makes the separation of the mixture more difficult. Endothermic mixing indicates the opposite: interaction between unlike components is energetically unfavorable (energy must be supplied to form the mixture) and the mixture they form is easier to separate. The work required for the separation of the components is equal to the lost work of mixing because, as we recall, lost work refers to the work needed to reverse a process (in this case, to separate the components). From eq. (6.34), this work is

$$W_{\text{lost}} = T_0 S_{\text{gen}}.$$

The entropy generation consists of the entropy change of the system, ΔS_{mix}, and the entropy change of the surroundings, $-\Delta H_{\text{mix}}/T$:[9]

$$S_{\text{gen}} = \Delta S_{\text{mix}} - \frac{\Delta H_{\text{mix}}}{T}.$$

---

9. For constant-pressure mixing, the amount of heat exchanged with the surroundings is $Q = \Delta H_{\text{mix}}$; accordingly, the entropy change of the surroundings at temperature T is

$$-\frac{Q}{T} = -\frac{\Delta H_{\text{mix}}}{T}.$$

Combining these results the lost work is,

$$W_{\text{lost}} = T_0 \left(\Delta S_{\text{mix}} - \frac{\Delta H_{\text{mix}}}{T} \right) = -\frac{T_0}{T} \Delta G_{\text{mix}}. \quad (9.23)$$

In the special case that the heat bath is at the temperature of the system ($T_0 = T$), the lost work simplifies to

$$W_{\text{lost}} = -\Delta G_{\text{mix}}.$$

As we recall, the lost work is also equal to the minimum amount of work that must be done to reverse the process, in this case, to separate the components. Since the Gibbs free energy of mixing is always negative, the work for separation is positive. The more important conclusion is that the Gibbs free energy of mixing is a direct measure of the energy released when the mixture is formed, also equal (in absolute value) to the amount of energy that must be consumed to separate the mixture.

9.5 Mixtures in the Ideal-Gas State

The ideal-gas state for a mixture is reached by reducing pressure at constant temperature and constant composition. When this state is reached, molecules behave as point masses that exert no interactions. Neither the chemical nature of the gas, nor its composition (if mixture) has any effect on its volumetric properties. Therefore, the equation of state of the ideal-gas mixture is the same as for a pure fluid in the ideal-gas state:

$$PV^{\text{igm}} = RT. \quad (9.24)$$

Here the superscript ^{igm} is used to indicate ideal-gas *mixture*. It follows from this equation that when we form an ideal-gas mixture by mixing the pure components in the ideal-gas state at constant temperature and pressure, the volume of the mixture formed is equal to the volume of the pure components before mixing:[10]

$$V^{\text{igm}} = \sum_i x_i V_i^{\text{ig}}. \quad (9.25)$$

---

10. The proof is straightforward: multiply eq. (9.24) by $n = n_1 + n_2 + \cdots$, to obtain

$$nV^{\text{igm}} = n_1 \frac{RT}{P} + n_2 \frac{RT}{P} + \cdots,$$

On the left-hand side we have the total volume of the mixture, and on the right-hand side we have the sum of the volumes of the pure components.

The enthalpy and entropy of the ideal-gas mixture are[11]

$$H^{\text{igm}} = \sum_i x_i H_i^{\text{ig}}, \qquad (9.26)$$

$$S^{\text{igm}} = \sum_i x_i S_i^{\text{ig}} - R \sum_i x_i \ln x_i. \qquad (9.27)$$

These results are given without proof (for the qualitative arguments that support these results, see the note later in this section). Accordingly, volume and enthalpy of mixing in the ideal-gas state are both zero:

$$\Delta V_{\text{mix}} = 0, \quad \Delta H_{\text{mix}} = 0.$$

The entropy of mixing is not zero but is given by

$$\Delta S_{\text{mix}}^{\text{igm}} = -R \sum_i x_i \ln x_i. \qquad (9.28)$$

This is always a positive quantity because the logarithm of all mole fractions is negative ($x_i < 1$) and the minus sign makes the right-hand side positive. According to this result, the mixing of components in the ideal-gas state is a spontaneous process and leads to entropy generation. The reverse process, separation, is not spontaneous and requires work.

From the above relationships we may easily identify the partial molar properties of mixture. Recall the basic relationship between molar and partial molar properties:

$$F = \sum_i x_i \overline{F}_i. \qquad [9.11]$$

Comparing this with eqs. (9.25), (9.26), and (9.27) for the volume, enthalpy and entropy of the ideal mixture we identify the respective partial molar properties as

$$\overline{V}_i^{\text{igm}} = V_i^{\text{ig}}, \qquad (9.29)$$

$$\overline{H}_i^{\text{igm}} = H_i^{\text{ig}}, \qquad (9.30)$$

$$\overline{S}_i^{\text{igm}} = S_i^{\text{ig}} - R \ln x_i. \qquad (9.31)$$

The partial molar volume and partial molar enthalpy are equal to the corresponding properties of the pure component. The partial molar entropy is higher than the entropy of the pure component by $-R \ln x_i$: a component in the mixture contribute more entropy per mole than the original entropy of the pure component before mixing. The properties of the ideal-gas state are summarized in Table 9-1.

---

11. We use the superscript ^{igm} rather than ^{ig} to remember that we are dealing with a mixture, not a pure component.

9.5 Mixtures in the Ideal-Gas State

Table 9-1: Properties of mixture in the ideal-gas state. All properties (mixture or pure component) are understood to be in the ideal-gas state. The superscripts ^{ig} and ^{igm} have been omitted for simplicity.

Property of ideal-gas mixture	Property of mixing	Partial molar
$V = \sum x_i V_i$	$\Delta V_{\text{mix}} = 0$	$\bar{V}_i = V_i$
$H = \sum x_i H_i$	$\Delta H_{\text{mix}} = 0$	$\bar{H}_i = H_i$
$S = \sum x_i S_i - R \sum x_i \ln x_i$	$\Delta S_{\text{mix}} = -R \sum x_i \ln x_i$	$\bar{S}_i = S_i - R_i \ln x_i$
$U = \sum x_i U_i$	$\Delta U_{\text{mix}} = 0$	$\bar{U}_i = U_i$
$G = \sum x_i G_i + RT \sum x_i \ln x_i$	$\Delta G_{\text{mix}} = RT \sum x_i \ln x_i$	$\bar{G}_i = G_i + RT \ln x_i$

Other Properties

Other properties of the ideal-gas mixture can be obtained easily. For example, for the internal energy we write

$$U^{\text{igm}} = H^{\text{igm}} - PV^{\text{igm}} = \sum_i x_i \left(H_i^{\text{ig}} - PV_i^{\text{ig}} \right),$$

where we used eqs. (9.26) and (9.25) for the enthalpy and volume of the ideal-gas mixture. The quantity in parenthesis on the right-hand side in the above equation is the ideal-gas internal energy. Therefore, the final result for internal energy is

$$U^{\text{igm}} = \sum_i x_i U_i^{\text{ig}}. \tag{9.32}$$

From this result it also follows that the internal energy of mixing is zero, and the partial molar internal energy is equal to the internal energy of the pure component:

$$\Delta U_{\text{mix}}^{\text{igm}} = 0, \tag{9.33}$$

$$\bar{U}_i = U_i. \tag{9.34}$$

The corresponding results for the Gibbs free energy are

$$G^{\text{igm}} = \sum_i x_i G_i^{\text{ig}} + RT \sum_i x_i \ln x_i, \tag{9.35}$$

$$\Delta G_{\text{mix}}^{\text{igm}} = RT \sum_i x_i \ln x_i, \tag{9.36}$$

$$\bar{G}_i = G_i + RT \ln x_i. \tag{9.37}$$

These are obtained from the relationship between G, H, and S and their derivation is left as an exercise.

NOTE

Enthalpy and Entropy of Ideal-Gas Mixture

The equations given for enthalpy and entropy of ideal-gas mixture were given here without proof. They can be proven using the tools of statistical mechanics, but this is beyond the scope of this book. Nonetheless, we can arrive at these equations by qualitative arguments. Since molecules in the ideal-gas state do not interact, the internal energy of the mixture is the same as the total internal energy of the pure components at same pressure and temperature; this means $\Delta U^{\text{igm}} = 0$. And since the volume of the mixture is the sum of the pure component volumes, we conclude the same for enthalpy, or $\Delta H^{\text{igm}} = 0$.

To obtain the entropy of the mixture, we imagine a component going from the pure state at P, T, into the mixture at same pressure and temperature (see Figure 9-2). This process increases the volume available to this component, from V_i (the molar volume of the pure component) to V (the molar volume of the mixture). This amounts to expanding this component isothermally from volume $V_i = x_i V$ to volume V. The entropy change for this process is

$$R \ln \frac{V}{V_i} = -R \ln x_i,$$

The entropy difference between the mixture and the pure components is obtained by summing the contribution of all components. To form 1 mol of mixture we need x_i mole of each component, therefore,

$$\Delta S_{\text{mix}}^{\text{igm}} = -R \sum_i x_i \ln x_i,$$

which is eq. (9.28) in the text.

Example 9.1: Non-Isothermal Mixing

A rigid insulated tank is divided into two parts, one that contains 50 mol of ethylene at 400 K, 1 bar, and the other contains 150 mol nitrogen at 300 K, 2 bar. The partition that divides the tank is removed and the contents are allowed to reach equilibrium. Determine the final pressure and temperature in the tank and determine the entropy generation. You may assume the pure gases and the final mixture to be in the ideal-gas state. The ideal-gas heat capacities of ethylene and nitrogen are 43 J/mol K and 29 J/mol K, respectively, and may be assumed independent of temperature.

Solution The tank as a whole is a closed system. The process is adiabatic and there is no PV or shaft work. The energy balance under these conditions is

$$\Delta U^{\text{tot}} = 0.$$

Since the tank is rigid, this is also a constant-volume process:

$$\Delta V^{\text{tot}} = 0.$$

9.5 Mixtures in the Ideal-Gas State

These two equations may be solved for the two unknowns, the final temperature and pressure in the tank.

Calculation of temperature. Using the subscripts $_1$ for ethylene and $_2$ for nitrogen, the internal energy at the initial state is

$$U_i^{\text{tot}} = n_1 U_1(T_1, P_1) + n_2 U_2(T_2, P_2),$$

where $U_1(T_1, P_1)$ is the internal energy of pure ethylene in the initial state, and $U_2(T_2, P_2)$ is the internal energy of pure nitrogen in its initial state. The internal energy of the mixture is given by eq. (9.32), which in this case takes the form,

$$U_f^{\text{tot}} = n_1 U_1(T_f, P_f) + n_2 U_2(T_f, P_f).$$

Here, $U_1(T_f, P_f)$ and $U_2(T_f, P_f)$ is the internal energy of pure components at the temperature and pressure of the mixture. Inserting these expressions into the energy balance we obtain the following expression

$$n_1 \Big(U_1(T_f, P_f) - U_1(T_1, P_1) \Big) + n_2 \Big(U_2(T_f, P_f) - U_2(T_2, P_2) \Big) = 0.$$

Each of the above terms contains the change in internal energy for the pure component from initial state (T_i, P_i) to final state (T_f, P_f). In the ideal-gas state, this change is $C_{V\,i}^{\text{ig}}$ $(T_f - T_i)$. Then, the above equation becomes

$$n_1 C_{V\,1}^{\text{ig}}(T_f - T_1) + n_2 C_{V\,2}^{\text{ig}}(T_f - T_2) = 0.$$

Solving for T_f the result is

$$T_f = \frac{n_1 C_{V\,1}^{\text{ig}} T_1 + n_2 C_{V\,2}^{\text{ig}} T_2}{n_1 C_{V\,1}^{\text{ig}} + n_2 C_{V\,2}^{\text{ig}}}.$$

By numerical substitution, using $C_{V\,i}^{\text{ig}} = C_{P\,i}^{\text{ig}} - R$, we find

$$T_f = 335.85 \text{ K}.$$

Calculation of pressure. The constant-volume condition may be written in the form,

$$n_1 V_1 + n_2 V_2 = (n_1 + n_2) V,$$

where $V_i = RT_i/P_i$ is the molar volume of the pure component i initially and $V = RT_f/P_f$ is the molar volume in the final state. Substituting the ideal-gas expressions for the molar volume, we obtain an equation in which the only unknown is P_f. Solving for the pressure we find

$$P_f = \frac{T_f}{x_1 T_1/P_1 + x_2 T_2/P_2},$$

where $x_i = n_i/(n_1 + n_2)$ is the mol fraction of component i in the final mixture. By numerical substitution,
$$P_f = 1.58 \text{ bar}.$$

Entropy generation. For an adiabatic process, the entropy generation is
$$S_\text{gen}^\text{tot} = S_f^\text{tot} - S_i^\text{tot}.$$

The initial entropy is the sum of the entropy of pure components at their initial state:
$$S_i^\text{tot} = n_1 S_1(T_1, P_1) + n_2 S_2(T_2, P_2).$$

The entropy of the final state is given by eq. (9.27), which gives
$$S_f^\text{tot} = n_1 S_1(T_f, P_f) + n_2 S_2(T_f, P_f) - R(n_1 \ln x_1 + n_2 \ln x_2).$$

Notice that in this expression the entropies of the pure components are at the pressure and temperature of the mixture. Substituting these expressions into the equation for the entropy generation we obtain
$$S_\text{gen} = n_1 \Delta S_1 + n_2 \Delta S_2 - R(n_1 \ln x_1 + n_2 \ln x_2),$$

where ΔS_i is the change of molar entropy of pure component from state (T_i, P_i) to state (T_f, P_f). This change is
$$\Delta S_i = C_{P\,i}^\text{ig} \ln \frac{T_f}{T_i} - R \ln \frac{P_f}{P_i} \quad \Rightarrow \quad \begin{cases} \Delta S_1 = -11.322 \text{ J/K mol} \\ \Delta S_2 = 5.231 \text{ J/K mol} \end{cases}$$

With these values, the entropy generation is
$$S_\text{gen}^\text{tot} = 1153.64 \text{ J/K}.$$

Example 9.2: Separation of Air

Calculate the minimum work required to separate air to its pure constituents at 1 bar, 300 K. The approximate composition of air by volume is: 78.09% nitrogen, 20.95% oxygen, 0.933% argon, and 0.03% carbon dioxide.

Solution Treating air at 1 bar, 300 K as an ideal gas, the enthalpy of mixing is zero. The entropy generation then is
$$S_\text{gen} = \Delta S_\text{mix}^\text{igm} = -R \sum_i x_i \ln x_i.$$

With $x_1 = 0.7809$, $x_2 = 0.2095$, $x_3 = 0.0093$ and $x_4 = 0.0003$, the entropy of mixing is

$$\Delta S_{\text{mix}} = 0.566625 R = 4.71 \text{ J/mol K}.$$

Comments Mixing is a spontaneous process and generates entropy. Accordingly, the separation of the mixture to its constituents is not a spontaneous process but requires work. The minimum amount of work to separate air into its components is

$$W = T_0 S_{\text{gen}} = (300 \text{ K})(0.566625 R) = 1413.28 \text{ J/mol}.$$

9.6 Equations of State for Mixtures

As with pure substances, the properties of mixtures can be tabulated for future reference. However, such tabulations quickly become impractical as the number of components increase. It is therefore of great practical value to be able to obtain the properties of mixture by calculation from an equation of state, as done for pure fluids. In general, equations of state developed for pure fluids may be extended to mixtures provided that composition is properly accounted for. Cubic equations such as the Soave-Redlich-Kwong and the Peng-Robinson equation require two parameters, a and b (see Chapter 2). For mixtures, these parameters are typically calculated from the corresponding parameters of the individual components. In this approach, the effect of composition is reflected in the mixture parameters a and b. Once these parameters have been determined, the calculation is identical to that for pure components. This includes the calculation of the compressibility factor, molar volume, residual enthalpy, and residual entropy.

Equations that prescribe how to obtain the mixture parameters a and b from those of the pure components are known as *mixing rules*. The following mixing rules are commonly used with cubic equations of state:

$$b = \sum_i x_i b_i, \tag{9.38}$$

$$a = \sum_i \sum_j x_i x_j (1 - k_{ij}) \sqrt{a_i a_j}, \tag{9.39}$$

with the summations going over all components of the mixture. Here a_i and b_i, are the parameters of the pure component i, x_i, $i = 1, \cdots, N$ are the mole fractions of the components, and k_{ij} is symmetric function of i and j with the properties,

$$k_{ij} = k_{ji}, \quad k_{ii} = 0.$$

This parameter reflects interactions between components but it should be treated as an adjustable parameter whose purpose is to improve the accuracy of the calculation.

Residual Properties The residual properties of mixture are given by the same equations as for pure components (see Table 5-3). The calculation also requires the derivative of the mixture parameter a with respect to temperature. This is obtained by differentiation of the mixing rule in eq. (9.39) and is given by

$$\frac{da}{dT} = \frac{1}{2} \sum_i \sum_j x_i x_j (1 - k_{ij}) \sqrt{a_i a_j} \left[\frac{1}{a_i} \frac{da_i}{dT} + \frac{1}{a_j} \frac{da_j}{dT} \right]. \tag{9.40}$$

where da_i/dT is the derivative of the parameter a_i of the pure component i.

Binary Mixture The equations given here apply to mixtures with any number of components. For a binary mixture in particular, they reduce to the following form:

$$b = x_1 b_1 + x_2 b_2, \tag{9.41}$$

$$a = x_1^2 a_1 + x_2^2 a_2 + 2 x_1 x_2 (1 - k_{12}) \sqrt{a_1 a_2}, \tag{9.42}$$

$$\frac{da}{dT} = x_1^2 \left(\frac{da_1}{dT} \right) + x_2^2 \left(\frac{da_2}{dT} \right)$$

$$+ x_1 x_2 (1 - k_{12}) \sqrt{a_1 a_2} \left(\frac{1}{a_1} \frac{da_1}{dT} + \frac{1}{a_2} \frac{da_2}{dT} \right). \tag{9.43}$$

The calculation of the mixture parameters adds one layer of computations to the overall calculation. Once the mixture values of a, b, and da/dT have been determined, the calculation of the compressibility factor, residual enthalpy, and residual entropy, proceeds in exactly the same way as for pure components. The procedure goes like this:

1. Calculate the pure component parameters a_i, b_i, and da_i/dT using the critical constants and acentric factor of the pure components. These equations were given in Chapter 2 and depend on the equation of state.

2. Calculate the mixture parameters using the mixing rules.

3. Calculate the dimensionless parameters A' and B', of the polynomial equation for Z:

$$A' = \frac{aP}{(RT)^2}, \qquad B' = \frac{bP}{RT}. \tag{2.36}$$

4. Solve the cubic polynomial for the compressibility factor. If there are more than one positive roots, choose according to the phase of the mixture:[12] for a liquid mixture choose the smallest root; for gas mixture, choose the largest.

5. Use the compressibility factor to calculate the residual enthalpy and entropy according to the equations in Table 5-3.

In working with eqs. (9.38)–(9.43) we follow the convention: a single subscript refers to pure component, whereas no subscript refers to *mixture*.

9.7 Mixture Properties from Equations of State

Residual properties are defined in the same manner as for pure components:

$$F = F^{\text{igm}} + F^R, \tag{9.44}$$

where F is the molar property of the mixture at a given state, F^{igm} is the same property in the hypothetical ideal-gas state, and F^R is the residual property. All terms in eq. (9.44) are at the same pressure, temperature, and composition. For the enthalpy of the mixture, this equation becomes

$$\boxed{H = \sum_i x_i H_i^{\text{ig}} + H^R.} \tag{9.45}$$

Here we have substituted for H^{igm} the expression from eq. (9.26), which gives the ideal-gas enthalpy of the mixture in terms of the ideal-gas enthalpy of the pure components. For entropy we obtain a similar result,

$$\boxed{S = \sum_i x_i S_i^{\text{ig}} - R \sum_i x_i \ln x_i + S^R.} \tag{9.46}$$

Here, the ideal-gas entropy of mixture is given by eq. (9.27) and includes the ideal entropy of mixing. Equations (9.45) and (9.46) reduce the calculation of mixture properties into a two-step calculation, one for the ideal-gas contributions, and one for the residual corrections.

To calculate the absolute enthalpy and entropy of mixture we need the absolute properties of the pure components, H_i^{ig} and S_i^{ig}. These must be calculated

---

12. The phase must be known ahead of time. We will learn in Chapter 10 that it is possible to determine the phase from the equation of state itself.

based on a reference state. We must specify one reference state for each component, not necessarily the same for all components, though this choice is convenient. If the reference state for component i is at temperature T_{0i} and pressure P_{0i}, its ideal-gas properties at temperature T and pressure P are

$$H_i^{ig} = H_{0i}^{ig} + \int_{T_{0i}}^{T} C_{P,i}^{ig} \, dT, \tag{9.47}$$

$$S_i^{ig} = S_{0i}^{ig} + \int_{T_{0i}}^{T} C_{P,i}^{ig} \frac{dT}{T} - R \ln \frac{P}{P_{0i}}, \tag{9.48}$$

where H_{i0}^{ig} and S_{i0}^{ig} are the ideal-gas enthalpy and entropy, respectively, of the component at its reference state. The value of these constants depends on the reference state adopted:

- *Actual enthalpy and entropy at P_0, T_0, are set to zero.* For this reference state we use,

$$H_{0i}^{ig} = -H_{0i}^R, \quad S_{0i}^{ig} = -S_{0i}^R, \tag{9.49}$$

where H_{0i}^R and S_{0i}^R are the residual properties of pure component at its reference state.[13]

- *Ideal-gas enthalpy and entropy at P_0, T_0, are set to zero.* In this case,

$$H_{0i}^{ig} = 0, \quad S_{0i}^{ig} = 0. \tag{9.50}$$

Once the reference state for each component is selected, H_{0i}^{ig} and S_{0i}^{ig} become numerical constants and the calculation proceeds as usual. Table 9-2 summarizes all the equations needed for the calculation of the enthalpy and entropy of a binary mixture based on the Soave-Redlich-Kwong equation of state. The calculation is somewhat involved but completely straightforward. Except for the calculation of the ideal-gas integrals, everything else involves serial algebraic calculations. The entire procedure can be coded easily for computer calculation. The required inputs are the critical properties, the ideal-gas heat capacities as functions of temperature, and the numerical values of the reference state values H_{0i}^{ig} and S_{0i}^{ig} of all components. The calculation should then be set up so that enthalpy and entropy are calculated for any user input of temperature, pressure, mol fractions and phase of the mixture. The phase is needed in order to select the proper root of the compressibility equation, if more than one real root is found.

---

13. Using $H_i = H_i^{ig} + H_i^R$ you should be able to confirm that this above reference state leads to $H_0 = 0$ and $S_0 = 0$.

9.7 Mixture Properties from Equations of State

Table 9-2: Properties of binary mixture using the Soave-Redlich-Kwong equation of state. Subscript i refers to pure component 1 or 2; the values of H_{0i}^{ig} and S_{0i}^{ig} are fixed by the reference state.

$H = x_1 H_1^{ig} + x_2 H_2^{ig} + H^R$	[9.45]
$S = x_1 S_1^{ig} + x_2 S_2^{ig} - R(x_1 \ln x_1 + x_2 \ln x_2) + S^R$	[9.46]
$H_i^{ig} = H_{0i}^{ig} + \int_{T_{0i}}^T C_{P,i}^{ig} dT$	[9.47]
$S_i^{ig} = S_{0i}^{ig} + \int_{T_{0i}}^T C_{P,i}^{ig} \dfrac{dT}{T} - R \ln \dfrac{P}{P_{0i}}$	[9.48]
$H^R = RT(Z-1) + \dfrac{T(da/dT) - a}{b} \ln \dfrac{Z+B'}{Z}$	[5.55]
$S^R = R \ln(Z - B') + \dfrac{da/dT}{b} \ln \dfrac{Z+B'}{Z}$	[5.56]
$Z^3 - Z^2 + (A' - B' - B'^2)Z - A'B' = 0$	[2.44]
$A' = \dfrac{aP}{(RT)^2}, \quad B' = \dfrac{bP}{RT}$	[2.36]
$b = x_1 b_1 + x_2 b_2$	[9.41]
$a = x_1^2 a_1 + x_2^2 a_2 + 2 x_1 x_2 (1 - k_{12}) \sqrt{a_1 a_2}$	[9.42]
$\dfrac{da}{dT} = x_1^2 \left(\dfrac{da_1}{dT}\right) + x_2^2 \left(\dfrac{da_2}{dT}\right) + x_1 x_2 (1 - k_{12}) \sqrt{a_1 a_2} \left(\dfrac{1}{a_1} \dfrac{da_1}{dT} + \dfrac{1}{a_2} \dfrac{da_2}{dT}\right)$	[9.43]
$b_i = 0.08664 \dfrac{RT_{ci}}{P_{ci}}$	[2.41]
$a_i = 0.42748 \dfrac{R^2 T_{ci}^2}{P_{ci}} \left[1 + \Omega_i \left(1 - \sqrt{T_{ri}}\right)\right]^2$	[2.42]
$\dfrac{da_i}{dT} = -0.42748 \dfrac{R^2 T_{ci}}{P_{ci}} \dfrac{\left(1 + \Omega_i \left(1 - \sqrt{T_{ri}}\right)\right) \Omega_i}{\sqrt{T_{ri}}}$	[5.57]
$\Omega_i = 0.480 + 1.574 \omega_i - 0.176 \omega_i^2$	[2.43]

Example 9.3: Properties of Real Mixture

Carbon dioxide and normal pentane at 4.5 °C, 16.15 bar form a liquid solution. Calculate the enthalpy and entropy of a solution of the two components that contains 32% by mole carbon dioxide. Use the SRK equation with $k_{12} = 0.12$. The reference state for both components is the ideal-gas state at 25 °C, 1 bar.

Solution We collect the critical parameters of the two components:

	T_c (K)	P_c (bar)	ω
CO_2 (1) :	304.2	73.74	0.225
Pentane (2) :	469.7	33.7	0.252

The ideal-gas heat capacities are both given by the polynomial expression

$$C_{P,i}^{ig}/R = a_0 + a_1 T + a_2 T^2 + a_3 T^3 + a_4 T^4,$$

with T in kelvin and the constants given by

	a_0	a_1	a_2	a_3	a_4
CO_2	3.259	0.001356	0.00001502	-2.374×10^{-8}	1.056×10^{-11}
Pentane	7.554	-0.000368	0.00011846	-1.4939×10^{-7}	5.753×10^{-11}

All equations are given in Table 9-2. Below we summarize the results.

Calculation of residual properties.

The parameters a, b, and da/dT of the pure component and of the mixture are:

	a (J m³/mol²)	b (m³/mol)	da/dT (J m³/mol² K)
CO_2	0.39863	2.97156×10^{-5}	-1.09172×10^{-3}
Pentane	2.78581	1.00397×10^{-4}	-5.56343×10^{-3}
Mixture	1.73256	7.77787×10^{-5}	-3.63996×10^{-3}

The parameters A and B of he mixture are

$$A = 0.525105, \quad B = 0.0544159.$$

The cubic equation for the compressibility factor is

$$Z^3 - Z^2 + 0.467728 Z - 0.028574 = 0,$$

and has one real root, which we accept as the compressibility factor of the solution:

$$Z = 0.0711422.$$

9.7 Mixture Properties from Equations of State

The residual enthalpy and entropy are

$$H^R = -22180.2 \text{ J/mol},$$

$$S^R = -60.5966 \text{ J/mol K}.$$

Calculation of ideal-gas properties.

The reference state is the ideal-gas state, therefore, H_{0i}^{ig} and S_{0i}^{ig} are zero for both components. With $T_0 = 298.15$ K, $P_0 = 1$ bar for both components, we obtain the following values for the ideal-gas enthalpy of pure component at $T = 277.65$ K, $P = 16.15$ bar:

$$H_1^{ig} = -749.987 \text{ J/mol}, \quad H_2^{ig} = -2402.78 \text{ J/mol}.$$

The ideal-gas enthalpy of the mixture is

$$H^{igm} = (0.32)(-749.987) + (0.68)(-2402.78) = -1873.89 \text{ J/mol}.$$

The ideal-gas entropy of pure components is

$$S_1^{ig} = -25.7346 \text{ J/mol K}, \quad S_2^{ig} = -31.4757 \text{ J/mol K}.$$

With these values we calculate the ideal-gas entropy of the mixture:

$$S^{igm} = (0.32)(-25.7346) + (0.68)(-31.4757) - (8.314)((0.32)\ln(0.32) + (0.68)\ln(0.68))$$
$$= -24.4268 \text{ J/mol K}.$$

Calculation of actual properties.

The actual properties of the mixture are calculated by adding the ideal gas and residual contributions:

$$H = (-1873.89) + (-22180.2) = -24054.1 \text{ J/mol},$$

$$S = (-24.4268) + (-60.5966) = -85.02 \text{ J/mol K}.$$

Example 9.4: Enthalpy of Mixing
Consider the mixing of carbon dioxide and normal pentane at 4.5 °C, 16.15 bar to form a solution that contains 32% by mol. Is this process endothermic or exothermic? Additional data: At this pressure and temperature pure carbon dioxide is vapor, and pure pentane is liquid.

Solution To answer the question we must calculate the enthalpy of mixing:

$$\Delta H_{\text{mix}} = H - x_1 H_1 - x_2 H_2,$$

where H is the enthalpy of the mixture at 4.5 °C, 16.15 bar, $x_1 = 0.32$, and H_1, H_2, are the enthalpies of the pure components at the same pressure and temperature. The enthalpy of mixture was calculated in the previous example; all that is needed is the calculation of the enthalpies of the pure components. These may be calculated by the same equations, with $x_1 = 1$ to obtain H_1, and $x_1 = 0$ to obtain H_2 (the same reference states must be used in these calculations as in the previous example). This calculation is left as an exercise. Instead, we will use an equivalent approach that requires fewer calculations.

We express the enthalpy of the mixture using the residual enthalpy:

$$H = x_1 H_1^{\text{ig}} + x_2 H_2^{\text{ig}} + H^R.$$

Similarly, the enthalpies of the pure components are

$$H_1 = H_1^{\text{ig}} + H_1^R,$$

$$H_2 = H_2^{\text{ig}} + H_2^R.$$

Substituting these expressions into the equation for ΔH_{mix}, the ideal-gas terms cancel out and the result is

$$\Delta H_{\text{mix}} = H^R - x_1 H_1^R - x_2 H_2^R,$$

where H^R is the residual enthalpy of the mixture, and H_1^R, H_2^R, are the residual enthalpies of the pure components at the same pressure and temperature. Therefore, only the residual properties of the pure components need to be calculated. Leaving the details of the calculation as an exercise, the final results are:

$$H^R = -22180.2 \text{ J/mol},$$

$$H_1^R = -795.0 \text{ J/mol},$$

$$H_2^R = -27892.9 \text{ J/mol}.$$

The enthalpy of mixing is

$$\Delta H_{\text{mix}} = (-22180.2) - (0.32)(-795.0) - (0.68)(-27892.9) = -2958.63 \text{ J/mol}.$$

The enthalpy of mixing is negative, therefore, this process is endothermic.

Comments In selecting the compressibility factor, the phase of the pure components must be taken into consideration: For carbon dioxide we select the largest root and for pentane the smallest.

9.7 Mixture Properties from Equations of State

Example 9.5: Compression of Real Mixture
A mixture of carbon dioxide (75% by mol) and pentane (25% by mol) is compressed from 1 bar, 250 K to 20 bar. Determine the work for compression the final temperature and the entropy generation, if the efficiency of the compressor is 80%. Use the SRK equation with $k_{12} = 0.12$.

Solution The procedure for the calculation of the compressor work is the same as for pure fluids (see Chapter 5). The calculation will be done in two steps, one for reversible operation and one for the actual process. The enthalpy and entropy of the mixture will be calculated as in Example 9.3, using the same reference states as in that example. First, we calculate the properties at the inlet state. These are summarized below:

$$T_1 = \text{K}, \quad P_1 = 1 \text{ bar}, \quad H_1 = -2750.35 \text{ kJ/kg}, \quad S_1 = -5.26747 \text{ kJ/kg K}.$$

For the subsequent calculations it will be useful to tabulate the properties of the compressed stream at various temperatures. This is done below.

T (K)	P (bar)	H (J/mol)	S (J/mol K)
375	20.	3692.95	−8.12694
400	20.	5517.11	−3.41831
425	20.	7383.45	1.10696
450	20.	9297.33	5.48221
475	20.	11261.5	9.72966

Reversible operation. For reversible operation, the exit state is defined by the conditions $P_2 = 20$ bar, $S_{2'} = S_1 = 3.92219$ J/mol K. By interpolation in the above table we find,

$$T_{2'} = 390.18, \quad H_{2'} = 4800.73 \text{ J/mol}.$$

Actual operation. The enthalpy of the actual outlet is

$$H_2 = H_1 + \frac{H_{2'} - H_1}{\eta} = (-2750.35) + \frac{4800.73 - (-2750.35)}{0.8} = 6688.5 \text{ J/mol}.$$

The work for compression is

$$W = H_2 - H_1 = (6688.5) - (-2750.35) = 9438.85 \text{ J/mol}.$$

The exit state is obtained by interpolation at $P_2 = 20$ bar, $H_2 = 6688.5$ J/mol. We find

$$T_2 = 415.7 \text{ K}, \quad S_2 = -0.5781 \text{ J/mol K}.$$

Entropy generation. The entropy generation is

$$S_{\text{gen}} = S_2 - S_1 = 4.689 \text{ J/mol K}.$$

Comments Energy balances involving mixtures are performed in the same manner as for pure fluids. All that is required is a method for the calculation of enthalpy and entropy at given pressure, temperature, and composition.

9.8 Summary

In dealing with multicomponent systems, the new element is composition, which is expressed through the number of moles (or mass) of each component. If composition remains constant during a process, a multicomponent mixture behaves as a pure fluid in all respects except when multiple phases are present. This observation makes the extension of the theoretical relationships from pure components to mixtures straightforward, though algebraically more involved. Of practical importance is the application of cubic equations to mixtures, which makes possible the calculation of volume, enthalpy and entropy of systems with several components. The new element is the introduction of mixing rules, which are recipes for the calculation of the parameters of the mixture EOS from those of the pure components. Like the equations themselves, mixing rules are *empirical*. The mathematical form of mixing rules is guided by theoretical considerations but ultimately these equations are based on trial and error.

Equations of state such as the Soave-Redlich-Kwong and the Peng-Robinson, are appropriate for like, nonpolar components. Components that interact strongly cannot be represented accurately by these equations. It is important then to establish whether a given equation of state is appropriate for the system of interest through validation of the results against known data. If the answer is yes, then the calculation of mixture properties follows the same procedure as for pure components.

9.9 Problems

Problem 9.1: A stream that contains a mixture of methane (25% by mol) and carbon monoxide is compressed from 1 bar, 35 °C to 12 bar. The compressor efficiency is 90%. Treating the mixture as an ideal gas, calculate the required work.
Additional data: $C_{PA} = 40.8$ J/mol K, $C_{PB} = 29.4$ J/mol K ($A =$ CH$_4$, $B =$ CO).

Problem 9.2: Air is compressed from 1 bar, 25°C to 50 bar, and subsequently is cooled to 300 K using cooling water. Assuming air to be an ideal-gas mixture and the compressor to be 100% efficient, calculate the work in the compressor and the heat removed in the heat exchanger per mol of air. Assume air to be ideal-gas mixture (79% N$_2$, 21% O$_2$) and use the heat capacities given in Perry's *Handbook*.

Problem 9.3: Streams A and B are mixed adiabatically to produce stream C. Stream A is at 1 bar, 50 °C, it contains pure methane and its flow rate is 0.2 mol/min. Stream B is at 1 bar, 100 °C, contains a methane-ethane mixture 50% in methane (by mol) and its flow rate is 0.8 mol/min. Stream C exits at 1 bar.

9.9 Problems

a) What is the composition in the final mixture?
b) What is the temperature and pressure at the end of the process?
c) Calculate the entropy generation.
d) What is the change in the total Gibbs energy of the system?
e) How much work is needed to separate the final stream into the two streams we had before mixing, at their respective temperature and pressure? Assume that the temperature of the surroundings is 20 °C.
f) What is the irreversible feature of this process?
Additional data: The ideal-gas C_P's of the pure components are: ethane: 10.3 R; methane: 6.2R; where R is the ideal-gas constant.

Problem 9.4: An insulated cylinder is divided by a partition into two compartments. Compartment A contains 0.2 mol of pure methane at 50 °C, 1 bar. Compartment B contains 0.8 mol of a methane-ethane mixture at 100 °C, 1 bar with $y_{\text{methane}} = 0.5$. The partition is removed and the system reaches equilibrium.
a) What is the composition of the final mixture?
b) What is the temperature and pressure at the end of the process?
c) Calculate the entropy generation.
d) What is the change in the total Gibbs energy of the system?
e) How much work is needed to separate the final mixture into A and B at the conditions before mixing? Assume that the temperature of the surroundings is 20 °C.
f) Which of the above answer would change (and by how much), if both compartments contained methane? All other conditions are the same.
Additional data: The ideal-gas C_P's of the pure components are: ethane: 10.3 R; methane: 6.2R; where R is the ideal-gas constant.

Problem 9.5: In Figure 9-3, all streams are mixtures of methane/carbon monoxide with molar compositions indicated on the flow chart. Calculate the

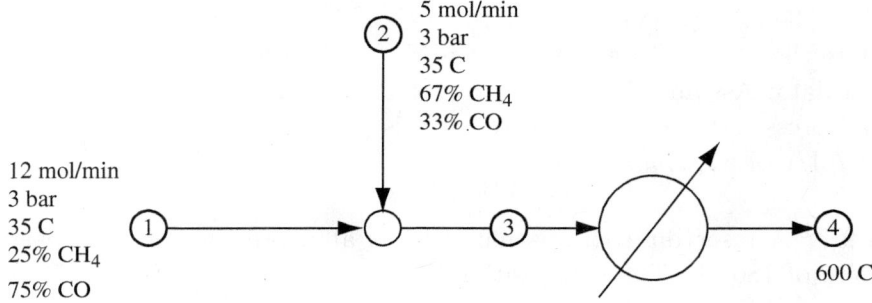

Figure 9-3: Flow chart for Problem 9.5.

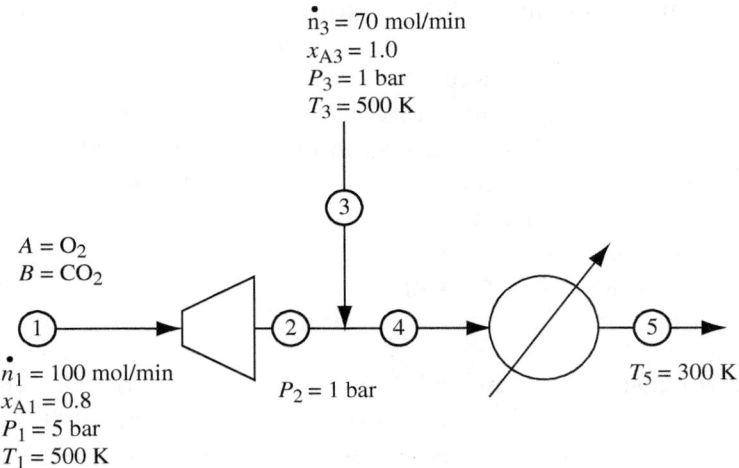

Figure 9-4: Flow chart for Problem 9.6.

material and energy balances and perform an entropy analysis of the mixing point, the heat exchanger, and of the entire process. The heat capacities of methane and carbon monoxide are $C_{PA} = 40.8$ J/mol K, and $C_{PB} = 29.4$ J/mol K, respectively.

Problem 9.6: A mixture of O_2/CO_2 (stream 1) expands in turbine (see Figure 9-4). The outlet (stream 2) is mixed with a stream of pure oxygen and the new stream (number 4) passes through a heat exchanger and exits as stream 5. The efficiency of the turbine is 80%. There are no heat losses and pressure drops in pipes and in the heat exchanger can be ignored.
a) Calculate the enthalpy and entropy of all streams. For the enthalpy and entropy of the pure components the reference state is at 5 bar, 500 K.
b) Calculate the work in the turbine, and the heat load in the heat exchanger.
c) Calculate the entropy generation.
d) Calculate the ideal work and the lost work for the process.
Additional data: Assume the components of their mixtures to be at the ideal-gas state at all pressures and temperatures of this problem with ideal-heat capacities, $C_{PA} = 44.7$ J/mol K, $C_{PB} = 31.1$ J/mol K.

Problem 9.7: A reservoir contains natural gas at a pressure of 5000 psig and a temperature of 180 °F. The composition of the gas is 80% methane, 15% ethane, and 5% CO_2 (all fractions are by mole). Estimate the temperature of the gas when it emerges at the surface under the following alternative assumptions:

9.9 Problems

a) The flow to the surface is assumed to be reversible adiabatic expansion.
b) The flow to the surface is assumed to be a throttling process.
Take the natural gas to be an ideal-gas mixture with temperature-dependent heat capacities for each component given in Perry's *Handbook*. Is the ideal-gas assumption reasonable?

Problem 9.8: A flow process receives two streams (A and B) that contain mixtures of methane and propane and produce a stream of pure methane (C) and a stream of pure propane (D). All streams may be considered ideal-gas mixtures with heat capacities, 35 and 53 J/mol K, for methane and ethane, respectively. Information about the system is given in the table below.
a) Fill out the missing entries on this table. For the enthalpy and entropy, use as reference state the pure components in the ideal-gas state at 1 bar, 300 K.
b) Calculate the ideal work for this process.

	A	B	C	D
Molar flow rate of methane (mol/min)	10		80	0
Molar flow rate of ethane (mol/min)		20	0	75
Mole fraction of methane				
Mole fraction of ethane				
P (bar)	5	10	2	1
T (K)	250	600	300	350
H (kJ/min)				
S (kJ/K/min)				

Problem 9.9: A process stream contains a mixture of argon (1) and ethane (2) with the following properties: $P = 50$ bar, $T = -123$ °C, $x_1 = 0.75$ (by mole). Under these conditions the residual properties of the stream are, $H^R = -7211$ J/mol, $S^R = -36$ J/mol K.
a) Calculate the enthalpy of the mixture.
b) Calculate the entropy of the mixture.
c) This stream is to be separated into its pure components in a flow process from which the two pure streams exit at 1 bar, 25 °C. Assuming the surroundings to be at 25 °C, what is the ideal work for this process? Is this work produced or absorbed by the process?
Additional information: The reference state for both components is the ideal-gas state at $P_{\text{ref}} = 1$ bar, $T_{\text{ref}} = -123$ °C. The ideal-gas heat capacities of the pure components are $C_P^{\text{ig}}(\text{argon}) = 3.5R$, $C_P^{\text{ig}}(\text{ethane}) = 5R$ (where R is the ideal-gas constant) and may be assumed independent of temperature. At 25 °C, 1 bar, argon, and ethane may be assumed to be in the ideal-gas state.

Problem 9.10: A flow process is used to split a mixture of carbon monoxide (1) and ethane (2) (stream A) into two streams, B and C. Information about the inlet and outlet streams is shown in the table below.
a) Fill out the table.
b) Calculate the entropy change for this process.

	A (inlet)	B (outlet)	C (outlet)
Molar flow (mol/s)	100		
Mol fraction of CO	0.5	0.1	0.9
P (bar)	100	50	1
T (°C)	25	0	25
S_1^{ig} (J/mol K)			
S_2^{ig} (J/mol K)			
S^R (J/mol K)	-3.1	-30	0
S (J/mol K)			

Reference states for both components: ideal-gas state at 50 bar, 25 °C. Ideal-gas heat capacities: $C_{P1}^{ig} = 30$ J/mol K; $C_{P2}^{ig} = 60$ J/mol K.

Problem 9.11: A batch of natural gas contains 93% by mol methane with the rest being mostly ethane. Use the SRK equation to estimate the volume of tank needed to store 100 kg of this natural gas in liquid form at 1 bar, at -165 °C.

Problem 9.12: a) Use the SRK equation to calculate the residual properties of a $N_2(1)/CH_4(2)$ mixture at 120 K, 10 bar, with $x_1 = 0.5$. Under the conditions, the mixture is in the liquid phase.
b) Calculate the enthalpy and entropy of the mixture in the previous part using as reference for both components the ideal-gas state at 25°C, 1 bar. The ideal-gas heat capacities of the two components are:

$$C_{P1} = 29 + 0.0005768\,T, \quad C_{P2} = 24 + 0.03926\,T$$

with T in K and C_P in J/mol/K.

Problem 9.13: In an air-liquefaction process, air is compressed from 1 bar, 25 °C, to 180 bar. To avoid overheating, compression is done in stages so that the exit temperature does not exceed 200 °C. The stream exiting a stage is cooled to 25 °C before it is led to the next stage. If the efficiency of each stage is 80%, determine the number of stages, and the total amount of compression work.

Problem 9.14: Use the SRK equation to do the following calculations:
a) The enthalpy (J/mol) of a vapor mixture at 1.2 bar, 130 °C, that contains 25% (by mol) n-hexane in n-octane.

9.9 Problems

b) The enthalpy of a liquid solution at 1.2 bar 90 °C that contains 25% (by mol) n-hexane in n-octane.

c) A mixture of n-hexane (25% by mole) in n-octane, initially at 1.2 bar, 130 °C, passes through a heat exchanger and exits as liquid at 90 °C. Assuming no pressure drop in the heat exchanger, calculate the amount of heat removed.

Additional data: Assume for simplicity that the ideal-gas heat capacities are constant: n-hexane = 176 J/mol K; n-octane = 247 J/mol K.

The reference state for both pure components is the ideal-gas state at 1 bar, 300 K.

Chapter 10

Theory of Vapor-Liquid Equilibrium

When a fluid is brought into the vapor-liquid region, it forms two coexisting phases, each with its own molar properties (volume, enthalpy, entropy, etc.). If the fluid is a mixture of components, then each phase also has its own composition. This fundamental property of mixtures is the basis of separation processes. A major goal of chemical engineering thermodynamics is to provide computational methodologies for the calculation of phase diagrams in systems with many components. This requires the determination of the precise conditions that lead to phase separation, the number of phases that form, and their composition. The thermodynamic property that holds the answers to these problems is the Gibbs free energy.

In this chapter we will learn how to:

- Develop the fundamental equations of phase equilibrium in terms of the chemical potential and fugacity.
- Calculate fugacity from an equation of state.
- Calculate vapor-liquid equilibrium of a mixture using the equation of state.

10.1 Gibbs Free Energy of Mixture

In Chapter 4 we determined that the equilibrium state of a closed system at constant temperature and pressure corresponds to conditions that minimize the Gibbs free energy:

$$\left(\Delta G \leq 0\right)_{T,P,n}. \qquad [4.45]$$

For a mixture that contains n_i moles of component i with $i = 1, 2, \cdots N$, this inequality remains valid provided that the moles of all components are held constant:[1]

$$\boxed{\left(\Delta G \leq 0\right)_{T,P,n_i}.} \qquad (10.1)$$

---

1. In our notational convention, subscript n_i indicates that all n_i, with $i = 1, 2, \cdots, N$, are held constant.

Therefore,

> *the equilibrium state of a multicomponent system whose temperature, pressure, and number of moles of all components are fixed, is a state that minimizes the Gibbs free energy.*

This is a statement of fundamental importance in thermodynamics and the starting point for the discussion of phase equilibrium. To apply it we must first develop expressions for the Gibbs free energy of multicomponent systems. We begin by writing the differential of G in the standard form,

$$dG^{\text{tot}} = \left(\frac{\partial G^{\text{tot}}}{\partial T}\right)_{P,n_i} dT + \left(\frac{\partial G^{\text{tot}}}{\partial P}\right)_{T,n_i} dP + \sum_i \left(\frac{\partial G^{\text{tot}}}{\partial n_i}\right)_{P,T,n_j} dn_i. \quad (10.2)$$

The derivatives with respect to temperature and pressure were obtained previously in Chapter 5:

$$\left(\frac{\partial G^{\text{tot}}}{\partial T}\right)_{P,n_i} = -S^{\text{tot}}, \quad \left(\frac{\partial G^{\text{tot}}}{\partial P}\right)_{T,n_i} = V^{\text{tot}}. \quad [5.18]$$

These were originally derived for a pure component, but they apply to *all systems of constant composition*, of which the pure component is a special case.[2] The partial derivatives with respect to n_i on the right-hand side of eq. (10.2) are the partial molar Gibbs free energies ($\overline{G}_i$) with respect to each component. The partial molar Gibbs free energy is a very important property. For this reason it is given a special name, *chemical potential*, and a special symbol, μ_i:

$$\boxed{\mu_i \equiv \overline{G}_i = \left(\frac{\partial G^{\text{tot}}}{\partial n_i}\right)_{P,T,n_j}.} \quad (10.3)$$

Using this definition and eqs. (5.18), the differential of the Gibbs free energy of mixture is written in the more compact form,

$$\boxed{dG^{\text{tot}} = -S^{\text{tot}} dT + V^{\text{tot}} dP + \sum_i \mu_i dn_i.} \quad (10.4)$$

This is the fundamental equation for the Gibbs free energy of mixture in terms of temperature, pressure, and number of moles of all components. An alternative form of this equation, which will be useful in subsequent developments, is obtained by developing an expression for the differential of the ratio G^{tot}/RT, which is dimensionless. We start by calculating the differential of this ratio:

$$d\left(\frac{G^{\text{tot}}}{RT}\right) = \frac{dG^{\text{tot}}}{RT} - \frac{G^{\text{tot}}}{RT^2} dT.$$

---
2. See also relevant discussion in Chapter 9.

10.1 Gibbs Free Energy of Mixture

For dG^{tot} we substitute eq. (10.4), for G^{tot} we substitute $H^{\text{tot}} - TS^{\text{tot}}$, and with some straightforward manipulation the result is

$$d\left(\frac{G^{\text{tot}}}{RT}\right) = -\frac{H^{\text{tot}}}{RT^2}dT + \frac{V^{\text{tot}}}{RT}dP + \sum_i \frac{\mu_i}{RT}dn_i. \quad (10.5)$$

This expression is equivalent to eq. (10.4). It has the advantage that it is in dimensionless form and involves the enthalpy, rather than the entropy of the mixture.

Multicomponent Equilibrium

When a mixture is brought into the two-phase region, it splits into two phases, each with its own composition. The situations is shown schematically in Figure 10-1. On a microscopic basis, molecules of both components pass continuously from one phase to the other. It is this molecular transfer that allows the system to find and maintain its equilibrium composition. When this composition is reached, the net transfer between phases is zero. Molecules continue to cross the interface in such way that the rate of transfer in one direction matches the rate in the reverse direction, so that the average (macroscopic) composition of each phase remains constant. We analyze this situation as follows. We form a mixture by mixing n_1 moles of component 1 with n_2 mols of component 2, and fix the temperature and pressure of the system. Next, we take an arbitrary number n_1^V moles of component 1 and n_2^V moles of component 2 and place them in the vapor phase, with the remaining moles placed in the liquid. Out of the infinitely many ways that the two components can be partitioned between the two phases, only one corresponds to the true equilibrium state. To identify the equilibrium state we apply the Gibbs inequality in eq. (10.1): since the Gibbs free energy must be at a minimum, if we transfer a small amount δn_i of component i from one phase to the other, the Gibbs energy must remain unchanged. That's because this change takes place at the bottom of the Gibbs curve (see Figure 4-7) where its derivative is flat. Suppose that we transfer δn_i mole of component i from the liquid to the vapor. The change in the Gibbs energy of the liquid is $-\mu_i^L \delta n_i$

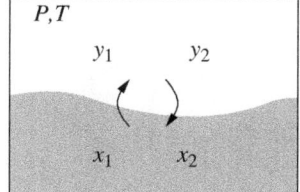 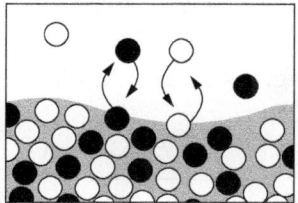

Figure 10-1: Macroscopic and molecular view of vapor-liquid equilibrium.

while the change in the vapor is $+\mu_i^V \delta n_i$. For the total Gibbs energy to remain constant, we must then have

$$\boxed{\mu_i^L = \mu_i^V.} \tag{10.6}$$

This expresses a very important result: *in equilibrium between phases, the chemical potential of each component is the same in both phases.*

Gibbs's Phase Rule

Equation (10.6) is a general criterion that applies not just to vapor-liquid systems, but to systems with any number of phases. As an example, at the triple point of a pure substance where liquid, vapor, and solid coexist, the chemical potential is the same in all three phases:

$$\mu^S = \mu^L = \mu^V.$$

A pure substance can simultaneously exist in three phases at most. With mixtures it is possible to observe coexistence of more phases. For example, two partially miscible liquids form two separate liquid phases. At its boiling point, such system consists of three phases, two liquids plus vapor; if a solid is added at amounts that exceed its solubility limit in both phases, the undissolved solid constitutes an additional phase. In general, by increasing the number of components we gain more degrees of freedom and it is possible to form more coexisting phases. A systematic way to count the degrees of freedom is provided by Gibbs's phase rule.

We consider a system with N components distributed among π phases. To specify the intensive properties of this system we need to know pressure, temperature, and the mol fractions in each phase. There are $N-1$ mole fractions in each phase,[3] or $(N-1)\pi$ in total. Along with pressure and temperature, the total number of unknowns is

$$2 + \pi(N-1).$$

The equations that are available to solve for these unknowns are given by the equilibrium condition: for each component i, the equilibrium criterion demands equality of the chemical potentials in each phase. Thus we have the following equations

$$\mu_i^{\text{phase 1}} = \mu_i^{\text{phase 2}} = \cdots = \mu_i^{\text{phase } \pi}; \quad (i = 1, 2, \cdots N).$$

This gives $\pi - 1$ equations[4] for each component, or $N(\pi - 1)$ equations in total. The number of degrees of freedom is the difference between the unknowns and the equations available:

$$\mathcal{F} = 2 + \pi(N-1) - N(\pi - 1),$$

---

3. The Nth mole fraction is obtained from the normalization condition, $x_1 + x_2 + \cdots = 1$.
4. In a series of equalities of the form $a = b = \cdots = z$, in which each term appears only once, the number of independent equations is equal to the number of equal signs.

10.2 Chemical Potential

or
$$\mathcal{F} = 2 + N - \pi. \qquad (10.7)$$

This equation is known as Gibbs's phase rule. One application of the phase rule is to determine the number of variables that must be specified to fully define the state of the system. For a single-phase pure component ($N = 1$, $\pi = 1$) the phase rule gives $\mathcal{F} = 2$, namely, that to fix the state we must specify two variables (pressure and temperature, or some other combination). Increasing the number of phases decreases the degrees of freedom. For a two-phase pure substance, only one variable is needed; indeed at saturation we may specify the pressure of the system or its temperature, but not both, since knowing one fixes the other. Another application of the phase rule is to determine the maximum number of phases that are possible. This corresponds to setting the degrees of freedom to zero (the degrees of freedom cannot be negative). For a single component we find $\pi_{\max} = 3$ while a binary mixture, $\pi_{\max} = 4$.

10.2 Chemical Potential

The chemical potential is an important property in phase equilibrium, its importance arising from the equilibrium criterion, eq. (10.6). As a partial molar property, it satisfies eqs. (9.10) and (9.11), which now become

$$G^{\text{tot}} = \sum_i n_i \mu_i, \qquad (10.8)$$

and

$$G = \sum_i x_i \mu_i. \qquad (10.9)$$

The chemical potential represents the molar contribution of a component to the Gibbs energy of the mixture. For a pure component ($x_i = 1$, $x_{j \neq i} = 0$), the chemical potential is equal to the molar Gibbs energy

$$\mu_i \Big|_{\text{pure}} = G_i = H_i - TS_i. \qquad (10.10)$$

To make equations more readable we will use the symbol G_i to indicate the chemical potential of *pure* component, and μ_i for the chemical potential of *component in a mixture*.

NOTE

Gibbs-Duhem equation revisited

As a partial molar property, the chemical potential obeys the Gibbs-Duhem equation. This equation was derived in Chapter 9 (see eq. 9.12). Applied to the chemical potential it gives

$$\sum_i x_i d\mu_i = 0, \quad (\text{constant } P, T). \qquad (10.11)$$

In a two component mixture this becomes (see eq. 9.13)

$$x_1 \frac{d\mu_1}{dx_1} + x_2 \frac{d\mu_2}{dx_1} = 0.$$

This equation says that the chemical potential of components in a mixture do not change independently of each other with changing composition.

Chemical Potential in Ideal Gas and in Real Mixture

In Chapter 9 we obtained the following equation for the Gibbs free energy of a mixture in the ideal-gas state:

$$G^{\text{igm}} = \sum_i x_i G_i^{\text{ig}} + RT \sum_i x_i \ln x_i. \qquad [9.35]$$

Comparing this with eq. (10.8), we identify the chemical potential μ_i^{igm}, of component i in the ideal-gas mixture as

$$\boxed{\mu_i^{\text{igm}} = G_i^{\text{ig}} + RT \ln x_i,} \qquad (10.12)$$

where G_i^{ig} is the Gibbs free energy (chemical potential) of the pure component in the ideal-gas state.

To obtain the chemical potential of component in a real mixture, we write the Gibbs free energy in terms of the residual Gibbs energy:

$$G = G^{\text{igm}} + G^R,$$

and write the corresponding relationship among the partial molar properties of each term:

$$\mu_i = \mu_i^{\text{igm}} + \overline{G}_i^R, \qquad (10.13)$$

where $\overline{G}_i^R$ is the partial molar residual Gibbs energy,

$$\overline{G}_i^R = \left(\frac{\partial n G^R}{\partial n_i}\right)_{T,P,n_j}. \qquad (10.14)$$

Thus the calculation of the chemical potential is reduced into a calculation of the residual partial molar Gibbs energy.

10.2 Chemical Potential

Example 10.1: Chemical Potential in Air
Calculate the chemical potential of the components in air at 1 bar, 25 °C. Assume the composition of air by volume to be: 78.09% nitrogen, 20.95% oxygen, 0.933% argon, and 0.03% carbon dioxide. The reference state for each component is the pure component at 25 °C, 1 bar.

Solution It is reasonable to treat air and its pure components at 1 bar, 25 °C as ideal gases. Then, the chemical potential of component i is given by eq. (10.12).

$$\mu_i = G_i^{ig} + RT \ln x_i,$$

where G_i^{ig} is the molar Gibbs energy of pure component at 1 bar, 25 °C. The reference state for each pure component specifies $H_i = 0$ and $S_i = 0$. from this we obtain, $G_i = H_i - TS_i = 0$. The chemical potential then simplifies to

$$\mu_i = RT \ln x_i.$$

The results are summarized below:

	x_i	μ_i^{igm} (J/mol)
N_2	0.7809	-613.0
O_2	0.2095	-3874.5
Ar	0.00933	-11587.3
CO_2	0.0003	-20107.5

Let us compare the combined Gibbs energy of the pure components to that of the mixture. For the pure components at 25 °C, 1 bar, it is

$$G_{\text{pure comp.}} = \sum_i x_i G_i = 0.$$

The Gibbs energy of the mixture is

$$G_{\text{air}} = \sum_i x_i \mu_i = -1404.56 \text{ J/mol}.$$

Consider the separation of air (25 °C, 1 bar) into its pure components (at same temperature and pressure). The change in Gibbs energy for this process is

$$G_{\text{pure comp.}} - G_{\text{air}} = (0) - (-1404.56) = 1404.56 \text{ J/mol}.$$

This change is *positive* and takes the system in the opposite direction of equilibrium. Therefore it will not occur spontaneously. It is possible of course to separate air into its pure components but this process requires the input of work.

Exercise: Calculate the amount of work that is needed to separate air into pure oxygen and nitrogen at 25 °C, 1 bar.

Example 10.2: Chemical Potential in Air

Repeat the calculations of the previous example if the reference state for all components is the pure component at $P_0 = 0.5$ bar, $T_0 = 800$ K.

Solution We outline the solution and leave the calculations as an exercise. The chemical potential of component i is
$$\mu_i = G_i^{\text{ig}} + RT \ln x_i.$$

Therefore, the only additional task is the calculation of the Gibbs energy of pure component. To calculate the Gibbs energy of pure component at the temperature and pressure of the mixture, T and P, we first calculate the enthalpy and entropy at this state. To do this we use the ideal-gas equations

$$H_i^{\text{ig}} = H_{0i} + \int_{T_0}^{T} C_{P\,i}^{\text{ig}} \, dT,$$

$$S_i^{\text{ig}} = S_{0i} + \int_{T_0}^{T} C_{P\,i}^{\text{ig}} \frac{dT}{T} - R \ln \frac{P}{P_0}.$$

Both H_{0i} and S_{0i} are equal to zero but we leave them in the equations until numerical substitution. The Gibbs energy of pure component at T, P, is

$$G_i^{\text{ig}} = H_i^{\text{ig}} - TS_i^{\text{ig}} = G_{i0}^{\text{ig}} + \int_{T_0}^{T} C_{P\,i}^{\text{ig}} \, dT - T \int_{T_0}^{T} C_{P\,i}^{\text{ig}} \frac{dT}{T} + RT \ln \frac{P}{P_0}.$$

The only piece of information needed to perform this calculation is the ideal-gas heat capacity as a function of temperature. If the heat capacity may be assumed to be independent of temperature in the interval T_0, to T, the result is simplified to

$$G_i^{\text{ig}} = C_P^{\text{ig}}(T - T_0) - T C_P^{\text{ig}} \ln \frac{T}{T_0} + RT \ln \frac{P}{P_0}.$$

Comments Notice that the value of ΔG_{mix} is not affected by the change of the reference states. The Gibbs energy of the mixture by the new reference state is

$$G_{\text{mixture}} = \sum_i x_i G_i^{\text{ig}} + RT \sum_i x_i \ln x_i,$$

and the Gibbs energy of the pure components at the temperature and pressure of the mixture is

$$G_{\text{pure comp.}} = \sum_i x_i G_i^{\text{ig}}.$$

Upon taking the difference, the G_i^{ig} terms cancel and the result is the familiar Gibbs energy of mixing.

10.3 Fugacity in a Mixture

In Chapter 7 we introduced fugacity and the fugacity coefficient as auxiliary variables related to the Gibbs energy of pure component. Here we extend these definitions to mixtures. The fugacity of component in mixture, f_i, is defined as

$$f_i = x_i \phi_i P, \tag{10.15}$$

where x_i is the mole fraction of component in the mixture, P is pressure, and ϕ_i is the *fugacity coefficient* of component, defined as

$$\ln \phi_i = \frac{\overline{G}_i^R}{RT}, \tag{10.16}$$

where $\overline{G}_i^R$ is the residual partial molar Gibbs energy. These definitions revert to those of the pure component by setting $x_i = 1$, as one can easily verify. The ideal-gas limit of these properties is obtained by noting that the residual partial molar Gibbs energy in the ideal-gas state is zero:

$$\phi_i^{\text{igm}} = 1, \quad f_i^{\text{igm}} = x_i P. \quad \text{(ideal-gas state)} \tag{10.17}$$

The fugacity of component in mixture is closely related to the chemical potential. Given two states of a mixture at the same temperature (but not necessarily at the same pressure or composition), the relationship between the chemical potentials and the fugacities in the two states is (see Example 10.3 for the derivation)

$$\mu_i^A - \mu_i^B = RT \ln \frac{f_i^A}{f_i^B}, \quad \text{(constant } T\text{)}, \tag{10.18}$$

where A and B refer to two states of component i at the same temperature (the derivation is given as an example below). This is an important equation that establishes the relationship between the two properties and shows that fugacity is nothing but a mathematical transformation of the chemical potential. An important consequence of this result is obtained by applying eq. (10.18) to vapor-liquid equilibrium. Taking one state to be the liquid and the other the vapor, both states are at the same temperature, same pressure, but different composition. Since the chemical potential of a component is the same in both phases, eq. (10.18) states that the same is true for the fugacity:

$$f_i^L = f_i^V. \tag{10.19}$$

Therefore, the criterion for phase equilibrium may be expressed either in terms of chemical potentials, or in terms of fugacity. Both approaches are equivalent and all problems in phase equilibrium may be solved using either one. We prefer to work with fugacity because its calculation does not require a reference state and because of its straightforward ideal-gas limit. Using eq. (10.15) to express fugacity in terms of the fugacity coefficient, and noting that pressure is the same in both phases, the equilibrium criterion becomes

$$x_i \phi_i^L = y_i \phi_i^V, \qquad (10.20)$$

where x_i and y_i are the mole fractions of component i in the liquid and the vapor respectively, and ϕ_i^L, ϕ_i^V, are its fugacity coefficients in the two phases. In the special case of pure component ($x_i = y_i = 1$), the fugacity coefficients in each phase are equal. In the general case, the fugacity coefficients are not equal but satisfy eq. (10.20).

Example 10.3: Relationship Between Chemical Potential and Fugacity
Derive eq. (10.18) between the chemical potential and fugacity of a mixture in two states at the same temperature.

Solution We start with eq. (10.13) and use eq. (10.12) for μ_i^{igm}:

$$\mu_i = G_i^{\text{ig}} + RT \ln x_i + RT \ln \phi_i.$$

Combining the logarithms and using $x_i \phi_i = f_i/P$, this becomes

$$\mu_i = G_i^{\text{ig}} + RT \ln \frac{f_i}{P}.$$

Consider now two states that are both at the same temperature, though not necessarily at the same pressure or composition. Taking the difference of the chemical potentials we find

$$\mu_i - \mu_i' = G_i^{\text{ig}} - G_i^{\text{ig}'} + RT \ln \frac{f_i}{f_i'} - RT \ln \frac{P}{P'}.$$

The term $G_i^{\text{ig}} - G_i^{\text{ig}'}$ is the change of the ideal-gas Gibbs free energy of pure component between two pressures on the same isotherm. This is given by

$$G_i^{\text{ig}} - G_i^{\text{ig}'} = RT \ln \frac{P}{P'}.$$

Combining the last two results we obtain eq. (10.18).

10.3 Fugacity in a Mixture

Example 10.4: Fugacity from Experimental Data
The system 1,4,dioxane/2-propanol at 1 atm, 83.5 °C, forms a vapor-liquid mixture with the following compositions (in mol %):

	Liquid	Vapor
1,4,dioxane	40.5	27.4
2-propanol	59.5	72.6

Estimate the fugacity and fugacity coefficient of each component in the liquid and the vapor. Data from B. Choffe, M. Cliquet, and S. Meunier, Rev. Inst. Franc. Petrole. 15, no. 6 (1960), p. 1051.

Solution The starting point is the equilibrium criterion in eq. (10.20), which relates the fugacity coefficients of a component to the mole fraction of the component in each phase. However, one coefficient must be known for the other to be calculated from this equation. In the absence of additional information, we must make some suitable approximations. We will assume that the vapor phase is ideal and take $\hat{\phi}_i^V \approx 1$ for both components. Then, the fugacity coefficient in the liquid phase is

$$\hat{\phi}_i^L = \frac{y_i}{x_i}\hat{\phi}_i^V \approx \frac{y_i}{x_i}.$$

Using 1 for 1,4,dioxane and 2 for 2-propanol, we find:

$$\hat{\phi}_1^L \approx \frac{0.274}{0.405} = 0.676543$$

$$\hat{\phi}_2^L \approx \frac{0.726}{0.595} = 1.22017.$$

The fugacities are calculated from eq. (10.15):

$$\begin{aligned}
f_1^L &= x_1\hat{\phi}_1^L P = (0.405)(0.676543)(1.013) &= 0.2776 \text{ bar} \\
f_1^V &= y_1\hat{\phi}_1^V P = (0.274)(1)(1.013) &= 0.2776 \text{ bar} \\
f_2^L &= x_2\hat{\phi}_2^L P = (0.595)(1.22017)(1.013) &= 0.7356 \text{ bar} \\
f_2^V &= y_2\hat{\phi}_2^V P = (0.726)(1)(1.01325) &= 0.7356 \text{ bar}
\end{aligned}$$

The fugacity of each component is the same in both phases, as it should.

Comments The assumption of ideality in the gas phase is reasonable but should be checked. This can be done by checking the pure components against a generalized compressibility graph. If the individual components are sufficiently close to the ideal-gas state at the pressure and temperature of the calculation, their mixture is likely to be in the ideal-gas state as well.

10.4 Fugacity from Equations of State

In Chapter 9 we discussed the use of equations of state for mixtures. Given the equation of state we may calculate all the residual properties of the mixture, including the residual Gibbs free energy. Since the fugacity coefficient is defined in terms of the residual partial molar Gibbs free energy, the equation of state may be used to calculate fugacity coefficients in gas and liquid mixtures. The calculation requires the application of the partial molar derivative to the residual Gibbs energy. Applying this mathematical operation to the composition-dependent terms of the residual Gibbs energy, we ultimately obtain an equation for the fugacity coefficient of the component. The equations below summarize the results for the two commonly used cubic equations of state:

- *Soave-Redlich-Kwong*

$$\ln \phi_i = \frac{b_i}{b}(Z-1) - \ln(Z-B') - C'_i \ln \frac{Z+B'}{Z}. \tag{10.21}$$

- *Peng-Robinson*

$$\ln \phi_i = \frac{b_i}{b}(Z-1) - \ln(Z-B') - \frac{C'_i}{2\sqrt{2}} \ln \frac{Z+(\sqrt{2}+1)B'}{Z-(\sqrt{2}-1)B'}. \tag{10.22}$$

In both equations the dimensionless parameter C'_i is defined as

$$C'_i = \frac{A'}{B'}\left(-\frac{b_i}{b} + \frac{2}{a}\sum_{j=1}^{N} x_j \sqrt{a_i a_j}(1-k_{ij})\right). \tag{10.23}$$

In writing these equations we follow the convention that parameters of the cubic equation with a single subscript (a_i, b_i) refer to pure components while unsubscripted parameters (a, b, A', B') refer to the mixture. Parameter C'_i is introduced for convenience and combines parameters of the pure component and of the mixture. In the case of binary mixture, this parameter simplifies to:

$$C'_1 = \frac{A'}{B'}\left(-\frac{b_1}{b} + \frac{2}{a}(x_1 a_1 + x_2\sqrt{a_1 a_2}(1-k_{12}))\right),$$

$$C'_2 = \frac{A'}{B'}\left(-\frac{b_2}{b} + \frac{2}{a}(x_2 a_2 + x_1\sqrt{a_1 a_2}(1-k_{12}))\right).$$

The numerical steps for the calculation of the fugacity coefficient are analogous to those for the residual properties of mixture:

1. The required input is temperature, pressure, composition, and phase of the system (in case the compressibility equation has three real roots).

10.4 Fugacity from Equations of State

2. Calculate the parameters a_i, b_i, of the pure components.

3. Calculate the parameters a, b, of the mixture using the mixing rules from eqs. (9.38) and (9.39).

4. Calculate the parameters A' and B' of the mixture using eq. (2.36), and the parameter C_i for each component from eq. (10.23).

5. Solve for the roots of the compressibility factor: if there are three real roots, choose based on the phase of the system.

6. Calculate the fugacity coefficients from eqs. (10.21) or (10.22) using the compressibility factor of the mixture.

7. Calculate the fugacity of each component from eq. (10.15).

This calculation is demonstrated with a numerical example below.

Example 10.5: Fugacity in Mixture Using the SRK Equation
A liquid mixture of carbon dioxide(1)/n-pentane(2) contains 32% by mole carbon dioxide. Calculate the fugacity of the two components at 4.5 °C, 16.15 bar using the SRK equation with $k_{12} = 0.12$.

Solution The critical parameters of the pure components are:

	T_c (K)	P_c (bar)	ω
CO_2 (1):	304.2	73.74	0.225
Pentane (2):	469.7	33.7	0.252

The various SRK parameters of the pure components and of the mixture are:

	a (J m^3/mol^2)	b (m^3/mol)	C_i (−)
CO_2	0.39863	2.97156×10^{-5}	4.75869
n-Pentane	2.78581	1.00397×10^{-4}	11.9516
Mixture	1.73256	7.77787×10^{-5}	

Using these values we find

$$A' = 0.525105, \quad B' = 0.0544159.$$

The cubic equation for the compressibility factor is

$$Z^3 - Z^2 + 0.467728 Z - 0.028574 = 0,$$

which has one real root:

$$Z = 0.0711422.$$

The fugacity coefficients of the two components are

$$\phi_1 = 2.80821, \quad \phi_2 = 0.02029.$$

Finally, the fugacity of the component in the solution is

$$f_1 = (0.32)(2.80821)(16.15) = 14.5 \text{ bar},$$
$$f_2 = (0.68)(0.02029)(16.15) = 0.223 \text{ bar}.$$

10.5 VLE of Mixture Using Equations of State

We now have all the necessary ingredients to compute the phase diagram of a mixture based entirely on an equation of state. The starting point is the equilibrium condition, which for a binary system gives

$$x_1 \phi_1^L = y_1 \phi_1^V, \tag{10.24}$$
$$x_2 \phi_2^L = y_1 \phi_2^V. \tag{10.25}$$

The variables that define a tie line are pressure, temperature, and the mole fractions in the two phases. This makes for a total of four unknowns (only one mole fraction is needed per phase as the other is calculated from the normalization condition). Given four variables and two equations, two variables must be specified in order to solve for the rest.[5] Depending on which variables are specified and which are unknown, VLE problems are classified according to the following scheme:

- *Bubble P.* Temperature and liquid-phase composition are specified; solve for bubble pressure and vapor mol fractions.

- *Bubble T.* Pressure and liquid-phase composition are specified; solve for bubble temperature and vapor mol fractions.

- *Dew P.* Temperature and vapor-phase composition are specified; solve for dew pressure and liquid mol fractions.

- *Dew T.* Pressure and vapor-phase composition are specified; solve for dew temperature and liquid mol fractions.

- *Flash.* Pressure and temperature are specified; solve for the compositions of the two phases.

---
5. We obtain the same result by the phase rule: with two components ($N = 2$) and two phases ($\pi = 2$), the number of degrees of freedom is $N + 2 - \pi = 2$.

10.5 VLE of Mixture Using Equations of State

The bubble and dew problems are named according to the unknowns, for example, in the *bubble P* problem we are solving for bubble pressure and composition of the vapor in equilibrium with a known liquid. The flash problem is analogous to a flash separator in which pressure and temperature are fixed. To solve any of these problems all that is needed is an equation for the fugacity coefficient of the form,

$$\phi_i = \phi_i(P, T, \text{compositions}, \text{phase}),$$

such as eqs. (10.21) and (10.22). In computer language, the above expression represents a procedure that returns the fugacity of component i given temperature, pressure, composition, and phase (liquid or vapor). The phase must be indicated because it is needed for the selection of the proper root for the compressibility factor.

For a binary system, all of the above problems reduce to two equations that must be solved for two unknowns. However, the nonlinear character of the equations and the discontinuous character of the isotherms require a trial-and-error methodology. We outline below an iterative method for the solution of the bubble P problem. The calculation is streamlined by introducing the K-factors defined in Chapter 8,

$$K_i \equiv \frac{y_i}{x_i} = \frac{\phi_i^L}{\phi_i^V}. \tag{10.26}$$

The first equality is the definition of K_i, given in eq. (8.4); the second equality follows from the definition of fugacity, eq. (10.15). Using the K factors, the mole fractions in the vapor are

$$y_i = K_i x_i \tag{10.27}$$

and obey the normalization condition

$$K_1 x_1 + K_2 x_2 = 1. \tag{10.28}$$

Equations (10.27) and (10.28) are solved iteratively as follows:

1. Start with a guess for the bubble pressure and the gas-phase mole fractions.

2. Calculate the fugacity coefficients of the liquid.

3. Calculate the fugacity coefficients of the vapor and the K factors of each component.

4. Obtain a refined estimate of the mole fractions using eq. (10.27):

$$y_i = \frac{K_i x_i}{K_1 x_1 + K_2 x_2}, \quad (i = 1, 2).$$

Here, division by $K_1 x_1 + K_2 x_2$ is needed because the mole fractions calculated directly from eq. (10.27) will not obey the normalization condition except at the bubble pressure.

5. Check the value of $K_1 x_1 + K_2 x_2$: if it has changed from the previous iteration, return to step 3 and repeat the calculation with the new vapor mole fractions until this quantity does not change any more.

6. If the converged value of $K_1 x_1 + K_2 x_2$ is 1, we have obtained the solution; if not, refine the estimate for the pressure, return to step 2 and repeat. To obtain a new guess for the pressure, use the quantity $K_1 x_1 + K_2 x_2$ as a guide: if this quantity is larger than one, pressure is too high and should be decreased; if less than one, pressure is too low and should be increased. One way to adjust pressure systematically is to calculate the new guess using the equation below:

$$P_{\text{new}} = \frac{P_{\text{old}}}{K_1 x_1 + K_2 x_2}.$$

Such numerical details may be varied according to the system to improve the speed of convergence.

The calculation described here consists of two nested loops, an inner loop that refines the mole fractions of the vapor, and an outer loop that refines the estimate for pressure. Care must be exercised to identify trivial solutions. A trivial solution is one for which the liquid and vapor phases have the same composition and same compressibility factor. In this case the fugacity coefficients of a component are the same in the two phases (i.e., $K_i = 1$ for all components) and all equations are satisfied. Such solution is physically acceptable only at the critical point, but spurious solutions of this type may appear at other pressures and temperatures.

By this calculation we obtain one tie line on the phase diagram, defined by the values of pressure, temperature, and composition of the two phases:

$$\{T, P, x_1, y_1\}.$$

This tie line may be plotted on a Pxy or Txy graph. The entire phase diagram can be constructed in this manner by repeating the calculation at different conditions. Figure 10-2 shows the Pxy graph of the system carbon dioxide/n-pentane calculated by the Soave-Redlich-Kwong equation of state (see Example 10.6 for additional details of this calculation). The calculated phase diagram is in very good agreement with experimental data for this system. At 277.65 K, both components are below their critical temperature and the Pxy graph extends over the entire range of compositions. At 344.32 K, carbon dioxide is above its critical temperature (304.2 K) and the phase envelop is detached from the axis that represents pure CO_2. The SRK calculation is very good in this case as well (close to the critical point of the mixture the numerical calculation converges very slowly and these points are not shown).

10.5 VLE of Mixture Using Equations of State

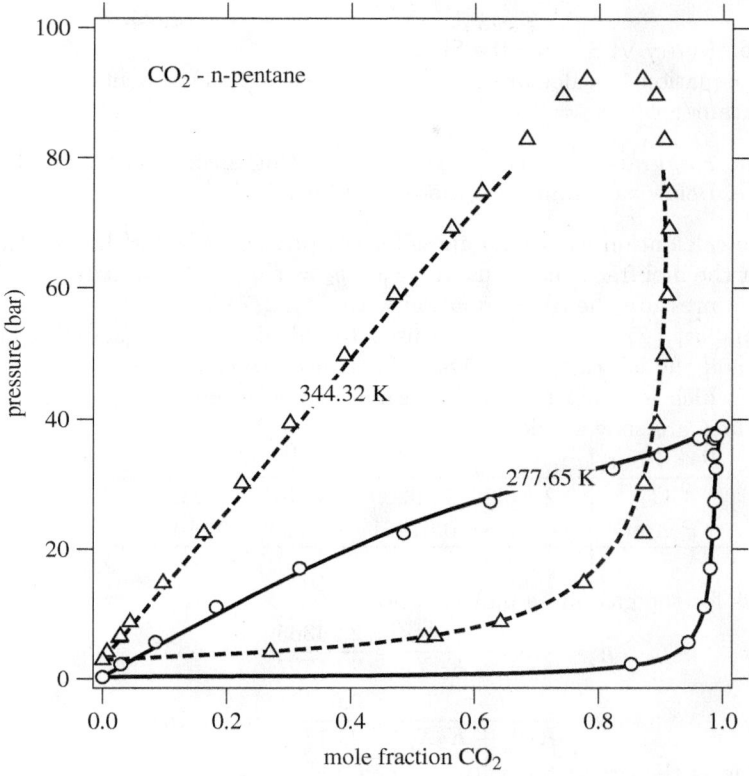

Figure 10-2: *Pxy* graph of carbon dioxide-*n* pentane calculated using the SRK equation with $k_{12} = 0.12$.

The Interaction Parameter

The interaction parameter is a required input for this calculation. It is an empirical parameter that must be obtained by fitting the calculated values from the equation of state to experimental data. For the calculations shown in Figure 10-2 the same value of interaction parameter was used in both temperatures. It is possible to obtain a better fit by allowing the interaction parameter to vary with temperature. For many systems, however, the interaction parameter is taken to be independent of temperature. A listing of k_{12} values for some binary systems is given in Sandler (Ref. [3]) for use with the Peng-Robinson equation. Though these values are specific to the Peng-Robinson equation, they may be used as a starting guess in the Soave-Redlich-Kwong equation. In general, before a serious calculation is undertaken based on an equation of state, the interaction parameter should be validated by comparison to data as extensively as possible.

Example 10.6: Binary VLE Using the SRK
Use the SRK equation to calculate the Pxy graph of $CO_2(1)/n$-pentane at 277.65 K. The interaction parameter is $k_{12} = 0.12$.

Solution The Pxy graph is constructed by calculating a series of tie line by the method outlined above. Below we summarize the results for the tie line at $x_1 = 0.2$.

To initiate the calculation we need a guess for the pressure and for the mol fractions in the gas phase. For the mol fractions we use $y_1 = x_1$, $y_2 = x_2$. For the pressure we pick $P = 4$ bar because at this pressure the SRK equation with $T = 277.65$ K and $x_1 = 0.2$ has three real roots. The smallest ($Z_L = 0.0190462$) is used to calculate the fugacity coefficients in the liquid phase, and the largest ($Z_V = 0.832693$) is used to calculate the fugacity coefficients in the vapor (which is taken to have the same composition as the liquid). The results of these calculations are shown below:

$P = 4$ bar	x_i	y_i	ϕ_i^L	ϕ_i^V	K_i	$K_i x_i$
CO_2	0.2	0.2	11.9269	1.06317	11.2183	2.24365
n-Pentane	0.8	0.8	0.07461	0.81303	0.09176	0.07341
	1.0	1.0				2.31707

The new guess for the gas-phase mole fractions is

$$y_1 = \frac{K_1 x_1}{K_1 x_1 + K_2 X_2} = \frac{2.24365}{2.31707} = 0.968317,$$

$$y_2 = \frac{K_2 x_2}{K_1 x_1 + K_2 X_2} = \frac{0.07341}{2.31707} = 0.03168.$$

The calculation is then repeated with the new mole fractions (all else remains the same) until the quantity $K_1 x_1 + K_2 X_2$ converges to a constant value. The pressure guess is then refined and the trial-end error method is repeated until the quantity $K_1 x_1 + K_2 X_2$ converges to 1. These iterations (not shown) converge to the final result for this tie line

$P = 10.7083$ bar	x_i	y_i	ϕ_i^L	ϕ_i^V	K_i	$K_i x_i$
CO_2	0.2	0.96956	4.53038	0.934523	4.8478	0.96956
n-Pentane	0.8	0.03044	0.02888	0.759036	0.0381	0.03044
	1.0	1.00000				1.00000

The bubble pressure is 10.7083 bar and the mole fraction of carbon dioxide in the vapor is $y_1 = 0.96956$. The complete Pxy graph is obtained by repeating the calculation at several liquid compositions between $x_1 = 0$ and $x_1 = 1$, as summarized in the table below:

$T = 277.65$ K

P (bar)	x_1	y_1	P (bar)	x_1	y_1
0.29	0	0	27.39	0.6	0.9869
5.58	0.1	0.9444	30.19	0.7	0.9884
10.71	0.2	0.9696	32.61	0.8	0.9900
15.57	0.3	0.9781	35.24	0.9	0.9925
20.04	0.4	0.9824	39.44	1	1.0000
24.01	0.5	0.9851			

10.6 Summary

The *Pxy* graph is shown in Figure 10-2. The data points are the experimental results of Besserer and Robinson, *Journal of Chemical and Engineering Data*, 18, no. 4 (1973). The agreement is very good. Results are also shown at 344.32 K. This temperature is above the critical temperature of CO_2, which causes the VLE region to detach from the vertical axis at $x_1 = 1$ and to exhibit a critical point. Numerical convergence near the critical point is more difficult. For this reason, the SRK calculation at 344.32 K was conducted up to a maximum pressure of about 80 bar.

NOTE

On Equations of State

The appeal of equations of state is that they can be used to obtain the entire phase behavior of a system over a wide range of pressures and temperature using tabulated values (critical properties, acentric factor) and a one adjustable parameter (interaction parameter). There are, however, two important limitations. The first one is that equations of state are generally applicable to small molecules that are chemically similar and interact via weak, nonpolar forces. While this encompasses many hydrocarbon systems that are industrially important, it excludes a much wider class of mixtures of strongly interacting molecules, including systems that contain water and other polar molecules. The second limitation is that these calculations quickly become cumbersome as the number of components increases. For this reason, alternative methods have been developed that can handle a broader variety of systems. The most widely used methodology is based on the notion of the ideal solution and makes use of activity coefficients. These topics are discussed in Chapters 11 and 12.

10.6 Summary

In this chapter we developed the criteria for phase equilibrium in a multicomponent mixture and showed how to construct the entire phase diagram if the equation of state is known. The most important theoretical result is equilibrium criterion which we expressed in the following equivalent forms:

$$\mu_i^L = \mu_i^V, \qquad [10.6]$$

$$f_i^L = f_i^V, \qquad [10.19]$$

$$x_i \phi_i^L = y_i \phi_i^V. \qquad [10.20]$$

The fundamental result is eq. (10.6), which equates the chemical potentials of a component in every phase the component is present. Equations (10.19)–(10.20) express the equilibrium criterion in terms of fugacity (or fugacity coefficient) and represent equivalent forms of eq. (10.6). Another important result is a relationship between chemical potential and fugacity:

$$\mu_i - \mu_i' = RT \ln \frac{f_i}{f_i'}. \qquad [10.18]$$

This equation relates the chemical potential and fugacity at two states at the same temperature. This relationship allows us to use fugacity interchangeably with chemical potential to solve problems in phase equilibrium.

In general, calculations of phase equilibrium are reduced to calculations of fugacity and since fugacity can be calculated from an equation of state, the vapor-liquid problem of mixture can be solved if such equation is available. The required inputs are (1) an appropriate equation of state, (2) suitable mixing rules, and (3) the ideal-heat capacities of the pure components. The appeal of equations of state is that they require little more than tabulated values (critical properties, acentric factor) of the pure components. This convenience comes with a price: the method is not guaranteed to work for arbitrary mixtures. Generally, it works very well for similar compounds that are weakly interacting, usually mixtures of short hydrocarbon chains and other non-polar molecules, but fails with many other systems, including many aqueous solutions. This problem requires the development of alternative methodologies that do not suffer from this weakness. These are discussed in Chapters 11 and 12.

10.7 Problems

Problem 10.1: a) How many degrees of freedom are there at the triple point of a pure substance?
b) What is the fugacity of ice at the triple point? The triple point of pure water is at 0.611 kPa and $T = 0.01$ °C.
c) Calculate the Gibbs energy of ice at the triple point.
d) Calculate the fugacity and the fugacity coefficient of liquid water at 0.01 °C and 300 bar.

Problem 10.2: a) First define the fugacity coefficient, ϕ, of the *mixture* as

$$\ln \phi = \frac{G^R}{RT},$$

10.7 Problems

where G^R is the residual Gibbs energy of the entire mixture. Then show that the fugacity coefficient of a species, ϕ_i, is related to the fugacity coefficient of the mixture as follows:

$$\ln \phi_i = \left(\frac{\partial n \ln \phi}{\partial n_i}\right)_{P,T,n_j}.$$

In other words, the logarithm of the fugacity coefficient—rather than the coefficient itself—is a partial molar property.

b) Based on your previous result, outline the procedure for the calculation of the fugacity coefficient of a component in a mixture using an equation of state such as the SRK.

Problem 10.3: At 50 °C, 1.2 bar, the system n-pentane (1), n-hexane (2), and n-heptane (3) exists in vapor-liquid equilibrium. The mole fractions in the two phases are: $x_1 = 0.6$, $x_2 = 0.2$, $x_3 = 0.2$ in the liquid; and $y_1 = 0.87$, $y_2 = 0.1$, $y_3 = 0.03$ in the vapor.
a) Calculate the fugacity and the fugacity coefficient of each component in the vapor and in the liquid.
b) Calculate the chemical potential of n-heptane in the liquid. The reference state is the ideal-gas state at 200 K, 35 bar.
Additional information: Ideal-gas heat capacity of n-heptane: $C_P = 200$ J/mole K.

Problem 10.4: Your company faxed you some results, shown below, for the system hydrogen (1) / methane (2). Use these data to answer the following:
a) Because the fax machine malfunctioned, some entries were missing. Supply the missing numbers and explain your calculations.
b) What is the saturation pressure of methane at 250 K?
c) Calculate the fugacity of saturated liquid methane at 250 K.
d) Calculate the fugacity of methane at 250 K, 300 bar.
e) A solution with composition $x_1 = 0.2$ is to be stored at 250 K. It is important to avoid the formation of vapor. What is the minimum pressure required?
f) Calculate the volume of the tank needed to store a saturated liquid mixture of the two components with composition $x_1 = 0.2$ at 250 K. Report the result in cubic meters per million of mol.

1 = hydrogen, 2 = methane

P (bar)	T (K)	x_1	y_1	ϕ_1^L	ϕ_1^V	ϕ_2^L	ϕ_2^V	Z_V	Z_L
0.074	250	0	0.0000	438.1532	1.0018	0.9953	0.9953	0.9953	0.0004
6.148	250	0.2	0.9876		0.9479	0.0126		0.9433	0.0310

Problem 10.5: Calculate the ideal work for the separation of the following mixtures into their pure components. In all cases initial mixture as well as the purified components are at 10 °C, 1 bar. You may take the surroundings to be at 25 °C.
a) Gas mixture containing 25% Ne (1) + 75% H_2 (2).
b) Gas mixture containing 25% methane (1) + 75% ethane (2).
c) Solution containing 25% ethanol (1) + 75% water (2). Additional data for this system at 10 °C, 1 bar are given below:

$\bar{H}_1$	=	−570.85 J/mole	$\bar{H}_2$ =	167.86 J/mole
H_1	=	0 J/mole	H_2 =	0 J/mole
$\bar{S}_1$	=	4.540 J/mole K	$\bar{S}_2$ =	1.586 J/mole K
S_1	=	0 J/mole K	S_2 =	0 J/mole K

d) Compare the answers and suggest reasons why some of these results are the same and some are not.

Problem 10.6: At 89.8 °C, 0.75 bar, water (1) and pyridine form a vapor-liquid mixture with mol fractions $x_1 = 0.3114$, $y_1 = 0.5411$.
a) Calculate the fugacity of component in the vapor.
b) Calculate the fugacity coefficient of the two components in the liquid.
c) Calculate the molar Gibbs energy of the vapor and of the liquid. The reference state for each component is the ideal-gas state at 89.8 °C, 0.75 bar.
You may assume the gas phase to be in the ideal-gas state.

Problem 10.7: A box is divided into two compartments via a partition. Part A contains a mixture of nitrogen and oxygen in molar ratio $N_2 : O_2 = 2 : 1$. The second compartment is three times as big in volume as compartment A and contains pure oxygen. Both compartments are at 1 bar, 300 K, and the entire box is insulated from the surroundings. The partition is removed and the system is allowed to reach equilibrium.
a) Calculate the chemical potential of each component before and after the removal of the partition. As reference state, use the pure components at 1 bar, 300 K.
b) Calculate the fugacity of each component before and after the removal of the partition.
You may assume the components and their mixtures to be in the ideal-gas state at the pressures and temperatures of this problem. Is this a reasonable assumption?

10.7 Problems

Problem 10.8: The following table is a set of VLE data for methanol(1)/water(2) at 333.15 K.

P/kPa	x_1	y_1	P/kPa	x_1	y_1
19.953	0.0000	0.0000	60.614	0.5282	0.8085
39.223	0.1686	0.5714	63.998	0.6044	0.8383
42.984	0.2167	0.6268	67.924	0.6804	0.8733
48.852	0.3039	0.6943	70.229	0.7255	0.8922
52.784	0.3681	0.7345	72.832	0.7776	0.9141
56.652	0.4461	0.7742	84.562	1.0000	1.0000

a) What is the saturation pressure of methanol at 333.15 K?
b) What is the phase of the mixture that contains 50% (by mole) methanol at 333.15 K and 40 kPa?
c) At $z_1 = 0.2$ and 30 kPa, the system is present as two phases. What is the chemical potential of methanol and water in the vapor phase using an ideal gas reference state at 300 K and 101 kPa? Is the assumption of treating the vapor phase as an ideal gas appropriate?
Additional information: The C_P's of methanol and water vapor are 49.0 and 33.8 J/mol K, respectively.

Problem 10.9: Use the SRK equation with $k_{12} = 0$ to compute the phase diagram of the system normal hexane (1)/normal octane (2) at the following conditions:
a) Pxy graph at 350 K.
b) Txy graph at 1 bar.
Make graphs and annotate them properly.

Problem 10.10: a) At 542 K, 21.7 bar, the system Ethylbenzene(1)/Benzene(2) is in vapor-liquid equilibrium. Available data for this system are given in the table below. Fill out the missing information on the table.

	x_i	y_i	ϕ_i^L	ϕ_i^V	f_i^L (bar)	f_i^V (bar)
Ethylbenzene(1)			0.531			7.150
Benzene(2)				0.844		9.686

b) Calculate the entropy of a vapor mixture with $y_1 = 0.3$ at 600 K, 12 bar. The reference state for each pure component is the ideal-gas state at 600 K, 12 bar.

Additional information for part b: The residual entropy of the mixture at 600 K, 12 bar, is -1.8 J/mol K; the ideal-gas heat capacities of the pure components are $C_{P1} = 120$ J/mol K, $C_{P2} = 150$ J/mol K.

Problem 10.11: The binary system CO_2 (1) normal pentane (2) can be described by the SRK equation with $k_{12} = 0.12$. Use this equation of state to do the following:
a) Compute and plot the Pxy graph of this system at 290 K. Annotate the graph properly. Show the tie line at 40 bar.
b) What is the phase of a mixture that contains 80% CO_2 by mol at 40 bar? If a two-phase system, report the composition (x_1, y_1) and molar fractions (V, L) of the two phases.
c) Compute and plot the Txy graph for this system at 40 bar. Show the tie line at 290 K.
d) How much heat (kJ/mol) is needed to boil a saturated liquid with $x_1 = 0.8$ under constant pressure $P = 40$ bar to produce saturated vapor?
e) A flow stream ($x_1 = 0.8$) is compressed in an adiabatic compressor whose efficiency is 100%. The inlet is at 1 bar, 290 K and the outlet is delivered at 40 bar. Calculate the work for the compression (kJ/mol) as well as the temperature of the exit stream.

Problem 10.12: Use the SRK equation to calculate the Pxy graph of carbon dioxide (1)/normal pentane at 250 K. For this system $k_{12} = 0.12$.
a) Plot the Pxy graph at 250 K.
b) A stream contains a mixture of the two components with overall composition $z_1 = 50\%$. The conditions are 15 bar, 250 K. What is the phase of the system?
c) The above stream is flashed isothermally to a pressure such that the vapor contains 90% carbon dioxide. Determine the pressure and the concentration of CO_2 in the liquid.

Problem 10.13: Use the SRK equation to compute a Txy graph for the system carbon dioxide (1)/normal pentane at 25 bar. Assume that $k_{12} = 0.12$ and independent of temperature.
a) Plot the Txy graph.
b) A saturated vapor-liquid stream at 25 bar with the overall composition $z_1 = 0.5$ contains 50% vapor. What are the compositions of the two phases?
c) The stream passes through a heat exchanger where it is heated until the dew point is reached. What is the temperature?
d) How much heat is exchanged in the previous step?

10.7 Problems

Problem 10.14: The table below gives data for the system hydrogen sulfide (1) in benzene (2) at 150 °C, using Peng-Robinson equation with $k_{12} = 0.015$.
a) Do you expect this system to exhibit a critical point at 150 °C?
b) Determine whether the solution behavior of benzene may be considered ideal in the concentration range $x_1 = 0$ to 0.30.
c) Explain why the fugacity coefficients of benzene in the liquid and in the vapor are equal at 5.76 bar but different at 36.99 bar.

x_1	y_1	P (bar)	ϕ_1^L	ϕ_1^V	ϕ_2^L	ϕ_2^V	Z^L	Z^V
0.00	0.0000	5.76	15.9442	1.0288	0.9002	0.9002	0.0171	0.8901
0.30	0.8133	36.99	2.4569	0.9062	0.1558	0.5843	0.0981	0.8105

Chapter 11

Ideal Solution

In the previous chapter we developed the general criteria for phase equilibrium and showed how the phase diagram and all other properties of a mixture can be calculated if the equation of state is known. In all property calculations that are based on equations of state we are starting from the ideal-gas state and use the equation of state to calculate residual properties, in effect, to extrapolate from the ideal gas to the actual state of the system. This approach works quite well for mixtures of nonpolar molecules whose interactions are captured accurately by cubic equations of state. Mixtures of strongly interacting molecules cannot be described by these methods with sufficient accuracy.[1] A different approach is needed to do calculations with these systems. Instead of using the ideal-gas state as the starting point to describe a liquid mixture, we use a state closer to the liquid itself: the ideal solution. In this chapter we define the ideal solution, identify systems that can be described under the simplified assumptions of the ideal-solution theory, and calculate their phase diagram.

The learning objectives for this chapter are to:

1. Define the ideal solution.

2. Obtain equations for the primary properties (volume, enthalpy, entropy) of ideal solution.

3. Obtain the fugacity of component in ideal solution.

4. Calculate the phase diagram of ideal solutions using Raoult's law.

11.1 Ideality in Solution

In a pure liquid all interactions are between molecules of the same kind. In solution, molecules are surrounded by a mix of similar and dissimilar molecules. The behavior of the solution is ultimately determined by the balance of self- and cross-interactions. A special case of practical and theoretical importance is when

---

[1]. Several adaptations of equations of state are now available that can be applied to strongly interacting systems. These are beyond the scope of this book. The interested reader may consult Ref. [3].

cross- and self-interactions are identical in magnitude. This situation arises when molecules are very similar in chemical nature and size, as in solutions of neighboring members of a homologous series. We call this special system an *ideal solution*. The ideal solution is defined by the following properties:

$$V^{\text{id}} = \sum x_i V_i \qquad \Delta V_{\text{mix}}^{\text{id}} = 0, \qquad (11.1)$$

$$H^{\text{id}} = \sum x_i H_i \qquad \Delta H_{\text{mix}}^{\text{id}} = 0, \qquad (11.2)$$

$$S^{\text{id}} = \sum x_i S_i - R \sum x_i \ln x_i \qquad \Delta S_{\text{mix}}^{\text{id}} = -R \sum_i x_i \ln x_i, \qquad (11.3)$$

where the superscript ^{id} is used to refer to "ideal solution." Because molecules that form ideal solutions interact with each other in the same manner as among themselves, there is no change of volume or enthalpy when the pure components are mixed to form the solution. The entropy, on the other hand, does not remain constant but increases by exactly the same amount as when ideal gases mix. There is a close relationship between ideal solutions and ideal mixtures, but an important difference as well: in the ideal-gas mixture, self-interactions and cross-interactions are both zero; in ideal solution, self- and cross-interactions are of equal strength, but they are *not* zero. In fact, they are quite strong, as a result of the close proximity of molecules in the liquid. The ideal-gas mixture, therefore, should be thought of as a special case of ideal solution. It has the properties of the ideal-solution, eqs. (11.1)–(11.3) but also additional special properties (for example, it obeys the ideal-gas law) that the ideal solution does not have.

Equations (11.1)–(11.3) give the primary properties of mixing. All other properties may be obtained by their relationship to those given above. In particular, for the internal energy and Gibbs free energy we obtain

$$U^{\text{id}} = \sum_i x_i U_i, \qquad (11.4)$$

$$G^{\text{id}} = \sum x_i G_i + RT \sum x_i \ln x_i, \qquad (11.5)$$

which are obtained by writing $U^{\text{id}} = H^{\text{id}} - PV^{\text{id}}$, and $G^{\text{id}} = H^{\text{id}} - TS^{\text{id}}$. From these equations we may easily identify the corresponding partial molar properties:

$$\bar{V}_i = V_i, \qquad (11.6)$$

$$\bar{H}_i = H_i, \qquad (11.7)$$

$$\bar{S}_i = S_i - R \ln x_i. \qquad (11.8)$$

The most important partial molar property is the chemical potential. It is the partial molar Gibbs free energy and is given by

$$\boxed{\mu_i^{\text{id}} = G_i + RT \ln x_i.} \qquad (11.9)$$

11.1 Ideality in Solution

Here, μ_i^{id} is the chemical potential of component in ideal solution, and G_i is the chemical potential (molar Gibbs free energy) of the pure liquid at the same temperature and pressure.

NOTE

The Practical Significance of the Ideal Solution
According to eqs. (11.1)–(11.5), the calculation of properties in ideal solution requires nothing more than the properties of the pure liquids at the pressure and temperature of the solution, and the composition. Since the properties of the pure components are generally known, the calculation of mixture properties is reduced to a very simple algebraic calculation. Compared to the methods of Chapter 10, this represents a significant simplification with respect to the amount of calculations involved.

Example 11.1: Ideal Entropy of Mixing
Express the entropy in terms of an ideal gas and a residual contribution to show that the ideal-solution entropy is given by eq. (11.3).

Solution We form a solution by mixing the pure components under constant temperature T and constant pressure P. For the entropy of pure component i we write

$$S_i = S_i^{\text{ig}} + S_i^R,$$

where S_i^{ig} is the ideal-gas part and S_i^R is the residual correction. The contribution of component i to the entropy of the mixture is given by the partial molar entropy. This too may be expressed in terms of an ideal-gas contribution and a residual correction:

$$\bar{S}_i = \bar{S}_i^{\text{igm}} + \bar{S}_i^R = S_i^{\text{ig}} - R \ln x_i + \bar{S}_i^R.$$

Consider now the residual correction in the two expressions. The term S_i^R measures deviations from ideal-gas behavior due to interactions in the pure liquid of component i. The residual term $\bar{S}_i^R$ measures deviations from ideal-gas behavior due to interactions of component i in the solution. Since the interaction of component i with other molecules is identical in the pure liquid and in the solution, these residual terms are equal:

$$\bar{S}_i^R = S_i^R.$$

For the difference $\bar{S}_i - S_i$ we then find

$$\bar{S}_i - S_i = (S_i^{\text{ig}} - R \ln x_i + \bar{S}_i^R) - (S_i^{\text{ig}} + S_i^R) = -R \ln x_i.$$

or

$$\bar{S}_i = S_i - R \ln x_i.$$

Finally, the entropy of the solution is obtained by summing the partial molar entropies in proportion to the mole fraction of each component

$$S = \sum_i \bar{S}_i = \sum_i x_i S_i - R \sum_i x_i \ln x_i.$$

The last result is eq. (11.3). This ideal-solution entropy reflects the fact that a solution can exist in more microscopic states than a pure fluid because in addition to arranging molecules in space and assigning energy to them, we have the additional freedom to assign a molecule to species i or j. As we discussed in Chapter 4, more microstates means higher entropy.

Example 11.2: Separation Work
Calculate the amount of work needed to separate into pure components a solution that contains n-heptane (1) and n-octane (2) with $x_1 = 0.3$, at 40 °C, 5 bar.

Solution The ideal work for separation is equal to the lost work of mixing:

$$W_{\text{ideal}} = T_0 \Delta S_{\text{mix}} - \Delta H_{\text{mix}}.$$

We treat the solution as ideal since the components are chemically similar. The enthalpy of mixing in this case is 0. The entropy of mixing of the ideal solution is $\Delta S_{\text{mix}} = -R(x_1 \ln x_1 + x_2 \ln x_2)$. With $x_1 = 0.3$, $x_2 = 0.7$, we find $\Delta S_{\text{mix}} = +0.611R = +5.08$ J/mol K. For ideal solutions the entropy of mixing is always positive. The lost work then is

$$W_{\text{ideal}} = (300)(5.08) - 0 = 1523.6 \text{ J/mol}.$$

We notice that the temperature (40 °C) and pressure (5 bar) of the system do not enter in the calculation. This is because the entropy and enthalpy of mixing in ideal solutions are independent of temperature and pressure.

11.2 Fugacity in Ideal Solution

The key problem is to calculate the phase diagram of ideal solution and for this we first need an expression for fugacity. We start with the basic relationship between fugacity and chemical potential in eq. (10.18). Taking state A to be component i in ideal solution, and state B the pure component at the same temperature and pressure, we have

$$\ln \frac{f_i^{\text{id}}}{f_{i,\text{pure}}} = \frac{\mu_i^{\text{id}} - G_i}{RT} = \ln x_i,$$

11.2 Fugacity in Ideal Solution

where we use eq. (11.9) to write the result on the far right. Solving for f_i we obtain the following simple expression:

$$f_i^{\mathrm{id}} = x_i f_{i,\mathrm{pure}}. \qquad (11.10)$$

This states that the fugacity of component in ideal solution is the product of the pure-component fugacity, and the mol fraction of component in solution. This equation is known as the Lewis-Randall rule and is the basis for VLE calculations with ideal solutions.

For the fugacity of pure liquid at the temperature and pressure of the solution we use the Poynting equation, which relates this fugacity to the fugacity of the saturated liquid at same temperature:

$$f_{i,\mathrm{pure}} = \phi_i^{\mathrm{sat}} P_i^{\mathrm{sat}} \exp\left(\frac{P - P_i^{\mathrm{sat}}}{RT} V_i^L\right), \qquad [7.16]$$

where P_i^{sat} is the saturation pressure, ϕ_i^{sat} is the fugacity coefficient of the saturated pure component, and V_i^L is the molar volume of the pure liquid. Then, the fugacity in ideal solution takes the form,

$$f_i^{\mathrm{id}} = x_i \phi_i^{\mathrm{sat}} P_i^{\mathrm{sat}} \exp\left(\frac{P - P_i^{\mathrm{sat}}}{RT} V_i^L\right). \qquad (11.11)$$

Here, except for the mol fraction, all other properties in the right-hand side refer to pure liquid i.

Example 11.3: Fugacity in Ideal Solution
Calculate the fugacity of water in a water/acetic acid solution at 25 °C, 1 bar, $x_w = 0.42$. Assume that these components form an ideal solution.

Solution From the Lewis-Randall rule, $f_w = x_w f_w$, where f_w is the fugacity of pure water at 25 °C, 1 bar. At this state, water is a compressed liquid. We calculate the fugacity using the Poynting factor:

$$f_w = P_w^{\mathrm{sat}} \phi_w^{\mathrm{sat}} \exp\left[\frac{P - P_w^{\mathrm{sat}}}{RT} V^L\right].$$

The saturation pressure of water at 25 °C is $P_w^{\mathrm{sat}} = 0.03166$ bar. This value is so much smaller than the critical pressure that the fugacity coefficient ϕ_w^{sat} can be safely approximated as unity. Using the molar volume of water, $V^L = 1.003 \times 10^{-3}$ m^3/kg $= 18.05 \times 10^{-6}$ m^3/mol, the Poynting factor is 1.00071. The fugacity of pure water is $f_w = 0.03166$ bar and its fugacity in the solution is

$$f_w^{\mathrm{id}} = x_w f_w = (0.42)(0.03166) = 0.0133 \text{ bar}.$$

Comments The calculation demonstrates that the Poynting correction is generally negligible for pressures around ambient.

11.3 VLE in Ideal Solution–Raoult's Law

To perform VLE calculations with ideal solutions, we begin with the equilibrium condition that requires the fugacity of component i to be the same in the liquid and in the vapor:
$$f_i^V = f_i^L.$$
For the fugacity of the liquid we use eq. (11.11); for the fugacity of the component in the vapor we use the general expression, $f_i^V = y_i \phi_i^V P$. Combining these expressions, the equilibrium criterion now takes the form,
$$y_i \phi_i^V P = x_i \phi_i^{\text{sat}} P_i^{\text{sat}} \exp\left[\frac{P - P_i^{\text{sat}}}{RT} V_i^L\right]. \tag{11.12}$$
Here, ϕ_i^V is the fugacity coefficient of component i in the vapor, not to be confused with ϕ_i^{sat}, which refers to the *saturated pure* component i at the same temperature. This result takes a much simpler form if pressure is relatively low. Then, the Poynting correction can be neglected and the fugacity coefficients may be set to 1. With these simplifications, eq. (11.12) becomes
$$\boxed{y_i P = x_i P_i^{\text{sat}}.} \tag{11.13}$$
This is known Raoult's law and requires only the saturation pressure of the pure components. The assumptions that make this simple equation valid are that the total pressure and the saturation pressure of component i be sufficiently low. Raoult's law should be understood as a shortcut for quick calculations.

In a binary mixture, Raoult's law gives
$$y_1 P = x_1 P_1^{\text{sat}}, \tag{11.14}$$
$$y_2 P = x_2 P_2^{\text{sat}}. \tag{11.15}$$
These two equations, along with the normalization conditions for the mol fractions,
$$x_1 + x_2 = 1,$$
$$y_1 + y_2 = 1,$$
constitute four equations that relate the following six variables:
$$\{x_1, x_2, y_1, y_2, P, T\}.$$
Temperature appears implicitly in the saturation pressure of pure component, which is assumed to be known as a function of temperature, usually via the Antoine equation, or some similar temperature-dependent expression. The six unknowns and four equations result in two degrees of freedom.[2] The typical problems are

---

2. The same result is reached through application of the phase rule: with two components ($N = 2$) and two phases ($\pi = 2$), the degrees of freedom are $\mathcal{F} = N + 2 - \pi = 2$.

11.3 VLE in Ideal Solution–Raoult's Law

Table 11-1: Classification of VLE Problems

	Knowns	Unknowns
Bubble P	T, x_i	P, y_i
Bubble T	P, x_i	T, y_i
Dew P	T, y_i	P, x_i
Dew T	P, y_i	T, x_i
Flash	P, T	x_i, y_i

classified as "bubble T or P," "dew T or P," or "flash." These classifications were introduced in Chapter 10 and are summarized again in Table 11-1. Solution strategies for performing these calculations are outlined below.

Bubble P Calculation Following the designation of VLE problems introduced in Chapter 10, in the bubble P problem we know temperature and the composition of the liquid and seek pressure and the composition of the vapor. Adding eqs. (11.14) and (11.15) to eliminate the unknown vapor fractions we obtain the bubble pressure of the solution:

$$P = x_1 P_1^{\text{sat}} + x_2 P_2^{\text{sat}}. \tag{11.16}$$

The vapor mol fractions are then calculated from eqs. (11.14) and (11.15):

$$y_1 = \frac{x_1 P_1^{\text{sat}}}{P}, \quad y_2 = \frac{x_2 P_2^{\text{sat}}}{P}, \tag{11.17}$$

where P is the bubble pressure calculated above. Notice that according to eq. (11.16), the bubble pressure of an ideal solution is an average of the saturation pressures of the pure components.

Bubble T Calculation In this case we know liquid composition and total pressure, and seek to calculate vapor composition and temperature. Again we add eqs. (11.14) and (11.15) to eliminate the unknown vapor fractions:

$$P = x_1 P_1^{\text{sat}} + x_2 P_2^{\text{sat}}. \qquad [11.16]$$

In this equation, the unknown is temperature, which appears implicitly in the saturation pressure through an equation such as the Antoine. Because this is a nonlinear equation in T, it requires a numerical solution. Once temperature is known, the saturation pressures are calculated and the vapor mole fractions are obtained as before, using

$$y_1 = \frac{x_1 P_1^{\text{sat}}}{P}, \quad y_2 = \frac{x_1 P_2^{\text{sat}}}{P}. \qquad [11.17]$$

Dew P Calculation Here the knowns are temperature and the composition of the vapor, and we seek to calculate pressure and the composition of liquid. We begin by solving eqs. (11.14) and (11.15) for the unknown mol fractions:

$$x_1 = y_1 P/P_1^{\text{sat}}, \tag{11.18}$$

$$x_2 = y_2 P/P_2^{\text{sat}}. \tag{11.19}$$

Adding these equations we eliminate the unknown compositions and obtain an equation that is solved for the dew pressure:

$$P = \frac{1}{y_1/P_1^{\text{sat}} + y_2/P_2^{\text{sat}}}. \tag{11.20}$$

Once the dew pressure is known, the liquid mole fractions are obtained from eqs. (11.18) and (11.19).

Dew T Calculation In this problem we know pressure and the composition of the vapor, and seek temperature and the composition of the liquid. As in the previous case, we first solve Raoult's law for the liquid mol fractions to obtain eqs. (11.18) and (11.18), then add them to obtain

$$1 = \frac{y_1 P}{P_1^{\text{sat}}} + \frac{y_2 P}{P_2^{\text{sat}}}. \tag{11.21}$$

This equation is solved for the unknown dew temperature. Once T is known, the saturation pressures are calculated and the mole fractions of the liquid are obtained by substitution into Raoult's law.

Flash Calculation In flash separations, a mixture of known overall composition is brought into the two-phase region and the resulting liquid and vapor are separated into two streams whose composition is given by the tie line that corresponds to the temperature and pressure in the separator. Calculations are streamlined by making use of the K factors. These were defined in Chapter 8 as $K_i = y_i/x_i$, and in the case of ideal solution are given by

$$K_i = \frac{P_i^{\text{sat}}}{P}. \tag{11.22}$$

Using the K factors, Raoult's law is written as

$$y_1 = K_1 x_1, \tag{11.23}$$

$$y_2 = K_2 x_2. \tag{11.24}$$

In addition, we have the mass balance between feed and the liquid and vapor streams at the exit:

$$z_1 = x_1 L + y_1 V, \tag{11.25}$$

$$1 = L + V, \tag{11.26}$$

11.3 VLE in Ideal Solution–Raoult's Law

where L and V are the fractions of liquid and vapor. Equations (11.23)–(11.26), along with the normalization conditions for the mole fractions, constitute the available equations of the flash problem. The corresponding variables are:

$$z_i, x_i, y_i, L, V, P, T.$$

For a binary system this leads to three degrees of freedom, which means that three independent variables must be specified for the problem to be well-defined.[3] The solution procedure will then depend on which variables are known and which must be solved for.

Example 11.4: Bubble and Dew Pressure
Calculate the bubble and dew pressure at 80 °C of a solution that contains 30% by mol acetonitrile (1) and 70% by mol nitromethane (2). Assume that the components form ideal solution.

Solution We need the saturation pressure of the pure components. The following equations are obtained from the NIST Chemistry WebBook:

$$\log_{10} \frac{P_1^{\text{sat}}}{\text{bar}} = 4.27873 - \frac{1355.37}{T/K - 37.853},$$

$$\log_{10} \frac{P_2^{\text{sat}}}{\text{bar}} = 4.40542 - \frac{1446.2}{T/K - 45.633},$$

with pressure in bar and temperature in kelvin. At 80 °C we find

$$P_1^{\text{sat}} = 0.955 \text{ bar}, \quad P_2^{\text{sat}} = 0.504 \text{ bar}$$

Bubble pressure. At the bubble point we have a liquid with composition $x_1 = 0.3$ and $x_2 = 0.7$. By Raoult's law,

$$y_1 P = x_1 P_1^{\text{sat}},$$
$$y_2 P = x_2 P_2^{\text{sat}}.$$

Adding the two equations we obtain the bubble pressure

$$P = x_1 P_1^{\text{sat}} + x_2 P_2^{\text{sat}} = (0.3)(0.955) + (0.7)(0.504) = 0.639 \text{ bar}.$$

---

3. Including the three normalization conditions (for z_i, x_i, and y_i) we have seven equations and 10 unknowns, or three degrees of freedom. In this case, in addition to the two degrees for vapor-liquid equilibrium of binary mixture we have one additional degree of freedom for the material balance between inlet and outlet of the separator.

The vapor compositions are

$$y_1 = \frac{x_1 P_1^{\text{sat}}}{P} = 0.448,$$

$$y_2 = 1 - y_1 = 0.552.$$

Dew pressure. In this case the given mole fractions give the composition of the vapor. Solving Raoult's law for the liquid mole fractions we obtain

$$x_1 = \frac{y_1 P}{P_1^{\text{sat}}},$$

$$x_2 = \frac{y_2 P}{P_2^{\text{sat}}}.$$

Adding the two equations we obtain an equation for the unknown pressure:

$$P = \frac{1}{\frac{y_1}{P_1^{\text{sat}}} + \frac{y_2}{P_2^{\text{sat}}}} = 0.587 \text{ bar}.$$

The corresponding composition of the liquid is

$$x_1 = \frac{y_1 P}{P_1^{\text{sat}}} = 0.185,$$

$$x_2 = 1 - x_1 = 0.815.$$

Comments Each of the above calculations has resulted in the points that define one tie line:

Bubble P: $T = 80\,°C$ $P = 0.639$ bar $x_1 = 0.3$ $y_1 = 0.448$,
Dew P: $T = 80\,°C$ $P = 0.587$ bar $x_1 = 0.185$ $y_1 = 0.3$.

These tie lines, which are shown on the Pxy graph in Figure 11-1, correspond to the same overall composition $z_1 = 0.3$. In the bubble P calculation, this refers to the composition of the liquid; in the dew P calculation, it refers to the composition of the vapor.

Example 11.5: Raoult's Law and Fugacity
Calculate the fugacity and fugacity coefficient of the components in the liquid and the vapor for a vapor-liquid system of acetonitrile and nitromethane at 80 °C that contains 30% by mol acetonitrile in the liquid.

Solution The system is at the bubble point of the solution, which was determined in the previous example:

$$T = 80\,°C, \quad P = 0.639 \text{ bar}, \quad x_1 = 0.3, \quad y_1 = 0.448.$$

11.3 VLE in Ideal Solution—Raoult's Law

The fugacity in the liquid phase is given by the Lewis-Randall rule, which under the assumptions of Raoult's law becomes,

$$f_i^L = x_i P_i^{\text{sat}}.$$

By numerical substitution,

$$f_1^L = (0.3)(0.955) = 0.285 \text{ bar},$$
$$f_2^L = (0.7)(0.504) = 0.353 \text{ bar}.$$

The fugacity coefficients are

$$\phi_i^L = \frac{f_i^L}{x_i P}$$

or,

$$\phi_1^L = \frac{(0.285 \text{ bar})}{(0.3)(0.639 \text{ bar})} = 1.494,$$

$$\phi_2^L = \frac{(0.353 \text{ bar})}{(0.7)(0.639 \text{ bar})} = 0.788.$$

The vapor phase in Raoult's law is treated as an ideal gas. The fugacity coefficient of the vapor components is 1, and the fugacity is

$$f_i^V = y_i P.$$

By numerical substitution,

$$f_1^V = (0.448)(0.639) = 0.285 \text{ bar},$$
$$f_2^V = (0.552)(0.639) = 0.353 \text{ bar}.$$

The fugacity coefficients in the vapor are given by

$$\phi_i^V = \frac{f_i^V}{P y_i},$$

and we find:

$$\phi_1^V = \frac{(0.285 \text{ bar})}{(0.448)(0.639 \text{ bar})} = 1,$$

$$\phi_2^V = \frac{(0.353 \text{ bar})}{(0.552)(0.639 \text{ bar})} = 1.$$

Comments The calculations confirm that the fugacity of component is the same in the liquid and in the vapor. We have also found that the fugacity coefficients in the vapor are 1. These results are not coincidental, they were both used, indirectly, in the calculation of the bubble pressure and vapor phase composition.

NOTE

Hypothetical Liquid State
At the conditions of Example 11.5 ($T = 80$ °C, $P = 0.639$ bar), only nitromethane ($P^{sat} = 0.504$ bar) exists as liquid in pure form. Pure acetonitrile ($P_2^{sat} = 0.955$ bar) is a vapor at this temperature. Ideal solution theory requires the properties of the pure *liquids* at the pressure and temperature of the solution. In the case of acetonitrile, this is a *hypothetical* liquid state, and the calculation of its properties requires extrapolation from the real liquid. For fugacity, the extrapolation is done according to the Poynting equation:

$$f_i^{id} = x_i \phi_i^{sat} P_i^{sat} \exp\left(\frac{P - P_i^{sat}}{RT} V_i^L\right), \qquad [11.11]$$

which we apply whether $P \geq P_i^{sat}$ (real liquid) or $P < P_i^{sat}$ (hypothetical liquid). Using the usual low pressure approximations ($\phi_i^{sat} = 1$, Poynting correction ≈ 1) this reduces to

$$f_i^{id} = x_i P_i^{sat},$$

which is the equation we used in Example 11.5. Other properties of the hypothetical liquid such as enthalpy or entropy, must also be extrapolated from those of the saturated liquid (see Example 11.8).

Example 11.6: Bubble and Dew Temperature
Calculate the bubble and dew temperature at 1 bar of a mixture that contains 30% by mol acetonitrile (1) and 70% by mol nitromethane (2).

Solution *Bubble temperature.* Solving Raoult's law for the vapor mol fractions and adding the resulting equations we obtain

$$P = x_1 P_1^{sat} + x_2 P_2^{sat}.$$

The composition of the liquid is $x_1 = 0.3$, $x_2 = 0.7$ and the only unknown is the bubble temperature, which appears implicitly in the saturation pressure via the Antoine equation. Several numerical alternatives exist for the solution of this equation. A simple approach is to write the equation as

$$\frac{x_1 P_1^{sat} + x_2 P_2^{sat}}{P} = 1,$$

and tabulate the left-hand side as a function of temperature:

T (K)	P_1^{sat} (bar)	P_2^{sat} (bar)	$(x_1 P_1^{sat} + x_2 P_2^{sat})/P$
...	...	...	...
360	1.178	0.638	0.80048
370	1.578	0.885	1.09282
380	2.077	1.203	1.46504
...	...	...	...

11.3 VLE in Ideal Solution–Raoult's Law

The solution is between $T = 360$ K and 370 K. By further refining the interval we find

$$T_{\text{bubble}} = 367.086 \text{ K} = 93.93 \,°\text{C}.$$

The saturation pressures at this temperature are calculated from the Antoine equation:

$$P_1^{\text{sat}} = 1.452 \text{ bar}, \quad P_2^{\text{sat}} = 0.806 \text{ bar}.$$

The gas-phase composition is obtained from Raoult's law:

$$y_1 = \frac{x_1 P_1^{\text{sat}}}{P} = 0.436,$$

$$y_2 = 1 - y_2 = 0.564.$$

Dew T calculation. We solve Raoult's law for the liquid mol fractions and add the resulting equations:

$$1 = \frac{y_1 P}{P_1^{\text{sat}}} + \frac{y_2 P}{P_2^{\text{sat}}}.$$

With $y_1 = 0.3$, $y_2 = 0.7$, the only unknown is temperature. By numerical solution we find

$$T = 369.388 \text{ K} = 96.23 \,°\text{C}.$$

The saturation pressures at this temperature are

$$P_1^{\text{sat}} = 1.551 \text{ bar}, \quad P_2^{\text{sat}} = 0.868 \text{ bar}.$$

The liquid-phase mol fractions are finally calculated from Raoult's law:

$$x_1 = \frac{y_1 P}{P_1^{\text{sat}}} = 0.193, \quad x_2 = 1 - x_1 = 0.807.$$

Comments Each of the above calculations produces a tie line:

Bubble *T*: $T = 93.93 \,°\text{C} \quad P = 1 \text{ bar} \quad x_1 = 0.3 \quad y_1 = 0.436,$

Dew *T*: $T = 96.23 \,°\text{C} \quad P = 1 \text{ bar} \quad x_1 = 0.193 \quad y_1 = 0.3.$

These tie-lines correspond to the same overall composition $z_1 = 0.3$ (see the *Txy* graph in Figure 11-1). In the bubble *T* calculation, this composition refers to the liquid; in the dew *T* calculation, it refers to the vapor.

Example 11.7: Phase Diagram
Calculate the *Pxy* graph of nitromethane/acetonitrile at 80 °C, and the *Txy* of the same system at 1 bar.

Solution The *Pxy* graph can be calculated by repeating the bubble *P* or the dew *P* calculation at various mole fractions in the range 0 to 1, at constant temperature 80 °C. The *Txy* graph may be constructed by repeating either the bubble *T* or the dew *T* calculation at constant pressure of 1 bar. Here we choose to solve the bubble problem for both the *Pxy* and *Txy* graphs. The results of these calculations are summarized in the table below:

	$T = 80$ °C			$P = 1$ bar	
x_1	y_1	P (bar)	x_1	y_1	T (°C)
0.0	0.000	0.504	0.0	0.000	100.76
0.1	0.174	0.549	0.1	0.165	98.36
0.2	0.321	0.594	0.2	0.309	96.09
0.3	0.448	0.639	0.3	0.436	93.94
0.4	0.558	0.685	0.4	0.547	91.89
0.5	0.654	0.730	0.5	0.646	89.94
0.6	0.740	0.775	0.6	0.734	88.08
0.7	0.815	0.820	0.7	0.812	86.31
0.8	0.883	0.865	0.8	0.882	84.63
0.9	0.945	0.910	0.9	0.944	83.02
1.0	1.000	0.955	1.0	1.000	81.47

These are plotted in Figure 11-1.

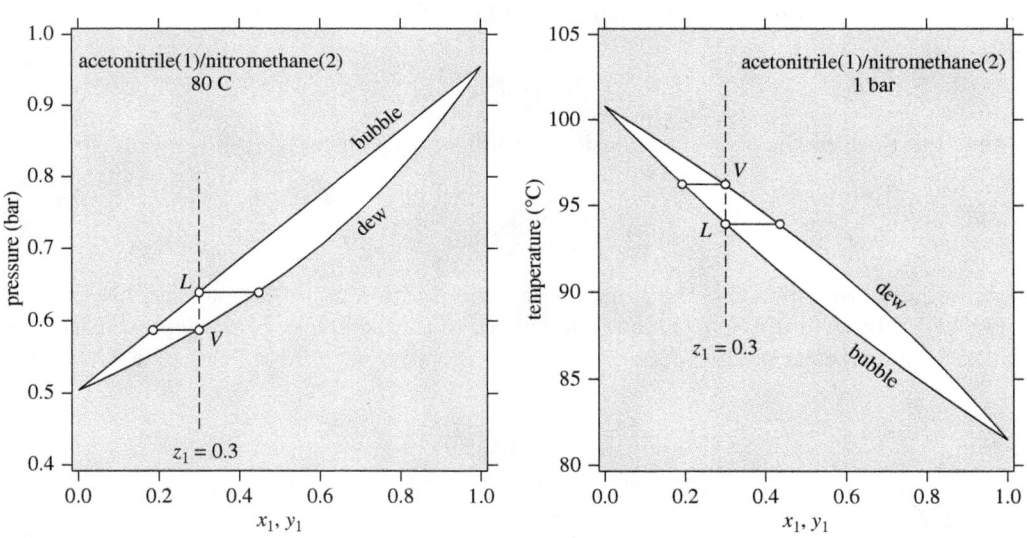

Figure 11-1: *Pxy* graph (left) and *Txy* graph (right) of the system acetonitrile (1)-nitromethane (2) calculated using Raoult's law.

11.4 Energy Balances

Energy balances involving ideal solutions require the internal energy (closed system) or the enthalpy (open system), both of which are given by straightforward expressions in terms of the corresponding properties of the pure liquids at the pressure and temperature of the solution:

$$U^{\text{id}} = \sum_i x_i U_i, \qquad [11.4]$$

$$H^{\text{id}} = \sum_i x_i H_i. \qquad [11.2]$$

For processes that do not involve change of composition, the change of the solution enthalpy is

$$\Delta H^{\text{id}} = \sum_i x_i \Delta H_i,$$

where ΔH_i is the enthalpy change of the pure liquid between the pressure and temperature of the initial and final state. If the process involves no phase change, the term ΔH_i may be calculated by integration of the liquid heat capacity between the initial (T_1) and final (T_2) temperatures:[4]

$$\Delta H_i = \int_{T_1}^{T_2} C_{P_i}^L \, dT.$$

Alternatively, ΔH_i may be calculated from tabulated values, or from an appropriate equation of state. The internal energy of the solution is calculated in a similar manner.[5]

For processes that involve phase transformation, the general approach is to use the ideal-solution equation for the enthalpy of the liquid and treat the vapor phase as an ideal-gas mixture. This reduces the problem to a calculation of the enthalpies of pure liquid and pure vapor components. If the calculation involves states near the phase boundary, hypothetical states may be involved, whose properties must be calculated by extrapolation from known real states. As an example, consider the constant-pressure heating of a solution that contains 30% acetonitrile

---

4. Since pressure makes negligible contribution to the enthalpy of liquids, this equation may be used even if the pressure changes between initial and final states.

5. Alternatively, once the enthalpy change is known, the internal energy of component i is

$$\Delta U_i = \Delta H_i - \Delta(PV_i) \approx \Delta H_i,$$

where $\Delta(PV_i)$ is the difference in the product PV_i between the initial and final state and is usually small enough to be negligible.

in nitromethane, at 1 bar. This is shown by the line LV on the Txy graph in Figure 11-1. The enthalpy change for this process is

$$\Delta H = H_{\text{dew}} - H_{\text{bubble}}, \tag{11.27}$$

where H_{dew} is the enthalpy of the vapor at the dew point and H_{bubble} is the enthalpy of the liquid in the bubble point. The enthalpy of the liquid calculated by the ideal-solution equation,

$$H_{\text{bubble}} = z_1 H_{L1} + z_2 H_{L2},$$

where z_i is the composition of the liquid and H_{Li} is the enthalpy of pure liquid i at the bubble temperature and pressure P, the fixed pressure of the process. Treating the vapor as an ideal-gas mixture, its enthalpy is

$$H_{\text{dew}} = z_1 H_{V1} + z_2 H_{V2},$$

where H_{Vi} is the enthalpy of pure vapor of component i at the dew temperature and pressure P. Taking the difference between the two expressions, the enthalpy change between the bubble and dew point is

$$\Delta H = z_1 (H_{V1} - H_{L1}) + z_2 (H_{V2} - H_{L2}). \tag{11.28}$$

The problem, therefore, is reduced to the calculation of enthalpy changes of pure components. Notice that two of the states involved in this equation are hypothetical: pure acetonitrile at the bubble temperature of the solution is a vapor; and pure nitromethane is liquid at the dew temperature of the mixture. Accordingly, H_{L1} refers to the enthalpy of a hypothetical liquid at T_{bubble}, P, and H_{V2} refers to a hypothetical vapor at T_{dew}, P. The properties of these hypothetical states may be approximated as those of the corresponding saturated phases: H_{L1} is taken to be the enthalpy of the pure saturated liquid at T_{bubble}, and H_{V2} as the enthalpy of the pure saturated vapor at T_{dew}.

NOTE

Hypothetical States

Hypothetical states appear because the ideal solution (and the ideal-gas) equations call for the properties of the pure *liquid* (and pure ideal-*gas* component) at the pressure and temperature of the solution. Specifying pressure, temperature *and* phase for a pure component represents an over-specification of state and thus may lead to conflict between the actual and specified state. The same is true with the ideal-mixture equations, which require the properties of pure components as gases at the temperature and pressure of the mixture. These hypothetical states are mathematical, not physical, states and their properties are calculated by the equations that apply to each phase.

The situation can be analyzed more clearly on the enthalpy/pressure chart of pure component (Figure 11-2). The more volatile component (acetonitrile in the previous example) is a hypothetical liquid at the bubble temperature. It is shown by point L_1, located on the liquid portion of the isotherm at T_{bubble} but extrapolated into the vapor-liquid region. Since isotherms in the compressed liquid region are almost vertical, the enthalpy of state L_1 is to a very good

11.4 Energy Balances

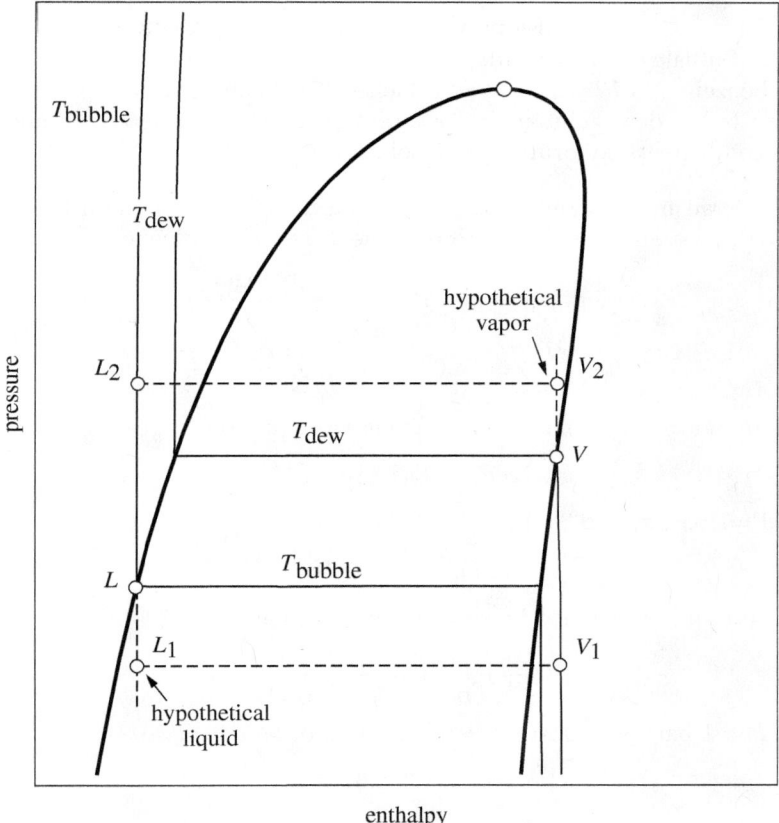

Figure 11-2: Enthalpy changes of pure component between bubble and dew temperature of solution.

approximation equal to that of saturated liquid, state L on the same isotherm. Using the notation on Figure 11-2,

$$H_{L1} \approx H_L.$$

Pure nitromethane vapor at the dew temperature corresponds to point V_2, which lies on the vapor portion of the isotherm at T_{dew} but is extrapolated into the vapor-liquid region. Because enthalpy in the ideal-gas state is independent of pressure, this portion of the isotherm is also vertical. Accordingly, the enthalpy of state V_2 is to a very good approximation the same as the enthalpy of saturated vapor at T_{dew}, shown by point V:

$$H_{V2} \approx H_V.$$

These approximations allow us to obtain the enthalpy of the hypothetical states from the corresponding values on the saturation line. At higher pressures close to the critical point these approximations break down because the isotherms are no longer vertical but bend significantly towards the saturation line.

Example 11.8: Enthalpy of Vaporization
A solution of benzene (23% by mole) and toluene (77% by mole) is brought from its bubble point at 1 bar to its dew point under constant pressure. Calculate the amount of heat assuming the components to form an ideal solution.

Solution First we must calculate the bubble and the dew temperature. The Antoine equations for this system are (1 refers to benzene, 2 refers to toluene):

$$\log_{10} \frac{P_1^{\text{sat}}}{\text{bar}} = 4.72583 - \frac{1660.65}{T/K - 1.461},$$

$$\log_{10} \frac{P_2^{\text{sat}}}{\text{bar}} = 4.07827 - \frac{1343.94}{T/K - 53.773}.$$

The bubble temperature is calculated by solving the equation

$$P = x_1 P_1^{\text{sat}}(T_B) + x_2 P_2^{\text{sat}}(T_B),$$

for T_B, with P=1 bar, $x_1 = 0.23$, $x_2 = 0.77$. We find

$$T_B = 373.57 \text{ K}.$$

The dew temperature is obtained by solving the equation

$$\frac{y_1 P}{P_1^{\text{sat}}(T_D)} + \frac{y_2 P}{P_2^{\text{sat}}(T_D)} = 1,$$

for T_D, with $P = 1$ bar, $y_1 = 0.23$, $y_2 = 0.77$. We find,

$$T_D = 378.32 \text{ K}.$$

Next, we collect the properties of the pure components at 378.32 K and 373.57 K, at a pressure of 1 bar. These are shown in the table below, along with the enthalpy of the saturated phases at the two temperatures.

		Benzene (1)					
T (K)	P (bar)	H (kJ/mol)	phase	T (K)	P (bar)	H (kJ/mol)	phase
373.57	1	32.943	vapor	378.32	1	33.451	vapor
373.57	1.8228	32.697	sat. vap.	378.32	2.0683	33.141	sat. vap.
373.57	1.8228	3.0667	sat. liq.	378.32	2.0683	3.7992	sat. liq.

		Toluene (2)					
T (K)	P (bar)	H (kJ/mol)	phase	T (K)	P (bar)	H (kJ/mol)	phase
373.57	0.75203	31.982	sat. vap.	378.32	0.86631	32.564	sat. vap.
373.57	0.75203	−1.8604	sat. liq.	378.32	0.86631	−0.99652	sat. liq.
373.57	1	−1.8589	liquid	378.32	1	−0.99572	liquid

Enthalpy at bubble point. At the bubble point, the system is liquid and its enthalpy is

$$H_B = z_1 H_{L1} + z_2 H_{L2},$$

11.4 Energy Balances

where H_{Li} is the enthalpy of the pure liquids at 373.57 K, 1 bar. For toluene, $H_{L2} = 1.8589$ kJ/mol. For benzene, this is a hypothetical liquid state; for H_{L1} we use the enthalpy of saturated liquid at 373.57 K, $H_{L1} = 3.0667$ kJ/mol. The enthalpy of the liquid at the bubble point is

$$H_B = (0.23)(3.0667) + (0.77)(-1.8589) = -0.7260 \text{ kJ/mol}.$$

Enthalpy at the dew point. Treating the vapor as an ideal-gas mixture, its enthalpy is

$$H_D = z_1 H_{V1} + z_2 H_{V2},$$

where H_{Vi} is the enthalpy of pure component i as vapor at 378.32 K, 1 bar. For benzene from the above table, $H_{V1} = 33.451$ kJ/mol. Toluene at this state is actually a liquid; the enthalpy of the hypothetical vapor is $H_{V2} = 32.564$ kJ/mol. The enthalpy of the vapor is

$$H_D = (0.23)(33.451) + (0.77)(32.564) = 32.768 \text{ kJ/mol}.$$

Finally, the enthalpy change between bubble and dew point is

$$\Delta H = (32.768) - (-0.7260) = 33.494 \text{ kJ/mol}.$$

This is the amount of heat that must be provided in order to boil the liquid until complete evaporation.

Comments The states of the pure liquid and vapor components are shown in Figure 11-2. Benzene at the bubble temperature corresponds to point L_1 (hypothetical liquid), and at the dew temperature to point V_1 (true vapor). Toluene at the bubble temperature is shown by point L_2 (true liquid), and at the dew temperature by point V_2 (hypothetical vapor). For demonstration purposes these states are shown on the same P-H graph. In reality, each component must be plotted on its own phase diagram.

Example 11.9: Ideal Solution in Heat Exchanger
The mixture of the previous example is heated in a heat exchanger from 25 °C to 150 °C at constant pressure of 1 bar. Calculate the amount of required heat.

Solution The initial temperature is below the saturation temperature of both components. Both pure components, therefore, exist as pure liquids at 25 °C, 1 bar. The final temperature is above the saturation temperature of both components. Both exist as vapor in the pure state. Using the ideal solution enthalpy for the initial state and the ideal-gas mixture equation for the vapor, the enthalpies at the two states are

$$H_a = z_1 H_{L1} + z_2 H_{L2},$$
$$H_b = z_1 H_{V1} + z_2 H_{V2},$$

and the amount of heat is

$$Q = \Delta H_{ab} = z_1(H_{V1} - H_{L1}) + z_2(H_{V2} - H_{L2}),$$

where $H_{Vi} - H_{Li}$ is the enthalpy change of pure component i from 1 bar, 25 °C, to 1 bar, 150 °C. These enthalpies may be calculated by any acceptable method. Using tabulated values from the NIST Chemistry WebBook:

P (bar)	T (°C)	H_1 (kJ/mol)	H_2 (kJ/mol)
1	25	-7.6978	-14.572
1	150	38.547	38.903

$$Q = (0.23)(38.547 - (-7.6978)) + (0.77)(38.903 - (-14.572)) = 51.81 \text{ kJ/mol}.$$

Comments When no hypothetical states are involved, the calculation of energy balances is straightforward.

11.5 Noncondensable Gases

When only a single component is present (pure fluid), below the boiling temperature the system is liquid, and above it vapor. When two liquids are mixed to form a solution, then a component produces a vapor over a range of temperatures, from the bubble point to the dew point. This range is expanded even further if the second component is a gas above its critical temperature. We refer to such a gas as *noncondensable* to emphasize the fact that it does not undergo a vapor-liquid transition as long as temperature remains above the critical point. For example, pure water[6] at 1 atm cannot exist as vapor below 100 °C but in the presence of air it produces vapor even at room temperature. The presence of a noncondensable gas alters the phase behavior of a system, even when the gas itself has negligible solubility in the liquid. In reality, gases have a finite solubility in the liquid and they participate in phase equilibrium by partitioning between both phases. This problem will be discussed in Chapter 13. However, under pressures and temperatures near ambient, the solubility of most gases in liquids is low and it is an acceptable approximation to assume that the liquid phase contains only the condensable species.

Suppose that an ideal solution of two components ($i = 1, 2$) is in the presence of a noncondensable gas (subscript g). Neglecting the solubility of the gas in the liquid, the liquid contains only the liquid components while the gas contains the

---

6. By "pure water" we mean a system that contains only water but no other substance. We may visualize such experiment by placing liquid water in a cylinder with a piston and moving the piston to touch the free surface of the liquid so that no gas is present.

11.5 Noncondensable Gases

noncondensable gas well as vapors of the liquid components. The vapor/liquid equilibrium of the condensable components is described by Raoult's law:

$$y_1 P = x_1 P_1^{\text{sat}}, \tag{11.29}$$

$$y_2 P = x_2 P_2^{\text{sat}}, \tag{11.30}$$

along with the normalization conditions

$$x_1 + x_2 = 1, \tag{11.31}$$

$$y_1 + y_2 + y_g = 1. \tag{11.32}$$

The only difference here is that the normalization condition of the gas-phase mole fractions must include the noncondensable species. It is the presence of the noncondensable species that allows the condensable components to form a vapor and establish vapor/liquid equilibrium. If it is absent, vapor/liquid equilibrium can be established only when the following equation is satisfied:

$$P = x_1 P_1^{\text{sat}} + x_2 P_2^{\text{sat}}.$$

When the noncondensable gas is present, this equation becomes[7]

$$P = x_1 P_1^{\text{sat}} + x_2 P_2^{\text{sat}} + y_g P,$$

and can be satisfied even when $x_1 P_1^{\text{sat}} + x_2 P_2^{\text{sat}} < P$ because air "fills in" the difference.

Applications–Humidification

In humidification we are dealing with a single condensable species (water) in the presence of a noncondensable gas (air). As a condensable species, water obeys Raoult's law:[8]

$$x_w P_w^{\text{sat}} = y_w P.$$

Since the solubility of air is neglected, the mol fraction of water in the liquid is 1 and the above equation simplifies to

$$P_w^{\text{sat}} = y_w P. \tag{11.33}$$

This equation states that the partial pressure of water vapor in the presence of liquid water is equal to the saturation pressure of pure water at this temperature. Equation 11.33, along with the normalization condition,

$$y_w + y_a = 1, \tag{11.34}$$

---

7. Multiply eq. (11.32) by the total pressure P, then substitute eqs. (11.29), (11.30) into the result.
8. A pure liquid is a special case of an ideal solution since all interactions are of the same strength (self-interactions).

fully specify the problem (y_a stands for the mole fraction of air). The relative humidity, RH, is defined as the ratio of the partial pressure of water vapor in air relative to the saturation pressure of water at the same temperature.

$$\text{RH} = \frac{y_w P}{P_w^{\text{sat}}(T)}. \tag{11.35}$$

If RH < 1, the air is *unsaturated* in water vapor. If RH $= 1$, it is saturated, that is, it holds the maximum amount of water vapor that is possible at the given pressure and temperature. The relative humidity of unsaturated air can be increased by manipulating any of the factors in the above equation: by bringing more water vapor into the system (increasing y_w), by increasing pressure, or by decreasing temperature (causes the saturation pressure to decrease). Once the system becomes saturated (RH $= 1$), it reaches the dew point. Attempting to increase the relative humidity past this point causes enough vapor to condense so as to keep the relative humidity equal to 1. The dew point, of the gas is defined by the condition RH $= 1$, or

$$\frac{y_w P}{P_w^{\text{sat}}(T)} = 1, \tag{11.36}$$

which in thermodynamic language expresses the fact that fugacity of water in the liquid (P_w^{sat}) is equal to the fugacity of water in the vapor ($y_w P$). The solution of problems based on these equations is demonstrated in the examples that follow. Although the discussion is for water in air, the same ideas apply to any condensable species in the presence of a noncondensable gas.

Example 11.10: Relative Humidity
The weather report calls for a temperature of 75 °F pressure 1 atm, and relative humidity 80%. What is the mol fraction of water vapor in air?

Solution From the steam tables at $T = 75$ °F $= 24$ °C we find the saturation pressure of water, $P_w^{\text{sat}} = 0.02982$ bar. The mole fraction of water vapor in air at saturation is

$$y_w^{\text{sat}} = \frac{0.02982}{1.013} = 0.02944.$$

The actual mole fraction is 80% of that at saturation:

$$y_w = (0.8)(0.02944) = 0.02355.$$

Notice that pure water (in the absence of air) would be liquid under these conditions. At 1 atm, water would produce vapor only at 100 °C. The presence of a new component, air, allows some water to evaporate at temperatures much lower than the boiling point of water.

11.5 Noncondensable Gases

Example 11.11: Dew Point
What is the dew point for the weather conditions of the previous problem?

Solution Condensation will occur when the relative humidity becomes 100%. As the temperature drops, the saturation pressure of water decreases and so does y_w^{sat}. The dew point is defined by the condition

$$\frac{y_w}{y_w^{\text{sat}}(T_{\text{dew}})} = 1 \quad \Rightarrow \quad y_w^{\text{sat}}(T_{\text{dew}}) = 0.02355.$$

We also have

$$y_w^{\text{sat}}(T_{\text{dew}}) = \frac{P^{\text{sat}}(T_{\text{dew}})}{P},$$

therefore,

$$P^{\text{sat}}(T_{\text{dew}}) = y_w^{\text{sat}}(T_{\text{dew}})P = (0.02355)(1.013) = 0.02386 \text{ bar.}$$

By interpolation in the steam tables we find $T_{\text{dew}} \approx 20.5\ °\text{C} = 68.9\ °\text{F}$. If the temperature drops below 70 °F you should carry an umbrella.

Example 11.12: Relative Humidity-2
If the temperature in the previous problem increases to 100 °F, what is the relative humidity?

Solution At 100 °F ≈ 38 °C, the saturation pressure is 0.06624 bar. The saturation mole fraction of water vapor at this temperature is

$$y_w^{\text{sat}} = \frac{0.06624}{1.013} = 0.06539.$$

Since the actual amount of vapor is the same as before, $y_w = 0.02355$, the relative humidity is

$$\text{RH} = \frac{0.02355}{0.06539} = 36\%.$$

The relative humidity decreases with temperature and increases with pressure.

Example 11.13: Dew Point-2
What is the dew point under the conditions of the previous example?

Solution The dew point depends on the actual mole fraction of vapor in the gas. This has not changed, so the dew temperature is the same, $T_{\text{dew}} \approx 20.5\ °\text{C} = 68.9\ °\text{F}$.

11.6 Summary

The ideal solution is a useful idealization that simplifies VLE calculations of systems composed of molecules whose cross interactions are nearly identical to their self interactions. It is an idealization because such similarity is encountered only in an approximate sense. It is very useful because several real systems are actually well approximated by ideal solutions. When this is the case, a property of the mixture can be calculated *solely* from the properties of the pure components and the amounts of components. The VLE problem, in particular, is very straightforward and reduces to Raoult's law. Compared to calculations that are based on equations of state, calculations with ideal solutions are very simple and require no interaction parameters or mixing rules. This simplicity comes with the limitation that ideal-solution theory works with a small number of systems whose self- and cross-interactions are similar. The true value of the ideal solution is that it provides the basis for treating more complex solutions through the introduction of excess properties. This approach is discussed in the next chapter.

11.7 Problems

Problem 11.1: A vapor mixture of acetone (1) and nitromethane (2) ($y_1 = 0.7$) initially at $P = 1$ bar and 110 °C is to be liquified by compression under constant temperature.
a) At what pressure does the system fully become a liquid?
b) Calculate the composition of the last bubble that condenses.
c) What is the fugacity of nitromethane in the liquid when the system is at the conditions of part (a)?
Assume that the system behaves according to Raoult's law. The saturation pressures of pure acetone and nitromethane at 110 °C are 463 kPa, and 132.4 kPa respectively.

Problem 11.2: A mixture contains n-pentane, n-hexane, and n-heptane at equal mole fractions. The temperature is 55 °C.
a) Calculate the bubble and dew pressure at 55 °C.
b) What is the phase of the system at 1.7 bar, 55 °C? If a two-phase system, calculate the amount and composition of each phase.
c) Repeat the previous part at 0.5 bar.
d) The pressure is adjusted so that 75% of the mixture is vapor while the temperature remains at 55 °C. Determine the pressure and the compositions of the two phases.
Additional data: At 55 °C the saturation pressures of n-C_5, n-C_6, and n-C_7 are 1.903 bar, 0.644 bar, and 0.231 bar respectively.

11.7 Problems

Problem 11.3: a) Calculate the enthalpy, entropy, and volume of a solution that contains 37.5% by mole heptane in decane, at 15 °C, 1 bar.
b) Calculate the amount of heat needed to raise the temperature of the solution to 40 °C under constant pressure of 1 bar.
c) Calculate the entropy change of the solution in part b.
Additional data: You may assume that the components form an ideal solution. The following properties of the pure components at 15 °C, 1 bar are known:

	C_7	C_{10}
ρ (mol/liter)	6.7823	5.1062
H (J/mol)	-17596	-52923
S (J/mol/K)	-52.632	-142.73
C_P (J/mol/K)	224.5	311.96

Problem 11.4: Assuming normal heptane and normal octane to form ideal solutions, do the following:
a) Calculate the chemical potential of n-C_7(1) in solution with n-C_8(2) at 40 °C, 1 bar, $x_1 = 0.45$. The reference state is the pure liquid at 40 °C, 1 bar.
b) Calculate the fugacity of normal heptane at the conditions of part (a).
c) Calculate the bubble pressure of the solution ($x_1 = 0.45$) at 40 °C.
d) Calculate the dew pressure of the solution at 40 °C.
e) Calculate the bubble temperature at 1 bar.
f) Calculate the dew temperature at 1 bar.
g) Calculate the amount and composition of vapor and liquid when a solution with $z_1 = 0.45$ is flashed to 40 °C, 0.065 bar.

Problem 11.5: A tank that contains a mixture of normal heptane (1)/normal octane is delivered to you. The composition of the mixture is unknown but the tank labels state that the bubble temperature of the mixture at 1 bar is 103 °C.
a) What is the composition of the mixture?
b) What is the dew temperature at 1 bar?
c) The mixture is to be flashed at 120 °C, 1.5 bar. Determine the recovery of normal heptane in the vapor stream.
d) You are requested to flash the mixture so that the liquid stream is 85% of the inlet (by mol). If the pressure is 1.5 bar, what temperature should be used?
Additional data: The saturation pressures of the two components are given by the Antoine equation:
$$\ln P_i^{\text{sat}} = A_i - B_i/(C_i + t)$$
where P^{sat} is in mm Hg, t is in Kelvin, and the parameters A_i, B_i, C_i, are

	A	B	C
nC_8	15.9426	3120.29	-63.63
nC_7	15.8737	2911.32	-56.51

Problem 11.6: a) A stream contains a mixture of 40% normal heptane (by mol) and 60% toluene. Calculate the bubble and dew pressure of the mixture at 120 °C.
b) You are requested to adjust the volatility of the mixture so that the bubble pressure at 120 °C is 4 bar. You may add either normal pentane or normal octane. Which component will you add and in what amount? Report the amount as mol of additive per mole of the original solution.
c) Regardless of your answer in part (b), your supervisor instructs you to add 0.2 moles of normal octane per mol of solution. This new stream is flashed and the vapor and liquid streams are separated. The available information for the streams in the separator are shown below. Fill out the missing entries and determine the pressure:

	Feed	Vapor	Liquid
Mol fraction of n-heptane		0.3728	0.2527
Mol fraction of n-octane		0.1609	
Mol fraction of toluene			
Mol/min in stream	100		

Additional information: The saturation pressures at 120 °C are: $P_{pentane} = 9.2$ bar, $P_{heptane} = 1.8$ bar, $P_{octane} = 1.1$ bar, $P_{toluene} = 1.0$ bar.

Problem 11.7: Assuming toluene (1) and heptane (2) to form ideal solutions, answer the following:
a) A solution of the two components that contains 62% mol toluene is slowly heated at constant pressure. When the solution begins to boil the temperature reads 120 °C. What is the pressure?
b) The bubble pressure at 120 °C of a mixture of the two components is 2 bar. What is the dew pressure of this mixture at 120 °C?
c) A mixture of toluene and heptane that contains 61% by mole toluene is at 120 °C, 2.45 bar. What is the fugacity of toluene in this mixture?
Additional data: The saturation pressures of the pure components at 120 °C are: $P_1^{sat} = 3.0$ bar, $P_2^{sat} = 1.81$ bar.

Problem 11.8: You have just been delivered a tank containing a mixture of n-butane (83.3% by mol) and n-octane at 25 °C and 1 bar. Your boss asks you to measure the viscosity of the mixture, so you have to take a sample for analysis. One technician (twice your age, and smiling) says you should collect the sample in a test tube because the contents are in the liquid phase. Another technician (three times your age, also smiling) says you should use a balloon because the contents of the tank are in the vapor phase. A third technician is looking at you waiting for your instructions. This is your first day on the job. What do you do?

11.7 Problems

State all your assumptions and justify your reasoning with calculations—this is no time for wrong decisions!

Additional data: The saturation pressures of the components at 25 °C are: n-butane: 2.34 bar; n-octane: 0.0175 bar

Problem 11.9: Assuming the system methanol(1)/ethanol(2) to form ideal solutions, do the following:

a) Calculate the bubble and dew pressure at 100 °C of a mixture that contains 20% methanol by mol.
b) Construct a Pxy graph at 100 °C.
c) Construct a Txy graph at 1.2 bar.
d) A stream that contains a methanol/ethanol mixture with 32% mol methanol is flashed to 100 °C to a pressure such that the molar rate of the vapor stream is 75% of the feed. Determine the temperature and the composition of the two phases.

The Antoine equation for the pure components are given below:

$$\text{MeOH:} \quad \ln \frac{P^{sat}}{\text{mmHg}} = 18.5875 - \frac{3626.55}{T/K - 34.29},$$

$$\text{EtOH:} \quad \ln \frac{P^{sat}}{\text{mmHg}} = 18.9119 - \frac{3803.98}{T/K - 41.68}.$$

Problem 11.10: a) A solution contains 25% by mol methanol in ethanol. The pressure is reduced at constant temperature $T = 85$ °C until the first bubble appears. What is the composition of that bubble?

b) We continue to reduce the pressure until the last drop evaporates. What is the composition of that drop?

Use the data in Problem 11.9.

Problem 11.11: A mixture contains 82% by mol nitromethane in acetonitrile. Determine the phase of the system at the conditions given below. If the system is in the vapor-liquid region, report the fraction of each phase.

a) 85 °C, 1 bar.
b) 120 °C, 2.7 bar.

Additional data: Assume the components to form an ideal solution. The saturation pressures for the pure components are given below.

$$\ln \frac{P_1^{sat}}{\text{kPa}} = 14.2742 - 2945.47/(t/°C + 224),$$

$$\ln \frac{P_2^{sat}}{\text{kPa}} = 14.2043 - 2972.64/(t/°C + 209).$$

Problem 11.12: a) Assuming that methanol (1), ethanol (2), acetonitrile (3), and nitromethane (4), form ideal solutions, calculate the bubble and dew pressure at 120 °C of a mixture that contains equal moles of each of the four components.
b) What is the phase of this mixture at 1 bar, 120 °C?
Additional data: the saturation pressures are: $P_1 = 6.43239$ bar, $P_2 = 4.34246$ bar, $P_3 = 3.02411$ bar, $P_4 = 1.75725$ bar.

Problem 11.13: The exhaust stream from a combustion process contains 10% water vapor (by mol). The pressure of the stream is 2 bar and the temperature 100 °C.
a) What is the relative humidity of the stream?
b) What is its dew temperature?
c) The stream is to be cooled to 25 °C at constant pressure. To avoid condensation, the exhaust stream is mixed with dry air prior to cooling. How much air is needed? Report the result in moles of dry air per mole of gas.

Problem 11.14: The pressure is 1 bar, the relative humidity is 75%, and the dew temperature is 25 °C.
a) What is the mol fraction of water vapor in the air?
b) What is the temperature?
c) Your dehumidifier removes water vapor from air by cooling the air to 12 °C. What fraction of water vapor is removed from the air?
d) What is the dew temperature of the air at the exit of the dehumidifier?
e) State and justify all your assumptions clearly.

Problem 11.15: a) A solution that contains nitromethane in acetonitrile is heated at constant pressure of 1 bar until the first bubble appears. If the temperature at this point is 85 °C, what is the composition of the original solution?
b) You are delivered a mixture of nitromethane in acetonitrile. The composition is not known but you are told that the dew temperature at 1 bar is 93.5 °C. What is the phase at 111 °C, 2 bar?

Problem 11.16: At 80 °C, 1.32 bar the system methanol(1)/ethanol(2) is in vapor-liquid equilibrium. The composition of the two phases is $x_1 = 0.25$, $y_1 = 0.37$. Calculate the fugacity, fugacity coefficient of each component in each phase. You may assume that the vapor phase is in the ideal-gas state.

Chapter 12

Nonideal Solutions

The ideal solution assumes equal strength of self- and cross-interactions between components. When this is not the case, the solution deviates from ideal behavior. Deviations are simple to detect: upon mixing, nonideal solutions exhibit volume changes (expansion or contraction) and exhibit heat effects that can be measured. Such deviations are quantified via the *excess* properties. An important new property that we encounter in this chapter is the *activity* coefficient. It is related to the excess Gibbs free energy and is central to the calculation of the phase diagram.

In this chapter you will learn how to:

1. Obtain excess properties (volume, enthalpy, Gibbs free energy) from experimental data.

2. Calculate activity coefficients using an appropriate model.

3. Calculate the phase diagram of nonideal mixtures

12.1 Excess Properties

An excess property is the difference between the property of solution and the same property calculated by the ideal-solution equations:

$$F^{\mathrm{E}} = F - F^{\mathrm{id}}. \tag{12.1}$$

For the primary properties of interest, volume, enthalpy, and entropy, this definition produces the following results:

$$V^{\mathrm{E}} = V - \sum_i x_i V_i = \Delta V_{\mathrm{mix}}, \tag{12.2}$$

$$H^{\mathrm{E}} = H - \sum_i x_i H_i = \Delta H_{\mathrm{mix}}, \tag{12.3}$$

$$S^{\mathrm{E}} = S - \sum_i x_i S_i + R \sum_i x_i \ln x_i. \tag{12.4}$$

Expressions for other properties are obtained similarly. For internal energy we have

$$U^{\mathrm{E}} = U - \sum_i x_i U_i = \Delta U_{\mathrm{mix}}. \tag{12.5}$$

Similarly, for the Gibbs free energy the result is

$$G^{\mathrm{E}} = G - \sum_i x_i G_i - RT \sum_i x_i \ln x_i. \qquad (12.6)$$

By definition, the excess properties of ideal solution are zero. We may view the excess properties as corrections that must be added to the ideal-solution equations in order to obtain the correct property of solution.

A special case is the limit of pure component, reached when the amount (moles) of component i in the solution is increased indefinitely while keeping all other components constant. In this case, $x_i \to 1$ for component i and $x_j \to 0$ for all others. According to (12.2)–(12.6), in this limit all excess properties reduce to zero.[1] This means that a pure liquid is a special case of ideal solution.[2] In a binary solution, a graph of any excess property as a function of x_1 meets the zero axis both at $x_1 = 1$ (pure component 1) and at $x_1 = 0$ (pure component 2).

NOTE

Excess Properties as a Mathematical Operator
It is convenient to think of the excess property as a mathematical operator that removes the ideal-solution part from a thermodynamic property. It is a linear operator and can be combined with other operators, such as the partial molar differentiation. Expressions that can be written between regular properties may be written for the excess and for the partial molar excess properties. For example, starting with the fundamental relationship

$$dG = -S dT + V dP,$$

which applies to systems of constant composition, we may immediately write the corresponding relationships for the excess and the excess partial molar properties:

$$dG^{\mathrm{E}} = -S^{\mathrm{E}} dT + V^{\mathrm{E}} dP,$$

$$d\overline{G}_i^{\mathrm{E}} = -\overline{S}_i^{\mathrm{E}} dT + \overline{V}_i^{\mathrm{E}} dP.$$

In the last equation we have applied both the excess and the partial molar operator. The order in which they are applied does not matter.

Excess Partial Molar Properties

Excess partial molar properties are obtained by applying the partial molar operator on an excess property. Recall that the definition of the partial molar operator is

$$\overline{F}_i = \left(\frac{\partial F^{\mathrm{tot}}}{\partial n_i} \right)_{P,T,n_j},$$

---

1. The summation $\sum_i x_i \ln x_i$ goes to zero when $x_i = 1$, $x_{j \neq i} = 0$.
2. In a pure liquid the only type of interaction is self-interaction. In this respect, the pure liquid satisfies the ideal-solution condition that all interactions be of equal strength.

12.1 Excess Properties

and involves the differentiation of extensive property F^{tot} with respect to n_i. It is convenient to express partial molar property $\bar{F}_i$ as a derivative in terms of x_i. In the case of a binary solution, this leads to simple results, as we will see. We begin by obtaining the relationship between derivatives in n_i and in x_i. Starting with

$$x_1 = \frac{n_1}{n_1 + n_2},$$

we take the derivative with respect to n_1 at constant n_2:

$$\left(\frac{\partial x_1}{\partial n_1}\right)_{n_2} = \left[\frac{\partial}{\partial n_1}\left(\frac{n_1}{n_1 + n_2}\right)\right]_{n_2} = \frac{n_2}{(n_1 + n_2)^2} = \frac{x_2}{n_1 + n_2}. \quad (12.7)$$

We return to the definition of partial molar and expand the right-hand side by applying the chain rule:

$$\bar{F}_1 = \left(\frac{\partial(n_1 + n_2)F}{\partial n_1}\right)_{n_2} = \left(\frac{\partial(n_1 + n_2)}{\partial n_1}\right)_{n_1} F + n\left(\frac{\partial F}{\partial n_1}\right)_{P,T,n_2}$$

$$= F + n\left(\frac{\partial F}{\partial x_1}\right)_{P,T}\left(\frac{\partial x_1}{\partial n_1}\right)_{n_2}.$$

Using eq. (12.7) the above results becomes

$$\boxed{\bar{F}_1 = F + (1 - x_1)\left(\frac{\partial F}{\partial x_1}\right)_{P,T}.} \quad (12.8)$$

The corresponding expression for the partial molar property of component 2 is obtained by symmetry: exchanging subscripts 1 and 2 we obtain

$$\bar{F}_2 = F + (1 - x_2)\left(\frac{\partial F}{\partial x_2}\right)_{P,T}.$$

Using $dx_2 = -dx_1$ we obtain $\bar{F}_2$ in the equivalent form

$$\boxed{\bar{F}_2 = F - x_1\left(\frac{\partial F}{\partial x_1}\right)_{P,T}.} \quad (12.9)$$

Here, both partial molar properties are given as derivatives with respect to the mol fraction of component 1. Equations (12.8) and (12.9) apply to any partial molar properties, including excess partial molar. There is a simple graphical interpretation of these equations, as shown in Figure 12-1. In a plot of property F against x_1, we draw a tangent line at a mol fraction of interest. The zero intercept of this line is the partial molar property of component 2 and the intercept at x_1 is the partial

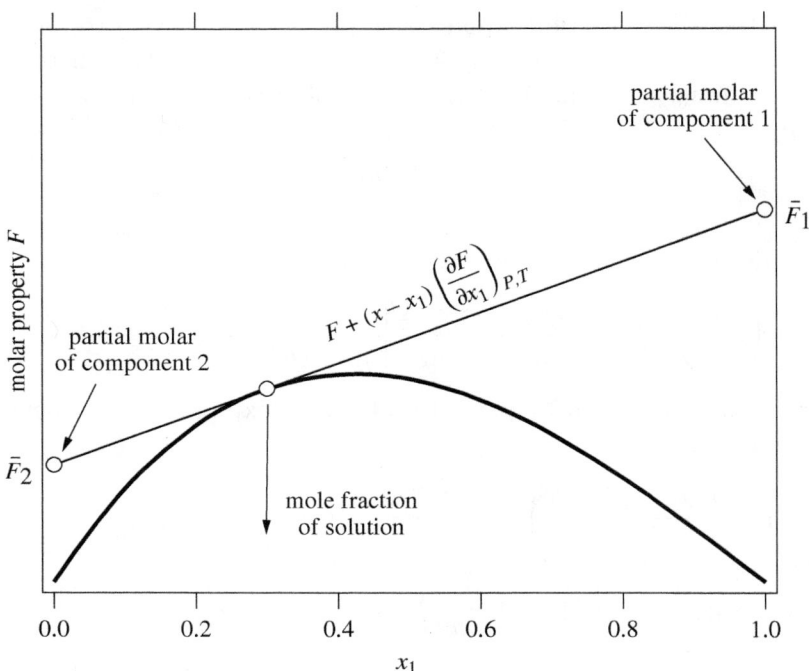

Figure 12-1: Graphical interpretation of partial molar properties in binary mixture.

molar of component 1. This is easy to prove if we notice that the tangent line has the equation

$$y = F + (x - x_1)\left(\frac{\partial F}{\partial x_1}\right)_{P,T},$$

where F and $\partial F/\partial x_1$ are the values of property F and its derivative at the point of the tangent. At $x = 0$ it reduces to eq. (12.9), which gives $\bar{F}_2$; at $x = 1$ it reduces to eq. (12.8), which gives $\bar{F}_1$. This graphical construction is useful in determining the partial molar excess properties from experimental data, as demonstrated with the examples below.

Example 12.1: Volume of Mixing
The data below give the density of ethanol-water solutions as a function of the weight fraction of ethanol. Determine the volume of solution that is formed when 0.1 m³ of ethanol is mixed with 0.1 m³ of water and report the excess volume in m³/mol.

12.1 Excess Properties

<table>
<tr><th colspan="4">Densities of ethanol/water solutions at 20 °C.[†]</th></tr>
<tr><th>Ethanol conc.
(weight fraction)</th><th>ρ
g/cm³</th><th>Ethanol conc.
(weight fraction)</th><th>ρ
g/cm³</th></tr>
<tr><td>0.00</td><td>1.0000</td><td>0.54</td><td>0.9049</td></tr>
<tr><td>0.10</td><td>0.9819</td><td>0.60</td><td>0.8911</td></tr>
<tr><td>0.20</td><td>0.9687</td><td>0.70</td><td>0.8676</td></tr>
<tr><td>0.30</td><td>0.9539</td><td>0.80</td><td>0.8436</td></tr>
<tr><td>0.40</td><td>0.9352</td><td>0.90</td><td>0.8180</td></tr>
<tr><td>0.50</td><td>0.9139</td><td>1.00</td><td>0.7893</td></tr>
</table>

[†]CRC Handbook of Chemistry and Physics, 60th ed. (New York: CRC Press, 1979), p. D-236.

Solution The masses of ethanol and water in the solution are

$$M_E = V_E^{\text{tot}} \rho_E = (0.1 \text{ m}^3)(789.3 \text{ kg/m}^3) = 78.93 \text{ kg},$$

$$M_W = V_W^{\text{tot}} \rho_W = (0.1 \text{ m}^3)(1000 \text{ kg/m}^3) = 100 \text{ kg}.$$

The mass fraction of ethanol is

$$w_E = \frac{78.93}{78.93 + 100} = 0.441.$$

By interpolation in the above table, the density of the solution at $w_1 = 0.441$ is

$$\rho = 0.9267 \text{ g/cm}^3 = 926.7 \text{ kg/m}^3,$$

and the volume of the solution is

$$V^{\text{tot}} = \frac{M}{\rho} = \frac{178.93 \text{ kg}}{926.7 \text{ kg/m}^3} = 0.1931 \text{ m}^3.$$

Upon mixing the volume changes by:

$$\Delta V^{\text{tot}} = 0.1931 - 0.1 - 0.1 = -0.00692 \text{ m}^3.$$

To express this on a molar basis, we calculate the number of moles in the solution:

$$n_E = \frac{78.93 \text{ kg}}{46.07 \times 10^{-3} \text{ kg/mol}} = 1713.26 \text{ mol},$$

$$n_W = \frac{100 \text{ kg}}{18.01 \times 10^{-3} \text{ kg/mol}} = 5552.47 \text{ mol}.$$

The molar volume of mixing, also equal to the excess molar volume, is

$$V^E = \frac{-0.00692 \text{ m}^3}{(1713.26 + 5552.47) \text{ mol}} = -0.952 \times 10^{-6} \text{ m}^3/\text{mol} = -0.952 \text{ cm}^3/\text{mol}.$$

Comments When ethanol is mixed with water at 25 °C, the volume of the solution contracts relative to the volume of the pure components. In this example, the volume contracts by about 3.5%.

Example 12.2: Excess Volume
With the data of the previous example, plot the excess volume as a function of the mol fraction of ethanol and determine the partial molar volumes of the two components at ethanol mol fraction of 0.4.

Solution For every entry of the table in the previous example we calculate the molar volume of solution and the corresponding mol fraction of ethanol. The calculation is facilitated by choosing a basis of 1 kg of solution. The moles of each component are

$$n_E = \frac{w_E}{M_{mE}}, \quad n_W = \frac{1 - w_E}{M_{mW}},$$

where w_E is the weight faction of ethanol, and $M_{mE} = 46.07 \times 10^{-3}$ kg/mol and $M_{mW} = 18.01 \times 10^{-3}$ kg/mol are the molar masses of ethanol and water, respectively. The molar volume of the solution is

$$V = \frac{1}{\rho}\left(\frac{1 \text{ kg}}{n_E + n_W}\right).$$

Finally, the excess volume is calculated as

$$V^E = V - x_E V_E - (1 - x_E) V_W,$$

where $V_E = 0.05837$ m^3/mol and $V_W = 0.01801$ m^3/mol are the molar volumes of pure ethanol and water, calculated from the given data. These calculations are summarized in the table below:

x (−)	ρ (kg/m^3)	V (10^{-6}m^3/mol)	V^E (10^{-6}m^3/mol)	x (−)	ρ (kg/m^3)	V (10^{-6}m^3/mol)	V^E (10^{-6}m^3/mol)
0.000	1000	18.01	0.0000	0.315	905	29.66	−1.0481
0.042	982	19.53	−0.1584	0.370	891	31.85	−1.0774
0.089	969	21.17	−0.4323	0.477	868	36.19	−1.0755
0.143	954	23.10	−0.6998	0.610	844	41.64	−0.9891
0.207	935	25.46	−0.8926	0.779	818	48.73	−0.7078
0.281	914	28.34	−1.0167	1.000	789	58.37	0.0000

These results are plotted in Figure 12-2. The excess partial molar volumes are indicated by the arrows and may be obtained by the graphical method. For a more accurate determination, the experimental excess volume is fitted to a polynomial, and the excess partial molar volumes are obtained by application of eqs. (12.8) and (12.9). The fitted polynomial for V^E is

$$V^E = c_4 x_1^4 + c_3 x_1^3 + c_2 x_1^2 + c_1,$$

with

$$c_1 = -6.19944 \times 10^{-6}, c_2 = 1.1561 \times 10^{-5}, c_3 = -8.6877 \times 10^{-6}, c_4 = 3.32617 \times 10^{-6}.$$

12.1 Excess Properties

Using the fitted equation we find the following values at $x_1 = 0.4$:

$$V^E = -1.101 \times 10^{-6} \text{ m}^3/\text{mol},$$

$$\overline{V}_E^E = -1.262 \times 10^{-6} \text{ m}^3/\text{mol},$$

$$\overline{V}_W^E = -9.932 \times 10^{-7} \text{ m}^3/\text{mol}.$$

These correspond to the three points on the tangent line in Figure 12-2.

NOTE

Fitting Excess Properties—Redlich-Kister Expansion

All excess properties of binary solutions must go to zero at $x_1 = 0$ and $x_1 = 1$. One equation that satisfies this condition is

$$f = x_1(1 - x_1)\left(c_0 + c_1 x_1 + c_2 x_2^2 + \cdots\right),$$

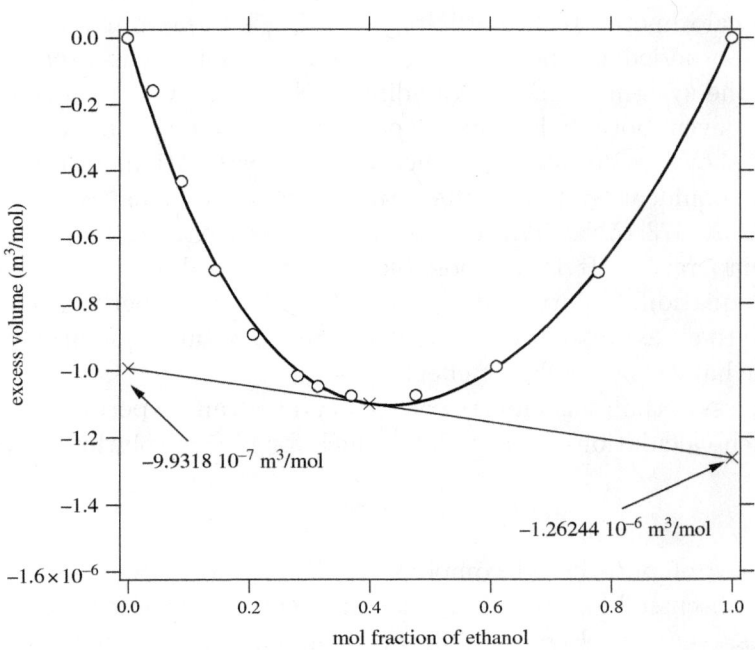

Figure 12-2: Excess volume of ethanol/water solutions at 20 °C (see Example 12.2).

where c_i are coefficients, to be obtained by fitting. This polynomial can be written in the more symmetric form as

$$f = x_1 x_2 \Big(a_0 + a_1(x_1 - x_2) + a_2(x_1 - x_2)^2 + \cdots \Big) \tag{12.10}$$

where a_i is a new set of coefficients that are related to the coefficients c_i ($c_0 = a_0 - a_1 + a_2 - \cdots$, $c_1 = 2a_1 - 4a_2 + \cdots$, and so on). In the above form, the polynomial is known as the *Redlich-Kister* expansion and is commonly used to fit experimental data of excess properties, not only volume, but also enthalpy and Gibbs free energy. This is not the only mathematical form used to fit excess properties. Nonpolynomial forms are often necessary with systems that are highly nonideal.

12.2 Heat Effects of Mixing

The mixing of components that form nonideal solution is accompanied by heat effects originating from the fact that the enthalpy of the solution is generally different than the sum of enthalpies of the pure components before mixing. The excess enthalpy of solution is the same as the enthalpy of mixing. It is determined experimentally by calorimetry. If the enthalpy of mixing is positive mixing is *endothermic* (heat must be added to the solution); if negative mixing is *exothermic* (heat is rejected by the system to the surroundings). Nonideal systems may exhibit either behavior, or even both behaviors, depending on temperature and composition. Figure 12-3 shows smoothed experimental data for solutions of ethanol in water. This highly nonideal system shifts from exothermic behavior at 323.15 K to endothermic at 373.15 K. The excess enthalpy is a strong function of composition and temperature. It is not possible generally to describe these dependences by a single equation. The typical approach is to fit the experimental data at each temperature to an expression with enough terms to obtain good accuracy. Pressure, on the other hand, has negligible effect.

If the excess enthalpy is known, energy balances can be performed. The starting point for such calculations is eq. (12.3), which for binary solution becomes

$$H = x_1 H_1 + x_2 H_2 + H^E.$$

The enthalpies of pure liquid components, H_1 and H_2, are calculated by any of the methods discussed for pure components; the excess enthalpy is calculated either from graphs, such as Figure 12-3, or from equations fitted to experimental data.

12.2 Heat Effects of Mixing

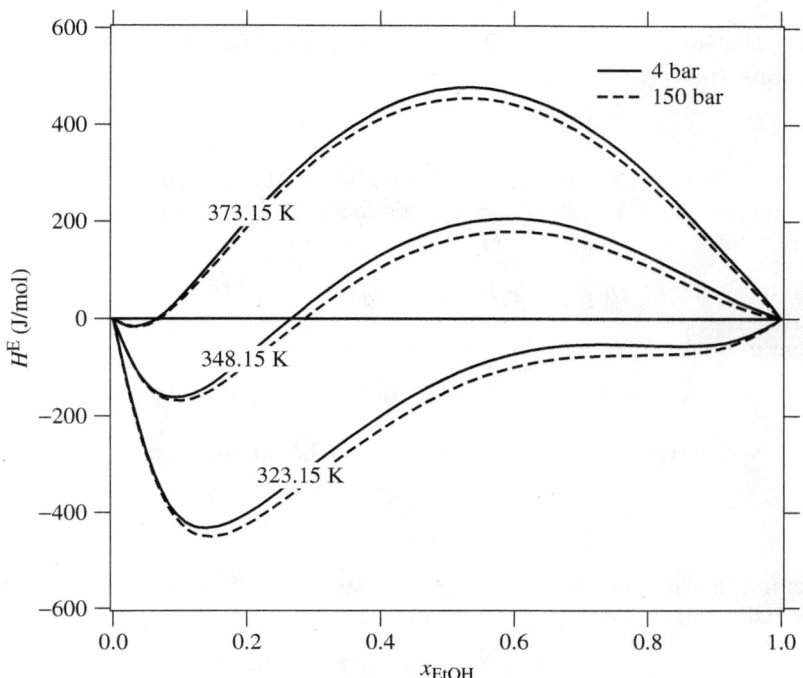

Figure 12-3: Excess enthalpy of ethanol(1)-water(2) solutions (from Ott et al., *J. Chem. Thermodynamics*, **18**[1986], pp. 867–875).

Example 12.3: Isothermal and Adiabatic Mixing
Determine the amount of heat exchanged with the surroundings when 0.2 mol of ethanol is mixed isothermally with 0.8 mol of water at 323.15 K. If both pure components are initially at 323.15 K and mixing is adiabatic, what is the temperature of the solution?

Solution *Isothermal mixing.* For isothermal mixing, the amount of heat is equal to the excess enthalpy of mixing. From Figure 12-3, the excess enthalpy at 323.15 K, $x_{\text{ethanol}} = 0.2$ is
$$Q = H^{\text{E}} = -403 \text{ J/mol}.$$
Therefore, the solution transfers to the bath 403 kJ/mol of heat.

Adiabatic mixing. If mixing is adiabatic, the energy balance reads
$$Q = 0 = \Delta H,$$

where ΔH is the enthalpy change between the final state (solution at temperature T') and the initial state (pure components at $T = 323.15$ K):

$$\Delta H = H(T') - x_1 H_1(T) - x_2 H_2(T) = 0. \qquad [\text{A}]$$

Here, we use subscript 1 for ethanol and 2 for water. In this equation the only unknown is the final temperature T'. To recognize this fact, we must express the enthalpy at T' in terms of enthalpies at T. Using eq. (12.3), the solution enthalpy is

$$H(T') = x_1 H_1(T') + x_2 H_2(T') + H^{\text{E}}(T').$$

Substituting into eq. [A] we obtain

$$x_1(H_1(T') - H_1(T)) + x_2(H_2(T') - H_2(T)) - H^{\text{E}}(T'). \qquad [\text{B}]$$

We express the difference $H_i(T') - H_i(T)$ using the liquid heat capacity of pure component,

$$H_i(T') - H_i(T) = \int_T^{T'} C_{Pi}^L dT \approx C_{Pi}(T' - T).$$

The approximation on the right-hand side is acceptable if the temperature change is not very large. Using this result in eq. [B] we obtain,

$$(x_1 C_{P1} + x_2 C_{P2})(T' - T) + H^{\text{E}}(T') = 0. \qquad [\text{C}]$$

Here, the only unknown is the final temperature T'. Still, to obtain a numerical solution we must express the excess enthalpy as a function of temperature, while the data are available in graphical form and at a few temperatures only. We accomplish this by linearizing the excess energy around its value at $T = 323.15$ K, by writing

$$H^{\text{E}}(T') = H^{\text{E}}(T) + \left(\frac{\partial H^{\text{E}}}{\partial T}\right)_{P,x}(T' - T), \qquad [\text{D}]$$

and approximate the derivative of excess enthalpy with temperature as

$$\left(\frac{\partial H^{\text{E}}}{\partial T}\right)_{P,x} \approx \frac{H^{\text{E}}(348.15 \text{ K}) - H^{\text{E}}(323.15 \text{ K})}{(348.15 - 323.15) \text{ K}}.$$

This approximation amounts to obtaining H^{E} in eq. [D] by linear interpolation between the known values 323.15 and 348.15 K. Substituting eq. [D] into [C] and solving for the unknown temperature we obtain the final result,

$$T' - T = \frac{-H^{\text{E}}(T)}{x_1 C_{P1} + x_2 C_{P2} + \left(\dfrac{\partial H^{\text{E}}}{\partial T}\right)_{P,x}}. \qquad [\text{E}]$$

Numerical substitutions. The heat capacities of the pure liquids at $T = 323.15$ K are

$$C_{P1} = 123.1 \text{ J/mol K}, \quad C_{P2} = 75.2 \text{ J/mol K}.$$

The excess enthalpies at 323.15 and 348.15 K are

$$H^{\mathrm{E}}(323.15\ \mathrm{K}) = -403\ \mathrm{J/mol}, \quad H^{\mathrm{E}}(348.15\ \mathrm{K}) = -76\ \mathrm{J/mol},$$

from which we find

$$\left(\frac{\partial H^{\mathrm{E}}}{\partial T}\right)_{P,x} = \frac{(-76) - (-403)}{348.15 - 323.15} = 13\ \mathrm{J/mol\ K}.$$

Substitution into eq. [E] gives

$$\Delta T = \frac{-(-403)}{(0.2)(123.1) + (0.8)(75.2) + (13)} = 4.1\ \mathrm{K}.$$

Comments Since mixing is exothermic, under isothermal conditions we observe an exchange of heat from the system to the surroundings; under adiabatic conditions we observe a temperature rise. The temperature rise in this problem is about 4 K, small enough that the approximation of constant heat capacities is acceptable.

Example 12.4: Excess Heat Capacity
Calculate the excess constant-pressure heat capacity and the actual constant-pressure of a solution of ethanol in water at 323 K that contains 20% ethanol by mol and use this result to obtain the constant pressure heat capacity of the solution.

Solution The enthalpy of nonideal solution is

$$H = H^{\mathrm{id}} + H^{\mathrm{E}}.$$

Differentiation with respect to temperature at constant pressure and composition gives

$$C_P = C_P^{\mathrm{id}} + \left(\frac{\partial H^{\mathrm{E}}}{\partial T}\right)_{P,x}.$$

Here the heat capacity is given as the sum of the ideal-solution heat capacity plus a correction. We identify the correction term as the excess heat capacity:

$$C_P^{\mathrm{E}} = \left(\frac{\partial H^{\mathrm{E}}}{\partial T}\right)_{P,x}.$$

The above derivative was estimated in the previous example using a finite difference between the known values of the excess enthalpy at 323.15 and 348.15 K, where we obtained the value 13.051 J/mol K. Therefore, the excess heat capacity is,

$$C_P^{\mathrm{E}} = 13.0\ \mathrm{J/mol\ K}.$$

The heat capacity of the solution is

$$C_P = C_P^{\text{id}} + C_P^{\text{E}}.$$

In ideal solution, the enthalpy is given by

$$H^{\text{id}} = x_1 H_1 + x_2 H_2.$$

Taking the derivative with respect to temperature while holding pressure and composition held constant, we have

$$C_P^{\text{id}} = x_1 C_{P1} + x_2 C_{P2}. \qquad \text{[A]}$$

That is, the heat capacity of ideal solution is the average heat capacity of the pure components. The heat capacity of the nonideal solution is

$$C_P = x_1 C_{P1} + x_2 C_{P2} + C_P^{\text{E}}.$$

Using $x_1 = 0.2$, $C_{P1} = 123.1$ J/mol K, $C_{P2} = 75.2$ J/mol K,

$$C_P = (0.2)(123.1) + (0.8)(75.2) + (13.0) = 97.8 \text{ J/mol K}.$$

Comments Recognizing the denominator of eq. [E] in the previous example as the heat capacity of the solution, the temperature rise for adiabatic mixing can be written in the simpler form,

$$\Delta T = \frac{-H^{\text{E}}}{C_P},$$

where C_P is the constant-pressure heat capacity of the solution. This heat capacity contains contributions from the pure components as well as from their nonideal interaction. In this particular example, nonideal interactions account for about 13% of the solution heat capacity.

Enthalpy Charts

Solution enthalpies are typically presented graphically in the form of excess enthalpies plotted as a function of composition and temperature, such as those in Figure 12-3. From such data we can obtain the enthalpy of solution by adding the excess enthalpy to the ideal-solution enthalpy:

$$H = x_1 H_1 + x_2 H_2 + H^{\text{E}}.$$

This equation may be used to calculate energy balances, as was done in Example 12.3. The same equation may be used to tabulate the enthalpy of the solution as a function of temperature and composition. Combining with an ideal-mixture treatment of the vapor phase, it is possible then to construct the enthalpy

12.2 Heat Effects of Mixing

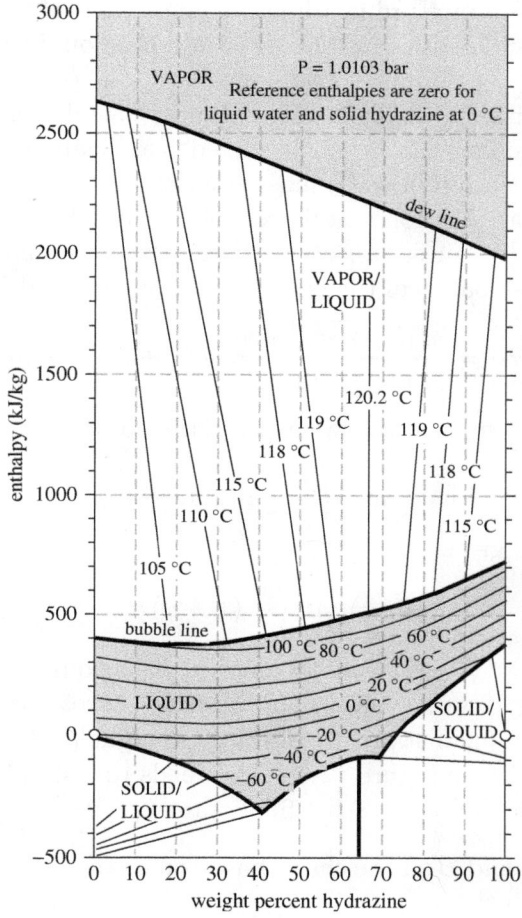

Figure 12-4: Enthalpy of hydrazine/water mixtures (adapted from Tyner, *AIChE J.* **1**, no. 1 [1955], p. 87).

chart over the entire phase region of the mixture. One example of such graph is given in Figure 12-4, which shows the enthalpy of hydrazine/water mixtures as a function of composition over a temperature range that includes solid, liquid, and vapor phases. There is a solid phase near the bottom, a liquid region, shown by the gray area in the lower part of the graph, a vapor region, shown by the gray area at the top, and a vapor-liquid coexistence region in between. Isotherms in the vapor-liquid region are straight lines that connect points on the bubble line (upper boundary of the liquid region) to the dew line (lower boundary of the vapor region). The linearity in this region is a direct consequence of the lever rule, according to which, the enthalpy of a two-phase mixture is a linear combination of the molar enthalpies in each phase. In the liquid, isotherms are curved due to nonideal interactions between the components.[3]

Enthalpy-composition charts simplify enthalpy calculations considerably. Adiabatic mixing, in particular, can be easily solved graphically in such charts. Suppose that a mass m_A of solution A with mass fraction x_A of component 1 (this is the component plotted on the horizontal axis of the enthalpy-composition chart) is mixed adiabatically with mass m_B of solution B that contains mass fraction x_B of component 1. By mass balance, the composition of the resulting mixture (C) is

$$x_C = \frac{m_A x_A + m_B x_B}{m_A + m_B} = w x_A + (1-w) x_B, \tag{12.11}$$

where $w = m_A/(m_A + m_B)$ is the weight fraction of solution A. The energy balance is

$$(m_A + m_B) H_C = m_A H_A + m_B H_B,$$

which, solved for H_C, gives

$$H_C = w H_A + (1-w) H_B. \tag{12.12}$$

Equations (12.11) and (12.12) are parametric equations of a linear relationship between H_C and x_C.[4] If we let w vary (by varying the relative amounts of solutions A and B) and plot H_C against x_C, we obtain a straight line that passes from points (x_A, H_A) and (x_B, H_B) (see Figure 12-5). This leads to a simple graphical procedure for calculating the temperature of the final state:

1. Define points A and B on the chart, based on the composition and temperature of the initial solutions.

2. Calculate the composition x_C of the final solution.

3. The state of the final solution lies on the straight line AB, at the composition x_C calculated above.

4. The temperature of the final solution corresponds to the isotherm that passes through point C.

This procedure also allows us to determine the phase of the final mixture. If point C lies in the two-phase region, then x_C is understood to refer to the overall composition of the mixture. The compositions of the individual phases are obtained

---

3. If the solution were ideal, isotherms in the liquid would be straight lines.
4. To prove this, eliminate w between eqs. (12.11) and (12.12) to obtain an equation between H_C and x_C.

12.2 Heat Effects of Mixing

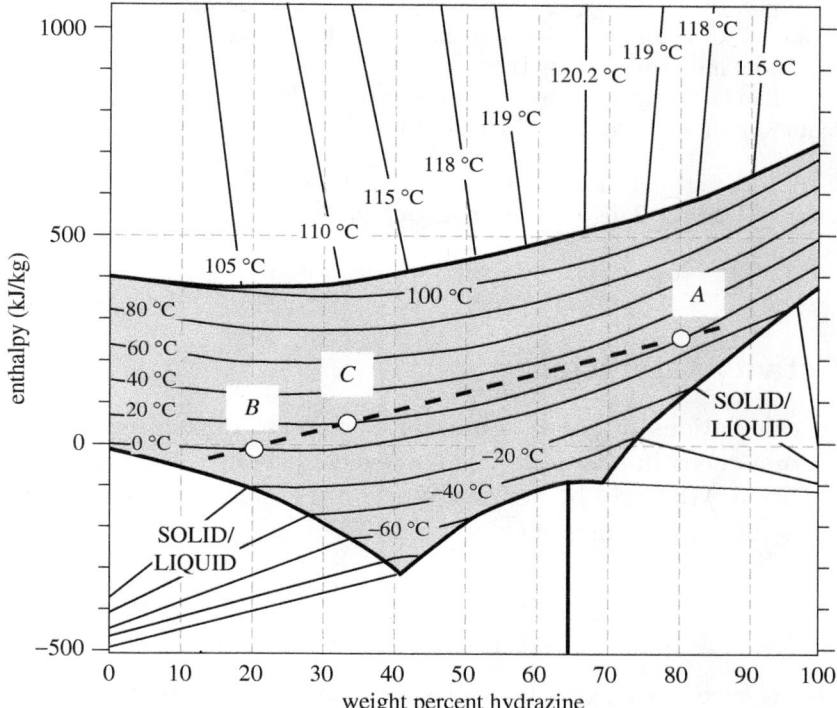

Figure 12-5: Adiabatic mixing of solutions on the enthalpy-composition chart. The initial solutions (A and B) and the solution they produce (C) lie on a straight line. (See Example 12.5.)

from the tie line and once these compositions are known, the amounts of each phase may be calculated using the lever rule. The same procedure works if enthalpy and compositions are given on a molar rather than mass basis.

Example 12.5: Using Enthalpy Charts
One kilogram of solution that contains 80% by weight hydrazine in water at 20 °C is mixed with 3.8 kg of a solution that contains 20% by weight hydrazine at 0 °C. If mixing is adiabatic, what is the temperature of the final solution?

Solution Using the subscript $_A$ for the initial hydrazine solution, $_B$ for water, and $_C$ for the final solution, the weight fraction of hydrazine in the final solution is

$$w_C = \frac{m_A w_A + m_B w_B}{m_A + m_B} = \frac{(1)(0.8) + (3.8)(0.2)}{(1) + (3.8)} = 0.325.$$

> States A, B, and C all lie on the same straight line. This line is defined by points A and B, whose composition and temperature is known. State C is then obtained at the intersection of AB with the line of constant composition $w_C = 0.325$. From Figure 12-5, we see that this state approximately lies on the 20 °C, therefore, the final temperature is $T_C \approx 20$ °C.
>
> **Comments** Even though solution A (20 °C) is mixed with a fairly cold solution B, the final temperature is still 20 °C. This is because mixing in this case is quite exothermic.

12.3 Activity Coefficient

To perform calculations of phase equilibrium we must obtain expressions for the fugacity of component in solution. The key excess property here is the Gibbs free energy of solution. We begin by defining a new dimensionless property, the *activity coefficient* (γ_i) of component i in solution:

$$\boxed{\ln \gamma_i = \frac{\overline{G}_i^{\mathrm{E}}}{RT},} \qquad (12.13)$$

where $\overline{G}_i^{\mathrm{E}}$ is the excess partial molar Gibbs free energy. Next, we obtain the chemical potential of component in solution. We begin by writing the solution Gibbs free energy in terms of the ideal-solution and the excess free energy,

$$G = G^{\mathrm{id}} + G^{\mathrm{E}}.$$

We apply the partial molar operator and notice that on the left-hand side we obtain the chemical potential of component in the solution. The first term on the right-hand side is the chemical potential of the component in ideal solution, and the second term is the excess partial molar Gibbs free energy:

$$\mu_i = \mu_i^{\mathrm{id}} + \overline{G}_i^{\mathrm{E}}. \qquad (12.14)$$

The chemical potential of component in ideal solution is given by eq. (11.9)

$$\mu_i^{\mathrm{id}} = G_i + RT \ln x_i, \qquad [11.9]$$

where G_i is the molar Gibbs free energy of pure liquid component i, and x_i is the mol fraction of component in solution. Combining this result, and making use of the definition of the activity coefficient, the chemical potential of component i in solution takes the form

$$\boxed{\mu_i = G_i + RT \ln \gamma_i x_i.} \qquad (12.15)$$

12.3 Activity Coefficient

To calculate the fugacity, we return to eq. (10.18),

$$\mu_i^A - \mu_i^B = RT \ln \frac{f_i^A}{f_i^B}, \qquad [10.18]$$

which provides the relationship between fugacity and chemical potential. Taking state A to be component i in solution ($\mu_i^A = \mu_i$, $f_i^A = f_i$) and B to be component i as the pure liquid ($\mu_i^B = G_i$, $f_i^B = f_{i,\text{pure}}$), we obtain

$$RT \ln \gamma_i x_i = RT \ln \frac{f_i}{f_{i,\text{pure}}},$$

or

$$\boxed{f_i = \gamma_i x_i f_{i,\text{pure}}.} \qquad (12.16)$$

The usefulness of the activity coefficient should be clear now: it acts as a correction factor in the chemical potential and fugacity and accounts for nonideal interactions between components. Before we study the activity coefficient in more detail, we examine some important limits:

- *Ideal Solution.* By definition in ideal solution all excess properties are zero. Accordingly, $\overline{G}_i^{\text{E}} = 0$ and the activity coefficient of all components at all compositions are 1. Equations (12.15) and (12.16) then revert to the correct results for ideal solution, eqs. (11.9) and (11.10), respectively.

- *Pure Liquid.* In the pure liquid all excess properties are zero. It follows that the activity coefficient of a pure liquid is 1. This must also understood to apply in the limiting sense, when the mol fraction of a component is made to approach 1 (by adding large amounts of that component, for example):

$$\gamma_i \to 1 \quad \text{as} \quad x_i \to 1. \qquad (12.17)$$

- *Infinite Dilution.* This is the limit approached by the minority species when the mol fraction of the majority species approaches 1. The activity coefficient at this limit is known as the activity coefficient at infinite dilution:

$$\gamma_i \to \gamma_i^\infty \quad \text{as} \quad x_i \to 0. \qquad (12.18)$$

The value of γ_i^∞ depends on component i as well as on all other components that are present at this limit.

Activity Coefficient and Excess Gibbs Energy

We return to the definition of the activity coefficient and solve for the excess partial molar Gibbs free energy:

$$\overline{G}_i^{\text{E}} = RT \ln \gamma_i.$$

The molar excess Gibbs free energy of solution is obtained by summing up the partial molar contributions of all components:

$$G^{\mathrm{E}} = RT \sum_i x_i \ln \gamma_i. \qquad (12.19)$$

This equation provides a direct relationship between activity coefficients and excess Gibbs free energy. We return now to eq. (10.5), which we apply to a system of constant composition[5]

$$d\left(\frac{G}{RT}\right) = -\frac{H}{RT^2} dT + \frac{V}{RT} dP, \quad \text{(constant composition)}.$$

Next we apply the partial molar and the excess operator to write

$$d\left(\frac{\overline{G}_i^{\mathrm{E}}}{RT}\right) = -\frac{\overline{H}_i^{\mathrm{E}}}{RT^2} dT + \frac{\overline{V}_i^{\mathrm{E}}}{RT} dP, \quad \text{(constant composition)},$$

and recognize the right-hand side as the differential of $\ln \gamma_i$:

$$d \ln \gamma_i = -\frac{\overline{H}_i^{\mathrm{E}}}{RT^2} dT + \frac{\overline{V}_i^{\mathrm{E}}}{RT} dP, \quad \text{(const. composition)}. \qquad (12.20)$$

This last result gives the differential of $\ln \gamma_i$ with respect to temperature and pressure at constant composition and leads to the following identifications,

$$\left(\frac{\partial \ln \gamma_i}{\partial T}\right)_{P,x} = -\frac{\overline{H}_i^{\mathrm{E}}}{RT^2}, \qquad (12.21)$$

$$\left(\frac{\partial \ln \gamma_i}{\partial P}\right)_{T,x} = \frac{\overline{V}_i^{\mathrm{E}}}{RT}. \qquad (12.22)$$

These partial derivatives quantify the sensitivity of the activity coefficient on temperature and pressure. Since the volume of mixing is generally very small, the effect of pressure on the activity coefficient is small and under typical conditions negligible. Sensitivity on temperature depends on the magnitude of the partial molar enthalpy. For small temperature differences it is common to neglect the effect of temperature and assume the activity coefficient to be constant. This approximation however is inappropriate for systems that are strongly endothermic or exothermic.

---

5. Since all n_i are constant, we have chosen the total number of moles to be 1. Then the superscript ^{tot} can be dropped since we are referring to 1 mol of solution.

12.4 Activity Coefficient and Phase Equilibrium

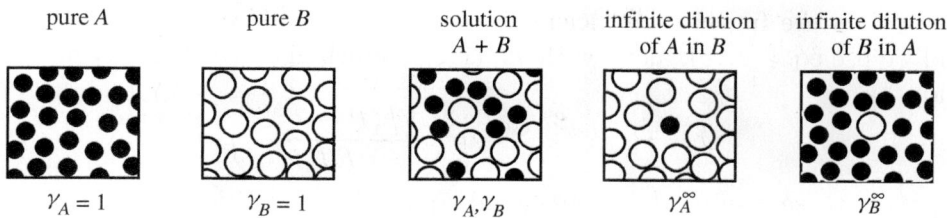

Figure 12-6: Molecular view of solution.

NOTE
Molecular View of Nonideal Effects

In a solution of two components, a molecule is surrounded by a mix of similar and dissimilar molecules and experiences a combination of self- and cross-interactions, as shown schematically in Figure 12-6. The total energy of interaction depends on the average number of neighbors and on how many of them are of the same kind. If self and cross interactions are identical in strength and the two components have similar molecular sizes, we obtain an ideal solution: a molecule is surrounded by the same number of neighbors on average, whether in pure liquid or in solution, and its total interaction with its neighbors does not depend on how many of them are of similar or dissimilar kind. If self and cross interactions are different, the solution environment surrounding a molecule is quite different from the environment in the pure liquid. This causes deviations from ideal-solution behavior. Difference in molecular size is also a source of nonidealities because the arrangement of neighbors around a molecule is affected. In the limit $x_i \to 1$, component i approaches ideal-solution behavior because it, interacts almost exclusively with molecules of the same type (component i is the majority component at this limit). In this solution the minority component is found at its infinite dilution limit: a molecule of the minority component is completely surrounded by dissimilar molecules and experience cross interactions only. The activity coefficient of the minority component shows its maximum deviation from 1 at the infinite dilution limit.

Contributions to nonideality that are caused by differences in molecular size are called *combinatorial* and are related to the arrangement of molecules in space. Contributions that are caused by molecular interactions are called *residual*. The goal of solution theories is to provide suitable approximations for these two sources of nonideal behavior.

12.4 Activity Coefficient and Phase Equilibrium

To perform VLE calculations, the starting point is the equality of the fugacity of component in the two phases:

$$f_i^V = f_i^L.$$

For the fugacity of the vapor we use the general expression

$$f_i = y_i \phi_i^V P,$$

where ϕ_i^V is the fugacity coefficient of component in the vapor mixture. For the liquid we use eq. (12.16), along with eq. (7.16), which gives the fugacity of the pure liquid:
$$f_i^L = \gamma_i x_i \phi_i^{\text{sat}} P_i^{\text{sat}} \exp\left(\frac{V_i^L(P - P_i^{\text{sat}})}{RT}\right).$$
Combining these results, the vapor-liquid equilibrium condition becomes
$$y_i \phi_i^V P = \gamma_i x_i \phi_i^{\text{sat}} P_i^{\text{sat}} \exp\left(\frac{V_i^L(P - P_i^{\text{sat}})}{RT}\right). \tag{12.23}$$

This equation is sometimes referred to as the "gamma-phi" method because it uses the fugacity coefficient for the vapor phase and the activity coefficient in the liquid phase. Except for ϕ_i^V and γ_i, all other properties in this equation are those of the pure component. At low pressures we may drop the fugacity coefficients (vapors are assumed to be in the ideal-gas state) and the Poynting correction, which is generally negligible. Under these conditions eq. (12.23) simplifies to
$$\boxed{y_i P = \gamma_i x_i P_i^{\text{sat}}.} \tag{12.24}$$

Setting $\gamma_i = 1$ (ideal solution) this reverts to Raoult's law. We may view then the activity coefficient as a correction to Raoult's law that accounts for deviations from ideal-solution behavior. Equation (12.24) is the basis of VLE calculations, provided that pressure is sufficiently low. It is also used to extract activity coefficients from experimental data, as we will see with examples below.

Positive and Negative Deviations from Raoult's Law The activity coefficient appears as a correction factor to Raoult's law. If $\gamma_i > 1$, the partial pressure of components above the solution are higher than those predicted by Raoult's law and the solution is said to exhibit *positive* deviations from ideal-solution behavior. The excess Gibbs free energy is positive in the case (see eq. 12.19). If $\gamma_i < 1$, partial pressures are lower than those predicted by Raoult's law, the excess Gibbs free energy is negative and the solution is said to exhibit *negative* deviations. The type of deviation may also be inferred by the general shape of the *Pxy* graph: under positive deviations, the bubble line lies above the straight line that connects the saturation pressures of the pure components; conversely, under negative deviations the bubble pressure lies below this line. In the special case of ideal solution, the bubble line is a straight line between the saturation pressures. Figure 12-7 shows the *Pxy* graph, the activity coefficients, and the excess Gibbs free energy for two systems, one that exhibits positive deviations (1,4, dioxane/methanol), and one that exhibits negative deviations (acetone/chloroform). The acetone/chloroform system also exhibits an azeotrope. This is not a necessary feature of systems with negative deviations from ideality.

12.4 Activity Coefficient and Phase Equilibrium

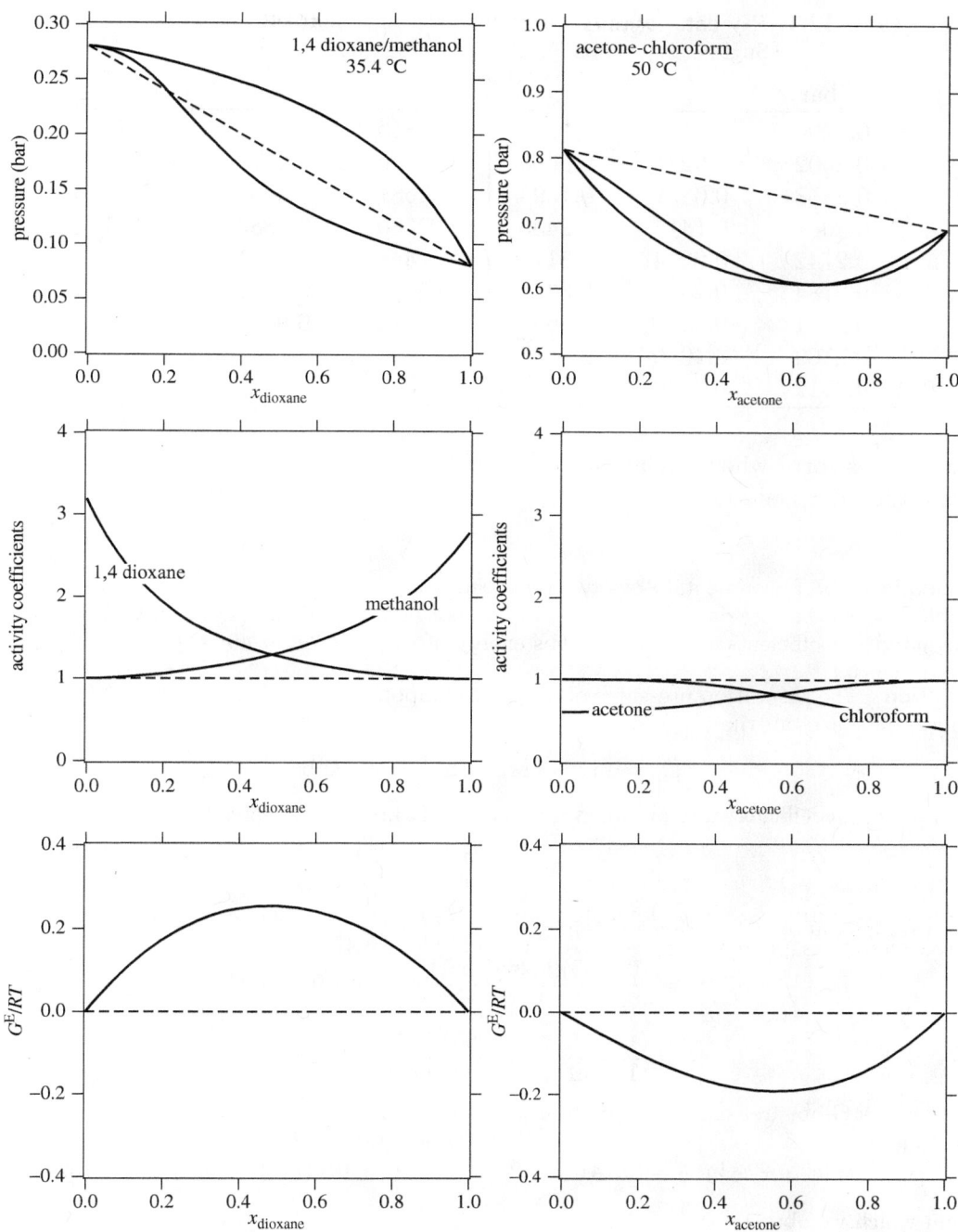

Figure 12-7: Positive deviations (left) and negative deviations (right) from ideal-solution behavior.

Table 12-1: Pxy data for the system ethanol(1)/acetonitrile(2) at 40 °C (from Sugi H., Katayama T., *J. Chem. Eng. Jpn.*, **11**, 167, (1978).)

P (bar)	x_1	y_1	P (bar)	x_1	y_1
0.2290	0	0	0.2810	0.4813	0.4376
0.2402	0.0281	0.0663	0.2766	0.5623	0.4753
0.2558	0.0831	0.1596	0.2681	0.6965	0.5344
0.2661	0.1415	0.2435	0.2580	0.7885	0.5987
0.2742	0.2314	0.3117	0.2418	0.8681	0.6687
0.2784	0.2884	0.3442	0.2189	0.9252	0.7708
0.2801	0.3330	0.3684	0.2012	0.9659	0.8664
0.2809	0.4040	0.4115	0.1799	1	1
0.2812	0.4140	0.414			

Azeotropes form when deviations from ideality are strong enough, whether these are positive or negative.

Example 12.6: Experimental Activity Coefficients

Table 12-1 shows Pxy data for the system ethanol (1)/acetonitrile (2) at 40 C. Determine the activity coefficients and excess Gibbs energy of the solution at 40 °C.

Solution The saturation pressures of the pure components correspond to $x_1 = 1$ (ethanol) and $x_1 = 0$ (acetonitrile):

$$P_1^{\text{sat}} = 0.1799 \text{ bar}, \quad P_2^{\text{sat}} = 0.2290 \text{ bar}.$$

The activity coefficients are obtained from eq. (12.24), which we solve for γ_i:

$$\gamma_i = \frac{y_i P}{x_i P_i^{\text{sat}}}.$$

For example, at $x_1 = 0.1415$ we obtain

$$\gamma_1 = \frac{(0.2435)(0.2661 \text{ bar})}{(0.1415)(0.1799 \text{ bar})} = 2.54619,$$

$$\gamma_2 = \frac{(1 - 0.2435)(0.2661 \text{ bar})}{(1 - 0.1415)(0.2290 \text{ bar})} = 1.02378.$$

The excess Gibbs free energy of the solution at this composition is

$$\frac{G^{\text{E}}}{RT} = x_1 \ln \gamma_1 + x_2 \ln \gamma_2 = (0.1415) \ln(2.54619) + (1 - 0.1415) \ln(1.02378) = 0.1524,$$

from which we obtain

$$G^{\text{E}} = (8.314 \text{ J/mol K})(313.15 \text{ K})(0.1524) = 396.78 \text{ J/mol}.$$

12.4 Activity Coefficient and Phase Equilibrium

Since temperature is constant, it is more convenient to work with the dimensionless ratio G^E/RT rather than G^E itself.

Repeating these calculations we obtain the results shown below.

x_1	γ_1	γ_2	G^E/RT	x_1	γ_1	γ_2	G^E/RT
0	–	1	0	0.4813	1.4208	1.3304	0.3171
0.0281	3.1517	1.0077	0.0397	0.5623	1.3002	1.4479	0.3096
0.0831	2.7321	1.0238	0.1051	0.6965	1.1438	1.7957	0.2712
0.1415	2.5462	1.0238	0.1524	0.7885	1.0891	2.1371	0.2279
0.2314	2.0540	1.0722	0.2202	0.8681	1.0358	2.6521	0.1592
0.2884	1.8473	1.1201	0.2577	0.9252	1.0141	2.9286	0.0933
0.3330	1.7230	1.1580	0.2790	0.9659	1.0034	3.4413	0.0454
0.4040	1.5909	1.2110	0.3017	1.0000	1	–	0
0.4140	1.5634	1.2276	0.3052				

Notice that the activity coefficients at infinite dilution ($x_1 = 0$, $x_1 = 1$) cannot be determined by direct calculation because the ratios y_1/x_1 and $(1-y_1)/(1-x_1)$ are indeterminate in this limit. As a consequence, the values of G^E/RT cannot be obtained by direct calculation either. However, these values are known to be zero and for this reason they are shown in the table. The results of these calculations are shown in Figure 12-8.

Comments The excess Gibbs free energy is zero at both ends ($x_1 = 0$, $x_1 = 1$), as is true for all excess properties. The activity coefficients have the expected general behavior: at the pure limit the activity coefficient approaches 1; at the infinite dilution limit it reaches the corresponding infinite dilution value. In this system the activity coefficients are larger than 1, that is, the system exhibits *positive* deviations from Raoult's law.

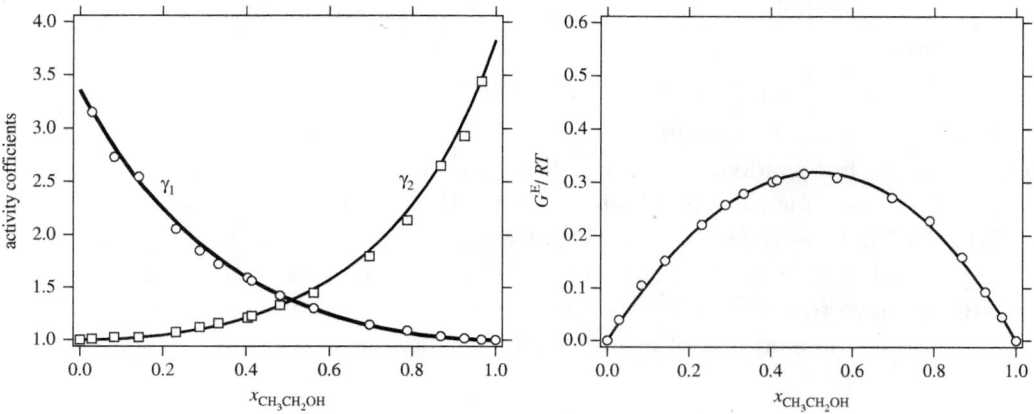

Figure 12-8: Activity coefficients and excess Gibbs energy of ethanol (1)/acetonitrile (2) solutions at 40 °C.

12.5 Data Reduction: Fitting Experimental Activity Coefficients

An equation for the activity coefficient can be obtained by fitting an appropriate equation to the experimental data. Rather than fitting each individual activity coefficient to its own function, the preferred procedure is to fit the excess Gibbs energy as a function x_1. The activity coefficients are obtained from this fit by noting from eq. (12.13) that $\ln \gamma_i$ is in fact a partial molar property; specifically, it is the partial molar property of the ratio G^E/RT. As such, $\ln \gamma_i$ can be obtained from the corresponding molar property of solution by applying eqs. (12.8) and (12.9) with F replaced by G^E/RT:

$$\ln \gamma_1 = \left(\frac{G^E}{RT}\right) + (1 - x_1)\frac{\partial}{\partial x_1}\left(\frac{G^E}{RT}\right), \qquad (12.25)$$

$$\ln \gamma_2 = \left(\frac{G^E}{RT}\right) - x_1 \frac{\partial}{\partial x_1}\left(\frac{G^E}{RT}\right). \qquad (12.26)$$

In this approach, a single fitted equation for the excess Gibbs free energy is used to obtain both activity coefficients. This procedure ensures that the fitted activity coefficients are consistent and satisfy the Gibbs-Duhem equation. The challenge is fitting an appropriate mathematical form to the excess Gibbs free energy. From a purely numerical standpoint, the fitted function must be of the form,

$$\frac{G^E}{RT} = x_1 x_2 f(x_1, x_2),$$

which satisfies the condition that the excess Gibbs free energy be zero at $x_1 = 0$ and $x_1 = 1$. The choice of $f(x_1, x_2)$ is open. One approach is adopt a Redlich-Kister polynomial,

$$f(x_1, x_2) = a_0 + a_1(x_1 - x_2) + a_2(x_1 - x_2)^2 + \cdots,$$

with enough terms to obtain an accurate fit. This approach is satisfactory with some systems but inadequate with others. One of the goals of theory is to provide informed guesses for this unknown function. Based on the assumptions employed, several models have been developed that lead to specific forms for the excess Gibbs energy of solution and the activity coefficients, as discussed in the next section. Before we move to this discussion, we demonstrate the numerical fitting procedure by continuing the example of ethanol/acetonitrile.

12.5 Data Reduction: Fitting Experimental Activity Coefficients

Example 12.7: Fitting Experimental Activity Coefficients
Use the data in Example 12.6 to fit the excess Gibbs free energy of the system ethanol/acetonitrile to a Redlich-Kister polynomial with two parameters and obtain equations for the activity coefficients. Use these equations to obtain the activity coefficients at infinite dilution and to construct the Pxy graph of this system at 40 °C.

Solution We fit the ratio G^{E}/RT to a two-parameter Redlich-Kister polynomial

$$\frac{G^{\mathrm{E}}}{RT} = x_1 x_2 (a_0 + a_1 (x_1 - x_2)).$$

By least-squares fitting we find

$$a_0 = 1.27729, \quad a_1 = 0.06540.$$

The activity coefficients are obtained by applying eqs. (12.25) and (12.26) to the fitted expression:

$$\ln \gamma_1 = (x_1 - 1)^2 (a_0 - a_1 + 4 a_1 x_1)$$
$$= 1.21188 - 2.16215 x_1 + 0.688648 x_1^2 + 0.261617 x_1^3,$$
$$\ln \gamma_2 = x_1^2 (a_0 - 3 a_1 + 4 a_1 x_1)$$
$$= 1.08107 x_1^2 + 0.261617 x_1^3.$$

The solid lines in Figure 12-8 were calculated using these coefficients. As we see, the fitted expressions provide a very accurate description of the experimental data, both for the activity coefficients and the excess Gibbs free energy.

The activity coefficients at infinite dilution are obtained by setting $x_1 = 0$ and $x_2 = 0$:

$$x_1 = 0: \quad \ln \gamma_1^\infty = a_0 - a_1 = 1.2119$$
$$x_2 = 0: \quad \ln \gamma_2^\infty = a_0 + a_1 = 1.3427$$

Calculation of Pxy graph. The starting point is eq. (12.24), which we apply to each component:

$$y_1 P = \gamma_1 x_1 P_1^{\mathrm{sat}}$$
$$y_2 P = \gamma_2 x_2 P_2^{\mathrm{sat}}.$$

(It is left as an exercise to show that the assumptions that allow us to use the simplified form of the equilibrium criterion are valid in this case). Adding these equations we obtain

$$P = \gamma_1 x_1 P_1^{\mathrm{sat}} + \gamma_2 x_2 P_2^{\mathrm{sat}}.$$

This gives the bubble pressure of a solution with composition $\{x_1, x_2\}$. The composition of the vapor is obtained from eq. (12.24), which we solve for y_1:

$$y_1 = \frac{\gamma_1 x_1 P_1^{\text{sat}}}{\gamma_1 x_1 P_1^{\text{sat}} + \gamma_2 x_2 P_2^{\text{sat}}}.$$

As an example, we show the results at $x_1 = 0.1415$. Using the fitted equations, the activity coefficients are

$$\gamma_1 = 2.51046, \quad \gamma_2 = 1.02264.$$

The bubble pressure of the solution is

$$P = (2.51046)(0.1415)(0.1799 \text{ bar}) + (1.02264)(1 - 0.1415)(0.2290 \text{ bar}) = 0.2650 \text{ bar}.$$

The vapor-phase compositions are

$$y_1 = \frac{(2.51046)(0.1415)(0.1799 \text{ bar})}{0.2650 \text{ bar}} = 0.2411, \quad y_2 = 1 - y_1 = 0.7589.$$

These calculations are summarized below and are compared to the experimental values:

	γ_1	γ_2	P (bar)	y_1
Fitted	2.5105	1.0226	0.2650	0.2411
Experimental	2.5462	1.0238	0.2661	0.2435
% difference	-1.4	-0.12	-0.41	-1.0

By repeating this calculations at liquid compositions between $x_1 = 0$ and $x_1 = 1$, we obtain the complete *Pxy* graph. This is shown in Figure 12-9. The calculated *Pxy* graph is in very good agreement with the experimental data, even though the system shows strong nonideality that results in an azeotrope.

Comments The particular expressions fitted to the excess Gibbs free energy and the resulting activity coefficients are known as the *Margules* model. This and other models for the activity coefficient are discussed in Section 12.6.

NOTE

Data Reduction

The procedure in Examples 12.6 and 12.7 is an example of *data reduction*: a large set of experimental data (*Pxy* data) is reduced into an equation for the activity coefficients with two parameters (a_0 and a_1). The entire *Pxy* graph may then be reconstructed from these two parameters.

12.6 Models for the Activity Coefficient

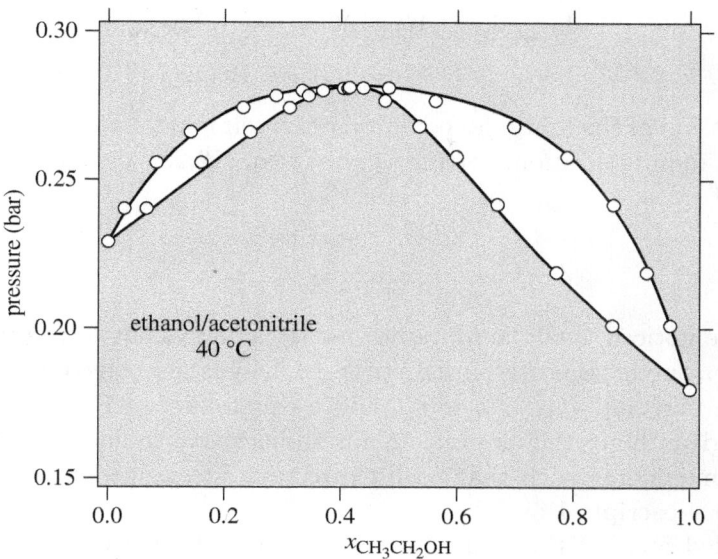

Figure 12-9: Experimental (symbols) and calculated (line) *Pxy* graph of ethanol/acetonitrile at 40 °C (see Example 12.7).

12.6 Models for the Activity Coefficient

Several equations exist for the activity coefficients in nonideal solution. These are based on models for the excess Gibbs free energy of the solution and they are known by the name of the scientists who developed them or by the theory used to model nonideal interactions. The literature on this topic is extensive and remains the subject of research. Here we review some of the most common models that have been used to relate the activity coefficient to composition.

Margules Equation

The Margules equation models the excess Gibbs free energy by a two-parameter Redlich-Kister polynomial. The excess Gibbs energy and the activity coefficients are given by the following equations:

$$\frac{G^{\mathrm{E}}}{RT} = x_1 x_2 \left[A_{21} x_1 + A_{12} x_2 \right], \tag{12.27}$$

$$\ln \gamma_1 = x_2^2 \left[A_{12} + 2(A_{21} - A_{12})x_1 \right], \qquad (12.28)$$

$$\ln \gamma_2 = x_1^2 \left[A_{21} + 2(A_{12} - A_{21})x_2 \right]. \qquad (12.29)$$

The two constants of the Margules equation bear a simple relationship to the activity coefficients at infinite dilution. Setting $x_1 = 0$ in eq. (12.28) and $x_2 = 0$ in eq. (12.29) we obtain,

$$\ln \gamma_1^\infty = A_{12},$$

$$\ln \gamma_2^\infty = A_{21}.$$

You may have noticed that the Margules model is equivalent to the two-parameter Redlich-Kister polynomial used in Example 12.7. This may be confirmed by setting $A_{12} = a_0 - a_1$ and $A_{21} = a_0 + a_1$ to the above equations (see Example 12.8, below). In the form given here, the expressions for the activity coefficients are symmetric in the two components such that each expression is obtained from each other by switching the subscripts 1 and 2.

A simplified form of the Margules equation is obtained if the two parameters are equal. Setting $A_{12} = A_{21} = A$, eqs. (12.27)–(12.29) become

$$\frac{G^{\mathrm{E}}}{RT} = A x_1 x_2,$$

$$\ln \gamma_1 = A x_2^2,$$

$$\ln \gamma_2 = A x_1^2$$

and the activity coefficients at infinite dilution are

$$\ln \gamma_1^\infty = \ln \gamma_2^\infty = A.$$

From theoretical considerations, this model describes nonideal interactions between molecules of similar molecular size.

Example 12.8: Margules Equation
Obtain the Margules constants for the system ethanol(1)/acetonitrile, using the data in Example 12.6.

Solution The excess Gibbs free energy in Example 12.7 was fitted to the equation,

$$\frac{G^{\mathrm{E}}}{RT} = x_1 x_2 (a_0 + a_1(x_1 - x_2)).$$

To see that this is the same as the Margules expression, we set $x_2 = 1 - x_1$ in eq. (12.27):

$$\frac{G^{\mathrm{E}}}{RT} = x_1 x_2 \Big(A_{12} + (A_{21} - A_{12})x_1 \Big).$$

12.6 Models for the Activity Coefficient

Equating the coefficients of equal powers in x_1 between the two expressions we find

$$A_{12} = a_0 - a_1,$$
$$A_{21} - A_{12} = 2a_1,$$

which we solve for the Margules parameters to obtain

$$A_{21} = a_0 - a_1,$$
$$A_{12} = a_0 + a_1.$$

Using the parameters a_0 and a_1 obtained in Example 12.6 ($a_0 = 1.27729$, $a_1 = 0.06540$), we have

$$A_{12} = 1.2119, \quad A_{21} = 1.3427.$$

Comments The general procedure for obtaining the Margules parameters is to fit eq. (12.27) to experimental data for the excess Gibbs free energy and obtain the parameters A_{12}, A_{21} from this fit. The fitting procedure can be simplified by first writing the excess Gibbs free energy of the Margules model as

$$\frac{G^{\mathrm{E}}}{x_1 x_2 RT} = A_{21} x_1 + A_{12} x_2 = A_{12} + (A_{21} - A_{12}) x_1.$$

This equation suggests that a graph of the ratio $G^{\mathrm{E}}/x_1 x_2 RT$ against x_1 must yield a straight line whose values at $x_1 = 0$ and $x_1 = 1$ are A_{12} and A_{21}, respectively. This graph is shown in Figure 12-10. While this linearization makes the fitting procedure very simple, it comes with some potential pitfalls. Data points near $x_1 = 0$ and $x_1 = 1$ are subject to numerical noise because the product $x_1 x_2$ in the denominator goes to infinity. As a result, any experimental error is magnified significantly. This is quite obvious in the first two points in Figure 12-10. These points affect the fitted parameters rather strongly, even though the excess Gibbs energy at these mole fractions is nearly zero. It is acceptable to reject such points and not use them in the numerical fitting. The line that is shown on Figure 12-10 is based on the values of A_{12} and A_{21} obtained from a_0, a_1, which were based in a fit of G^{E}/RT. A least squares fit of $G^{\mathrm{E}}/RTx_1 x_2$ would yield somewhat different values, depending on which experimental points are kept in the fit and which are rejected.

Example 12.9: Reconstructing the Pxy Graph from Limited Data

The system benzene(1)-acetonitrile(2) at 45 °C forms an azeotrope with the composition $x_1 = 0.53$ and pressure 0.372 bar. Calculate the Pxy graph at this temperature using the simplified one-parameter Margules equation. Is this equation appropriate?

Additional data The saturation pressures are $P_1^{\mathrm{sat}} = 0.298$ bar and $P_2^{\mathrm{sat}} = 0.278$ bar.

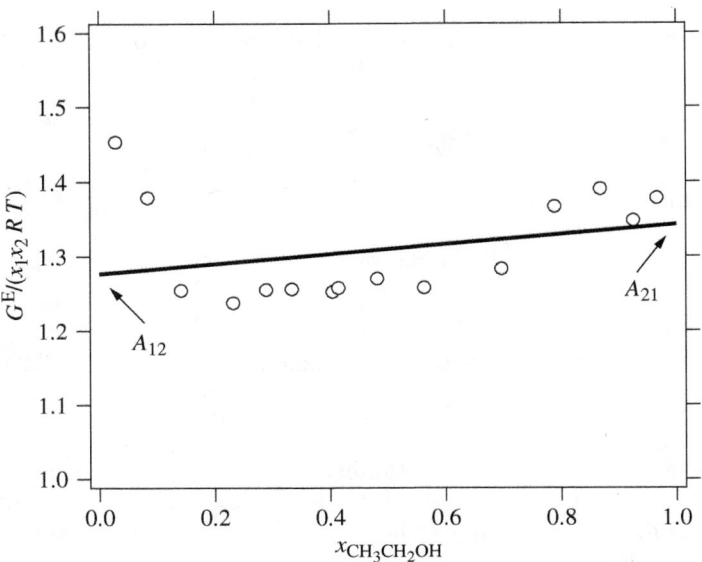

Figure 12-10: Linearized form of excess Gibbs free energy in the Margules equation for the system ethanol (1)/acetonitrile (2) at 40 °C (see Example 12.8).

Solution To calculate the value of A in the one-parameter form of the Margules equation we need one experimental value of the activity coefficient. This is obtained from the known azeotropic point. With $y_1^{az} = x_1^{az}$ we find

$$y_1^{az} P^{az} = \gamma_1^{az} x_1^{az} P_1^{sat} \quad \Rightarrow \quad \gamma_1^{az} = \frac{P^{az}}{P_1^{sat}} = \frac{0.372}{0.298} = 1.2483.$$

The parameter A is

$$A = \frac{\ln \gamma_1^{az}}{(x_2^{az})^2} = \frac{\ln(1.2483)}{(1 - 0.53)^2} = 1.004.$$

Notice that A may also be obtained from the activity coefficient of the second component; if the simple one-parameter Margules is appropriate, the same value of A should be obtained from either calculation:

$$\gamma_2^{az} = \frac{P^{az}}{P_2^{sat}} = \frac{0.372}{0.278} = 1.3381,$$

which leads to the following value for A:

$$A = \frac{\ln \gamma_2^{az}}{(x_1^{az})^2} = \frac{\ln(1.3381)}{(0.53)^2} = 1.0369.$$

The difference between the two values is about 4%, which implies that the one-parameter Margules is probably acceptable for approximate calculations.

12.6 Models for the Activity Coefficient

Comments The advantage of the one-parameter Margules is numerical convenience due to the simpler expression of the activity coefficient. Nonetheless, we are given enough information to obtain the values of both parameters of the original Margules using $x_1 = 0.53$, $x_2 = 0.47$, $\gamma_1 = 1.2483$, and $\gamma_2 = 1.3381$, eqs. 12.28 and (12.29) give

$$0.2218 = 0.234154 A_{21} - 0.013254 A_{12},$$

$$0.2913 = 0.264046 A_{12} + 0.016854 A_{21},$$

which we solve for the two parameters to obtain, $A_{12} = 1.0389$ and $A_{21} = 1.00605$.

van Laar Equation

This model is based on the following equations:

$$\frac{G^{\mathrm{E}}}{RT} = \frac{B_{12} B_{21} x_1 x_2}{B_{12} x_1 + B_{21} x_2}, \tag{12.30}$$

$$\ln \gamma_1 = B_{12} \left[1 + \frac{B_{12} x_1}{B_{21} x_2}\right]^{-2}, \tag{12.31}$$

$$\ln \gamma_2 = B_{21} \left[1 + \frac{B_{21} x_2}{B_{12} x_1}\right]^{-2}. \tag{12.32}$$

Notice that in this model the excess Gibbs free energy is not a polynomial in the mol fraction. The van Laar equation is a two-parameter model and the constants B_{12} and B_{21} are related to the activity coefficients at infinite dilution:

$$\ln \gamma_1^\infty = B_{12}, \quad \ln \gamma_2^\infty = B_{21}.$$

In the special case that $B_{12} = B_{21}$, the van Laar model reduces to the same expression as the one-parameter Margules equation ($A_{12} = A_{21}$).

Wilson Equation

The Wilson equation is based on local composition theories and accounts for intermolecular interactions between a molecule and its immediate neighbors. The Wilson expression for the excess Gibbs energy and the resulting equations for the activity coefficients are

$$\frac{G^{\mathrm{E}}}{RT} = -x_1 \ln(x_1 + \Lambda_{12} x_2) - x_2 \ln(x_2 + \Lambda_{21} x_1), \tag{12.33}$$

$$\ln \gamma_1 = -\ln(x_1 + x_2 \Lambda_{12}) + x_2 \left(\frac{\Lambda_{12}}{x_1 + x_2 \Lambda_{12}} - \frac{\Lambda_{21}}{x_2 + x_1 \Lambda_{21}} \right), \tag{12.34}$$

$$\ln \gamma_2 = -\ln(x_2 + x_1 \Lambda_{21}) + x_1 \left(\frac{\Lambda_{21}}{x_2 + x_1 \Lambda_{21}} - \frac{\Lambda_{12}}{x_1 + x_2 \Lambda_{12}} \right). \tag{12.35}$$

The two parameters Λ_{12} and Λ_{21} are related to molecular properties of the components:

$$\Lambda_{12} = \frac{V_2}{V_1} \exp\left(-\frac{\lambda_{12} - \lambda_{11}}{RT}\right), \tag{12.36}$$

$$\Lambda_{21} = \frac{V_1}{V_2} \exp\left(-\frac{\lambda_{21} - \lambda_{22}}{RT}\right), \tag{12.37}$$

where V_i is the molar volume of the pure component i, and λ_{ij} is the interaction energy between components i and j (cross interaction if $i \neq j$, self-interaction if $i = j$). In the Wilson model, nonidealities arise from interactions of unequal strength ($\lambda_{ii} \neq \lambda_{ij}$) and from unequal sizes ($V_i \neq V_j$). In addition to connecting the excess Gibbs free energy to molecular properties, Wilson's equation also introduces a temperature dependence of the activity coefficient. The interaction parameters are not known and are usually treated as adjustable parameters, to be obtained by numerical fitting.

Non Random Two Liquid Model (NRTL)

This equation is also based on local composition theories. The expression for the Gibbs free energy is

$$\frac{G^{\mathrm{E}}}{RT} = x_1 x_2 \left(\frac{\tau_{21} G_{21}}{x_1 + x_2 G_{21}} + \frac{\tau_{12} G_{12}}{x_2 + x_1 G_{12}} \right), \tag{12.38}$$

with

$$\tau_{12} = \frac{g_{12} - g_{22}}{RT}, \qquad G_{12} = \exp(-\alpha \tau_{12}), \tag{12.39}$$

$$\tau_{21} = \frac{g_{21} - g_{11}}{RT}, \qquad G_{21} = \exp(-\alpha \tau_{21}). \tag{12.40}$$

The corresponding activity coefficients are given by

$$\ln \gamma_1 = x_2^2 \left[\tau_{21} \left(\frac{G_{21}}{x_1 + x_2 G_{21}} \right)^2 + \frac{\tau_{12} G_{12}}{(x_2 + x_1 G_{12})^2} \right], \tag{12.41}$$

$$\ln \gamma_2 = x_1^2 \left[\tau_{12} \left(\frac{G_{12}}{x_2 + x_1 G_{12}} \right)^2 + \frac{\tau_{21} G_{21}}{(x_1 + x_2 G_{21})^2} \right]. \tag{12.42}$$

In this equation, g_{ij} is the interaction energy (analogous to λ_{ij} of the Wilson equation). The parameter α arises from the fact that locally the composition of the solution may differ from the bulk concentration giving rise to nonrandom arrangement of molecules. It is usually set to a constant value between 0.2 and 0.3. Treating α as a constant, the NRTL model contains two adjustable parameters, τ_{12} and τ_{21} (notice that the individual values of the g_{ij} are not needed for the calculation).

Flory-Huggins Model

Polymer-solvent systems differ from usual solutions in that one component (polymer) is much larger than the other (solvent). This leaves less space for the solvent, limits the ways that a chain can stretch in solution and creates a situation where segments of a long polymer chain may be interacting with other segments of the same chain. The Flory-Huggins model accounts for these effects.[6] The excess Gibbs energy is given by the expression

$$\frac{G^{\mathrm{E}}}{RT} = \left(x_1 \ln \frac{\Phi_1}{x_1} + x_2 \ln \frac{\Phi_2}{x_2} \right) + \chi (x_1 + rx_2) \Phi_1 \Phi_2. \tag{12.43}$$

Here, 1 refers to the solvent and 2 to the polymer; Φ_i is the volume fraction of component i, and is defined as

$$\Phi_i = \frac{x_i V_i}{x_1 V_1 + x_2 V_2},$$

V_i is the molar volume of component i, r is the ratio of the molecular volumes,

$$r = \frac{V_2}{V_1} > 1,$$

and χ is a measure of the interaction energy between polymer and solvent. The corresponding activity coefficients are

$$\ln \gamma_1 = \ln \frac{\Phi_1}{x_1} + \left(1 - \frac{1}{r}\right) \Phi_2 + \chi \Phi_2^2, \tag{12.44}$$

$$\ln \gamma_2 = \ln \frac{\Phi_2}{x_2} + (r-1) \Phi_1 + r\chi \Phi_1^2. \tag{12.45}$$

The first term in parenthesis on the right-hand side of eq. (12.43) reflects entropic effects that arise from the number of possible ways that macromolecules and solvent

---

6. Paul J. Flory received the 1974 Nobel Prize in Chemistry in recognition of his contributions to polymer science.

can be arranged in space; this term is also known as the combinatorial contribution. The second term on the right-hand side is the enthalpic contribution and arises from differences between polymer-polymer and polymer-solvent interactions; this term is also referred to as the residual contribution (not to be confused with the residual properties introduced earlier, which measure deviations from the ideal-gas state). Even if this term is zero (i.e., $\chi = 0$), the solution is nonideal due to the size difference between polymer and solvent.

UNIQUAC

The name of this model is an acronym of *universal quasichemical*, the name of the theory used to drive it. Like the Flory-Huggins model, UNIQUAC separates nonideal contributions to the excess Gibbs free energy into a combinatorial and a residual term:

$$\frac{G^{\mathrm{E}}}{RT} = g^R + g^C. \tag{12.46}$$

The combinatorial term accounts for differences in size and shape of the molecules and depends entirely on properties of the pure components. The residual term accounts for binary interactions and contains parameters characteristic for a pair of components. For binary solution these are given by the expressions

$$g^C = x_1 \ln \frac{\Phi_1}{x_1} + x_2 \ln \frac{\Phi_2}{x_2} + \frac{z}{2}\left(x_1 q_1 \ln \frac{\theta_1}{\Phi_1} + x_2 q_2 \ln \frac{\theta_2}{\Phi_2}\right), \tag{12.47}$$

$$g^R = -q_1 x_1 \ln(\theta_1 \tau_{11} + \theta_2 \tau_{21}) - q_2 x_2 \ln(\theta_2 \tau_{12} + \theta_2 \tau_{22}). \tag{12.48}$$

The activity coefficients are also expressed as a sum of residual and combinatorial contributions,

$$\ln \gamma_i = \ln \gamma_i^C + \ln \gamma_i^R, \tag{12.49}$$

These terms are given by the following expressions:

$$\ln \gamma_1^C = \ln \frac{\Phi_1}{x_1} + 1 - \frac{\Phi_1}{x_1} - 5 q_1 \left(\ln \frac{\Phi_1}{\theta_1} + 1 - \frac{\Phi_1}{\theta_1}\right), \tag{12.50}$$

$$\ln \gamma_1^R = q_1 \left(1 - \ln(\theta_1 + \theta_2 \tau_{21}) - \frac{\theta_1}{\theta_1 + \theta_2 \tau_{21}} - \frac{\theta_2 \tau_{12}}{\theta_1 \tau_{12} + \theta_2}\right), \tag{12.51}$$

with

$$\Phi_i = \frac{x_i r_i}{x_1 r_1 + x_2 r_2}, \tag{12.52}$$

$$\theta_i = \frac{x_i q_i}{x_1 q_1 + x_2 q_2}, \tag{12.53}$$

$$\tau_{ij} = \exp(-a_{ij}/T). \tag{12.54}$$

12.6 Models for the Activity Coefficient

The corresponding expressions for γ_2 are obtained by interchanging the subscripts 1 and 2. The method requires the following parameters: r_i, which is a measure of the molecular size of component i, and q_i, which represents the surface area of the component, and two binary parameters a_{ij} which represent interactions. The parameters r_i and q_i characterize the pure components and can be calculated from tabulated values. The binary interaction parameters a_{ij} represent the difference between the cross-interaction energy, u_{ij} and the self-interaction, u_{jj}:

$$a_{ij} = u_{ij} - u_{jj},$$

(notice that $a_{ij} \neq a_{ji}$). The two a_{ij} constants are treated as adjustable parameters and are obtained by fitting eq. (12.46) to experimental data. Assuming a_{ij} to be independent of temperature, a single set of these parameters may be used to calculate activity coefficients at other temperatures. UNIQUAC works well with a large variety of nonelectrolyte solutions, including polar components such as water, as well as nonpolar molecules such as hydrocarbons.

Group Contribution Method for r_i, q_i The parameters r_i and q_i are obtained by the *group contribution* method, a general approach for the estimation of pure component properties that is based on the premise that a given molecular group makes the same contribution to a property regardless of the molecule to which it is attached. For example, the molecular group CH_3 is assumed to make the same contribution to a property (enthalpy, entropy, surface area, etc.) whether it is part of ethane, toluene, or any other molecule. Since there are far fewer molecular groups than there are molecules, the required tabulations are significantly reduced. A number of molecular units, or subgroups, have been identified and their r and q values have been tabulated (see Table D-1 in the appendix). In the nomenclature of UNIFAC, a structural unit is a "subgroup." Subgroups are grouped together in "main groups." For instance, CH_3, CH_2, CH, and C are all subgroups of the main group "CH_2." Each subgroup has been assigned an index value, k, that is unique for that subgroup. For example, $k = 9$ identifies the subgroup $C=C$ (carbon double bond). To obtain r and q of a molecule, the molecule is broken down in its constituent subgroups and the r and q values are computed as follows:

$$r_i = \sum_k \nu_k^{(i)} R_k,$$

$$q_i = \sum_k \nu_k^{(i)} Q_k,$$

where $\nu_k^{(i)}$ is the number of subgroups of type k that are present in component i, and Q_k, R_k, are the corresponding values of the subgroup and are obtained from

Table D-1 (in the appendix). Calculations based on the UNIQUAC equation are more involved compared to other methods but the method is appealing because it works well with many nonideal systems. The calculation of the activity coefficients and of the Pxy graph are demonstrated in the example below.

Example 12.10: Calculation of Pxy Graph from UNIQUAC
Use the UNIQUAC equation to calculate the Pxy graph of ethanol(1)/acetonitrile(2) at 45 °C. The interaction parameters are

$$a_{12} = 294.5 \text{ K}, \quad a_{21} = -41.3 \text{ K}.$$

Solution Ethanol consists of three subgroups, CH_3, CH_2, and OH. Acetonitrile consists of a single group, CH_3CN. The parameters of these subgroups from Table D-1 in the appendix are

Subgroup	k	R_k	Q_k
CH_3	1	0.9011	0.848
CH_2	2	0.6744	0.540
OH	14	1.0000	1.200
CH_3CN	40	1.8701	1.724

The parameters of ethanol are

$$r_1 = (1)(0.9011) + (1)(0.6744) + (1)(1.000) = 2.5755,$$
$$q_1 = (1)(0.848) + (1)(0.540) + (1)(1.200) = 2.588,$$

and for acetonitrile,

$$r_2 = (1)(1.8701) = 1.8701,$$
$$q_2 = (1)(1.724) = 1.724.$$

A sample calculation is shown at $x_1 = 0.8$:

$$\Phi_1 = \frac{(0.8)(2.5755)}{(0.8)(2.5755) + (0.2)(1.8701)} = 0.846362,$$

$$\Phi_2 = \frac{(0.2)(1.8701)}{(0.8)(2.5755) + (0.2)(1.8701)} = 0.153638,$$

$$\theta_1 = \frac{(0.8)(2.588)}{(0.8)(2.588) + (0.2)(1.724)} = 0.857237,$$

$$\theta_2 = \frac{(0.2)(1.724)}{(0.8)(2.588) + (0.2)(1.724)} = 0.142763,$$

12.6 Models for the Activity Coefficient

$$\tau_{12} = \exp(-294.5/318.15) = 0.390454,$$
$$\tau_{12} = \exp(-(-41.3)/318.15) = 1.14098.$$

The combinatorial and residual terms are

$$\ln \gamma_1^C = -0.000566751 \qquad \ln \gamma_1^R = 0.0595364,$$
$$\ln \gamma_2^C = -0.00809672 \qquad \ln \gamma_2^R = 0.830032.$$

Finally, the activity coefficients are

$$\gamma_1 = \exp(-0.000566751 + 0.0595364) = 1.0607,$$
$$\gamma_2 = \exp(-0.00809672 + 0.830032) = 2.2749.$$

We construct the Pxy by preforming a bubble P calculation:

$$P = \gamma_1 x_1 P_1^{\text{sat}} + \gamma_2 x_2 P_2^{\text{sat}}$$
$$= (1.0607)(0.8)(0.1799 \text{ bar}) + (2.2749)(0.2)(0.2290 \text{ bar}) = 0.281 \text{ bar}$$
$$y_1 = \frac{\gamma_1 x_1 P_1^{\text{sat}}}{P} = \frac{(1.0607)(0.8)(0.1799)}{0.281 \text{ bar}} = 0.4482.$$

The complete Pxy graph is obtained by repeating the calculation at various liquid compositions. The phase diagram that is calculated in this manner is in very good agreement with the experimental measurements, as shown in Figure 12-11.

Comments In this example the interaction parameters, a_{12} and a_{21}, were given. These were actually calculated by fitting the above experimental data to eq. 12.46.

UNIFAC

This method deserves special mention because, unlike all of the previous methods, it allows the prediction of activity coefficients based entirely on tabulated parameters i.e., no fitting of parameters is necessary. It builds on UNIQUAC and is based on the premise that a solution may be regarded as a mixture of *structural units* rather than of chemical species. For example, a mixture of n-pentane and n-heptane is considered as a mixture of CH_2 and CH_3 subgroups and so is a mixture of cyclohexane and ethane. In this approach, interaction parameters are determined between a finite number of subgroups and are tabulated. It is then possible to calculate activity coefficients for any solution, binary or multicomponent, from a relatively small number of tabulated values. This is the main advantage of the method. Its

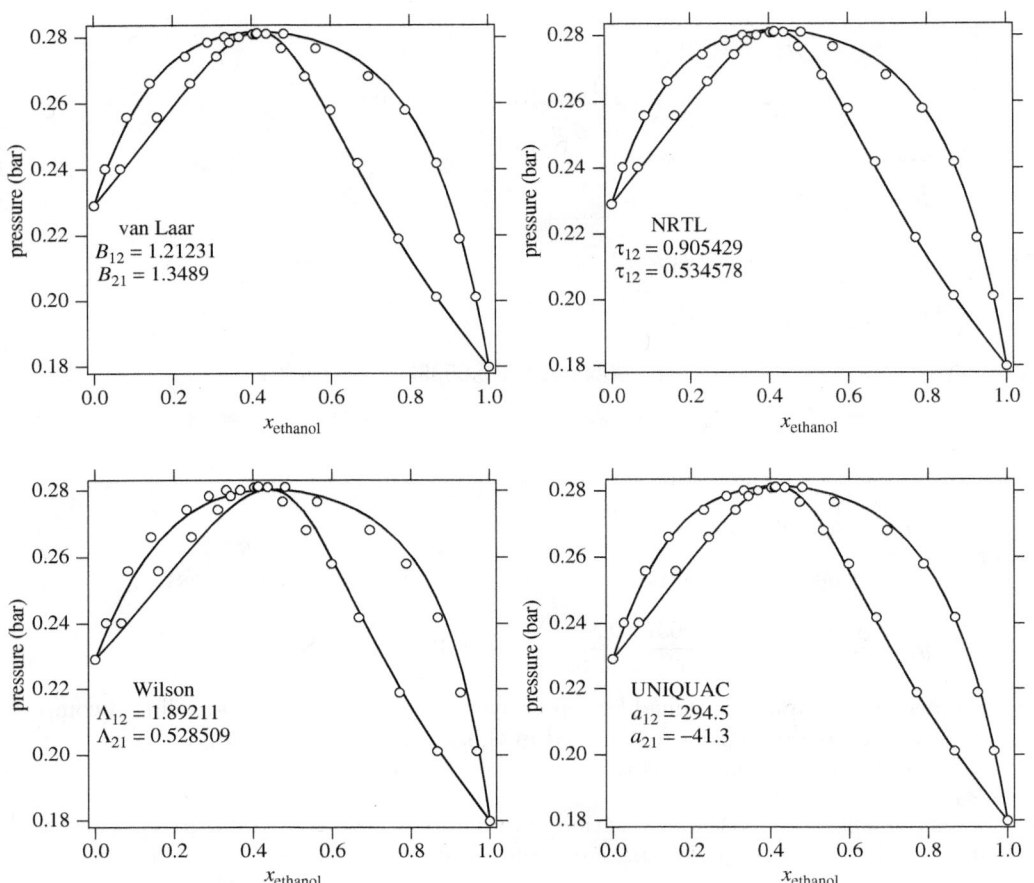

Figure 12-11: Calculated *Pxy* graphs for the system ethanol/acetonitrile at 40 °C using various activity coefficient models. The parameters have been fitted to the experimental points for the excess Gibbs free energy of solution.

applicability is limited to components that are liquid at 25 °C. Parameters for the UNIFAC equation have been determined by fitting a large body of experimental values so as to obtain good agreement for as many experimental systems as possible. As a result, some systems are predicted better than others. At the moment, however, this is the method most extensively used that is capable of predicting the activity coefficient.

UNIFAC (Universal Functional Activity Coefficient) applies the UNIQUAC equation to a solution of subgroups to calculate the combinatorial and the residual contributions. In UNIFAC, these represent interactions among the *subgroups* rather

12.6 Models for the Activity Coefficient

than among the molecules themselves. The method requires a set of size and surface area constants R_k and Q_k for each subgroup k and a set of interaction parameters a_{mk} for each pair of subgroups. The size and surface-area tables are the same as those used by UNIQUAC.[7] Table D-1 in the appendix shows these values for 44 standard subgroups. Interactions are tabulated by groups, rather than subgroups (i.e., all subgroups within a main group have the same interaction energy). As in UNIQUAC, the interaction parameter a_{mn} is not equal to a_{nm}.

To perform a UNIFAC calculation we must first determine $\nu_k^{(i)}$, the number of functional subgroups of type k that appear in the chemical formula of species i. Also needed are the mole fractions in the solution and temperature. The activity coefficient in UNIFAC is calculated by adding the contributions of the combinatorial term, (γ^C), and the residual term, (γ^R):

$$\boxed{\ln \gamma_i = \ln \gamma_i^C + \ln \gamma_i^R}. \tag{12.55}$$

These terms are calculated by the following set of equations:

$$\ln \gamma_i^C = 1 - J_i + \ln J_i - 5 q_i \left(1 - \frac{J_i}{L_i} + \ln \frac{J_i}{L_i} \right), \tag{12.56}$$

$$\ln \gamma_i^R = q_i (1 - \ln L_i) - \sum_k \left(\theta_k \frac{s_{ki}}{\eta_k} - G_{ki} \ln \frac{s_{ki}}{\eta_k} \right), \tag{12.57}$$

$$J_i = \frac{r_i}{\sum_j r_j x_j}, \tag{12.58}$$

$$L_i = \frac{q_i}{\sum_j q_j x_j}, \tag{12.59}$$

$$r_i = \sum_k \nu_k^{(i)} R_k, \tag{12.60}$$

$$q_i = \sum_k \nu_k^{(i)} Q_k, \tag{12.61}$$

$$G_{ki} = \nu_k^{(i)} Q_k, \tag{12.62}$$

$$\theta_k = \sum_i G_{ki} x_i, \tag{12.63}$$

---

7. In UNIQUAC, these tables are used to calculate the size and area parameters of the components that form the solution; in UNIFAC, these values are used directly in the calculation of the activity coefficients.

$$s_{ki} = \sum_{m} G_{mi} \tau_{mk}, \qquad (12.64)$$

$$\eta_k = \sum_{i} s_{ki} x_i, \qquad (12.65)$$

$$\tau_{mk} = \exp\left[-\frac{a_{mk}}{T}\right]. \qquad (12.66)$$

The various terms are explained below:

i	identifies component i of the solution
j	summation index running through all components
k	identifies a subgroup
m	summation index running through all subgroups present in the solution
$\nu_k^{(i)}$	number of subgroups k in component i
R_k	relative volume of subgroup k, tabulated
Q_k	relative surface area of subgroup k, tabulated
a_{mk}	interaction energy between subgroups m and k, in units of kelvin, tabulated

The above equations may be used with any number of components.[8] The procedure is tedious for hand calculations but easily automated for computer calculations. Tables listing several subgroups and the corresponding R_k and Q_k values are given in Table D-1 in the appendix. The table should be consulted before preparing the input for the UNIFAC calculation since some molecules (e.g., water, methanol, acetonitrile) constitute a single subgroup by themselves. The calculation is illustrated in the example below.

Example 12.11: A Detailed Calculation using UNIFAC
Calculate the Pxy graph of ethanol/acetonitrile at 40 °C and compare with the data for this system given in Table 12-1.

Solution The steps of the calculation will be demonstrated for $x_1 = 0.1$. The structural subgroups for this system were determined in Example 12.10:

Subgroup	k	R_k	Q_k	$\nu_k^{(1)}$	$\nu_k^{(2)}$
CH_3	1	0.9011	0.848	1	0
CH_2	2	0.6744	0.540	1	0
OH	14	1.0000	1.200	1	0
CH_3CN	40	1.8701	1.724	0	1

---

8. The extension of UNIFAC to multicomponent solutions is based on the assumption that interaction between multiple components may be obtained from pair interactions between all possible pairs. This means that no additional tabulations are needed for the calculation of multicomponent solutions.

12.6 Models for the Activity Coefficient

The interaction parameters between the subgroups are

	Interaction energy, a_{mk}			
	$k=1$	$k=2$	$k=14$	$k=40$
$m=1$	0	0	986.5	597
$m=2$	0	0	986.5	597
$m=14$	156.4	156.4	0	6.712
$m=40$	24.82	24.82	185.4	0

For example, the interaction between subgroup "CH$_3$" ($k=1$) and subgroup "OH" ($k=14$) is $a_{1,14} = 986.5$ K.

This is all that is required from the UNIFAC tables. Below we summarize the results for the various intermediate variables ($i=1$ refers to ethanol, $i=2$ refers to acetonitrile):

	$(i=1)$	$(i=2)$
r	2.5755	1.8701
q	2.588	1.724
J	1.32714	0.963651
L	1.42952	0.952276

The τ_{mk} parameters are calculated as $\exp(-a_{mk}/T)$:

	τ_{mk}			
	$k=1$	$k=2$	$k=14$	$k=40$
$m=1$	1.	1.	0.0428415	0.148609
$m=2$	1.	1.	0.0428415	0.148609
$m=14$	0.60687	0.60687	1.	0.978794
$m=40$	0.9238	0.9238	0.553193	1.

All other parameters are summarized below:

G_{ki}		s_{ki}		θ_k	η_k
$(i=1)$	$(i=2)$	$(i=1)$	$(i=2)$		
0.848	0	2.11624	1.59263	0.0848	1.64499
0.54	0	2.11624	1.59263	0.054	1.64499
1.2	0	1.25946	0.953705	0.12	0.98428
0.	1.724	1.38082	1.724	1.5516	1.68968

The combinatorial contributions are

$$\ln \gamma_1^C = -0.00925364, \quad \ln \gamma_2^C = -0.000066914,$$

and the residual contributions are

$$\ln \gamma_1^R = 0.7086, \quad \ln \gamma_2^R = 0.00920115.$$

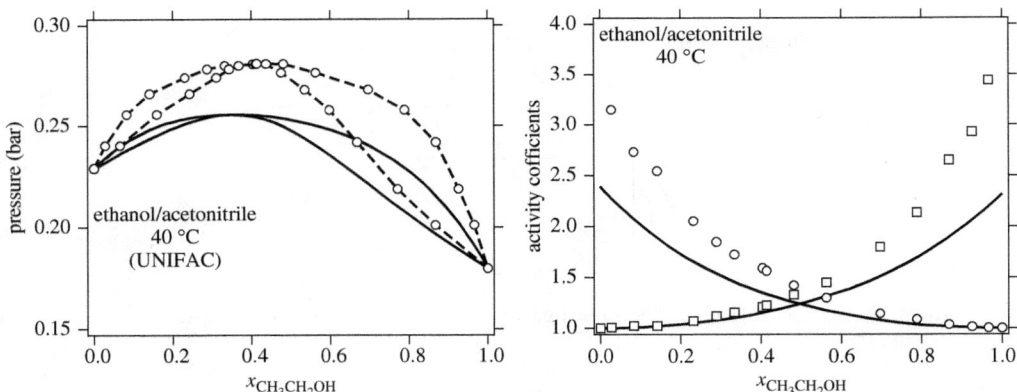

Figure 12-12: *Pxy* graph and activity coefficients for the system ethanol/acetonitrile at 40 °C calculated from UNIFAC.

Finally, the activity coefficients are

$$\gamma_1 = 2.01244,$$
$$\gamma_2 = 1.00918.$$

The *Pxy* graph is obtained by repeating this calculation at various liquid compositions in the range $x_1 = 0$ to $x_1 = 1$. The results are shown in Figure 12-12.

Comments The agreement with the experimental data is not very good. UNIFAC indeed predicts an azeotrope at $x_1 = 0.33$ (compared to $x_1 = 0.414$ from the experimental data) but its pressure is under-predicted by about 0.025 bar. We must note, however, that as a predictive model, UNIFAC has no adjustable parameters. The next example demonstrates a case in which UNIFAC produces a much better prediction of phase behavior.

Example 12.12: Using UNIFAC (2)
Use UNIFAC to calculate the *Pxy* graph for the system n-pentane/1-butanol at 30 °C.

Solution N-pentane contains two "CH_3" and three "CH_2" subgroups. 1-butanol contains one "CH_3", three "CH_2" and one "OH". The R and Q parameters are summarized below.

Subgroup	k	R_k	Q_k	$\nu_k^{(1)}$	$\nu_k^{(2)}$
CH_3	1	0.9011	0.848	2	1
CH_2	2	0.6744	0.540	3	3
OH	14	1	1.200	0	1

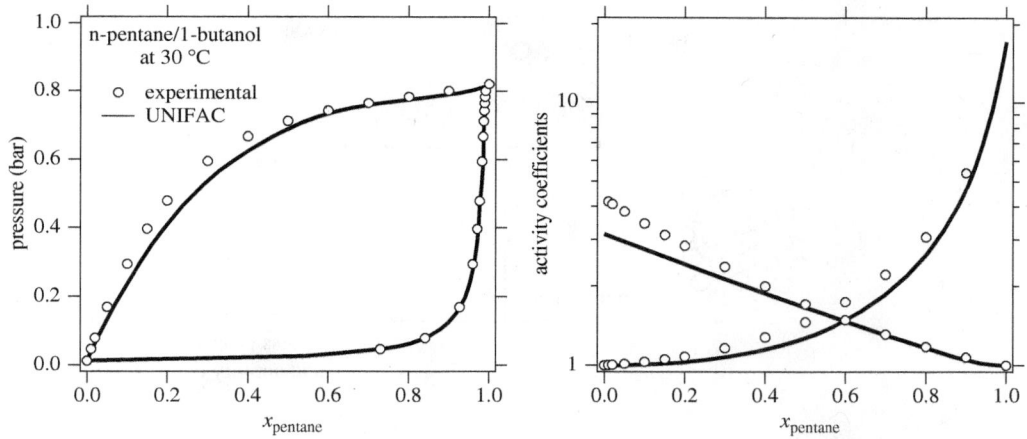

Figure 12-13: *Pxy* graph and activity coefficients for the system pentane/1-butanol at 30 °C calculated from UNIFAC (data from Ronc and Ratcliff, *Can. J. Chem. Eng.* **54**, no. 326 [1976]).

The calculation is done as in the previous example and is left as an exercise. The *Pxy* graph is shown in Figure 12-13. In this case the agreement is very good.

12.7 Summary

Activity coefficients offer an alternative to equations of state for the calculation of solution properties and vapor-liquid equilibrium. The activity coefficient is closely related to the excess Gibbs free energy, and quantifies deviations from ideal-solution behavior. There is a certain similarity to the fugacity coefficient, which, as we recall, is based on the residual Gibbs free energy and quantifies deviations from the ideal-gas state. The similarity is not coincidental. In both cases we are applying the same basic methodology:[9] we adopt an idealized state as our reference state (ideal-gas state, ideal solution) and use appropriate corrections (residual properties, excess properties) to quantify systems that deviated from this idealization. This relationship is depicted schematically in Figure 12-14 and highlights the similarities and differences between calculations based on equations of state and those based on

---

[9]. The Gibbs free energy is the work that must be done to transfer a molecule from one state to another at constant temperature and pressure. For example, $\overline{G}_i^R$ is the work required to transfer a molecule from an isolated environment of no neighbors (ideal-gas state) into a solution with other molecules.

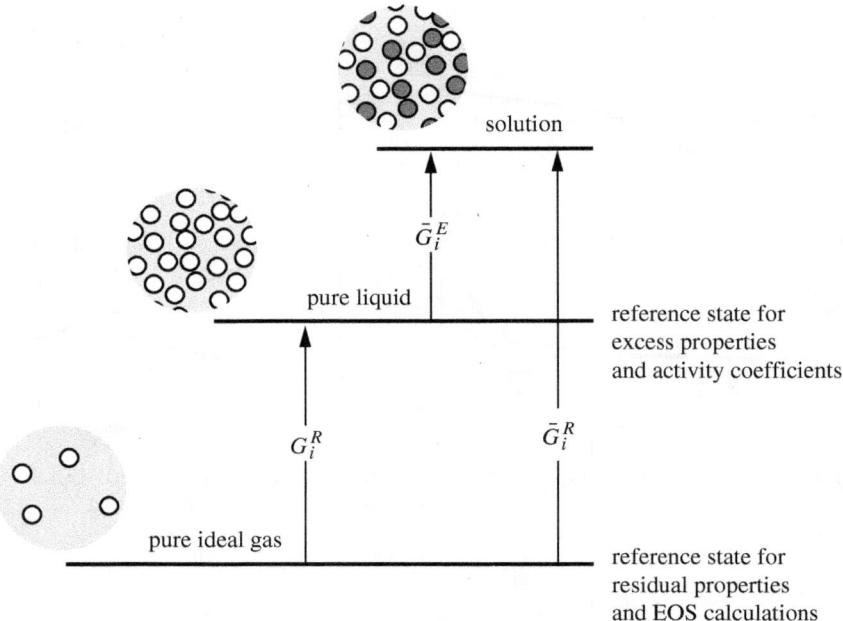

Figure 12-14: Qualitative representation of VLE approach based on equations of state (using residual properties) and approach based on activity coefficients (using excess properties).

activity coefficients.[10] Starting with the ideal-gas state as a reference, deviations are measured via residual properties (partial molar residual, if the target state is the component in solution). These residuals require an equation of state for their calculation. If the reference state is the pure liquid, then deviations are quantified by the excess property (partial molar excess, if the target state is the component in solution). We can now see why the approach based on activity coefficients is so successful: It starts from a reference state that is much closer to the target state. This makes the necessary corrections smaller in magnitude and easier to model. The downside is that these corrections cannot be calculated via purely predictive equations but require input from experiments.

The activity coefficient is a correction factor in Raoult's law that accounts for deviations from ideal-solution behavior. If the activity coefficient is known, VLE calculations can be performed and the phase diagram of the system can be constructed.

---

10. The arrows in Figure 12-14 are drawn qualitatively. In reality, the magnitude of the various Gibbs free energies is in most cases negative.

The activity coefficient depends on temperature, pressure and composition but the most critical variable is composition. The main challenge is how to obtain the activity coefficient for a given system, if experimental data at the desired conditions are not available. Models such as the Margules, van Laar, and the like, provide a correlation between activity coefficients and composition via an equation with a small number of adjustable parameters. These parameters are fitted using experimental data and the agreement between the resulting correlation and the original data is generally very good. UNIFAC contains no adjustable parameters but uses values that have been obtained by regressing a large number of experimental data for several different mixtures. As a result, the accuracy of its predictions varies from system to system. For example, the system n-pentane/1-butanol is predicted fairly accurately, even though it is quite strongly no-ideal (notice the magnitude of the activity coefficients at infinite dilution). By contrast, the system ethanol/acetonitrile is predicted less accurately. The predictive advantage of UNIFAC comes at the expense of accuracy, which is not guaranteed. In the literature we find recommendations that provide decision trees for selecting the appropriate model based on the chemical character of the components. Unfortunately, there is no foolproof method by which to determine which model is best for a given system. The best approach is try various models and validate them against available data before a final choice is made.

12.8 Problems

Problem 12.1: A binary solution is described by the one-parameter equation

$$\frac{G^{\mathrm{E}}}{RT} = A(x_1 - x_2^2),$$

where A is constant. At 50 °C and $x_1 = 0.30$ the partial pressures of the two components in the vapor are $P_1 = 105.3$ mm Hg, $P_2 = 288$ mm Hg respectively. The saturation pressure of pure components at 50 °C: $P_1 = 228$ mm Hg, $P_2 = 380$ mm Hg. Calculate the Pxy graph of this system at 50 °C.

Problem 12.2: The table below gives the enthalpy of benzene(1)-heptane(2) solutions at 20 °C 1 bar as a function of the mol fraction of benzene.
a) Calculate and plot the excess enthalpy at 20 °C, 1 bar.
b) Calculate the partial molar excess enthalpy at $x_1 = 0.2$, 20 °C, 1 bar.
c) Calculate the heat that is released or absorbed when 10 mol of benzene is mixed with 40 mol of heptane isothermally at 20 °C, 1 bar.

d) Calculate the temperature change when 10 mol of benzene is mixed with 40 mol of heptane adiabatically, 1 bar, if the initial temperature of the pure components is 20 °C. The heat capacity of the pure liquids are $C_{P1} = 134$ J/mol K, $C_{P2} = 222$ J/mol K.

x_1	H (J/mol)	x_1	H (J/mol)
0	635.00	0.50	1229.51
0.14	956.91	0.58	1185.85
0.20	1035.06	0.65	1110.76
0.22	1066.37	0.71	1005.74
0.30	1174.31	0.83	688.42
0.33	1184.04	0.88	537.71
0.37	1203.61	1	−50.00
0.44	1237.33		

Problem 12.3: The table below gives the excess enthalpy for the system acetone(1)-water(2) at 25 °C, 1 bar.

a) Calculate the partial molar enthalpy of each component at $x_1 = 0.5$, 25 °C, 1 bar. The reference state for each component is the pure liquid at 25 °C, 1 bar.

b) Repeat the previous part with the reference state for water changed to the pure liquid at 2 bar,

c) A solution that contains 50% by mole acetone is mixed with equal moles of a solution that contains 80% by mole acetone. The mixing takes place in a heat bath at 25 °C. Determine whether any amount of heat is exchanged between the system and the surroundings. If so, calculate that heat and report whether the mixing is endo- or exothermic.

x_{acetone}	H^{E} (J/mol)	x_{acetone}	H^{E} (J/mol)
0.0000	0.00	0.5000	−167.57
0.0424	−388.11	0.6271	70.27
0.1186	−630.27	0.7246	200.00
0.1314	−638.92	0.7542	234.60
0.2458	−612.97	0.8771	277.84
0.2542	−591.35	0.9534	234.60
0.3729	−414.05	1.0000	0.00

Problem 12.4: a) Based on Figure 12-5, is the mixing of hydrazine and water an endothermic or exothermic process?

b) Calculate the excess enthalpy for a solution that contains 60% by mol hydrazine at 20 °C.

12.8 Problems

Problem 12.5: Determine the phase of a mixture of hydrazine and water at 1 bar, 110 °C if the overall mol fraction of hydrazine is 20%. If the state is a mixture of vapor and liquid, report the amounts and compositions in each phase.
(Use Figure 12-5 to answer these questions.)

Problem 12.6: Pure hydrazine is mixed with pure water to form a solution that contains 40% hydrazine (by mol). Both components are initially at 20 °C.
a) Determine the temperature of the final solution if mixing is adiabatic.
b) Determine the amount of heat exchanged with the surroundings if mixing is isothermal.
(Use Figure 12-5 to answer these questions.)

Problem 12.7: The activity coefficients of the system benzene(1)/acetonitrile(2) are given by
$$\ln \gamma_1 = A x_2^2, \quad \ln \gamma_2 = A x_1^2.$$
At 45 °C, $A = 1$.
a) Obtain the activity coefficients of the two components at infinite dilution.
b) Does this system form an azeotrope at 45 °C? If so, report the pressure and the azeotropic composition.
c) Calculate the bubble and dew pressure of a system containing 30% benzene by mole at 45 °C.
Additional data. The saturation pressures of components 1 and 2 at 45 °C are 223.7 and 208.4 Torr, respectively.

Problem 12.8: The data in the table below are for the system benzene (1)-diethylamine at 55 °C (Letcher, T. M. and Bayles, J.W., *J. Chem. Eng. Data*, **16**, 266 (1971)):

x_1	y_1	P (bar)
0.000	0.000	1.0018
0.125	0.066	0.9389
0.159	0.085	0.9205
0.286	0.157	0.8533
0.430	0.252	0.7737
0.493	0.301	0.7387
0.538	0.338	0.7133
0.620	0.415	0.6662
0.646	0.441	0.6519
0.802	0.633	0.5601
0.845	0.698	0.5336
0.862	0.725	0.5241
0.912	0.812	0.4930
0.953	0.894	0.4668
1.000	1.000	0.4365

a) Fit the excess Gibbs energy to a polynomial expression of the form

$$\frac{G^{ex}}{RT} = x_1(1-x_1)(a_0 + a_1 x_1 + a_2 x_1^2)$$

and obtain the values of a_0, a_1 and a_2.
b) Express the activity coefficients in terms of the above parameters.
c) Make a graph of the Pxy graph calculated from the fitted activity coefficients and compare with the experimental data.
d) Instead of this three parameter fit it has been suggested to use the Margules equation. What is your recommendation?

Problem 12.9: Use the data below for the system benzene(1)/acetonitrile(2) at 45 °C to answer the following questions:
a) What is the saturation pressure of the pure components at 45 °C?
b) What is the phase of the pure components at 45 °C and 34 kPa?
c) Calculate the fugacity of the pure components at the conditions of part (b).
d) Two moles of benzene are mixed with 3 moles of acetonitrile under constant temperature and pressure at 45 °C and 33 kPa. What is the phase of the mixture (vapor, liquid, or mixed)?
e) What is bubble pressure of the mixture at 45 °C?
f) Mark on the graph the points before and after mixing.
g) Calculate the fugacity of the two components at the composition of the azeotrope.
h) Using data from the graph, estimate the activity coefficient of benzene at infinite dilution.

P (bar)	x_1	y_1	P (bar)	x_1	y_1
0.2778	0.0000	0.0000	0.3717	0.5500	0.5431
0.3022	0.0500	0.1243	0.3710	0.6000	0.5687
0.3208	0.1000	0.2124	0.3692	0.6500	0.5960
0.3351	0.1500	0.2784	0.3662	0.7000	0.6260
0.3459	0.2000	0.3302	0.3617	0.7500	0.6601
0.3542	0.2500	0.3725	0.3554	0.8000	0.7002
0.3604	0.3000	0.4081	0.3467	0.8500	0.7488
0.3650	0.3500	0.4392	0.3350	0.9000	0.8098
0.3682	0.4000	0.4671	0.3192	0.9500	0.8899
0.3704	0.4500	0.4932	0.2982	1.0000	1.0000
0.3715	0.5000	0.5182			

Problem 12.10: The activity coefficients for the system carbon methyl acetate (1)/methanol (2) at 60 °C can be described by the equation

$$\gamma_1 = e^{Ax_2^2}, \quad \gamma_2 = e^{Ax_1^2}.$$

12.8 Problems

The saturation pressures of the pure components at 60 °C are:

$$P_1^{\text{sat}} = 1.126 \text{ bar},$$

$$P_2^{\text{sat}} = 0.847 \text{ bar}.$$

It is also known that at 60 °C the two liquids are fully miscible and that $\gamma_1^\infty = 2.89$.
a) Does this system exhibit positive or negative deviations from ideality? How can you tell?
b) Calculate the bubble pressure and the composition of the vapor at 60 °C of a solution that contains 63.4% methyl acetate.
c) Calculate the Pxy graph of this system at 60 °C.

Problem 12.11: The laboratory division of your company has sent you the following data for the system n-pentane (1)/propionaldehyde (2) at 35 °C: The activity coefficients of the two components at infinite dilution are 4.0 and the saturation pressures of the pure components are 0.7 and 1 bar, respectively.
a) Does this system exhibit positive, negative, or no deviations from ideal-solution behavior?
b) Assuming the validity of the Margules equation, determine the constants A_{12} and A_{21}.
c) You suspect that this system forms an azeotrope. What is the composition of the two phases and the pressure at the azeotrope?
d) Draw a qualitative Pxy graph at 35 °C. On the graph show all the information that you have for this system.
e) Your are designing a single flash separation process for a stream that contains 80% n-pentane (the rest is propionaldehyde). The mixture is to be flashed at 35 °C and you must determine the pressure so that the purity of n-pentane is 95%. What is this pressure?

Problem 12.12: At 50 °C the system acetone(1) and chloroform(2) forms an azeotrope with composition $x_1 = 0.416$. The activity coefficients are given by the equation

$$\ln \gamma_1 = A x_2^2, \quad \ln \gamma_2 = A x_1^2,$$

where A is constant.
a) What is the pressure at the azeotropic composition at 50 °C?
b) One mole of acetone is mixed with one mole of chloroform at 50 °C, 0.5 bar. What is the phase of the pure components before mixing? What is the phase of the system after mixing?
Additional information: The saturation pressures of the pure components at 50 °C are $P_{\text{acetone}} = 0.81$ bar, $P_{\text{chloroform}} = 0.41$ bar.

Problem 12.13: a) Use the data for the system ethyl-methyl ketone(1)-toluene(2) given below (T = 50 °C) to obtain the parameters of the van Laar equation.
b) Reconstruct the Pxy using the van Laar equation and compare with the data.

x_1	y_1	P (bar)
0.0000	0.0000	0.1230
0.0895	0.2716	0.1551
0.1981	0.4565	0.1861
0.3193	0.5934	0.2163
0.4232	0.6815	0.2401
0.5119	0.7440	0.2592
0.6096	0.8050	0.2796
0.7135	0.8639	0.3012
0.7934	0.9048	0.3175
0.9102	0.9590	0.3415
1.0000	1.0000	0.3609

Problem 12.14: The following data are for the system water(1)/diethylamine(2).
a) Do the components form an ideal solution?
b) Calculate the fugacity and fugacity coefficient of water in a solution that contains 20% water by mole, at the bubble point at 311.5 K.

P (Pa)	T (K)	x_1	y_1
54382.19	311.50	0.0000	0.0000
48955.97	311.50	0.2000	0.0530
6759.44	311.50	1.0000	1.0000

Problem 12.15: A stream (1) that carries a mixture of heptane (60% by mol) and benzene (40%) is compressed to unspecified pressure (stream 2) in a compressor whose efficiency is 80%. Stream 2 is cooled to 152 °C and the resulting vapor-liquid mixture (stream 3) is separated in a vapor-liquid separator. One of the two streams that exit the separator contains 65% heptane. The activity coefficients for this system are given by the Margules equation with $A_{12} = 3$, $A_{21} = 1$, where the subscripts 1 and 2, refer to heptane and benzene, respectively. The saturation pressures of heptane and benzene at 152 °C are $P_1^{sat} = 3.9$ bar, and $P_2^{sat} = 2.5$ bar, respectively; the ideal heat capacities are $C_{P1} = 208$ J/mol K and $C_{P2} = 130$ J/mol K. The gas phase may be assumed ideal.
a) Determine the pressure in all streams.
b) Calculate the molar rates and composition (mol fraction of heptane) in all streams.
c) Calculate the work in the compressor and the temperature of stream 2.

12.8 Problems

d) Calculate the heat load in the heat exchanger.
e) Calculate the entropy generation for the entire process.

Problem 12.16: The data below are for a vapor/liquid system of two unnamed components:

x_1	y_1	P (Torr)	x_1	y_1	P (Torr)
0	0	120	0.6	0.9375	1108.9
0.1	0.7385	415.3	0.7	0.9439	1147
0.2	0.8485	649.3	0.8	0.9503	1175.9
0.3	0.892	828.3	0.9	0.9621	1208.6
0.4	0.9149	959.1	1	1	1250
0.5	0.9286	1049.7			

a) Calculate the activity coefficients of the two components and plot them against x_1. Your graph should be done on the computer and it should be properly annotated.
b) Does this system exhibit positive or negative deviations from ideality?
c) Determine whether the system is best described by the Margules or the van Laar equation and determine the parameters of the corresponding equation.

Problem 12.17: The system water (1)/phenol (2) is described by the van Laar equation with $B_{12} = 0.83$, $B_{21} = 3.22$.
a) Construct the Pxy graph of this system at 100°C.
b) If a solution that contains 65% (mol) water is flashed to 0.5 bar, 100 °C, what is the composition and relative amounts of the phases that are formed?
The saturation pressure of phenol at 100 °C is 0.055 bar.

Problem 12.18: Assuming that the system ethyl methyl ketone (1)/ toluene (2) at 50 °C is described by the Margules equation with $A_{12} = 0.3681$, $A_{21} = 0.2046$, do the following:
a) Calculate the bubble pressure at 50 °C of a mixture with $z_1 = 0.3$.
b) Calculate the dew pressure at 50 °C of a mixture with $z_1 = 0.3$.
c) A mixture with $z_1 = 0.3$ is flashed to 50 °C at pressure such that a stream is received that contains 88% toluene by mol. Determine the pressure and the percent recovery of toluene in the toluene-rich stream.
Additional information: $P_1^{sat} = 0.3609$ bar, $P_2^{sat} = 0.123$ bar.

Problem 12.19: The following information is available for the system tetrachloromethane (1) acetic acid (2) at 25 °C:

$$P_1^{sat} = 0.1213 \text{ bar} \quad P_2^{sat} = 0.0157 \text{ bar,}$$
$$\gamma_1^\infty = 3.0 \quad \gamma_2^\infty = 4.0.$$

a) Assuming that the activity coefficients are given by the equations below,

$$\ln \gamma_1 = x_2^2 \left[A + 2(B - A)x_1 \right],$$
$$\ln \gamma_2 = x_1^2 \left[B + 2(A - B)x_2 \right],$$

obtain the constants A and B.

b) What is the phase of a system that contains 30% tetrachloromethane, at $P = 0.7$ bar, $T = 25\,°\mathrm{C}$?

c) Calculate the fugacity of tetrachloromethane at the conditions of part (b).

d) Calculate the *fugacity coefficient* of tetrachloromethane at the conditions of part (a).

e) A mixture of the two components is flashed to conditions that produce a vapor-liquid mixture. The mol fraction of tetrachloromethane in the liquid is $x_1 = 0.6$ and the temperature of the two-phase system is 25 °C. What is the pressure?

f) Calculate the composition of the vapor in the previous part.

g) Calculate the vapor and liquid fractions formed in part (e) if the mol fraction of tetrachloromethane in the initial mixture (before flashing) is 80%.

Problem 12.20: At 70 °C, 1 bar, the system acrylonitrile(1)/water(2) forms two phases with composition, $x_{1A} = 0.968$, $x_{1W} = 0.073$, where the subscripts A and W refer to the acrylonitrile-rich and water-rich phases, respectively. Assuming that the Margules equation is appropriate for this system:

a) Calculate the activity coefficients of the two components at the given conditions.

b) Calculate the bubble pressure of the two-phase system at 70 °C and the composition of the vapor.

c) Calculate the Pxy phase diagram at 70 °C.

Additional information: The saturation pressures of the pure components at 70 °C are

$$P_1^{\mathrm{sat}} = 0.791 \text{ bar}, \quad P_2^{\mathrm{sat}} = 0.312 \text{ bar}.$$

Problem 12.21: a) Use the SRK equation to calculate the activity coefficient of oxygen in air (gas) at the dew pressure of air at 120 K.

b) Use the SRK equation to calculate the activity coefficient of oxygen in liquid air at the bubble pressure of air at 120 K.

Assume that air is a mixture of oxygen (21%) and nitrogen (79%) and that $k_{12} = 0$.

Problem 12.22: The table below shows the results from a VLE calculation for the system normal heptane (1)/normal decane (2) but because someone spilled coffee, some entries are missing.

12.8 Problems

a) Calculate the activity coefficient of n-decane at $P = 17.22$ bar, $T = 570$ K, $x_{\text{decane}} = 0.7$.
b) Is Raoult's law valid for this system at 570 K? (Answers without justification will not count.)
c) Fill out the missing information in the table. You may not fill out the table by interpolation!
d) Draw a qualitative Pxy graph for this system at 570 K.
e) State and justify your assumptions.

Component 1 = n-heptane; component 2 = n-decane

P (bar)	T (K)	x_1	y_1	ϕ_1^L	ϕ_1^V	ϕ_2^L	ϕ_2^V	Z_V	Z_L
11.02	570	0.00	0.000	1.912	0.938		0.765	0.685	0.093
12.99	570	0.10		1.627	0.911	0.659	0.723	0.664	0.109
13.50	570		0.217		0.905	0.638	0.713	0.659	0.114
17.22	570	0.30	0.431	1.237	0.861	0.520	0.640	0.611	0.148

Problem 12.23: The partial molar enthalpies of species 1 and 2 in a binary mixture at 25 °C are given by

$$\overline{H}_1 = 100(1 + 0.08x_2^2); \quad \overline{H}_2 = 80(1 + 0.1x_1^2),$$

where the enthalpy is in kJ/mol of mixture.
a) Calculate the molar enthalpies of the pure components.
b) Calculate the partial molar enthalpies of the two components at infinite dilution.
c) Determine the excess partial molar enthalpies of the two components as functions of the composition and calculate their values at infinite dilution.
d) Calculate the molar enthalpy of an equimolar solution.
e) Calculate the enthalpy change when the two components are mixed at constant temperature and pressure to produce an equimolar solution.
f) Is this an endothermic or exothermic system?

Problem 12.24: Use the UNIFAC equation to estimate the enthalpy of mixing of a solution that contains 37% (mol) acetic acid in water at 20 °C, 1 bar.

Problem 12.25: The infinite dilution activity coefficients of the system water/acetic acid at 100 °C are

$$\gamma_w^\infty = 5.5, \quad \gamma_{ac}^\infty = 2.57.$$

a) Obtain the parameters of the Margules equation for this system.
b) Calculate the bubble pressure at 100 °C of a solution with $x_w = 0.63$.

c) Calculate the dew pressure at 100 °C of a solution with $x_w = 0.63$.
Additional data: The Antoine parameters for the two components are:

	A	B	C
Water	18.3036	3816.44	−46.13
Acetic acid	16.8080	3405.57	−56.34

to be used with

$$\ln P^{\text{sat}} = A - \frac{B}{T+C},$$

with T in K and P^{sat} in Torr.

Problem 12.26: a) You just finished an SRK calculation for the system carbon dioxide (1) / normal pentane (2) at 290 K when a virus wiped out your hard disk. Fortunately you had printed a copy (see table below), but in your panic to save it, you spilled coffee and can't make some of the numbers. Fill in the missing entries.
b) How big of a container (cm^3) is needed to store 1000 moles of a solution that contains 50% by mole carbon dioxide at 290 K, 30.1 bar?
c) What is the fugacity of carbon dioxide in solution with n-pentane, at $x_1 = 0.5$, $T = 290$ K, $P = 30.1$ bar?
d) Calculate the Poynting factor of carbon dioxide in solution with n-pentane, at $x_1 = 0.5$, $T = 290$ K, $P = 30.1$ bar.
e) Calculate the activity coefficient of carbon dioxide in the liquid at 50% mole fraction.
f) Does this system obey Raoult's law at 290 K?

P (bar)	x_1	y_1	ϕ_1^L	ϕ_1^V	ϕ_2^L	ϕ_2^V	Z^V	Z^L
0.5	0	0.000	128.211	1.009	0.979	0.979	0.979	0.003
30.1	0.5	0.978	1.645	0.841	0.021	0.481	0.799	0.116
34.8	0.6	0.980		0.817	0.021		0.761	0.125
42.5			0.955	0.777	0.026	0.329	0.692	0.134
53.5	1	1.000	0.716	0.716	0.052	0.221	0.578	0.147

Problem 12.27: Use the UNIFAC equation to do the following calculations for the system methanol(1)-acetic acid(2):
a) Calculate the Pxy graph for this system at 80 °C.
b) Do you expect the volume of the system to expand or to contract upon mixing the pure components?
c) Calculate the fugacity of methanol in mixture with acetic acid at $x_{\text{MeOH}} = 0.8$, 80 °C, 0.4 bar.
d) Repeat the previous calculation at the bubble pressure of the solution instead of 0.4 bar.

12.8 Problems

e) Calculate the chemical potential of methanol in mixture with acetic acid at $x_{\text{MeOH}} = 0.8$, 80 °C, 0.4 bar. The reference state is the pure liquid at 80 °C, 0.4 bar. How is your answer affected by the fact that methanol is in the *vapor* phase at 80 °C, 0.4 bar?

f) Repeat the previous part if the reference state is the pure liquid at saturation at 80 °C.

Additional data: The saturation pressures of methanol and acetic acid at 80 °C are 1.81 and 0.2763 bar, respectively.

Problem 12.28: Use the UNIFAC method to calculate the following phase diagrams for the system water(1)-ethanol(2):
a) Pxy graph at 100 °C.
b) Txy graph at 1 bar.

Problem 12.29: Use the UNIFAC equation to calculate the Pxy graph for the system water/acetic acid at 100 °C. The saturation pressure of acetic acid at 100 °C is 0.57 bar.

Chapter 13

Miscibility, Solubility, and Other Phase Equilibria

In the previous chapters we applied the theory of phase equilibrium to a very important and very common problem in chemical engineering: vapor-liquid equilibrium. This is not the only type of phase equilibrium that is encountered in practice. Some liquids have limited miscibility in each other. When mixed, they form two liquid phases and give rise to liquid-liquid equilibrium (LLE). When such a system is brought to boil, it forms a third phase, vapor, and the thermodynamic problem is one of vapor-liquid-liquid equilibrium (VLLE). Limited solubility is also encountered in mixtures of gases with liquids (oxygen in water, for example) and of solids in liquids (glucose in water). Another problem of industrial and biological relevance is osmosis. In this case partial equilibrium is established between two liquids via a semipermeable membrane that restricts the passage of one component. These problems may seem unrelated but they have a common thread: they are all governed by the basic principle that requires the chemical potential (or the fugacity) of a component distributed in various phases to be the same in all phases. In this chapter we apply the principles of thermodynamics to such problems. The objectives in this chapter are to:

1. Apply the minimization of the Gibbs free energy as a criterion to determine the limits of mutual miscibility of liquids.

2. Use Henry's law to calculate the fugacity of a dissolved gas and calculate phase equilibrium in gas-liquid systems.

3. Calculate equilibrium through a semipermeable membrane and analyze separation processes that are based on osmotic effects.

13.1 Equilibrium between Partially Miscible Liquids

Liquids exhibit partial miscibility when their interactions show strong positive deviations from ideality. This indicates that cross interactions are unfavorable to such an extent that full miscibility is not possible. When a mixture of two liquids forms

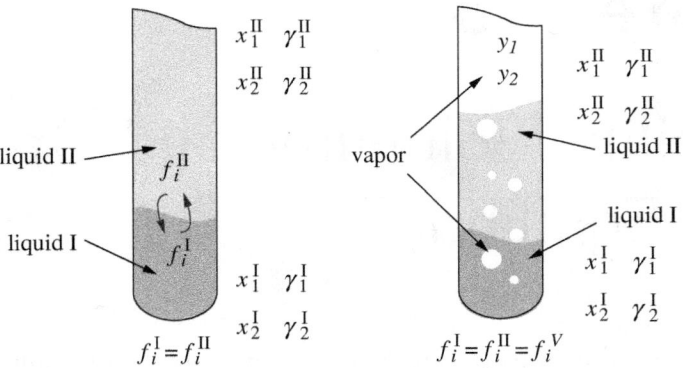

Figure 13-1: Schematic of liquid-liquid equilibrium (LLE) and vapor-liquid-liquid (VLLE) equilibrium between partially miscible liquids.

two separate liquid phases, each phase contains both components, though the composition is generally different in each phase. The situation is schematically shown in Figure 13-1, in which the two phases have been identified by the superscripts I and II. Each phase is saturated in both components so that if small additional amounts are added of either component, these will be distributed among the two liquids in such a way that the composition in each liquid will remain unchanged. The distribution of components is governed by the general-phase equilibrium criterion that requires the fugacity of a component to be the same in both phases. Using the low-pressure approximation of the fugacity of component in solution, the equilibrium criterion becomes:

$$\begin{aligned} f_1^I &= f_1^{II} \\ f_2^I &= f_2^{II} \end{aligned} \quad \Rightarrow \quad \begin{aligned} x_1^I \gamma_1^I P_1^{\text{sat}} &= x_1^{II} \gamma_1^{II} P_1^{\text{sat}}, \\ x_2^I \gamma_2^I P_2^{\text{sat}} &= x_2^{II} \gamma_2^{II} P_2^{\text{sat}}, \end{aligned}$$

where x_i^I is the mol fraction of component i in phase I, γ_i^I is its activity coefficient, P_i^{sat} is the saturation pressure at the temperature of the system, and similarly for phase II. Because the compositions in each liquid are different, the activity coefficients are also different. The saturation pressures are of course the same since both phases are at the same temperature. Thus the equilibrium criterion simplifies to

$$\begin{aligned} x_1^I \gamma_1^I &= x_1^{II} \gamma_1^{II}, \\ x_2^I \gamma_2^I &= x_2^{II} \gamma_2^{II}. \end{aligned} \tag{13.1}$$

These equations express the conditions for liquid-liquid equilibrium (LLE). If the activity coefficients are known, the compositions of the two-liquid phases can be obtained by solving the system of the two equations.

13.1 Equilibrium between Partially Miscible Liquids

When a two-liquid system is brought to boil, the vapor constitutes a third phase, which coexists with the two liquids. The equilibrium criterion at this point applies to all three phases, and thus we have,

$$\begin{aligned} f_1^I &= f_1^{II} = f_1^V \\ f_2^I &= f_2^{II} = f_2^V \end{aligned} \Rightarrow \begin{aligned} x_1^I \gamma_1^I P_1^{sat} &= x_1^{II} \gamma_1^{II} P_1^{sat} = y_1 P, \\ x_2^I \gamma_2^I P_2^{sat} &= x_2^{II} \gamma_2^{II} P_2^{sat} = y_2 P. \end{aligned} \qquad (13.2)$$

These equations fix the equilibrium compositions of the three phases. As a consequence of these relationships, and as long as both liquid phases are present, the system boils as an azeotrope: the boiling temperature and the composition of all phases remain constant and the only change is that the vapor phase increases at the expense of the liquid. When one of the two liquids boils off completely, the system consists of a single liquid and a vapor and behaves as usual, namely, its boiling temperature and the composition of the two phases vary continuously as more liquid is converted into vapor.

To perform calculations with partially miscible liquids we need expressions for the activity coefficients. These may be calculated by any of the models discussed in Chapter 12, with the notable exception of the Wilson model, which is not capable of describing partially miscible liquids. If the activity coefficients are known, LLE and VLLE calculations are fundamentally no different from the VLE calculations discussed in the previous chapter.

Example 13.1: Solubility Limits Using Activity Coefficients

Methanol and carbon disulfide at 10 °C exhibit partial miscibility. The activity coefficients may be represented by the NRTL equation with $\alpha = 0.2$ and $\tau_{12} = 49.62/T$, $\tau_{21} = 940.7/T$, where T is temperature in K. Determine the composition of the equilibrium phases at 10 °C.

Solution Using $\alpha = 0.2$ and the values of τ_{ij} given in the problem statement we find

$$G_{12} = 0.96556, \quad G_{21} = 0.514543.$$

The corresponding NRTL expressions for the activity coefficients are

$$\ln \gamma_1 = \frac{0.870055(1-x_1)^2}{(0.503811(1-x_1)+x_1)^2} + \frac{0.222994(1-x_1)^2}{(1-0.0456573 x_1)^2},$$

$$\ln \gamma_2 = \frac{1.72695 x_1^2}{(0.503811(1-x_1)+x_1)^2} + \frac{0.212813 x_1^2}{(1-0.0456573 x_1)^2}.$$

To obtain the mutual solubility of methanol and carbon disulfide, we solve eq. (13.1) for x_1^I and x_1^{II} using these activity coefficients. This requires a numerical solution and it is best done using a mathematical package. The solution is

$$x_1^I = 0.040724, \quad x_1^{II} = 0.742787.$$

To confirm that these satisfy the equilibrium criterion we calculate the activity coefficients of methanol in the two phases:

$$\ln\gamma_1^{I} = 19.9129, \quad \ln\gamma_1^{II} = 1.09174,$$
$$\ln\gamma_2^{I} = 1.01025, \quad \ln\gamma_2^{II} = 3.76774.$$

These indeed satisfy eq. (13.1):

methanol: $(0.040724)(19.9129) = (0.742787)(1.09174) = 0.810932,$
carbon disulfide: $(1 - 0.040724)(1.01025) = (1 - 0.742787)(3.76774) = 0.969113.$

Therefore, the solubility of methanol in carbon disulfide at 10 °C is

$$x_1^{I} = 0.0407 = 4.07\% \text{ by mol},$$

and the solubility of carbon disulfide in methanol is

$$x_2^{II} = 1 - 0.743 = 0.257 = 25.7\% \text{ by mol}.$$

Comments The numerical solution of eq. (13.1) requires some attention. These equations are always satisfied at $x_1^{I} = x_1^{II}$ but such solutions are trivial and do not represent the composition of true coexisting phases. Trivial solutions must be identified and rejected, whenever the numerical method happens to converge to one of them.

13.2 Gibbs Free Energy and Phase Splitting

In practice it is useful to know before mixing whether we will obtain a homogeneous solution, or whether phase splitting will occur. One way to answer this question is suggested in the previous example: if eq. 13.1 has a nontrivial solution, the system phase separates into two liquids whose compositions are given by the solution to these equations. If no solution exists, then the liquids form a homogenous solution at all compositions. Alternatively, the determination of phase splitting can be done by examining the Gibbs energy of mixing. Recall that the equilibrium state of a system at fixed pressure and temperature has the minimum possible Gibbs free energy. With this in mind, we analyze the problem of phase splitting as follows. We form 1 mol of solution by mixing x_1 mol of liquid component 1 and x_2 mol of liquid component 2 ($x_1 + x_2 = 1$). If components are fully miscible at this composition, the result is a single liquid with composition (x_1, x_2). The Gibbs free energy of this single-phase system is

$$G = G^{\mathrm{id}} + G^{\mathrm{E}} = \underbrace{x_1 G_1 + x_2 G_2 + RT(x_1 \ln x_1 + x_2 \ln x_2)}_{G^{\mathrm{id}}} + \underbrace{RT(x_1 \ln \gamma_1 + x_2 \ln \gamma_2)}_{G^{\mathrm{E}}}$$

$$= x_1 G_1 + x_2 G_2 + \underbrace{RT\big(x_1 \ln(x_1 \gamma_1) + x_2 \ln(x_2 \gamma_2)\big)}_{\Delta G_{\mathrm{mix}}}.$$

13.2 Gibbs Free Energy and Phase Splitting

Choosing the reference state for each component to be the pure liquid (i.e., $G_1 = G_2 = 0$), the Gibbs energy of a single phase system simplifies to

$$G = \Delta G_{\text{mix}} = RT\bigl(x_1 \ln(x_1 \gamma_1) + x_2 \ln(x_2 \gamma_2)\bigr). \tag{13.3}$$

If the liquids are only partially miscible, then two phases are formed, phase I with composition (x_1^I, x_2^I), and phase II with composition (x_1^{II}, x_2^{II}). The number of moles in each phase are related by the mass-balance equations in the form of the familiar lever rule:

$$n^I = \frac{x_1 - x_1^{II}}{x_1^I - x_1^{II}}, \quad n^{II} = \frac{x_1^I - x_1}{x_1^I - x_1^{II}}.$$

Notice that $n^I + n^{II} = 1$ because we are working with a total of 1 mol. As an extensive property, the Gibbs free energy of the two-phase system is the sum of the two phases:

$$G^{I+II} = n^I G^I + n^{II} G^{II} = n^I G^I + (1 - n^I) G^{II}, \tag{13.4}$$

where G^I and G^{II} are the molar Gibbs free energies of each liquid phase. These are given by eq. (13.3) using the composition of the corresponding phase.

To decide whether the system forms a single phase or two liquid phases we compare the Gibbs free energy for each case:

- If $G < G^{I+II}$, the single-phase system is more stable;
- If $G > G^{I+II}$, the two-phase system is more stable.

To demonstrate the application of this criterion, we reanalyze the system methanol/carbon disulfide of Example 13.1. The Gibbs free energy of the solution is calculated from eq. (13.3) using the NRTL equation with the constants given in Example 13.1 and is plotted in Figure 13-2 as a function of the mol fraction of methanol.

Suppose that we mix 0.3 mol of methanol with 0.7 mol of carbon disulfide. If the system forms a single phase, its state would be represented by point A on the Gibbs line. If it forms two phases, say, liquids B' and C', the overall state is represented by point A'. Points B' and C' lie on the Gibbs curve (both states are a single-phase liquid) and their composition satisfies the condition $x_{B'} < x_A < x_{C'}$. Point A' lies on the straight line that connects points B' and C'. The molar Gibbs energy that corresponds to this point (read off the vertical axis at point A') is the overall molar Gibbs energy of the two-phase system.[1] Comparing the two states, state A' has lower Gibbs energy. According to the Gibbs inequality, state A' is more stable than A. Liquid states B' and C' were selected arbitrarily in this trial. Many other

---

1. To convince yourself of this, look at eq. (13.4) and notice that G^{I+II} moves on a straight line between G^I and G^{II} as n^I is varied between 1 and 0.

550   Chapter 13   Miscibility, Solubility, and Other Phase Equilibria

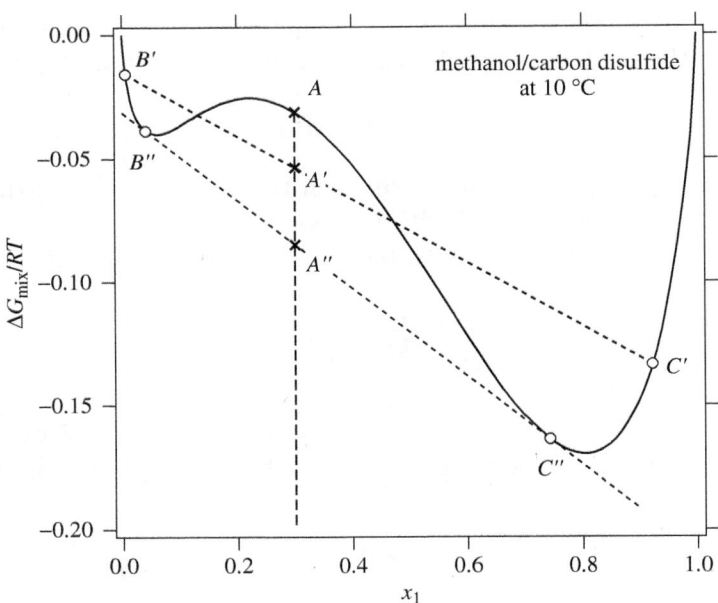

Figure 13-2: Graphical determination of phase splitting. Of all states along the dashed vertical line, A'' is the most stable one because it has the lowest possible Gibbs free energy. This state corresponds to a mixture of two liquid phases, B'' and C''.

two-liquid states can be constructed in this manner in such a way that the overall state has even lower Gibbs energy than point A'. This can be done systematically by moving point B' and C' lower. This procedure converges to points B'' and C'' whose distinguishing feature is that they form a straight line that is tangent to the Gibbs curve at both contact points. All other lines that connect any two points to the left and to the right of point A lie above line $B''C''$ and thus correspond to *higher* molar Gibbs energy. We conclude then that state A'', consisting of a mixture of liquids B'' and C'', is the *most stable* among all single or two-phase systems that can be constructed. Accordingly, state C'' is identified as liquid phase II (methanol saturated in carbon disulfide) and state B'' with liquid phase I (carbon disulfide saturated in methanol).

The procedure described here represents the graphical solution of eq. (13.1) (the compositions of points B'' and C'' are the same as those obtained by solving eq. [13.1]). Its main advantage is that the graphical procedure allows us to determine whether phase splitting takes place by simply inspecting the *shape* of the Gibbs free energy: phase splitting occurs only when the Gibbs free energy contains a *concave* segment because it is then possible to connect two points of the curve with a straight line that lies *below* the curve. By contrast, in a convex curve, any straight line between two points lies above the curve, producing a two-phase system with higher Gibbs energy (and thus less favorable) than the single-phase liquid. Therefore, the presence of a concave portion is necessary and sufficient condition for phase splitting. The compositions of the coexisting liquids are then identified by drawing a double-tangent line, that is, one that is tangent to the Gibbs free energy at both contact points.

Equilibrium and Stability Having identified points B'' and C'' as the miscibility limits, the Gibbs graph must be corrected by removing the portion of the curve that represents nonequilibrium states. To do that, we erase the portion of the curve between points B'' and C'' and connect the two points with a straight tie line. The resulting Gibbs free energy is shown in Figure 13-3 by the solid line with points B'' and C'' now relabeled as I and II, respectively. The system consists of a single liquid in the regions $0 \leq x_1 \leq x_1^I$ and $x_1^{II} \leq x_1 \leq 1$; between x_1^I and x_1^{II} the system forms two liquid phases and the Gibbs free energy of this two-phase mixture lies on the straight tie-line that connects points I and II. The portion of the Gibbs energy between points I and II contains one concave region near the center (shown by large dashes in Figure 13-3) with two convex portions to each side. The convex parts represent metastable states and may be observed under special

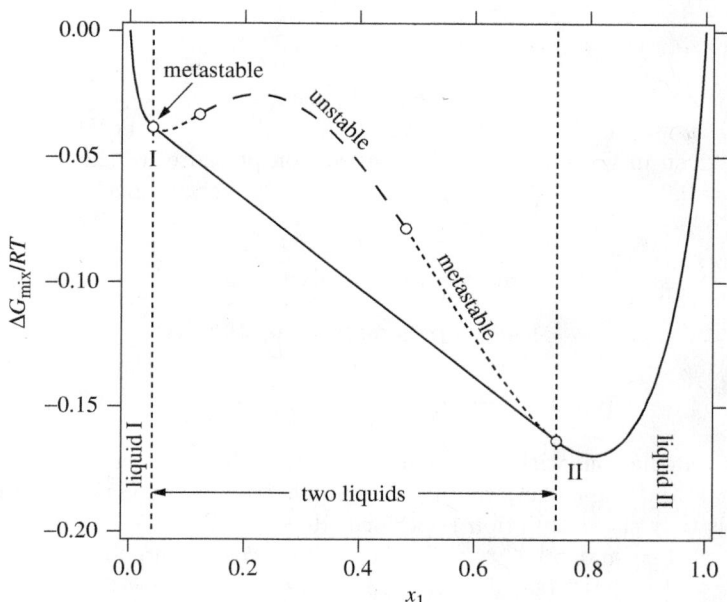

Figure 13-3: Stable (solid lines), metastable (dotted line), and unstable (large dashes) portions of the Gibbs free energy of solution. Concave parts on the Gibbs curve are unstable; convex parts are stable or metastable.

experimental conditions.[2] States on the concave portion are *unstable* and cannot exist as a single liquid.

NOTE

Mathematical Conditions for Stability

The stability criteria can be put in mathematical form. Convexity is indicated by positive values of the second derivative. Therefore, stable and metastable states must satisfy the condition

$$\left(\frac{\partial G^2}{\partial x_1}\right)_{P,T} > 0, \quad \text{stability.} \tag{13.5}$$

When the second derivative is zero, we reach the limit of stability:

$$\left(\frac{\partial G^2}{\partial x_1}\right)_{P,T} = 0, \quad \text{limit stability.} \tag{13.6}$$

Beyond this point the second derivative is negative, indicating that the Gibbs energy is a concave function of x_1 and thus *unstable*.

A note on the Wilson equation: This equation results in a convex graph for the Gibbs free energy of mixing for all values of the parameters Λ_{12} and Λ_{21}; therefore, it is not capable of predicting phase splitting. All other models discussed in Chapter 12 can produce concave shapes and thus may be used to describe partially miscible systems.

Example 13.2: *Pxy* Graph of Partially Miscible Liquids

Construct the *Pxy* graph of methanol/disulfide at 10 °C using the NRTL equation with the parameters given in Example 13.1. The saturation pressures of the two components at 10 °C are, $P_1^{\text{sat}} = 0.0741$ bar, $P_2^{\text{sat}} = 0.264$ bar, where 1 refers to methanol and 2 to carbon disulfide.

Solution The *Pxy* is calculated most easily by solving the bubble-*P* problem:

$$P = x_1 \gamma_1 P_1^{\text{sat}} + (1 - x_1) \gamma_2 P_2^{\text{sat}},$$

$$y_1 = \frac{x_1 \gamma_1 P_1^{\text{sat}}}{P}.$$

As in usual VLE calculations with a single liquid, the graph is constructed by calculating the bubble pressure *P* and the corresponding vapor mole fraction at various values of x_1. The only difference is that the calculation is performed *only within the range of full miscibility*, namely, for $0 \leq x_1 \leq x_1^{\text{I}}$ and $x_1^{\text{II}} \leq x_1 \leq 1$. The activity coefficients are calculated using the NRTL equation with $\alpha = 0.2$ (see Example 13.1):

---

2. Metastable states have a strong tendency to revert to the more stable states. To observe them experimentally one must avoid impurities and surface imperfections of the containing vessels, which tend to act as nucleation sites of the more stable phases.

13.2 Gibbs Free Energy and Phase Splitting

$$\ln \gamma_1 = \frac{0.870055(1-x_1)^2}{(0.503811(1-x_1)+x_1)^2} + \frac{0.222994(1-x_1)^2}{(1-0.0456573x_1)^2},$$

$$\ln \gamma_2 = \frac{1.72695 x_1^2}{(0.503811(1-x_1)+x_1)^2} + \frac{0.212813 x_1^2}{(1-0.0456573x_1)^2}.$$

The calculations are summarized in the table shown below:

x_1	γ_1	γ_2	y_1	P(bar)	x_1	γ_1	γ_2	y_1	P(bar)
0.000	32.838	1.000	0.000	0.264	0.743[‡]	1.092	3.768	0.190	0.316
0.005	30.717	1.000	0.042	0.274	0.775	1.067	4.045	0.203	0.301
0.010	28.770	1.001	0.077	0.283	0.807	1.048	4.341	0.221	0.283
0.015	26.979	1.002	0.105	0.290	0.839	1.032	4.658	0.245	0.262
0.020	25.330	1.003	0.129	0.297	0.871	1.020	4.996	0.280	0.235
0.025	23.810	1.004	0.148	0.303	0.904	1.011	5.356	0.332	0.204
0.031	22.408	1.006	0.165	0.308	0.936	1.005	5.740	0.417	0.167
0.036	21.112	1.008	0.179	0.312	0.968	1.001	6.149	0.579	0.124
0.041[†]	19.913	1.010	0.190	0.316	1.000	1.000	6.585	1.000	0.074

[†]saturated carbon disulfide phase (x_1^I); [‡]saturated methanol phase (x_1^{II});

The Pxy graph is shown in Figure 13-4.

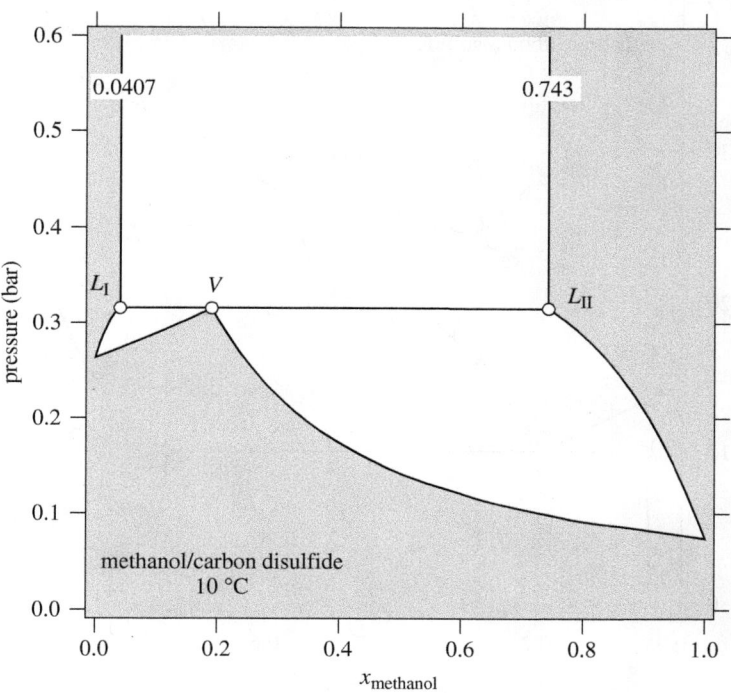

Figure 13-4: Pxy graph of methanol/carbon disulfide at 10 °C (see Example 13.2).

Comments The above calculation produces the vapor-liquid boundary of the system. To obtain the liquid-liquid boundary, we assume that the mutual solubility of the components is independent of pressure. This is a very good assumption because pressure has negligible effect on the solubility of liquids. On the Pxy graph, this is indicated by drawing the liquid-liquid boundaries as vertical lines starting at points L_I and L_{II}, and moving upwards.

Example 13.3: *Txy* Graph of Partially Miscible Liquids
Calculate the Txy graph of methanol(1)/carbon disulfide(2) at 0.3155 bar. The activity coefficients are given by the NRTL equation with $\alpha = 0.2$ and with τ_{ij} given by the following expressions as a function of temperature:

$$\frac{\tau_{12}}{R} = 0.009134 \ T^2 - 6.502 \ T + 1158.35,$$

$$\frac{\tau_{21}}{R} = -0.16377 \ T^2 + 91.5487 \ T - 11851.2,$$

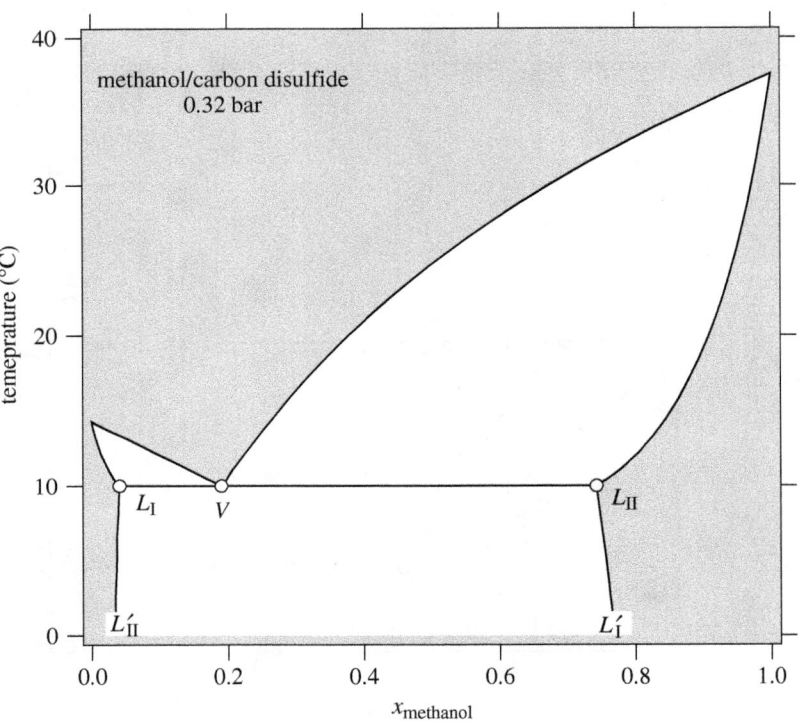

Figure 13-5: *Txy* graph of methanol/carbon disulfide at 0.32 bar.

13.2 Gibbs Free Energy and Phase Splitting

with T in Kelvin. The saturation pressures of the pure components are given by the Antoine equation with the coefficients given below:

$$\log_{10} \frac{P_1^{\text{sat}}}{\text{bar}} = 5.20409 - \frac{1581.34}{T - 33.5},$$

$$\log_{10} \frac{P_2^{\text{sat}}}{\text{bar}} = 4.06683 - \frac{1168.62}{T - 31.616},$$

with T in Kelvin.

Solution The solution will be easier to follow by first examining the Txy graph in Figure 13-5. Temperature affects mutual solubility and the shape of the liquid-liquid phase boundary. Since we have temperature-dependent activity coefficients, it possible compute the liquid-liquid boundary as a function of temperature. This calculation will be performed up to the bubble temperature. Above the bubble temperature the calculation will be done by solving the bubble T problem.

Txy below the bubble temperature: In this part of the calculation we establish the phase boundary between the two liquids provided that the system contains no vapor, i.e., it is below the bubble point. This refers to lines $L_I - L_I'$ and $L_{II} - L_{II}'$ in Figure 13-5. The calculation is performed as follows: we pick a temperature and we solve eq. (13.1) for x_1^I and x_1^{II}. Once the compositions of the two liquids are known, we compute the bubble pressure of the two-phase system. If it is below the system pressure of 0.3155 bar, the system is below the bubble point. We then increment the temperature and repeat the calculation, until the bubble point is reached. This calculation produces the liquid-liquid boundary of the phase diagram.

As a demonstration, we show the calculation at $T = 5\,°\text{C} = 278.15$ K. Following the same procedure as in Example 13.1 we find:

$$T = 278.15 \text{ K}, \quad x_1^I = 0.0361579, \quad x_1^{II} = 0.757001.$$

The bubble pressure of the two-liquid system is calculated as

$$P = x_1^I \gamma_1^I P_1^{\text{sat}} + (1 - x_1^I) \gamma_2^I P_2^{\text{sat}} = x_1^{II} \gamma_1^{II} P_1^{\text{sat}} + (1 - x_1^{II}) \gamma_2^{II} P_2^{\text{sat}}.$$

Either phase I or II may be used in this calculation. Using phase I and $P_1^{\text{sat}} = 0.0550051$ bar, $P_2^{\text{sat}} = 0.212145$ bar, we find

$$P = (0.0361579)(22.6526)(0.0550051) + (0.963842)(1.00843)(0.212145) = 0.25125 \text{ bar}.$$

This pressure is below the pressure of the system; therefore, the system exists as two liquids. We repeat this calculations at various temperatures and summarize the results below:

T (°C)	x_1^I	x_1^{II}	γ_1^I	γ_2^I	P_1^{sat} (bar)	P_2^{sat} (bar)	P_{bubble} (bar)
5	0.0362	0.7570	22.65	1.0084	0.0550	0.2121	0.2513
6	0.0369	0.7544	22.14	1.0087	0.0584	0.2217	0.2632
7	0.0378	0.7517	21.61	1.0091	0.0621	0.2316	0.2755
8	0.0387	0.7488	21.06	1.0094	0.0659	0.2419	0.2884
9	0.0397	0.7459	20.49	1.0098	0.0699	0.2525	0.3017
10[†]	0.0407	0.7428	19.91	1.0103	0.0741	0.2635	0.3155

[†] bubble temperature

We find from this table that the bubble temperature of the two-phase system is 10 °C since at this temperature the bubble pressure is equal to the system pressure.

Txy above the bubble temperature. To calculate the Txy graph above the bubble temperature, we solve the bubble T problem:

$$P = x_1\gamma_1(T)P_1^{\text{sat}}(T) + x_2\gamma_2(T)P_2^{\text{sat}}(T),$$

$$y_1 = \frac{x_1\gamma_1(T)P_1^{\text{sat}}(T)}{P},$$

where $P = 0.3155$ bar. We fix x_1, solve the first equation for T, and use the second equation to calculate y_1. As with the Pxy graph, the calculation is done only within the region of full miscibility at the conditions of the bubble point ($T_{\text{bubble}} = 10$ °C), namely,

$$0 \leq x_1 \leq 0.0407, \quad \text{and} \quad 0.7428 \leq x_1 \leq 1.$$

The calculations are summarized in the table below.

T (°C)	x_1	y_1	γ_1	γ_2	T (°C)	x_1	y_1	γ_1	γ_2
14.3	0.0000	0.0000	31.1681	1.0000	10.0	0.7428‡	0.1905	1.0917	3.7677
13.4	0.0051	0.0428	29.5237	1.0002	11.1	0.7749	0.2067	1.0674	4.0334
12.6	0.0102	0.0776	27.9308	1.0007	12.5	0.8071	0.2290	1.0476	4.3126
11.9	0.0153	0.1060	26.4037	1.0015	14.3	0.8392	0.2602	1.0319	4.6043
11.4	0.0204	0.1294	24.9503	1.0026	16.7	0.8714	0.3051	1.0197	4.9068
11.0	0.0255	0.1490	23.5744	1.0041	19.8	0.9035	0.3722	1.0108	5.2170
10.6	0.0305	0.1653	22.2771	1.0058	24.0	0.9357	0.4784	1.0046	5.5298
10.3	0.0356	0.1790	21.0574	1.0079	29.6	0.9678	0.6597	1.0011	5.8367
10.0	0.0407†	0.1905	19.9129	1.0103	37.5	1.0000	1.0000	1.0000	6.1240

†saturated carbon disulfide phase (x_1^{I}); ‡saturated methanol phase (x_1^{II});

This table corresponds to the part of the Txy graph from the tie-line of the triple point and above (see Figure 13-5).

Comments The calculation is more involved than that of the Pxy graph because temperature affects the liquid-liquid boundary while pressure does not. To establish the liquid-liquid boundary, it is important to have an activity coefficient model with temperature dependent coefficients. The coefficients used in this example were taken from the DECHEMA series (Sørensen and Artl, *Liquid-Liquid Equilibrium Data Collection*, Chemistry Data Series, vol. V, Part 1).

13.3 Liquid Miscibility and Temperature

At fixed pressure the liquid-liquid boundary ceases to exist above the bubble point (in the case of Figure 13-5, 10 °C). If the system pressure is increased, the bubble

13.3 Liquid Miscibility and Temperature

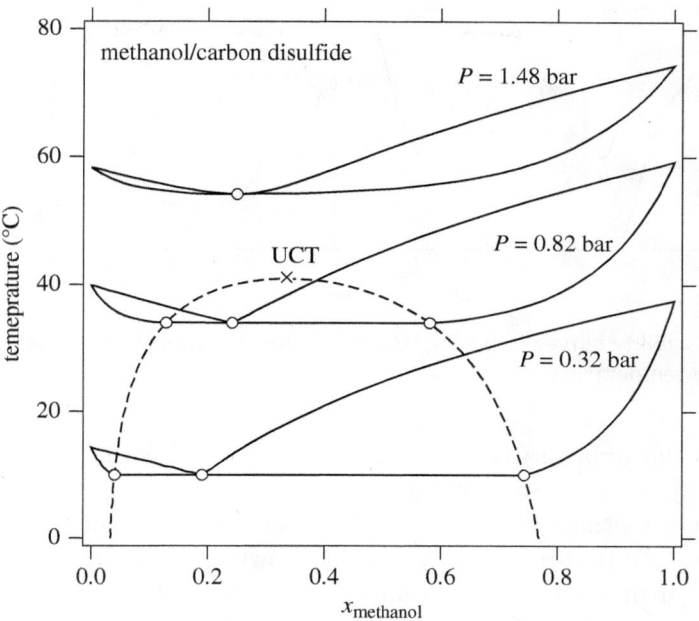

Figure 13-6: *Txy* graphs of methanol/carbon disulfide at various pressures. Above the upper consolute temperature (UCT) the system exhibits full miscibility (see Example 13.3).

temperature increases as well, and a similar calculation would reveal that the liquid-liquid boundary extends to even higher temperatures, until the new bubble point is reached. This behavior is seen in Figure 13-6, which shows the *Txy* graph of methanol/carbon disulfide at three different pressures. For this system, increasing temperature results in increased mutual solubility, as indicated by the convergence of the two liquid branches (dashed line). At the point where the two meet, full miscibility is restored: above the temperature of this point the two liquids are miscible at all compositions. This temperature is called *upper consolute temperature* (UCT) and is very much analogous to the critical point of a vapor/liquid mixture in the sense that the two phases become indistinguishable at this point. Above the UCT the two components form a homogeneous minimum boiling azeotrope. The relationship between partial miscibility and minimum-temperature azeotropy is not coincidental. Nonideal systems that exhibit positive deviations from ideality indicate unfavorable cross-interaction between components. When mixed they form solutions that have a higher tendency to erase the interaction by forming a vapor at lower temperature than the boiling point of the pure components. If the self-interaction is sufficiently strong, then the system exhibits phase separation,

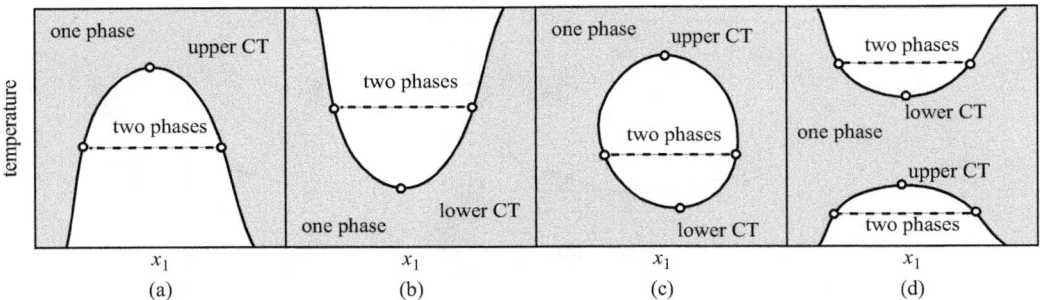

Figure 13-7: Miscibility curves in binary liquids. Dashed lines are tie lines between two liquids. (CT = consolute temperature).

which disrupts the unfavorable interaction. Figure 13-6 demonstrates this gradual transition as a function of temperature.

In the system methanol/carbon disulfide, the lower part of the solubility curve is terminated at the freezing boundary (not shown). Many systems exhibit similar behavior with an upper consolute temperature. Other behaviors are encountered as well, as shown in Figure 13-7. Some systems exhibit a lower consolute temperature, with full miscibility at low temperatures and partial miscibility above. It is also possible, though less common, to observe an upper and a lower consolute temperature (case c in Figure 13-7). In some cases, there are two disjoint two-phase regions. Such systems exhibit partial miscibility at low temperature, full miscibility at intermediate temperatures, and partial miscibility again at higher temperatures.

13.4 Completely Immiscible Liquids

When the mutual solubility of the liquids is so low that they may be regarded as completely immiscible, certain simplifications are possible so that we may construct the phase diagrams without the need for activity coefficients. These simplifications arise from the fact that the two-liquid phases each practically consist of the pure components.

Before we show how to calculate these phase diagrams we will examine their features first. Figure 13-8 shows the Txy graph of water(1) / toluene(2) at 1 bar. This system phase-separates and because the mutual solubility is very low, the two phases can be considered pure. In the water-rich phase we have $x_w^\alpha = 1$ and in the toluene-rich phase $x_w^\beta = 0$. As a result, the pure liquid phases have been reduced to a vertical line, one at $x_w = 0$ and one at $x_w = 1$. The region marked $W + T$ indicates the two-liquid region. A point in this region splits into two-liquid phases that are essentially pure water and toluene, respectively. The region marked

13.4 Completely Immiscible Liquids

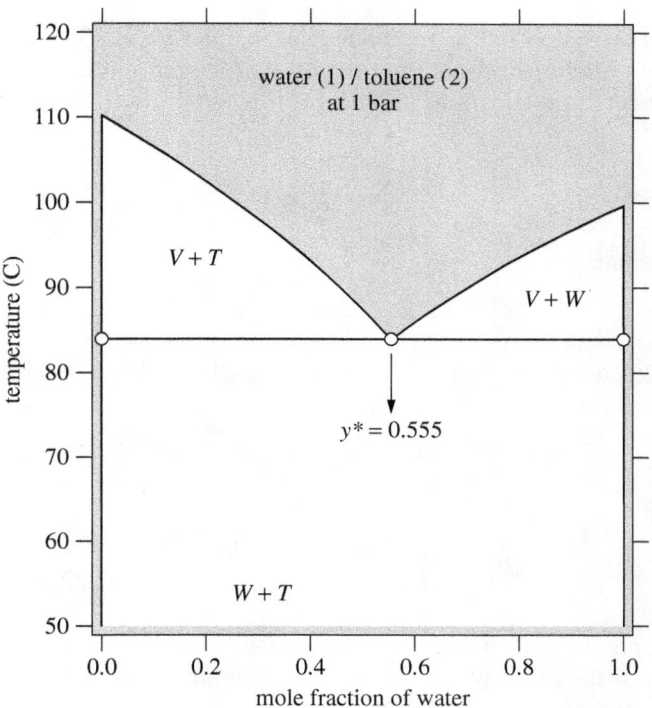

Figure 13-8: *Txy* graph of water-toluene assuming complete immiscibility.

$V + W$ represents equilibrium between water-rich liquid (which we have assumed to be pure water) and vapor that contains both components. The region marked $V + T$ represents equilibrium between toluene-rich liquid (which we have assumed to be pure toluene) and vapor containing both components. We will demonstrate the calculations of the phase diagram with an example.

Example 13.4: *Txy* Graph of Immiscible Liquids
Calculate the Txy of the system water(1) / toluene(2) at 1 bar assuming complete immiscibility. The Antoine constants are given below:

$$\ln \frac{P_w^{sat}}{\text{Torr}} = 18.3036 - \frac{3816.44}{T(K) - 46.13},$$

$$\ln \frac{P_t^{sat}}{\text{Torr}} = 16.0137 - \frac{3096.52}{T(K) - 53.67},$$

where subscript w refers to water and subscript t to toluene.

Solution To calculate the Txy graph we first determine y^*, the composition of the vapor at the triple point. All three phases are present at this point. With the simplification that each liquid phase contains only one component, the equilibrium conditions are

$$y_w^* P = P_w^{\text{sat}}, \tag{13.7}$$

$$y_t^* P = P_t^{\text{sat}}. \tag{13.8}$$

Adding the two equations we obtain

$$P = P_w^{\text{sat}} + P_t^{\text{sat}}.$$

Here, the only unknown is temperature, which appears implicitly in the saturation pressures. Solving by trial and error we find

$$T^* = 84\ °C, \quad P_w^{\text{sat}} = 0.555\ \text{bar}, \quad P_t^{\text{sat}} = 0.445\ \text{bar}.$$

The mole fraction y_w^* is

$$y_w^* = \frac{P_w^{\text{sat}}(T^*)}{P} = \frac{0.555}{1} = \quad \Rightarrow \quad y_w^* = 0.555.$$

For the construction of the Txy graph we must consider the ranges $y_w = 0$ to y_w^*, and y_w^* to 1 separately. This is dictated by the fact that a different liquid is present in each of these two ranges.

Region $y_w \leq y_w^*$. In this region the liquid is pure toluene. No water is contained in the liquid phase. Therefore, the equilibrium condition is written for toluene only:

$$y_t P = P_t^{\text{sat}}.$$

This equation defines the dew line. The simplest way to calculate the dew line is to set the temperature and solve for y_t. Notice that the temperature in this range of the graph must be between T^* and T_t^{sat}. The mole fraction of water is then obtained as $y_w = 1 - y_t$. For example, at 90 °C we find

$$P_t^{\text{sat}} = 0.5421\ \text{bar},$$

$$y_t = \frac{P_t^{\text{sat}}}{P} = 0.5421,$$

$$y_w = 1 - 0.5421 = 0.4579.$$

The procedure is continued until the entire range $0 \leq y_w \leq y_w^*$ is covered.

Region $y_w \geq y_w^*$. In this region the liquid is pure water and the equilibrium condition is written for water only:

$$y_w P = P_w^{\text{sat}}. \tag{13.9}$$

13.4 Completely Immiscible Liquids

This equation describes the dew line to the right of point y_w^*. The procedure is the same as above. We fix the value of the temperature to a value between T^* and T_w^{sat} and solve for y_w. For example, at 90 °C we find

$$P_w^{sat} = 0.7010,$$

$$y_w = \frac{P_w^{sat}}{P} = 0.701,$$

$$y_t = 1 - 0.701 = 0.299.$$

The complete Txy graph is obtained by repeating the calculation for various temperatures in the range T^* to T_w^{sat}.

Example 13.5: *Pxy* Graph of Immiscible Liquids
Calculate the *Pxy* graph of the system water(1)/toluene(2) at 84 °C.

Solution Again the first step is to calculate the conditions at the triple point. The two components are distributed among three phases. The equilibrium conditions are given by eqs. (13.7) and (13.8):

$$y_w^* P = P_w^{sat}$$
$$y_t^* P = P_t^{sat}$$

and by addition,

$$P = P_w^{sat} + P_t^{sat},$$

At 84 °C, we find

$$P_w^{sat} = 0.5551, \quad P_t^{sat} = 0.4444.$$

Therefore,

$$P^* = 0.5551 + 0.4444 = 1 \text{ bar},$$

$$y_w^* = \frac{P_w^{sat}}{P} = 0.5551.$$

(These values are the same as in the previous example because the temperature given here happens to be the same as the temperature at the triple point at 1 bar.) Having determined the conditions at the triple point we calculate the graph by considering the regions below and above y_w^* separately.

In the region below y_w^* the liquid is pure toluene. No water is contained in the liquid phase. Therefore, the equilibrium condition should be written for toluene only:

$$y_t P = P_t^{sat}.$$

The simplest way to calculate this dew line is to fix the mole fraction and calculate the dew pressure from the above equation. For example, with $y_w = 0.1$ we find

$$y_t = 1 - 0.1 = 0.9,$$

$$P = \frac{P_t^{\text{sat}}}{y_t} = \frac{0.4444}{0.9} = 0.494 \text{ bar}.$$

In the region above y_w^* the equilibrium condition is written for water only:

$$y_w P = P_w^{\text{sat}}.$$

The dew pressure is calculated by fixing the value of y_w and solving for P. For example, at $y_w = 0.6$ we have

$$P = \frac{P_w^{\text{sat}}}{y_w} = \frac{0.5551}{0.6} = 0.925 \text{ bar}.$$

The complete Pxy graph constructed in this manner is shown in Figure 13-9.

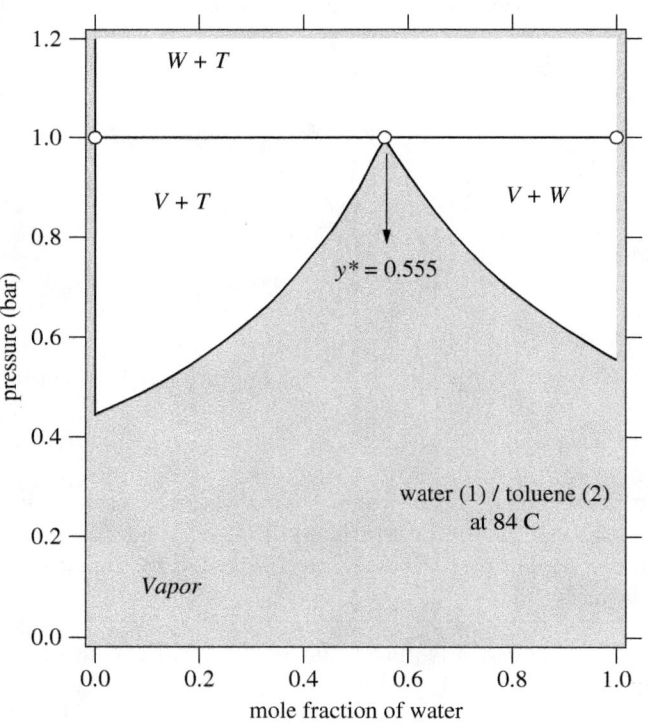

Figure 13-9: Pxy graph of immiscible liquids.

NOTE

On Complete Immiscibility

It is important to understand the limitations of these simplified calculations. While we have assumed complete immiscibility, in reality there is *some* amount, however small, of toluene dissolved in water, and of water in toluene. The assumption of complete immiscibility allows us to calculate the bubble and dew points quite accurately. This simplification, however, may not be appropriate for other calculations. For example, the presence of toluene in drinking water is regulated by EPA to a maximum of 1 mg/L, corresponding to a mol fraction of approximately 2×10^{-7}. If the purpose of the calculation is to determine compliance with regulations, the assumption of complete immiscibility would be inappropriate. An accurate activity coefficient model is needed for this calculation.

13.5 Solubility of Gases in Liquids

In the examples considered so far, both components are liquids at near ambient conditions. This allows us to calculate the fugacity in solution using the generalized form of the Lewis-Randall equation, eq. (12.16), which requires the fugacity of the pure *liquid*. Many systems of industrial and biological significance involve a component that at ambient conditions is a gas above its critical temperature. The system water/oxygen is one such example. The solubility of oxygen, an important process for sustaining underwater life, is a case of vapor-liquid equilibrium in which one component (oxygen, $T_c = 154.58$ K) is a gas above its critical temperature.

The solubility of a gas solute in a solvent is a problem of phase equilibrium. Suppose that we contact oxygen with water at 25 °C. Some oxygen dissolves in water, some water evaporates into the gas phase, and an equilibrium state is established between the two phases. This state is governed by the equality of fugacities: the fugacity of dissolved oxygen is equal to the fugacity of oxygen in the gas, and the fugacity of water in the liquid is equal to that in the vapor. The fugacity of water can be calculated by the general form of the Lewis-Randall equation. For the oxygen, however, we need a different approach, since the component is above its critical temperature. This is done using Henry's law, a relationship that gives the fugacity of a solute in the limit of infinite dilution:

$$\boxed{f_i = k_i^H x_i, \quad \text{when } x_i \to 0.} \tag{13.10}$$

Here, f_i is the fugacity of the solute, x_i is the mol fraction of the solute, and k_i^H is Henry's law constant. Henry's law represents the limiting behavior of fugacity as the concentration of solute approaches zero. This situation is demonstrated in Figure 13-10. As the mol fraction of solute is reduced to zero, fugacity also

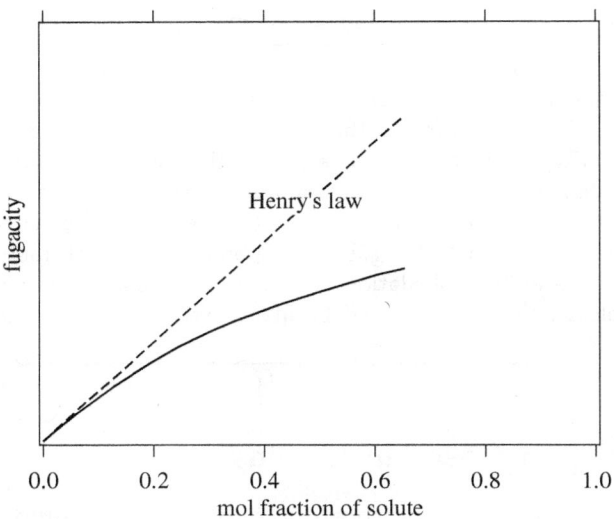

Figure 13-10: Fugacity of solute in the limit of infinite dilution.

approaches zero. Henry's law constant is the slope of the fugacity at $x_i = 0$. Henry's law in eq. (13.10) represents a linear extrapolation of fugacity from point $x_i = 0$ into the region $x_i > 0$. Within a small range of compositions, the linear extrapolation is essentially indistinguishable from the actual fugacity. In this region, Henry's law provides an accurate estimate of the fugacity of solute. At higher concentrations the linear extrapolation and the actual fugacity diverge and Henry's law breaks down. For gases with very low solubility (for example, most gases in water at near room temperature), eq. (13.10) provides an acceptable approximation.

VLE Using Henry's Law

Calculations of vapor-liquid equilibrium involving a gas dissolved in a liquid are performed using the Lewis-Randall rule for the fugacity of the liquid component and Henry's law for the vapor component. Using the subscript s for the solvent, and i for the gas, the equilibrium criterion for the two components is,

$$y_s P = \gamma_s x_s P_s^{\text{sat}}, \tag{13.11}$$

$$y_i P = x_i k_i^H, \tag{13.12}$$

where P is the total pressure, and $y_{s,i}$, $x_{s,i}$ are the mole fractions in each phase ($y_s + y_i = x_s + x_i = 1$). The first equation expresses the equality of fugacities of the solvent in the gas and in the liquid, and the second equation expresses the equality of

13.5 Solubility of Gases in Liquids

fugacities of the gas species in the gas phase and in the solution. From eq. (13.12), the solubility of the gas in the liquid is

$$x_i = \frac{y_i P}{k_i^H}. \tag{13.13}$$

This result states that the solubility (mol fraction of gas in the liquid) is proportional to the partial pressure of the gas above the liquid and inversely proportional to the value of k_i^H. That is, a high value of Henry's law constant indicates low solubility, and vice versa.

If Henry's law constant is known, eqs. (13.11) and (13.12) may be used to calculate the composition of phases at equilibrium. In many cases of practical interest the solubility of the gas in the liquid is so low that the liquid phase is nearly the pure solvent. This allows us to write $\gamma_s \approx 1$, which eliminates the need for an activity coefficient model and simplifies the VLE equations to

$$y_s P = x_s P_s^{\text{sat}}, \tag{13.14}$$

$$y_i P = x_i k_i^H. \tag{13.12}$$

Any of the standard VLE calculations can be performed but the simplest one is the bubble-P calculation. First, we add the above equations to eliminate the gas-phase compositions,

$$P = x_s P_s^{\text{sat}} + x_i k_i^H.$$

Next we set $x_s = 1 - x_i$ and solve for the concentration of the solute in the liquid:

$$x_i = \frac{P - P_s^{\text{sat}}}{k_i^H - P_s^{\text{sat}}}. \tag{13.15}$$

This gives the solubility of the gas at total pressure P. The remaining compositions are calculated easily by back substitution.

Other Units for Henry's Law Constant

Solving eq. (13.12) for k_i^H we have

$$k_i^H = \frac{y_i P}{x_i} = \frac{P_i}{x_i}, \tag{13.16}$$

where P_i is the partial pressure of the gas above the saturated liquid and x_i is the mol fraction of the gas in the liquid. This equation expresses Henry's law constant as the ratio of the partial pressure of the gas to the solubility of the gas in the liquid. Henry's law constant can be measured experimentally by application of this equation: the solvent of interest is brought into contact with the gas of interest and the value of k_i^H is obtained as the ratio of the partial pressure of the gas over the solubility. Because solubility can be expressed in different units, Henry's law constants are often reported in various forms. Common in the chemistry literature is the use of *molality* for concentration. Molality is the amount of solute in mol divided by the mass of the solvent, in kg. Molality can be converted into mol fraction by simple mass balance equations. The relationship between the two is

$$c_i = \frac{x_i}{(1-x_i)M_s}, \quad x_i = \frac{c_i}{c_i + 1/M_s}, \quad (13.17)$$

where c_i is the concentration in units of molality (mol/kg) and M_s is the molar mass of the solvent (kg/mol). Since Henry's law is usually applied in the dilute limit ($x_i \ll 1$), the relationship between molality and mol fraction can be approximated by a simple proportionality,

$$c_i \approx \frac{x_i}{M_s}. \quad (13.18)$$

In molality units, Henry's law is written as

$$f_i = c_i k_i^{H'}, \quad (13.19)$$

where $k_i^{H'}$ is Henry's law constant when liquid concentrations are expressed in molality and has units of bar kg/mol. Comparing with eq. (13.10), the relationship between k_i^H and $k_i^{H'}$ is

$$k_i^{H'} = k_i^H \frac{x_i}{c_i} = k_i^H M_s, \quad (13.20)$$

in which we used $x_i/c_i = M_s$, from eq. (13.18).

NOTE

Alternative (and Potentially Confusing) Forms of Henry's Law Constant

Henry's law constant as given in eq. (13.10) is the standard definition in the chemical engineering literature. In other branches of science various alternative definitions are encountered. Such definitions can be reconciled with the definition given here but, in what is a potential source of confusion (and of serious error in the calculation), these other forms are also commonly referred to as "Henry's law constants." In the environmental chemistry literature, the definition that seems to be in common use is the *reciprocal* of the definition given here. Using the symbol k_H, this constant is defined as

$$k_{H,i} = \frac{c_i}{P_i},$$

where c_i is the molality of the dissolved gas and P_i is its partial pressure. This is the form of Henry's law constant that is reported in the NIST WebBook. Another expression is in the form of the ratio

$$k'_{H,i} = \frac{c_i^L}{c_i^G},$$

13.5 Solubility of Gases in Liquids

where c_i^L and c_i^G is the concentration of species i in the liquid and in the gas phase, respectively. This has the form of a partition (or distribution) coefficient and is dimensionless. When obtaining values of Henry's law constant from the literature it is important to be clear about the definition associated with the reported values.

Example 13.6: Henry's Law
Calculate the amount (mol fraction and molality) of oxygen dissolved in water equilibrated in pure oxygen at 1.013 bar, 25 °C. Henry's law constant at 25 °C is 42735 bar.

Solution We will calculate the amount of oxygen by neglecting the dissolution of nitrogen or any other gases present in air. At 25 °C the saturation pressure of water is 0.03175 bar. Applying eq. (13.15) we have

$$x_i = \frac{(1.013 - 0.03175) \text{ bar}}{(42735 - 0.03175) \text{ bar}} = 2.296 \times 10^{-5}.$$

Comments The mol fraction of dissolved oxygen is very low. This justifies the two major assumptions of the calculation, namely, that Henry's law is appropriate for this system, and that the activity coefficient of water can be neglected.

A further observation has to do with the relative magnitudes of the various terms in eq. (13.15). For a solvent of low volatility, that is, at low temperature compared to its boiling point, P_s^{sat} is small, both compared to the total pressure P and Henry's law constant, k_i^H. Dropping the saturation pressure from the numerator and denominator of eq. (13.15) we obtain the simplified expression,

$$x_i = \frac{P}{k_i^H}.$$

Notice that the total pressure P is also equal to the partial pressure P_i of the gas (in our approximation the gas phase contains only component i, since the vapor pressure of the solvent has been neglected).

Example 13.7: Simultaneous Dissolution of Two Gases
Determine the amount of oxygen and nitrogen dissolved in water in equilibrium with air at 25 °C, 1.013 bar. Henry's law constants for oxygen and nitrogen at 25 °C are 42735 bar and 92593 bar, respectively.

Solution We take air to be a mixture of oxygen (21% by mole) and nitrogen (79%). We further assume that the solubility of each gas is not affected by the presence of the other. This is a three-component system that contains oxygen, nitrogen, and water. Using subscripts $_1$ and $_2$ for oxygen and nitrogen, respectively, and w for water, the equilibrium conditions are

$$y_1 P = x_1 k_1^H,$$
$$y_2 P = x_2 k_2^H,$$
$$y_w P = x_w P_w^{\text{sat}},$$

along with the mol fraction normalizations,

$$y_1 + y_2 + y_w = 1, \quad x_1 + x_2 + x_w = 1.$$

The unknowns are the six mol fractions. One additional equation is needed before the problem can be fully specified. Although the mol fractions of oxygen and nitrogen in dry air are known, their equilibrium values will be different as different amounts of each gas will dissolve in the liquid. Nonetheless, the actual amounts dissolved are very small, and if the total amount of air is much larger than the amounts of oxygen and nitrogen dissolved, dissolution will not affect the gas phase mol fractions significantly. Thus, a reasonable assumption is that the ratio of oxygen to nitrogen in air at equilibrium will be the same as in dry air, namely,

$$\frac{y_1}{y_2} = \frac{y_1^*}{y_2^*} = \frac{0.21}{0.79}.$$

Here we set the ratio of the mol fractions, rather than the mol fractions themselves, equal to the dry value to allow for the presence of water vapor in the gas. With this additional equation we have six equations, which we now solve for the six mol fractions. Solving for the liquid mole fractions we find

$$x_1 = \frac{y_1^* (P - P_w^{\text{sat}})}{k_1^H - P_w^{\text{sat}} \left(y_1^* - y_2^* \frac{k_1^H}{k_2^H}\right)},$$

$$x_2 = \frac{y_2^* (P - P_w^{\text{sat}})}{k_2^H - P_w^{\text{sat}} \left(y_2^* - y_1^* \frac{k_2^H}{k_1^H}\right)}.$$

By numerical substitution with $P_w^{\text{sat}} = 0.03175$ bar we find

$$x_1 = 4.82 \times 10^{-6}, \quad x_2 = 8.372 \times 10^{-6}.$$

The gas-phase mol fractions are obtained by back substitution into the equilibrium equations:

$$y_1 = \frac{x_1 k_1^H}{P} = \frac{(4.82 \times 10^{-6})(42735 \text{ bar})}{1.013 \text{ bar}} = 0.203,$$

$$y_2 = \frac{x_2 k_2^H}{P} = \frac{(1.81 \times 10^{-5})(92593 \text{ bar})}{1.013 \text{ bar}} = 0.765.$$

13.5 Solubility of Gases in Liquids

> The mol fraction of water is given by the balance in each phase:
>
> $$x_w = 0.999987, \quad y_w = 0.0313421.$$
>
> **Comments** If the volatility of the solvent is neglected (this is equivalent to setting $P_w^{\text{sat}} = 0$ in the above equations), then $y_1 = y_1^*$, $y_2 = y_2^*$ and the solubility of gas component i is
>
> $$x_i = \frac{y_i P}{k_i^H}.$$
>
> In this approximation, the mol fraction of gas component i in the liquid is equal to the partial pressure of the component in the gas divided by its Henry's law constant.

Temperature and Pressure Effects on Henry's Law Constant

For a given system of solute/solvent, Henry's law constant is a function of temperature and total pressure. The effect of pressure is small and can be neglected because pressure has generally little effect on properties in the liquid phase. Temperature is a more sensitive variable and must be accounted for. Commonly, literature data on Henry's law constant are presented in the form of a temperature dependent equation of the form,

$$\ln \frac{k_i^H(T)}{k_i^H(T_0)} = A \left(\frac{1}{T} - \frac{1}{T_0} \right), \tag{13.21}$$

where $k_i^H(T_0)$ is the value of Henry's law constant at some known temperature T_0, often chosen as $T_0 = 298.15$ K. The parameter A (with units of K) is obtained through numerical fitting of data and depends on the nature of the solute and solvent. Occasionally, more complex equations are used to capture more accurately the effect of temperature.[3] For many gases at low pressures, increasing temperature increases solubility. This trend is often reversed at higher temperatures.

NOTE

Henry's Law and Formal Thermodynamics
Henry's law owes its name to William Henry, the British chemist who reported the linearity between solubility and partial pressure in the early 1800s. The empirical observation of linearity was made independently of thermodynamics and took the force of a "physical law." It is not a new physical principle, however, and the constant it introduces is fully accounted for by thermodynamics. To establish the relationship between formal quantities introduced earlier and Henry's law constant, we return to the general expression for the fugacity of a species in a mixture in eq. (10.15). Applying this equation to the infinite dilution limit, we have

$$f_i = x_i \phi_i^\infty P, \tag{13.22}$$

---

3. See, for example, Sandler [3].

where ϕ_i^∞ is the fugacity coefficient of component at the infinite dilution limit ($x_i \to 0$). Comparing eq. (13.22) with eq. (13.10), we obtain a relationship between Henry's law constant and the fugacity coefficient at infinite dilution,

$$k_i^H = \phi_i^\infty P. \tag{13.23}$$

Using $\ln \phi = \bar{G}_i^R/RT$, we further relate k_i^H to the residual partial Gibbs energy:

$$\ln \frac{k_i^H}{P} = \ln \phi_i^\infty = \left(\frac{\bar{G}_i^R}{RT}\right)^\infty. \tag{13.24}$$

Recall that the fugacity coefficient can be calculated from an equation of state, as discussed in Chapter (10). Therefore, Henry's law constant can be obtained from the equation of state as well.

The dependence of Henry's law constant on pressure is obtained from the known dependence of the residual Gibbs energy. The partial derivative with respect to temperature is

$$\left(\frac{\partial \ln(k_i^H/P)}{\partial(1/T)}\right)_P = \frac{\left(\bar{H}_i^R\right)^\infty}{R}, \tag{13.25}$$

and with respect to pressure is[4]

$$\left(\frac{\partial \ln(k_i^H/P)}{\partial P}\right)_T = \frac{\left(\bar{V}_i^R\right)^\infty}{RT}, \tag{13.26}$$

where $\left(\bar{H}_i^R\right)^\infty$ is the residual partial molar enthalpy and $\left(\bar{V}_i^R\right)^\infty$ the residual partial molar volume, both in the limit of infinite dilution. Since molar (and partial molar) volumes in the liquid phase are small, the right-hand side of eq. (13.26) is a small number and makes a contribution only if pressure varies enormously. Returning to eq. (13.25), we integrate with respect to temperature between temperatures T_0 and T. The result is

$$\ln \frac{k_i^H(T)}{k_i^H(T_0)} = \frac{\left(\bar{H}_i^R\right)^\infty}{R}\left(\frac{1}{T} - \frac{1}{T_0}\right).$$

This is the same as eq. (13.21). That is, the empirical fit in eq. (13.21) is of the form expected from the theoretical considerations under the assumption that the residual partial molar enthalpy of solute at infinite dilution is constant. In practice, the factor $\left(\bar{H}_i^R\right)^\infty/R$ is treated as an adjustable parameter and is determined from measurements of Henry's law constant at two temperatures.

Figure 13-11 shows an SRK calculation of Henry's law constant as a function of temperature and pressure (see Example 13.9). As we expect, the effect of temperature is much stronger than that of pressure.

---

[4]. For a given system of solute/solvent, Henry's law constant is a function of pressure and temperature but not of composition. This is because it refers to a very specific composition of the solution, the infinite dilution limit.

13.5 Solubility of Gases in Liquids

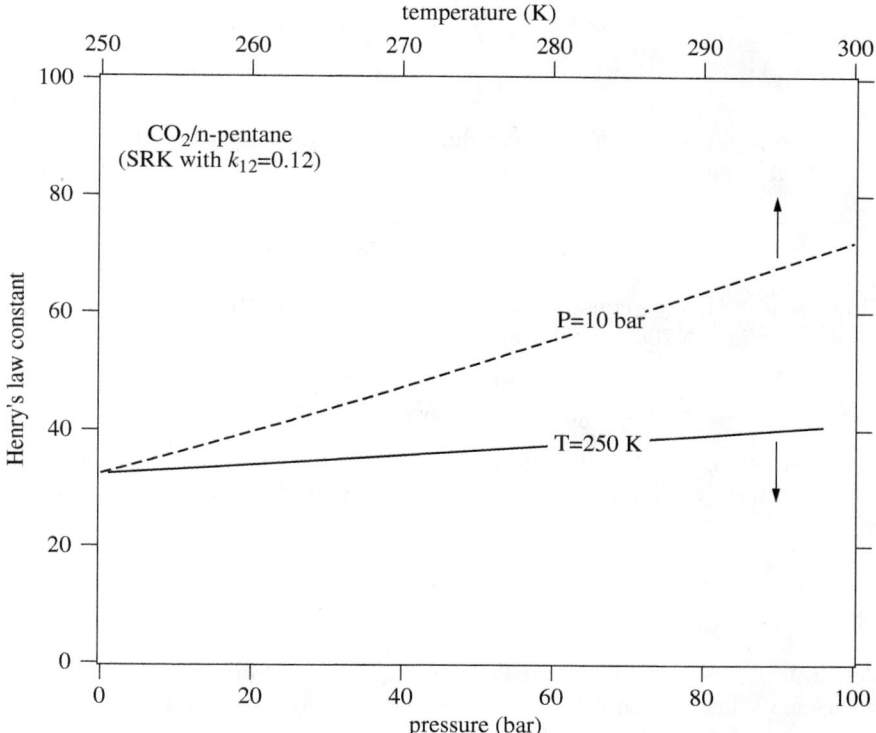

Figure 13-11: SRK calculation of Henry's law constant for carbon dioxide in n-pentane. The solid line is plotted against pressure (bottom axis) at constant temperature; the dashed line is plotted against temperature (top axis) at constant pressure.

Example 13.8: Carbonated Soda
Estimate the amount of carbonation in a can of soda at 2 °C. Henry's law constant for carbon dioxide in water is given by the equation,

$$k^H = 1585.98 \exp\left[-2400\left(\frac{1}{T} - \frac{1}{298.15}\right)\right],$$

with k_i^H in bar and T in kelvin. Make any suitable assumptions as needed.

Solution We will assume that soda is an equilibrium mixture of water and carbon dioxide and will ignore the effect of preservatives, sweeteners, and color agents on the thermodynamic properties. A good estimate for the pressure in the can is about 2 bar. The equilibrium conditions for water and carbon dioxide are:

$$y_w P = x_w P_w^{\text{sat}},$$

$$y_c P = x_c k^H,$$

where the subscript c stands for carbon dioxide. We add the two equations and solve for the mol fraction of carbon dioxide in the liquid:

$$x_c = \frac{P - P_w^{\text{sat}}}{k^H - P_w^{\text{sat}}}.$$

At $2\,°C$ the value of Henry's constant is 809.229 bar and from the steam tables the saturation pressure of water is 0.00706 bar. By numerical substitution,

$$x_c = \frac{2 - 0.00706}{809.229 - 0.00706} = 2.46 \times 10^{-3}.$$

Comments As long as the can contains some gas above the liquid, the assumption of equilibrium is valid, provided that the can has been stored at $2\,°C$ for a sufficient period of time.

Example 13.9: Henry's Law Constant From an Equation of State
Calculate Henry's law constant for carbon dioxide in n-pentane at 10 bar, 250 K using the SRK equation of state with $k_{12} = 0.12$.

Solution The fugacity coefficients are calculated from eq. (10.21) with all the necessary parameters given in Table 9-2. With $T = 250$ K, $P = 10$ bar, $x_1 = 0$, $x_2 = 1$, we calculate the following values for the various parameters of pure component and mixture:

	Mixture	CO_2	n-pentane
a	2.94591	0.43021	2.94591
b	0.000100397	0.0000297156	0.000100397
da/dT	−0.00602914	−0.00119522	−0.00602914
C'_i		5.31651	14.1173

The parameter A' and B' of the mixture are

$$A' = 0.681898, \quad B' = 0.0483024.$$

The cubic equation for the compressibility factor is

$$Z^3 - Z^2 + 0.631263 Z - 0.0329373 = 0,$$

which has one real root:

$$Z = 0.0570364.$$

13.5 Solubility of Gases in Liquids

Using the above results, the fugacity coefficient of carbon dioxide is

$$\phi_1 = 3.31944 = \phi_1^\infty.$$

Since the calculations was done at $x_1 = 0$, this is the fugacity coefficient at infinite dilution. Henry's law constant is

$$k_1^H = \phi_1^\infty P = (3.31944)(10 \text{ bar}) = 33.2 \text{ bar}.$$

Comments This calculation may be repeated at other pressures and temperatures. Figure 13-11 shows a graph of Henry's law constants for this system calculated by the SRK equation.

Infinite Dilution and Ideal Solution as Reference States

Henry's law was introduced as a way of calculating the fugacity of a component in solution when the component is above its critical temperature at the temperature of the solution. Nonetheless, Henry's law may be used even when the component is below its critical point. There is a certain symmetry between the Lewis-Randall rule, which applies in the limit $x_i \to 1$, and Henry's law, which applies in the limit $x_i \to 0$. The relationship is demonstrated in Figure 13-12, which shows the fugacity of carbon dioxide in n-pentane, plotted as a function of the mol fraction of carbon dioxide at constant temperature. In this case carbon dioxide is below its critical temperature (304.2 K) and forms a liquid solution at all compositions between 0 and 1. The Lewis-Randall rule gives the fugacity of component by the linear relationship

$$f_i = x_i f_i^{\text{pure}} \qquad (x_i \to 1).$$

This is a straight line that is *tangent* to the true fugacity at $x_i = 1$. As a tangent line, it provides a good approximation of the true fugacity over some range of compositions near $x_i = 1$. Beyond this range the linear form diverges from the true fugacity and the Lewis-Randall rule is corrected through the use of the activity coefficient:

$$f_i = \gamma_i x_i f_i^{\text{pure}}.$$

Henry's law gives the fugacity of component by the linear relationship

$$f_i = k_i^H x_i, \qquad (x_i \to 0).$$

This straight line is tangent to the true fugacity at the opposite corner, at $x_i = 0$. As with the Lewis-Randall rule, it provides a good representation of fugacity within

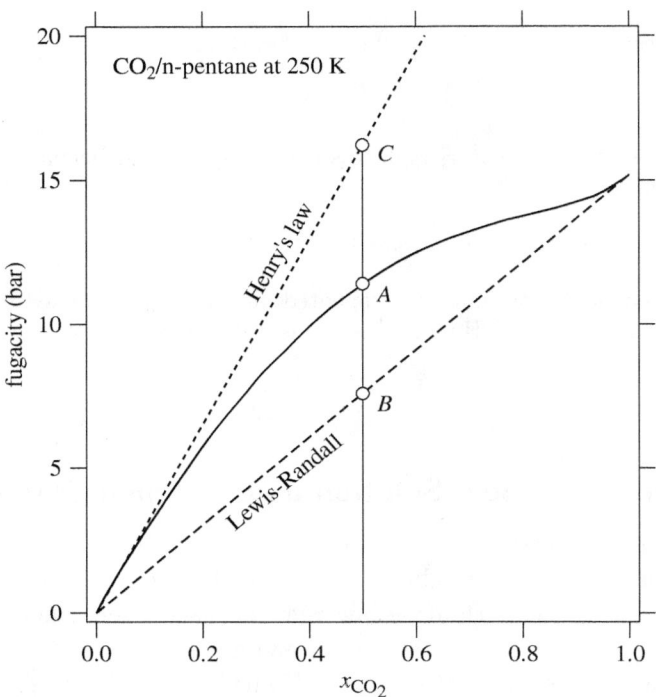

Figure 13-12: Fugacity of carbon dioxide in *n*-pentane (saturated liquid at 250 K).

a limited region near $x_i = 0$ but breaks down at higher concentrations. This failure can be corrected using a new activity coefficient, γ_i^H:

$$f_i = \gamma_i^H k_i^H x_i.$$

Figure 13-12 provides a graphical interpretation of these activity coefficients. When the real fugacity of component i is given by point A, the Lewis-Randall calculation gives point B, and Henry's law gives point C. The respective activity coefficients are the factors by which we must multiply the fugacity at B and C to bring these points right on A.

Both the Lewis-Randall rule and Henry's law represent limiting behaviors where the fugacity of component may be calculated using simple relationships. A system that behaves according to the Lewis/Randall rule is called "ideal in the Lewis-Randall sense" (or simply, ideal solution), and a system that obeys Henry's law is called "ideal in the sense of Henry's law." And in both cases, activity coefficients are introduced to account from deviations from the "ideal" behavior.

13.6 Solubility of Solids in Liquids

The dissolution of solids in liquids arises often in practice because the liquid phase provides a more homogeneous environment for contact between components as well as for chemical reactions. Solids generally have a finite solubility in a liquid solvent. Exceeding the solubility limit produces a two-phase system, a solid in contact with the solution. This problem of solid liquid equilibrium (SLE) is treated by the same general thermodynamics tools developed so far. When the liquid is saturated in the solid component (*solute*), the solute satisfies the equilibrium criterion,

$$f_i^{\text{pure},S} = f_i^L = \gamma_i x_i f_i^{\text{pure},L}. \tag{13.27}$$

Here, $f_i^{\text{pure},S}$ is the fugacity of the component in the solid, and f_i^L its fugacity in the liquid, which can be expressed, using an activity coefficient, in terms of the mole fraction of the dissolved solid and the fugacity of the solute as a pure *liquid* at the temperature of the solution. Since the solute does not exist as a pure liquid at the temperature of the solution, $f_i^{\text{pure},L}$ refers to a *hypothetical* liquid and must be calculated by extrapolation from the liquid phase. The situation is demonstrated in Figure 13-13, which shows the pressure-enthalpy graph of the pure solute. The pure solute at temperature T of the solution is shown by point S, which lies in the solid region. The hypothetical liquid at T is shown by point L. This point lies on the dashed line $L'L$, which represents the hypothetical liquid isotherm at T, that is, the isotherm that is obtained by assuming that the liquid at L' does not solidify but continues to exist as a liquid. The fugacity of point S is $f_i^{S,\text{pure}}$ and of point L is $f_i^{L,\text{pure}}$. For each of these fugacities we apply eq. (10.4)

$$\left(\frac{\partial \ln f_i^{L,\text{pure}}}{\partial T}\right)_P = \left(\frac{\partial (G^L/RT)}{\partial T}\right)_P = -\frac{H^L}{RT^2},$$

$$\left(\frac{\partial \ln f_i^{S,\text{pure}}}{\partial T}\right)_P = \left(\frac{\partial (G^S/RT)}{\partial T}\right)_P = -\frac{H^S}{RT^2},$$

and upon taking the difference we have

$$\left[\frac{\partial}{\partial T}\left(\ln \frac{f_i^{L,\text{pure}}}{f_i^{S,\text{pure}}}\right)\right]_P = -\frac{\Delta H_{\text{fus}}}{RT^2},$$

where $\Delta H_{\text{fus}} = H_L - H_S$ is the heat of fusion for the hypothetical melting $S \to L$ at T. Next we integrate the ratio of fugacities with respect to temperature from the normal melting point (states S_n, L_n) to the hypothetical melting point (states S, L) noting that the fugacities of the solid and liquid at the normal melting point are

576   Chapter 13   Miscibility, Solubility, and Other Phase Equilibria

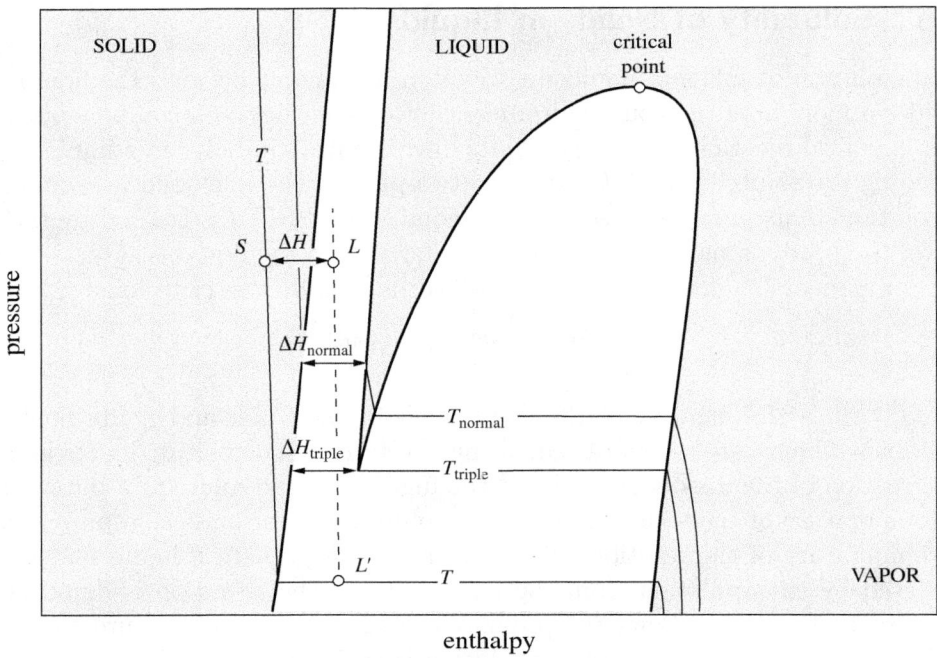

Figure 13-13: Hypothetical liquid state *L* of pure solute for the calculation of solubility. Dashed line *LL'* is a hypothetical liquid isotherm at the same temperature as solid-state *S*.

equal to each other since this is an equilibrium state.[5] Finally, we assume the heat of fusion to be constant and equal to its value at the normal melting point. The result of the integration is

$$\ln \frac{f_i^{L,\text{pure}}}{f_i^{S,\text{pure}}} = \frac{\Delta H_{\text{fus}}}{RT_n}\left(\frac{T_n}{T} - 1\right), \tag{13.28}$$

where T_n is the normal melting point. Combining this result with eq. (13.27) we obtain the final expression for the solubility of the solute:

$$\ln \gamma_i x_i = -\frac{\Delta H_{\text{fus}}}{RT_n}\left(\frac{T_n}{T} - 1\right). \tag{13.29}$$

If solute and solvent are similar in chemical structure and interaction, their solution is approximately ideal ($\gamma_i \approx 1$). In this case we obtain a simpler relationship for the *ideal solubility* of solid solute in a solvent at temperature T:

---

5. The fugacities at S and at the hypothetical liquid L are *not* equal because this is not a pair of coexisting phases.

13.6 Solubility of Solids in Liquids

$$\boxed{\ln x_i = -\frac{\Delta H_{\text{fus}}}{RT_n}\left(\frac{T_n}{T} - 1\right).} \qquad (13.30)$$

It is interesting to observe that the result depends on the heat of fusion. This alludes to the fact that the dissolution of a solid is akin to melting, a process that disrupts the bonding between molecules in the solid phase. In the case of dissolution this disruption is caused not by temperature but by the presence of the solvent. Even in the absence of specific interaction ($\gamma_i = 1$), the solvent is capable of bringing a certain amount of solute into the solution.

Before we demonstrate the application of this equation, a few comments on the derivation. The use of the normal melting point as a starting point for the integration of the fugacity ratio is done for convenience, since the melting temperature and heat of fusion at 1 atm are usually available from tables. Since the heat of fusion does not change much with temperature, other temperatures where the heat of fusion is known can be used.[6] The derivation assumes that the heat of fusion is independent of temperature. More accurate expressions require the heat capacity of the solid and liquid (see, for example, Sandler [3] or Prausnitz [4]). Another assumption that was made but not explicitly stated is that the solid phase is pure solute, that is, no solvent is dissolved into the solid. This is usually an acceptable assumption.

Example 13.10: Solubility of Naphthalene and Benzene
The melting point of naphthalene is 80.09 °C and of benzene is 5.48 °C. Calculate the solubility of naphthalene in benzene at 40 °C, and that of benzene in naphthalene at 0 °C. The heat of fusion of naphthalene and benzene is

$$\Delta H_N = 19061 \text{ J/mol}, \quad \Delta H_B = 9866.3 \text{ J/mol},$$

where N refers to naphthalene and B to benzene.

Solution *Naphthalene in benzene.* Assuming the two components to form an ideal solution, the solubility (mole fraction) of naphthalene at 40 °C is calculated from eq. (13.30):

$$\ln x_N = -\frac{19061 \text{ (J/mol)}}{(8.314 \text{ J/mol K})(353.24 \text{ K})}\left(\frac{353.24 \text{ K}}{313.15 \text{ K}} - 1\right) = -0.830893,$$

from which we find,

$$x_N = 0.43566.$$

---

6. In some texts the result is given in terms of the melting temperature and heat of fusion at the triple point. For most solids the temperature of the triple point is fairly close to the normal melting point so that either choice gives essentially the same result.

Benzene in naphthalene. The calculation is done in the same manner:

$$\ln x_B = -\frac{9866.3 \text{ (J/mol)}}{(8.314 \text{ J/mol K})(278.63 \text{ K})}\left(\frac{278.63 \text{ K}}{273.15 \text{ K}} - 1\right) = -0.0854468,$$

from which we find

$$x_B = 0.918.$$

Comments Even though naphthalene and benzene in pure form at 0 °C are both solid, upon mixing them they form a liquid that can contain up to 91.8% benzene by mol. This is demonstration of freezing-point depression and is discussed further in the next section.

Phase Diagram and Freezing Point Depression

The calculation of Example 13.10 can be repeated at other temperatures to obtain the solubility of naphthalene and benzene in each other. Figure 13-14 shows the calculated solubility lines plotted against the mole fraction of naphthalene (the solubility of benzene is plotted against $x_N = 1 - x_B$). Point A corresponds to the solubility of naphthalene in benzene at 40 °C and point A' to the solubility of benzene in naphthalene at 0 °C. Both solubilities increase with increasing temperature, as eq. (13.30) indicates. The solubility lines also indicate the phase boundary of the system. To see why this is so, let us examine point A, which gives the solubility (maximum mole fraction) of naphthalene in benzene at 40 °C. To the left of this point we have a homogenous solution because the mol fraction of naphthalene is less

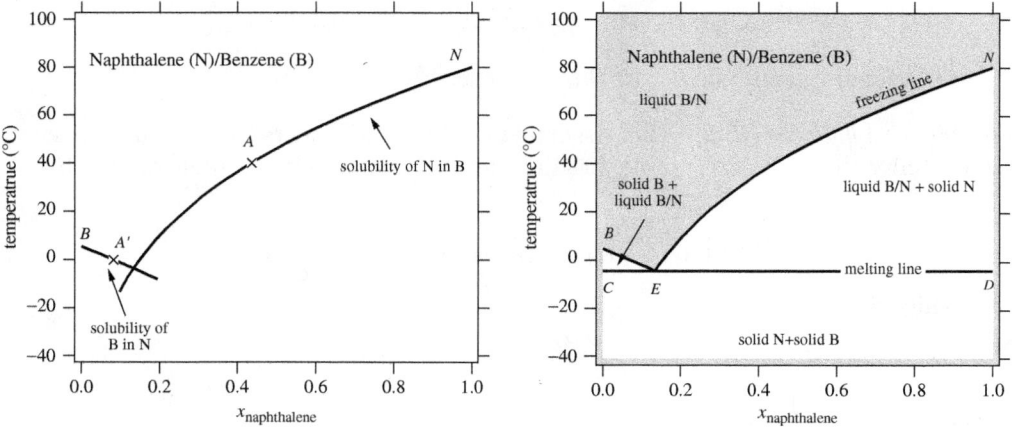

Figure 13-14: Ideal solubility between naphthalene (N) and benzene (B). Left: solubility curves; right: phase diagram. (See Example 13.10.)

13.6 Solubility of Solids in Liquids

than the solubility limit. To the right, we have a heterogeneous mixture that consists of a liquid that is saturated in naphthalene, along with a solid phase of undissolved naphthalene, since the amount present is above the solubility limit. Therefore, the solubility curve of naphthalene marks the boundary between the homogenous liquid (above) and the heterogeneous mixture of liquid with undissolved naphthalene, below. Similarly, the solubility curve of benzene marks the boundary between the homogeneous solution (above) and a liquid mixed with undissolved benzene, below. The two curves meet at point E, which marks the end of each solubility curve. Line BEN is formed by applying eq. (13.30) for benzene from $x_N = 0$ ($x_B = 1$) to x_N^E (eutectic point), and the same equation for naphthalene from $x_N = x_N^E$ to $x_N = 1$. To complete the phase diagram we draw the tie line that passes through point E (see graph on the right pane of Figure 13-14). The graph represents the phase diagram of two completely immiscible solids whose liquid is completely miscible at all compositions. It is very similar to the Txy graph of two completely immiscible liquids (see Figure 13-8) and it is read in a similar manner. The region above line BEN is a homogeneous liquid. All other regions are a mixture of two phases, two solids (below CED), or a liquid and a solid (regions between lines CED and BEN). Line BEN is the freezing line of the mixture: approached from above (liquid), it represents the point where a solid phase forms. Line CED is the melting line: approached from below (mixture of two solids), it represents the temperature where a liquid forms. In metallurgy, the freezing and melting lines are known as *solidus* and *liquidus*, respectively.

A characteristic of such systems is that the freezing line is curved downwards, that is, mixtures of the two components have a lower freezing point than either component. This is known as freezing point depression. A common application of this phenomenon is in deicing. By mixing ice with a nonvolatile solute, the ice-solute mixture becomes liquid even at temperatures where pure water would be solid.

Example 13.11: Eutectic Point
Determine the eutectic temperature and composition in the system naphthalene/benzene.

Solution According to Figure 13-14, the eutectic point is the temperature where the two solubilities meet. Mathematically, this point is defined by the following conditions:

$$\ln x_N = -\frac{\Delta H_N}{RT_N}\left(\frac{T_N}{T_E} - 1\right),$$

$$\ln x_B = -\frac{\Delta H_B}{RT_B}\left(\frac{T_B}{T_E} - 1\right),$$

$$x_N + x_B = 1.$$

These equations are solved simultaneously for the eutectic composition, x_N, x_B, and the eutectic temperature, T_E. Using the parameters in Example 13.10 we find

$$T_E = 269.6 \text{ K} = -3.58 \text{ °C},$$

$$x_N = 0.133,$$

$$x_B = 0.867.$$

13.7 Osmotic Equilibrium

Osmosis is a process in which a system establishes equilibrium between two regions separated by a partition that is permeable to some but not all components. A typical situation is that of a mixture of two components one of which is small enough to pass through the membrane but the other is too big to fit. Osmosis is an important process in biology, as most living organisms depend on membranes to regulate selective transport through them. It is also the basis of reverse osmosis, a separation process that employs a semipermeable membrane to purify a component from a mixture. The basic osmotic experiment consists of a solution that contains a large solute and which is brought into contact with the pure solvent via a membrane that is permeable to the solvent but not to the solute (Figure 13-15).[7] Initially both compartments are at the same temperature and pressure. In the equilibrated system, the pressure is higher in the compartment that contains both components due to

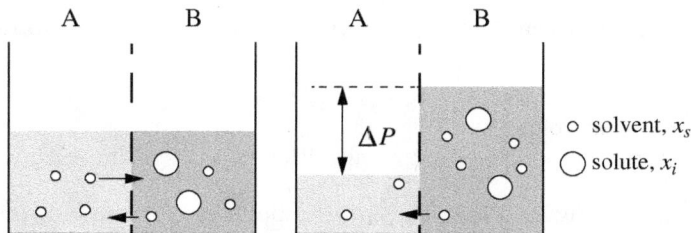

Figure 13-15: The osmotic process: Two regions establish equilibrium via a semipermeable membrane. At equilibrium, pressure is higher in the compartment where both components are present.

---

7. In the case demonstrated in Figure 13-15, compartment A contains solvent only. This is not necessary. If both compartments initially contain solvent and solute, but at different concentrations, solvent will be transferred from the compartment with the higher concentration to the one with lower concentration.

13.7 Osmotic Equilibrium

the transfer of solvent from compartment A. In *reverse osmosis*, pressure is applied to compartment B causing the transfer of solvent into compartment A. This process produces pure solvent from a solvent/solute mixture and is used commercially in water purification.

In the basic osmotic process, the equilibrium state is one in which pressure and composition in the two compartments are unequal. This is a case of constrained equilibrium: The construction of the membrane prevents the equilibration of pressures; it also prevents the equilibration of the chemical potential of the solute, since it cannot pass through the membrane. On the other hand, it permits the equilibration of the solvent, which is free to pass between both sides, and the equilibration of temperature, since molecular transport, even of one species only, necessarily transports energy between the two sides. The equilibrium conditions for this situation are,

$$T^A = T^B,$$

$$f^A_{\text{solv}} = f^B_{\text{solv}},$$

where A and B indicate the compartment (A is pure solvent, B is solvent plus solute). Fugacity f^A_{solv} refers to pure solvent at pressure P_A; fugacity f^B_{solv} refers to mixed solvent with mol fraction x_s at pressure $P_B = P_A + \Pi$, where Π is the osmotic pressure and is equal to the pressure difference between the two compartments at equilibrium. For the fugacities we write,

$$f^A_{\text{solv}} = f^{\text{pure}}_{\text{solv}}(T, P_A),$$

$$f^B_{\text{solv}} = \gamma^B_{\text{solv}} x^B_{\text{solv}} f^{\text{pure}}_{\text{solv}}(T, P_B),$$

where x^B_{solv} and γ^B_{solv} is the mol fraction and activity coefficient of the solvent in the compartment that contains the solute. The fugacities of pure solvent that appear in these two equations are related via the Poynting equation:

$$\frac{f^{\text{pure}}_{\text{solv}}(T, P_B)}{f^{\text{pure}}_{\text{solv}}(T, P_A)} = \exp\left[\frac{\Pi V_{\text{solv}}}{RT}\right]. \tag{13.31}$$

where V_{solv} is the molar volume of the solvent. Combining these results and solving for the osmotic pressure we obtain,

$$\Pi = -\frac{RT}{V_{\text{solv}}} \ln \gamma^B_{\text{solv}} x^B_{\text{solv}}. \tag{13.32}$$

As we see, the osmotic pressure of the solution depends on the activity of the solvent, $a^B_{\text{solv}} = \gamma^B_{\text{solv}} x^B_{\text{solv}}$, in the solution that contains the solute. A simpler expression is obtained in the limit of infinite dilution, where the activity coefficient of the solvent is approximately unity. Using the approximation $-\ln x_{\text{solv}} \approx 1 - x_{\text{solv}} = x_i$, which is valid when $x_{\text{solv}} \approx 1$, eq. (13.32) becomes,

$$\boxed{\Pi = x_i \frac{RT}{V_{\text{solv}}}, \quad (x_i \to 0),} \tag{13.33}$$

where $x_i = 1 - x_{\text{solv}}$ is the mol fraction of the *solute*. Noting that $x_i/V_{\text{solv}} \approx c_i$, where c_i is the molar concentration of the solute (in mol per unit volume), another form of eq. (13.33) is[8]

$$\Pi = c_i RT, \quad (c_i \to 0). \tag{13.34}$$

The general result is eq. (13.32) while eqs. (13.33) and (13.34) are approximations when the concentration of the solute is low. To apply the general form of the equation we must know the activity coefficient of the solvent, which depends on concentration. This dependence is not always known and to circumvent this difficulty, an alternative expression for the osmotic pressure is developed that is easier to accommodate experimental data. In this form, the osmotic pressure is expressed as a series expansion in c_i:

$$\boxed{\frac{\Pi}{c_i RT} = 1 + B(T) c_i + C(T) c_i^2 + \cdots.} \tag{13.35}$$

This is known as *osmotic virial expansion* and is analogous to the virial expansion of the compressibility factor. At very low solute concentrations, the linear and higher-order terms in c_i on the right-hand side are negligible and eq. (13.35) reverts to (13.34). The coefficients $B(T)$, $C(T)$, and others are the *osmotic virial coefficients*, and for a given solute/solvent system they are functions of temperature only. These coefficients can be obtained experimentally by fitting experimental values of the osmotic pressure to a polynomial function of concentration. The higher coefficients are difficult to obtain accurately and in practice the series is usually truncated past the second or third term.

NOTE

Chemical Potential as Driving Force for Diffusion
The osmotic experiment demonstrates that the driving force for diffusion is chemical potential, *not* concentration. In the initial state, the chemical potential of the solvent is different in the

---

8. To obtain this result, write $x_i = n_i/(n_i + n_s)$ and work out the ratio x_i/V_{solv}:

$$\frac{x_i}{V_{\text{solv}}} = \frac{n_i}{(n_i + n_s) V_{\text{solv}}}.$$

For low concentrations of solute, the molar volume of the solvent is equal to that of the solution, that is, $(n_i + n_s) V_{\text{solv}} \approx V^{\text{tot}}$, and the above becomes equal to the molar concentration of the solute.

13.7 Osmotic Equilibrium

two compartments. It is higher in compartment A (pure solvent), and lower in compartment B (mixed solvent). This is seen by writing

$$\mu_s^A - \mu_s^B = RT \ln \frac{f_{solv}^A}{f_{solv}^B},$$

where $f_{solv}^A = f_{solv}^{pure}$ is the fugacity of pure solvent, and $f_{solv}^B = x_B f_{solv}^{pure}$ is the fugacity of mixed solvent. Since $x_B < 1$, the chemical potential of the solvent in A is higher and causes the transfer of solvent molecules from A to B. The same arguments apply to the solute except that the semipermeable membrane does not allow this component to equilibrate (this analogous to an adiabatic wall that prevents the equilibration of temperature between two regions). The identification of the chemical potential as the driving force for molecular transport is important because it helps us understand the conditions of thermal, mechanical, and chemical equilibrium better. Pressure is the driving force for bulk (convective) motion; mechanical equilibrium requires uniformity of pressure. Temperature is the driving force for the transfer of heat; thermal equilibrium requires uniformity of temperature. Chemical potential is the driving force for molecular transport (diffusion); chemical equilibrium requires uniformity of the chemical potential of any species.

This discussion brings us to a question that may still be lingering: *Why do we call it "chemical potential"?* Just as electrons move in response to an electric potential, chemical species move in response to their *chemical potential*. This motion is from regions of high potential to regions of low potential. The imbalance allows the transport of solvent against a pressure gradient: the chemical potential "pushes" the solvent with stronger force than the force due to pressure. In other words, chemical potential can produce a real mechanical force. Equilibrium with respect to molecular transport is established when the chemical potential is uniform in all regions where a species can wander. In the osmotic experiment, the chemical potential in the side of the pure solvent remains constant. On the solute side it is lower but increases continuously with molecular transport because both the mole fraction of the solvent and pressure increase. At equilibrium, solvent molecules continue to pass freely from one side to the other but there is no net change to either side because the chemical potential is uniform. At this point, the force due to chemical potential balances exactly the mechanical force exerted by the pressure difference.

Reverse Osmosis

In the osmotic process, shown in Figure 13-15, pure solvent is brought into contact with a solvent/solute solution resulting in elevated pressure on the side of the solution. If the solution is subjected to pressure, this will cause a *reverse* transfer of solvent from the solution to the pure solvent because pressure increases the chemical potential of the solvent in the solution and forces it pass on the other side. This process is called *reverse osmosis* or hyperfiltration, and may be used to purify a solvent. One of the most prominent applications is in desalination of seawater. In desalination by reverse osmosis, sea water (containing approximately 3.5% wt of salts of mostly Na^+ and Cl^- ions) is compressed to a value above its osmotic pressure and is brought into contact with a semipermeable membrane that allows water

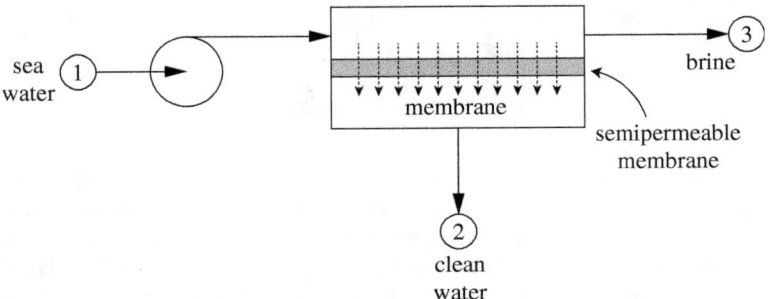

Figure 13-16: Schematic of desalination process by reverse osmosis.

to pass but not any of the ions. The process is shown schematically in Figure 13-16. Effectively, pressure causes water to pass through the membrane, which acts as a filter that keeps out the ions, hence *hyperfiltration*. However, there are important differences between regular filtration and reverse osmosis. In filtration, the applied pressure is needed to overcome the resistance to the flow of the liquid through the filter; even small pressure will produce a trickle of flow. In reverse osmosis, pressure establishes a gradient of chemical potential from the seawater side to the freshwater side that squeezes freshwater out of the water/salt solution. For this process to work, the pressure on the seawater side must be at least equal to the osmotic pressure of the seawater solution. If it is less, freshwater will pass into the seawater by forward osmosis. In desalination, the goal is to produce a stream of pure solvent (water). The same process, however, can be used to concentrate a dilute solution of a solute. An important advantage of reverse osmosis is that it requires no heat (its energy input is entirely in the form of work), which makes it possible to operate at room temperature. For this reason reverse osmosis finds application in the separation of temperature-sensitive products such as proteins.

Example 13.12: Osmotic Pressure
Determine the mol fraction of sugar in water at 25 °C that would produce an osmotic pressure of 10 bar.

Solution The mol fraction of solute that produces osmotic pressure Π is found by solving eq. (13.32) for $x_i = 1 - x_{\text{solv}}$. Taking $\gamma_{\text{solvent}} = 1$ we find

$$x_i = 1 - \exp\left[-\frac{\Pi V_w}{RT}\right].$$

13.7 Osmotic Equilibrium

With $\Pi = 10$ bar $= 10^6$ Pa, $V_w = 1.8 \times 10^{-5}$ m^3/mol, $T = 298.15$ K, we find

$$x_i = 1 - \exp\left[-\frac{(10^6 \text{ Pa})(1.8 \times 10^{-5} \text{ m}^3/\text{mol})}{(8.314 \text{ J/mol K})(298.15 \text{ K})}\right] = 7.24 \times 10^{-3}.$$

If we use eq. (13.33) instead, the result is

$$x_i = \frac{(10^6 \text{ Pa})(1.8 \times 10^{-5} \text{ m}^3/\text{mol})}{(8.314 \text{ J/mol K})(298.15 \text{ K})} = 7.26 \times 10^{-3},$$

which is essentially the same result.

Comments A pressure of 10 bar corresponds to the hydrostatic pressure of approximately 10 m of water and is produced by less than 1% by mol of solute! The calculation demonstrates that it takes a very small mole fraction of solute to produce a fairly large osmotic pressure. The result may seem counterintuitive, but it is just another manifestation of the fact the pressure has a small effect on liquid properties: it takes a fairly large osmotic pressure to change the fugacity of the mixed solvent enough to make it equal to that of the pure solvent. The mathematical explanation is that the molar volume of liquids, which appears in the denominator of eq. (13.32), is very small compared to RT.

Example 13.13: Desalination

Determine the operating pressure to produce 1 m^3 of freshwater from 2 m^3 seawater (3.5% NaCl wt) by the reverse-osmosis process shown in Figure 13-16. Assume that the water in the brine that exits the unit is at equilibrium with the clean water in the filtrate and neglect any nonideal effects. The feed is at ambient conditions (25 °C, 1 bar).

Solution Since stream 3 exits in equilibrium with pure water, its pressure is equal to its osmotic pressure. First, we must calculate the amount of salt in stream 3. Taking the density of seawater to be essentially that of pure water, the process produces 1 kg of water from 2 kg of seawater. Since no salt passes through the membrane, the concentrated brine solution contains a mass fraction of salt,

$$w_3 = \frac{m_1 w_1}{m_2} = \frac{(2 \text{ kg})(0.035)}{1 \text{ kg}} = 0.07 = 7\% \text{ wt},$$

where m_i refers to the mass of stream i and w_i to the mass fraction of salt in that stream. We now construct the material balance table shown below (molar mass of NaCl: 58.44×10^{-3} kg/mol):

	1	2	3
m (kg)	2	1	1
w	0.035	0.	0.07
n (mol)	108.42	55.56	52.86
x	0.0110	0	0.0227

The mol fraction of NaCl in stream 3 is 0.0227. Sodium chloride dissociates fully into Na^+ and Cl^- ions, both of which contribute to the osmotic pressure of the solution. The mol fraction of "solute" in this case is twice that of the undissociated salt. Using eq. (13.32), the osmotic pressure of stream 3 is

$$\Pi_3 = -\ln(1 - 2 \times 0.0227)\frac{(8.314 \text{ J/mol K})(298.15 \text{ K})}{18.15 \times 10^{-6} \text{ m}^3/\text{mol}} = 64.3 \text{ bar}.$$

Comments The osmotic pressure of seawater (stream 1) is 31 bar; the osmotic pressure of stream 3 is twice that, because the extraction of 1 m³ of water concentrates the salt by a factor of 2. The pressure requirement will be lower if we accept a lower amount of freshwater per kg of seawater, but it must exceed the osmotic pressure of seawater.

13.8 Summary

The processes discussed in this chapter demonstrate the great variety of phase equilibrium that can arise beyond the basic vapor-liquid problems discussed in most of the previous chapters. Many other systems could be included: The adsorption of gases onto solids (used in the removal of pollutants from air), the distribution of detergents in water/oil systems, the wetting of solid surface by a liquid, the formation of an electrochemical cell when two metals make contact are all examples of multiphase/multicomponent equilibrium. They all share one important common element: their equilibrium state is determined by the requirement that the chemical potential of any species must be the same in any phase where the species can be found. These problems are beyond the scope of this book. The important point is this: The mathematical development of equilibrium (Chapter 10) is extremely powerful and encompasses any system whose behavior is dominated by equilibrium.

13.9 Problems

Problem 13.1: Determine the solubility of hexane in methanol and decane at 25 °C based on the available activity coefficients at infinite dilution:
a) Hexane in methanol: $\gamma_H^\infty = 18.97$, $\gamma_M^\infty = 21$
b) Hexane in decane: $\gamma_H^\infty = 0.89$, $\gamma_D^\infty = 0.91$.
Hint: Assume that the Margules equation is valid for these systems.

Problem 13.2: A tank contains a mixture of water and carbon dioxide. The mol fraction of CO_2 in the liquid is 5.0×10^{-4} and the temperature is 70 °C.

13.9 Problems

a) Calculate the total pressure as well as the composition of the vapor phase.
b) To what temperature should you bring the system in order to increase the mol fraction of CO_2 in the liquid by a factor of 5, if the total pressure is to remain constant?
c) Calculate the composition of the vapor phase for the conditions of part (b). Henry's constant of CO_2 is

$$k_i^H = 9.12 \times 10^4 e^{-2400/T} \text{ (bar kg/mol)},$$

with T in kelvin. Notice that Henry's law constant is expressed in units of molality.

Problem 13.3: The fugacity coefficient of CO_2 in liquid n-pentane at 344.15 K, 2.93 bar approaches the value $\phi = 39.2$ as the mol fraction of CO_2 approaches zero. Use this information to calculate Henry's law constant for this system.

Problem 13.4: A chemical process produces a waste stream that is mostly air but contains dimethlyamine (DMA) at a mol fraction of 0.01. To satisfy emission standards, the gas stream must be purified to contain no more than 100 ppm of DMA. To achieve this concentration, the gas is passed through a liquid scrubber in which water is sprayed from the top of the tank while the gas rises. During the contact between the drops and the gas, some DMA is transferred to the liquid phase. The inlet water contains no DMA and the exiting streams may be assumed to be in equilibrium with each other. The process is operated at 25 °C, 1 bar.
a) Calculate the required flow rate of water to achieved the desired purification. Report the result in kg of water per kg of air.
b) If the water flow rate must remain less than 50 kg water/kg of air, what is the required pressure?
c) Critique the assumption that the exit streams are at equilibrium.
Additional data: Henry's law constant for DMA at 25 °C is 1.84 bar.

Problem 13.5: The VLE data in the table below were obtained using the SRK equation. The system under consideration is carbon dioxide (1) and cyclohexane (2). Using only data given in this problem, answer the following questions:
a) What is the saturation pressure of cyclohexane at 250 K?
b) Calculate the solubility (mole fraction) of CO_2 in cyclohexane at 250 K under total pressure of 1 bar.
c) A vapor mixture of carbon dioxide/cyclohexane is to be separated by partial condensation in a vapor-liquid separator at 250 K so that carbon dioxide is received at a purity of 80%. Determine the pressure and the mole fraction of CO_2 in the liquid.

d) If the feed stream in the previous question contains 30% CO_2 by mole, calculate the recovery of carbon dioxide (i.e., moles of CO_2 in the purified stream as a percentage of moles CO_2 in the feed).
e) Clearly state and justify your assumptions.

P bar	T (K)	x_1	y_1	ϕ_1^L	ϕ_1^V	ϕ_2^L	ϕ_2^V	Z^V	Z^L
0.0403	250	0	0	374.04	1.0005	0.9977	0.9977	0.9977	0.0002

Problem 13.6: Pure water (stream A) and H_2S (stream B) are brought into contact in a bubbler where they reach equilibrium at 1 bar, 50 °C. The gas stream that leaves the bubbler is then compressed to 5 bar and subsequently passes through a heat exchanger which cools the compressed stream to 25 °C.
a) Determine the composition of streams D and C.
b) Determine whether stream G contains any liquid and if so, calculate the fraction of the liquid.
Additional data: The saturation pressure of water (P_w) and Henry's law constant for H_2S in water (k^H) are given below as a function of T:

$$\ln P_w = -37.224 + 0.16686 \times T - 0.00017985 \times T^2,$$

$$\ln k^H = -14.13 + 0.11365 \times T - 0.00015146 \times T^2.$$

In the above, T is in kelvin while both P_w^{sat} and K_H are in bar.

Problem 13.7: An air stream that contains 8% ammonia (by mol) is treated in absorption tower that removes 95% of the ammonia. In this unit, the gas stream is brought into contact with freshwater at 1 bar, 20 °C. In a simplified treatment of the process, assume that the liquid that exits the tower is in equilibrium with the exiting air stream. Determine the required flow rate of water per mol of the air stream entering the unit. The following equilibrium data are available:

x	0.0206	0.0310	0.0407	0.0502	0.0735	0.0962
y	0.0158	0.0240	0.0329	0.0418	0.0660	0.0920

Data from Wark, Warner, and Davis, *Air Pollution: Its Origin and Control*, 3rd ed., (Boston: Addison-Wesley, 1998), p. 329.

Problem 13.8: Water and normal heptane are essentially immiscible.
a) What is the bubble temperature of a liquid mixture that contains 50% by mol normal heptane at 1 bar?
b) What is the dew temperature of a vapor mixture with 50% normal heptane at 1 bar?
c) What is the dew pressure of an equimolar vapor mixture at 50 °C?

13.9 Problems

Problem 13.9: Water and normal octane are practically *immiscible* in each other.
a) Calculate the dew temperature at 2 bar of a vapor mixture that contains 65% water and 35% octane (by mol). What is the composition of the first liquid to condense?
b) A mixture that contains 75% water and 25% octane (by mol) is brought to 115 °C, 2 bar. Which phases are present?
c) 100 mol of vapor mixture that contains 75 mol water and 25 mol octane is cooled at constant $P = 2$ bar until 10 mol of liquid octane have been collected. How many moles of water have condensed at that point?
The saturation pressures of the two components are given by the equations below (P in bar, T in C):

$$P_W = e^{11.6832 - 3816.44/(T+227.02)}, \quad P_O = e^{9.3222 - 3120.29/(T+209.52)}$$

where W stands for water and O stands for octane.

Problem 13.10: Water and normal octane are essentially immiscible in each other. Consider a solution of the two components that contains 32% water by mol. In all of the following the temperature is 50 °C.
a) At what pressure does the liquid begin to boil?
b) What is the composition of the first bubble?
c) Which phase boils off first?
d) What is the pressure when the first liquid phase boils off?
e) At what pressure does the second liquid phase disappear?
f) What is the composition of the vapor at that point?
Additional data: You may use Antoine equations given in Problem 13.9.

Problem 13.11: Water and normal octane are essentially immiscible in each other. Consider a solution of water and normal octane containing 80% water by mol. In all of the following the pressure is 1 bar.
a) At what temperature does the liquid begin to boil?
b) What is the composition of the first bubble?
c) Which phase boils off first?
d) What is the temperature when the first liquid phase boils off?
e) At what temperature does the second liquid phase disappear?
f) What is the composition of the vapor at that point?
Antoine parameters are given in the Problem 13.9.

Problem 13.12: At 100 °C water (w) and nitrobenzene (n) are only partially miscible. At this temperature, the solubility of nitrobenzene in water is 0.147 mol %, while the solubility of water in nitrobenzene is 8.3 mol %.

a) Assuming that each liquid phase behaves ideally with respect to the concentrated species (that is, $\gamma_w = 1$, in the water-rich phase and $\gamma_n = 1$ in the nitrobenzene-rich phase), calculate the activity coefficient of each component at infinite dilution.
b) Show that if boiling occurs under constant pressure, the boiling temperature must remain constant until one of the two-liquid phases completely evaporates.
c) 100 mol of the water-rich phase are mixed with 100 mol of the nitrobenzene-rich phase at 100 °C and the pressure is adjusted until boiling starts. What is the pressure?
d) Calculate the composition of the vapor phase in the previous part.
e) If boiling continues indefinitely, which liquid phase will disappear first?
f) Draw a qualitative Pxy graph for this system at 100 °C. Show all the important features on the graph. The saturation pressure of nitrobenzene at 100 °C is 21 Torr.

Problem 13.13: The activity coefficients for the system hexane/ethanol at 85 °C are given by

$$\ln \gamma_1 = 2.8 x_2^2, \quad \ln \gamma_2 = 2.8 x_1^2.$$

a) Construct the Pxy graph for this system at 85 °C.
b) Determine the bubble pressure of a mixture with the overall composition $x_{hex} = 0.5$ and report the composition of all the phases present.
Additional data: The saturation pressures of the pure components are

$$P_{\text{hex}}^{\text{sat}} = 1.64 \text{ bar}, \quad P_{\text{ethanol}}^{\text{sat}} = 1.31 \text{ bar}.$$

Problem 13.14: Benzene and water are essentially immiscible in each other. Consider a liquid produced by mixing 25 moles of benzene with 75 moles of water at 1 bar, 25 °C:
a) At what temperature does the liquid begin to boil?
b) What is the composition of the first bubble?
c) Which phase boils off first?
d) What is the temperature when the first liquid boils off?
e) At what temperature does the second liquid phase disappear?
f) What is the composition of the vapor at that point?
Additional data: The Antoine equations for the two components are given below (temperature in K, pressure in Torr):

$$\ln P_W^{\text{sat}}(\text{Torr}) = 18.3036 - \frac{3816.44}{T(K) - 46.13},$$

$$\ln P_B^{\text{sat}}(\text{Torr}) = 15.9080 - \frac{2788.51}{T(K) - 52.36}.$$

13.9 Problems

Problem 13.15: Butanol and water are partially miscible liquids. At 1.013 bar, the bubble point of the two-phase system is at 93 °C, and the mol fraction of butanol in the two liquids is 4% and 40%, respectively. Determine the activity coefficients of the two components in the two liquids at 93 °C. Make any suitable assumptions. Additional data: The saturation pressures of butanol and water at 93 °C are 0.387 bar, 0.7849 bar, respectively.

Problem 13.16: A desalination process such as the one in Figure 13-16 is used to produce freshwater from seawater that contains 3.25% wt salts. If the operating pressure is 42 bar, determine the amount (kg) of freshwater that is produced per kg of seawater. Assume seawater to be a solution of NaCl and take its density to be that of freshwater.

Chapter 14

Reactions

Up to this point we have concentrated exclusively on nonreacting systems, namely, systems in which the molecular identity of all species is preserved. In reactions, chemical identity is not preserved, as a reactant species is converted into a product species. Even so, the *atomic* identity of the constituent elements *is* preserved. In fact, a chemical reaction is simply a rearrangement of atoms into new molecules, and in this respect, is very similar to the passages of molecules from one phase into another. There are more analogies between chemical reactions and mixing of two liquids, for example. In both cases we observe the transition from an initial state to a final one that is accompanied by heat effects. A chemical reaction may not go to completion, just as two liquids may only be partially miscible. Chemical equilibrium refers to a state where no net interconversion is observed between products and reactants. The situation is again analogous to phases at equilibrium with no net mass transfer between them. In this qualitative picture, products and reactants can be viewed as analogous to phases, with matter being transferred from the reactant side to the product side until equilibrium is achieved. With this analogy in mind, it should not be surprising that the general concepts developed in phase equilibrium apply to reacting species as well. Thus, we will be talking about fugacity, chemical potential, and the Gibbs energy.

In this chapter, you will learn to:

1. Calculate the standard enthalpy and Gibbs free energy of reaction from tabulated values.

2. Calculate the equilibrium constant of a reaction.

3. Perform energy and material balances in reacting systems.

14.1 Stoichiometry

For the purposes of thermodynamic calculations, a chemical reaction represents a process in which, reactants, on the left-hand side of the reaction equation, are converted into products on the right-hand side. For example, for the formation of ammonia by reaction between nitrogen and hydrogen we write,

$$\frac{3}{2}H_2 + \frac{1}{2}N_2 = NH_3. \tag{14.1}$$

The stoichiometric coefficients that appear in the above equation ensure that the reaction is balanced. We will represent the stoichiometric coefficients of species i by ν_i and will adopt the convention by which, the stoichiometric coefficients of the reactants are negative, and of the products positive. For the ammonia reaction,

$$\nu_{H_2} = -\frac{3}{2}, \quad \nu_{N_2} = -\frac{1}{2}, \quad \nu_{NH_3} = +1.$$

Stoichiometric coefficients represent the ratios by which reactant and product species participate in the chemical reaction, and may be fractional, as in the above example. Since ratios are preserved if all stoichiometric coefficients are multiplied by the same number, the stoichiometry of a chemical reaction can be represented in many equivalent ways. It is, important, therefore, to clearly specify the stoichiometry that is adopted in a particular calculation and use it consistently throughout. The sum of the stoichiometric coefficients,

$$\nu = \sum \nu_i, \tag{14.2}$$

is the change in the total number of moles when the reaction proceeds from left to right according to the indicated stoichiometry. For the ammonia example, $\nu = -3/2 - 1/2 + 1 = -1$, i.e., for each mole of ammonia produced there is a net decrease in the total number of moles by 1 mole.

When reactants are converted into products the moles of all species change but these changes are not independent of each other because they are interrelated through stoichiometry. For each 3/2 moles of hydrogen reacted, we have 1/2 moles of nitrogen consumed, and 1 mole of ammonia produced. If δn_{H_2} moles of hydrogen are consumed, then

$$|\delta n_{N_2}| = \left(\tfrac{1}{2}\right)\frac{|\delta n_{H_2}|}{\left(\tfrac{3}{2}\right)}, \quad |\delta n_{NH_3}| = (1)\frac{|\delta n_{H_2}|}{\left(\tfrac{3}{2}\right)}.$$

These results can be combined into a simpler relationship,[1]

$$\frac{\delta_{H_2}}{\nu_{H_2}} = \frac{\delta n_{N_2}}{\nu_{N_2}} = \frac{\delta n_{NH_3}}{\nu_{NH_3}}.$$

The common value of the ratio $\delta n_i/\nu_i$ is called *extent of reaction*. For a general reaction with any number of species, the reaction extent ξ is defined as

$$\xi = \frac{n_i - n_{i0}}{\nu_i}, \tag{14.3}$$

---

1. Keep in mind that δn_{H_2} and δn_{N_2} are negative (consumed) while δn_{NH_3} is positive.

14.1 Stoichiometry

where $n_i - n_{i0} = \delta n_i$ is the change in the moles of species i by reaction from its initial number n_{i0}. Solving for n_i, the moles of species i after reaction is

$$n_i = n_{i0} + \xi \nu_i. \tag{14.4}$$

The total number of moles is calculated by summing this equation over all species:

$$n = n_0 + \xi \nu, \tag{14.5}$$

where n_0 and n are the total moles before and after the reaction, and ν is the sum of the stoichiometric coefficients, defined in eq. (14.2). Finally, the mole fraction of species i after the reaction is

$$x_i = \frac{n_i}{n} = \frac{n_{i0} + \xi \nu_i}{n_0 + \xi \nu}. \tag{14.6}$$

These equations demonstrate the usefulness of the extent of reaction. While the number of moles of all species changes during reaction, these changes are interrelated through a common variable, the extent of reaction. If the extent of reaction is known, all compositions can be calculated. The extent of reaction is a measure of how far the reaction has proceeded. If it is positive, it indicates that the reaction has advanced from left to right; if negative, it indicates that the reaction has advanced from right to left. Although we will be referring to the species on the left as "reactants" and those on the right as "products," these terms should be understood to be by convention. The actual direction of the chemical reaction would be indicated by the sign of the reaction extent.

Example 14.1: Extent of Reaction
A reactor contains an equimolar mixture of hydrogen, nitrogen, and ammonia under conditions that reaction (14.1) can take place. Determine the possible values of the extent of reaction and the composition (mole fractions) of the mixture when the reaction has produced 0.5 mol of ammonia.

Solution We construct the stoichiometric table below:

	H_2	N_2	NH_3
ν_i	$-\frac{3}{2}$	$-\frac{1}{2}$	1
n_{i0}	1	1	1
n_i	$1 - \frac{3\xi}{2}$	$1 - \frac{\xi}{2}$	$\xi + 1$

If the reaction proceeds to completion from left to right, the maximum value of ξ is

$$\xi_{\max} = \frac{2}{3},$$

because at this point one of the reactants (hydrogen) is fully consumed. If the reaction proceeds to completion from right to left, the reaction will stop when all ammonia is consumed, that is,
$$\xi_{\min} = -1.$$
Therefore, the range of possible values of ξ is
$$-1 \leq \xi \leq 2/3.$$
If the reaction proceeds from left to right until the reactor has produced 0.5 mol of ammonia, then
$$n_{\mathrm{NH}_3} = \xi + 1 = 1.5 \quad \Rightarrow \quad \xi = 0.5.$$
From the stoichiometric table,
$$n_{\mathrm{H}_2} = 0.25, \quad n_{\mathrm{N}_2} = 0.75, \quad n_{\mathrm{H}_2} = 1.5,$$
and the total number of moles is 2.5. The corresponding mole fractions are
$$x_{\mathrm{H}_2} = 0.1, \quad x_{\mathrm{N}_2} = 0.3, \quad x_{\mathrm{H}_2} = 0.6.$$

Comments The value of the reaction extent depends on the chosen stoichiometry but the mol fractions are independent of that choice. You should confirm this by repeating the calculation with different stoichiometric coefficients.

14.2 Standard Enthalpy of Reaction

A chemical reaction that proceeds to completion is a process whose initial state consists of all reactants and its final state of all products. This process is accompanied by a change in enthalpy, which is given by the difference between products and reactants:
$$\Delta H_{\mathrm{rxn}} = H_{\mathrm{products}} - H_{\mathrm{reactants}}.$$
The precise value of this difference depends on various process details such as the temperature of reactants, whether reactants are mixed before entering the reactor or fed as pure components, whether the reaction product is delivered as a mixture or is separated into its pure components, etc. To avoid such ambiguities and to facilitate tabulations of reaction properties, we adopt a *standard* state for each component in the reaction and report all reaction properties based on these states. The following standard states are in use:

> *Standard state for gases (g):* It is defined as the pure substance *in the ideal-gas state* at 1 bar.

14.2 Standard Enthalpy of Reaction

Standard state for liquids (l): It is defined as the pure substance in the liquid phase at 1 bar.

Standard state for solids (s): It is defined as the pure substance in the solid phase at 1 bar.[2]

Standard state for aqueous solutes (aq): The standard state of a dissolved solute is defined as a solution at 1 bar that obeys Henry's law at concentration of unit molality, that is, 1 mol of solute per kg of solvent.

The standard state specifies the pressure (always 1 bar) and purity of component. With the exception of the standard state for aqueous solutes, components are taken to be pure. Temperature is not specified but is fixed by the user according to the problem at hand. Since pressure and composition (purity) are fixed, *all properties of the standard state are functions of temperature only.*

Standard enthalpies of reaction are tabulated for the *formation reaction* of a species. The formation reaction is a balanced chemical equation that produces one mole of the species in its standard state from the pure constituent elements, also at their standard states.[3] By convention, the standard enthalpy of pure elements is zero. Tabulated values of the standard enthalpy of formation at 25 °C are available for a large number of components in various handbooks (selected values are given in Appendix C.) These values are accompanied by an indication of the standard state to which the reported value refers.

The enthalpy of formation allows the calculation of the standard enthalpy of any reaction. For the reaction

$$\nu_1 A + \nu_2 B + \cdots \rightarrow \nu_i C + \nu_{i+1} D + \cdots$$

it is calculated as the algebraic sum of the standard enthalpies of formation of all species:

$$\Delta H^\circ = \sum_i \nu_i H_i^\circ, \qquad (14.7)$$

where ΔH° is the standard enthalpy of the reaction, and H_i° is the standard formation enthalpy of species i. Since the coefficients of the reactants are negative, this is the difference between the formation enthalpies of reactants and products. The reaction is *exothermic* if $\Delta H^\circ < 0$; in this case the enthalpy of the products is lower than that of the reactants and the difference is released as heat. If $\Delta H^\circ > 0$, the reaction is *endothermic* and requires heat to proceed as written.

---
2. This standard state is sometimes denoted by (c) for *crystalline*.
3. The stoichiometry of the formation reaction is fixed by the requirement that the reaction produces one mole of the species. Accordingly, the species appears on the product side with stoichiometric coefficient $\nu_i = +1$.

Example 14.2: Standard Enthalpy of Reaction
Steam reforming is an important industrial reaction in which natural gas, which is mostly methane, is used to produce hydrogen by reaction with water:

$$\mathrm{CH_4(g) + H_2O(g) = CO(g) + 3H_2(g)}.$$

The reaction requires high temperatures ($\sim 800\ ^\circ$C) and the use of catalysts. Calculate the standard enthalpy of reaction at 25 °C and determine whether it is exothermic or endothermic.

Solution We construct the table below using tabulated values of the standard enthalpy of formation

	$\mathrm{CH_4(g)}$	$\mathrm{H_2O(g)}$	$\mathrm{CO(g)}$	$\mathrm{H_2(g)}$	
ν_i	-1	-1	1	3	
H_i°	$-74{,}520$	$-241{,}818$	$-110{,}525$	0	(J/mol)

The standard enthalpy of the reaction at 25 °C is

$$\Delta H^\circ = (-1)(-74{,}520) + (-1)(-241{,}818) + (1)(-110{,}525) + (3)(0) = 205{,}813\ \mathrm{J}.$$

This value is positive, therefore, the reaction is endothermic.

Comments The value of the standard enthalpy of reaction depends on the stoichiometry adopted, and for this reason, the stoichiometry should be indicated when the enthalpy is reported. To avoid ambiguities, the result could be reported as "205,813 J per mole of methane reacted," or as "68,604 J per mol of hydrogen produced."

The formation enthalpy of water is usually reported for both the gas and the liquid standard state. The correct value is the one that matches the standard state in the reaction as given in the problem statement, in this case, *g*. Notice that this is a *hypothetical* state since pure water at 1 bar, 25 °C is liquid.

NOTE

Standard Enthalpy of Reaction

The standard enthalpy of reaction should be understood as a calculation between two very specific states, that of the pure reactants before reactions, and the pure products after reaction. The calculation can be visualized as a process, as shown schematically in Figure 14-1 for the methane reforming reaction. The pure reactants, methane and water in the ideal-gas state at 25 °C and 1 bar, are mixed, possibly compressed, and heated to reaction temperature. The products are cooled and separated to be delivered as pure gases at 25 °C, 1 bar. The standard enthalpy corresponds to the difference between states *A* and *G*. Since we are dealing with a state function, the enthalpy of reaction depends only on the end states but is independent of the internal details of the process.

14.2 Standard Enthalpy of Reaction

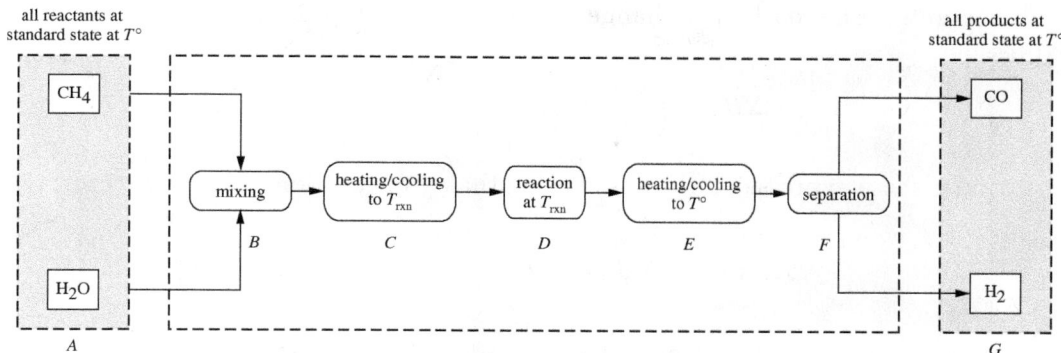

Figure 14-1: Path for the calculation of standard enthalpy of reaction.

Effect of Temperature on Enthalpy of Reaction

As we have seen, the standard enthalpy of any reaction at 25 °C may be calculated from tabulated values of the enthalpy of formation. If the enthalpy is needed at a different temperature, it can be easily calculated from its value at 25 °C. The calculation is done according to the following path:

1. Begin with all reactants at their standard state at T and calculate the enthalpy change to bring each reactant to its standard state at 25 °C (ΔH_1).

2. Calculate the standard enthalpy of reaction at 25 °C (ΔH°_{298}). This calculation delivers the products in their standard state at 25 °C.

3. Calculate the enthalpy change to bring the products from their standard state at 25 °C to their standard state at temperature T (ΔH_3).

The total change is

$$\Delta H^\circ(T) = \Delta H_1 + \Delta H^\circ_{298} + \Delta H_3.$$

The enthalpy change for step 1 is[4]

$$\Delta H_1 = \left(\sum_i \int_T^{298} |\nu_i| C^\circ_{Pi} dT \right)_{\text{reactants}} = \left(\sum_i \int_{298}^T \nu_i C^\circ_{Pi} dT \right)_{\text{reactants}}.$$

where C°_{Pi} is the heat capacity of species i in its standard state, $|\nu_i|$ is the absolute value of its stoichiometric coefficient, and the summation runs only through

---

4. To obtain the result in the far right, change the order of the limits in ΔH_1. This brings a minus sign out which we use to write $-|\nu_i| = \nu_i$, since ν_i for the reactants is negative.

the reactants. The enthalpy change for the third step is given by a similar expression,

$$\Delta H_3 = \left(\sum_i \int_{298}^{T} \nu_i C_{Pi}^{\circ} dT \right)_{\text{products}}.$$

Combining these expressions the standard enthalpy of reaction at temperature T is

$$\Delta H^{\circ}(T) = \Delta H_{298}^{\circ} + \int_{298}^{T} \left(\sum_i \nu_i C_{Pi}^{\circ} \right) dT, \quad (14.8)$$

where we exchanged the order of the integral and the summation. To write the result in simpler form, we define the heat capacity difference between reactants and products,

$$\Delta C_P^{\circ} = \sum_i \nu_i C_{Pi}^{\circ}. \quad (14.9)$$

Using this definition, we obtain the final result in the form

$$\boxed{\Delta H^{\circ}(T) = \Delta H_{298}^{\circ} + \int_{298}^{T} \Delta C_P^{\circ} dT.} \quad (14.10)$$

In summary, the only additional step in the calculation of the standard heat of reaction at temperature T is the integration of ΔC_P° from 298 K (the temperature of tabulated formation standard enthalpies) to T. The superscript $^{\circ}$ indicates that all heat capacities are those of the standard state. For example, if the standard state is notated as (g), the *ideal-gas* heat capacity must be used, regardless of whether the real state under the conditions of the reaction is indeed an ideal gas or not.

Example 14.3: Standard Heat of Reaction at Other Temperatures
Calculate the standard heat of the methane reforming reaction at 800 °C.

$$CH_4(g) + H_2O(g) = CO(g) + 3H_2(g).$$

Solution The standard heat of the reaction at 25 °C was calculated in Example 14.2 where we found

$$\Delta H_{298}^{\circ} = 205,813 \text{ J}.$$

All species are in the ideal-gas state. Their heat capacities are given in the form

$$\frac{C_P^{\circ}}{R} = c_0 + c_1 T + c_2 T^2 + c_3 T^3 + c_4 T^4,$$

14.3 Energy Balances in Reacting Systems

with T in kelvin and with coefficients c_i given in the table below:

Species	ν_i	c_0	c_1	c_2	c_3	c_4
CH_4	1	4.568	-0.008975	0.00003631	-3.407×10^{-8}	1.091×10^{-11}
H_2O	1	4.395	-0.004186	0.00001405	-1.564×10^{-8}	6.32×10^{-12}
CO	1	3.912	-0.003913	0.00001182	-1.302×10^{-8}	5.15×10^{-12}
H_2	3	2.883	0.003681	-7.72×10^{-6}	6.920×10^{-9}	-2.13×10^{-12}

Since all heat capacities are given by the same mathematical form, a fourth-order polynomial in T, ΔC_P can also be expressed in the same polynomial form with coefficients c'_i calculated as

$$c'_i = \sum_k \nu_k c_i^{(k)}.$$

Here $c_i^{(k)}$ is the coefficient of T^i in the heat capacity of species k. Applying this formula we obtain

$$\frac{\Delta C_P^\circ}{R} = 3.598 + 0.020291\,T - 0.0000617\,T^2 + 5.745 \times 10^{-8}\,T^3 - 1.847 \times 10^{-11}\,T^4$$

in which all coefficients are calculated by the same equation as ΔC_P°.

The standard heat of reaction at 800 °C is

$$\Delta H_{1073}^\circ = \Delta H_{298}^\circ + \int_{T_0}^{T} \Delta C_P^\circ \, dT.$$

The enthalpy of reaction at 298 K was obtained in the previous example where we found $\Delta H_{298}^\circ = 205{,}813$ J/mol. The integral of ΔC_P° is

$$(8.314 \text{ J/mol K}) \int_{298.15}^{1073.15} \left(3.598 + 0.020291\,T - 0.0000617\,T^2 + 5.745 \times 10^{-8}\,T^3 - 1.847 \times 10^{-11}\,T^4 \right) dT = 19{,}820 \text{ J}$$

Therefore, the enthalpy of reaction at 1073.15 K is

$$\Delta H_{1073}^\circ = (205{,}813) + (19{,}820) = 225{,}633 \text{ J per mol of methane reacted.}$$

This is about 10% higher than the value at 25 °C.

14.3 Energy Balances in Reacting Systems

A chemical reactor is a vessel in which a reaction takes place. Industrial reactors come in a variety of designs, shapes and sizes. Liquid-phase reactors are usually

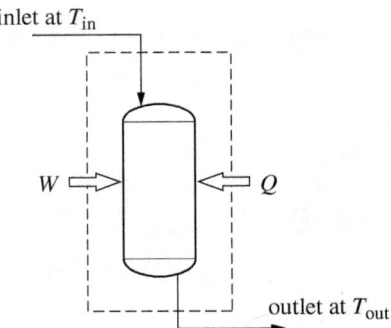

Figure 14-2: Setup for energy balance of reactor.

stirred tanks whereas gas-phase reactions often take place inside tubular reactors. In general, chemical reactions operate at elevated temperatures to achieve practical rates of conversion. Additional heating/cooling is supplied to account for endothermic or exothermic effects. Chemical reactors may operate in batch mode (all chemicals loaded and the reaction is allowed to proceed for a certain time), semi-batch (some chemicals are loaded initially, others are continuously added), or continuous. As with other unit operations considered in previous chapters, the internal details of chemical reactors are not important when performing overall material and energy balances, as long as the inlet and outlet states are specified. For the purposes of illustration we will consider a flow reactor, as in Figure 14-2.

The steady-state energy balance around the reactor is

$$\Delta(\dot{n}H) = \dot{Q} + \dot{W}_s, \tag{14.11}$$

Unless pumps, stirrers, or other mechanical devices are part of the reactor, the shaft work is zero. Typically, a reactor will exchange heat with the surroundings but it is also possible that operation may be adiabatic. The inlet stream contains the reactants, not necessarily in stoichiometric form, and possibly other species that do not participate in the reaction.[5] The outlet contains the products plus any leftover reactants. The energy balance under these conditions is

$$\Delta(\dot{n}H) = \dot{Q},$$

or,

$$\Delta(nH) = Q. \tag{14.12}$$

---

5. The feed may also contain some product species, usually in small amounts that are left in the stream after a separation process.

14.3 Energy Balances in Reacting Systems

The last result expresses the energy balance over a period of time δt. For the calculation of the enthalpy change we adopt the following path:

1. Inlet stream is separated to its pure components and each component is brought to its standard state at $T_0 = 25\,°\text{C}$.

2. Reaction progresses to extent ξ at T_0 to produce all species at their standard state at the same temperature.

3. The species after the reaction are mixed and brought to the pressure and temperature of the outlet.

This procedure can be applied to any reacting system to obtain the enthalpy change. A general expression cannot be written for all cases because the details of the process (phase of streams, enthalpy of mixing) can vary between cases. A simple result will be given here under the following simplifying assumptions: (i) neglect all mixing effects on enthalpy during mixing and separation of components, and (ii) assume that no phase transitions take place during the heating/cooling steps between T_0 and the temperature of the inlet and outlet streams. With these assumptions the enthalpy change is

$$\Delta(nH) = \xi \Delta H^\circ_{\text{rxn}}(T_0) + \sum_{\text{inlet}} \int_{T_{\text{in}}}^{T_0} n_{i0} C_{Pi}\, dT + \sum_{\text{outlet}} \int_{T_0}^{T_{\text{out}}} n_i C_{Pi}\, dT. \quad (14.13)$$

Here, $\Delta H^\circ_{\text{rxn}}(T_0)$ is the standard heat of reaction at T_0, n_i are the moles of species i in the outlet stream, n_{i0} are the moles of species i in the inlet, and C_{Pi} is its heat capacity. Notice that the moles in the inlet, the moles in the outlet, and the extent of reaction are interrelated by stoichiometry:

$$n_i = n_{i0} + \nu_i \xi.$$

The general procedure for the calculation is to first obtain the mole balance through stoichiometry, and then to calculate the energy balance via eqs. (14.12) and (14.13). If the simplifying assumptions are not valid, the enthalpy must be calculated by applying steps 1 to 3 of the calculation path to the inlet and outlet streams.

Example 14.4: Heat of Combustion
Methane is mixed with air at 20% excess over the stoichiometric requirement and undergoes combustion in a burner operating at 2 bar. The inlet gases are at 40 °C and the effluent stream is at 1000 °C. Determine the amount of heat assuming complete oxidation of methane

to carbon dioxide and water. For simplicity, assume the heat capacities of the species to be constant and equal to the values given below (in J/mol K):

$$C_{P,CH_4} = 55.42, \quad C_{P,O_2} = 32.53, \quad C_{P,N_2} = 30.37, \quad C_{P,CO_2} = 48.65, \quad C_{P,H_2O} = 36.94.$$

Solution The combustion reaction of methane is

$$CH_4 + 2O_2 = CO_2 + 2H_2O.$$

For stoichiometric combustion we need 2 mol of oxygen. The actual amount of oxygen is 20% above the stoichiometric requirement, or, $2 + (0.2)(2) = 2.4$ mol. Assuming air to be 21% oxygen and 79% nitrogen, the inlet stream contains nitrogen in the amount (79/21) times the amount of oxygen:

$$\frac{79}{21}(2.4) = 9.02857.$$

Even though nitrogen does not participate in the reaction, it is a component of the mixture and must be included in the mole and energy balance. We now construct the stoichiometric table:

	CH_4	O_2	N_2	CO_2	H_2O	
ν_i	-1	-2	0	1	2	
n_{i0}	1	2.4	9.02857	0	0	mol
n_i	$1-\xi$	$2.4-2\xi$	9.02857	ξ	2ξ	mol
H°_{298}	-75520	0	0	-393509	-241818	J/mol
C_{Pi}	55.42	32.53	30.37	48.65	36.94	J/mol K

For complete combustion of methane, $\xi = 1$.

The assumptions behind eq. (14.13) are acceptable in this problem: mixing effects can be neglected as products and reactants are close to the ideal-gas state at 2 bar and at the temperatures of this problem; in addition, no phase changes take place in bringing the inlet species from inlet conditions (40 °C, 2 bar) to standard conditions at 25 °C, 1 bar, or the outlet species from standard conditions to 1000 °C, 2 bar. Therefore, eq. (14.13) may be used. With $T_{in} = 40\,°C = 313.15$ K, $T_{out} = 1000\,°C = 1273.15$ K, $T_0 = 25\,°C = 298.15$ K, we have:

$$\Delta H^\circ_{rxn}(T_0) = (-1)(-75,520) + (-2)(0) + (1)(-393,509) + (2)(-241,818)$$
$$= -801,625 \text{ J},$$

$$\sum_{inlet} \int_{T_{in}}^{T^\circ} n_{i0} C_{Pi} dT = \Big((1)(55.42) + (2.4)(32.53) + (9.02857)(30.37)\Big)(298.15 - 313.15)$$
$$= -6,115.35 \text{ J},$$

$$\sum_{outlet} \int_{T^\circ}^{T_{out}} n_i C_{Pi} dT = \Big((0.4)(32.53) + (9.02857)(30.37) + (1)(48.65) + (2)(36.94)\Big) \times$$
$$(1,273.15 - 298.15) = 399,496. \text{ J}.$$

14.3 Energy Balances in Reacting Systems

The overall balance is

$$Q = (-801,625) + (-6,115) + (399,496) = -408,244. \text{ J}.$$

Comments The combustion of hydrocarbons is highly exothermic and an important source of energy. In this case, the amount of heat that is released is less than that of the enthalpy of reaction. This is because a substantial amount of enthalpy is carried out by the hot effluent stream.

We have treated all species as ideal gases. Pressure does not enter the calculation because it has no effect on enthalpy in the ideal-gas state.

To simplify the calculations we used a constant C_P for all species. The calculation is only moderately more involved if temperature-dependent values are used.

Example 14.5: Adiabatic Flame Temperature
Adiabatic flame temperature is the temperature of the effluent gases that is obtained if no heat is exchanged between the flame and its surroundings. In this case the heat of reaction is consumed to increase the temperature of the effluent gases.

Calculate the adiabatic flame temperature of methane in 20% excess air at 2 bar.

Solution For adiabatic conditions the energy balance is

$$Q = 0 = \xi \Delta H_{\text{rxn}}^{\circ}(T_0) + \sum_{\text{inlet}} \int_{T_{\text{In}}}^{T_0} n_{i0} C_{Pi} \, dT + \sum_{\text{outlet}} \int_{T_0}^{T_{\text{out}}} n_i C_{Pi} \, dT.$$

In this equation the only unknown is the effluent temperature, T_{out}. If the heat capacities are taken to be constant, the above equation becomes,

$$\xi \Delta H_{\text{rxn}}^{\circ}(T_0) + \sum_{\text{inlet}} n_{i0} C_{Pi}(T_0 - T_{\text{in}}) + \sum_{\text{outlet}} n_i C_{Pi}(T_{\text{outlet}} - T_0) = 0,$$

which is solved for T_{out} to give

$$T_{\text{out}} = T_0 + \frac{-\xi \Delta H_{\text{rxn}}^{\circ}(T_0) - \sum_{\text{inlet}} n_{i0} C_{Pi}(T_0 - T_{\text{in}})}{\sum_{\text{outlet}} n_i C_{Pi}}.$$

Using

$$\xi = 1,$$
$$\Delta H_{\text{rxn}}^{\circ}(T_0) = -801,625 \text{ J},$$

$$\sum_{\text{inlet}} n_{0i} C_{Pi}(T_0 - T_{\text{in}}) = -6115.35 \text{ J},$$

$$\sum_{\text{outlet}} n_i C_{Pi} = 409.74 \text{ J/K},$$

we obtain

$$T_{\text{out}} = 2269.5 \text{ K} = 1996 \text{ °C}.$$

Comments The adiabatic flame temperature depends on the fuel (in this case methane) but also on the composition of the reacting mixture. The highest temperature is achieved when combustion is done in pure oxygen under stoichiometric conditions (why?). This calculation is left as an exercise.

14.4 Activity

Calculations of chemical equilibrium, which will be the topic of the next section, are facilitated through the introduction of the *activity*, a property closely related to fugacity and chemical potential. The activity of a component i in mixture is defined as the ratio of its fugacity over the fugacity of the same component at its *standard state*:

$$\boxed{a_i = \frac{f_i}{f_i^\circ}.} \tag{14.14}$$

Activity is dimensionless fugacity, normalized by the fugacity at the standard state. An immediate consequence of the definition of activity is its relationship to the chemical potential. First, recall that the standard state, which was introduced in Section 14.2, is at the same temperature as the state of interest. We now return to eq. (10.18), which relates fugacities and chemical potentials between two states at the same temperature:

$$\ln \frac{f_i^B}{f_i^A} = \frac{\mu_i^B - \mu_i^A}{RT}. \tag{10.18}$$

We apply this equation with B referring to the state of interest and A the standard state. The result is

$$\mu_i - \mu_i^\circ = RT \ln \frac{f_i}{f_i^\circ}. \tag{14.15}$$

Solving for the chemical potential we obtain the desired relationship:

$$\boxed{\mu_i = \mu_i^\circ + RT \ln a_i.} \tag{14.16}$$

14.4 Activity

If the activity is known, the chemical potential can be immediately calculated, and vice versa. As we will see with examples, the definition of the standard states provides all the necessary information for the calculation of the activity.

NOTE

Activity and Chemical Potential
Equation (14.16) is very similar in form to eq. (10.12) for the chemical potential of component in ideal-gas mixture,

$$\mu_i^{igm} = G_i^{ig} + RT \ln x_i, \qquad [10.12]$$

and eq. (12.15) for the chemical potential in ideal solution,

$$\mu_i^{id} = G_i + RT \ln x_i. \qquad [12.15]$$

In both equations the dependence of the chemical potential on the mol fraction of component i is contained in the term $RT \ln x_i$. Compare this with eq. 14.16, which contains the term $RT \ln a_i$. In light of this observation, we view activity as the *effective* concentration of a component in a mixture, in the sense that when it is used in eq. (10.12) or (12.15) in place of x_i it returns the chemical potential of the component. In nonideal systems, the activity of a component depends not only on the mol fraction of the component of interest, but also on the composition of the entire mixture. This dependence will become more clear in the following section where we develop expressions for the activity for different standard states.

Activity of Gas

The standard state for gases, denoted by (g), is specified as *the pure gas in the ideal-gas state, at the temperature of the system and at* $P° = 1$ *bar.* Accordingly, the fugacity is calculated using the ideal-gas relation $f_i° = \phi_i° P° = (1) \cdot P° = P°$. The fugacities at the actual and at the standard state are

$$f_i = y_i \phi_i P,$$
$$f_i° = P°,$$

and the corresponding activity is

$$\boxed{a_i = y_i \phi_i \frac{P}{P°}.} \qquad (14.17)$$

In general, the activity of gas components will require the calculation of the fugacity coefficient using the methods discussed in Chapter 10. If the ideal-gas approximation is applicable, then we may set $\phi_i = 1$, and in this case the activity reduces to $a_i = y_i P/P°$.

TIP

Working with P°

It may be tempting to set $P^\circ = 1$ in the above equation and write $a_i = y_i \phi_i P$. Don't do it! This substitution produces an equation that is marginally simpler, but dimensionally incorrect, as the left-hand side is dimensionless whereas the right-hand side has units of pressure. Such equation is valid *only* if special units are used (bar, in this case) and this can lead to confusion and errors in the calculation. To avoid these problems, retain P° in your formulas until final numerical substitution. This recommendation applies to all formulas that involve P° or other constants of the reference state.

Activity of Liquid

To obtain an equation for the activity based on the liquid standard state, we use eq. (12.16) for the fugacity of component in solution,

$$f_i = \gamma_i x_i f_{i,\text{pure}}, \qquad [12.16]$$

to write

$$a_i = \frac{f_i}{f_i^\circ} = \frac{\gamma_i x_i f_{i,\text{pure}}}{f_i^\circ}, \qquad (14.18)$$

where $f_{i,\text{pure}} = f_{i,\text{pure}}(P, T)$ is the fugacity of the pure species i and $f_i^\circ = f_i(P^\circ, T)$ is the fugacity of the pure liquid at the standard state. The actual and the standard state are both at the same temperature but different pressures. Accordingly, the ratio of the corresponding fugacities is given by the Poynting factor,

$$\frac{f_{i,\text{pure}}}{f_i^\circ} = \exp\left(\frac{P - P^\circ}{RT} V_i\right), \qquad [7.16]$$

where V_i is the liquid molar volume of pure component i. Combining these results the activity takes the form

$$\boxed{a_i = \gamma_i x_i \exp\left(\frac{P - P^\circ}{RT} V_i\right).} \qquad (14.19)$$

This general equation can be put into simpler form in some special cases: *Low pressure*: At pressures not much higher than P°, the Poynting correction is negligible and the activity simplifies to

$$a_i = \gamma_i x_i. \qquad (14.20)$$

Low pressure, ideal solution. With the additional assumption of ideal solution ($\gamma_i = 1$), the previous result gives

$$a_i = x_i. \qquad (14.21)$$

14.4 Activity

Low pressure, pure liquid. For a pure liquid, $\gamma_i = 1$ and $x_i = 1$, and the activity becomes

$$a_i = 1. \tag{14.22}$$

Activity of Solute

The fugacity of solute with molality c_i is given by eq. (13.19),

$$f_i = c_i k^{H'}, \qquad [13.19]$$

where k'^H is Henry's law constant in terms of molality. In this expression we assume that the concentration of solute is low enough that the activity coefficient is unity. The standard state is defined as a solution of molality $c° = 1$ mol/kg "that obeys Henry's law." This specification means that the fugacity in the standard state is

$$f_i° = c° k^{H'}.$$

Combining these results we obtain the activity as

$$\boxed{a_i = \frac{c_i}{c°},} \tag{14.23}$$

where c_i is the molality in the solution and $c° = 1$ mol/kg is the standard molality. In other words, the activity is numerically equal to the concentration of the solute expressed in mol per kg of solvent. It is recommended to retain the term $c°$ in the formulas until final substitution in order to avoid potential confusion with units.

Activity of Solid

Solids generally do not mix intimately with other substances and even when they are part of a multicomponent system, they remain essentially pure. This is certainly true in the case of mechanical mixtures of bulk or powdered solids, for example, sugar and sodium chloride mixed together. There are instances in which solid components intermix to form a solid solution, as in alloys. In most cases of interest, however, solid components may be treated as fully immiscible, and therefore pure. Under these conditions the fugacity of solid in a mixture is equal to the fugacity of the pure solid at the same temperature and pressure:

$$f_i(x_i, T, P) = f_{i\text{pure}}(T, P).$$

The fugacity at the standard state is

$$f_i° = f_{i\text{pure}}(T, P°).$$

The two fugacities are related through the Poynting factor,[6]

$$f_i = f_i^\circ \exp\left(\frac{P - P^\circ}{RT} V_i\right), \qquad [7.15]$$

where V_i is the molar volume of the solid. Using the above results, the activity of the solid is

$$\boxed{a_i = \exp\left(\frac{P - P^\circ}{RT} V_i\right).} \qquad (14.24)$$

Unless pressure is substantially higher than 1 bar, the Poynting factor can be neglected and the activity simplifies to

$$a_i \approx 1. \qquad (14.25)$$

This result suffices for most practical situations.

NOTE

Activity and Standard State

The standard state should be viewed as a set of precise and unambiguous instructions on how to calculate the absolute properties of a species. For the activity, these instructions are summarized below:

Standard state	Activity	
(g)	$a_i = y_i \phi_i \dfrac{P}{P^\circ},$	[14.17]
(l)	$a_i = \gamma_i x_i \exp\left(\dfrac{P - P^\circ}{RT} V_i\right),$	[14.19]
(aq)	$a_i = \dfrac{c_i}{c^\circ},$	[14.23]
(s)	$a_i = \exp\left(\dfrac{P - P^\circ}{RT} V_i\right).$	[14.24]

When these instructions are put into words, they often involve states that are not real but hypothetical. This is of no consequence because the sole purpose of the standard state is to specify the mathematical formulas for the calculation of absolute properties. For example, eq. (14.17) reads: "The activity of a component in a gas mixture is equal to its fugacity in the mixture, divided by the standard pressure." This is more conventionally expressed by saying that the denominator is "the activity of pure component in the ideal-gas state at pressure P°". This wordier expression, which makes reference to the standard state, has the advantage that it provides instructions for the calculation of any property, not just activity.

---

6. Solids are incompressible; therefore, the Poynting factor can be used to relate the fugacity at different pressures on the same isotherm.

14.4 Activity

Example 14.6: Activity in the Gas Phase
Calculate the activity of CO_2 and normal pentane at 1.2 bar, $y_{CO_2} = 0.8$.

Solution Assuming the mixture to be in the ideal-gas state, the activity is calculated using (14.17) with $\phi_i = 1$:

$$a_{CO_2} = \frac{(0.8)(1.0)(1.2 \text{ bar})}{1.0 \text{ bar}} = 0.96,$$

$$a_{nC_5} = \frac{(0.2)(1.0)(1.2 \text{ bar})}{1.0 \text{ bar}} = 0.24.$$

Comments To determine whether the ideal-gas assumption is appropriate we use the SRK with $k_{12} = 0.12$, to calculate the fugacity coefficients. Performing the calculation as in Example 10.5 we find

$$\phi_{CO_2} = 0.994 \quad \phi_{nC_5} = 0.962.$$

The ideal-gas assumption is excellent for carbon dioxide and accurate to within $\sim 3\%$ for normal pentane.

Example 14.7: Activity of Liquid
Calculate the activity of ethanol in solution with water at 50 °C, $P = 1$ bar, $x_{\text{ethanol}} = 0.25$.

Solution Water/ethanol solutions are nonideal and the activity coefficient must be calculated. Using the UNIFAC method we find

$$\gamma_{\text{ethanol}} = 1.865.$$

Neglecting the Poynting factor, the activity is

$$a_{\text{ethanol}} = (1.865)(0.25) = 0.466.$$

Comments As the actual pressure is equal to the standard pressure, the Poynting factor is exactly equal to 1.

Example 14.8: Activity of Solute
Calculate the activity of oxygen in water at 25 °C, 5 bar, $x_1 = 3 \times 10^{-5}$, using the standard state for aqueous solutes.

Solution The concentration of oxygen expressed as molality is

$$c_g = \frac{x_g}{x_w \mathrm{MW}_w} = \frac{3 \times 10^{-5}}{(1)(18 \times 10^{-3})} = 1.67 \times 10^{-3} \text{ mol/kg}$$

Applying eq. (14.23), the activity of dissolved oxygen is

$$a_g = 1.67 \times 10^{-3}.$$

Comments Let us calculate the solubility of oxygen in water at 25 °C, at the standard pressure $P° = 1$ bar. This is given by eq. (13.19), which we solve for the concentration of the solute:

$$c_g = \frac{P°}{k_g^{H'}}.$$

From the NIST WebBook at 25 °C, $k_g^{'H} = 769.2$ bar kg/mol. With this value we find

$$c_g = \frac{1 \text{ bar}}{769.2 \text{ bar kg/mol}} = 0.0013 \text{ mol/kg}.$$

The standard molality (1 mol/kg) exceeds the solubility of oxygen in water at the temperature of this problem. The standard state in this case is a *hypothetical* state.

Example 14.9: Activity of Solid
Calculate the activity of sand at a depth of 200 m below the surface of the ocean where the temperature is 5 °C. Make appropriate assumptions.

Solution We assume the sand to be silica (SiO$_2$, $\rho = 2.65$ g/cm^3, $M_m = 60.08 \times 10^{-3}$ mol/kg) and the density of seawater to be 1 g/cm^3 (in reality a bit higher). The pressure at depth $h = 200$ m is

$$P = P_0 + \rho g h = 1 \text{ bar} + (1000 \text{ kg/m}^3)(9.81 \text{ m/s}^2)(200 \text{ m})(10^{-5} \text{ bar/Pa}) = 20.6 \text{ bar}.$$

The activity at $P = 20.6$ bar, $T = 278.15$ K, is calculated from eq. (14.24). Using

$$V_S = \frac{M_m}{\rho_S} = \frac{60.08 \times 10^{-3} \text{ mol/kg}}{2650 \text{ kg/m}^3} = 2.267 \times 10^{-5} \text{ m}^3/\text{mol},$$

we find

$$a_{\text{sand}} = \exp\left[\frac{P - P°}{RT} V_i\right]$$

$$= \exp\left[\frac{(20.6 - 1) \text{ bar}}{(8.314 \text{ J/mol K})(278.15 \text{ K})} \left(2.267 \times 10^{-5} \frac{\text{m}^3}{\text{mol}}\right)\left(\frac{10^5 \text{ Pa}}{\text{bar}}\right)\right]$$

$$= e^{0.01923} = 1.02.$$

Comments The Poynting factor is very close to unity and we could have assumed $a_{\text{sand}} \approx 1$.

14.4 Activity

Example 14.10: Relationship between Standard States

Tables often include formation properties of a species in more than one standard state. Develop a relationship between the Gibbs free energy of formation in the liquid and in the gas standard state.

Solution The problem really asks for the calculation of the difference in the free energy of formation between two standard states. The gas standard state is in the hypothetical ideal-gas state at $T°$, $P°$, and the liquid is in the pure liquid state at the same temperature and pressure. We construct the following path: from the hypothetical ideal-gas state at $P°$, $T°$ (g), to the saturated vapor at $T°$, to the saturated liquid at $T°$, to the pure liquid at $T°$, $P°$. To obtain $G°(l)$ we add corresponding changes to $G°(g)$:

$$G°(l) = G°(g) + RT° \ln \frac{\phi° P°}{P°} + RT° \ln \frac{\phi^{sat} P^{sat}}{\phi° P°} + (P° - P^{sat}) V_L.$$

The last term on the right-hand side is the Poynting correction from saturated liquid at P^{sat}, $T°$, to compressed liquid at $P°$, $T°$. Before that is the change from gas at $P°$, $T°$, to saturated vapor, and before that is the change from hypothetical ideal-gas state at $P°$, $T°$ to real state at the same conditions. Accordingly, $\phi°$ is the fugacity coefficient at $P°$, $T°$, ϕ^{sat} is the fugacity coefficient of saturated vapor, and V_L is the molar volume of the liquid. After cancellations the result is

$$G°(l) = G°(g) + RT° \ln \frac{\phi^{sat} P^{sat}}{P°} + (P° - P^{sat}) V_L.$$

If the saturation pressure is near standard pressure, the fugacity coefficient and the Poynting factor can be dropped to obtain the simpler result,

$$\boxed{G°(l) \approx G°(g) + RT° \ln \frac{P^{sat}}{P°}.} \qquad (14.26)$$

This equation relates the free energy of formation in the liquid and gas standard states, provided that the low-pressure approximations apply.

Numerical example For water, the saturation pressure at $T° = 298.15$ K is $P^{sat} = 0.03169$ bar. Using eq. (14.26), the difference between the free energies of formation in the liquid and gas standard states is

$$G°(l) - G°(g) = RT° \ln \frac{P^{sat}}{P°} = (8.314 \text{ J/mol K})(298.15 \text{ K}) \ln \frac{0.03169 \text{ bar}}{1 \text{ bar}} = -8555.73 \text{ J/mol}.$$

This should be compared with the difference that is obtained from the tabulated values, which are

$$G°(g) = -228,572 \text{ J/mol}, \quad G°(l) = -237,129 \text{ J/mol}.$$

From these we calculate,

$$G°(l) - G°(g) = -237,129 - (-228,572) = -8557 \text{ J/mol}.$$

This result is in excellent agreement with that obtained from eq. (14.26).

14.5 Equilibrium Constant

In principle all reactions are reversible: just as reactants have a tendency to combine and form products, products have the tendency to recombine and form the initial reactants. At equilibrium, the forward rate is balanced by the reverse rate, all net conversion ceases, and the composition of the system becomes constant in time. Suppose we load a closed reactor with a mixture that contains arbitrary amounts of the reactant and product species, and initiate the reaction while maintaining constant temperature and pressure. If we monitor the progress to equilibrium through the extent of reaction, we will observe it to increase in the positive or negative direction, indicating that the reaction progresses in the forward or reverse direction, until equilibrium is reached. Since temperature and pressure are held constant, the equilibrium state corresponds to conditions that minimize the Gibbs free energy. This condition allows us to obtain precise mathematical relationships for the equilibrium constant of the reaction. As an example, consider the ammonia synthesis reaction,

$$\frac{3}{2}H_2 + \frac{1}{2}N_2 = NH_3,$$

and suppose that it has reached equilibrium. We perturb the reaction by causing it to proceed by a small $\delta\xi$ to right. This results to the consumption of $3\,\delta\xi/2$ moles of hydrogen, $\delta\xi/2$ moles of nitrogen, and the production of $\delta\xi$ moles of ammonia. The corresponding change in the Gibbs energy of the mixture is

$$\Delta G = -\frac{3\,\delta\xi}{2}\mu_{H_2} - \frac{\delta\xi}{2}\mu_{N_2} + \delta\xi\,\mu_{NH_3} = \left(-\frac{3}{2}\mu_{H_2} - \frac{1}{2}\mu_{N_2} + \mu_{NH_3}\right)\delta\xi.$$

Before the perturbation the reaction was at equilibrium, that is, the state was located at the bottom of the Gibbs free energy, as shown in Figure 14-3. The small perturbation moves the system along the tangent line at the minimum, leaving the Gibbs free energy unchanged, meaning $\Delta G = 0$, or

$$-\frac{3}{2}\mu_{H_2} - \frac{1}{2}\mu_{N_2} + \mu_{NH_3} = 0.$$

For a reaction with general stoichiometric coefficients ν_i the corresponding result is

$$\sum_i \nu_i \mu_i = 0.$$

The chemical potentials in this expression are given by eq. (14.16). With this substitution, the result is

$$\underbrace{\sum_i \nu_i \mu_i^\circ}_{\Delta G^\circ} + RT \underbrace{\sum_i \ln a_i^{\nu_i}}_{\ln\left(a_1^{\nu_1} a_2^{\nu_2}\cdots\right)} = 0, \qquad (14.27)$$

14.5 Equilibrium Constant

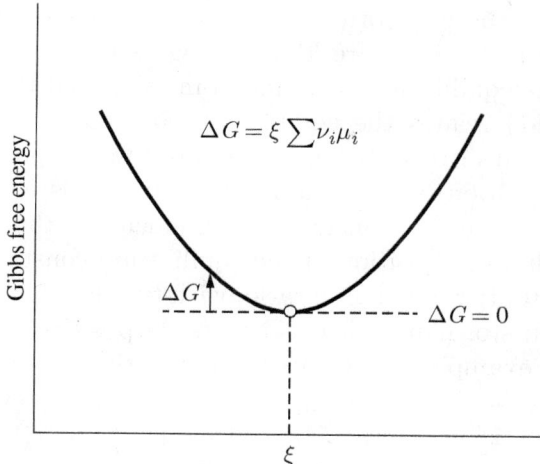

Figure 14-3: Determination of reaction equilibrium.

where μ_i° is the chemical potential of species i in its standard state and a_i is its activity based on that standard state. The first term on the right-hand side is the standard Gibbs energy of the reaction at temperature T:

$$\sum_i \nu_i \mu_i^\circ = \Delta G^\circ. \tag{14.28}$$

By the properties of logarithms, the summation of the logarithms in eq. (14.27) is the log of the product of the activities, each raised to the corresponding stoichiometric coefficient:

$$\sum_i \ln a_i^{\nu_i} = \ln \left(a_1^{\nu_1} a_2^{\nu_2} \cdots \right) = \ln \left(\prod_i a_i^{\nu_i} \right).$$

Using these results we rewrite eq. (14.27) in the form

$$-\frac{\Delta G^\circ}{RT} = \ln \left(\prod_i a_i^{\nu_i} \right). \tag{14.29}$$

We now define the equilibrium constant of chemical reaction as

$$\boxed{K = \exp\left(-\frac{\Delta G^\circ}{RT}\right).} \tag{14.30}$$

Combining this definition with eq. (14.29) we obtain a second expression for the equilibrium constant:

$$\boxed{K = \prod_i a_i^{\nu_i} = a_1^{\nu_1} a_2^{\nu_2} \cdots.} \tag{14.31}$$

Equation (14.30) gives the equilibrium constant in terms of the standard Gibbs energy of the reaction at temperature T. As we will see, this equation allows us to obtain the value of the equilibrium constant from tabulated thermodynamic properties. Equation (14.31) relates the equilibrium constant to the activities of the participating species and ultimately to the equilibrium composition, since activity depends on the composition of the reacting mixture. This equation allows us to obtain the equilibrium from experimental measurements of the equilibrium composition, or to predict that composition, if the equilibrium constant is known.

The product of activities is in fact a ratio, with the reaction products appearing in the numerator (their stoichiometric coefficients are positive), and the reactants in the denominator. For example, for the ammonia reaction this term takes the form

$$\prod_i a_i^{\nu_i} = a_{H_2}^{-3/2} a_{N_2}^{-1/2} a_{NH_3}^{1} = \frac{a_{NH_3}}{a_{H_2}^{3/2} a_{N_2}^{1/2}}.$$

Since activity is related to concentration, eq. (14.31) can be viewed as a relationship between composition at equilibrium, and standard Gibbs free energy of reaction.

TIP

Equilibrium Constant and Stoichiometry
According to eq. (14.31), the numerical value of the equilibrium constant depends on the stoichiometric coefficients. A simple relationship exists between the values of the equilibrium constant obtained with different stoichiometric coefficients. If the stoichiometric coefficients ν_i are multiplied by a factor λ, then the new equilibrium constant is

$$K' = K^\lambda, \qquad (14.32)$$

where K is the equilibrium constant with stoichiometric coefficients ν_i and K' is the constant for stoichiometric coefficients $\nu'_i = \lambda \nu_i$. This follows easily from eq. (14.31).

Example 14.11: Calculation of Equilibrium Constant at 25 °C
Use tabulated values of the Gibbs energy of formation to calculate the equilibrium constant of the reaction

$$2NO(g) + O_2(g) = 2NO_2(g),$$

at 25 °C.

Solution We set up the stoichiometric table as follows:

	NO	O_2	NO_2	
ν_i	-2	-1	$+2$	
$\Delta H^\circ_{f,i}$	90,250	0	31,180	J/mol
$\Delta G^\circ_{f,i}$	86,550	0	51,310	J/mol

14.5 Equilibrium Constant

The standard Gibbs energy of reaction is

$$\Delta G^\circ = (-2)(86,550) + (-1)(0) + (2)(51,310) = -70,480 \text{ J}.$$

The equilibrium constant is

$$K = \exp\left[-\frac{-70,480}{(8.314)(298.15)}\right] = 2.23 \times 10^{12}.$$

Comments The numerical value of the equilibrium constant depends on the stoichiometry. If the above reaction is written as

$$\text{NO(g)} + \frac{1}{2}\text{O}_2(\text{g}) = \text{NO}_2(\text{g}),$$

the corresponding standard Gibbs energy is half of the previous value,

$$\Delta G^\circ = \frac{-70,480 \text{ J}}{2} = -35,240 \text{ J},$$

and the equilibrium constant is equal to the square root of the previous value:

$$K = \left(2.23 \times 10^{12}\right)^{1/2} = 1.49 \times 10^6.$$

The stoichiometry must be indicated when the equilibrium constant is reported.

Equilibrium Constant and Temperature

The equilibrium constant at 25 °C is calculated directly from tabulations of the Gibbs free energy of formation. Once this value is known, the equilibrium constant can be calculated at any other temperature. To obtain the equation that governs the variation of the equilibrium constant with temperature, the starting point is eq. (10.5), which provides the relationship between the Gibbs free energy, temperature, pressure, and composition:

$$d\left(\frac{G^{\text{tot}}}{RT}\right) = -\frac{H^{\text{tot}}}{RT^2}dT + \frac{V^{\text{tot}}}{RT}dP + \sum_i \frac{\mu_i}{RT}dn_i. \qquad [10.5]$$

To apply this equation to ΔG°, we replace enthalpy on the right-hand side with ΔH°. Noting that the standard Gibbs free energy is a function of temperature only, the right-hand side reduces to the temperature term alone:

$$d\left(\frac{\Delta G^\circ}{RT}\right) = -\frac{\Delta H^\circ}{RT^2}dT.$$

Using $\Delta G°/RT = -\ln K$, we obtain

$$\frac{d\ln K}{dT} = \frac{\Delta H°}{RT^2}. \qquad (14.33)$$

This is known as the *van't Hoff* equation and gives the variation of the equilibrium constant with temperature. If K is known at a temperature T_0, it is obtained at any other temperature by integration:

$$\ln \frac{K(T)}{K(T_0)} = \int_{T_0}^{T} \frac{\Delta H°}{RT^2} dT. \qquad (14.34)$$

The integration is simplified if the standard enthalpy is assumed constant:

$$\int_{T_0}^{T} \frac{\Delta H°}{RT^2} dT \approx -\frac{\Delta H°}{R}\left(\frac{1}{T} - \frac{1}{T_0}\right), \qquad (14.35)$$

and the equilibrium constant at temperature T from eq. (14.34) becomes

$$K(T) \approx K(T_0) \exp\left[-\frac{\Delta H°}{R}\left(\frac{1}{T} - \frac{1}{T_0}\right)\right]. \qquad (14.36)$$

Since small errors in $\Delta H°$ are amplified inside the exponential term, the proper procedure is to use eq. (14.8) to express $\Delta H°$ in terms of T, then integrate the resulting expression in terms of temperature.

Example 14.12: Equilibrium at Other Temperatures
Calculate the equilibrium constant of the reaction

$$2\mathrm{NO}(g) + \mathrm{O}_2(g) = 2\mathrm{NO}_2(g),$$

at 1200 °C assuming the enthalpy of reaction to be constant.

Solution The equilibrium constant for this reaction at 25 °C was calculated in Example 14.11 and was found to be
$$K_{298} = 2.23 \times 10^{12}.$$

The standard enthalpy of the reaction at 25 °C is calculated from the tabulated standard enthalpies of formation:
$$\Delta H°_{298} = -118,140 \text{ J}.$$

14.5 Equilibrium Constant

Using eq. (14.36), the equilibrium constant at 1200 °C (1473.15 K) is

$$\ln \frac{K(T)}{K_{298}} = -\frac{(-118,140)}{8.314}\left(\frac{1}{1473.15} - \frac{1}{298.15}\right) = -36.7 \Rightarrow$$

$$K(1473\text{ K}) = K_{298} \times e^{-36.7} = 2.50 \times 10^{-4}.$$

Comments Notice that the equilibrium constant decreases (and substantially so) with temperature. Indeed, from eq. (14.33) we expect that the equilibrium constant must decrease with temperature if the reaction is exothermic ($\Delta H < 0$).

Example 14.13: Heat of Reaction as a Function of Temperature
In the previous example, we assumed that $\Delta H°$ is constant in the range 25 °C to 1200 °C. Is this a good approximation?

Solution To answer this question we will calculate the equilibrium constant by taking into account the variation of the standard heat of reaction. First, we will do this by an approximation, then by the most accurate calculation.

Approximate calculation. As an approximation, we will calculate the standard heat of reaction at 1200 °C and we will apply eq. (14.36) using an average value of $\Delta H°$ between 25 °C and 1200 °C. We collect the heat capacities of the reactants, which are given by the polynomial form,

$$\frac{C_P^{ig}}{R} = c_0 + c_1 T + c_2 T^2 + c_3 T^3 + c_4 T^4,$$

with T in kelvin given below:

	c_0	c_1	c_2	c_3	c_4
NO	4.534	−0.007644	0.00002066	-2.156×10^{-8}	8.06×10^{-12}
O_2	3.63	−0.001794	6.58×10^{-6}	-6.01×10^{-9}	1.79×10^{-12}
NO_2	3.2973	0.0033374	3.234×10^{-6}	-5.6799×10^{-9}	2.0855×10^{-12}

The term $\Delta C_P°$ is calculated from the above expressions. Since all heat capacities are given by the same polynomial form, $\Delta C_P°$ is also given as a fourth-order polynomial in T with coefficients c_i' given by

$$c_i' = \sum_j \nu_j c_i^{(j)}$$

where $c_i^{(j)}$ is the corresponding coefficient of species j and ν_j is its stoichiometric coefficient. Applying this equation to the values in the above table we find

$$\Delta C_P° = -50.7437 + 0.197514\, T - 0.000344466\, T^2 + 3.14021 \times 10^{-7} T^3 - 1.14226 \times 10^{-10} T^4.$$

The standard heat of the reaction at 1200 °C (1473.15 K) is calculated from eq. (14.10):

$$\Delta H^\circ_{1473} = \Delta H^\circ_{298} + \int_{298}^{1473.15} \Delta C^\circ_P \, dT = (-7458.62) + (-114,140) = -121,599 \text{ J/mol}$$

We notice that the standard enthalpy of reaction increases in absolute value by about 6%. Its average value between the two temperatures is

$$\Delta H^\circ = \frac{(-114,140) + (-121,599)}{2} = -117,869 \text{ J/mol}.$$

Using this value in eq. (14.36) we find

$$-\frac{\Delta H^\circ}{R}\left(\frac{1}{T} - \frac{1}{T_0}\right) = -\frac{(-117,869)}{(8.314)}\left(\frac{1}{1473.15} - \frac{1}{298.15}\right) = -37.9269,$$

and

$$K_{1473} = K_{298} e^{-37.9269} = 7.53 \times 10^{-5}.$$

Most accurate calculation. For the most accurate calculation, the standard enthalpy of reaction must be expressed in terms of temperature and the resulting expression must be integrated according to eq. (14.34). The starting expression is eq. (14.10). Using the expression obtained above for ΔC°_P, the standard enthalpy of reaction as a function of T is

$$\Delta H^\circ_T = \Delta H^\circ_{298} + \int_{298}^{T} \Delta C^\circ_P \, dT =$$
$$- 105313 - 50.7437\,T + 0.098757\,T^2 - 0.000114822\,T^3 + 7.85054 \times 10^{-8}\,T^4$$
$$- 2.28452 \times 10^{-11}\,T^5.$$

The expression to be integrated in eq. (14.34) is

$$\frac{\Delta H^\circ}{RT^2} = 0.0118784 - 12666.9/T^2 - 6.1034/T - 0.0000138107\,T + 9.44255 \times 10^{-9}\,T^2$$
$$- 2.7478 \times 10^{-12}\,T^3.$$

Performing the integration we find

$$\int_{298.15}^{1473.15} \frac{\Delta H^\circ}{RT^2} dT = -37.3027.$$

The equilibrium constant is finally obtained from eq. (14.34) as

$$K_{1473} = K_{298}\, e^{-37.3027} = 1.406 \times 10^{-4}.$$

Comments The three methods give quite different results. Taking the last method to be the most accurate, the first method gives an error of 78% and the second method −46%. To understand what is going on, we summarize the results in the table below:

Method	K_{1473}	$\ln(K_{1473}/K_{298})$
$\Delta H^\circ \approx \Delta H^\circ_{298}$	2.500×10^{-4}	−36.7269
$\Delta H^\circ \approx 0.5(\Delta H^\circ_{298} + \Delta H^\circ_{1473})$	7.530×10^{-5}	−37.9269
$\Delta H^\circ = $ function of T	1.406×10^{-4}	−37.3027

14.5 Equilibrium Constant

Each of the three methods uses a different approach for the calculation of the quantity

$$\ln \frac{K_{1473}}{K_{298}},$$

which is then used to calculate the value of K_{1473}. This quantity is calculated fairly accurately by all three methods, with an error that is less than 2%. Yet, the error in the equilibrium constant is substantially larger. This is a common numerical pitfall that one should be aware of: when e is raised to a logarithm, small errors in the logarithm are amplified exponentially. If the quantity of interest is not the logarithm itself but e raised to that value, one must obtain the most accurate value possible.

Other Forms of the Equilibrium Constant

The fundamental relationship between equilibrium constant and composition is via the product of the activities of species raised to the corresponding stoichiometric coefficients:

$$K = \prod_i a_i^{\nu_i}. \tag{14.37}$$

If reactants and products are in the gas phase, all activities are given by

$$a_i = \frac{\phi_i y_i P}{P^\circ} = \frac{\phi_i P_i}{P^\circ},$$

where y_i is the mole fraction and $P_i = y_i P$ is the partial pressure of component i. The equilibrium constant in eq. (14.37) may then be expressed as

$$K = K_\phi K_y \left(\frac{P}{P^\circ}\right)^\nu = K_\phi K_P (P^\circ)^{-\nu},$$

where

$$K_y = \prod_i y_i^{\nu_i}, \quad K_P = \prod_i P_i^{\nu_i}, \quad K_\phi = \prod_i \phi_i^{\nu_i},$$

and ν is the sum of stoichiometric coefficients. Similarly, if all species are aqueous solutes, the equilibrium constant is

$$K = K_c (c^\circ)^{-\nu}$$

with

$$K_c = \prod_i c_i^{\nu_i}.$$

The expressions, K_y, K_P, K_c, all have the mathematical form of the equilibrium constant in eq. (14.37), with activity replaced by y_i, P_i, or c_i and are often called "equilibrium constants." As we see here, they are indeed related to the equilibrium constant. However, the term *equilibrium constant* should be strictly reserved for the expression that gives the equilibrium constant in terms of activities. This constant is dimensionless and a function of temperature only. The derivative forms, K_y, K_P, may have units, and are functions of pressure, if ν is not zero. These expressions may be used in calculations but should not be confused with the thermodynamic quantity that we call equilibrium constant.

14.6 Composition at Equilibrium

A common problem is to calculate the composition of a reacting mixture at equilibrium at a specified temperature. To do this, it is always easier if we start with the stoichiometric table of the reaction. The first step is to express all the concentrations in terms of the extent of reaction, ξ. We then calculate the activity of each species and finally, we equate the product of activities to the equilibrium constant. This produces an equation where the only unknown is ξ. Once the extent of reaction is known, all the mole fractions can be computed from the stoichiometric table. If the temperature of the calculation is at 25 °C, the equilibrium constant is obtained directly from tabulated values of the standard Gibbs free energy of formation. To calculate the equilibrium constant at another temperature, an additional step is neeed to obtain the heat of reaction and the Gibbs energy at the desired temperature. This procedure is demonstrated with examples below.

Example 14.14: Equilibrium Composition
NO, oxygen and NO_2 react according to the reaction
$$NO(g) + \frac{1}{2}O_2(g) = NO_2(g)$$
at 1200 °C, 3 bar. The reaction mixture initially contains equal amounts of all three species. Determine the equilibrium composition.

Solution In Example 14.13 we calculated the equilibrium constant for this reaction at 1200 °C (1474 K) and found it to be 1.406×10^{-4}, however, the stoichiometry in that example was different than the one used here. Applying eq. (14.32) with $\lambda = 1/2$ we obtain the equilibrium constant for the new stoichiometry:
$$K_{1473} = \left(1.406 \times 10^{-4}\right)^{1/2} = 0.0118575.$$
Next, we construct the stoichiometric table on the basis of 1 mol of NO:

	NO	O_2	NO_2	
ν	-1	$-\frac{1}{2}$	1	
n_{0i}	1	1	1	mol
n_i	$1-\xi$	$1-\frac{\xi}{2}$	$1+\xi$	mol

14.6 Composition at Equilibrium

The total number of moles is

$$n = 1 - \xi + 1 - \frac{\xi}{2} + 1 + \xi = 3 - \frac{\xi}{2},$$

and the corresponding mole fractions are

$$y_{NO} = \frac{1-\xi}{3-\xi/2}, \quad y_{O_2} = \frac{1-\xi/2}{3-\xi/2}, \quad y_{NO_2} = \frac{1+\xi}{3-\xi/2}.$$

Treating all species in the ideal-gas state, the equilibrium constant for the reaction is

$$K_{1473} = \frac{y_{NO_2}}{y_{NO} y_{O_2}^{1/2}} \cdot \left(\frac{P^\circ}{P}\right)^{1/2}.$$

Substituting the expressions for the mole fractions, this equation becomes

$$K_{1473} \left(\frac{P}{P^\circ}\right)^{1/2} = \frac{(1+\xi)(3-\xi/2)^{1/2}}{(1-\xi)(1-\xi/2)^{1/2}}. \tag{14.38}$$

The only unknown here is the extent of reaction. Because of the nonlinear nature of this equation, a trial-and-error method will be needed. Using $K_{1473} = 0.0118575$, $P = 3$ bar, $P^\circ = 1$ bar and solving for ξ we find

$$\xi = -0.9735.$$

By back substitution in the stoichiometric table we obtain the mole fractions at equilibrium:

$$y_{NO} = 0.566, \quad y_{O_2} = 0.426, \quad y_{NO_2} = 0.00759.$$

The negative sign indicates that the reaction proceeds from right to left.

Comments The stoichiometry of the reaction affects the value of the equilibrium constant and of the extent of reaction but not the final compositions. You should be able to confirm this by repeating these calculations using the stoichiometry of Example 14.13.

Example 14.15: Effect of Pressure
Repeat the calculation of Example 14.14 at 0.01 bar.

Solution Following the steps of Example 14.14 we find

$$\xi = -0.9984,$$
$$y_{NO} = 0.571,$$
$$y_{O_2} = 0.428,$$
$$y_{NO_2} = 0.000443.$$

Notice that the reaction is shifted even more to the left and that the concentration of NO_2 decreases by about 95% compared to its value in the previous example.

Comments This example demonstrates the effect of pressure on the composition of the equilibrium mixture. The equilibrium constant itself is independent of pressure but the equilibrium *composition* is affected by pressure. This dependence comes through the term (on the left-hand side in eq. [14.38])

$$\left(\frac{P}{P^\circ}\right)^{-\nu} K,$$

where ν is the sum of the stoichiometric coefficients of the gas-phase species. If ν is negative, increasing pressure enhances the effect of the equilibrium constant and shifts the reaction to the right. If ν is positive, then pressure has the opposite effect and shifts the equilibrium composition towards the left. In the special case that $\nu = 0$, pressure has no effect on the equilibrium composition.

14.7 Reaction and Phase Equilibrium

The equilibrium constant in eq. (14.31) represents a relationship between the fugacities of the participating species. As such, it is an additional constraint that reduces the degrees of freedom of a thermodynamic system. Specifically, the degrees of freedom are reduced by one for each reaction that takes place. We determine the degrees of freedom by counting unknowns and available equations. In a system with N components distributed among π phases, the unknowns are $N-1$ mole fractions per phase[7] plus pressure and temperature. This makes for $(N-1)\pi + 2$ unknowns. The available equations are, $\pi - 1$ phase equilibrium conditions per component, plus one equilibrium condition per reaction. This makes for $N(\pi - 1) + R$ equations. The degrees of freedom is the difference between the number of unknowns and the number equations that are available to solve for them:

$$\mathcal{F} = N + 2 - \pi - R. \tag{14.39}$$

This equation gives the number of variables that we must specify in order to fix the state of the system. In the absence of reactions ($R = 0$) this reverts to the Gibbs's

---

7. Since the sum of the mol fractions add up to 1, we only have $N-1$ unknown fractions in each phase.

14.7 Reaction and Phase Equilibrium

phase rule obtained in Chapter 10. The presence of reactions reduces the degrees of freedom. Suppose that nitrogen and hydrogen react to form ammonia in the gas phase. In this case we have three components (N_2, H_2, and NH_3), one phase and one reaction, which leads to three degrees of freedom. To fix the equilibrium state of this system we must specify three intensive variables, for example, pressure, temperature, and mol fraction of one component; or we could specify the mole fraction of two components and temperature or pressure (but not both). If a system of three reactants forms a two-phase system, then the phase equilibrium conditions decrease the degrees of freedom to 2. In this case, fixing the pressure and temperature are sufficient to fully specify the state. The example below demonstrates the calculation of chemical equilibrium in the presence of two phases, in this case a liquid and a vapor.

Example 14.16: Chemical Equilibrium and VLE
Ethylene can be produced from the dehydration of ethanol according to the reaction,

$$CH_3CH_2OH = C_2H_4 + H_2O.$$

This is an alternative method for producing ethylene, an important chemical feedstock, from bioethanol. The reaction is reversible and can also be used to produce ethanol from ethylene. Determine the equilibrium composition between the three components at 120 °C and 7 bar.

Solution At 7 bar, 120 °C, both ethanol and water are liquid, but ethylene is a gas. We anticipate the formation of two phases, a vapor and a liquid. With $N = 3$, $\pi = 2$, and $R = 1$, the number of degrees of freedom for this problem is

$$\mathcal{F} = 3 + 2 - 2 - 1 = 2.$$

Since pressure and temperature are given, the state of the system is fully specified. First we choose the standard states for the components in the equilibrium constant. We are free to choose them at will and we will take them to be all in the gaseous (g) standard state. The equilibrium condition then becomes

$$\left(\frac{y_2 y_3}{y_1}\right)\left(\frac{\phi_2 \phi_3}{\phi_1}\right)\left(\frac{P}{P^\circ}\right) = K,$$

with the subscripts 1, 2, and 3 referring to ethanol, ethylene, and water, respectively. The vapor-liquid equilibrium conditions for the three components are:

$$y_1 \phi_1 P = \gamma_1 x_1 f_1^{\text{pure}},$$
$$y_2 \phi_2 P = k_2^H x_2,$$
$$y_3 \phi_3 P = \gamma_3 x_3 f_3^{\text{pure}}.$$

Here we have used activity coefficients for the fugacity of ethanol and water in the liquid, and Henry's law for the fugacity of ethylene in the liquid. In addition we have the two normalization conditions,

$$x_1 + x_2 + x_3 = 1,$$
$$y_1 + y_2 + y_3 = 1.$$

These six equations are to be solved for the six unknown mol fractions. To continue we will make a number of simplifying assumptions:

(a) We will take the gas phase to be ideal, which makes all the fugacity coefficients equal to 1.

(b) We will neglect the Poynting correction so that the fugacity of pure ethanol and pure water is simply equal to the saturation pressure:

$$f_i^{\text{pure}} = P_i^{\text{sat}} \qquad (i = 1, 3).$$

(c) We will neglect the solubility of ethylene in the liquid. This replaces the equilibrium condition for ethylene by the condition,

$$x_2 = 0.$$

Notice that this does *not* imply $y_2 = 0$ but rather that $k_2^H = \infty$, so that the product $k_2^H x_2$ has a finite value, equal to the fugacity in the gas phase. With these simplifications the equations to be solved are:

$$\frac{y_2 y_3}{y_1} = K \frac{P^\circ}{P}, \qquad [A]$$

$$y_1 P = \gamma_1 x_1 P_1^{\text{sat}}, \qquad [B]$$

$$y_3 P = \gamma_3 x_3 P_3^{\text{sat}}, \qquad [C]$$

$$x_1 + x_3 = 1, \qquad [D]$$

$$y_1 + y_2 + y_3 = 1. \qquad [E]$$

Equilibrium constant. We collect the formation enthalpies and free energies of the components:

	1 ethanol (g)	2 ethylene (g)	3 water (g)	
ν	-1	1	1	
H_{298}°	-234.95	52.5	-241.81	kJ/mol
G_{298}°	-167.73	68.48	-228.42	kJ/mol

The free energy and the enthalpy of reaction are

$$\Delta G_{298}^\circ = 7790 \text{ J/mol}, \quad \Delta H_{298}^\circ = 45640 \text{ J/mol},$$

14.7 Reaction and Phase Equilibrium

and the equilibrium constant at $T^\circ = 298$ K is

$$K_{298} = 0.04092.$$

To calculate the equilibrium constant at 120 °C (393.15 K) we will use the shortcut equation (14.36), which assumes the standard heat of reaction to be independent of temperature and equal to ΔH°_{298}. We find

$$K = (0.04092) \exp\left[-\frac{7790}{(8.314 \text{ J/mol K})}\left(\frac{1}{393.15 \text{ K}} - \frac{1}{298}\right)\right] = 3.6929.$$

Liquid-phase fugacities. To proceed we need a method to calculate the activity coefficients in ethanol/water solutions. We will use UNIFAC for this purpose. We also need the saturation pressures of ethanol and water at 120 °C. We use the following values calculated form the Antoine equation:

$$P_1^{\text{sat}} = 4.325 \text{ bar}, \quad P_3^{\text{sat}} = 1.917 \text{ bar}.$$

Numerical procedure. The activity coefficients depend on the unknown mole fractions in the liquid, which complicates the direct solution of the above equations. We adopt a trial-and-error approach. For the first pass we assume the activity coefficients to be 1. Equations [A]–[E] are then solved and the following values are obtained:

$$y_1 = 0.2093, y_2 = 0.6097, y_3 = 0.1811, x_1 = 0.3387, x_2 = 0, x_3 = 0.6613.$$

With the mol fractions in the liquid known, we calculate the activity coefficients in the liquid using UNIFAC:

$$x_1 = 0.3387, x_3 = 0.6613 \quad \Rightarrow \quad \gamma_1 = 1.554, \gamma_2 = 1.259.$$

Using the activity coefficients calculated above we solve eqs. [A]–[E] again and use the new liquid compositions to refine the estimate of the activity coefficients. The procedure is repeated until the solution converges and the mol fractions do not change any more. A sample of the iterations are shown in the table below:

iteration	liquid				vapor		
	γ_1	γ_3	x_1	x_3	y_1	y_2	y_3
1	1.0000	1.0000	0.3387	0.6613	0.2093	0.6097	0.1811
2	1.5544	1.2593	0.2535	0.7465	0.2435	0.4991	0.2574
3	1.8787	1.1635	0.2084	0.7916	0.2419	0.5060	0.2522
⋮	⋮	⋮	⋮	⋮	⋮	⋮	⋮
20	2.7281	1.0629	0.1429	0.8571	0.2409	0.5096	0.2494
21	2.7282	1.0629	0.1429	0.8571	0.2409	0.5096	0.2494

Notice that the gas-phase compositions converge quickly but the liquid phase compositions require more iterations to converge.

Comments In this problem we chose the standard states to be in the gas phase. This choice is made arbitrarily and independently of whether the reaction actually takes place in the liquid or in the gas phase. We could have chosen other standard states, for example, liquid for water and/or ethanol and gas or aqueous for ethylene. The first consideration is whether the formation values for the state of choice are available or not (in this problem, for example, liquid and gas state values for ethanol and water can be found in the appendix, but for ethylene the only values listed are for the gas standard state). Once the standard state is fixed, then the activities to be used in the equilibrium constant must match the standard states. In principle, the final answer must be the same regardless of the choice of the standard state, except for possible inaccuracies in the tabulated values.

If the pressure is too high for the gas phase to be assumed ideal, we must then calculate the fugacity coefficients by an appropriate method, for example, using an equation of state. Since fugacity coefficients depend on the composition of the mixture, the calculation must proceed by a trial-and-error method similar to that of this example: starting with a guess for all activity and fugacity coefficients, calculate the mole fractions and use them refine the guesses of ϕ_i and γ_i.

NOTE

Phase Equilibrium as a Chemical Reaction

The equilibrium conditions in the previous example (eqs. [A]–[E] in Example 14.16) express relationships between mole fractions. In the case of chemical equilibrium, this is a relationship between the mol fractions of products and reactants. In the case of phase equilibrium, it is a relationship between the mol fraction of the same component in two different phases. The similarity between phase and chemical equilibrium can be made more clear by treating the transfer of a species between phases as a chemical reaction. For example, the evaporation-condensation of water can be represented as

$$H_2O(g) = H_2O(l),$$

with equilibrium constant

$$K = \frac{a(l)}{a(g)} = \frac{x_w \gamma_w e^{(P-P^\circ)V_w/RT}}{y_w \phi P/P^\circ} \approx \frac{x_w \gamma_w}{y_w} \frac{P^\circ}{P},$$

where we have used eq. (14.19) for the activity of the liquid and eq. (14.17) for the activity of the vapor and applied the low-pressure approximations to obtain the result on the far right. For the equilibrium constant of this process we also have $K = \exp(-\Delta G^\circ/RT)$, with ΔG° given in eq. (14.26), which leads to

$$K = \exp\left[-\frac{G^\circ(l) - G^\circ(g)}{RT}\right] = \frac{P^\circ}{P^{\text{sat}}}.$$

Combining these results we obtain the following relationship between the mol fractions in the two phases:

$$\frac{x_w \gamma_w}{y_w} = \frac{P}{P^{\text{sat}}}.$$

14.8 Reaction Equilibrium Involving Solids

We recognize this result as the standard condition for phase equilibrium (to see this more clearly write it as $y_w P = x_w \gamma_w P^{\text{sat}}$) but we may also view it as the equilibrium constant for the equilibrium between phases. That is, whether we treat the problem as a phase equilibrium process, or a reaction equilibrium process, we obtain the same answer. The "reaction" here involves not a change in the chemical nature of the species but a change of the state of "aggregation," from a vapor into a liquid. This relationship between phase and chemical equilibrium should not come as a surprise. In both cases we are applying the same principle, namely that the equilibrium state at fixed temperature and pressure is the state that minimizes the Gibbs free energy of the entire system. This principle leads to the equality of fugacities in phase equilibrium, and in the introduction of the equilibrium constant in reaction equilibrium.

14.8 Reaction Equilibrium Involving Solids

When a gas-phase or liquid-phase system involves a solid as a reactant or product, we have a multi-phase system but this calculation is simplified enormously by the fact that solid phases can be treated as pure. If the reaction involves a gas or liquid phase, only the moles of those species that are present in the gas or liquid participate in the equation that determines the equilibrium of the system. The fact that the activity of solids does not depend on the amount of the solid also implies that it is possible for equilibrium to fully consume a solid reactant. This is not possible with reactions that do not involve solids unless the equilibrium is shifted very strongly towards the products. The calculation of chemical equilibrium is demonstrated with the example below.

Example 14.17: Solid-Gas Reaction
A gas stream that contains 15% (mol) H_2, 70% CH_4 and 15% N_2 is to be heated but there is concern that this may lead to deposition of carbon according to the reaction

$$CH_4 = C + 2H_2.$$

Determine whether deposition will occur at 600 K, and at 800 K, if the pressure is 2 bar.

Solution We construct the stoichiometric table below:

	$CH_4(g)$	$C(s)$	$H_2(g)$	$N_2(g)$
ν	-1	1	2	0
H°_{298} (kJ/mol)	-74.52	0	0	0
G°_{298} (kJ/mol)	-50.45	0	0	0
n_{i0}	0.7	0	0.15	0.15
n_i	$0.7-\xi$	ξ	$0.15+2\xi$	0.15
y_i	$\dfrac{0.7-\xi}{1+\xi}$	0	$\dfrac{0.15+2\xi}{1+\xi}$	$\dfrac{0.15}{1+\xi}$

The last row gives the mole fractions in the *gas* phase. Since carbon is present only in the solid phase, its moles are not included in the determination of the total moles of gas-phase species:

$$n_{\text{tot}}^{\text{gas}} = 0.7 - \xi + 0.15 + 2\xi + 0.15 = 1 + \xi.$$

This is the number of moles used in the calculation of the gas-phase mol fractions. The equilibrium condition involves only the gas-phase reactants, methane and hydrogen. Neglecting the fugacity coefficients, the equilibrium reads

$$\frac{y_{H_2}^2}{y_{CH_4}} \left(\frac{P}{P^\circ}\right) = K(T).$$

Expressing the mol fractions in terms of ξ, this equation becomes

$$\frac{(0.15 + 2\xi)^2}{(0.7 - \xi)(1 + \xi)} = K(T) \left(\frac{P^\circ}{P}\right),$$

and by expanding we obtain a quadratic equation in ξ:

$$0.0225 - \frac{0.7 P_0}{P} K(T) + \xi \left(0.6 + \frac{0.3 P_0}{P} K(T)\right) + \xi^2 \left(4 + \frac{P_0}{P} K(T)\right) = 0 \qquad [A]$$

This equation is to be solved for the unknown extent of the reaction.

Numerical substitutions. The standard enthalpy and Gibbs free energy of the reaction are $\Delta H_{298}^\circ = 74,520$ J/mol and $\Delta G_{298}^\circ = 50,450$ J/mol, respectively. The equilibrium constant at 298.15 K is

$$K_{298} = 1.44894 \times 10^{-9}.$$

The equilibrium constant at T will be calculated by the simplified expression in eq. (14.36). At 600 K and at 800 K we find,

$$K_{600} = 5.362 \times 10^{-3}, \quad K_{800} = 0.2245.$$

Calculation at 600 K. With $K = 5.362 \times 10^{-3}$ eq. [A] becomes

$$0.0187465 + 0.601609\xi + 4.00536\xi^2 = 0,$$

whose two roots are

$$\xi = -0.10608, \quad \xi = -0.044121.$$

Both roots are negative, which implies the reverse reaction is taking place. However there is no carbon present initially, so the reverse reaction cannot actually occur. We conclude that no carbon will be deposited at 600 Kelvin.

Calculation at 800 K. With $K = 0.2245$ eq. [A] becomes

$$-0.134674 + 0.66736\xi + 4.22453\xi^2 = 0,$$

14.8 Reaction Equilibrium Involving Solids

whose roots are
$$\xi = -0.274224, \quad \xi = 0.116252.$$

The positive root corresponds to forward reaction and we conclude that carbon deposition will occur at 800 K. The resulting gas-phase mol fractions are

$$y_{CH_4} = 0.523, \quad y_C = 0, \quad y_{H_2} = 0.343, \quad y_{N_2} = 0.134.$$

Comments At 600 K there is no deposition of carbon. This corresponds to the reaction going quantitatively from right (products) to left (reactants). Normally, a reversible reaction will shift from left to right until the mol fractions adjust to the values that satisfy the condition for chemical equilibrium. However, when a solid is involved, its mol fraction does not appear in the equilibrium constant, and this makes it possible to reach an equilibriums state that the solid is fully consumed.

Example 14.18: Phase Diagram for Carbon Deposition
Use the data of Example 14.17 to create a pressure-temperature graph and show the region where carbon deposition is expected according to the reaction

$$CH_4 = C + 2H_2.$$

Solution In the previous example we obtained the following equation for the extent off reaction:

$$0.0225 - \frac{0.7 P_0}{P} K(T) + \xi \left(0.6 + \frac{0.3 P_0}{P} K(T) \right) + \xi^2 \left(4 + \frac{P_0}{P} K(T) \right) = 0 \quad [A]$$

Deposition occurs if this equation has a positive root. Inspecting the coefficients of this quadratic polynomial we notice that the sum of the roots is always negative. This means that if the quadratic equation has real roots, at least one is always negative. Carbon deposition requires the presence of a positive root. In order to have a positive root, the constant term of the quadratic must be negative:

$$0.0225 - \frac{0.7 P_0}{P} K(T) \leq 0 \quad \Rightarrow \quad \boxed{P \leq 31.111 P_0 K(T).}$$

The last inequality defines the region of the PT plane where deposition of carbon is thermodynamically feasible. This is plotted in Figure 14-4 with the shaded region marking the range of pressure and temperatures where deposition can occur.

Comments These results are specific to the composition of the methane/hydrogen/nitrogen mixture. If this composition changes, the phase diagram must be recalculated by repeating the calculations in Example 14.17.

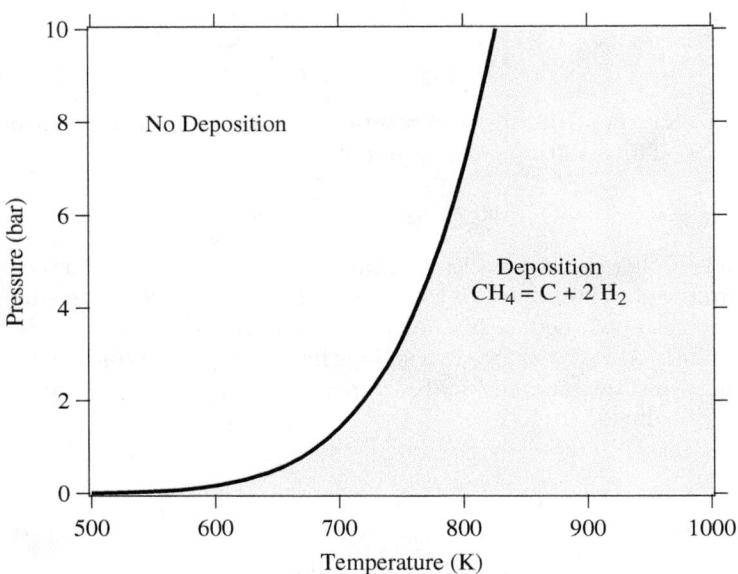

Figure 14-4: Phase diagram for carbon deposition according to the reaction $CH_4 = C + 2H_2$ (see Example 14.18).

14.9 Multiple Reactions

When multiple reactions take place, each reaction provides an equilibrium condition that may be used in the calculation of the final composition. However, it is not always obvious how many independent reactions are present. For example, in the reforming of methane by steam we may write the following four reactions that produce a mixture of carbon monoxide, carbon dioxide and hydrogen:

$$CH_4 + H_2O = CO + 3H_2, \qquad [R1]$$

$$CH_4 + 2H_2O = CO_2 + 4H_2, \qquad [R2]$$

$$CH_4 + CO_2 = 2CO + 2H_2, \qquad [R3]$$

$$CO + H_2O = CO_2 + H_2. \qquad [R4]$$

Notice that reaction R4 can be obtained by subtracting reaction R1 from R2. This means that at least one reaction in this system is not independent. A systematic method to determine the number of independent reactions is to begin with the matrix of stoichiometric coefficients of all components that appear in these reactions.

14.9 Multiple Reactions

We then proceed to combine rows by linear transformations with the goal of eliminating stoichiometric coefficients by turning them into zeros, one coefficient at a time. When no more eliminations are possible, we count the number of rows that contain nonzero coefficients. This gives the number of independent reactions. The process is demonstrated in the example that follows.

Example 14.19: Determining the Number of Independent Reactions
Determine the number of independent reactions in the system of the four reactions R1 through R4, above.

Solution We begin by constructing the matrix of the stoichiometric coefficients:

CH_4	H_2O	CO	CO_2	H_2
-1	-1	1	0	3
-1	-2	0	1	4
-1	0	2	-1	2
0	-1	-1	1	1

Each row in the matrix represents a reaction, and each column the stoichiometric coefficients of the components in the corresponding reaction. Next we replace a row r_i by the linear combination of row i with row j, namely,

$$r'_i = ar_i + br_j,$$

with the coefficients a and b chosen so as to eliminate a stoichiometric coefficient in r'_i. For example, by replacing the first row with the difference between the first and second row we turn the first element of the first row into zero:

CH_4	H_2O	CO	CO_2	H_2
0	1	1	-1	-1
-1	-2	0	1	4
-1	0	2	-1	2
0	-1	-1	1	1

Replace the second row (r_2) with $r_2 - r_3$:

CH_4	H_2O	CO	CO_2	H_2
0	1	1	-1	-1
0	-2	-2	2	2
-1	0	2	-1	2
0	-1	-1	1	1

Replace r_1 with $2r_1 + r_2$:

CH$_4$	H$_2$O	CO	CO$_2$	H$_2$
0	0	0	0	0
0	1	1	−1	−1
−1	0	2	−1	2
0	−1	−1	1	1

Replace r_2 with $r_2 + r_3$:

CH$_4$	H$_2$O	CO	CO$_2$	H$_2$
0	0	0	0	0
0	0	0	0	0
0	1	1	−1	−1
−1	0	2	−1	2

No additional eliminations can be done past this point. We are left with two rows of nonzero coefficients; therefore, the number of independent reactions is 2.

Comments The elimination process can be done in many different ways and the surviving rows may contain different coefficients from ones obtained above, but the *number* of nonzero rows will be the same. This method comes from linear algebra and amounts to the determination of the *rank* of the matrix of the stoichiometric coefficients.

The reactions that correspond to the last table of stoichiometric coefficients are

$$CO_2 + H_2 = CO + H_2O \qquad [R5]$$

$$CH_4 + CO_2 = 2CO + 2H_2 \qquad [R5]$$

However, *any two independent* reactions between the five components may be used to calculate the equilibrium composition.

Example 14.20: Equilibrium Composition with Multiple Reactions
Methane is reformed by reaction with steam according to reactions R1 – R4 on page 632. Determine the equilibrium composition at 600 K, 2 bar, of a reaction mixture whose feed consists of a methane-water mixture at molar ratio 1:2.

Solution We determined that only two of the four reactions are independent. We may choose any two independent reactions to describe the system, namely, any two of the four in R1–R4, or any linear combination of these, such as reactions R5 and R6 in Example 14.19. We choose reactions R1 and R2 and build the stoichiometric table shown below using the gas standard state for all components.

14.9 Multiple Reactions

	CH_4	H_2O	CO	CO_2	H_2
H°_{298} (kJ/mol)	-74.52	-241.81	-110.53	-393.51	0
G°_{298} (kJ/mol)	-50.45	-228.42	-137.16	-394.38	0
ν_1	-1	-1	1	0	3
ν_2	-1	-2	0	1	4
n_{i0}	1	2	0	0	0
n_i	$1-\xi_1-\xi_2$	$2-\xi_1-2\xi_2$	ξ_1	ξ_2	$3\xi_1+4\xi_2$

The mol fraction of component i in the equilibrium mixture is

$$y_i = \frac{n_i}{n_{\text{tot}}} = \frac{n_i}{3+2\xi_1+2\xi_2},$$

with n_i for the above table. Neglecting the fugacity coefficients, the two equilibrium conditions are

$$\frac{\xi_1(3\xi_1+4\xi_2)^3}{(2-\xi_1-2\xi_2)(1-\xi_1-\xi_2)(3+2\xi_1+2\xi_2)^2}\left(\frac{P}{P^\circ}\right)^2 = K_1,$$

$$\frac{\xi_2(3\xi_1+4\xi_2)^4}{(2-\xi_1-2\xi_2)^2(1-\xi_1-\xi_2)(3+2\xi_1+2\xi_2)^2}\left(\frac{P}{P^\circ}\right)^2 = K_2.$$

These are to be solved for the two unknown extents of reaction ξ_1 and ξ_2.

Numerical substitutions. We calculate the equilibrium constant of the two reactions using the shortcut equation (14.36). The results of these calculations are summarized below:

Reaction #	ΔH°_{298} (J/mol)	ΔG°_{298} (J/mol)	K_{298}	K_{600}
R1	205,800	141,710	0.0026906	0.20306
R2	164,630	112,910	0.0089575	0.28466

With $K_1 = 0.20306$, $K_2 = 0.28466$, and $P = 2$ bar, the two equilibrium equations are solved to give $\xi_1 = 0.182$ and $\xi_2 = 0.233$. The composition of the equilibrium mixture is summarized in the table below:

	CH_4	H_2O	CO	CO_2	H_2
n_i	0.5848	1.3516	0.1820	0.2332	1.4789
y_i	0.1527	0.3529	0.0475	0.0609	0.3861

Comments It is left as an exercise to show that the same result is obtained if a different set of reactions is used to compute equilibrium.

14.10 Summary

As with all equilibrium, the final state in a reacting mixture is such that the Gibbs free energy is at a minimum. In this respect, chemical equilibrium requires no new theories beyond those developed in Chapter 10. However, the fact that the chemical nature of species changes during the reaction requires the introduction of consistent reference states. This is done by the adoption of the standard states for gases (g), liquids (l), solids (s) and aqueous solutes (aq). The specification of the standard state must be understood as a set of rules for the unambiguous calculation of three main properties of a reacting system: the activity of components, the enthalpy of reaction, and the Gibbs free energy of the reaction. Although activity was introduced in the context of reactions, its usefulness is general and can be traced to eq. (14.16)

$$\mu_i = \mu_i^\circ + RT\ln a_i, \qquad [14.16]$$

which shows that activity allows us to obtain the chemical potential of a component in any mixture, reacting or not. In this textbook we did not introduce activity until the discussion of reactions, mainly because we have chosen fugacity, rather than chemical potential, as the primary property in phase equilibrium.

Most problems in chemical equilibrium are solved using the two fundamental equations for the equilibrium constant:

$$K = \exp\left(-\frac{\Delta G^\circ}{RT}\right), \qquad [14.30]$$

and

$$K = \prod_i a_i^{\nu_i}. \qquad [14.31]$$

The first equation expresses the equilibrium constant in terms of quantities that can be obtained from tabulated values (formation enthalpy and Gibbs free energy, heat capacities). The second equation expresses the equilibrium constant in terms of the activity of species, and ultimately, in terms of mol fractions. We must remember that activities and formation properties of each species must refer to the same standard state.

If multiple equilibria take place, all equilibrium conditions must be satisfied. This is true whether we are dealing with multiple reactions, or with simultaneous reaction *and* phase equilibrium. The most important consideration in these types of problems is the determination of the number of independent equilibrium conditions that can be written. Specifically, we must write one equilibrium condition for each independent chemical reaction in addition to any phase equilibria that may be present.

14.11 Problems

Some attention must be paid if one of the reactants is in the solid phase. Equilibrium calculations are simplified in this case because the activity of the solid does not appear in the equilibrium constant (recall from eq. [14.24] that activity of the solid is practically 1 unless pressure is high). This creates the interesting situation that a reversible reaction involving a solid may proceed to completion. This is very similar to the familiar dissolution of solids in water and other solvents. If an amount of a soluble solid (e.g., salt or sugar) is mixed with a liquid, some of the solid dissolves until the solubility limit is reached, at which point we have equilibrium between the undissolved and dissolved fractions of the solid. If, however, the amount of solid is below the solubility limit, the entire amount dissolves. This is an example of a reversible reaction that goes to completion. Such process is possible only if it involves a solid, because the concentration of the solid does not appear in the equilibrium constant. With components in any other phase, the mol fraction of a reactant appears in the denominator of the equilibrium constant, and this prevents a reversible reaction from reaching completion unless the equilibrium constant is a very large number.

14.11 Problems

Problem 14.1: Hydrogen peroxide reacts with sodium thiosulfate according to the reaction,

$$Na_2S_2O_3 + \alpha H_2O_2 \to \beta Na_2S_3O_6 + \gamma Na_2SO_4 + \delta H_2O.$$

a) Determine the stoichiometric coefficients.
b) A reactor is initially loaded with an aqueous solution that contains 10% hydrogen peroxide (by mole), 10% sodium thiosulfate, and the rest is water. Assuming the reaction goes to completion, what is the composition (mole fractions) of the final solution? The reaction takes place at 25 °C.
c) Determine the amount of heat that is exchanged with the surroundings.

Problem 14.2: A reactor is loaded with equal amounts of hydrogen, nitrogen, ammonia, and helium. If the reaction is

$$3H_2 + N_2 = 2NH_3,$$

do the following:
a) Calculate the minimum and maximum values of ξ.
b) Calculate all mole fractions when the mole fraction of helium is 0.3. What is the conversion of hydrogen?
c) Calculate all mole fractions when the mole fraction of helium is 0.215. What is the conversion of hydrogen?

Problem 14.3: Calculate the adiabatic flame temperature of methane in 20% excess air at 1 bar. Both methane and air are initially at 25 °C, 1 bar.
Additional data. Assume that the reaction is complete and that products are carbon dioxide and water.
The ideal-gas C_P's (in J/mol K) of the reactants and products as a function of temperature are given below (T must be in K):

$$C_P^{ig} = a + bT$$

	a (J/mol K)	b (J/mol K^2)
Methane	44.2539	0.02273
CO_2	44.3191	0.00730
Oxygen	30.5041	0.00349
Nitrogen	29.2313	0.00307
Water	32.4766	0.00832

Note: First calculate the temperature assuming the C_p's to be constant and equal to their value at 25 °C; then repeat using the temperature-dependent heat capacities given above.

Problem 14.4: a) Calculate the standard heat of reaction for the complete combustion of methane at 800 °C.
b) Calculate the amount of heat that is released in a furnace that burns methane in 20% excess air, if the furnace temperature is 800 °C and the pressure is 2 bar. Assume that the inlet gases are already at 800 °C as they enter the furnace.
c) Repeat part b, but this time the inlet gases are at 40 °C, 2 bar.
Additional data. Assume that the reaction is complete and that products are carbon dioxide and water.
The heat capacities of the gases are given in Problem 14.3.

Problem 14.5: a) Determine the heat of vaporization of benzene at 25 °C using tabulated heats of formation.
b) Determine Henry's law constant of ammonia in water at 25 °C from the tabulated (g) and (aq) Gibbs energies of formation.
c) Generate an entry for the Gibbs energy of formation of CO based on the standard state for aqueous solutes (aq).
d) N_2O gas is bubbled through water at 12 bar, 10 °C. Calculate the activity of N_2O (aq) and of water (l) in the liquid based on the indicated standard states.

Problem 14.6: The standard enthalpy of formation of water in the gas standard state is $H^\circ_{298}(g) = -241,818$ J/mol. Use this value and information from the steam tables to obtain the enthalpy of formation in the liquid standard state and compare with the tabulated value $H^\circ_{298}(l) = -285,830$ J/mol.

14.11 Problems

Problem 14.7: The equilibrium constant of the gas-phase reaction $2A(g) + 3B(g) = 2C(g)$ is experimentally found to be

$$K = ae^{b/T},$$

with $b = 1200$ K, $a = 10^{-5}$.

a) Calculate $\Delta G°$, $\Delta H°$, and $\Delta S°$, at 25 °C. Is the reaction exothermic, or endothermic?

b) A constant-volume reactor is loaded with an equimolar mixture of A, B, and C at 12 bar. The reaction is allowed to take place at constant temperature and when equilibrium is reached, the pressure in the reactor is 15 bar. Unfortunately, due to a malfunction in the thermometer, the actual temperature is not known. What is the temperature of the reactor?

Additional data. All three gases may be assumed ideal.

Problem 14.8: Ammonia is produced by the gas-phase synthesis

$$3H_2 + N_2 = 2NH_3.$$

The reactor in your plant can operate from 1 to 20 bar, and from 25 to 1000 °C. What conditions would you choose for maximum yield?

Problem 14.9: a) Calculate the equilibrium constant at 1000 °C for ammonia synthesis in the stoichiometry written below:

$$3H_2 + N_2 = 2NH_3.$$

b) An equimolar mixture of the three species enters a reactor at 1000 °C, 10 bar. Determine the composition at equilibrium, if temperature and pressure remain constant.

c) Determine the amount of heat that must be added to, or removed from the reactor to maintain the temperature at 1000 °C. Report the result in J per mol of hydrogen in the feed.

Additional data. The ideal-gas C_P (in J/mol K) of hydrogen, nitrogen, and ammonia are 29.52, 30.58, and 45.4, respectively.

Problem 14.10: You are asked to carry out the gas-phase hydrogenation of benzene to cyclohexane.

a) What conditions of pressure and temperature (high/low) would result in maximum conversion?

b) Determine the equilibrium composition of a mixture that contains 30% benzene (by mole) and 70% hydrogen at 200 °C, 5 bar.

Problem 14.11: Benzene is formed from the gas-phase dehydrogenation of cyclohexane according to the reaction:

$$C_6H_{12}(g) = C_6H_6(g) + 3H_2(g).$$

a) Calculate the standard heat of reaction and the equilibrium constant at 25 °C.
b) A stream that contains 50% by mole cyclohexane in nitrogen is fed to a reactor that operates a constant pressure and temperature. The reactor pressure is 5 bar and the desired conversion of cyclohexane is 85%. Determine the amount of heat per mol in the feed that must be added or removed from the reactor in order to maintain the temperature constant.
c) Determine the composition of all species at the exit of the reactor.
d) Assuming the reaction reaches equilibrium, determine the reactor temperature.
e) It has been suggested that conversion could be increased by operating at different pressure. You may vary the reactor pressure between 1 and 10 bar. If temperature and inlet conditions remain the same, at what pressure would you operate the reactor? Explain your answer.
f) List all of your assumptions and explain where each assumption was used.

Problem 14.12: Isopropyl alcohol (IPA) is dehydrogenated to produce propionaldehylde (PA) according to the reaction

$$C_3H_8O(g) = C_3H_6O(g) + H_2(g).$$

The equilibrium constant for this reaction is

$$K(T) = e^{-29.5 + 6673 K/T},$$

with T in K.
a) Is the reaction endothermic, or exothermic?
b) A reactor is fed with a mixture of IPA and nitrogen that contains 70% IPA by mol. The reaction takes place at $T = 500$ K, and $P = 3$ bar. Determine the conversion of IPA if the reaction reaches equilibrium.

Problem 14.13: Acetaldehyde can be produced from the dehydrogenenation of ethanol according to the following reaction:

$$CH_3CH_2OH(g) = CH_3CHO(g) + H_2(g).$$

Additional data. The ideal-gas heat capacities of the pure species in this reaction are: ethanol: 118 J/mole K; acetaldehyde: 90 J/mole K; hydrogen: 30 J/mole K.

14.11 Problems

a) Calculate the equilibrium constant at 25 °C.
b) Calculate the heat of reaction at 25 °C. Is the reaction exothermic or endothermic?
c) Assuming that the heat of reaction does not change much with temperature, calculate the equilibrium constant at 550 °C.
d) Is it an acceptable assumption that the heat of reaction does not change much with temperature? (Answers without justification do not count.)
e) The reactor initially contains one mole of ethanol and nothing else. The reaction is run at 550 °C and 2 bar until equilibrium is reached. What is the mole fraction of species at equilibrium?
f) It is desirable to react 98% of the initial amount of ethanol. If the temperature is to remain at 550 °C, at what pressure should you run the reaction?

Problem 14.14: The gas-phase dehydrogenation of propane is carried out according to the reaction

$$C_3H_8 \text{ (g)} = C_3H_6 \text{ (g)} + H_2 \text{ (g)}.$$

The reaction takes place in the presence of a catalyst in an isothermal reactor that is maintained at 5 bar.
a) Calculate the equilibrium constant at 25 °C.
b) Calculate the standard enthalpy of reaction at 25 °C. Is the reaction exothermic or endothermic?
c) The reactor feed contains pure propane. If the desired conversion of propane is 95%, at what temperature should you run the reaction? Assume the reaction reaches equilibrium.
d) Due to upstream process modifications, the reactor feed has now changed from pure propane into a mixture that contains 50% nitrogen (by mol). If the reactor pressure remains at 5 bar, how should you adjust the temperature to achieve 95% conversion?
e) Just as you finished adjusting the temperature to handle the propane/nitrogen feed of the previous part, the compressor malfunctions and the reactor pressure drops to 1 bar. How should you adjust the temperature to achieve 95% conversion?
f) Clearly state all the assumptions.

Problem 14.15: The standard Gibbs energy of formation of a substance based on one standard state (s_1) may be related to any other standard state (s_2) according to the following equation:

$$\Delta G^\circ(s_2) - \Delta G^\circ(s_1) = \mu(s_2) - \mu(s_1)$$

where $\mu(s_i)$ is the chemical potential of the substance at the standard state s_i (s_i could be g, l, etc.)

a) Show that

$$\Delta G°(s_2) - \Delta G°(s_1) = RT \ln \frac{f(s_2)}{f(s_1)},$$

where $f(s_i)$ is the fugacity of the substance at the standard state s_i.

b) For CO_2, $\Delta G(g) = 394.6$ kJ/mol and $\Delta G(aq) = 386.5$ kJ/mol. Use these data to calculate Henry's constant for CO_2 in water at 25 °C.

c) Calculate the equilibrium constant at 25 °C for the reaction

$$CO(aq) + H_2O(l) = CO_2(aq) + H_2(aq).$$

Henry's constants for CO and H_2 in water at 25 °C are 57089 bar and 71673 bar, respectively.

Problem 14.16: A major air pollution problem in combustion is the formation of NO from the reaction between O_2 and N_2:

$$N_2 + O_2 = 2NO.$$

a) Calculate the equilibrium constant and the heat of reaction at 25 °C for this reaction. Is the reaction exothermic, or endothermic?

b) Determine the effect of pressure and temperature on the amount of NO formed from this reaction. Explain your answer.

c) A high-temperature furnace operates at 5 bar with temperatures in the range 1,000 to 2,000 K by burning methane in 20% excess air. You are concerned that the air in the furnace may produce unacceptable amounts NO. By regulation, the concentration of NO in the gases leaving the furnace may not exceed 500 ppm by mole. At what temperatures can you operate the furnace without violating the regulation?

Note: A concentration of 1 ppm (parts per million) corresponds to mole fraction of 10^{-6}. You may assume that the heat of reaction is independent of temperature.

Problem 14.17: The gas phase reaction oxidation of SO_2 to SO_3 is carried out at a pressure of 1 bar with 20% excess air in an adiabatic reactor. The reactants enter at 25 °C and equilibrium is obtained at the exit of the reactor.

a) Write the balanced equation for the formation of one mole of sulfur trioxide.

b) What is the heat of reaction of the equation in part (a)?

c) Determine the composition and the temperature of the product stream from the reactor.

14.11 Problems

The heat capacities of the components are given by the equation

$$\frac{C_P}{R} = A + BT + CT^2 + \frac{D}{T^2},$$

with T in kelvin and with parameters given below:

	A	$10^3 \times B$	$10^6 \times C$	$10^{-5} \times D$
SO_2	5.699	0.801	–	-1.1015
O_2	3.639	0.506	–	-0.227
N_2	3.280	0.593	–	0.040
SO_3	8.060	1.056	–	-2.028

Problem 14.18: Consider the esterification reaction

$$CH_3COOH(l) + CH_3CH_2OH(l) = CH_3COOCH_2CH_3(l) + H_2O(l).$$

a) Calculate the equilibrium constant at 25 °C.
b) A reactor is loaded with a solution of acetic acid in ethanol containing 80% acetic acid by mole. The reaction is allowed to proceed to equilibrium at constant temperature of 25 °C. Assuming the components to form an ideal solution, calculate the composition of the equilibrium mixture.
c) Repeat the calculation using activity coefficients calculated from UNIFAC.

Problem 14.19: The water gas shift reaction

$$CO(g) + H_2O(g) = CO_2(g) + H_2(g),$$

is used to produce high-purity hydrogen. The reaction is carried out in a 10 m³ reactor that contains a copper catalyst and is operated at 1000 K and 1.5 bar. The equilibrium constant is given by

$$K(T) = e^{-5.057 + 4951.4/T},$$

where T is in K.
a) Is the reaction exothermic or endothermic?
b) Calculate the value of K_P at 1000 K, with P in bar. What are the units of K_P?
c) The composition of the reactor feed is 1 mole of CO and 5 moles of steam. Assuming the reaction to reach equilibrium, what is the composition (in terms of mole fraction) of the stream exiting the reactor?
d) After reading your report, your boss points out that for fuel cell applications, the mol fraction of CO must not exceed 5×10^{-3} in order to avoid poisoning of

the fuel cell anode. A colleague in your team suggests running the reaction at a different pressure; a young intern suggests running the reaction at a different temperature. Which suggestion do you adopt and why? Determine the new pressure or new temperature (depending on your decision) that meets the CO requirement.

Problem 14.20: Consider the water-gas shift reaction,

$$CO(g) + H_2O(g) = CO_2(g) + H_2(g).$$

A reactor initially contains one mole of each of the four species. The temperature is 900 K and the pressure is 1 bar.
a) What is the minimum and maximum possible value of the extent of reaction for this system? What does it mean if the reaction coordinate is negative?
b) Calculate the extent of reaction at equilibrium. What percentage of the initial amount of CO has reacted when equilibrium is achieved?
c) The same reaction now takes place in the presence of 5 moles of nitrogen. (The initial amount of the four reacting species is the same as before). What is the percent conversion of CO at equilibrium? Compare your answer with the result in part (b): is it higher, lower, or the same? Why?
d) Estimate the amount of heat per mole of CO that must be added or removed from the reactor to keep the temperature constant.
e) At what temperature should you run the reaction in order to react 50% of the initial amount of CO?
f) If you don't want to change the temperature, what else can you do to increase the conversion of CO?
g) State your assumptions.

Problem 14.21: An open bottle in the lab contains crystals of Na_2SO_4. The temperature is 25 °C, the pressure 1 bar, and the relative humidity 85%. Determine whether the formation of the decahydrate according to the reaction below,

$$Na_2SO_4 + 10H_2O = Na_2SO_4 \cdot 10H_2O,$$

can take place under these conditions. What is the minimum relative humidity that will allow the reaction to proceed in the forward direction?

Problem 14.22: A reversible reaction is conducted in a steady-state flow reactor. To improve conversion, a junior member of the design team suggests that you recycle a fraction of the reactor outlet back into the reactor. What is your opinion about this suggestion? *Hint:* Use the reaction $A(g) = B(g)$ as an example and show that the recycle stream has no effect whatsoever.

14.11 Problems

Problem 14.23: SO_3 is produced by oxidation of SO_2 in the presence of air. SO_2 is mixed with air so that the molar ratio of SO_2 to oxygen is 1:1.2 and is passed through a reactor at 1 bar, 850 K, until equilibrium is established. To improve conversion, SO_3 is separated out of the product mixture, while a portion of the other gases (SO_2, O_2, and N_2) is recycled back into the reactor with the rest being used elsewhere in the plant. Assuming that all of the SO_3 is obtained at the exit of the separator in pure form (i.e., no oxygen or nitrogen are present), determine the flow rate of the recycle stream (per 100 mol in the feed) that is required to produce an overall conversion of 95% and report the per-pass conversion of the reactor.

Problem 14.24: a) Calculate the solubility of ammonia in water at 25 °C if the partial pressure of ammonia in the gas phase is 2 bar.
b) Calculate Henry's law constant for ammonia in water at 25 °C.
Additional data:
ΔG°_{298} for $NH_3(g) = -16450$ J/mol,
ΔG°_{298} for $NH_3(aq) = -26500$ J/mol.

Bibliography

[1] E. W. Lemmon, M. O. McLinden, and D. Friend, *NIST Chemistry WebBook, NIST Standard Reference Database Number 69*, ch. Thermophysical Properties of Fluid Systems. NIST, 2010.

[2] R. H. Perry, D. W. Green, and J. O. Maloney, eds., *Perry's Chemical Engineering Handbook*, 7th ed. New York: McGraw-Hill, 1997.

[3] S. I. Sandler, *Chemical, Biochemical, and Engineering Thermodynamics*. Hoboken, NJ: Wiley, 2006.

[4] J. M. Prausnitz, R. N. Lichtenhaler, and E. G. de Azevedo, *Molecular Thermodynamcis of Fluid-Phase Equilibria*, 3rd ed. Upper Saddle River, NJ: Prentice Hall PTR, 1999.

[5] D. Chandler, *Introduction to Modern Statistical Mechanics*. Oxford University Press, 1987.

[6] A. Z. Panagiotopoulos, *Essential Thermodynamics*. Princeton, NJ: Drios Press, 2011.

[7] B. Poling, J. M. Prausnitz, and J. P. O'Connell, *The Properties of Gases and Liquids*, 5th ed. New York: McGraw-Hill, 2007.

[8] J. M. Smith and H. C. V. Ness, *Introduction to Chemical Engineering Thermodynamics*, 6th ed. New York: McGraw-Hill, 1996.

Appendix A

Critical Properties of Selected Compounds

Table A-1: Critical and related properties of selected species (sorted alphabetically by chemical formula).

Name	Formula	Molar mass (g/mol)	T_C (K)	P_C (bar)	V_C (cm^3/mol)	Z_C —	ω —
Argon	Ar	39.948	150.86	48.98	74.57	0.291	−0.002
Bromine	Br$_2$	159.908	584.10	103.00	135.00	0.269	0.119
Carbon tetrachloride	CCl$_4$	153.822	556.30	45.57	276.00	0.271	
Chloroform	CHCl$_3$	119.377	536.50	55.00	240.00	0.296	
Dichloromethane	CH$_2$Cl$_2$	84.932	510.00	61.00			
Methyl chloride	CH$_3$Cl	50.488	416.20	66.80	143.00	0.276	0.151
Methane	CH$_4$	16.043	190.56	45.99	98.60	0.286	0.011
Methanol	CH$_4$O	32.042	512.64	80.97	118.00	0.224	0.565
Methylamine	CH$_5$N	31.057	430.00	74.20	125.00	0.259	0.283
Carbon monoxide	CO	28.010	135.85	34.94	93.10	0.292	0.045
Carbon dioxide	CO$_2$	44.010	304.12	73.74	94.07	0.274	0.225
Acetylene	C$_2$H$_2$	26.038	308.30	61.14	112.20	0.268	0.189
R-134a	C$_2$H$_2$F$_4$	102.032	374.26	40.59	200.80	0.262	0.326
R-134	C$_2$H$_2$F$_4$	102.032	391.74	46.40	190.40	0.271	0.293
Ethylene	C$_2$H$_4$	28.054	282.34	50.41	131.10	0.282	0.087
Acetic acid	C$_2$H$_4$O$_2$	60.053	594.45	57.90	171.00	0.2	0.445
Ethyl chloride	C$_2$H$_5$Cl	64.514	460.30	53.00	199.00	0.276	
Ethane	C$_2$H$_6$	30.070	305.32	48.72	145.50	0.279	0.099
Ethanol	C$_2$H$_6$O	46.069	513.92	61.48	167.00	0.240	0.649
Ethylamine	C$_2$H$_7$N	45.084	456.40	56.30	181.80	0.267	0.276
Propylene	C$_3$H$_6$	42.081	364.90	46.00	184.60	0.280	0.142
Acetone	C$_3$H$_6$O	58.080	508.10	47.00	209.00	0.233	0.307
Methyl acetate	C$_3$H$_6$O$_2$	74.079	506.80	46.90	228.00	0.254	

continued on next page

Appendix A Critical Properties of Selected Compounds

Name	Formula	Molar mass (g/mol)	T_C (K)	P_C (bar)	V_C (cm^3/mol)	Z_C –	ω –
Propane	C_3H_8	44.097	369.83	42.48	200.00	0.276	0.152
1-Propanol	C_3H_8O	60.096	536.78	51.75	219.00	0.254	0.629
2-Propanol	C_3H_8O	60.096	508.30	47.62	220.00	0.248	0.665
1,3-Butadiene	C_4H_6	54.092	425.00	43.20	221.00	0.270	0.195
1-Butene	C_4H_8	56.108	419.50	40.20	240.80	0.278	0.194
cis-2-Butene	C_4H_8	56.108	428.60	41.00	237.70	0.276	0.218
trans-2-Butene	C_4H_8	56.108	435.50	42.10	233.80	0.269	0.203
Methyl ethyl ketone	C_4H_8O	72.107	536.80	42.10	267.00	0.252	0.322
Tetrahydrufurane	C_4H_8O	72.107	540.20	51.90	224.00	0.259	
Ethyl acetate	$C_4H_8O_2$	88.106	523.20	38.30	286.00	0.252	0.361
n-Butane	C_4H_{10}	58.123	425.12	37.96	255.00	0.274	0.200
Isobutane	C_4H_{10}	58.123	407.85	36.40	262.70	0.278	0.186
1-Butanol	$C_4H_{10}O$	74.123	563.05	44.23	275.00	0.260	0.590
Diethyl ether	$C_4H_{10}O$	74.123	466.70	36.40	280.00	0.263	0.281
Cyclopentane	C_5H_{10}	70.134	511.60	45.08	260.00	0.276	
n-Pentane	C_5H_{12}	72.150	469.70	33.70	311.00	0.268	0.252
Chlorobenzene	C_6H_5Cl	112.558	632.40	45.20	308.00	0.265	0.251
Benzene	C_6H_6	78.114	562.05	48.95	256.00	0.268	0.210
Phenol	C_6H_6O	94.113	694.25	61.30	229.00	0.243	0.442
Cyclohexane	C_6H_{12}	84.161	553.50	40.73	308.00	0.273	0.211
n-Hexane	C_6H_{14}	86.177	507.60	30.25	368.00	0.264	0.300
Toluene	C_7H_8	92.141	591.75	41.08	316.00	0.264	0.264
n-Heptane	C_7H_{16}	100.204	540.20	27.40	428.00	0.261	0.350
Ethylbenzene	C_8H_{10}	106.167	617.15	36.09	374.00	0.263	0.304
o-Xylene	C_8H_{10}	106.167	630.30	37.32	370.00	0.263	0.312
m-Xylene	C_8H_{10}	106.167	617.00	35.41	375.00	0.259	0.327
p-Xylene	C_8H_{10}	106.167	616.20	35.11	378.00	0.259	0.322
n-Octane	C_8H_{18}	114.231	568.70	24.90	492.00	0.259	0.399
Isooctane	C_8H_{18}	114.231	543.90	28.70	469.70	0.266	0.304
Cumene	C_9H_{12}	120.194	631.00	32.09	437.70	0.261	0.326
n-Nonane	C_9H_{20}	128.258	594.60	22.90	555.00	0.257	0.445
n-Decane	$C_{10}H_{22}$	142.285	617.70	21.10	624.00	0.256	0.490
Biphenyl	$C_{12}H_{10}$	154.211	773.00	33.80	497.00	0.261	0.404
Chlorine	Cl_2	70.905	417.00	77.00	124.00	0.275	
Hydrogen	H_2	2.016	32.98	12.93	64.20	0.303	-0.217
Water	H_2O	18.015	647.14	220.64	55.95	0.229	0.344

continued on next page

Appendix A Critical Properties of Selected Compounds

Name	Formula	Molar mass (g/mol)	T_C (K)	P_C (bar)	V_C (cm^3/mol)	Z_C —	ω —
Hydrogen sulfide	H_2S	344.250	572.00	59.10	150.00	0.186	
Ammonia	NH_3	17.031	405.40	113.53	72.47	0.255	0.257
Helium	He	4.003	5.19	2.27	57.30	0.301	−0.39
Krypton	Kr	83.800	209.40	55.00	91.20	0.288	
Nitric oxide	NO	30.006	180.00	64.80	58.00	0.251	0.582
Nitrogen	N_2	28.014	126.20	33.98	90.10	0.289	0.037
Nitrous oxide	N_2O	44.013	309.60	72.55	97.00	0.273	
Oxygen	O_2	31.999	154.58	50.43	73.37	0.288	
Sulfur dioxide	SO_2	64.065	430.80	78.84	122.00	0.269	
Sulfur trioxide	SO_3	80.064	490.90	82.10	126.50	0.254	
Xenon	Xe	131.290	289.74	58.40	118.00	0.286	

Note: Data from Poling, Prausnitz, and O'Connell, *The Properties of Gases and Liquids*, 5th ed. (New York: McGraw-Hill, 2001). Small discrepancies beween sources are common.

Appendix B

Ideal-Gas Heat Capacities

Table B-1: Ideal-gas heat capacity of selected substances according to the equation

$$\frac{C_P^{ig}}{R} = a_0 + a_1 T + a_2 T^2 + a_3 T^3 + a_4 T^4$$

where R is the ideal-gas constant and T is in kelvin. The range of validity is in kelvin.

Name	Formula	Range	a_0	$a_1 \times 10^3$	$a_2 \times 10^5$	$a_3 \times 10^8$	$a_4 \times 10^{11}$
Argon	Ar		2.5	0	0	0	0
Bromine	Br_2		3.212	7.16	−1.528	1.445	−0.499
Carbon tetrachloride	CCl_4	50–1000	2.518	41.882	−7.16	5.739	−1.756
Chloroform	$CHCl_3$	200–1000	2.389	26.218	−3.145	1.857	−0.423
Dichloromethane	CH_2Cl_2	200–1000	2.71	11.561	0.324	−1.37	0.662
Methyl chloride	CH_3Cl	200–1000	3.578	−1.75	3.071	−3.714	1.408
Methane	CH_4	200–1000	4.568	−8.975	3.631	−3.407	1.091
Methanol	CH_4O	50–1000	4.714	−6.986	4.211	−4.443	1.535
Methylamine	CH_5N	50–1000	4.193	−2.122	4.039	−4.738	1.757
Carbon monoxide	CO	50–1000	3.912	−3.913	1.182	−1.302	0.515
Carbon dioxide	CO_2	50–1000	3.259	1.356	1.502	−2.374	1.056
Acetylene	C_2H_2	50–1000	2.41	10.926	−0.255	−0.79	0.524
R-134a	$C_2H_2F_4$	50–1000	3.064	25.42	0.586	−3.339	1.176
R-134	$C_2H_2F_4$	50–1000	3.084	32.841	−2.425	0.488	0.162
Ethylene	C_2H_4	50–1000	4.221	−8.782	5.795	−6.729	2.511
Acetic acid	$C_2H_4O_2$	50–1000	4.375	−2.397	6.757	−8.764	3.478
Ethyl chloride	C_2H_5Cl	50–1000	3.029	9.885	2.967	−4.55	1.871
Ethane	C_2H_6	200–1000	4.178	−4.427	5.66	−6.651	2.487
Ethanol	C_2H_6O	50–1000	4.396	0.628	5.546	−7.024	2.685
Ethylamine	C_2H_7N	50–1000	4.64	2.069	5.797	−7.659	3.043
Propylene	C_3H_6	50–1000	3.834	3.893	4.688	−6.013	2.283
Acetone	C_3H_6O	50–1000	5.126	1.511	5.731	−7.177	2.728
Methyl acetate	$C_3H_6O_2$	200–1000	4.242	14.388	3.338	−4.93	1.931
Propane	C_3H_8	298–1000	3.847	5.131	6.011	−7.893	3.079
1-Propanol	C_3H_8O	50–1000	4.712	6.565	6.31	−8.341	3.216
2-Propanol	C_3H_8O	50–1000	3.334	18.853	3.644	−6.115	2.543
1,3-Butadiene	C_4H_6	50–1000	3.607	5.085	8.253	−12.371	5.321
1-Butene	C_4H_8	50–1000	4.389	7.984	6.143	−8.197	3.165
cis-2-Butene	C_4H_8	50–1000	5.584	−4.89	9.133	−10.975	4.085

continued on next page

Name	Formula	Range	a_0	$a_1 \times 10^3$	$a_2 \times 10^5$	$a_3 \times 10^8$	$a_4 \times 10^{11}$
trans-2-Butene	C_4H_8	50–1000	3.689	19.184	2.23	−3.426	1.256
Methyl ethyl ketone	C_4H_8O	50–1000	6.349	11.062	4.851	−6.484	2.469
Tetrahydrufurane	C_4H_8O	200–1000	5.171	−19.464	16.46	−20.42	8.000
Ethyl acetate	$C_4H_8O_2$	50–1000	10.228	−14.948	13.033	−15.736	5.999
n-Butane	C_4H_{10}	298–1000	5.547	5.536	8.057	−10.571	4.134
Isobutane	C_4H_{10}	200–1000	3.351	17.883	5.477	−8.099	3.243
1-Butanol	$C_4H_{10}O$	50–1000	4.467	16.395	6.688	−9.69	3.864
Diethyl ether	$C_4H_{10}O$	50–1000	4.618	37.492	−1.87	1.316	−6.98
Cyclopentane	C_5H_{10}	100–1000	5.019	−19.734	17.917	−21.696	8.215
n-Pentane	C_5H_{12}	50–1000	7.554	−0.368	11.846	−14.939	5.753
Chlorobenzene	C_6H_5Cl	200–1000	0.104	38.288	1.808	−5.732	2.718
Benzene	C_6H_6	200–1000	3.551	−6.184	14.365	−19.807	8.234
Phenol	C_6H_6O	50–1000	2.582	17.501	8.894	−14.435	6.317
Cyclohexane	C_6H_{12}	50–1000	4.035	−4.433	16.834	−20.775	7.746
n-Hexane	C_6H_{14}	100–1000	8.831	−0.166	14.302	−18.314	7.124
Toluene	C_7H_8	200–1000	3.866	3.558	13.356	−18.659	7.69
n-Heptane	C_7H_{16}	50–1000	9.634	4.156	15.494	−20.066	7.77
Ethylbenzene	C_8H_{10}	200–1000	4.544	10.578	13.644	−19.276	7.885
o-Xylene	C_8H_{10}	50–1000	3.289	34.144	4.989	−8.335	3.338
m-Xylene	C_8H_{10}	50–1000	4.002	17.537	10.59	−15.037	6.008
p-Xylene	C_8H_{10}	50–1000	4.113	14.909	11.81	−16.724	6.736
n-Octane	C_8H_{18}	50–1000	10.824	4.983	17.751	−23.137	8.98
Isooctane	C_8H_{18}	200–1000	0.384	77.059	0.665	−5.565	2.619
Cumene	C_9H_{12}	200–1000	2.985	34.196	11.938	−20.152	8.923
n-Nonane	C_9H_{20}	50–1000	12.152	4.575	20.416	−26.777	10.465
n-Decane	$C_{10}H_{22}$	200–1000	13.467	4.139	23.127	−30.477	11.97
Biphenyl	$C_{12}H_{10}$	200–1000	−0.843	61.392	6.352	−13.754	6.169
Chlorine	Cl_2	200–1000	3.056	5.3708	−0.8098	0.5693	−0.15256
Hydrogen	H_2	50–1000	2.883	3.681	−0.772	0.692	−0.213
Water	H_2O	50–1000	4.395	−4.186	1.405	−1.564	0.632
Hydrogen sulfide	H_2S	50–1000	4.266	−3.438	1.319	−1.331	0.488
Ammonia	NH_3	50–1000	4.238	−4.215	2.041	−2.126	0.761
Helium	He	50–1000	2.5	0	0	0	0
Krypton	Kr		2.5	0	0	0	0
Nitric oxide	NO		4.534	−7.644	2.066	−2.156	0.806
Nitrogen	N_2	50–1000	3.539	−0.261	0.007	0.157	0.099
Nitrous oxide	N_2O	50–1000	3.165	3.401	0.989	−1.88	0.89
Oxygen	O_2	50–1000	3.63	−1.749	0.658	0.601	0.179
Sulfur dioxide	SO_2	50–1000	4.417	−2.234	2.344	−3.271	1.393
Sulfur trioxide	SO_3	50–1000	3.426	6.479	1.691	−3.356	1.59
Xenon	Xe	50–1000	2.5	0	0	0	0

Note: Data from Poling, Prausnitz, and O'Connell, *The Properties of Gases and Liquids*, 5th ed. (New York: McGraw-Hill, 2001).

Appendix C

Standard Enthalpy and Gibbs Free Energy of Reaction

Table C-1: Standard enthalpy and Gibbs free energy of formation of selected species.

Name	Formula	State	$H°_{298}$ (kJ/mol)	$G°_{298}$ (kJ/mol)
Argon	Ar	g	0	0
Bromine	Br$_2$	g	0	0
Carbon tetrachloride	CCl$_4$	l	−95.81	−53.53
Chloroform	CHCl$_3$	l	−102.93	−70.09
Dichloromethane	CH$_2$Cl$_2$	l	−95.40	−68.84
Methyl chloride	CH$_3$Cl	g	−81.96	−58.42
Methane	CH$_4$	g	−74.52	−50.45
Methanol	CH$_4$O	l	−200.94	−162.24
Methylamine	CH$_5$N	l	−22.53	32.73
Carbon monoxide	CO	g	−110.53	−137.16
Carbon dioxide	CO$_2$	g	−393.51	−394.38
Acetylene	C$_2$H$_2$	g	190.92	201.30
R-134a	C$_2$H$_2$F$_4$	g	−907.10	−838.40
R-134	C$_2$H$_2$F$_4$	g	−892.40	−824.60
Ethylene	C$_2$H$_4$	g	52.50	68.48
Acetic acid	C$_2$H$_4$O$_2$	l	−432.25	−374.27
Ethyl chloride	C$_2$H$_5$Cl	l	−112.26	−60.43
Ethane	C$_2$H$_6$	g	−83.82	−31.86
Ethanol	C$_2$H$_6$O	l	−234.95	−167.73
Ethylamine	C$_2$H$_7$N	g	−47.47	36.28
Propylene	C$_3$H$_6$	g	20.00	62.50

continued on next page

Appendix C Standard Enthalpy and Gibbs Free Energy of Reaction

Name	Formula	State	$H°_{298}$ (kJ/mol)	$G°_{298}$ (kJ/mol)
Acetone	C_3H_6O	l	−217.10	−152.60
Methyl acetate	$C_3H_6O_2$	l	−408.80	−321.40
Propane	C_3H_8	g	−104.68	−24.29
1-Propanol	C_3H_8O	l	−255.20	−159.81
2-Propanol	C_3H_8O	l	−272.70	−173.32
1,3-Butadiene	C_4H_6	g	110.00	150.60
1-Butene	C_4H_8	g	−0.54	70.37
trans-2-Butene	C_4H_8	g	−11.00	63.34
cis-2-Butene	C_4H_8	g	−7.40	65.46
Methyl ethyl ketone	C_4H_8O	l	−238.60	−146.50
Tetrahydrufurane	C_4H_8O	l	−184.18	−79.57
Ethyl acetate	$C_4H_8O_2$	l	−444.50	−328.00
n-Butane	C_4H_{10}	g	−125.79	−16.57
Isobutane	C_4H_{10}	g	−134.99	−21.44
1-Butanol	$C_4H_{10}O$	l	−274.60	−150.17
Diethyl ether	$C_4H_{10}O$	l	−250.80	−120.70
Cyclopentane	C_5H_{10}	l	−77.10	38.92
n-Pentane	C_5H_{12}	l	−146.76	−8.65
Chlorobenzene	C_6H_5Cl	l	51.09	98.36
Benzene	C_6H_6	l	82.88	129.75
Phenol	C_6H_6O	l	−96.40	−32.55
Cyclohexane	C_6H_{12}	l	−123.10	32.26
n-Hexane	C_6H_{14}	l	−166.92	0.15
Toluene	C_7H_8	l	50.17	122.29
n-Heptane	C_7H_{16}	l	−187.80	8.20
Ethylbenzene	C_8H_{10}	l	29.92	130.73
o-Xylene	C_8H_{10}	l	19.08	122.05
m-Xylene	C_8H_{10}	l	17.32	118.89
p-Xylene	C_8H_{10}	l	18.03	121.48
n-Octane	C_8H_{18}	l	−208.75	16.27
Isooctane	C_8H_{18}	l	−224.01	14.21
Cumene	C_9H_{12}	l	4.00	
n-Nonane	C_9H_{20}	l	−228.86	25.00
n-Decane	$C_{10}H_{22}$	l	−249.53	33.30
Biphenyl	$C_{12}H_{10}$	l	182.42	281.08
Chlorine	Cl_2	g	0	0
Hydrogen	H_2	g	0	0
Water	H_2O	g	−241.81	−228.42
Water	H_2O	l	−285.83	−237.13
Hydrogen sulfide	H_2S	g	−20.63	−33.43

continued on next page

Appendix C Standard Enthalpy and Gibbs Free Energy of Reaction

Name	Formula	State	H°_{298} (kJ/mol)	G°_{298} (kJ/mol)
Ammonia	NH_3	g	−45.94	−16.41
Helium	He	g	0	0
Krypton	Kr	g	0	0
Nitric oxide (NO)	NO	g	90.25	86.58
Nitrogen	N_2	g	0	0
Nitrous oxide (N_2O)	N_2O	g	82.05	104.18
Oxygen	O_2	g	0	0
Sulfur dioxide	SO_2	g	−296.81	−300.14
Sulfur trioxide	SO_3	l	−395.72	−370.93
Xenon	Xe	g	0	0

Note: Data from Poling, Prausnitz, and O'Connell, *The Properties of Gases and Liquids*, 5th ed. (New York: McGraw-Hill, 2001). Depending on the source, descrepancies of the order of about ±0.200 kJ/mol are not uncommon.

Appendix D

UNIFAC Tables

Table D-1: Parameters for the UNIFAC Equation. "Main" refers to the index of the main group; k refers to the index of the subgroup.

Main	k	Name	R_k	Q_k	Example
1	1	CH_3	0.9011	0.848	n-hexane: 4 CH_2 + 2 CH_3
1	2	CH_2	0.6744	0.540	isobutane: 1 CH, 3 CH_3
1	3	CH	0.4469	0.228	neopentane: 1 C, 4 CH_3
1	4	C	0.2195	0.000	2,2,4,trimethylpentane, 5CH_3, 1 CH_2, 1CH,C
2	5	CH_2=CH	1.3454	1.176	1-hexene: 1 CH_2=CH, 3 CH_2, 1 CH_3
2	6	CH=CH	1.1167	0.867	2-hexene: 1 CH=CH, 2 CH_3, 2 CH_2
2	7	CH_2=C	1.1173	0.988	2-methyl-1-butene: 2 CH_3, 1 CH_2, 1 CH=C
2	8	CH=C	0.8886	0.676	2-methyl-2-butene: 3 CH_3, 1 CH_2=C
2	70	C=C	0.6605	0.485	2,3-dimethylbutene: 4 CH_3, 1 C=C
3	9	ACH	0.5313	0.400	benzene: 6 ACH
3	10	AC	0.3652	0.120	styrene: 1 CH_2=CH, 5 ACH, 1 AC
4	11	$ACCH_3$	1.2663	0.968	toluene: 5 ACH, 1 $ACCH_3$
4	12	$ACCH_2$	1.0396	0.660	ethylbenzene: 5 ACH, 1 $ACCH_2$, 1 CH_3
4	13	ACCH	0.8121	0.348	cumene: 2 CH_3, 5 ACH, 1 ACCH
5	14	OH	1.0000	1.200	n-propanol: 1 OH, 1 CH_3, 2 CH_2
6	15	CH_3OH	1.4311	1.432	methanol: 1 CH_3OH
7	16	H_2O	0.9200	1.400	water: 1 H_2O
8	17	ACOH	0.8952	0.680	phenol: 1 ACOH, 5 ACH
9	18	CH_3CO	1.6724	1.488	dimethylketone: 1 CH_3CO, 1 CH_3
9	19	CH_2CO	1.4457	1.180	diethylketone: 1 CH_2CO, 2 CH_3, 1 CH_2
10	20	CHO	0.9980	0.948	ethanal: 1 CHO, 1 CH_3
11	21	CH_3COO	1.9031	1.728	methyl acetate: 1 CH_3COO, 1 CH_3
11	22	CH_2COO	1.6764	1.420	methyl propanate: 1 CH_2COO, 2 CH_3
12	23	HCOO	1.2420	1.188	methyl formate: 1 HCOO, 1 CH_3
13	24	CH_3O	1.1450	1.088	dimethyl ether: 1 CH_3, 1 CH_3O
13	25	CH_2O	0.9183	0.780	ethyl ether: 1 CH_2O, 1 CH_3, 1 CH_2
13	26	CHO	0.6908	0.468	diisopropyl ether: 4 CH_3, 1 CH, 1 CHO
13	27	THF	0.9183	1.100	tetrahydrofuran : 3 CH_2, 1 THF
14	28	CH_3NH_2	1.5959	1.544	methylamine: 1 CH_3NH_2
14	29	CH_2NH_2	1.3692	1.236	propylamine: 1 CH_2NH_2, 1 CH_3, 1 CH_2
14	30	$CHNH_2$	1.1417	0.924	isopropylamine: 2 CH_3, 1 $CHNH_2$

continued on next page

Main	k	Name	R_k	Q_k	Example
15	31	CH_3NH	1.4337	1.244	dimethyl amine: 1 CH_3, 1 CH_3NH
15	32	CH_2NH	1.2070	0.936	diethylamine: 1 CH_2NH, 2 CH_3, 1 CH_2
15	33	CHNH	0.9795	0.624	diisopropyl amine: 4 CH_3, 1 CH, 1 CHNH
16	34	CH_3N	1.1865	0.940	trimethylamine: 2 CH_3, 1 CH_3N
16	35	CH_2N	0.9597	0.632	triethylamine: 1 CH_2N, 2 CH_2, 3 CH_3
17	36	$ACNH_2$	1.0600	0.816	aniline: 1 $ACNH_2$, 5 ACH
18	37	C_5H_5N	2.9993	2.113	pyridine: 1 C_5H_5N
18	38	C_5H_4N	2.8332	1.833	methylpyridine: 1 C_5H_4N, 1 CH_3
18	39	C_5H_3N	2.6670	1.553	2,2-dimethylpyridine, 2 CH_3, 1 C_5H_3N
19	40	CH_3CN	1.8701	1.724	acetonitrile: 1 CH_3CN
19	41	CH_2CN	1.6434	1.416	propionnitrile: 1 CH_2CN, 1 CH_3
20	42	COOH	1.3013	1.224	acetic acid: 1 CH_3, 1 COOH
20	43	HCOOH	1.5280	1.532	formic acid: 1 HCOOH
21	44	CH_2Cl	1.4654	1.264	1-chlorobutane: 1 CH_3, 2 CH_2, 1 CH_2Cl
21	45	CHCl	1.2380	0.952	chloroethane: 1 CH_2Cl, 1 CH_3
21	46	CCl	1.0060	0.724	2-chloro-2-methylpropane: 3 CH_3, 1 CCl
22	47	CH_2Cl_2	2.2564	1.988	dichloromethane: 1 CH_2Cl_2
22	48	$CHCl_2$	2.0606	1.684	1,1-dichloroethane: 1 CH_3, 1 $CHCl_2$
22	49	CCl_2	1.8016	1.448	2,2-dichloropropane: 1 $CHCl_2$, 2 CH_3
23	50	$CHCl_3$	2.8700	2.410	chloroform: 1 $CHCl_3$
23	51	CCl_3	2.6401	2.184	1,1-trichloroethane, 1 CH_3, 1 CCl_3
24	52	CCl_4	3.3900	2.910	tetrachloromethane: CCl_4
25	53	ACCl	1.1562	0.844	carbon tetrachloride
26	54	CH_3NO_2	2.0086	1.868	nitromethane: 1 CH_3NO_2
26	55	CH_2NO_2	1.7818	1.560	1-nitropropane: 1 CH_2NO_2, 2 CH_3, 1 CH_2
26	56	$CHNO_2$	1.5544	1.248	2-nitropropane: 1 $CHNO_2$, 2 CH_3
27	57	$ACNO_2$	1.4199	1.104	nitrobenzene: 1 $ACNO_2$, 5 ACH
28	58	CS_2	2.0570	1.650	carbon disulfide: 1 CS_2
29	59	CH_3SH	1.8770	1.676	methanethiol: 1 CH_3SH
29	60	CH_2SH	1.6510	1.368	ethanethiol: 1 CH_3, CH_2SH
30	61	furfural	3.1680	2.481	furfural
31	62	$(CH_2OH)_2$	2.4088	2.248	ethylene glycol: 1 $(CH_2OH)_2$
32	63	I	1.2640	0.992	iodoethane: 1 CH_3, 1 CH_2, 1 I
33	64	Br	0.9492	0.832	bromomethane: 1 CH_3, 1 Br
34	65	CH≡C	1.2920	1.088	1-hexyne: 1 CH_3, 2 CH_2, 1 CH≡C
34	66	C≡C	1.0613	0.784	2-hexyne: 2 CH_3, 2 CH_2, 1 C≡C
35	67	Me_2SO	2.8266	2.472	dimethylsulfoxide: 1 Me_2SO
36	68	ACRY	2.3144	2.052	acrylonitrile: 1 ACRY
37	69	Cl–(C=C)	0.7910	0.724	trichlorethylene: 1 CH=C, 3 Cl–(C=C)
38	71	ACF	0.6948	0.524	fluorobenzene: 5 ACH, 1 ACF
39	72	DMF	3.0856	2.736	N,N-dimethylformamide: 1 DMF
39	73	$HCON(CH_2)_2$	2.6322	2.120	N,N-diethylformamide: 2 CH_3, 1 $HCON(CH_2)_2$

continued on next page

Appendix D UNIFAC Tables

Main	k	Name	R_k	Q_k	Example
40	74	CF_3	1.4060	1.380	perfluoroethane: 2 CF_3
40	75	CF_2	1.0105	0.920	
40	76	CF	0.6150	0.460	
41	77	COO	1.3800	1.200	butylacetate: 2 CH_3, 3 CH_2, 1 COO
42	78	SiH_3	1.6035	1.263	methylsilane: 1 CH_3, 1 SiH_3
42	79	SiH_2	1.4443	1.006	diethylsilane: 2 CH_3, 2 CH_2, 1 SiH_2
42	80	SiH	1.2853	0.749	trimethysilane: 3 CH_3, 1 SiH
42	81	Si	1.0470	0.410	tetramethylsilane: 4 CH_3, 1 Si
43	82	SiH_2O	1.4838	1.062	
43	83	$SiHO$	1.3030	0.764	
43	84	SiO	1.1044	0.466	hexamethyldisiloxane: 6 CH_3, 1 Si, 1 SiO
44	85	NMP	3.9810	3.200	N-methylpyrrolidone: 1 NMP
44	86	CCl_3F	3.0356	2.644	trichlorofluoromethane: 1 CCl_3F
45	87	CCl_2F	2.2287	1.916	tetrachloro-1,2-difluoroethane: 2 CCl_2F
45	88	$HCCl_2F$	2.4060	2.116	dichlorofluoromethane: 1 $HCCl_2F$
45	89	$HCClF$	1.6493	1.416	2-chloro-2-fluoroethane: 1 CH_3, 1 $HCClF$
45	90	$CClF_2$	1.8174	1.648	2-chloro-2,2-difluoroethane, 1 CH_3, 1 $CClF_2$
45	91	$HCClF_2$	1.9670	1.828	chlorodifluoromethane: 1 $HCClF_2$
45	92	$CClF_3$	2.1721	2.100	chlorotrifluoromethane: 1 $CClF_3$
45	93	CCl_2F_2	2.6243	2.376	dichlorodifluoromethane: 1 CCl_2F_2
46	94	$CONH_2$	1.4515	1.248	acetamide: 1 CH_3, 1 $CONH_2$
46	95	$CONHCH_3$	2.1905	1.796	N-methylamide: 1 CH_3, 1 $CONHCH_3$
46	96	$CONHCH_2$	1.9637	1.488	N-ethylamide: 2 CH_3, 1 $CONHCH_2$
46	97	$CON(CH_3)_2$	2.8539	2.428	N,N-dimethylacetamide: 1 CH_3, 1 $CON(CH_3)_2$
46	98	$CONCH_3CH_2$	2.6322	2.120	N,N-methylethylacetamid: 2 CH_3, 1 $CONCH_3CH_2$
46	99	$CON(CH_2)_2$	2.4054	1.812	N,N-diethylacetamide: 3 CH_3, 1 $CON(CH_2)_2$
47	100	$C_2H_5O_2$	2.1226	1.904	2-ethoxyethanol: 1 CH_3, 1 CH_2, 1 $C_2H_5O_2$
47	101	$C_2H_4O_2$	1.8952	1.592	2-ethoxy-1-propanol: 2 CH_3, 1 CH_2, 1 $C_2H_4O_2$
48	102	CH_3S	1.6130	1.368	dimethylsulfide: 1 CH_3, 1 CH_3S
48	103	CH_2S	1.3863	1.060	diethylsulfide: 2 CH_3, 1 CH_2, 1 CH_2S
48	104	CHS	1.1589	0.748	diisopropylsulfide: 4 CH_3, 1 CH, 1 CHS
49	105	MORPH	3.4740	2.796	morpholine: 1 MORPH
50	106	C_4H_4S	2.8569	2.140	thiophene: 1 C_4H_4S
50	107	C_4H_3S	2.6908	1.860	2-methylthiophene: 1 CH_3, 1 C_4H_3S
50	108	C_4H_2S	2.5247	1.580	2,3,dimethylthiophene: 2 CH_3, 1 C_4H_2S

Note: Data from Poling, Prausnitz, and O'Connell, *The Properties of Gases and Liquids*, 5th ed. (New York: McGraw-Hill, 2001).

Appendix E

Steam Tables

The tables that follow give the properties of water in the following units:

P is in bar
T is in °C
V is in m³/kg
U is in kJ/kg
H is in kJ/kg
S is in kJ/kg K

The reference state is the saturated liquid at the triple point:

at $P_0 = 0.006117$ bar, $T_0 = 0.01$ °C : $H_L = 0$ kJ/kg and $S_L = 0$ kJ/kg/K

Table E-1: Saturated Steam (part 1 of 2)

T	P^{sat}	V^L	V^V	U^L	U^V	H^L	H^V	S^L	S^V
0.01	0.006117	0.001000	206.00	0.00	2374.9	0.00	2500.9	0.0000	9.1555
1	0.006571	0.001000	192.44	4.18	2376.3	4.18	2502.7	0.0153	9.1291
2	0.007060	0.001000	179.76	8.39	2377.7	8.39	2504.6	0.0306	9.1027
3	0.007581	0.001000	168.01	12.60	2379.0	12.60	2506.4	0.0459	9.0765
4	0.008135	0.001000	157.12	16.81	2380.4	16.81	2508.2	0.0611	9.0506
5	0.008726	0.001000	147.02	21.02	2381.8	21.02	2510.1	0.0763	9.0249
6	0.009354	0.001000	137.64	25.22	2383.2	25.22	2511.9	0.0913	8.9994
7	0.01002	0.001000	128.93	29.42	2384.5	29.43	2513.7	0.1064	8.9742
8	0.01073	0.001000	120.83	33.62	2385.9	33.63	2515.6	0.1213	8.9492
9	0.01148	0.001000	113.31	37.82	2387.3	37.82	2517.4	0.1362	8.9244
10	0.01228	0.001000	106.31	42.02	2388.7	42.02	2519.2	0.1511	8.8998
12	0.01403	0.001001	93.724	50.41	2391.4	50.41	2522.9	0.1806	8.8514
14	0.01599	0.001001	82.798	58.79	2394.1	58.79	2526.5	0.2099	8.8038
16	0.01819	0.001001	73.291	67.17	2396.7	67.17	2530.2	0.2390	8.7571
18	0.02065	0.001001	65.003	75.55	2399.6	75.55	2533.8	0.2678	8.7112
20	0.02339	0.001002	57.761	83.92	2402.4	83.92	2537.5	0.2965	8.6661
22	0.02645	0.001002	51.422	92.29	2405.1	92.29	2541.1	0.3250	8.6218
24	0.02986	0.001003	45.863	100.65	2407.8	100.66	2544.7	0.3532	8.5783
26	0.03364	0.001003	40.977	109.02	2410.5	109.02	2548.4	0.3813	8.5355
28	0.03783	0.001004	36.675	117.38	2413.2	117.38	2552.0	0.4091	8.4934
30	0.04247	0.001004	32.882	125.74	2415.9	125.75	2555.6	0.4368	8.4521
32	0.04759	0.001005	29.529	134.10	2418.7	134.11	2559.2	0.4643	8.4115
34	0.05325	0.001006	26.562	142.46	2421.4	142.47	2562.8	0.4916	8.3715
36	0.05947	0.001006	23.932	150.82	2424.0	150.82	2566.4	0.5187	8.3323
38	0.06632	0.001007	21.595	159.18	2426.7	159.18	2570.0	0.5457	8.2936
40	0.07384	0.001008	19.517	167.53	2429.4	167.54	2573.5	0.5724	8.2557
42	0.08209	0.001009	17.665	175.89	2432.1	175.90	2577.1	0.5990	8.2183
44	0.09112	0.001009	16.013	184.25	2434.8	184.26	2580.7	0.6255	8.1816
46	0.1010	0.001010	14.535	192.61	2437.4	192.62	2584.2	0.6517	8.1454
48	0.1118	0.001011	13.213	200.96	2440.1	200.98	2587.8	0.6778	8.1099
50	0.1235	0.001012	12.028	209.32	2442.8	209.34	2591.3	0.7038	8.0749
52	0.1363	0.001013	10.964	217.68	2445.4	217.70	2594.8	0.7296	8.0405
54	0.1502	0.001014	10.007	226.04	2448.0	226.06	2598.4	0.7552	8.0066
56	0.1653	0.001015	9.1454	234.41	2450.7	234.42	2601.9	0.7807	7.9733
58	0.1817	0.001016	8.3688	242.77	2453.3	242.79	2605.4	0.8060	7.9405
60	0.1995	0.001017	7.6677	251.13	2455.9	251.15	2608.8	0.8312	7.9082
62	0.2187	0.001018	7.0338	259.50	2458.5	259.52	2612.3	0.8563	7.8764
64	0.2394	0.001019	6.4601	267.87	2461.1	267.89	2615.8	0.8811	7.8451
66	0.2618	0.001020	5.9402	276.24	2463.7	276.27	2619.2	0.9059	7.8142
68	0.2860	0.001022	5.4684	284.61	2466.3	284.64	2622.7	0.9305	7.7839
70	0.3120	0.001023	5.0397	292.99	2468.9	293.02	2626.1	0.9550	7.7540
72	0.3400	0.001024	4.6498	301.36	2471.4	301.40	2629.5	0.9793	7.7245
74	0.3701	0.001025	4.2947	309.74	2474.0	309.78	2632.9	1.0035	7.6955
76	0.4024	0.001026	3.9709	318.13	2476.5	318.17	2636.3	1.0276	7.6669
78	0.4370	0.001028	3.6754	326.51	2479.0	326.56	2639.7	1.0516	7.6388
80	0.4741	0.001029	3.4053	334.90	2481.6	334.95	2643.0	1.0754	7.6110
82	0.5139	0.001030	3.1582	343.29	2484.1	343.34	2646.4	1.0991	7.5837
84	0.5564	0.001032	2.9319	351.69	2486.6	351.74	2649.7	1.1227	7.5567
86	0.6017	0.001033	2.7244	360.09	2489.0	360.15	2653.0	1.1461	7.5301
88	0.6502	0.001035	2.5341	368.49	2491.5	368.56	2656.3	1.1694	7.5039
90	0.7018	0.001036	2.3591	376.90	2494.0	376.97	2659.5	1.1927	7.4781
92	0.7568	0.001037	2.1983	385.31	2496.4	385.38	2662.8	1.2158	7.4526
94	0.8154	0.001039	2.0502	393.72	2498.8	393.81	2666.0	1.2387	7.4275
96	0.8777	0.001040	1.9138	402.14	2501.2	402.23	2669.2	1.2616	7.4027
98	0.9439	0.001042	1.7880	410.56	2503.6	410.66	2672.4	1.2844	7.3782
100	1.014	0.001043	1.6719	418.99	2506.0	419.10	2675.6	1.3070	7.3541
102	1.089	0.001045	1.5645	427.43	2508.4	427.54	2678.7	1.3296	7.3303
104	1.168	0.001047	1.4653	435.87	2510.7	435.99	2681.8	1.3520	7.3068
106	1.251	0.001048	1.3734	444.31	2513.1	444.44	2684.9	1.3743	7.2836
108	1.340	0.001050	1.2883	452.76	2515.4	452.90	2688.0	1.3965	7.2607
110	1.434	0.001052	1.2094	461.21	2517.7	461.36	2691.1	1.4187	7.2380
112	1.533	0.001053	1.1362	469.67	2519.9	469.83	2694.1	1.4407	7.2157
114	1.637	0.001055	1.0681	478.14	2522.2	478.31	2697.1	1.4626	7.1937
116	1.748	0.001057	1.0049	486.61	2524.4	486.80	2700.1	1.4844	7.1719
118	1.864	0.001059	0.9461	495.09	2526.7	495.29	2703.0	1.5062	7.1504
120	1.987	0.001060	0.8913	503.57	2528.9	503.78	2705.9	1.5278	7.1291
122	2.116	0.001062	0.8403	512.07	2531.0	512.29	2708.8	1.5494	7.1081
124	2.252	0.001064	0.7927	520.56	2533.2	520.80	2711.7	1.5708	7.0873
126	2.395	0.001066	0.7483	529.07	2535.3	529.32	2714.5	1.5922	7.0668
128	2.545	0.001068	0.7068	537.58	2537.4	537.85	2717.3	1.6134	7.0465
130	2.703	0.001070	0.6681	546.10	2539.5	546.39	2720.1	1.6346	7.0264

Table E-1: Saturated Steam, cont. (part 2 of 2)

T	P^{sat}	V^L	V^V	U^L	U^V	H^L	H^V	S^L	S^V
132	2.868	0.001072	0.6318	554.63	2541.6	554.93	2722.8	1.6557	7.0066
134	3.042	0.001074	0.5979	563.16	2543.6	563.49	2725.5	1.6767	6.9869
136	3.224	0.001076	0.5662	571.70	2545.7	572.05	2728.2	1.6977	6.9675
138	3.415	0.001078	0.5364	580.25	2547.6	580.62	2730.8	1.7185	6.9483
140	3.615	0.001080	0.5085	588.81	2549.6	589.20	2733.4	1.7393	6.9293
142	3.824	0.001082	0.4823	597.38	2551.6	597.79	2736.0	1.7600	6.9105
144	4.043	0.001084	0.4577	605.95	2553.5	606.39	2738.5	1.7806	6.8918
146	4.272	0.001086	0.4346	614.54	2555.4	615.00	2741.0	1.8011	6.8734
148	4.511	0.001088	0.4129	623.13	2557.2	623.62	2743.5	1.8216	6.8551
150	4.761	0.001091	0.3925	631.73	2559.0	632.25	2745.9	1.8420	6.8370
152	5.022	0.001093	0.3733	640.34	2560.9	640.89	2748.3	1.8623	6.8191
154	5.294	0.001095	0.3552	648.97	2562.6	649.55	2750.6	1.8825	6.8014
156	5.578	0.001097	0.3381	657.60	2564.4	658.21	2752.9	1.9027	6.7838
158	5.873	0.001100	0.3220	666.24	2566.1	666.89	2755.2	1.9228	6.7664
160	6.181	0.001102	0.3068	674.89	2567.8	675.57	2757.4	1.9428	6.7491
162	6.502	0.001104	0.2925	683.56	2569.4	684.28	2759.6	1.9627	6.7320
164	6.836	0.001107	0.2789	692.23	2571.1	692.99	2761.7	1.9826	6.7150
166	7.184	0.001109	0.2662	700.92	2572.6	701.71	2763.8	2.0025	6.6982
168	7.545	0.001112	0.2541	709.61	2574.2	710.45	2765.9	2.0222	6.6815
170	7.921	0.001114	0.2426	718.32	2575.7	719.21	2767.9	2.0419	6.6649
172	8.311	0.001117	0.2318	727.05	2577.2	727.97	2769.9	2.0616	6.6485
174	8.716	0.001119	0.2215	735.78	2578.7	736.75	2771.8	2.0811	6.6322
176	9.137	0.001122	0.2118	744.53	2580.1	745.55	2773.6	2.1007	6.6161
178	9.573	0.001125	0.2026	753.28	2581.5	754.36	2775.4	2.1201	6.6000
180	10.03	0.001127	0.1939	762.06	2582.8	763.19	2777.2	2.1395	6.5841
182	10.50	0.001130	0.1856	770.84	2584.2	772.03	2778.9	2.1589	6.5682
184	10.98	0.001133	0.1777	779.64	2585.4	780.89	2780.6	2.1782	6.5525
186	11.49	0.001136	0.1702	788.46	2586.7	789.76	2782.2	2.1974	6.5369
188	12.01	0.001139	0.1631	797.29	2587.9	798.66	2783.8	2.2166	6.5214
190	12.55	0.001141	0.1564	806.13	2589.1	807.57	2785.3	2.2358	6.5060
192	13.11	0.001144	0.1500	814.99	2590.2	816.49	2786.8	2.2549	6.4907
194	13.69	0.001147	0.1438	823.87	2591.3	825.44	2788.2	2.2739	6.4755
196	14.29	0.001150	0.1380	832.76	2592.3	834.40	2789.5	2.2929	6.4603
198	14.91	0.001153	0.1325	841.67	2593.3	843.39	2790.8	2.3119	6.4453
200	15.55	0.001157	0.1272	850.60	2594.3	852.39	2792.1	2.3308	6.4303
205	17.24	0.001164	0.1151	872.99	2596.5	874.99	2794.9	2.3779	6.3932
210	19.07	0.001173	0.1043	895.49	2598.4	897.73	2797.4	2.4248	6.3565
215	21.06	0.001181	0.09469	918.12	2600.0	920.61	2799.4	2.4714	6.3202
220	23.19	0.001190	0.08610	940.88	2601.4	943.64	2801.1	2.5178	6.2842
225	25.49	0.001199	0.07841	963.78	2602.4	966.84	2802.3	2.5641	6.2485
230	27.97	0.001209	0.07151	986.83	2603.0	990.21	2803.0	2.6102	6.2131
235	30.62	0.001219	0.06530	1010.0	2603.3	1013.8	2803.3	2.6561	6.1777
240	33.47	0.001229	0.05971	1033.4	2603.2	1037.5	2803.1	2.7019	6.1425
245	36.51	0.001240	0.05466	1057.0	2602.8	1061.5	2802.3	2.7477	6.1074
250	39.76	0.001252	0.05009	1080.7	2601.9	1085.7	2801.0	2.7934	6.0722
255	43.23	0.001264	0.04594	1104.7	2600.5	1110.1	2799.1	2.8391	6.0370
260	46.92	0.001276	0.04218	1128.7	2598.8	1134.8	2796.6	2.8847	6.0017
265	50.85	0.001289	0.03875	1153.3	2596.5	1159.8	2793.5	2.9304	5.9662
270	55.03	0.001303	0.03562	1177.9	2593.7	1185.1	2789.7	2.9762	5.9304
275	59.46	0.001318	0.03277	1202.9	2590.3	1210.7	2785.1	3.0221	5.8943
280	64.16	0.001333	0.03015	1228.1	2586.3	1236.7	2779.8	3.0681	5.8578
285	69.15	0.001349	0.02776	1253.7	2581.7	1263.0	2773.7	3.1143	5.8208
290	74.42	0.001366	0.02556	1279.6	2576.0	1289.8	2766.6	3.1608	5.7832
295	79.99	0.001385	0.02353	1306.0	2570.4	1317.0	2758.6	3.2076	5.7449
300	85.88	0.001404	0.02166	1332.7	2563.5	1344.8	2749.6	3.2547	5.7058
305	92.09	0.001425	0.01994	1359.9	2555.8	1373.1	2739.4	3.3024	5.6656
310	98.65	0.001448	0.01834	1387.7	2547.0	1402.0	2727.9	3.3506	5.6243
315	105.6	0.001472	0.01686	1416.1	2537.2	1431.6	2715.1	3.3994	5.5816
320	112.8	0.001499	0.01548	1445.1	2526.0	1462.1	2700.7	3.4491	5.5373
325	120.5	0.001528	0.01419	1475.0	2513.5	1493.4	2684.5	3.4997	5.4911
330	128.6	0.001561	0.01298	1505.7	2499.3	1525.7	2666.2	3.5516	5.4425
335	137.1	0.001597	0.01185	1537.5	2483.1	1559.3	2645.6	3.6048	5.3910
340	146.0	0.001638	0.01078	1570.5	2464.6	1594.4	2622.1	3.6599	5.3359
345	155.4	0.001685	0.009770	1605.3	2443.2	1631.4	2595.0	3.7175	5.2763
350	165.3	0.001740	0.008801	1642.1	2418.2	1670.9	2563.6	3.7783	5.2109
355	175.7	0.001808	0.007866	1681.9	2388.2	1713.7	2526.4	3.8438	5.1377
360	186.7	0.001895	0.006945	1726.1	2351.3	1761.5	2481.0	3.9164	5.0527
365	198.2	0.002016	0.006004	1777.6	2303.0	1817.6	2422.0	4.0010	4.9482
370	210.4	0.002222	0.004946	1845.9	2229.4	1892.6	2333.5	4.1142	4.7996
373.946	220.64	0.003110	0.003110	2018.9	2018.9	2087.5	2087.5	4.4120	4.4120

Table E-2: Superheated Steam (part 1 of 5)

P in bar (T^{sat} in C)		L	V	50	100	150	200	250	300	350	400	450	500	550	600	700
0.05 (32.88)	V	0.00101	28.186	29.782	34.419	39.043	43.663	48.281	52.898	57.515	62.131	66.747	71.363	75.979	80.594	89.826
	U	137.76	2419.8	2444.4	2516.0	2588.2	2661.5	2736.2	2812.4	2890.1	2969.4	3050.3	3132.9	3217.3	3303.3	3480.8
	H	137.77	2560.8	2593.4	2688.0	2783.4	2879.8	2977.6	3076.9	3177.6	3280.0	3384.0	3489.7	3597.1	3706.3	3929.9
	S	0.4763	8.3939	8.4976	8.7700	9.0097	9.2251	9.4216	9.6027	9.7713	9.9293	10.078	10.220	10.354	10.483	10.725
0.1 (45.81)	V	0.00101	14.671	14.867	17.197	19.514	21.826	24.136	26.446	28.755	31.064	33.372	35.680	37.988	40.296	44.912
	U	191.80	2437.2	2443.3	2515.5	2587.9	2661.3	2736.1	2812.3	2890.0	2969.3	3050.2	3132.9	3217.2	3303.3	3480.8
	H	191.81	2583.9	2592.0	2687.4	2783.0	2879.6	2977.4	3076.7	3177.5	3279.9	3384.0	3489.7	3597.1	3706.3	3929.9
	S	0.6492	8.1489	8.1741	8.4488	8.689	8.905	9.101	9.283	9.451	9.609	9.758	9.900	10.034	10.163	10.405
0.2 (60.06)	V	0.00102	7.648		8.586	9.749	10.907	12.064	13.220	14.375	15.530	16.684	17.839	18.993	20.147	22.455
	U	251.38	2456.0		2514.5	2587.3	2661.0	2735.8	2812.1	2889.9	2969.2	3050.1	3132.8	3217.1	3303.2	3480.7
	H	251.40	2608.9		2686.2	2782.3	2879.1	2977.1	3076.5	3177.4	3279.8	3383.8	3489.6	3597.0	3706.2	3929.8
	S	0.8320	7.9072		8.1262	8.3680	8.5842	8.7811	8.9624	9.1311	9.2892	9.4383	9.5797	9.7143	9.8431	10.086
0.5 (81.32)	V	0.00103	3.240		3.4188	3.8899	4.3563	4.8207	5.2841	5.7470	6.2095	6.6718	7.1339	7.5959	8.0578	8.981
	U	340.42	2483.2		2511.5	2585.7	2660.0	2735.1	2811.6	2889.4	2968.8	3049.9	3132.6	3216.9	3303.1	3480.6
	H	340.48	2645.2		2682.4	2780.2	2877.8	2976.2	3075.8	3176.8	3279.3	3383.5	3489.2	3596.7	3706.0	3929.7
	S	1.0910	7.5930		7.6952	7.9412	8.1591	8.3568	8.5386	8.7076	8.8658	9.0150	9.1565	9.2912	9.4200	9.6625
1 (99.61)	V	0.00104	1.694		1.6960	1.9367	2.1725	2.4062	2.6389	2.8710	3.1027	3.3342	3.5656	3.7968	4.0279	4.490
	U	417.33	2505.5		2506.2	2582.9	2658.2	2733.9	2810.7	2888.7	2968.3	3049.4	3132.2	3216.6	3302.8	3480.4
	H	417.44	2674.9		2675.8	2776.6	2875.5	2974.5	3074.5	3175.8	3278.5	3382.8	3488.7	3596.3	3705.6	3929.4
	S	1.3026	7.3588		7.3610	7.6147	7.8356	8.0346	8.2171	8.3865	8.5451	8.6945	8.8361	8.9709	9.0998	9.3424
1.5 (111.35)	V	0.00105	1.1594			1.2856	1.4445	1.6013	1.7571	1.9123	2.0671	2.2217	2.3762	2.5305	2.6847	2.993
	U	466.92	2519.2			2580.1	2656.5	2732.7	2809.7	2888.0	2967.7	3048.9	3131.7	3216.3	3302.5	3480.2
	H	467.08	2693.1			2772.9	2873.1	2972.9	3073.3	3174.9	3277.8	3382.2	3488.2	3595.8	3705.2	3929.1
	S	1.4335	7.2229			7.4207	7.6447	7.8451	8.0284	8.1983	8.3571	8.5067	8.6484	8.7833	8.9123	9.1550
2 (120.21)	V	0.00106	0.8857			0.9599	1.0805	1.1989	1.3162	1.4330	1.5493	1.6655	1.7814	1.8973	2.0130	2.2444
	U	504.47	2529.1			2577.1	2654.7	2731.5	2808.8	2887.3	2967.1	3048.4	3131.3	3215.9	3302.2	3479.9
	H	504.68	2706.2			2769.1	2870.8	2971.3	3072.1	3173.9	3277.0	3381.5	3487.6	3595.4	3704.8	3928.8
	S	1.5301	7.1269			7.2809	7.5081	7.7100	7.8940	8.0643	8.2235	8.3733	8.5151	8.6501	8.7792	9.0220

Temperature (°C)

Appendix E Steam Tables

P in bar (T^{sat} in C)		50	100	150	200	250	300	350	400	450	500	550	600	700
2.5 (127.41)	V U H S	0.00107 535.08 535.35 1.6072	0.7187 2536.8 2716.5 7.0524	0.7644 2574.1 2765.2 7.1707	0.8621 2652.9 2868.4 7.4013	0.9574 2730.2 2969.6 7.6046	1.0517 2807.9 3070.8 7.7895	1.1454 2886.6 3172.9 7.9602	1.2387 2966.5 3276.2 8.1196	1.3317 3048.0 3380.9 8.2696	1.4246 3130.9 3487.1 8.4116	1.5174 3215.6 3594.9 8.5467	1.6101 3301.9 3704.4 8.6759	1.7952 3479.7 3928.5 8.9188
3 (133.53)	V U H S	0.00107 561.13 561.46 1.6718	0.6058 2543.2 2724.9 6.9916	0.6340 2571.0 2761.2 7.0791	0.7164 2651.0 2866.0 7.3132	0.7965 2729.0 2967.9 7.5181	0.8753 2807.0 3069.6 7.7037	0.9536 2885.9 3172.0 7.8749	1.0315 2966.0 3275.4 8.0346	1.1092 3047.5 3380.2 8.1848	1.1867 3130.5 3486.6 8.3269	1.2641 3215.2 3594.5 8.4622	1.3414 3301.6 3704.0 8.5914	1.4958 3479.5 3928.2 8.8344
3.5 (138.86)	V U H S	0.00108 583.93 584.31 1.7275	0.5242 2548.5 2732.0 6.9401	0.5408 2567.8 2757.1 7.0002	0.6124 2649.1 2863.5 7.2381	0.6815 2727.7 2966.3 7.4445	0.7494 2806.1 3068.4 7.6310	0.8167 2885.2 3171.0 7.8026	0.8836 2965.4 3274.6 7.9626	0.9503 3047.0 3379.6 8.1130	1.0168 3130.1 3486.0 8.2553	1.0832 3214.9 3594.0 8.3906	1.1495 3301.3 3703.6 8.5199	1.2819 3479.2 3927.9 8.7630
4 (143.61)	V U H S	0.00108 604.29 604.72 1.7766	0.4624 2553.1 2738.1 6.8954	0.4709 2564.4 2752.8 6.9305	0.5343 2647.3 2861.0 7.1724	0.5952 2726.5 2964.6 7.3805	0.6549 2805.2 3067.1 7.5677	0.7139 2884.4 3170.0 7.7398	0.7726 2964.8 3273.9 7.9001	0.8311 3046.5 3379.0 8.0507	0.8894 3129.7 3485.5 8.1931	0.9475 3214.5 3593.6 8.3286	1.0056 3301.0 3703.2 8.4579	1.1215 3479.0 3927.6 8.7012
4.5 (147.91)	V U H S	0.00109 622.73 623.22 1.8206	0.4139 2557.1 2743.4 6.8560	0.4164 2560.9 2748.3 6.8677		0.4736 2645.3 2858.5 7.1139	0.5281 2725.2 2962.8 7.3237	0.5814 2804.2 3065.9 7.5117	0.6341 2883.7 3169.0 7.6843	0.6863 2964.2 3273.1 7.8449	0.7384 3046.0 3378.3 7.9957	0.7902 3129.3 3484.9 8.1383	0.8420 3214.2 3593.1 8.2738	0.8936 3300.7 3702.8 8.4032	0.9968 3478.8 3927.3 8.6466
5 (151.84)	V U H S	0.00109 639.64 640.19 1.8606	0.3748 2560.7 2748.1 6.8206			0.4250 2643.4 2855.9 7.0611	0.4744 2723.9 2961.1 7.2726	0.5226 2803.3 3064.6 7.4614	0.5701 2883.0 3168.1 7.6345	0.6173 2963.6 3272.3 7.7954	0.6642 3045.6 3377.7 7.9464	0.7109 3128.9 3484.4 8.0891	0.7576 3213.9 3592.6 8.2247	0.8041 3300.4 3702.5 8.3543	0.8970 3478.6 3927.0 8.5977
5.5 (155.46)	V U H S	0.00110 655.27 655.88 1.8972	0.3426 2563.9 2752.3 6.7885			0.3853 2641.4 2853.3 7.0128	0.4305 2722.6 2959.4 7.2261	0.4745 2802.4 3063.3 7.4158	0.5178 2882.3 3167.1 7.5894	0.5608 2963.1 3271.5 7.7505	0.6035 3045.1 3377.0 7.9017	0.6461 3128.5 3483.9 8.0446	0.6885 3213.5 3592.2 8.1803	0.7308 3300.1 3702.1 8.3099	0.8153 3478.3 3926.8 8.5535
6 (158.83)	V U H S	0.00110 669.84 670.50 1.9311	0.3156 2566.8 2756.1 6.7592			0.3521 2639.4 2850.7 6.9684	0.3939 2721.3 2957.7 7.1834	0.4344 2801.4 3062.1 7.3740	0.4743 2881.5 3166.1 7.5480	0.5137 2962.5 3270.7 7.7095	0.5530 3044.6 3376.4 7.8609	0.5920 3128.1 3483.3 8.0039	0.6309 3213.2 3591.7 8.1398	0.6698 3299.8 3701.7 8.2694	0.7473 3478.1 3926.5 8.5131

Temperature (°C)

Table E-2: Superheated Steam, cont. (part 2 of 5)

P in bar (T^{sat} in °C)		L	V	\multicolumn{13}{c}{Temperature (°C)}

P in bar (T^{sat} in °C)		L	V	200	250	300	350	400	450	500	550	600	650	700	750	800
6.5 (161.99)	V	0.00110	0.293	0.3241	0.3629	0.4005	0.4374	0.4739	0.5102	0.5463	0.5822	0.6181	0.6539	0.6897	0.7254	0.7611
	U	683.50	2569.4	2637.4	2720.0	2800.5	2880.8	2961.9	3044.1	3127.7	3212.8	3299.5	3387.9	3477.9	3569.6	3663.0
	H	684.22	2759.6	2848.0	2955.9	3060.8	3165.1	3269.9	3375.7	3482.8	3591.3	3701.3	3812.8	3926.2	4041.1	4157.7
	S	1.9626	6.7321	6.9270	7.1439	7.3354	7.5099	7.6717	7.8233	7.9665	8.1024	8.2321	8.3564	8.4759	8.5911	8.7024
7 (164.95)	V	0.00111	0.2728	0.3000	0.3364	0.3714	0.4058	0.4398	0.4735	0.5070	0.5405	0.5738	0.6071	0.6403	0.6735	0.7066
	U	696.37	2571.8	2635.3	2718.7	2799.5	2880.1	2961.3	3043.6	3127.3	3212.5	3299.2	3387.6	3477.7	3569.4	3662.8
	H	697.14	2762.7	2845.3	2954.1	3059.5	3164.1	3269.1	3375.1	3482.3	3590.8	3700.9	3812.6	3925.9	4040.8	4157.5
	S	1.9921	6.7070	6.8884	7.1071	7.2995	7.4745	7.6366	7.7884	7.9317	8.0678	8.1976	8.3220	8.4415	8.5567	8.6680
7.5 (167.76)	V	0.00111	0.2555	0.2791	0.3133	0.3462	0.3784	0.4102	0.4417	0.4731	0.5043	0.5354	0.5665	0.5975	0.6285	0.6595
	U	708.55	2574.0	2633.2	2717.3	2798.6	2879.4	2960.7	3043.2	3126.9	3212.1	3298.9	3387.3	3477.4	3569.2	3662.7
	H	709.38	2765.6	2842.5	2952.3	3058.2	3163.1	3268.4	3374.4	3481.7	3590.4	3700.5	3812.2	3925.6	4040.6	4157.3
	S	2.0198	6.6835	6.8520	7.0727	7.2660	7.4415	7.6039	7.7559	7.8994	8.0355	8.1654	8.2898	8.4094	8.5246	8.6360
8 (170.41)	V	0.00111	0.2403	0.2609	0.2932	0.3242	0.3544	0.3843	0.4139	0.4433	0.4726	0.5019	0.5310	0.5601	0.5892	0.6182
	U	720.13	2576.0	2631.1	2716.0	2797.6	2878.6	2960.1	3042.7	3126.5	3211.8	3298.6	3387.1	3477.2	3569.0	3662.5
	H	721.02	2768.3	2839.8	2950.5	3056.9	3162.2	3267.6	3373.8	3481.2	3589.9	3700.1	3811.9	3925.3	4040.3	4157.0
	S	2.0460	6.6615	6.8176	7.0403	7.2345	7.4106	7.5733	7.7255	7.8690	8.0053	8.1353	8.2598	8.3794	8.4947	8.6060
9 (175.36)	V	0.00112	0.2149	0.2304	0.2596	0.2874	0.3145	0.3411	0.3675	0.3938	0.4199	0.4459	0.4718	0.4977	0.5236	0.5494
	U	741.72	2579.7	2626.7	2713.3	2795.7	2877.2	2959.0	3041.7	3125.7	3211.1	3298.0	3386.6	3476.7	3568.6	3662.1
	H	742.72	2773.0	2834.1	2946.9	3054.3	3160.2	3266.0	3372.5	3480.1	3589.0	3699.3	3811.2	3924.7	4039.8	4156.6
	S	2.0944	6.6212	6.7538	6.9806	7.1768	7.3538	7.5172	7.6698	7.8136	7.9501	8.0803	8.2049	8.3246	8.4399	8.5513
10 (179.89)	V	0.00113	0.1943	0.2060	0.2327	0.2580	0.2825	0.3066	0.3304	0.3541	0.3777	0.4011	0.4245	0.4478	0.4711	0.4944
	U	761.56	2582.8	2622.3	2710.5	2793.7	2875.7	2957.8	3040.8	3124.9	3210.4	3297.4	3386.1	3476.3	3568.2	3661.8
	H	762.68	2777.1	2828.3	2943.2	3051.7	3158.2	3264.4	3371.2	3479.0	3588.1	3698.6	3810.5	3924.1	4039.3	4156.1
	S	2.1384	6.5850	6.6955	6.9266	7.1247	7.3028	7.4668	7.6198	7.7640	7.9007	8.0309	8.1557	8.2755	8.3909	8.5024
11 (184.07)	V	0.00113	0.1774	0.1860	0.2107	0.2339	0.2563	0.2783	0.3001	0.3217	0.3431	0.3645	0.3858	0.4070	0.4282	0.4494
	U	779.95	2585.5	2617.6	2707.7	2791.8	2874.2	2956.6	3039.8	3124.1	3209.7	3296.9	3385.5	3475.8	3567.8	3661.4
	H	781.20	2780.7	2822.3	2939.5	3049.1	3156.2	3262.8	3369.9	3477.9	3587.2	3697.8	3809.9	3923.5	4038.8	4155.7
	S	2.1789	6.5520	6.6414	6.8772	7.0773	7.2564	7.4210	7.5745	7.7189	7.8558	7.9863	8.1111	8.2310	8.3465	8.4580
12 (187.96)	V	0.00114	0.1632	0.1693	0.1924	0.2139	0.2345	0.2548	0.2748	0.2946	0.3143	0.3339	0.3535	0.3730	0.3924	0.4118
	U	797.13	2587.9	2612.9	2704.8	2789.8	2872.7	2955.4	3038.8	3123.3	3209.0	3296.3	3385.0	3475.4	3567.4	3661.0
	H	798.50	2783.8	2816.1	2935.7	3046.4	3154.1	3261.2	3368.6	3476.8	3586.2	3697.0	3809.2	3922.9	4038.3	4155.2
	S	2.2163	6.5217	6.5908	6.8314	7.0336	7.2138	7.3791	7.5330	7.6777	7.8148	7.9454	8.0704	8.1904	8.3059	8.4175

Appendix E Steam Tables

Temperature (°C)

P in bar (T^{sat} in C)		L	V	200	250	300	350	400	450	500	550	600	650	700	750	800
13 (191.61)	V	0.00114	0.1512	0.1552	0.1769	0.1969	0.2161	0.2349	0.2534	0.2718	0.2900	0.3081	0.3262	0.3442	0.3621	0.3801
	U	813.28	2590.0	2607.9	2701.9	2787.8	2871.2	2954.2	3037.8	3122.5	3208.3	3295.7	3384.5	3474.9	3567.0	3660.7
	H	814.76	2786.5	2809.6	2931.8	3043.7	3152.1	3259.6	3367.3	3475.7	3585.3	3696.2	3808.5	3922.4	4037.8	4154.8
	S	2.2512	6.4936	6.5430	6.7888	6.9931	7.1745	7.3404	7.4947	7.6397	7.7770	7.9078	8.0329	8.1530	8.2686	8.3803
14 (195.05)	V	0.00115	0.1408	0.1430	0.1635	0.1823	0.2003	0.2178	0.2351	0.2522	0.2691	0.2860	0.3028	0.3195	0.3362	0.3529
	U	828.52	2591.8	2602.8	2699.0	2785.8	2869.7	2953.0	3036.9	3121.6	3207.7	3295.1	3384.0	3474.5	3566.6	3660.3
	H	830.13	2788.9	2803.0	2927.9	3041.0	3150.1	3258.0	3366.0	3474.7	3584.4	3695.4	3807.8	3921.8	4037.2	4154.3
	S	2.2839	6.4675	6.4975	6.7488	6.9553	7.1378	7.3044	7.4591	7.6045	7.7420	7.8729	7.9981	8.1183	8.2340	8.3457
15 (198.30)	V	0.00115	0.1317	0.1324	0.1520	0.1697	0.1866	0.2030	0.2192	0.2352	0.2510	0.2668	0.2825	0.2981	0.3137	0.3293
	U	842.99	2593.5	2597.4	2696.0	2783.7	2868.2	2951.8	3035.9	3120.8	3207.0	3294.5	3383.5	3474.0	3566.2	3659.9
	H	844.72	2791.0	2796.0	2924.0	3038.3	3148.0	3256.4	3364.7	3473.6	3583.5	3694.6	3807.2	3921.2	4036.7	4153.9
	S	2.3147	6.4431	6.4537	6.7111	6.9199	7.1035	7.2708	7.4259	7.5716	7.7093	7.8404	7.9657	8.0860	8.2018	8.3135
16 (201.38)	V	0.00116	0.1237		0.1419	0.1587	0.1746	0.1901	0.2053	0.2203	0.2352	0.2500	0.2647	0.2794	0.2940	0.3086
	U	856.76	2594.9		2692.9	2781.7	2866.6	2950.6	3034.9	3120.0	3206.3	3293.9	3382.9	3473.6	3565.8	3659.6
	H	858.61	2792.9		2919.9	3035.5	3146.0	3254.7	3363.3	3472.5	3582.6	3693.9	3806.5	3920.6	4036.2	4153.4
	S	2.3438	6.4200		6.6754	6.8865	7.0713	7.2392	7.3948	7.5407	7.6787	7.8099	7.9354	8.0557	8.1716	8.2834
17 (204.31)	V	0.00116	0.1167		0.1330	0.1489	0.1640	0.1786	0.1930	0.2072	0.2212	0.2352	0.2491	0.2629	0.2767	0.2904
	U	869.91	2596.2		2689.8	2779.6	2865.1	2949.4	3033.9	3119.2	3205.6	3293.3	3382.4	3473.1	3565.3	3659.2
	H	871.89	2794.5		2915.9	3032.7	3143.9	3253.1	3362.0	3471.4	3581.6	3693.1	3805.8	3920.0	4035.7	4153.0
	S	2.3715	6.3983		6.6413	6.855	7.0408	7.2094	7.3654	7.5117	7.6499	7.7813	7.9068	8.0273	8.1432	8.2551
18 (207.12)	V	0.00117	0.1104		0.1250	0.1402	0.1546	0.1685	0.1821	0.1955	0.2088	0.2220	0.2351	0.2482	0.2612	0.274
	U	882.51	2597.3		2686.7	2777.5	2863.6	2948.2	3032.9	3118.4	3204.9	3292.7	3381.9	3472.6	3564.9	3658.8
	H	884.61	2796.0		2911.7	3029.9	3141.8	3251.5	3360.7	3470.3	3580.7	3692.3	3805.1	3919.4	4035.2	4152.5
	S	2.3978	6.3776		6.6087	6.8247	7.0119	7.1812	7.3377	7.4842	7.6226	7.7542	7.8799	8.0004	8.1164	8.2284
20 (212.38)	V	0.00118	0.0996		0.1115	0.1255	0.1386	0.1512	0.1635	0.1757	0.1877	0.1996	0.2115	0.2233	0.2350	0.247
	U	906.27	2599.2		2680.3	2773.2	2860.5	2945.8	3031.0	3116.7	3203.5	3291.5	3380.9	3471.7	3564.1	3658.1
	H	908.62	2798.4		2903.2	3024.3	3137.6	3248.2	3358.1	3468.1	3578.9	3690.7	3803.8	3918.2	4034.2	4151.6
	S	2.4470	6.3392		6.5474	6.7685	6.9582	7.1290	7.2863	7.4335	7.5723	7.7042	7.8301	7.9509	8.0670	8.1791

Appendix E Steam Tables

Table E-2: Superheated Steam, cont. (part 3 of 5)

P in bar (T^{sat} in C)		L	V	225	250	300	350	400	450	500	550	600	650	700	750	800
22 (217.26)	V	0.00119	0.091	0.0931	0.1004	0.1134	0.1255	0.1371	0.1484	0.1595	0.1704	0.1813	0.1921	0.2028	0.2136	0.2242
	U	928.37	2600.7	2619.6	2673.6	2768.9	2857.3	2943.4	3029.0	3115.1	3202.1	3290.3	3379.8	3470.8	3563.3	3657.4
	H	930.98	2800.2	2824.5	2894.5	3018.5	3133.4	3244.9	3355.4	3465.9	3577.0	3689.1	3802.4	3917.1	4033.1	4150.7
	S	2.4924	6.3040	6.3531	6.4903	6.7168	6.9091	7.0813	7.2396	7.3873	7.5266	7.6588	7.7850	7.9059	8.0222	8.1344
24 (221.80)	V	0.00119	0.0832	0.0842	0.0911	0.1034	0.1146	0.1253	0.1357	0.1459	0.1560	0.1660	0.1760	0.1858	0.1957	0.2055
	U	949.09	2601.8	2610.0	2666.8	2764.5	2854.1	2940.9	3027.0	3113.4	3200.7	3289.1	3378.8	3469.9	3562.5	3656.7
	H	951.95	2801.5	2812.1	2885.5	3012.6	3129.1	3241.6	3352.7	3463.7	3575.2	3687.6	3801.1	3915.9	4032.1	4149.8
	S	2.5344	6.2714	6.2926	6.4365	6.6688	6.8638	7.0375	7.1967	7.3450	7.4848	7.6173	7.7437	7.8648	7.9813	8.0936
26 (226.05)	V	0.00120	0.0769		0.0833	0.0948	0.1053	0.1153	0.1250	0.1345	0.1439	0.1531	0.1623	0.1714	0.1805	0.1896
	U	968.62	2602.5		2659.7	2760.0	2850.9	2938.4	3025.0	3111.8	3199.3	3287.9	3377.7	3469.0	3561.7	3655.9
	H	971.74	2802.5		2876.2	3006.6	3124.8	3238.3	3350.0	3461.5	3573.3	3686.0	3799.7	3914.7	4031.1	4148.9
	S	2.5738	6.2411		6.3854	6.6238	6.8216	6.9968	7.1570	7.3060	7.4461	7.5790	7.7056	7.8269	7.9435	8.0559
28 (230.06)	V	0.00121	0.0714		0.0765	0.0875	0.0974	0.1068	0.1158	0.1247	0.1334	0.1420	0.1506	0.1591	0.1676	0.1760
	U	987.12	2603.0		2652.3	2755.5	2847.7	2935.8	3023.0	3110.1	3197.9	3286.7	3376.7	3468.1	3560.9	3655.2
	H	990.50	2803.0		2866.5	3000.5	3120.5	3234.9	3347.4	3459.3	3571.5	3684.4	3798.4	3913.5	4030.0	4147.9
	S	2.6107	6.2126		6.3365	6.5814	6.7821	6.9589	7.1200	7.2696	7.4102	7.5434	7.6703	7.7918	7.9085	8.0211
30 (233.86)	V	0.00122	0.0667		0.0706	0.0812	0.09056	0.0994	0.1079	0.1162	0.1244	0.1324	0.1405	0.1484	0.1563	0.1642
	U	1004.72	2603.3		2644.7	2750.8	2844.4	2933.4	3021.0	3108.5	3196.5	3285.5	3375.6	3467.1	3560.1	3654.5
	H	1008.37	2803.3		2856.5	2994.3	3116.1	3231.6	3344.7	3457.0	3569.6	3682.8	3797.0	3912.3	4029.0	4147.0
	S	2.6456	6.1858		6.2893	6.5412	6.7449	6.9233	7.0853	7.2356	7.3767	7.5102	7.6373	7.7590	7.8759	7.9885
32 (237.46)	V	0.00122	0.0625		0.0655	0.0756	0.08454	0.0929	0.1009	0.1088	0.1165	0.1240	0.1316	0.1390	0.1465	0.1539
	U	1021.53	2603.3		2636.7	2746.1	2841.1	2930.9	3019.0	3106.8	3195.1	3284.3	3374.6	3466.2	3559.2	3653.7
	H	1025.45	2803.2		2846.2	2988.0	3111.6	3228.2	3341.9	3454.8	3567.7	3681.2	3795.6	3911.2	4028.0	4146.1
	S	2.6787	6.1604		6.2434	6.5029	6.7097	6.8897	7.0527	7.2036	7.3451	7.4790	7.6064	7.7283	7.8453	7.9581
34 (240.90)	V	0.00123	0.0588		0.0609	0.0707	0.07923	0.0872	0.0948	0.1022	0.1095	0.1166	0.1237	0.1308	0.1378	0.1448
	U	1037.64	2603.2		2628.4	2741.3	2837.7	2928.4	3017.0	3105.1	3193.7	3283.1	3373.5	3465.3	3558.4	3653.0
	H	1041.83	2803.0		2835.3	2981.6	3107.1	3224.8	3339.2	3452.6	3565.9	3679.6	3794.3	3910.0	4026.9	4145.2
	S	2.7102	6.1362		6.1986	6.4662	6.6762	6.8579	7.0219	7.1735	7.3154	7.4496	7.5773	7.6993	7.8165	7.9294
36 (244.19)	V	0.00124	0.0554		0.0568	0.0663	0.07451	0.0821	0.0893	0.0964	0.1033	0.1101	0.1168	0.1234	0.1301	0.1367
	U	1053.12	2602.9		2619.7	2736.5	2834.3	2925.8	3014.9	3103.4	3192.3	3281.8	3372.5	3464.4	3557.6	3652.3
	H	1057.57	2802.5		2824.0	2975.1	3102.6	3221.3	3336.5	3450.3	3564.0	3678.0	3792.9	3908.8	4025.9	4144.3
	S	2.7403	6.1131		6.1545	6.4309	6.6443	6.8276	6.9927	7.1449	7.2873	7.4219	7.5498	7.6720	7.7893	7.9023

Temperature (°C)

Appendix E Steam Tables

Temperature (°C)

P in bar (T^{sat} in C)		L	V	225	250	300	350	400	450	500	550	600	650	700	750	800
38 (247.33)	V U H S	0.00125 1068.02 1072.76 2.7690	0.0525 2602.4 2801.8 6.0910		0.0531 2610.4 2812.1 6.1107	0.0624 2731.3 2968.4 6.3968	0.07028 2830.9 3098.0 6.6137	0.0775 2923.2 3217.9 6.7988	0.0844 3012.9 3333.7 6.9649	0.0911 3101.8 3448.1 7.1178	0.0977 3190.8 3562.1 7.2607	0.1042 3280.6 3676.4 7.3955	0.1105 3371.4 3791.5 7.5237	0.1169 3463.4 3907.6 7.6461	0.1232 3556.8 4024.8 7.7636	0.1294 3651.5 4143.4 7.8767
40 (250.36)	V U H S	0.00125 1082.42 1087.43 2.7967	0.0498 2601.8 2800.9 6.0697			0.0589 2726.2 2961.7 6.3638	0.06647 2827.4 3093.3 6.5843	0.0734 2920.6 3214.4 6.7712	0.0800 3010.8 3331.0 6.9383	0.0864 3100.1 3445.8 7.0919	0.0927 3189.4 3560.2 7.2353	0.0989 3279.4 3674.8 7.3704	0.1049 3370.4 3790.2 7.4989	0.1110 3462.5 3906.4 7.6215	0.1170 3556.0 4023.8 7.7391	0.1229 3650.8 4142.5 7.8523
45 (257.44)	V U H S	0.00127 1116.43 1122.14 2.8613	0.0441 2599.7 2798.0 6.0198			0.0514 2712.9 2944.1 6.2852	0.05842 2818.6 3081.5 6.5153	0.0648 2914.1 3205.6 6.7069	0.0708 3005.6 3324.0 6.8767	0.0765 3095.8 3440.2 7.0320	0.0821 3185.9 3555.5 7.1765	0.0877 3276.4 3670.8 7.3126	0.0931 3367.7 3786.7 7.4416	0.0985 3460.2 3903.4 7.5647	0.1038 3553.9 4021.2 7.6827	0.1092 3649.0 4140.2 7.7963
50 (263.94)	V U H S	0.00129 1148.07 1154.50 2.9207	0.0394 2597.0 2794.2 5.9737			0.0453 2698.9 2925.6 6.2109	0.05197 2809.4 3069.3 6.4515	0.0578 2907.4 3196.6 6.6481	0.0633 3000.4 3317.0 6.8208	0.0686 3091.6 3434.5 6.9778	0.0737 3182.3 3550.8 7.1235	0.0787 3273.3 3666.8 7.2604	0.0836 3365.1 3783.3 7.3901	0.0885 3457.9 3900.5 7.5137	0.0933 3551.8 4018.6 7.6321	0.0982 3647.1 4137.9 7.7459
55 (269.97)	V U H S	0.00130 1177.76 1184.92 2.9759	0.0356 2593.7 2789.7 5.9307			0.0404 2684.1 2906.2 6.1396	0.04668 2800.0 3056.8 6.3919	0.0522 2900.6 3187.5 6.5938	0.0572 2995.1 3309.9 6.7693	0.0621 3087.3 3428.7 6.9282	0.0668 3178.7 3546.0 7.0751	0.0714 3270.2 3662.8 7.2129	0.0759 3362.4 3779.8 7.3432	0.0803 3455.6 3897.5 7.4673	0.0848 3549.8 4016.0 7.5861	0.0891 3645.3 4135.6 7.7002
60 (275.59)	V U H S	0.00132 1205.82 1213.73 3.0274	0.0324 2589.9 2784.6 5.8901			0.0362 2668.3 2885.5 6.0702	0.0423 2790.3 3043.9 6.3356	0.0474 2893.6 3178.2 6.5431	0.0522 2989.8 3302.8 6.7216	0.0567 3082.9 3422.9 6.8824	0.0610 3175.1 3541.2 7.0306	0.0653 3267.2 3658.8 7.1692	0.0694 3359.8 3776.4 7.3002	0.0735 3453.2 3894.5 7.4248	0.0776 3547.7 4013.4 7.5439	0.0816 3643.4 4133.3 7.6583
65 (280.86)	V U H S	0.00134 1232.49 1241.17 3.0760	0.0297 2585.6 2778.8 5.8515			0.0326 2651.5 2863.5 6.0018	0.0385 2780.3 3030.6 6.2819	0.0434 2886.6 3168.7 6.4953	0.0479 2984.3 3295.5 6.6771	0.0521 3078.5 3417.1 6.8397	0.0561 3171.4 3536.4 6.9892	0.0601 3264.1 3654.7 7.1287	0.0640 3357.1 3772.9 7.2603	0.0678 3450.9 3891.5 7.3854	0.0716 3545.6 4010.7 7.5050	0.0753 3641.6 4131.0 7.6197

Table E-2: Superheated Steam, cont. (part 4 of 5)

| P in bar (T^{sat} in C) | | L | V | \multicolumn{14}{c|}{Temperature (°C)} |
				300	325	350	375	400	450	500	550	600	650	700	750	800
70 (285.83)	V U H S	0.00135 1258.0 1267.4 3.1220	0.027 2580.9 2772.6 5.8146	0.0295 2633.4 2839.8 5.9335	0.0326 2707.4 2935.5 6.0970	0.0353 2770.0 3016.8 6.2303	0.0377 2826.5 3090.4 6.3460	0.0400 2879.4 3159.1 6.4501	0.0442 2978.8 3288.2 6.6351	0.0482 3074.1 3411.3 6.7997	0.0520 3167.8 3531.5 6.9505	0.0557 3261.0 3650.6 7.0909	0.0593 3354.4 3769.4 7.2232	0.0628 3448.5 3888.5 7.3488	0.0664 3543.6 4008.1 7.4687	0.0698 3639.7 4128.7 7.5837
75 (290.54)	V U H S	0.00137 1282.4 1292.7 3.1658	0.0253 2575.8 2765.8 5.7792	0.0267 2613.7 2814.3 5.8644	0.0298 2693.6 2917.4 6.0407	0.0325 2759.3 3002.7 6.1805	0.0348 2817.8 3078.8 6.3002	0.0370 2872.0 3149.3 6.4070	0.0410 2973.3 3280.7 6.5954	0.0448 3069.7 3405.3 6.7620	0.0483 3164.1 3526.7 6.9141	0.0518 3257.9 3646.5 7.0555	0.0552 3351.7 3765.9 7.1885	0.0586 3446.2 3885.4 7.3145	0.0619 3541.5 4005.5 7.4348	0.0651 3637.8 4126.3 7.5502
80 (295.01)	V U H S	0.00138 1306.0 1317.1 3.2077	0.0235 2570.4 2758.6 5.7448	0.0243 2592.1 2786.4 5.7935	0.0274 2679.1 2898.3 5.9849	0.0300 2748.2 2988.1 6.1319	0.0323 2808.9 3066.9 6.2560	0.0343 2864.5 3139.3 6.3657	0.0382 2967.7 3273.2 6.5577	0.0418 3065.2 3399.4 6.7264	0.0452 3160.4 3521.8 6.8798	0.0485 3254.7 3642.4 7.0221	0.0517 3349.0 3762.4 7.1557	0.0548 3443.8 3882.4 7.2823	0.0579 3539.4 4002.9 7.4030	0.0610 3636.0 4124.0 7.5186
85 (299.27)	V U H S	0.00140 1328.8 1340.7 3.2478	0.0219 2564.6 2751.0 5.7115	0.0220 2568.1 2755.4 5.7193	0.0252 2663.7 2878.3 5.9294	0.0278 2736.8 2972.9 6.0845	0.0300 2799.7 3054.7 6.2132	0.0320 2856.9 3129.1 6.3259	0.0357 2962.0 3265.6 6.5216	0.0391 3060.7 3393.4 6.6925	0.0424 3156.7 3516.9 6.8473	0.0455 3251.6 3638.3 6.9905	0.0485 3346.3 3758.9 7.1248	0.0515 3441.5 3879.4 7.2519	0.0545 3537.3 4000.2 7.3730	0.0574 3634.1 4121.7 7.4889
90 (303.35)	V U H S	0.00142 1350.9 1363.7 3.2866	0.0205 2558.4 2742.9 5.6790		0.0233 2647.4 2857.0 5.8736	0.0258 2724.9 2957.2 6.0378	0.0280 2790.2 3042.3 6.1716	0.0300 2849.1 3118.8 6.2875	0.0335 2956.2 3257.9 6.4871	0.0368 3056.2 3387.3 6.6601	0.0399 3152.9 3511.9 6.8163	0.0429 3248.4 3634.2 6.9605	0.0458 3343.6 3755.4 7.0955	0.0486 3439.1 3876.6 7.2231	0.0514 3535.2 3997.6 7.3446	0.0541 3632.2 4119.4 7.4608
95 (307.25)	V U H S	0.00144 1372.1 1386.0 3.3240	0.0192 2552.0 2734.4 5.6472		0.0215 2630.1 2834.4 5.8170	0.0240 2712.5 2940.9 5.9917	0.0262 2780.5 3029.4 6.1309	0.0281 2841.1 3108.2 6.2502	0.0316 2950.3 3250.2 6.4538	0.0347 3051.6 3381.2 6.6291	0.0377 3149.2 3506.9 6.7867	0.0405 3245.3 3630.0 6.9319	0.0433 3340.9 3751.9 7.0676	0.0460 3436.7 3873.3 7.1957	0.0486 3533.1 3994.9 7.3176	0.0512 3630.4 4117.1 7.4341
100 (311.00)	V U H S	0.00145 1393.3 1407.9 3.3603	0.0180 2545.1 2725.5 5.6159		0.0199 2611.4 2810.2 5.7593	0.0224 2699.5 2924.0 5.9458	0.0246 2770.6 3016.2 6.0910	0.0264 2833.0 3097.4 6.2139	0.0298 2944.4 3242.3 6.4217	0.0328 3046.9 3375.1 6.5993	0.0357 3145.4 3501.9 6.7584	0.0384 3242.1 3625.8 6.9045	0.0410 3338.2 3748.3 7.0409	0.0436 3434.3 3870.3 7.1696	0.0461 3531.0 3992.3 7.2918	0.0486 3628.5 4114.7 7.4087
110 (318.08)	V U H S	0.00149 1433.9 1450.3 3.4300	0.0160 2530.5 2706.4 5.5545		0.0170 2569.1 2755.6 5.6373	0.0196 2671.9 2887.8 5.8541	0.0217 2749.7 2988.7 6.0129	0.0235 2816.2 3075.1 6.1438	0.0267 2932.4 3226.2 6.3605	0.0296 3037.6 3362.6 6.5430	0.0322 3137.8 3491.9 6.7050	0.0347 3235.7 3617.5 6.8531	0.0371 3332.7 3741.2 6.9910	0.0395 3429.5 3864.2 7.1207	0.0418 3526.8 3987.0 7.2437	0.0441 3624.7 4110.1 7.3612

Appendix E Steam Tables

P in bar (T^{sat} in C)		L	V	300	325	350	375	400	450	500	550	600	650	700	750	800
120 (324.68)	V	0.00153	0.0143		0.0143	0.0172	0.0193	0.0211	0.0242	0.0268	0.0293	0.0317	0.0339	0.0361	0.0383	0.0404
	U	1473.0	2514.4		2516.6	2641.3	2727.5	2798.6	2920.0	3028.0	3130.0	3229.2	3327.1	3424.7	3522.6	3621.0
	H	1491.3	2685.6		2688.4	2848.0	2959.5	3051.9	3209.8	3350.0	3481.7	3609.0	3734.1	3858.0	3981.6	4105.4
	S	3.4965	5.4941		5.4988	5.7607	5.9362	6.0762	6.3027	6.4902	6.6553	6.8055	6.9448	7.0756	7.1994	7.3175
130 (330.86)	V	0.00157	0.0128			0.0151	0.0173	0.0190	0.0220	0.0245	0.0269	0.0291	0.0312	0.0332	0.0352	0.0372
	U	1511.0	2496.7			2607.1	2703.7	2780.2	2907.2	3018.3	3122.2	3222.7	3321.6	3419.9	3518.3	3617.2
	H	1531.4	2662.9			2803.6	2928.3	3027.6	3192.9	3337.1	3471.4	3600.5	3726.9	3851.9	3976.3	4100.7
	S	3.5606	5.4339			5.6635	5.8600	6.0104	6.2475	6.4404	6.6087	6.7610	6.9018	7.0336	7.1583	7.2771
140 (336.67)	V	0.00161	0.0115			0.0132	0.0155	0.0172	0.0201	0.0225	0.0248	0.0268	0.0288	0.0308	0.0326	0.0345
	U	1548.3	2477.2			2567.7	2678.1	2760.9	2894.1	3008.4	3114.3	3216.1	3316.0	3415.1	3514.0	3613.4
	H	1570.9	2638.1			2752.9	2894.9	3002.2	3175.6	3324.1	3461.0	3591.9	3719.7	3845.7	3970.9	4096.0
	S	3.6230	5.3730			5.5595	5.7832	5.9457	6.1945	6.3931	6.5648	6.7192	6.8615	6.9944	7.1200	7.2393
150 (342.16)	V	0.00166	0.0103			0.0115	0.0139	0.0157	0.0185	0.0208	0.0229	0.0249	0.0268	0.0286	0.0304	0.0321
	U	1585.3	2455.8			2520.8	2650.3	2740.5	2880.7	2998.4	3106.3	3209.5	3310.4	3410.2	3509.8	3609.6
	H	1610.2	2610.9			2693.0	2858.9	2975.5	3157.8	3310.8	3450.5	3583.3	3712.4	3839.5	3965.6	4091.3
	S	3.6844	5.3108			5.4435	5.7049	5.8817	6.1433	6.3479	6.5230	6.6797	6.8235	6.9576	7.0839	7.2039
160 (347.36)	V	0.00171	0.0093			0.0098	0.0125	0.0143	0.0170	0.0193	0.0214	0.0232	0.0250	0.0267	0.0284	0.0301
	U	1622.3	2431.9			2460.7	2619.8	2719.0	2866.8	2988.1	3098.2	3202.8	3304.7	3405.3	3505.5	3605.7
	H	1649.7	2580.8			2617.0	2819.5	2947.5	3139.6	3297.3	3439.8	3574.6	3705.1	3833.3	3960.2	4086.6
	S	3.7457	5.2463			5.3045	5.6238	5.8177	6.0935	6.3045	6.4832	6.6422	6.7876	6.9228	7.0499	7.1706
170 (352.29)	V	0.00177	0.0084				0.0112	0.0130	0.0158	0.0180	0.0199	0.0218	0.0235	0.0251	0.0267	0.0282
	U	1660.0	2405.1				2586.0	2696.1	2852.6	2977.7	3090.0	3196.1	3299.0	3400.4	3501.1	3601.9
	H	1690.0	2547.4				2775.9	2917.8	3120.9	3283.6	3429.1	3565.9	3697.8	3827.8	3954.8	4081.9
	S	3.8077	5.1785				5.5384	5.7533	6.0449	6.2627	6.4451	6.6064	6.7534	6.8897	7.0178	7.1391
180 (356.99)	V	0.00184	0.0075				0.0100	0.0119	0.0147	0.0168	0.0187	0.0204	0.0221	0.0236	0.0251	0.0266
	U	1698.9	2374.6				2547.6	2671.8	2837.9	2967.1	3081.7	3189.3	3293.3	3395.4	3496.8	3598.1
	H	1732.0	2509.5				2726.9	2886.3	3101.7	3269.7	3418.3	3557.0	3690.4	3820.7	3949.4	4077.2
	S	3.8717	5.1055				5.4465	5.6881	5.9973	6.2222	6.4085	6.5722	6.7208	6.8583	6.9872	7.1091

Temperature (°C)

Table E-2: Superheated Steam, cont. (part 5 of 5)

P in bar

T^{sat} in C		L	V	375	400	425	450	475	500	550	600	650	700	750	800
200 (365.75)	V	0.00204	0.006	0.0077	0.0099	0.0115	0.0127	0.0138	0.0148	0.0166	0.0182	0.0197	0.0211	0.0225	0.0239
	U	1786.33	2294.2	2449.0	2617.8	2723.3	2807.1	2879.6	2945.3	3064.8	3175.5	3281.7	3385.5	3488.1	3590.4
	H	1827.10	2411.4	2602.6	2816.8	2952.9	3061.5	3155.8	3241.2	3396.2	3539.2	3675.6	3808.2	3938.5	4067.7
	S	4.0154	4.9299	5.2275	5.5525	5.7510	5.9041	6.0322	6.1445	6.3390	6.5077	6.6596	6.7994	6.9301	7.0534
220 (373.71)	V	0.00275	0.0036	0.0049	0.0083	0.0099	0.0111	0.0122	0.0131	0.0148	0.0163	0.0178	0.0191	0.0204	0.0216
	U	1961.4	2085.5	2246.8	2554.2	2680.6	2774.4	2852.9	2922.8	3047.6	3161.6	3270.0	3375.5	3479.4	3582.6
	H	2021.9	2164.2	2354.7	2735.8	2897.8	3019.0	3121.0	3211.8	3373.8	3521.2	3660.6	3795.5	3927.6	4058.2
	S	4.3109	4.5308	4.8251	5.4050	5.6417	5.8124	5.9511	6.0704	6.2736	6.4475	6.6029	6.7451	6.8776	7.0022
240	V			0.0021	0.0067	0.0085	0.0098	0.0108	0.0118	0.0134	0.0148	0.0161	0.0174	0.0186	0.0197
	U			1822.7	2475.8	2633.4	2739.4	2825.0	2899.4	3029.9	3147.4	3258.2	3365.4	3470.6	3574.8
	H			1872.2	2637.4	2837.4	2974.0	3084.8	3181.4	3350.9	3502.9	3645.6	3782.8	3916.7	4048.8
	S			4.0727	5.2366	5.5289	5.7212	5.8720	5.9991	6.2116	6.3910	6.5499	6.6946	6.8289	6.9549
260	V			0.0019	0.0053	0.0073	0.0086	0.0097	0.0106	0.0121	0.0135	0.0148	0.0160	0.0171	0.0182
	U			1782.8	2372.9	2580.6	2702.0	2795.6	2875.1	3011.8	3133.0	3246.2	3355.2	3461.7	3567.0
	H			1832.8	2510.3	2770.6	2926.1	3047.0	3150.2	3327.6	3484.4	3630.4	3770.0	3905.8	4039.3
	S			4.0059	5.0300	5.4106	5.6296	5.7942	5.9298	6.1523	6.3374	6.5000	6.6473	6.7833	6.9107
280	V			0.0018	0.0039	0.0062	0.0076	0.0087	0.0096	0.0111	0.0124	0.0136	0.0147	0.0158	0.0168
	U			1757.6	2226.3	2520.9	2661.8	2764.8	2850.1	2993.4	3118.4	3234.2	3344.9	3452.8	3559.2
	H			1809.3	2334.2	2695.8	2875.1	3007.7	3117.9	3303.9	3465.7	3615.1	3757.1	3894.8	4029.7
	S			3.9637	4.7550	5.2841	5.5367	5.7170	5.8621	6.0953	6.2863	6.4527	6.6026	6.7405	6.8693
300	V			0.0018	0.0028	0.0053	0.0067	0.0078	0.0087	0.0102	0.0114	0.0126	0.0137	0.0147	0.0156
	U			1738.4	2068.8	2452.8	2618.8	2732.6	2824.1	2974.6	3103.5	3222.0	3334.6	3443.9	3551.4
	H			1792.2	2152.8	2611.9	2820.9	2966.7	3084.8	3279.8	3446.9	3599.7	3744.2	3883.8	4020.2
	S			3.9316	4.4756	5.1473	5.4419	5.6402	5.7956	6.0403	6.2374	6.4077	6.5602	6.7000	6.8303
350	V			0.0017	0.0021	0.0034	0.0050	0.0061	0.0069	0.0083	0.0095	0.0106	0.0115	0.0124	0.0133
	U			1703.0	1914.6	2253.2	2497.4	2645.3	2755.3	2925.9	3065.7	3191.1	3308.6	3421.5	3531.7
	H			1762.5	1988.3	2373.4	2671.0	2857.3	2998.0	3218.1	3399.0	3560.9	3711.9	3856.3	3996.5
	S			3.8725	4.2139	4.7749	5.1945	5.4480	5.6331	5.9093	6.1229	6.3032	6.4625	6.6072	6.7411
400	V			0.0016	0.0019	0.0025	0.0037	0.0048	0.0056	0.0070	0.0081	0.0091	0.0099	0.0107	0.0115
	U			1676.9	1854.8	2097.1	2363.9	2549.6	2681.7	2875.2	3026.9	3159.6	3282.1	3398.9	3511.9
	H			1742.6	1931.2	2198.5	2511.6	2740.1	2906.7	3154.6	3350.4	3521.8	3679.4	3828.8	3972.8
	S			3.8290	4.1142	4.5036	4.9446	5.2555	5.4746	5.7859	6.0170	6.2079	6.3743	6.5239	6.6614

Temperature (°C)

Appendix E Steam Tables

P in bar (T^{sat} in C)	L	V	\	375	400	425	450	475	Temperature (°C) 500	550	600	650	700	750	800
450	V U H S			0.0016 1656.1 1727.9 3.7938	0.0018 1816.5 1897.6 4.0506	0.0022 2012.2 2110.6 4.3611	0.0029 2246.1 2377.3 4.7361	0.0038 2451.5 2623.4 5.0710	0.0046 2604.8 2813.4 5.3209	0.0059 2823.0 3090.2 5.6685	0.0070 2987.3 3301.5 5.9179	0.0079 3127.7 3482.5 6.1197	0.0087 3255.6 3647.0 6.2932	0.0095 3376.1 3801.3 6.4479	0.0102 3492.1 3949.3 6.5891
500	V U H S			0.0016 1638.5 1716.5 3.7640	0.0017 1787.8 1874.4 4.0029	0.0020 1959.7 2060.1 4.2737	0.0025 2160.0 2284.4 4.5892	0.0032 2361.3 2519.9 4.9095	0.0039 2528.0 2722.5 5.1759	0.0051 2769.8 3025.7 5.5566	0.0061 2947.2 3252.6 5.8245	0.0070 3095.6 3443.5 6.0372	0.0077 3228.9 3614.8 6.2180	0.0084 3353.3 3774.1 6.3777	0.0091 3472.3 3926.0 6.5226
600	V U H S			0.0015 1609.7 1699.9 3.7147	0.0016 1745.2 1843.2 3.9316	0.0018 1892.7 2001.6 4.1627	0.0021 2054.7 2179.8 4.4134	0.0025 2227.1 2375.4 4.6791	0.0030 2393.3 2570.4 4.9357	0.0040 2664.8 2902.1 5.3519	0.0048 2866.9 3157.0 5.6528	0.0056 3031.3 3366.8 5.8867	0.0063 3175.5 3551.4 6.0815	0.0069 3307.7 3720.6 6.2512	0.0075 3432.7 3880.2 6.4034
700	V U H S			0.0015 1586.2 1688.5 3.6742	0.0016 1713.3 1822.9 3.8777	0.0017 1847.8 1967.2 4.0881	0.0019 1990.9 2123.4 4.3080	0.0021 2141.7 2291.7 4.5368	0.0025 2293.9 2466.3 4.7663	0.0032 2569.4 2795.0 5.1786	0.0040 2789.3 3067.5 5.5003	0.0046 2968.2 3293.6 5.7522	0.0053 3122.8 3490.5 5.9600	0.0058 3262.7 3669.0 6.1390	0.0063 3393.6 3835.8 6.2982
800	V U H S			0.0014 1566.3 1680.4 3.6395	0.0015 1687.4 1808.7 3.8338	0.0016 1813.7 1944.1 4.0311	0.0018 1945.7 2087.6 4.2331	0.0020 2082.9 2239.5 4.4397	0.0022 2222.6 2397.6 4.6475	0.0028 2489.2 2710.0 5.0391	0.0034 2717.4 2988.1 5.3674	0.0040 2907.7 3225.7 5.6321	0.0045 3071.6 3432.9 5.8509	0.0050 3218.7 3619.7 6.0382	0.0055 3355.2 3793.3 6.2039
900	V U H S			0.0014 1548.9 1674.7 3.6090	0.0015 1665.6 1798.5 3.7965	0.0016 1786.1 1927.6 3.9847	0.0017 1910.5 2062.7 4.1748	0.0018 2038.7 2204.0 4.3669	0.0020 2169.1 2350.4 4.5593	0.0025 2424.0 2645.2 4.9289	0.0030 2653.5 2920.8 5.2540	0.0035 2851.0 3164.4 5.5255	0.0040 3022.6 3379.5 5.7526	0.0044 3176.1 3573.5 5.9470	0.0048 3317.8 3753.0 6.1184
1000	V U H S			0.0014 1533.5 1670.7 3.5815	0.0014 1646.7 1791.1 3.7638	0.0015 1762.8 1915.5 3.9452	0.0016 1881.8 2044.6 4.1267	0.0017 2003.4 2178.3 4.3086	0.0019 2126.9 2316.3 4.4900	0.0022 2371.0 2596.0 4.8406	0.0027 2597.8 2865.1 5.1580	0.0031 2799.2 3110.6 5.4316	0.0035 2976.1 3330.8 5.6640	0.0040 3135.4 3530.7 5.8644	0.0043 3281.6 3715.2 6.0405

Index

Note: page numbers with "f" indicate figures; those with "t" indicate tables.

Absolute temperature, 23–24
Absolute zero, 22t
Accumulation, 253
Acentric factor, 47–51
   defined, 47–48
   of simple, normal, and polar fluids, 50–51
Activity, 606–13
   chemical potential and, 607
   of gas, 607–8, 611
   of liquid, 608–9, 611
   $P°$, working with, 608
   of solid, 609–10, 612
   of solute, 609, 611–12
   standard state and, 610, 613
Activity coefficient, 504–11
   for data reduction, 512–15
   defined, 504
   deviations from Raoult's law, 508–10, 509f
   experimental, 510–15
   Gibbs free energy and, excess, 505–7, 510–14, 511f
   limits of, 505
   phase equilibrium and, 507–10
   solubility limits using, 547–48
   usefulness of, 505
Activity coefficient models, 515–31
   Flory-Huggins, 521–22
   Margules, 515–19
   NRTL, 520–21
   UNIFAC, 525–31
   UNIQUAC, 522–25
   vanLaar, 519
   Wilson, 519–20
Adiabatic flame temperature, 605–6
Adiabatic mixing
   heat effects of, 497–99
   of solutions on enthalpy charts, 503f
   in steady-state processes, 276–78, 277f
Adiabatic processes. *See* Reversible adiabatic processes
Adiabatic production of work, 261, 262
Adiabatic system, 13
Air, chemical potential in, 441–42

American Engineering system of units, 22
   pound-mol (lb-mol) in, 24
   pressure measured in, 23
   temperature measured in, 23
Ammonia refrigerator, 304–6
Analogous relationships, 215, 408
Antoine equation, 38–40
   Clausius-Clapeyron equation and, 343
   determining phase with, 39–40
   units in, 38–39
ASHRAE designation of refrigerants, 307t
Athermal process of mixing, 410
Avogadro's number, 22t
Azeotropes, 380–81, 380f

Balances in open systems, 251–335
   energy balance, 254f, 255–58
   entropy balance, 258–66
   flow streams, 252–53, 252f
   liquefaction, 309–15
   mass balance, 253–55, 254f
   power generation, 295–300
   problems, 324–35
   refrigeration, 301–9
   summary, 323–24
   thermodynamics of steady-state processes, 272–94
   unsteady-state balances, 315–23
Bar, 23
Bath
   change of, entropy, 159–61
   default, exergy and, 189–91
   heat, in entropy balance, 265–66
   stream acting as, in entropy balance, 264–65
Benedict-Webb-Rubin equation, 67–71, 70f, 71t
Benzene, solubility of, 577–78, 578f
Binary mixtures, 420–21, 423t
Binary VLE using Soave-Redlich-Kwong, 452–53
Boltzmann constant, 8, 22t
Bubble temperature (or point), 371
BWR equation. *See* Benedict-Webb-Rubin equation

Calorie, 101
Carbon dioxide (CO_2)
   density of liquid, 10
   ideal-gas state and, 10–11
   intermolecular potential of, 6
   Pxy graph of CO_2/n-pentane calculated using Soave-Redlich-Kwong, 451f, 452–53
Carnot cycle, 168–77
   defined, 169
   in entropy, 168–77, 169f
   first-law analysis of, 170
   ideal gas used in, 173–74
   internal combustion engine example of, 173
   second-law analysis of, 171
   steam used in, 175–76
   steps in, 169–70
   thermodynamic efficiency of, 171–73
   on TS graph, 169f
   work value of heat and, 182–83, 182f
Carnot efficiency, 171–72
Celsius scale, 23–24
Chemical engineer, 5
Chemical engineering thermodynamics, 5
Chemical equilibrium, 15
Chemical potential
   activity and, 607
   in air, 441–42
   in fugacity, 444
   in Gibbs free energy, 440–42
   in ideal-gas state, 440
   in real mixture, 440
   in theory of VLE, 439–42
Chemical thermodynamics, 3–5, 4f
Classical thermodynamics
   laws of, 11–12
   $vs.$ statistical thermodynamics, 11–13
Clausius, Rudolph, 178
Clausius-Clapeyron equation in VLE, 341–43
Clausius statement
   as alternative statement of second law, 177–82
   defined, 178
   as equivalent statement of second law, 180–82
Closed systems
   defined, 13
   first law for, 98–101
   second law in, 150–53
Combinatorial molecular interactions, 507
Completely immiscible liquids, 558–61
   calculations, limitations of, 563
   Pxy graph of, 561–62, 562f
   Txy graph of, 559–61, 559f
Composite system, 13

Compressed liquids, 292–94
   entropy of, 156
   Poynting equation for calculating, 345–47
   throttling of, 288–89, 290f
Compressibility factor, 43–45, 45f
   in BWR equation, 69
   in corresponding states, 45–47, 46f, 51, 52
   in Lee-Kesler method, 56, 68
   in Peng-Robinson equation, 67
   in Pitzer method, 48, 71
   in Rackett equation, 73
   roots in, acceptable, 63, 66, 67
   in Soave-Redlich-Kwong equation of state, 61
   in van der Waals equation, 58–60
   in virial equation, 53, 54, 56
   ZP graph of pure fluids and, 43–45, 45f, 56
Compressor
   graphical solution on Molliev chart, 293f
   steam/gas, 290–92
Condensed phases, 9
Constant composition, systems of, 407–8
Constant-pressure cooling, 104–5
Constant pressure entropy
   change at, 155
   no phase change in, 153–54
Constant-pressure heat capacity (C_P), 111–12
   $vs.$ constant-volume heat capacity, 111–12, 112f
   as state function, 113
   of steam, 114–15
Constant-pressure heating, 102–3
Constant-pressure (or isobaric) process, 18–19
Constants, 22t
Constant-temperature (or isothermal) process, 18–19, 107, 108
Constant-volume cooling, 102
Constant-volume heat capacity (C_V), 109
   $vs.$ constant-pressure heat capacity, 111–12, 112f
   as state function, 113
Constant-volume heating, 101–2
Constant-volume path entropy, no phase change in, 154
Constant-volume process, 18–19
Constrained equilibrium, 15–16
Conversions, 25t
Corresponding states
   in compressibility factor, 45–47, 46f, 51, 52
   in phase diagrams of pure fluids, 45, 46f
Cubic equations of state
   application to, 230–34
   fugacity calculated from, 352–53
   isotherm, unstable and metastable parts of, 62–63

Peng-Robinson, 61, 66–67
PVT behavior of, 61–63
roots of, 63–65
Soave-Redlich-Kwong, 30f, 60–61, 66
of state, residual properties from, 232t
Van der Waals, 58–60, 65–66
for volume, solving, 65–67
working with, 64

Data reduction, 514
Definitions, 13–17
  equilibrium, 15–16
  property, 16–17
  system, 13–14
Density
  empirical equations for, 72–77
  of liquid CO_2, 10
Derivatives with functions of multiple variables, 110
Desalination, 585–86
Dew point, 483
Diathermal system, 13
Differentials
  exact, 211–12
  form of, 209–11
  inexact, 212
  integration of, 213–14, 214f
Diffusion in osmotic equilibrium, 582–83
Disorder, defined, 196
Distillation, 378, 379f
Double interpolations, 41, 41f, 42

Elementary process, 18–19
Empirical equations, 57
  for density, 72–77
  summary, 78
Endothermic process, 410
Energy, 24–25, 25t
  forms of, 90
  heat and, 96–97
  ideal-gas state, 124–32
  internal, 96
  process requirements, calculating, 5
  storable, 89
  transfer, sign convention for direction of, 89
  work and, 88–89, 88f
Energy balance, 254f, 255–58
  in closed systems, 257–58
  in enthalpy of vaporization, 478–79
  entropy used in, 163–66
  in heat exchanger, 479–80
  in ideal solution, 475–80
  irreversible processes and, 133–38
  in open systems, 254f, 255–58

steady state, 258
using C_P and ΔH_{vap}, 122–24
using tables, 121–22
Energy balances in reacting systems, 601–6
  adiabatic flame temperature and, 605–6
  heat of combustion and, 603–5
  setup for energy balance of reactor, 602f
Enthalpy
  defined, 103
  enthalpy-entropy (Mollier) chart, 244–45, 245f
  equations for, 217–19
  excess, heat effects of mixing and, 496, 497f
  ideal-gas, 234
  of ideal-gas mixture, 416
  lever rule for, 119–20
  of mixing, 425
  pressure and temperature effect on, 220–22
  pressure-explicit relations, 229
  of reactions, standard, 596–601, 599f
  residual, 223, 236, 237f
  as state function, 103, 125–26
  of steam, 103–5
  temperature-enthalpy chart, 243–44, 244f
  of vaporization, 478–79
Enthalpy charts, 500–504
  adiabatic mixing of solutions on, 503f
  of hydrazine/water mixtures, 501f
  pressure, 242–43, 243f
  using, 503–4
Enthalpy-entropy (Mollier) chart, 244–45, 245f
  graphical solution of compressed liquids on, 293f
  in steam turbine, 285–86, 285f
Entropy
  calculation of (See Entropy, calculation of)
  Carnot cycle in, 168–77, 169f
  change of bath, 159–61
  in energy balance, 163–66
  energy balances using, 163–66
  equations for, 217–19
  equilibrium and, 191–95
  exergy and, 189–91
  generation, 167–68
  heat calculated using, 166
  ideal-gas, 234
  of ideal-gas mixture, 416
  ideal work in, 183–88
  of irreversibility, 168
  of liquids, 156, 157
  lost work in, 183–88, 191
  measurable quantities relating to, 151–53
  of mixing, in ideal solution, 463–64
  molecular view of, 195–99, 197f
  pressure and temperature's effect on, 220–22

Entropy (*Continued*)
  pressure-explicit relations, 229–30
  problems, 201–4
  residual, 224, 236, 238f
  reversibility and, 176–77
  second law and, 149–204
  of solids, 155–56, 157
  stability and, 191–95
  as state function, 153
  summary, 199–201
  temperature and, 153
  thermodynamic potentials, 194–95
  of universe, 168
  of vaporization, 157, 158
Entropy, calculation of, 153–63, 199
  basic formula, 154
  change at constant pressure, 155
  change of bath, 159–61
  compressed liquids, 156
  constant-pressure path, no phase change, 153–54
  constant-volume path, no phase change, 154
  energy balances with phase change, using C_P and ΔH_{vap}, 158–59
  ideal-gas state, 161–62, 161f
  phase change, 157–58, 157f
  solids, 155–56
  steam, 163
  tabulated values used for, 162
  vaporization, 158
Entropy balance, 258–66
  heat exchange between streams, 262–63
  in open systems, 258–66
  processes with entropy generation, 262f
  reversible adiabatic process, 260–61
  steady state, 260, 261
  stream acting as a bath, 264–65
Entropy generation
  cooling system, choosing, 269–71
  in power plant, 267
  processes with, 262f
Equations of state
  appeal of, 453
  binary VLE using Soave-Redlich-Kwong, 452–53
  BWR, 67–71, 70f, 71t
  in compressibility factor, 44
  cubic, 57–67
  defined, 29
  fugacity from, 446–48
  Henry's law from, 572–73
  ideal-gas law and, 34, 54
  interaction parameter, 451–53
  Lee-Kesler, 68–70
  mathematical representation of, 44
  for mixtures, 419–21
  other, 67–71
  phase diagrams from, 356–58
  Pitzer, 71
  properties from mixtures, 421–27
  saturation pressure from, 353–56
  summary, 77
  in theory of VLE, 448–53
  in truncated virial equation, 54
Equilibrium, 15–16, 545–91. *See also* Liquid miscibility
  chemical, 15
  conditions at constant temperature and pressure, 191–94
  conditions fulfilled by, 15
  constrained, 15–16
  entropy and, 191–95
  gas liquids, solubility of, 561–74
  liquid-liquid (LLE), 369
  liquid-liquid-vapor (LLVE), 369
  mechanical, 15
  osmotic, 580–86
  phase splitting and Gibbs free energy, 548–56
  problems, 586–91
  schematic of, 546f
  solids in liquids, solubility of, 575–80
  solids involved in, 629–32
  vs. steady state, 261
  summary, 586
  thermal, 15
  thermodynamic potentials and, 194–95
Equilibrium composition, 622–24
  with multiple reactions, 634–35
  pressure in, effect of, 623–24
Equilibrium constant, 614–20, 621
Ethylene, phase diagram of, 356–58
Euler's theorem, 406
Eutectic point, 579–80
Exact differential, 211–12
Excess properties, 489–96
  as mathematical operator, 490
  molar, 490–91, 492f
  in nonideal solutions, 489–96
  Redlich-Kister expansion, 495–96
  volume, 494–95, 495f
  in volume of mixing, 492–94
Exergy, 189–91
Exothermic process, 410
Expansion, PV work in, 92–93
Expansion and Compression of Liquids, 292–94
Experimental activity coefficient, 510–15
  fitting, 512–14

Index

Pxy graph of, 515f
Extensive property, 16–17

Fahrenheit, 23–24
First law
 Carnot cycle analyzed by, 170
 for closed system, 98–101
 defined, 87
 energy and, 87–148
 history of, 100–101
 Joule's experiment, 100–101
 operator Δ used in, 99–100
 paths in, 101–8
 problems, 140–48
 schematic illustration of, 98f
 summary, 139–40
Fixed property, 17
Flash separation, 375–76, 376f
Flash units, multiple, 377–78
Flory-Huggins equation, 521–22
Flow streams, 252–53, 252f
Force, 25t
Fugacity, 343–53
 chemical potential and, 444
 defined, 343–43
 from equations of state, 446–48
 from experimental data, 445
 Gibbs free energy and, 344–45
 Henry's law used to calculate, 573–74, 574f
 in ideal-gas state, 344
 in ideal solution, 464–65
 Lewis-Randall rule used to calculate, 471, 563, 564
 in mixture using Soave-Redlich-Kwong, 447–48
 of solubility of gases in liquids, 564f
 steam, 349
 from steam tables, 347
 in theory of VLE, 443–48
Fugacity, calculation of, 345–53
 compressibility factor used in, 348–49
 from cubic equations of state, 352–53
 from generalized graphs, 349–52, 350f, 351f
 Poynting equation for compressed liquids, 345–47
 saturated liquid, 346
 from Soave-Redlich-Kwong, 352, 353
 tabulated properties used in, 347–48
Function, differential and integral forms of, 209–11
Fundamental relationships
 calculating, 214–17
 change of variables, 217
 summary of, 216t

Gas
 activity of, 607–8, 611
 Linde process for liquefaction of, 310f
Gases in liquids, solubility of, 563–74. *See also* Henry's law
 fugate of, 564f
Generalized correlations, calculating, 235–36, 237–38f
Generalized graphs, 77
 fugacity calculated from, 349–52, 350f, 351f
Gibbs, J. Willard, 4–5, 4f
Gibbs-Duhem equation, 406–7, 439–40
Gibbs free energy, 5, 194–95, 195f
 calculating, 215, 225, 242
 chemical potential in, 440–42
 excess, activity coefficient and, 505–7, 510–14, 511f
 fugacity and, 344–45
 in Margules equation, linearized form of excess, 518f
 multicomponent equilibrium in, 437–38, 437f
 phase rule in, 438–39
 phase splitting and, 548–56
 in theory of VLE, 435–39
Gibbs's phase rule, 438–39

Heat, 96–97
 capacities, 109–19
 of combustion in reacting systems, 603–5
 direction of flow, second law to determine, 184–85
 entropy used to calculate, 166
 as path function, 105–6
 sensible, 109
 sign convention for, 97
 specific, 111
 transfer, in entropy generation, 188
 transfer, irreversible processes in, 133
 of vaporization, 119–24, 120f
 of vaporization using Soave-Redlich-Kwong, 338
 work and, shared characteristics with, 97
 work value of, Carnot cycle and, 182–83, 182f
Heat bath (or heat reservoir), 19
Heat capacity, 109–19
 constant-pressure heat capacity, 111–12, 112f
 constant-volume heat capacity, 109, 111–12, 112f
 defined, 112–13
 effect of pressure and temperature on, 113–19, 113f
 excess, 499–500

Heat effects of mixing, 496–504
   enthalpies, excess, 496, 497f
   enthalpy charts, 500–504
   excess heat capacity, 499–500
   isothermal, 497–99
Heat exchanger, 278–81, 279f
   in ideal solution, 479–80
Heat of vaporization (ΔH_{vap}), 119–24, 120f
   C_P and ΔH_{vap}, energy balances with phase change using, 122–24
   latent, 119
   Pitzer correlation for, 121–22
   tables, energy balances with phase change using, 121–22
Helmholtz free energy, 194, 216, 225, 242
Henry's law, 564–74
   alternative forms of, 566–67
   carbonated soda and, 571–72
   from equation of state, 572–73
   formal thermodynamics and, 569–70, 571f
   fugacity calculated with, 573–74, 574f
   vs. Lewis-Randall rule, 573–74
   other units for, 565–66
   in simultaneous dissolution of two gases, 567–69
   temperature and pressure effects on, 569
   VLE using, 564–65
Household refrigerator, 304
Humidification, 481–83
   dew point, 483
   relative humidity, 482, 483
Hypothetical states, 476–77

Ideal gas
   filling a tank with, 318–18
   throttling of, 288
   venting, 322–23
Ideal-gas constant, 22t
Ideal-gas law
   calculated isotherms using, 57f
   defined, 34
   directions to, 34
   equation of state and, 34, 54
   molar volume calculated with, example of, 55
   summary, 78
   vs. truncated viral equation, 54
Ideal-gas state, 10–11, 124–32
   calculating, 219–20
   Carnot cycle using, 173–75
   chemical potential in, 440
   equations, summary of, 219t
   fugacity in, 344
   hypothetical, 223

   irreversible expansion of, 137
   in isothermal compression of steam, 126–27
   log-x mean, 130
   mixtures in, 413–19
   paths in, 132f
   residual properties of, 225–26
   reversible adiabatic compression of nitrogen, 131–32
   reversible adiabatic process, 127–30, 131–32, 132f
   reversible adiabatic process in, 164–66
   triple-product rule in, 207
Ideal solution, 461–88
   defined, 462
   energy balances in, 475–80
   entropy of mixing in, 463–64
   fugacity in, 464–65
   ideality in, 461–63
   noncondensable gases in, 480–83
   practical significance of, 463
   problems, 484–88
   in separation of work, 464
   summary, 484
   VLE in, Raoult's law, 466–74
Ideal work, 183–88
   for heat transfer, 268–69
   in open systems, 266–71
Incompressible phases, 220–21
Inexact differential, 212
Integral form of a function, 209–11
Integrals with functions of multiple variables, 110
Integration of differentials, 213–14, 214f
Integration paths, 109–10
Intensive property, 16–17
Interaction parameter, 451–53
Intermolecular potential, 6–7, 7f
Internal energy, 96, 215, 219, 224–25, 241–42
Interpolations, 40–43, 41f
   accuracy in, 40
   example of, using steam tables, 41–43
   types of, 41f
Irreversible processes, energy balances and
   adiabatic mixing, 134–36
   heat transfer, 133
   of ideal gas, 137
   against pressure, 138
   against vacuum, 136–37
Isentropic step in Carnot cycle, 170
Isolated system, 13
Isothermal compression of steam
   in first law paths, 108
   in ideal-gas state, 126–27
   in PV work, 94–95, 96f
Isothermal mixing, heat effects of, 497–99

Index

Isothermal step in Carnot cycle, 169–70
Isotherms
  critical point and, 33
  of Lee-Kesler method, 68
  of modified BWR equation, 67, 70f
  pressure-volume, of acetone, 74–75, 76f
  on PV graph, 31, 33
  of Soave-Redlich-Kwong equation, 61–62, 62f
  at zero pressure, 44
  on ZP graph, 53, 57f

Joule, James Prescott, 100–101

Kelvin (K), 23
Kelvin-Plank statement, 178
K-factors, 382, 449, 468
Kinetic energy, 90

Latent heat of vaporization, 119
Laws of classical thermodynamics, 11–12
Lee-Kesler method
  compressibility factor in, 56, 68
  in phase diagrams of pure fluids, 48–49, 48f, 49f, 50f
  for residual enthalpy, 237f
  for residual entropy, 238f
Length, 25t
Lever rule, 34–36, 35f
  defined, 35
  for enthalpy, 119–20
  partially miscible liquids and, 388
  in phase behavior of mixtures, 372–73
  setup for application of, 35f
  state located with, steam table example, 43
  using, reasons for, 36
Lewis-Randall rule
  defined, 565
  fugacity in liquid phase given by, 471, 563, 564
  $vs.$ Henry's law, 573–74
Linde process, 310f
Linear interpolations, 41f, 71, 79
Liquefaction, 309–15
  of gas, Linde process for, 310f
  of nitrogen, 310f, 311–14
Liquid-liquid equilibrium (LLE), 369, 546f
Liquid-liquid-vapor equilibrium (LLVE), 369
Liquid miscibility
  complete, 558–61
  partial, equilibrium between, 545–48
  temperature and, 556–58
Liquids. $See\ also$ Compressed liquids; Vapor-liquid mixture
  activity of, 608–9, 611
CO_2, density of, 10
  constant-pressure cooling of, 115–17
  of enthalpy and entropy, pressure and temperature's effect on, 220–22
  entropy of, 156, 157
  expansion of, 292–94
  fugacity of saturated, 346
  gases in, solubility of, 563–74
  heat capacities, 114
  hypothetical state of, 472
  properties, compressed, 117–19, 118f
  pump, calculating, 294
  solids in, solubility of, 575–80
Log-x mean, 130
Lost work in open systems, 266–71

Magdeburg hemispheres, 27–28
Margules equation, 515–19
  Gibbs free energy in, 518f
  Pxy graph reconstructed from limited data, 517, 518–19
Mass, 25t
Mass balance, 253–55, 254f
Mathematical conditions for stability, 552
Maxwell-Boltzmann distribution, 8, 9f
Maxwell relationships, 214, 216
Measurable quantities relating to entropy, 151–53
Mechanical equilibrium, 15
Mechanically reversible process, 91–92, 92f
Methane, virial coefficient used for residual properties of, 227–28
Mixing, 411–13. $See\ also$ Mixtures
  athermal process of, 410
  endothermic process of, 410
  enthalpy of, 425
  exothermic process of, 410
  heat effects of, 496–504
  molecular view of, 412f
  non-isothermal, 416–18
  properties of, 409–11
  rules, 419
  separation and, 411–13
  separation of air, 418–91
  volume of, 409–10
  volume of, excess properties in, 492–95
Mixtures, 401–33. $See\ also$ Mixing
  binary, 420–21
  composition of, 402–4
  equations of state for, 419–21
  formation of, at constant pressure and temperature, 410f
  fugacity in, using Soave-Redlich-Kwong, 447–48

Mixtures (*Continued*)
  homogeneous properties of, 403–4
  in ideal-gas state, 413–19
  mathematical treatment of, 404–9
  phase behavior of, 369–400
  problems, 428–33
  properties from equations of state, 421–27
  real, compression of, 427
  real, properties of, 424–25
  residual properties of, 420, 421
  subscripts, 411
  summary, 428
  VLE of, using equations of state, 448–53
Mixtures, mathematical treatment of, 404–9
  Euler's theorem, 406
  Gibbs-Duhem equation, 406–7
  partial molar property, 405, 408–8
  systems of constant composition, 407–8
Molar mass, 23
Molar property, 16–17
  excess, in nonideal solutions, 490–91, 492f
  partial, 405, 408–9
  residual, 408–9
  residual partial, 408–9
Molar volume
  of compressed liquid, 73
  in compressibility factor, 44
  in corresponding states, 51, 52
  in cubic equations, 64, 65, 66, 67
  ideal-gas, 55
  liquid, in Rackett equation, 72–73
  mean intermolecular distance and, 29
  proper, selecting, 63
  in PT graph of pure fluid, 37, 37f
  in PV graph of pure fluid, 30f, 31–35
  in PVT behavior of pure fluid, 29–30
  in Soave-Redlich-Kwong equation, 62
  in van der Waals equation, 59
  in virial equation, 55, 56
Mole, 24
Molecular basis of thermodynamics, 5–11
  condensed phases, 9
  density of liquid CO_2, 10
  ideal-gas state, 10–11
  intermolecular potential, 6–7, 7f
  Maxwell-Boltzmann distribution, 8, 9f
  phase transitions, 8–9
  temperature and pressure, 7–8
Molecular collisions, 7–8, 97
Molecular view of entropy, 195–99, 197f
  equally probable microstates, 199
  probabilities and, 198–99
Mollier chart. *See* Enthalpy-entropy (Mollier) chart

Multicomponent mixture, state of, 18
Multiple reactions, 632–35
  equilibrium compositions with, 634–35
  number of, determining, 633–34

Naphthalene, solubility of, 577–78, 578f
Newton, 23
Nitrogen
  liquefaction of, 310f, 311–14
  reversible adiabatic compression of, 131–32
Nomenclature, 17
Noncondensable gases. *See also* Humidification
  in ideal solution, 480–83
Nonideal solutions, 489–543. *See also* Activity coefficient
  excess properties, 489–96
  heat effects of mixing, 496–504
  molecular view of, 507, 507f
  problems, 533–43
  summary, 531–33
Nonrandom two liquid model (NRTL). *See* NRTL equation
NRTL equation, 520–21

Open systems. *See also* Balances in open systems
  defined, 13
  ideal and lost work in, 266–71
Operator Δ, 99–100, 255
Osmotic equilibrium, 580–86, 580f
  desalination and, 585–86
  diffusion in, chemical potential and, 582–83
  pressure in, 584–85
  reverse osmosis, 583–84, 584f
Osmotic pressure, 584–85

$P°$, working with, 608
Partially miscible liquids, 384–89
  equilibrium between, 545–48
  lever rule and, 388
  Pxy graph of, 387f, 552–54, 553f
  qualitative-phase diagram of, 388–89, 389f
  Txy graph of, 385f, 554–56, 554f, 557f
Partial molar, 405, 408–9
Pascal, 23
Path, 18–19, 18f
  defined, 18
  illustration of, 18f
Paths in first law, 101–8
  alternate, 106–7, 107f
  constant-pressure cooling, 104–5
  constant-pressure heating, 102–3
  constant-temperature process, 107, 108
  constant-volume cooling, 102

Index

constant-volume heating, 101–2
enthalpy of steam, 103–5
heat as path function, 105–6
integration, 109–10
in isothermal compression of steam, 108
work as path function, 105–6
Peng-Robinson equation
    compressibility factor in, 67
    fugacity calculated from, 352, 353
    fugacity from, 446–47
    residual properties from, 232t
Perry's Chemical Engineers' Handbook
    (Perry), 220, 233, 234, 275, 338, 357, 362t
Phase behavior of mixtures, 369–400
    azeotropes, 380–81
    compositions at bubble and dew point, 374f
    distillation, 378, 379f
    flash separation, 375–76, 376f
    flash units, multiple, 377–78
    lever rule in, 372–73
    partially miscible liquids, 384–89
    problems, 394–400
    Pxy graph, 373–79
    summary, 393
    ternary systems, 390–93
    Txy graph, 370–73
    VLE at elevated pressures and temperatures, 383–84, 384f
    xy graph, 381–83, 382f
Phase change, entropy and, 153–58, 157f
    constant-pressure path, no phase change, 153–54
    constant-volume path, no phase change, 154
    energy balances with, using C_P and ΔH_{vap}, 158–59
Phase diagrams. *See also* Pure fluids, phase diagrams of
    of acetic acid/water/methylisobutyl ketone, 390f
    from equations of state, 356–58
    of ethylene, 356–58
    in Raoult's law, 473–74, 474f
Phase equilibrium
    chemical equilibrium and VLE, 625–28
    as chemical reaction, 628–29
    reactions and, 624–29
Phases, calculating incompressible, 220–21
Phase splitting Gibbs free energy and, 548–56, 550f
    equilibrium and stability in, 551–52, 551f
Phase transitions, 8–9
Pipe, flow through, 272–76
    energy losses in, schematic for calculating, 273f
    pressure drop, 275–76

Pitzer method
    compressibility factor in, 48, 71
    for heat of vaporization, 121–22
    in phase diagrams of pure fluids, 47, 48
Planck, Max, 178
Potential energy, 90
Pound-mol (lb-mol), 24
Power, 25t
Power generation in open systems, 295–300
Poynting equation, 345–47
PR equation. *See* Peng-Robinson equation
Pressure, 7–8, 22–23, 25t
    in equilibrium composition, effect of, 623–24
    Henry's law and, 569
    irreversible expansion against, 138
Pressure, effect on heat capacities, 113–19, 113f
    compressed liquid properties, 117–19, 118f
    constant-pressure cooling of liquid, 115–17
    heat capacity of steam, 114–15
    ideal-gas heat capacity, 114
    liquid heat capacities, 114
Pressure-enthalpy chart, 242–43, 243f
Pressure-explicit relations, 228–30
    enthalpy, 229
    entropy, 229–30
Pressurizing, 321
Processes, 18–19, 18f
    defined, 18
    energy requirements of, calculating, 5
    entropy and, 176–77
Properties, 16–17
    defined, 16–17
    extensive and intensive properties, 16–17
    fixed, 17
    molar, 16–17
    nomenclature and, 17
    of pure component, 17
    specific, 16–17
    tabulated values of, 40–43
Properties, calculating, 205
    cubic equations, application to, 230–34
    differentials, integration of, 213–14
    enthalpy, 217–19
    entropy, 217–19
    fundamental relationships, 214–17
    generalized correlations, 235–36, 237–38f
    ideal-gas state, 219–20
    in compressible phases, 220–21
    pressure-explicit relations, 228–30
    problems, 246–50
    of reference states, 236, 239–42
    residual properties, 222–28
    summary, 245–46

Properties, calculating (*Continued*)
  thermodynamic charts, 242–45
  thermodynamics, 205–12
Properties of Gases and Liquids, The (Polling, Prausnitz, and O'Connell), 38, 114, 121
PT graph of pure fluid, 36–40, 37f
Pure component
  multicomponent mixture, state of, 18
  property of, 17
  state of, 17–18
Pure fluid, 337–66
  fugacity, 343–53
  phase diagrams from equations of state, 356–58
  problems, 360–66
  saturation pressure from equations of state, 353–56
  summary, 358–60
  two-phase systems, 337–40
  VPE of, 340–43
Pure fluids, phase diagrams of, 29–85
  compressibility factor, 43–45, 45f
  corresponding states, 45, 46f
  empirical equations, 57
  PVT behavior of, 29–40
  tabulated values of properties, 40–43
  virial equation, 53
PV graph of pure fluid, 31–36, 31f
  critical point, 33
  ideal-gas state, 33–34
  two-phase region and lever rule, 34–36, 35f
PVT behavior
  boiling in open air, 32
  PT graph of pure fluid, 36–40, 37f
  of pure fluids, phase diagrams of, 29–40
  PV graph of pure fluid, 31–36, 31f
  PVT surface of pure fluid, 30f
PVT surface of pure fluid, 30f
PV work, 90–96, 91f
  equation for, obtaining, 90–91
  in expansion, 92–93
  in isothermal compression of steam, 94–95, 96f
  reversible, 91–92, 92f
  Soave-Redlich-Kwong equation used for, 93–94
Pxy graph, 373–79
  of CO_2/n-pentane calculated using Soave-Redlich-Kwong, 451f, 452–53
  of completely immiscible liquids, 561–62, 562f
  of ethanol/water at elevated temperatures, 384f
  of experimental activity coefficient, 515f
  of heptane and decane, 375f
  forisobutane/furfural system, 389f
  of partially miscible liquids, 387f, 552–54, 553f
  reconstructing from limited data, 517, 518–19
  for UNIFAC equation, 530f, 531f
  UNIQUAC equation used to calculate, 524–25, 526f

Quantities relating to entropy, measurable, 151–53
Quasi-static process, 19–22, 20f
  example of, 21–22
  explained, 21
  reversible, 20–21

Rackett equation, 73
Rankine (R), 23
Rankine power plant, 295–300, 295f
Raoult's law, 466–74
  bubble and dew pressure in, 469–70
  bubble T calculation in, 467
  deviations from, 508–10, 509f
  dew P calculation in, 467–68
  dew T calculation in, 468
  flash calculation in, 468–69
  fugacity and, 470–71
  hypothetical liquid state in, 472
  phase diagram in, 473–74, 474f
Reactions, 593–645
  activity, 606–13
  energy balances in, 601–6
  equilibrium composition, 622–24
  equilibrium constant, 614–20
  equilibrium involving solids, 629–32
  multiple, 632–35
  phase equilibrium and, 624–29
  problems, 637–45
  standard enthalpy of, 596–601, 599f
  stoichiometry, 593–96
  summary, 636–37
Real mixtures
  chemical potential in, 440
  compression of, 427
  properties of, 424–25
Redlich-Kister expansion, 495–96
Reference states
  calculating, 236, 239–42
  using, 240–41
Refrigerants, 307–9
  ASHRAE designation of, 307t
  environment and, 308–9
  saturation pressure of, 308f

Refrigeration, 301–9
  ammonia refrigerator, 304–6
  cycle on PH graph, 303f
  household refrigerator, 304
  refrigerants, 307–9
  thermodynamic analysis of, 302–6
  vapor-compression refrigeration cycle, 301f
Relative humidity (RH), 482, 483
Relative volatility, 382
Residual molar, 408–9
Residual molecular interactions, 507
Residual partial molar, 408–9
Residual properties, 222–28
  applications, 225–28
  from corresponding states, 235–36
  from cubic equations of state, 232–34, 232t
  defined, 22
  enthalpy, 223
  entropy, 224
  in ideal-gas state, 225–26
  of methane, using virial coefficient, 227–28
  of mixtures, 420, 421
  other, 224–25
  from truncated virial equation, 226–27
  volume, 224
Reverse osmosis, 583–84, 584f
Reversible PV work, 91–92, 92f
Reversibility, entropy and, 176–77
Reversible adiabatic processes, 127–30, 131–32, 132f
  in entropy balance, 260–61
  in ideal-gas state, 164–66
  in mixing, 134–36
  in nitrogen compression, 131–32
  in steam compression, 165
Reversible quasi-static process, 20–21
Roots
  in compressibility factor, acceptable, 63, 66, 67
  of cubic equation of state, 63–65

Saturated liquid, fugacity of, 346
Saturation pressure from equations of state, 353–56, 354f, 355f
Second law
  alternative statements of, 177–82
  Carnot cycle analyzed by, 171
  in closed system, 150–53
  consequences of, 180, 200–201
  defined, 149
  direction of heat flow determined by, 184–85
  entropy and, 149–204
  in feasibility of process, 185–87
Sensible heat, 109

Separation of work, ideal solution in, 464
Separation units, 4f, 5
Shaft work, 90, 91f
Sign convention
  for heat, 97
  for transfer energy, direction of, 89
  for work, 89
Simple interpolations, 41, 41f
Simple system, 13
SI system of units, 22
  energy measured in, 24
  mole, 24
  pressure measured in, 22–23
  temperature measured in, 23
Soave-Redlich-Kwong equation
  binary mixture properties using, 423t
  binary VLE using, 452–53
  calculation of Henry's law constant, 569–70, 571f
  CO_2/n-pentane calculated using, 451f, 452–53
  coefficient of thermal expansion using, 207–9
  compressibility factor in, 61
  fugacity from, 352, 353, 446–48
  heat of vaporization using, 338
  for PV work, 93–94
  residual properties from, 232–34, 232t
  vapor-liquid mixture calculated by, 339–40
Solids
  activity of, 609–10, 612
  entropy of, 155–56, 157
  in liquids, solubility of, 575–80
Solubility
  of benzene, 577–78, 578f
  eutectic point, 579–80
  of gases in liquids, 563–74
  limits using activity coefficients, 547–48
  of naphthalene, 577–78, 578f
  of solids in liquids, 575–80
Solute, activity of, 609, 611–12
Specific energy, 25t
Specific heat, 111
Specific property, 16–17
SRK equation. *See* Soave-Redlich-Kwong equation
Stability, 355–56
  entropy and, 191–95
  equilibrium and, in phase splitting, 551–52, 551f
  mathematical conditions for, 552
Standard state, activity and, 610, 613
State function, 17
  constant-volume heat capacity as, 109, 113
  enthalpy as, 103, 125–26
  entropy as, 153

State function (*Continued*)
   heat capacity as, 109
   internal energy as, 89, 96, 107
Statistical thermodynamics, 11–13
Steady state
   energy balance, 258
   entropy balance, 260, 261
   vs. equilibrium, 261
Steam
   in Carnot cycle, 175–76
   Carnot cycle using, 175–76
   compression of, 291–92
   entropy of, 163
   filling a tank with, 317–18
   fugacity of, 349
   tables, 663–75
   tables, fugacity from, 347
   venting, 322
Steam turbine, 282–87
   actual operation, 283
   ideal and lost work in, 286–87
   Mollier graph used in, 285–86, 285f
   reversible operation, 283
   steam tables used in, 283–84
Stoichiometry, 593–96
Storable energy, 89
Subscripts, 411
Successive interpolations, 42
Summary, 26
Surroundings, in system, 13
System, 13–14
   adiabatic, 13
   classifying, 13–14
   closed, 13
   composite, 13
   diathermal, 13
   equilibriumin, 15–16
   examples of, 14f
   isolated, 13
   open, 13
   simple, 13
   surroundings in, 13
   universe in, 13

Tabulated properties
   entropy calculated using, 162
   fugacity calculated using, 347–48
   interpolations, 40–43, 41f
Temperature
   adiabatic flame, 605–6
   defined, 8
   entropy and, 153
   in heat transfer, 20f
   Henry's law and, 569
   liquid miscibility and, 556–58
   measuring, 23–24
   in molecular collision, 7–8
   upper consolute, 557
Temperature, effect on heat capacities, 113–19, 113f
   compressed liquid properties, 117–19, 118f
   constant-pressure cooling of liquid, 115–17
   heat capacity of steam, 114–15
   ideal-gas heat capacity, 114
   liquid heat capacities, 114
Temperature-enthalpy chart, 243–44, 244f
Ternary systems, 390–93
Thermal equilibrium, 15
Thermal expansion coefficient using Soave-Redlich-Kwong, 207–9
Thermodynamic analysis
   of processes, 314–15
   of refrigeration, 302–6
Thermodynamic charts, 242–45
   enthalpy-entropy (Mollier chart), 244–45, 245f
   pressure-enthalpy, 242–43, 243f
   temperature-enthalpy, 243–44, 244f
Thermodynamic cycle, notion of, 169
Thermodynamic potential function, 195
Thermodynamic potentials, 194–95, 195f
Thermodynamics
   chemical, 3–5, 4f
   definitions, 13–17
   development of, 12–13
   how and *why* in, 12–13
   introduction, 3–5
   molecular basis of, 5–11
   multicomponent mixture, state of, 18
   problems, 26–28
   process and path, 18–19, 18f
   pure component, state of, 17–18
   quasi-static process, 19–22, 20f
   statistical *vs.* classical, 11–13
   summary, 26
   units, 22–25
Thermodynamics, calculus of, 205–12
   differential and integral forms of a function, 209–12
   thermal expansion coefficient using Soave-Redlich-Kwong, 207–9
   triple-product rule in ideal-gas state, 207
Thermodynamics of steady-state processes, 272–94
   adiabatic mixing, 276–78, 277f
   expansion and compression of liquids, 292–94
   flow through pipe, 272–76
   gas compression, 290–92

Index

heat exchanger, 278–81, 279f
power generation, 295–300
steam turbine, 282–87
throttling, 287–90
Thomson, William, 100, 178
Throttling, 287–90
  of compressed liquid, 288–89, 290f
  of ideal gas, 288
Time, 25t
Torr, 23
Triple-product rule in ideal-gas state, 207
Truncated virial equation, residual properties from, 226–27
Turbines, gas/steam, 282–87
Two-phase systems, pure fluid in, 337–40
  condensing a vapor-liquid mixture, 340
  heat of vaporization using Soave-Redlich-Kwong, 338
  vapor-liquid mixture calculated by Soave-Redlich-Kwong, 339–40
Txy graph, 370–73, 379f
  of completely immiscible liquids, 559–61, 559f
  of ethanol/water at elevated temperatures, 384f
  of heptane and decane, 370f
  of partially miscible liquids, 385f, 554–56, 554f, 557f

UNIFAC equation, 525–31
  detailed calculation using, 528–30
  Pxy graph for, 530f, 531f
  terms calculated by, 527–28
  using, 530, 532
UNIQUAC equation, 522–25
  group contribution method in, 523–24
  Pxy graph calculated from, 524–25, 526f
Units, 22–25
  constants, 22t
  conversions, 25t
  energy, 24–25, 25t
  mole, 24
  pressure, 22–23, 25t
  temperature, 23–24
Universal Functional Activity Coefficient (UNIFAC). See UNIFAC equation
Universe
  entropy of, 168
  in system, 13
Unsteady-state balances, 315–23
  filling a tank with ideal gas, 318–18
  filling a tank with steam, 317–18
  pressurizing a tank, 315–19, 315f
  venting a tank, 319–23

  venting ideal gas, 322–23
  venting steam, 322
Upper consolute temperature (UCT), 557

Vacuum, irreversible expansion against, 136–37
Van der Waals equation
  compressibility factor in, 58–60
  residual properties from, 232t
Van der Waals forces, 6
Van Laar equation, 519
Vaporization. See also Vapor-liquid mixture
  enthalpy of, 478–79
  entropy of, 157, 158
  heat of, using Soave-Redlich-Kwong, 338
Vapor-liquid equilibrium (VLE), 369. See also Henry's law; Vapor-liquid equilibrium (VLE)
  Antoine equation, 343
  binary, using Soave-Redlich-Kwong, 452–53
  chemical equilibrium and, 625–28
  chemical potential, 341
  Clausius-Clapeyron equation, 341–43
  elevated pressures and temperatures, 383–84, 384f
  at elevated pressures and temperatures, 383–84, 384f
  equations in, solving, 449–50, 451f
  in ideal solution, 466–74 (See also Raoult's law)
  problems in, classification of, 448–49
  of pure fluid, 340–43
Vapor-liquid equilibrium, theory of, 435–59
  chemical potential in, 439–42
  fugacity in, 443–48
  Gibbs free energy in, 435–39
  using equations of state, 448–53
Vapor-liquid-liquid equilibrium (VLLE), 546f
Vapor-liquid mixture
  condensing, 340
  Soave-Redlich-Kwong equation used to calculate, 339–40
Vapor-liquid separator, 4f, 5
Variables
  change of, in fundamental relationships, 217
  derivatives and integrals with functions of multiple, 110
Venting, 321
  ideal gas, 322–23
  steam, 322
  a tank, 319–23
Virial equation
  Benedict-Webb-Rubin equation and, 68
  calculated isotherms using, 57f

Virial equation (*Continued*)
   compressibility factor in, 53, 54, 56
   molar volume calculated with, example of, 55–56
   residual properties of methane using, 227–28
   second virial coefficient, 53–54, 55
   summary, 78
   truncated, 54, 226–27
Volume, 25t
   excess properties in, 494–95, 495f
   residual, 224
Volume of mixing, 409–10
Von Guericke, Otto, 27–28

Wilson equation, 519–20
Work, 88–89, 88f
   adiabatic production of, 261, 262
   heat and, shared characteristics with, 97
   ideal, 183–88
   lost, in entropy, 183–88, 191
   maximum, in feasibility design, 186–87
   as path function, 105–6
   PV, 90–96, 91f
   separation of, ideal solution in, 464
   shaft, 90, 91f
   sign convention for, 89
   value of heat, Carnot cycle and, 182–83, 182f

x, y, z, convention for, 372
xy graph, 381–83, 382f

ZP graph of pure fluids, 43–45, 45f, 56

REGISTER YOUR PRODUCT at informit.com/register
Access Additional Benefits and SAVE 35% on Your Next Purchase

- Download available product updates.
- Access bonus material when applicable.
- Receive exclusive offers on new editions and related products.
  (Just check the box to hear from us when setting up your account.)
- Get a coupon for 35% for your next purchase, valid for 30 days. Your code will be available in your InformIT cart. (You will also find it in the Manage Codes section of your account page.)

Registration benefits vary by product. Benefits will be listed on your account page under Registered Products.

InformIT.com–The Trusted Technology Learning Source

InformIT is the online home of information technology brands at Pearson, the world's foremost education company. At InformIT.com you can

- Shop our books, eBooks, software, and video training.
- Take advantage of our special offers and promotions (informit.com/promotions).
- Sign up for special offers and content newsletters (informit.com/newsletters).
- Read free articles and blogs by information technology experts.
- Access thousands of free chapters and video lessons.

Connect with InformIT–Visit informit.com/community
Learn about InformIT community events and programs.

Addison-Wesley • Cisco Press • IBM Press • Microsoft Press • Pearson IT Certification • Prentice Hall • Que • Sams • VMware Press

ALWAYS LEARNING PEARSON